THE SUPPORTING ROOTS OF TREES AND WOODY PLANTS: FORM, FUNCTION AND PHYSIOLOGY

Developments in Plant and Soil Sciences

VOLUME 87

The titles published in this series are listed at the end of this volume.

The Supporting Roots of Trees and Woody Plants: Form, Function and Physiology

Edited by

A. STOKES
Laboratoire de Rheologie du Bois de Bordeaux
BP 10
Domaine de l'Hermitage
33610 Cestas Gazinet
France

Partly reprinted from *Plant and Soil*, Volume 217, No. 1/2 (1999)

KLUWER ACADEMIC PUBLISHERS
Dordrecht / Boston / London

Library of Congress Cataloging-in-Publication Data

The supporting roots of trees and woody plants : form, function, and
physiology / edited by Alexia Stokes.
p. cm. -- (Developments in plant and soil sciences ; v. 87)
ISBN 0-7923-5964-X (hb : alk. paper)
1. Trees--Roots. 2. Woody plants--Roots. I. Stokes, Alexia.
II. Series.
QK644.S86 1999
575.5'4--dc21 99-44183

ISBN 0-7923-5964-X

Published by Kluwer Academic Publishers,
P.O. Box 17, 3300 AA Dordrecht, The Netherlands

Sold and distributed in North, Central and South America
by Kluwer Academic Publishers,
101 Philip Drive, Norwell, MA 02061, U.S.A.

In all other countries, sold and distributed
by Kluwer Academic Publishers,
P.O. Box 322, 3300 AH Dordrecht, The Netherlands

Printed on acid-free paper

Printed in the Netherlands

Contents

*Chapters indicated with an asterisk are reprinted from *Plant and Soil* Volume 217, No. 1/2 (1999).

Preface

This publication comprises the proceedings of the first International Conference devoted to the structural roots of trees and woody plants. *'The Supporting Roots – Structure and Function,'* 20–24 July 1998, Bordeaux, France. The meeting was held under the auspices of IUFRO WPS 2.01.13 'Root Physiology and Symbiosis,' and its aim was to bring together scientific researchers, foresters and arboriculturalists, to discuss current problems in structural root research and disseminate knowledge to an audience from a wide disciplinary background.

For the first time in an international conference, emphasis was placed on presenting recent reseach in the field of tree anchorage mechanics and root biomechanics. The way in which tree stability can be affected by root system symmetry and architecture was addressed, as well as how movement during wind sway can influence the development and shape of woody roots. The role of different nursery and planting techniques was discussed, in relation to effects on root system form and development. Root response to different environmental stresses, including water, temperature, nutrient and mechanical stress was addressed in detail. The structure and function of woody roots was also considered at different levels, from coarse to fine roots, with several papers discussing the interaction between roots and the rhizosphere.

One of the conference highlights was the presentation of new methods in root research, by a series of workshops held at LRBB-INRA, Pierroton, on the northern border of the Gascony forest. Various techniques were demonstrated, with workshops on tree stability, root-wood strength, root system excavation, 3D digitalising of root systems and state-of-the-art software to simulate root growth and architecture. The book cover is an example of the recent advances in this area and shows an image of a *Pinus pinaster* root system, measured by 3D digitalising and analysed with AMAPmod (Danjon *et al.*, this volume, image conception T. Fourcaud).

Some of the papers presented in this book of proceedings, are also included in a special issue of the journal Plant and Soil (Vol. 217, Nos. 1–2, 1999). I would like to than Prof. H. Lambers (Editor, Plant and Soil) for his help and co-operation in this project.

The '*Supporting Roots*' conference would not have been possible without financial help from INRA, the Aquitaine Region, University of Bordeaux I and the European Community, the latter assisting the visit of scientists from former Eastern European countries. I wish to express my thanks to the sponsors, as well as to all those who assisted throughout the conference. Sincere thanks are also due to L. Thorin and Y. Cazenave, for their help in preparing this book.

Alexia Stokes,
July 1999

ANCHORAGE AND BIOMECHANICS

A. Stokes (ed.), *The Supporting Roots of Trees and Woody Plants: Form, Function and Physiology*, 3–17.

The development of symmetry, rigidity and anchorage in the structural root system of conifers

M.P. Coutts[1,*], C.C.N. Nielsen[2] and B.C. Nicoll[1]
[1]*Forestry Commission Research Agency, Northern Research Station, Roslin, Midlothian, Scotland, EH25 9SY, UK and* [2]*Arboretum, Department of Botany, Dendrology and Forest Genetics, Royal Veterinary and Agricultural University, Kirkegardsveg 3A, DK-2970 Horsholm, Denmark*

Key words: anchorage, development, growth, root systems, symmetry, tree stability

Abstract

The stability of shallowly rooted trees can be strongly influenced by the symmetry of the 'structural' system of woody roots. Root systems of forest trees are often markedly asymmetric, and many of the factors affecting symmetry, including root initiation and the growth of primary and woody roots, are poorly understood. The internal and environmental factors that control the development, with respect to symmetry and rigidity, of shallow structural root systems are reviewed and discussed with particular reference to Sitka spruce (*Picea sitchensis* Bong. Carr.). Areas where there is insufficient knowledge are highlighted. A scheme is proposed that represents the root system as a set of spokes that are variable in number, size and radial distribution. Rigidity can vary between and along each of the spokes. The root system is presented as a zone of competition for assimilates, where allocation to individual roots depends upon their position and local variations in conditions. Factors considered include the production of root primordia of different sizes, effects of soil conditions such as the supply of mineral nutrients and water on growth of primary and woody roots, and the effect of forces caused by wind action on growth of the cambium, giving rise to roots which, in cross section, resemble I- or T-beams, and efficiently resist bending.

Introduction

An improved understanding of root development is desirable in both forestry and arboriculture. It is especially relevant to tree stability because the woody root system provides anchorage and structural support. Where rooting depth is limited by soil conditions, the radial symmetry of the woody root system becomes especially important (Coutts 1983). Field observations have shown that root systems are often markedly asymmetric; for example, they develop less in the direction of plough furrows (Coutts et al. 1990) or nearby tree stumps (Quine et al. 1991). Even where there is no physical obstruction and the site is relatively even, root systems often have remarkably uneven development (Nicoll et al. 1995).

The woody root system forms a stiffened base to support the tree, and where the main roots are poorly developed or absent at one side, stability will be reduced (Coutts 1983 and Nielsen 1990, 1991). Many internal and environmental factors influence root development. In this account we consider certain internal factors such as the production of root tips capable of forming structural roots, and environmental factors, including mineral nutrition and mechanical stress, which affect the growth of the woody roots. It has not been possible to consider all features of the root environment; notable omissions are temperature, drought and waterlogging.

Because of the many factors involved, the process of root system development is difficult to understand without an appropriate model. Modelling can be expected to improve our understanding and expose important areas where little is known. It may eventually be possible to link a root development model to other stability models, for example the root anchorage model described by Blackwell et al. (1990).

* FAX No: +44 131 445 5124.
Plant and Soil is the original source of publication of this article. It is recommended that this article is cited as: *Plant and Soil* **217**: 1–15, 1999.

The object of this paper is to collate information on root system architecture and growth that will be required for modelling root system development with respect to symmetry and rigidity. Most of the relevant work has been carried out on Sitka spruce (*Picea sitchensis* Bong. Carr.) because of problems of windthrow in that species in upland Britain and Ireland, but information on other species has been included where appropriate. Results of previous studies that might be used in modelling are briefly described, with some reworking of published data.

Symmetry

Concepts and methods of analysis

Asymmetry in a radially 'spoked' structure such as an idealised root system (Figure 1) can be of more than one type. In what we term Type I asymmetry, the individual spokes (or roots) can vary in diameter, and this can give an asymmetrical structure even if the arrangement is regular. In Type II asymmetry, the structure may have spokes (roots) which are not uniformly arranged although they may be more evenly sized. In tree root systems both types of asymmetry occur together. In addition to arrangement and diameter, rigidity can vary between and along spokes, through differences in cross sectional shape and branching, and variation in material properties.

The symmetry of root systems has previously been measured in terms of the centre of mass of all of the lateral roots (Coutts et al. 1990). Root mass can be represented by root diameter (Quine et al. 1991) or root cross sectional area (CSA) (Nicoll et al. 1995). The centre of mass (or of diameter or CSA) has a magnitude and a direction from the origin in the centre of the stump and can be used in analysis of variance to compare treatments. The 'centre of CSA' calculation gives greatest weighting to the largest structural roots, i.e. the roots that are most important for stability (Nicoll and Ray 1996). However, for all centre of 'mass' descriptions, magnitude varies with root system size and although scaling to stem diameter can be used (Quine et al. 1991), it is not an ideal way of expressing symmetry. New methods which relate more closely to the mechanics of anchorage are required.

In the root anchorage model (Blackwell et al. 1990) the distance from the tree to the point of bending (fulcrum or hinge) on the lee side of the root system can be varied. The larger the distance, (other things being equal) the stronger the anchorage. This distance to the fulcrum, *xy* (see Figure 2) is related to the stiffness of the roots on the lee side and to their azimuths. Therefore, as stiffness of a root with a circular cross-section is related to d^4 (Coutts 1983),

$$xy \infty \sum_{t=1}^{n} d_i^4 \, sin\theta_i$$

The distance *xy* is also proportional to root system size, and therefore to scale the system, we can represent d^4 for each root as a percentage of $\sum d^4$ for the whole root system. To find the minimum value for *xy*, the line *xy* is rotated through 360 ° in small intervals and recalculated at each position. The minimum value is related to the length of the lever arm on which forces act in the root system.

An increase in the length of the lever arm will amplify all main components of the anchorage, i.e. the forces that are required to cause the soil to fail under and around the root-soil plate, and those forces required to lift the mass of the root-soil plate and pull out or break the roots on the windward side (Deans and Ford 1983, Coutts 1986). For example, in a shallow root-soil plate, doubling the distance of the hinge from the base of the tree will approximately double the turning moment required for uprooting.

Data on root system symmetry

Data on Type I asymmetry (variable root size) are given in Figure 3, which shows results for 17 root systems of 18-year-old Sitka spruce (*Picea sitchensis* (Bong.) Carr.) trees established from transplants (see Coutts 1983). This analysis describes diameter growth differentiation between roots. The largest roots on each tree are ranked according to size. Fifty percent of the root material was present in the 1–4 largest roots and 80% in the 3–10 largest roots on these trees (Figure 3). Thus most material in the root system is allocated to a small number of large woody roots. The largest roots summing up to 80% of the total root mass are hereafter described as 'dominant roots'. An analysis of Type II asymmetry is presented by Henderson et al. (1983a). They measured roots on seven 16-year-old Sitka spruce trees and found that roots of more than 5 mm dia. showed significant regularity of distribution on three of them, significant clustering on one, and neither (i.e. randomness) on the remaining three. In a recent study on Scots pine (unpublished) the roots were found to be distributed at random, and a random distribution can be assumed for modelling purposes.

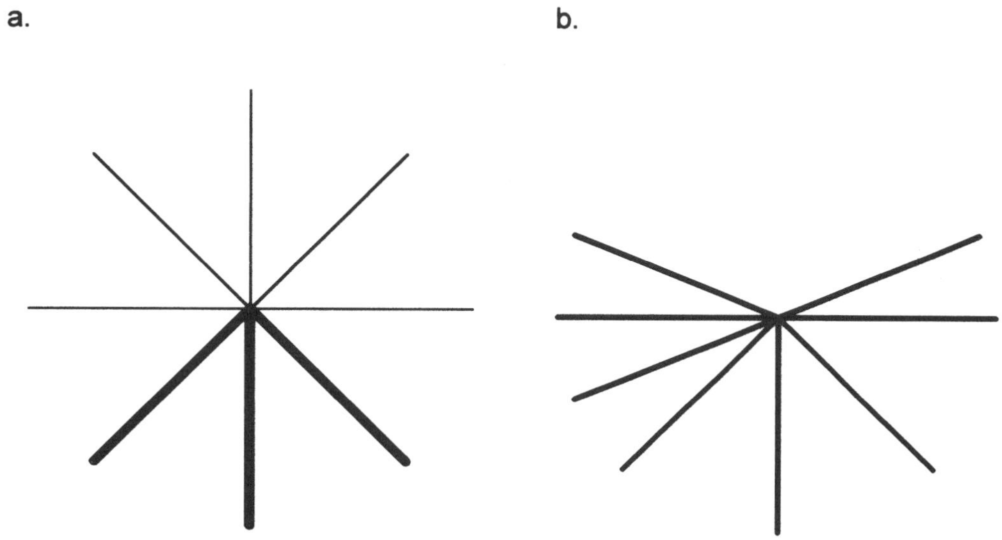

Figure 1. Types of asymmetry of a spoked system, such as a shallowly rooted tree root system; a. Type I, a regular arrangement with variation in spoke diameter, b. Type II, irregular arrangement with uniform spoke diameter.

Figure 2. Root system in plan view: *xy* is the distance from the stem centre to the fulcrum 'hinge', marked by a solid line on the lee side of the tree. Direction of tree fall is shown by an arrow.

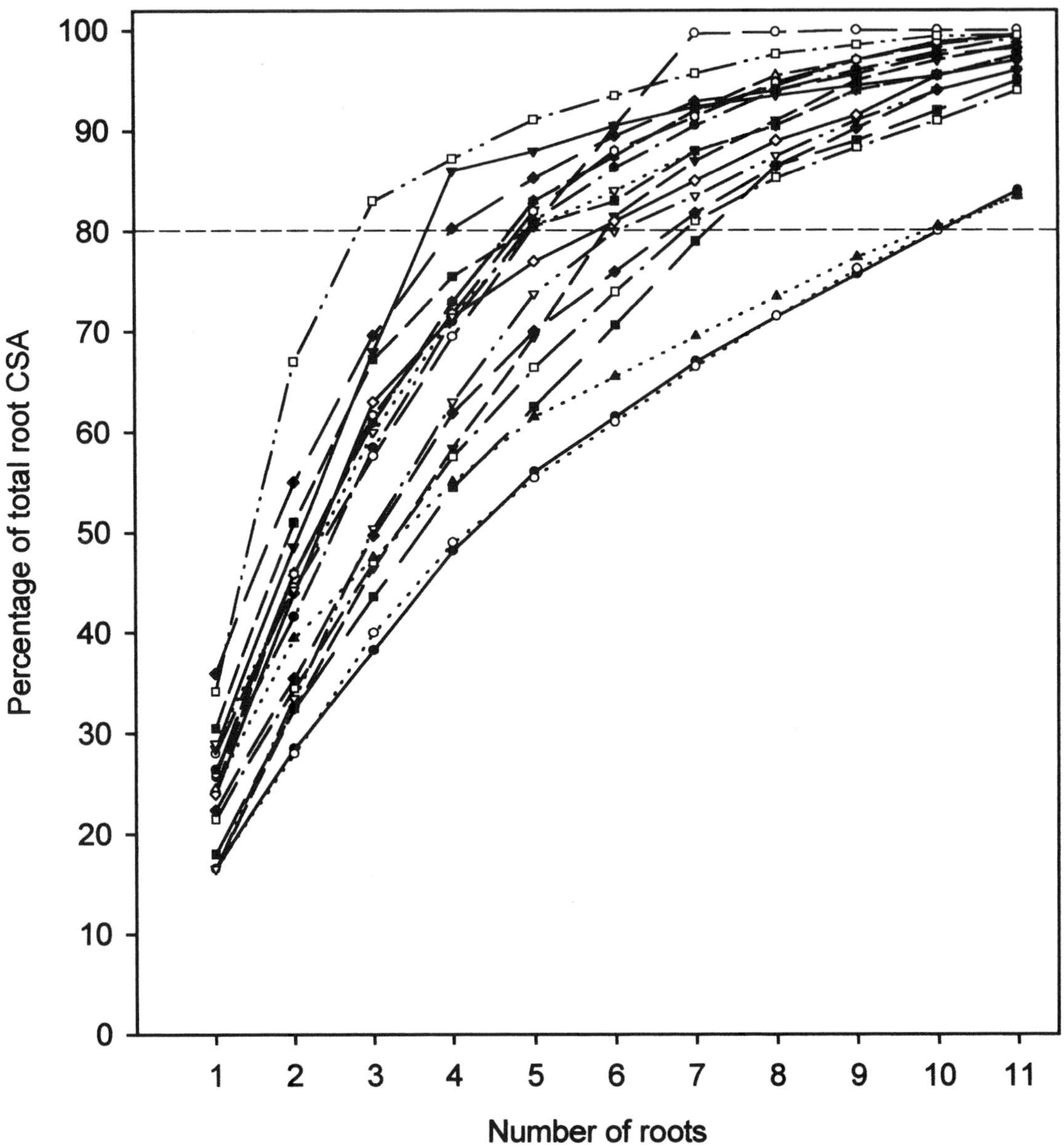

Figure 3. The proportion of the total cross sectional area (CSA) of the woody roots of Sitka spruce, in the largest root, in the largest + the second largest, the largest + second largest + third largest, and so on to the 11 largest roots. Each curve represents the root system of an individual tree measured 18 years after planting. The CSA of each root was measured at a point 20 cm from the centre of the stump. The largest roots that together make up at least 80% of the total CSA are classed as 'dominant roots: From Coutts (1983).

Branching and tapering of roots

Although the root system might be modelled at the outset as a set of spokes with different sizes and positions, a growth model relevant to stability should eventually include root branching. This is because, even if the total cross-sectional area is the same before and after a point of branching (Coutts 1983), stiffness of the root, being related to d^4, is greatly reduced beyond that point. In addition, the information on the ranking of root sizes in Figure 3 is based only on measurements at a fixed radius close to the stump. Further out, branching will change the relationships. Various models describing branching already exist (for example, Henderson et al. 1983b), but none deals with the growth process giving rise to major branches in woody roots. Obviously, taper of roots will also influence the lever arm. The stronger the taper, the shorter will be the lever arm (Nielsen 1990). Therefore, the root system may be better modelled as a set of spokes with rigidity that varies between and along the spokes.

Changes of symmetry with age

The largest roots in the first few years after sowing or out-planting are not necessarily the largest later on, but in a study of Sitka spruce by Coutts (1983) there was a clear distinction between the five largest roots and the remainder by eight years after planting (Figure 4), and Reynolds (1987) obtained comparable results with a number of other conifers. In undisturbed forest stands, the ranking of roots tends to be stable (Coutts 1983), but large changes in the environment like heavy thinnings, can influence distribution of growth between roots in such a way that Type I asymmetry decreases (Nielsen 1990).

Production of large root tips

The lateral roots originate from primordia on the tap root, on existing lateral roots, or (adventitiously) on the stem base. The seedling tap root produces a number of large lateral roots near its base, close to the junction with the hypocotyl. A radial arrangement is ensured by production of primordia adjacent to the more or less symmetrically arranged primary *xylem* elements. Large root primordia produce roots of large tip diameter. The positions of large primordia determine the starting points and general direction (azimuth) of the woody roots. The tip diameter of lateral roots is linearly related to extension growth (Wilcox 1968), and only the primary roots that have larger than average tip diameters produce strong secondary growth and become important structural roots (Horsley and Wilson 1971, Coutts 1983). Factors known to affect the production of large root tips were reviewed by Coutts (1987). Damage to the parent root apex or slowing down of its growth rate, deflection of the parent root from its original direction, and adventitious origin, all appeared to give rise to large lateral root tips. Nursery treatments such as root pruning, which could affect root tip size, may therefore affect later development. In planted trees the original arrangement of roots is disturbed and roots may be broken. Exactly how the tree recovers from this disturbance through production of new root apices and growth of existing ones, is not known. However, Coutts et al. (1990) described new roots on spruce grown from transplants that assumed an array of growth directions comparable to those of the primary lateral roots from direct sown trees.

Without a quantitative basis this information is inadequate for modelling but is a fairly tractable topic for research. For the present we will assume an arbitrary number of potential woody roots originating at random around the base of the tree. The production of root primordia and associated influences are shown in the schematic diagram of root development in Figure 5.

Growth rate and competition for assimilates

Little is known about allocation between root and shoot in mature trees. Although Johanson (1941) showed that under ‘poor’ growth conditions the root system acquired large amounts of assimilates relative to the stem; in young trees the root:shoot ratio increased in response to reduced nutrient supply, especially when phosphorus was limiting (Coutts and Philipson 1979). Wind movement of the tree is also known to increase the relative allocation of assimilates to the structural roots (Wilson 1975).

The main roots on a root system of Sitka spruce are often of comparable ages (Coutts 1983), so that the large differences in their sizes are mainly attributable to differences in growth rate. Individual roots compete for the C supply from the shoot system (Coutts 1987). Thus any enhanced growth of one part of the system is at the expense of the other parts, with roots destined to become dominant roots having relatively rapid thickening. The competitive process between thickening structural roots on a Sitka spruce tree is illustrated in Figure 4. Competition between the roots on the same root system may be increased when the total C supply from the shoot becomes limited. Root systems of trees growing under conditions which limit photosynthesis, such as shade or water stress, may therefore be less symmetrical than normal but there is no information on this.

Position of root origin

Differential growth is strongly influenced by the root environment (see later), but the depth of a root on the root system also appears to be an important influence on the growth rate of lateral roots. Roots originating at the top of the system, near the stem base, have an advantage over deeper roots, perhaps because they are the first to receive assimilates from the shoots. However, ‘internal’ effects of position have not been completely separated from those of environ-

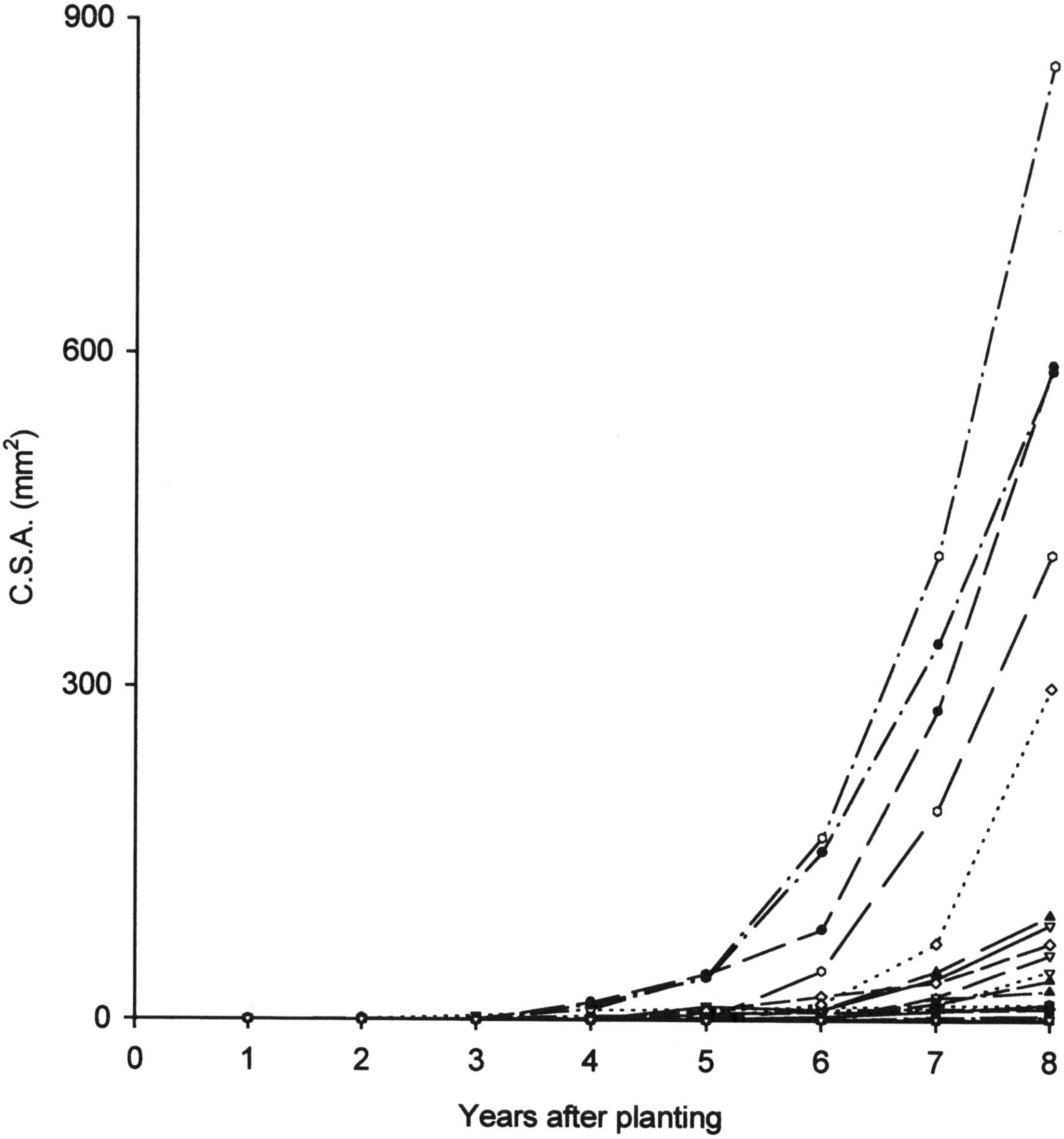

Figure 4. Cumulative cross sectional area (CSA) of the 20 largest woody roots on a Sitka spruce tree 8-years after planting, measured at a point 20 cm from the centre of the stump. Each line represents a different root. From Coutts (1983).

mental gradients such as soil temperature and nutrition and there are no data relating depth of origin to growth rate. Observations (e.g. Coutts and Nicoll 1991) suggest the relationship is non-linear, with much greater growth at the top than lower down.

In small trees where the phloem path is short, uneven assimilate supply from unbalanced crown development, or one-sided illumination, may have a small influence on the relative development of roots (Stokes et al. 1995), but in tall trees assimilates have the scope to move laterally around the stem and differential growth of the roots will be sink driven.

Mineral nutrition

When root systems are divided between different nutrient regimes, the part with a high nutrient supply grows faster than the part with a low supply (Coutts and Philipson 1976). The growth response (increase in dry weight) of roots to the concentration of nutrient

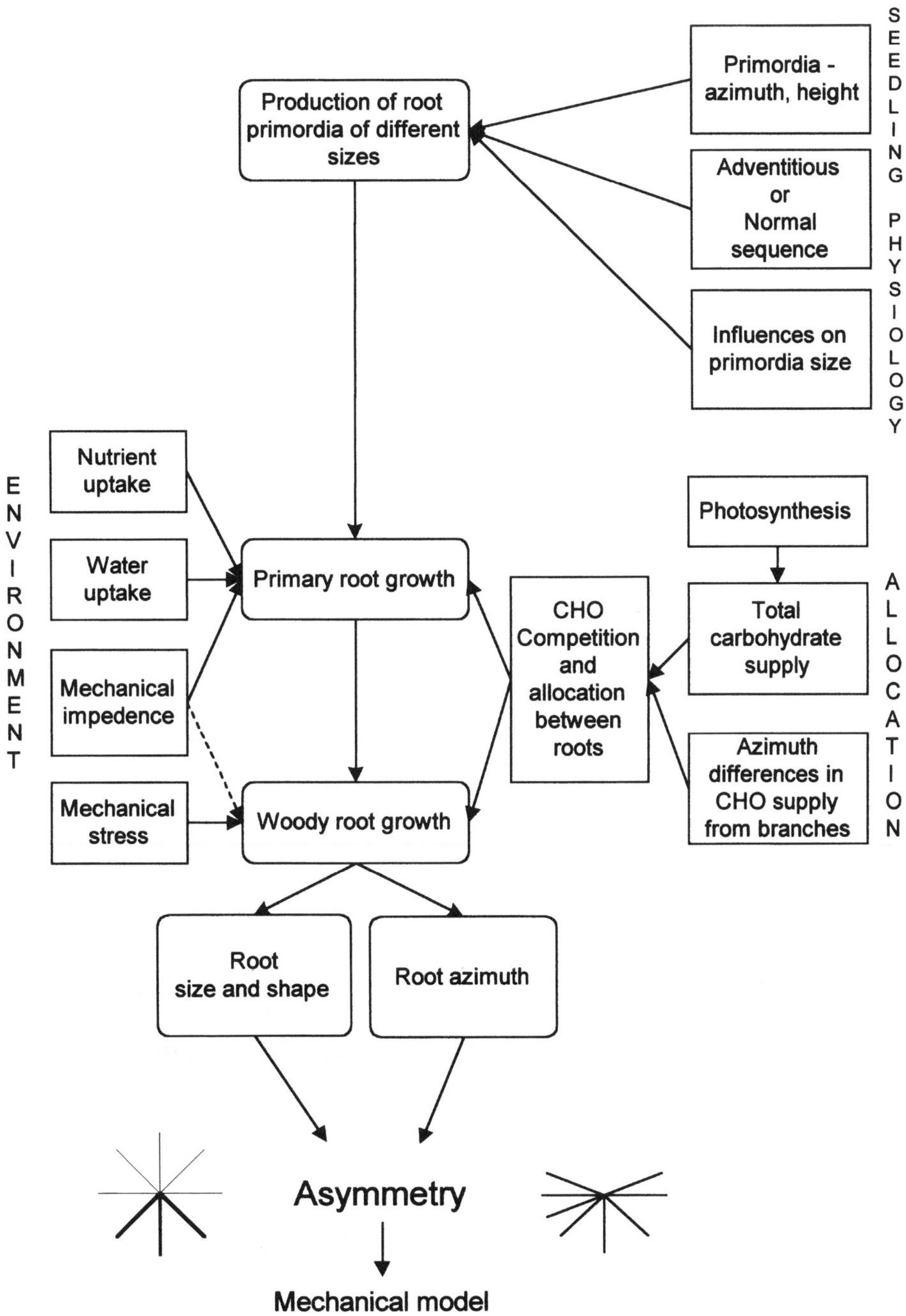

Figure 5. Scheme summarising the development of the structural root system in response to internal and external factors.

supplied to Sitka spruce showed a non-linear relationship (Figure 6). The difference in growth rate between the parts in high and low nutrient regimes increased with nutrient supply. Interpretation of the relationship shown should be made with care because the plants had only two main roots, and increasing the nutrient supply to one root increased the supply to the plant as a whole and stimulated total root and shoot growth. Data more appropriate to modelling could be obtained by arranging concentrations so that the total supply of nutrients to the plant remained constant.

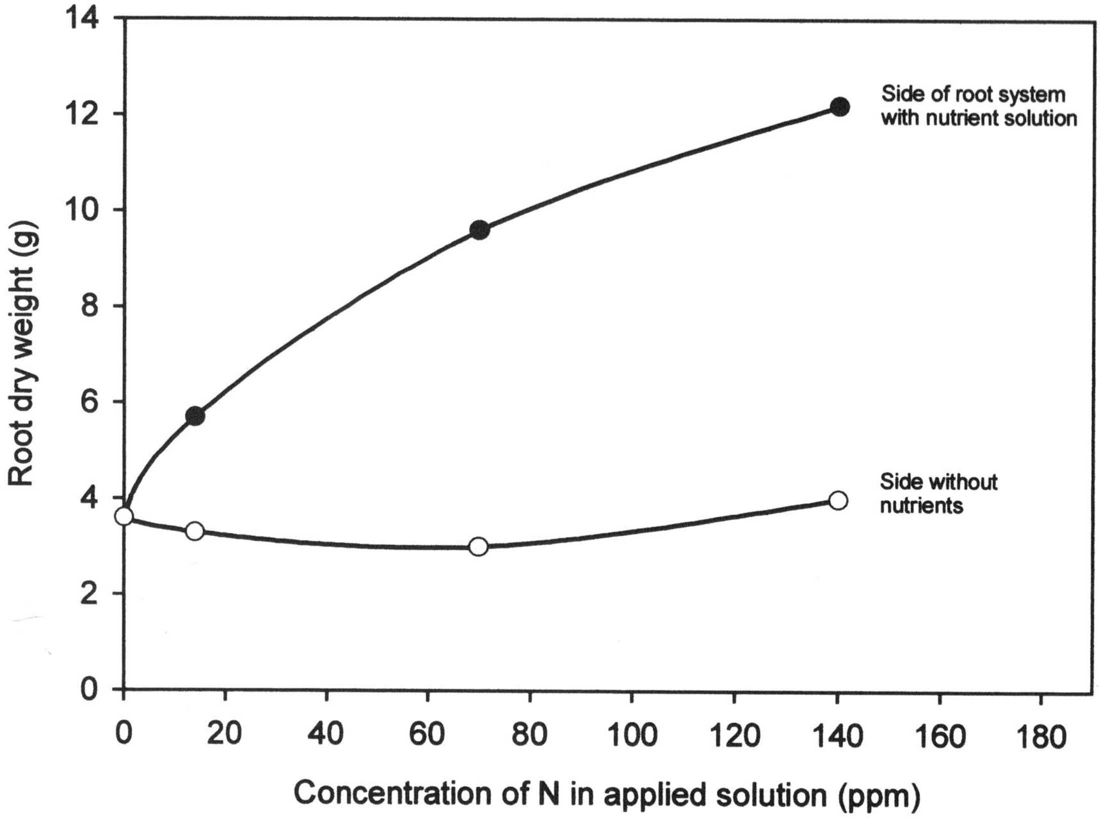

Figure 6. Effect of nutrient concentration on mean growth in dry weight by Sitka spruce roots. The roots were split between containers of soil, one of which was supplied with nutrient solution at various concentrations, while the other received water only. Data from Coutts and Philipson (1976).

If differential root growth is induced at an early stage of development by local variations in the environment, the growth differences may become exaggerated later, even in the absence of the original stimulus, if the more vigorous parts of the root system assume dominance over the remainder. This possibility was tested using lodgepole pine (*Pinus contorta* (Douglas ex Loudon) with roots divided between containers of culture solution giving high and low nutrient supply (Coutts and Philipson 1977). After differential growth had been induced (Figure 7) some plants were transferred to a uniformly high nutrient environment. This change caused an increase (relative to the high/low treatment) in growth of the 'previously low-nutrient' root, and a relative decrease in growth of the 'previously high-nutrient' root. However, the growth of the two halves of the root system continued to diverge. Calculation of relative growth rate (RGR) (Figure 8) shows that in the high/low nutrient regime the high-nutrient roots had a larger RGR than low-nutrient roots, and RGR in both halves showed the usual decrease with time. The change from high/low to uniform high/high nutrient at 56 days gave approximately equal RGRs for both halves of the root system at 84 days and 133 days (there were not enough plants to continue the original high/low treatment to 133 days). Thus growth had become related to the size of each half, and there was no indication of 'dominance' by the high-nutrient root. The short time period of the experiment leaves some uncertainty as to what the long-term outcome would have been; for clarification it will be necessary to grow 'induced' plants in uniform conditions for a longer period.

The experiments described were carried out by varying the concentrations of nutrient supplied, but growth in the field might be assumed to relate to the total uptake of nutrients over time. The concentration of available nutrients will vary with soil temperature and moisture content, and uptake will also be a func-

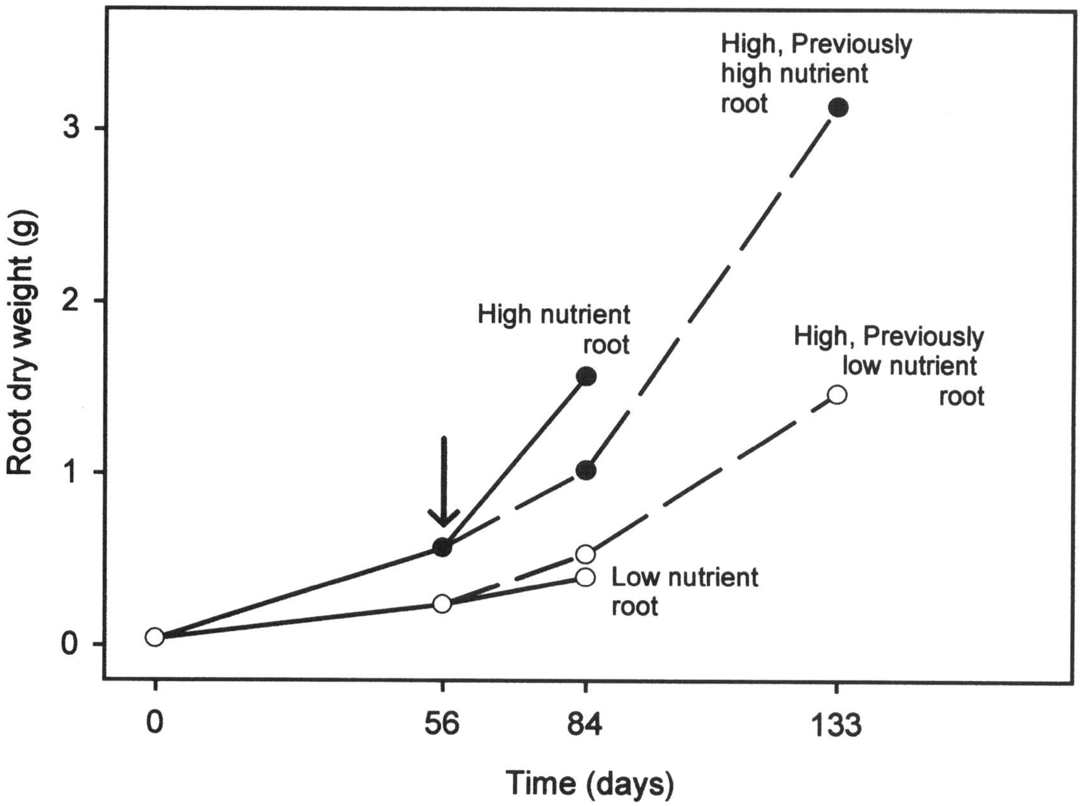

Figure 7. Time course of dry matter accumulation in lodgepole pine roots. The plants had divided roots growing in either an uneven (solid line) nutrient regime of high and low nutrient concentrations or in uniform high (dashed line) nutrient conditions. Certain plants were transferred from the uneven to the uniform regime at 56 days (arrowed). From Coutts and Philipson (1977).

tion of the total length of roots on any part of the root system.

Soil physical properties

Soil temperature, mechanical impedance and aeration can vary greatly with depth but are generally more uniform near the surface over the area occupied by a tree root system, and are therefore less likely to have important effects on root system symmetry. Waterlogging of the soil (Kozlowski 1982) and high mechanical impedance (Taylor and Gardner 1960) can inhibit root extension, but there is little information on how they affect secondary growth directly. The maximum radial pressure exerted by woody roots is around 800 kPa (MacLeod and Cram 1996), a value which would frequently be encountered at a depth in dense, compacted soils.

Primary–woody root relationships

When the primary roots (i.e. roots of primary structure) encounter conditions which stimulate their growth, there appears to be a feed-back system which causes a corresponding growth increase in the woody root from which they are growing. One would therefore expect to find a relationship between the sizes of the primary and woody parts of the root. Data from a study on the growth of Sitka spruce in response to nutrients (Coutts and Philipson 1976) indicated a linear relationship between primary and woody root growth (Figure 9).

Results similar to those obtained with nutrients were given when Sitka spruce roots were divided between containers of wet and dry soil (Coutts 1982). In the wet regime (matric potential –5 to –6 kPa) root dry weight was about six times larger than in the dry (–30 or –60 kPa) regime. A large part of this effect might be caused by reduced availability of nutrients in the dry soils, compounded by direct effects of water and associated changes in mechanical impedance on root extension. There is not enough information on soil water and differential root growth to reveal the shape of the relationship between them. Other factors

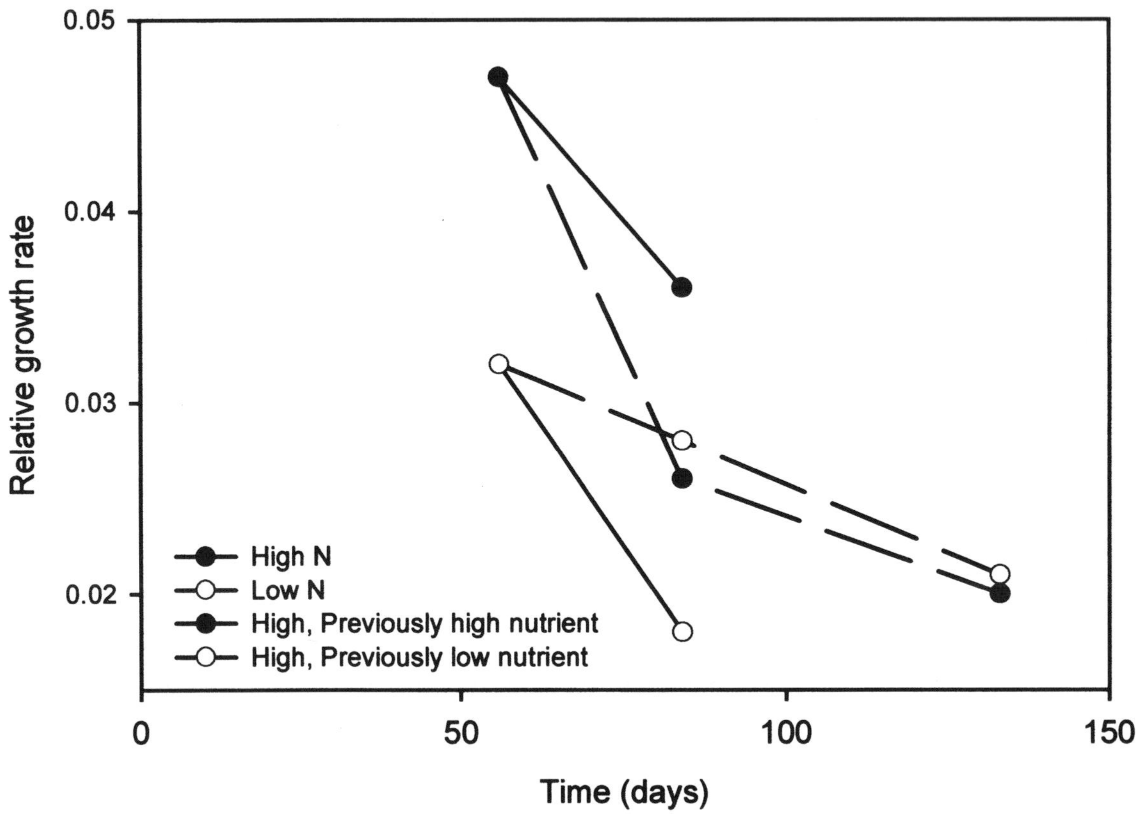

Figure 8. Relative growth rate $\frac{W_2 - logeW_1}{t_2 - t_1}$ of lodgepole pine roots, where W =dry weight and t = time. Calculated from data shown in Figure 7, from Coutts and Philipson (1977).

that adversely affect the growth of primary roots, such as poor aeration, would be expected to have similar results to water or nutrient supply. In the schematic diagram (Figure 5) the supply of water and nutrients is shown acting on primary root growth – which acts on woody growth – and on the process of competition.

Adaptive growth in response to mechanical stress

Growth stimulation

The effect of wind on above-ground growth and development of trees has been investigated for many years. For example, secondary thickening is known to be restricted in stem and branch by wind movement but accelerated in the stem base (see Telewski 1995), and resistance to wind bending is increased through greater thickening of the lower stem along the axis of the prevailing wind direction (Robertson 1991). These above-ground developmental responses counteract increasing movement as the tree grows and improve stability by allocating assimilate to parts of the tree where stress is greatest (see Morgan and Cannell 1994). Adaptive growth below-ground may be even more important as a compensatory mechanism, but there is only limited information on biomass allocation between root and shoot, and within the structural root system of large forest trees.

In a thinned stand of *Pinus strobus* L., Wilson (1975) found an increase in growth-ring width in the lower stem and in the base of structural roots corresponding with increased wind movement after the thinning. There was considerably less thickening on trees guyed after thinning. Urban et al. (1994) reported an immediate increase in thickening of structural roots but a 4 year delay in the increase of diameter growth in the stem in *Picea glauca* (Moench) exposed after removal of neighbouring trees. Similar differences in timing between stem and root thickening after thinning of *Pinus resinosa* Ait. were reported by Fayle (1983).

Adaptive growth of dominant woody roots has been documented by several authors. Experiments, where trees are prevented from swaying by guying

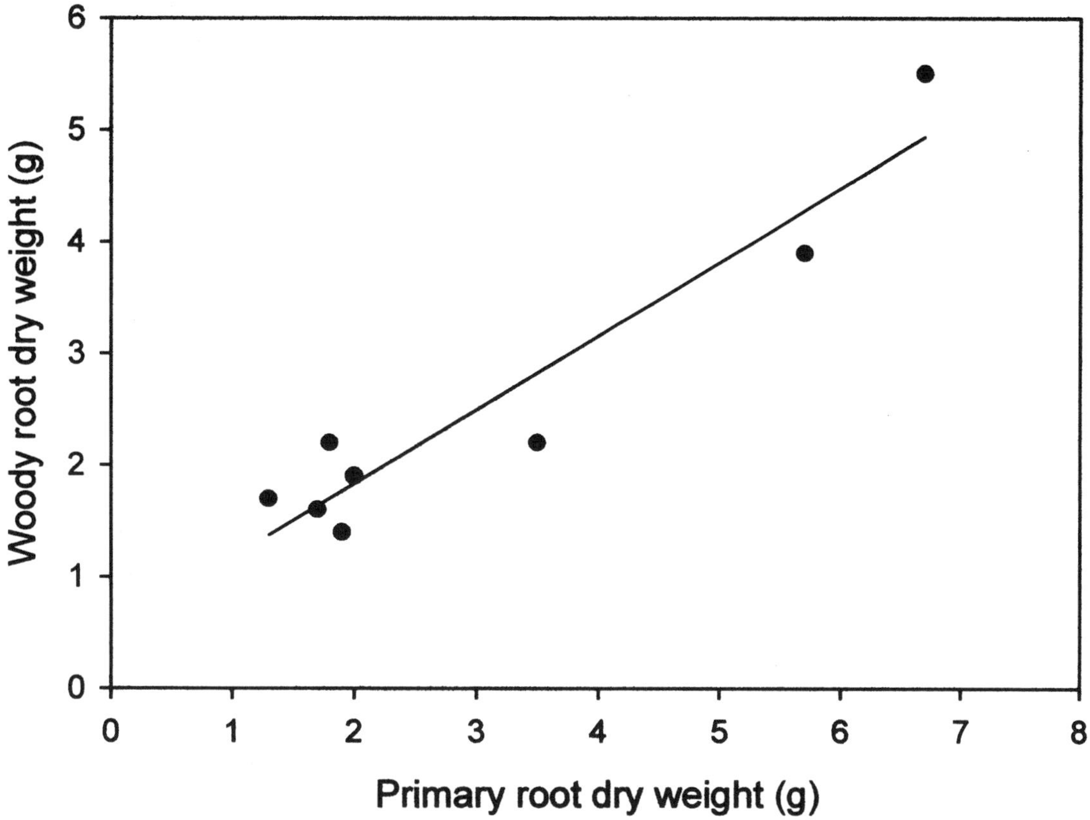

Figure 9. Primary and woody root dry weights of Sitka spruce growing in peat. Each point is a treatment mean from a split-root experiment ($n = 24$). Data from Coutts and Philipson (1976).

(e.g. Jacobs 1953) or are mechanically flexed (Stokes et al. 1997), and detailed field studies (Fayle 1968, Hintikka 1972, Deans 1981, Nielsen 1990 and Nicoll and Ray 1996) show that the allocation of assimilates within the structural root system is affected by distribution of mechanical stress.

As shallow rooted trees grow in height, their increasing mass and movement in the wind increase the flexing of structural roots near the soil surface. Trees employ various mechanical modifications during the development of root systems to limit this movement and to resist overturning. Nicoll and Ray (1996) and Ray and Nicoll (1998) found a direct relationship between root-soil plate area (in plan view) and depth in 46-year-old Sitka spruce on peaty gley soil, with the shallowest plates having the largest areas (Figure 10a). They also showed that anchorage was related to the rigidity of the plate, a factor which had an important effect on the resistance to failure in the soil. Rigidity also extends the distance from the stem to the point of bending on the lee side, conferring a mechanical advantage on the root system.

Although the shallow rooted trees had a smaller total mass than deeper rooted trees, they allocated a larger proportion of assimilate to their structural root systems. Therefore the shallow rooted trees – that would have experienced the most wind movement – had larger root:shoot ratios (Figure 10b). These results are comparable to those of Fritzsche (1933) and Nielsen (1990) who showed increasing root/stem-ratios with increasing wind load.

Roots on the leeward side of the tree in relation to the prevailing wind direction show higher diameter growth (Coutts et al. 1990, Nicoll et al. 1995, Nicoll and Ray 1996) and stronger taper (Nielsen 1990) than roots in other directions. Although roots all around the tree must act under tension and compression at different times, the larger bending stresses experienced by lee-side roots enhance radial growth, giving rise to buttress-like structures with the strength and stiffness required to resist uprooting. Supporting but-

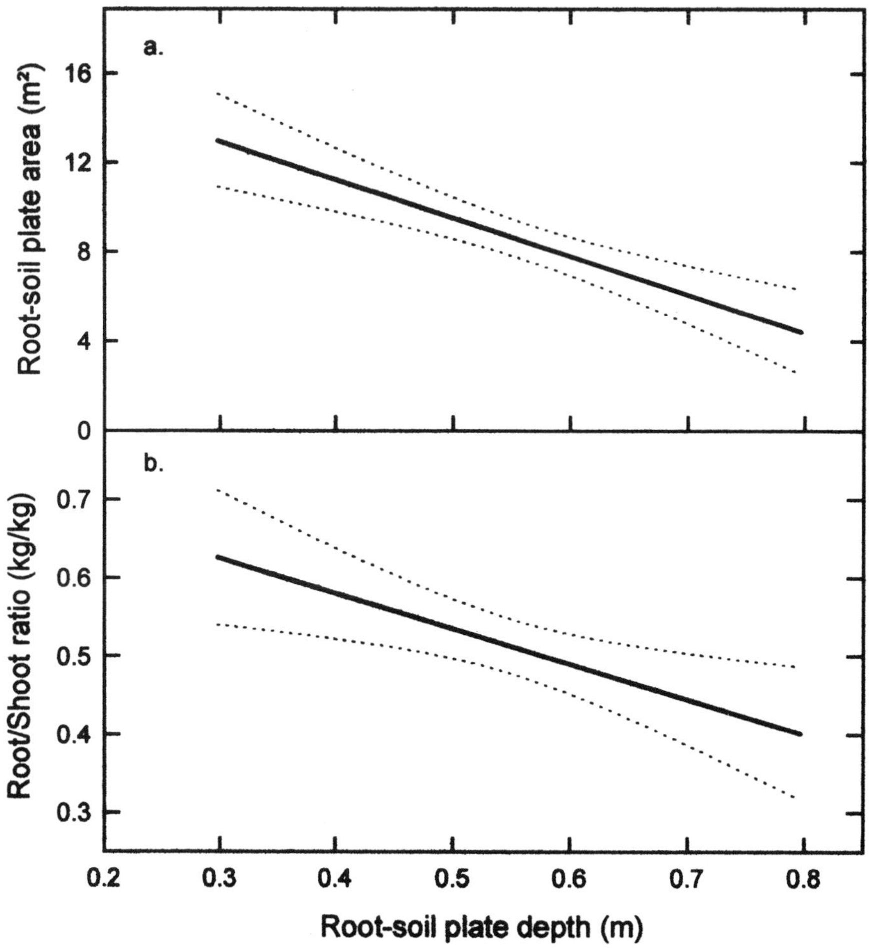

Figure 10. a. The relationship between root-soil plate area (normalised for tree size by dividing by stem mass) and plate depth. b. Root:shoot fresh mass ratio against root-soil plate thickness. Data from Nicoll and Ray (1996). Solid lines are regressions, dotted lines are 95% confidence intervals, $n = 39$.

tresses may reflect growth to equalise stress (Ennos 1995, Mattheck 1993), and also increase the leverage required for over-turning by moving the 'hinge' point of the root system further away from the base of the tree.

Cross sectional shapes of structural roots

In response to wind movement, trees with shallow structural roots commonly develop root cross-sectional shapes comparable in appearance to 'I-beams' or 'T-beams' (Büsgen et al. 1929, Rigg and Harrar 1931, Hintika 1972). These shapes (see Figure 11) appear to use a minimum of material to maximise resistance to bending or flexing within the soil (Wood 1995) and further increase rigidity of the root-soil plate (Nicoll and Ray 1996).

T-beam development was significantly greater on the side of the Sitka spruce root systems in the lee of the prevailing wind, compared to the windward side (Nicoll and Ray 1996). This adaptive growth strengthens the anchorage because the T-beam is particularly well designed to resist the compressive forces caused by wind action. Roots were more I-beam shaped on the windward side of the tree, where this shape effectively resists vertical flexing.

Change in adaptive growth with age

As the tree grows, symmetry of the structural root system may be increasingly influenced by adaptive secondary growth related to wind sway. Growth ring analysis by Nielsen (1990) indicated that the adaptive diameter growth of structural roots increased with

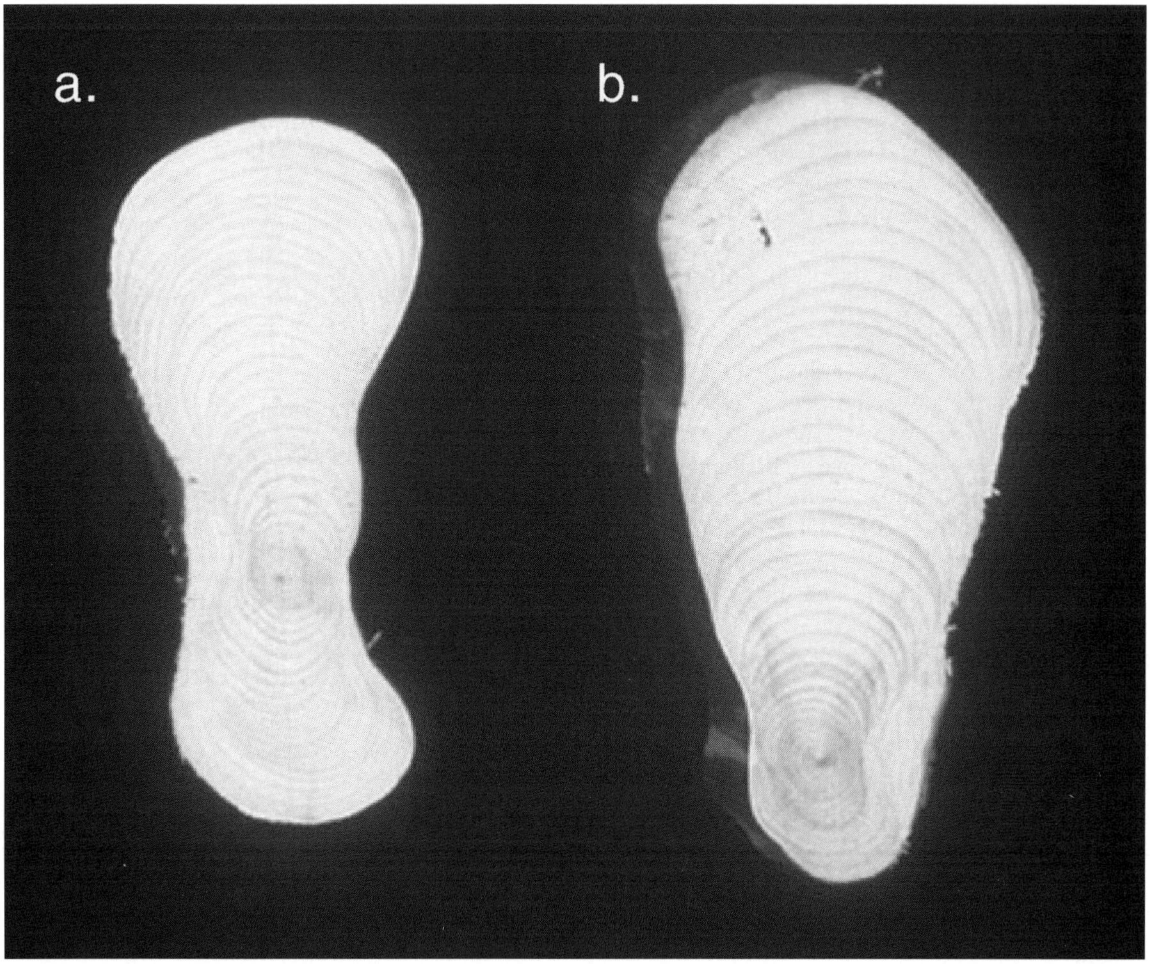

Figure 11. Cross sections of structural roots with a. 'I-beam shape', and b. 'T-beam shape'. The shape in a. is a well-developed example of a Sitka spruce root from the prevailing wind side of the tree; b. is more characteristic of the leeward side.

age, especially in thinned stands. Adaptive secondary root growth caused by mechanical stress appeared to compensate for the irregular arrangement of dominant roots, hence symmetry increased in older trees.

Modelling effects of adaptive growth

These results show that root system symmetry with respect to tree stability should not be considered only in terms of the allocation of biomass between roots and their distribution round the tree. Allocation within each individual root also varies, yielding shapes of significance to mechanical stability. In Figure 5 the mechanical stress is shown acting directly on woody root growth with subsequent effects on differential root size, shape and stiffness. In a root development model, root size and shape should be given so that stiffness can be calculated for the determination of the fulcrum position in the mechanical model (Blackwell et al. 1990).

Conclusions

It is clear that the development of a tree root system is a complex process involving many internal and environmental factors and their interactions. A greatly simplified scheme is shown in Figure 5, which relates the main influences on development to the sequence of events as the root system grows.

Quantitative information on all aspects of this development is poor. We know least about the earliest and latest stages. In particular, what factors determine the sizes and spatial distribution of lateral root primordia in young seedlings, and how much do changes in their environment modify the growth of individual roots and hence the symmetry of the structure in older trees? The quantitative influence of mechanical stress, which acts directly on the cambium, is more difficult to investigate, but becomes increasingly important as the tree grows taller and becomes more subject to wind action. Rather more information is available on the effects of nutrients and water on primary and secondary growth, where increased activity in primary roots appears to enhance secondary thickening.

A number of important features have been omitted from Figure 5 to avoid over-complicating the picture. For example disturbance to the tree at a young age (up to and including planting), and changes to the assimilate supply due to the varying demands of the shoot system, are not shown. Some of the changes which take place during the development of a forest stand, especially those made by management, such as thinning, will affect conditions both above and below-ground. For example, competition with neighbouring trees will alter crown development, wind loading and assimilate production; these changes will have repercussions below-ground in terms of assimilate supply and mechanical stress, together with changes in the competition between root systems for nutrients and water.

One benefit from attempts to model a complicated system is to highlight areas where research is most needed to further our understanding. The most productive areas for investigation in the present context are to improve our basic knowledge on the production and development of root primordia, to obtain quantitative data on the effects of mechanical stress on root diameter growth, shape and stiffness, and to measure the scale of changes in structure as the root systems of more mature trees respond to a changing environment.

References

Blackwell P G, Rennolls K and Coutts M P 1990 A root anchorage model for shallowly rooted Sitka spruce. Forestry 63, 73–91.

Büsgen M, Münch E and Thomson T 1929 The structure and life of forest trees. Chapman and Hall, London. 436 pp.

Coutts M P 1982 Growth of Sitka spruce seedlings with roots divided between soils of unequal matric potential. New Phytol. 92, 49–61.

Coutts M P 1983 Development of the structural root system of Sitka spruce. Forestry 56, 1–16.

Coutts M P 1986 Components of tree stability in Sitka spruce on peaty gley soil. Forestry 59, 173–197.

Coutts M P 1987 Developmental processes in the tree root systems. Can. J. For. Res. 17, 761–767.

Coutts M P and Nicoll B C 1991 Orientation of the lateral roots of trees I. Upward growth of surface roots and deflection near the soil surface. New Phytol. 119, 227–234.

Coutts M P and Philipson J J 1976 The influence of mineral nutrition on the root development of trees. I. The growth of Sitka spruce with divided root systems. J. Exp. Bot. 27, 1102–1111

Coutts M P and Philipson J J 1977 The influence of mineral nutrition on the root development of trees. III. Plasticity of root growth in response to changes in the nutrient environment. J. Exp. Bot. 28, 1071–1075.

Coutts M P and Philipson J J 1979 Mineral nutrition and tree root growth. *In* Mineral nutrition of fruit trees. Eds D Atkinson, J E Jackson, R O Sharples and W M Waller. pp 123–150, Butterworths, London.

Coutts M P, Walker C and Burnand A C 1990 Effects of establishment method on root form of lodgepole pine and Sitka spruce and on the production of adventitious roots. Forestry 63, 143–159.

Deans J D 1981 Dynamics of coarse root production in a young plantation of *Picea sitchensis*. Forestry 54, 139–155.

Deans J D and Ford E D 1983 Modelling root structure and stability. Plant Soil 71, 189–195.

Ennos A R 1995 Development of buttresses in rainforest trees: the influence of mechanical stress. *In* Wind and Trees. Eds M P Coutts and J Grace. pp 293–301. Cambridge University Press, Cambridge.

Fayle D C F 1968 Radial growth in tree roots. Technical report no. 9. Faculty of Forestry, University of Toronto.

Fayle D C F 1983 Differences between stem and root thickening at their junction in red pine. Plant Soil 71, 161–166.

Fritzsche K 1933 Sturmgefahr und Anpassung. Tharanther Forstliches Jahrbuch. 1–94.

Henderson R, Ford E D, Renshaw E and Deans J D 1983a Morphology of the structural root system of Sitka spruce. I. Analysis and quantitative description. Forestry 56, 121–135.

Henderson R , Ford E D and Renshaw E 1983b Morphology of the structural root system of Sitka spruce. II. Computer simulation of rooting patterns. Forestry 56, 137–153.

Hintikka V 1972 Wind-induced root movements in forest trees. Commun. Inst. For. Fenn. 76, 1–56.

Horsley S B and Wilson, B F 1971 Development of the woody portion of the root system of *Betula papyrifera*. Am. J. Bot. 58, 141–147.

Jacobs M R 1953 The effect of wind sway on the form and development of *Pinus radiata* D. Don. Aust. J. For. 2, 35–51.

Johanson S 1941 Tillväxtens lokalisiering hos träden efter huggningar. Skogen, Sweden 16 (28), 229.

Kozlowski T T 1982 Water supply and tree growth. Part II. Flooding. Forestry Abstracts 43, 145–161.

Mattheck C 1993 Design in der Natur: Der Baum als Lehrmeister. Rombach-Verlag, Freiburg.

MacLeod R D and Cram W J 1996 Forces exerted by tree roots. Arboricultural Research and Information Note 134/96/EXT. Arboricultural Advisory and Information Service, Farnham, UK.

Morgan J and Cannell M G R 1994 Shape of tree stems – a re-examination of the uniform stress hypothesis. Tree Physiol. 14, 49–62.

Nicoll B C and Ray D 1996 Adaptive growth of Sitka spruce root systems in response to wind action and site conditions. Tree Physiol. 16, 891–898.

Nicoll B C, Easton E P, Milner A D, Walker C and Coutts M P 1995 Wind stability factors in tree selection: distribution of biomass within root systems of Sitka spruce clones. *In* Wind and Trees. Eds M P Coutts and J Grace. pp 276–292. Cambridge University Press, Cambridge, UK.

Nielsen C C N 1990 Einflüsse von Pflanzenabstand und Stammzahlhaltung auf Wurzelform, Wurzelbiomasse, Verankerung sowie auf die Biomassenverteilung im Hinblick auf die Sturmfestigkeit der Fichte. University of Göttingen Faculty of Forestry and Niedersächsischen Forstlichen Versuchsanstalt, Thesis 100, J D Sauerländer's Verlag, Frankfurt am Main. p 279.

Nielsen C C N 1991 Zur verankerungsökologie der Fichte. Forst und Holz 46, 178–182.

Quine C P, Bernand A C, Coutts M P and Reynard B R 1991 Effects of mounds and stumps on the root architecture of Sitka spruce on a peaty gley restocking site. Forestry 64, 385–401.

Ray D and Nicoll B C 1998 The effect of soil water-table depth on root-plate development and stability of Sitka spruce. Forestry 71, 169–182.

Reynolds E R C 1987 Development of the root crown of some conifers. Plant Soil 98, 397–405.

Rigg G B and Harrar E S 1931 The root systems of trees growing in sphagnum. Am. J. Bot. 18, 391–397.

Robertson A 1991 Control of wood density, bole eccentricity, and tree ring width in relation to vector winds in wave forest. Can. J. For. Res. 21, 73–82.

Stokes A, Fitter A H and Coutts M P 1995 Responses of young trees to wind and shading: effects on root architecture. J. Expl. Bot. 46, 1139–1146.

Stokes A, Nicoll B C, Coutts M P and Fitter A H 1997 Responses of young Sitka spruce clones to mechanical perturbation and nutrition: effects on biomass allocation, root development and resistance to bending. Can. J. For. Res. 27, 1049–1057.

Taylor H M and Gardner H R 1960 Relative penetrating ability of different plant roots. Agron. Journal. 52, 579–581.

Telewski F W 1995 Wind-induced physiological and developmental responses in trees. *In* Wind and Trees. Eds M P Coutts and J Grace. pp 237–263. Cambridge University Press, Cambridge.

Urban S T, Lieffers V J and MacDonald S E 1994. Release in radial growth in the trunk and structural roots of white spruce as measured by dendrochronology. Can. J. For. Res. 24, 1550–1556.

Wilcox H E 1968 Morphological studies of root of red pine, *Pinus resinosa* I. Growth characteristics and patterns of branching. Am. J. Bot. 55, 247–254.

Wilson B F 1975 Distribution of secondary thickening in tree root systems. *In*: The Development and Function of Roots. Eds J G Torrey and D T Clarkson. pp 197–219. Academic Press, New York.

Wood C J 1995 Understanding wind forces on trees. *In* Wind and Trees. Eds M P Coutts and J Grace. pp 133–164. Cambridge University Press, Cambridge.

A. Stokes (ed.), *The Supporting Roots of Trees and Woody Plants: Form, Function and Physiology*, 19–29.

Strain distribution during anchorage failure of *Pinus pinaster* Ait. at different ages and tree growth response to wind-induced root movement

Alexia Stokes*
Laboratoire de Rhéologie du Bois de Bordeaux, CNRS/INRA/Université Bordeaux I, BP 10, Domaine de l'Hermitage, Cestas Gazinet 33610, France

Key words: adaptive growth, anchorage, maturation strain, mechanical stress, scaling, wood formation

Abstract

Winching tests were carried out on 5- 13- and 17-year-old tap rooted Maritime pine (*Pinus pinaster* Ait.) in order to determine how the mode of anchorage failure changes throughout the life of a tree. As trees were pulled sideways, strain along the lateral roots was recorded using strain gauges attached to a strain indicator. Measurements of strain in the root system, taken during winching, provide information about root movement when loaded by wind. The mode of mechanical failure appeared to depend on tree age. The youngest trees bent over completely during winching, but did not break due to the plasticity of their trunks. The 13-year-old trees either broke at the base of the tree (due to the presence of grafting scar tissue) or at the base of the tap-root. The oldest trees broke at the base of the tap-root and sounds of roots breaking were also heard. Strain was twice as great in the trunk compared to the roots in the 5- and 13-year-old trees and was three times greater in the compression roots of 17-year–old trees compared to that in the trunk. In one 17-year-old tree, strain was found to increase at a distance of 35 cm in tension roots before decreasing again. Although the mode of failure changed with tree size, anchorage strength increased proportionally with the third power of trunk diameter, therefore another reason why failure differs with tree age must exist. In order to determine if different types of wood were being laid down in the lateral roots in response to wind loading, maturation strains, indicating the existence of mechanical stress in developing wood cells, were measured at different points along the roots. A high correlation was found between maturation strain and strain measured during winching, in roots that lay in the wind direction only. Therefore, trees appear to be able to respond to external loading stress, even at a local level within a root.

Introduction

Increasing damage to forest and urban trees due to the frequency of catastrophic storms in recent years (Quine, 1995), has led to an interest in research into tree stability and anchorage mechanics (Coutts, 1983, 1986; Crook and Ennos, 1996; Ennos, 1993; Peltola and Kellomäki, 1993). Different mechanisms of anchorage failure have been identified, depending on root morphology and architecture (Crook et al., 1997; Ennos, 1989; Stokes et al., 1999). In order to resist mechanical loading by, e.g., the wind, trees must transfer external forces down the roots into the soil, as directly as possible. Root systems must therefore be large enough to transfer these forces into the ground and rigid enough to prevent rupture in highly stressed areas.

In young trees, a vertical tap root and many lateral roots are usually present, whereas in older trees, the lateral roots become relatively larger and the tap root may cease to grow (Köstler et al., 1968). In some species, e.g. *Pinus*, the tap-root continues growing with the rest of the tree and is an important component of anchorage. The tap-root anchors the tree centrally and the horizontal lateral roots act like guy ropes (Ennos, 1993). However, in the south-west region of France, originally a swampy area until the mid-1900s, when

* FAX No: +33 5 56 68 07 13.
E-mail: stokes@lrbb3.pierroton.inra.fr
Plant and Soil is the original source of publication of this article. It is recommended that this article is cited as: *Plant and Soil* **217:** 17–27, 1999.

it was forested with Maritime pine, the shallow sandy soil is subject to winter water-logging. In many parts of the region, a hard-pan can be found at a depth of 0.5–1.0 m. Both these factors hinder vertical root growth and contribute to the formation of a shallow root system, which has a large, but very short tap-root. Such types of root system are poorly anchored in the soil, especially in wet winters, when soil cohesive properties are reduced. Therefore, as the tree ages, or due to local environmental factors, the way in which it is anchored into the soil may alter, as the root system changes from a vertically anchored tap-root system into a less stable 'plate' root system.

In plate root systems, the most important component in resisting uprooting is that of the 'windward' roots which are pulled upwards during overturning (Coutts, 1983, 1986). Tensile and shearing forces are then present in the windward part of the root-system and must be transferred to the soil. Compressive forces are transferred to 'leeward' roots which are pushed downwards onto the load-bearing surface of the soil, and the point at which the root-soil plate is levered out of the ground (the point of rotation) occurs close to the stem. Lateral roots of such types of root systems will be very close to the soil surface and hence experience greater bending stresses than deeper growing roots which are stress protected by the soil.

Resistance to stem breakage seems to increase as a function of dbh^3 (dbh = diameter at breast height) whereas resistance to uprooting has been found to be a function of height $\times$ dbh^2 (Crook and Ennos, 1997; Gardiner et al., 1997; Nykänen et al., 1997). Therefore, these results suggest that the anchorage systems of larger trees become directly weaker compared with the structural strength of the trunk. If this is the case, the anchorage moment of Maritime pine, a tap-rooted tree, which grows in shallow soils, should also scale to the second power of tree trunk diameter.

The exact way in which roots move during wind sway has begun to be studied using strain gauges (Ennos, 1995). Strain gauges convert longitudinal deformations of a metal element into an electrical signal, thus indicating how a material is deformed under loading. Using such gauges, Crook et al. (1997) determined the mechanical behaviour of buttressed and non-buttressed trees during anchorage failure. Watson (1995, 1999) found that in *Pinus radiata*, leeward roots appear to be less sensitive than windward roots to changes in stem displacement. However, in these studies, strain gauges were not placed at regular intervals along the lateral roots, thus information about how roots move during wind loading is necessary for a better understanding of anchorage failure.

Tree roots are known to be very sensitive to mechanical loading, and respond to external stresses by laying down wood in areas susceptible to rupture (Mattheck, 1995; Nicoll and Ray, 1996; Stokes and Guitard, 1997). Mechanical properties of roots have been investigated and linked to the behaviour of roots under wind loading (Stokes and Mattheck, 1996; Stokes et al., 1997). However, little is known about the relationship between root movement and consequences for wood formation along the root. As wood is laid down within the tree, the new layer shrinks longitudinally during maturation. Lignin is deposited and microfibrils shorten. However, if the tree is leaning in one direction, reaction wood forms (Timell, 1986). During the maturation of compression wood in gymnosperms, the new wood layers expand longitudinally (Archer, 1986). However, these maturing cells cannot deform fully, as they are attached to older cells. Hence, these maturing cells exist in a state of mechanical stress. It is only when the wood is cut, that these 'maturation' stresses are released in the form of residual strains along the longitudinal axis. Longitudinal residual maturation strains (LRMS), indicating the existing mechanical stress in developing wood cells, have been extensively studied in the aerial parts of trees and have been implicated in the reorientation process of the trunk (Archer, 1986; Fournier et al., 1994). LRMS are also closely related to variations in elastic modulus, basic density, shrinkage and fibre length (Archer, 1986). Measurements can be made non-destructively on standing trees, therefore they can be good indicators of the formation of different types of wood in a growing tree. Until recently, information about maturation strains in tree root systems has been relatively scarce. A study by Gartner (1997) showed that LRMS are higher in stems than in roots, as stems experience greater stresses than root systems. Stokes et al. (1998) found that in Maritime pine, LRMS were positive in leeward roots and negative in windward roots, thus suggesting that wind loading can influence root-wood formation.

In order to determine how the mechanism of failure changes and how roots move during wind sway at different stages of a tree's life, winching tests were carried out on three age groups of Maritime pine. Strain distribution in the root system was measured using strain gauges and trees were winched over until anchorage failure occurred. Longitudinal, residual, maturation strains were then measured in the trunk and

Table 1. Tree DBH and height, distance between experimental and anchoring tree, height of winch cable and position of strain gauges along the lateral roots of Maritime pine trees of different ages. (Means ± standard error)

Tree age (years)	*n*	DBH (cm)	Tree height (m)	Distance of experimental tree from anchoring tree (m)	Height at which cable was attached in the experimental tree (m)	Distance of gauges along lateral roots from stem centre (m)
5	6	6.4 ± 0.5	3.2 ± 0.4	7.6 ± 0.7	2.0	0.10, 0.20, 0.30
13	10	15.2 ± 0.5	10.5 ± 0.4	13.9 ± 0.6	4.0	0.10, 0.30, 0.45, 0.65
17	3	24.3 ± 1.9	14.6 ± 0.3	16.7 ± 0.6	5.0	0.25, 0.35, 0.45

roots and related to root movement during winching and tree sway during wind loading.

Materials and methods

Study site and experimental layout

Maritime pine trees of different ages growing in a plantation in the Forêt de l'Hermitage, Gironde, south-west France (latitude 44° 44′ N, longitude 0° 46′ W, 58 m a.s.l. at 45 km from the Atlantic Ocean) were examined. The plantation was subjected to prevailing winds from the north-west. The mean windspeed in the region is 3.3 m s^{-1}, with maximum speeds reaching 33.0 m s^{-1} (Météo-France). A random selection of six 5-year-old, ten 13-year-old and three 17-year-old Maritime pine trees was made, with a mean diameter breast height (DBH) of 6.4 ± 0.5 cm, 15.16 ± 0.5 cm and 24.3 ± 1.87 cm respectively (Table 1). The trees were growing as a monoculture with 1 × 4 m between 5- and 13-year–old trees and 2 × 4 m between 17-year-old trees. Seven of the 13-year-old trees were stock which had been grafted when young, which may influence results. The scar from the graft could be seen at a height of 75 cm up the stem. The remaining three trees had not been grafted.

The trees were growing in a medium humid sandy podzol. A hard pan existed at a depth of 0.5–1.0 m into which vertically growing roots could not penetrate. The water table may rise close to ground level in very wet winters and fall to about 1.5m in mid-summer (Danjon et al., 1999). Soil samples were taken from the soil around each tree immediately after the winching tests had been carried out in order to determine soil water content. The samples were weighed when fresh and then oven-dried for three days at 80 °C and dry weight measured.

Winching tests

A hand winch was attached to the base of an anchoring tree at a distance of several metres from the experimental tree (Table 1). The winch cable was attached to the experimental tree in the crown at a height of a few metres (Table 1), rather than at the base of the tree so that strain in the tree during winching would imitate that under wind loading as best possible. The tree stem was protected with a rubber sling at the point of attachment to avoid damage by the metal cable during winching. A force transducer (sensitive to 0.05 kN) was attached between the winch and the winch cable and could measure forces applied to the tree up to 10 kN. The compass direction in which the tree was pulled was noted.

Two lateral roots per tree were excavated, one in the direction of the winch pull (roots in compression) and the second counter to the winch pull (roots in tension). In order to measure movement in these roots during mechanical loading, plastic backed strain gauges (Kyowa, Japan, KFG-10-120-C1-11, 10 mm gauge length, 120 ohm resistance) were used to estimate longitudinal strains. Strain gauges convert longitudinal deformations of a metal element into an electrical signal and must be connected to a strain indicator (Kyowa, Japan, SD-10) in quarter bridge mode, via a switch and balance unit (Vishay Measurements Group, North Carolina, U.S.A., SB-10), in order for the electrical signal to be converted into micro-deformations (μstrain). Bark was removed with a chisel at DBH and at three or four intervals along the length of a root, starting from the stem-root joint (Table 1). Care was taken not to damage the surface fibres of the wood during removal of the bark. Strain gauges were glued to the wood where the bark had been removed, using Loctite 401 multi-usage glue, which took approximately 15 minutes to dry. In the 17-year-old trees, strain gauges were also glued onto the underside of the lateral roots, at intervals corresponding to those on the upper side. A small hand mirror was used to help position the strain gauge longitudinally on the lower surface of the root.

The initial values of each strain gauge were recorded before winching commenced. The trees were then pulled sideways using increments of force of 0.05 kN for the 5-year-old trees, 0.25 kN for the 13-year-old trees and 0.5 kN for the 17-year-old trees. Strain was measured in each of the gauges after each increment of force had been applied. The tree was winched sideways until the trunk was plastically deformed or failure occurred.

The lateral roots were removed from each 5- and 17-year-old tree (mean diameter at root base was 2.41 ± 0.1 cm for the 5-year-old trees and 5.82 ± 1.77 cm for the 17-year-old trees). The azimuth (orientation from north), depth and diameter (horizontal and vertical) of each root were noted. After all tests had been carried out on the roots, cross-sections were cut at each of the sampling sites and the horizontal and vertical radii measured to give an indication of root growth and eccentricity (Stokes et al., 1998).

Measurement of longitudinal residual maturation strains (LRMS)

In order to determine if there are differences in wood type around the stem and between the first order lateral roots growing in different directions around the trunk, measurements of LRMS were made. Measurements were made on 5- and 17-year-old trees only, as it was found that after an analysis of LRMS in the roots of Maritime pine (Stokes et al., 1997, 1998) mechanical behaviour may influence wood formation (the 13-year-old trees had already been harvested). Measurements were made on the stem at DBH in the field and on roots at the laboratory. Roots were clamped in a vice at their base, and where each strain gauge had been attached to the stem or root, two 5 mm deep grooves were cut radially into the wood, 2 mm at each side of the gauge. The deformation of the wood underneath the gauge after the two cuts were made was recorded on the strain indicator. If the LRMS are positive, this means that cells expanded length wise on cutting the wood as happens with wood held in compression. If strains are negative, this means that the cells shrank when the stress was released and that the wood had developed under tensile stress.

Table 2. Water content in the soil samples and bending moment (M_B) around the base of the tree required to cause permanent stem displacement or tree failure. Mean longitudinal residual maturation strain values are given for all lateral roots tested. (Means ± standard error)

Tree age years	Soil water content (%)	Maximum M_B around the tree base (kNm)	Mean LRMS (μstrain) in all roots
5	13.0 ± 1.7	0.47 ± 0.03	61.8 ± 22.8
13	15.2 ± 2.2	6.55 ± 0.56	–
17	20.3 ± 4.5	12.75 ± 0.75	23.8 ± 77.9

Results

Winching tests

Soil water content was found to be similar on all days of the winching tests (Table 2) and therefore differences in soil shear strength were not considered to influence tree anchorage.

Mechanical failure by stem breakage or uprooting of the 5-year-old trees did not occur due to the plasticity of the stem, which remained irreversibly deformed after a mean bending moment (M_B) of 0.47 kNm had been reached (Table 2). In the 13-year-old trees, the seven grafted stems broke at a height of 50–100 cm and the remaining stems at the junction between the tap-root and the stem. Failure usually occurred first on the compression side of the stem, before then failing in tension (Figures 1, 2). After damage had occurred in compression, strain increased positively meaning that the tree could offer no further resistance to winching (Figure 1). Although the stem had been damaged, strain was still found to increase in the roots, as they were slowly being pulled out of the ground (Figures 1, 2). At a distance of 0.65 m along the compression root, strain was found to be positive, thus indicating that the root at this point was held in tension (Figure 1). This distance must therefore lie beyond the root hinge (Coutts, 1983, 1986), and the point at which the root is pushed into the ground. A mean M_B of 6.55 kNm, was needed to cause mechanical failure of the 13-year-old trees (Table 2). The 17-year-old trees were not completely pulled over for fear of felling neighbouring trees. However, noises of roots breaking were heard and failure began to occur at the base of the tap-root and lateral roots after a mean M_B of 12.75 kNm (Table 2) had been reached.

A regression of M_B against DBH of all trees showed that a significant relationship existed ($Y = 0.738 \times -4.44$, $P = 0.002$, $R^2 = 0.83$), therefore M_B

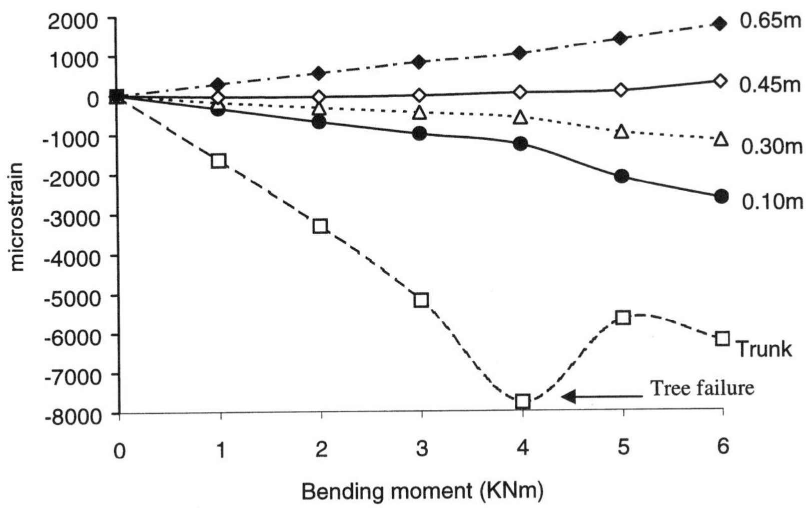

Figure 1. Strain in the compression side of the trunk and roots in 13-year-old Maritime pine during winching. Greatest strain occurs in the trunk, and even after failure occurs, deformation in the roots continues to increase.

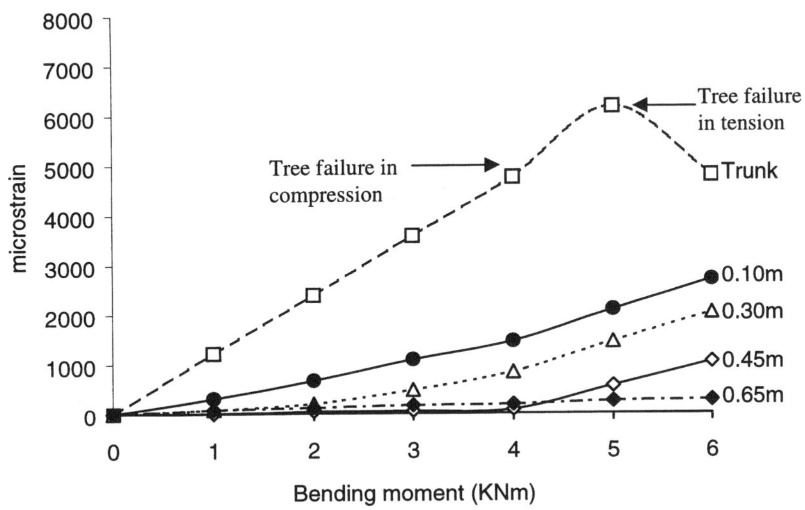

Figure 2. Strain in the tension side of the trunk and roots in 13-year-old Maritime pine during winching. Failure in the trunk occurs after that on the compression side of the tree.

is correlated with tree size. In order to determine how M_B scales with DBH, log M_B was regressed against log DBH which results in the following relationship:

$$\text{Log Bending Moment (kNm)} = 2.87 \log \text{DBH} - 5.52$$

Hence:

$$\text{Bending Moment around tree base (kNm)} = 4\times10^{-4}\,(\text{DBH})^{2.87}$$

Therefore, in these trees, the bending moment around the base of the tree scales to almost the third power of tree trunk diameter.

The pattern of strain within the root systems of Maritime pine varied with tree age. In the youngest trees, the greatest strain was found to be on the tension side of the trunk, and at the bases of the lateral roots. Strain then diminished along the lateral roots and was insignificant at a distance of 30 cm from the trunk in the tension roots. However, in the roots which were held in compression during winching, the strains were negative up to a distance of 30 cm from the stem

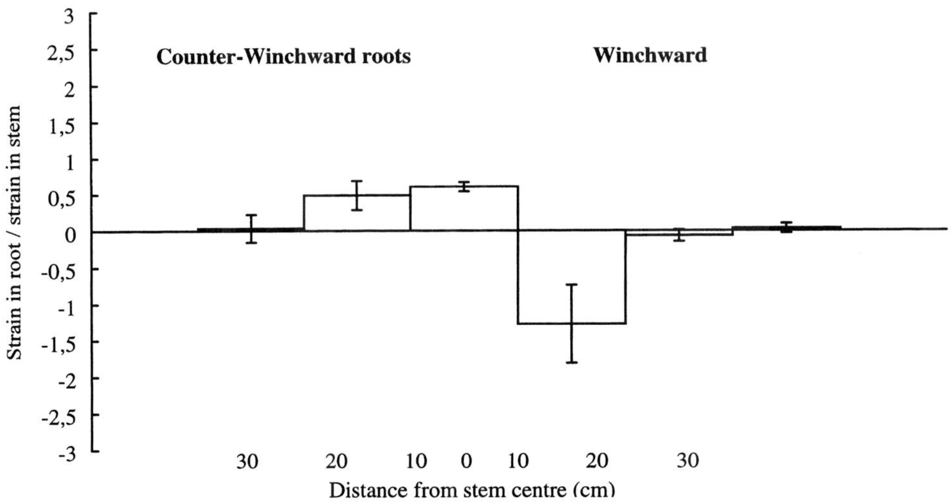

Figure 3. Mean strain is greater in the trunk than in the lateral roots of 5-year-old Maritime pine, and decreases along the roots. Each bar shows the mean strain in the root/mean strain in the stem at intervals along the root. Bars are ± standard error.

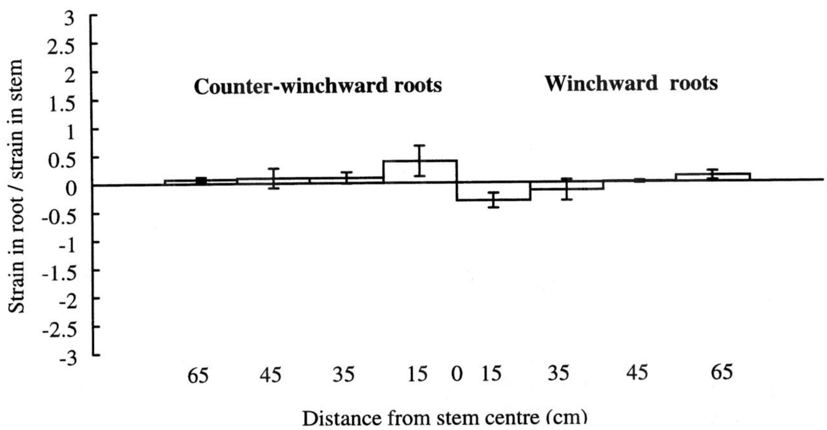

Figure 4. Mean strain is greater in the trunk than in the lateral roots of 13-year-old Maritime pine, and decreases along the roots. Each bar shows the mean strain in the root/mean strain in the stem at intervals along the root. Bars are ± standard error.

centre, after which the sign changed and the roots were then held in tension (Figure 3). In 13-year-old trees (Figure 4), less than half the strain was noted in the lateral roots compared to that in the trunk, but the pattern of decreasing strain with increasing distance from the stem was similar to that observed in 5-year-old trees (Figure 3).

Strain was found to be nearly three times greater at the surface of the base of compression lateral roots of 17-year-old trees, compared to that observed in the stem on the same side of the tree. The strain then rapidly diminished along the roots, becoming positive at a distance of 45 cm along the root (Figure 5). However, in one tree, (Figure 6), strain did not follow the same pattern as in the rest of the trees. Instead of decreasing along the tension roots, strain increased at a distance of 35 cm before decreasing again. This pattern was not repeated on the lower surface of the roots, where strain was maximal nearer the trunk. However, in compression roots, strain was greatest nearer the trunk on the upper surface, but on the lower surface, was found to be maximal again at a distance of 35 cm before decreasing again. In all the trees, strain on the underside of the roots was 1–1.5× greater than that in the stem up to a distance of 35 cm along the root, before becoming negligible. Approximately the same amount of strain was present on both surfaces of the tension roots as was found in the stem on the tension side of the tree. In general, strain was not found to diminish along the tension roots, as in the younger trees, but as

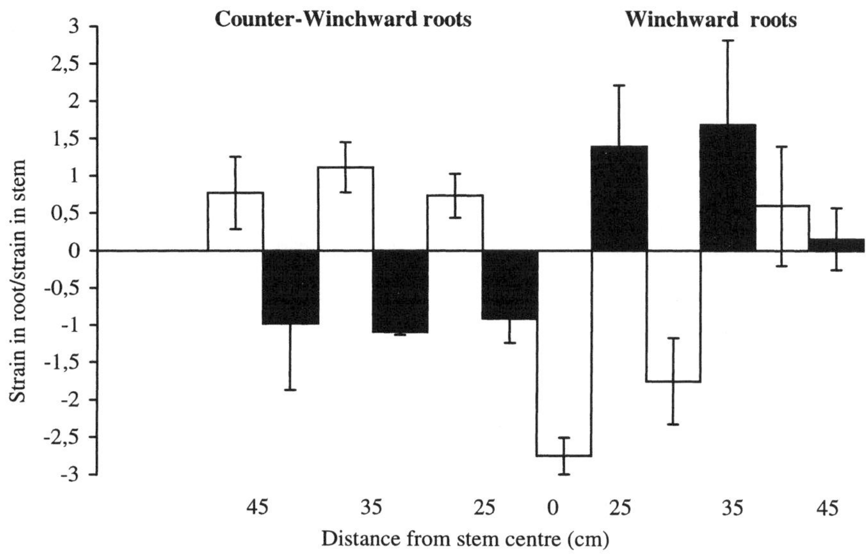

Figure 5. Mean strain is greater in the lateral roots of 17-year-old Maritime pine on both the upper (white bars) and lower (black bars) surfaces of the roots, compared to that in the trunk. Each bar shows the mean strain in the root/mean strain in the stem at intervals along the root. Bars are ± standard error.

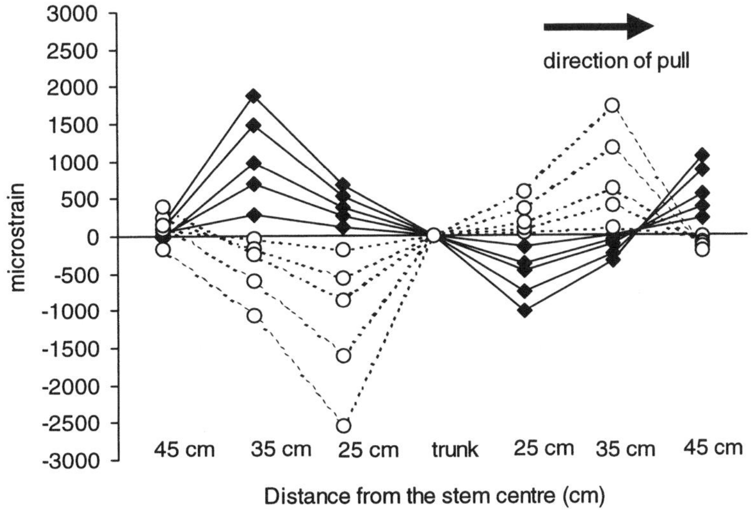

Figure 6. Strain distribution increases along the tension roots before decreasing again in one 17-year-old Maritime pine. Curves are strain at each interval along the root, for each increment of force of 0.5 kN applied during winching. At a distance of >35 cm from the stem centre, buckling in the roots can be seen to occur, due to the change in sign.

the gauges were only placed up to a distance of 45 cm, the decrease in strain probably occurred beyond this point and was not recorded (Figure 5).

Longitudinal residual maturation strains (LRMS)

LRMS in the trunk were highly variable and were not found to correlate with wind direction. A mean value of +40.2 ± 49.3 μstrain was found in the 5-year-old trees and -167 ± 881 μstrain in the 17-year-old trees. However, LRMS were significantly greater and positive in the sector underneath the trunk if it was leaning, than in the sector above the direction of lean. A complete analysis of results is presented in Stokes et al. (1998).

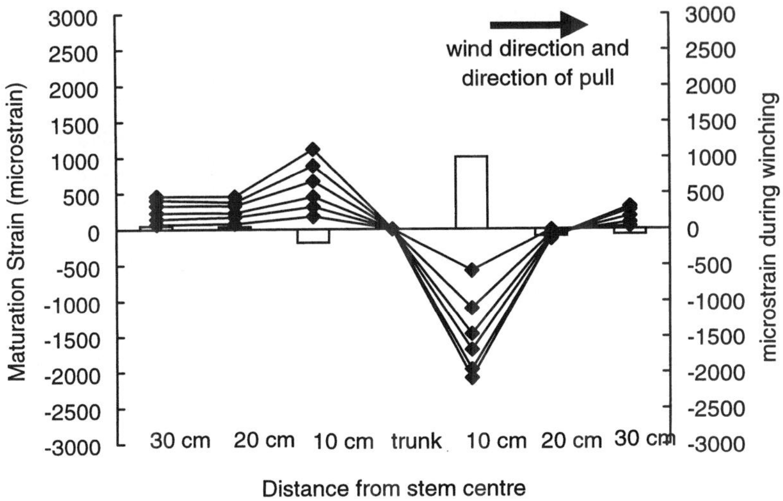

Figure 7. Longitudinal residual maturation strains (LRMS) in the lateral roots of one 5-year-old Maritime pine. The tree was pulled over in the same direction as the prevailing wind. LRMS were greatest in leeward roots close to the trunk. Curves are strain at each interval along the root, for each increment of force of 0.05 kN applied during winching.

In the root systems, strains varied both along a root and between roots, and regressions of root length, diameter and eccentricity were not significant, therefore variation was not due to root morphology. Initially, no correlation was found between the mean LRMS and the strains measured in roots during winching. However, when each individual tree was examined, taking into account the wind direction, it was found that the pattern of maturation strain was similar to the distribution of strain along the roots as the tree was being winched over, regardless of age (Figures 6, 7). Highest values of LRMS corresponded to areas of greatest strain during overturning, which would be similar to strains in the tree during wind sway. A significant positive increase in mean strain in leeward roots was also found in 5-year-old trees (Stokes et al., 1997, 1998).

Conclusion

Apart from the youngest trees which did not fail mechanically by stem breakage or uprooting, due to the plasticity of their trunks, trees failed first in compression and then in tension. Wood is more resistant to rupture in tension than in compression, therefore a wooden beam, analogous to a tree trunk, will always fail first in compression (Guitard, 1987). The seven 13-year-old trees which had been grafted when young, broke at a height of 50–100 cm which corresponded approximately to the height of the grafting scar. If these trees had not been grafted, thereby causing an area of mechanical weakness in the stem, they may have failed further down the stem, at the base of the tap-root, as in the three remaining trees which had not been grafted, or the 17-year-old trees. Very little strain was measured in the root systems of the 13-year-old trees compared to that observed in the trunk, which may be due to the presence of the weak grafted zone. As a tree is winched over, the direction of force is longitudinal and downwards. In the grafted area, scar tissue will exist, with fibres running both radially and longitudinally, therefore, there will be shearing stresses acting in different directions. The wood strength in some of these directions will not be as strong as that away from the scar, thus making this area more likely to fail mechanically. As the amount of strain is related to the zone of rupture (Stokes et al., 1999), strain will be greater in the stem than in the roots.

Older trees ruptured either in the roots or at the base of the tap-root, where it was joined to the trunk. According to Crook and Ennos (1997), the type of mechanical failure seen in these trees is due to the fact that as the tree trunk increases in size, the tap-root does not, possibly due to local environmental conditions. Therefore, young trees fail mechanically in their trunks and older trees in their root system. Hence, the anchorage strength of such trees becomes relatively weaker compared with the structural strength of the trunk (Gardiner et al., 1997).

However, unlike the tropical tap-rooted species that Crook and Ennos (1997) studied, the anchorage moment of Maritime pine scaled to the third power of trunk diameter, not the second power. However, if vertical root growth was affected by local soil conditions, the horizontal lateral roots may have increased in size in order to compensate for the lack of tap-root, thus affecting tree anchorage. An analysis of both above- and below-ground architecture would need to be made in order to provide further data concerning the scaling of anchorage moment with tree size. The point of winch-cable attachment in the trunk, as a function of tree height, also differed between different age classes, which may influence strain distribution and site of failure. Therefore, it would be necessary to verify these results by repeating the experiment on trees of different sizes, with the winch-cable at the same proportion of stem height for all trees. A further reason for the difference in scaling between different species studied, may be due to local soil conditions. Fraser and Gardiner (1967) found that the anchorage moment of *Picea sitchensis* for trees of different sizes on brown earth and peaty gley differed significantly. Trees growing on brown earth were much more resistant to overturning than those on peaty gley. These relationships differed depending on soil depth. Therefore, when considering the stability of a forest stand, it is necessary to take into consideration many different factors before being able to calculate wind stability factors.

Strains in the root system in young trees during overturning appear to be greatest at the base of the stem-root joint before decreasing along the root, as suggested by Ennos (1989) and Crook and Ennos (1996). Compression roots did not appear to be less sensitive to stem movement than tension roots, as suggested by Watson (1999). However, the strain gauges in Watson's studies were placed at a distance of 1.0–20 m along the lateral roots, where strain distribution will be different. The young trees were well-anchored into the ground by their tap-roots. As each tree was pulled over, it rotated about a point on the leeward side of the tree, possibly in the tap-root. The compression roots were pushed down into the soil before buckling, indicated by the change in sign of the strain gauge at a distance of >30 cm from the trunk. The tap-root pushed into the soil on the compression side of the tree, thus stretching the tension roots which were well-anchored into the soil at their extremities. These roots were thus held in tension and resisted being pulled up. In the 17-year-old trees, a similar pattern of strain distribution was seen, except in one tree, where strain was found to increase along the tension roots before decreasing again. This pattern was reversed on the lower surface of the roots, where strain was highest at 35 cm along the compression roots. Similar patterns of strain distribution have also been found in *Picea abies* (Stokes et al., 1999). As trees age, or in unfavourable environments, root systems develop laterally rather than vertically, leading to the development of plate-root systems, e.g. *Picea* sp. (Köstler et al., 1968). In such root systems of 34-year-old *Picea sitchensis*, it has also been observed that soil movement is greatest at a distance of approximately 50 cm along the tension (or windward) roots (Coutts, 1983, 1986; Hintikka, 1972; Stokes et al., 1999). As these trees were pulled over, they would have rotated about a point on the leeward side of the trunk. The compression roots would then be pushed down into the soil close to the trunk. Further along the root, buckling occurred, as in the younger trees. Roots growing away from the direction of pull would be held in tension, but as the tap-root was relatively smaller (or non-existent in the case of spruce) in these trees, the tension roots would be freer to move through the soil and would be subject to buckling at a short distance from the trunk.

It has been shown in some tree species, e.g. *Pinus sylvestris*, that both compression and bending strength decrease along the root, whereas in trees with plate root systems, e.g. *Picea* and *Populus*, strength increases at a distance of 0.5-1 m along the root before decreasing again (Stokes and Mattheck, 1996). Nicoll and Ray (1996) found that in leeward roots of 46-year-old *Picea sitchensis*, radial growth was more pronounced on the upper side of the biological centre, which increases the rigidity of the root in compression. However, on the windward side of the tree, roots developed eccentric cross-sectional shapes, comparable to 'I-beams' which would better resist vertical flexing. This I-beam shape was found to be most pronounced at a distance of 0.75 m from the stem. Therefore, if strain is maximal at a certain distance from the trunk in windward roots, adaptive growth may occur in this region to better resist possible rupture in the root.

In compression roots which lay on the lee side of the tree the LRMS were positive, i e. the cells expanded length wise on cutting the wood. This observation was more pronounced close to the trunk, where strains during winching were greatest. In tension roots on the windward side of the tree, LRMS were generally found to be negative, i.e. the cells shrunk when the stress was released. Negative LRMS indicate the presence of normal wood and suggest that the wood in these roots developed initially under tensile stresses, as confirmed by the strain values from the winching tests. Positive LRMS are usually associated with the formation of compression wood (Timell, 1986); however, compression wood in roots can only form if the root is exposed to sunlight (Fayle, 1968). All the roots examined were buried well in the soil, and compression wood had not been observed in any of the cross-sections (Stokes et al., 1998). Therefore, another kind of abnormal wood may have formed under wind loading, which may have similar properties to compression wood, such as increased cell density and percentage of lignin present. A study of the cellular characteristics along these roots would be needed to determine exactly how wind sway influences wood formation.

It appears that the mode of failure during mechanical loading changes throughout the life of a tree, and may depend on root system architecture. Further work is required on the relationship between root system architecture and anchorage strength in order to understand better how trees are anchored. It would be of value to measure strain during natural wind loading, and compare it to strain data obtained by winching. Trees appear to be able to grow adaptively to strain due to external stresses, even at a local level within the root system. Therefore, if a tree can respond to wind stress, foresters should use this knowledge when deciding where and how to plant, and more importantly, when to thin or harvest their plantation. The best time to thin or harvest would be at the beginning of the growing season, rather than just before winter, when growth ceases and trees cannot respond to the new wind environment.

Acknowledgements

The author is indebted to S. Berthier, G. Dumont, L. Lagunegrand, F. Martin and S. Sacriste for their assistance with fieldwork. This work was carried out with aid from the Institut National de Recherche Agronomique (INRA).

References

Archer R A 1986 Growth Stresses and Strains in Trees. Springer Series in Wood Science, Springer-Verlag, Berlin, Heidelberg.

Coutts M P 1983 Root architecture and tree stability. Plant Soil 171–88.

Coutts M P 1986 Components of tree stability in Sitka spruce on peaty, gley soil. Forestry 59, 173–97.

Crook M J and Ennos A R 1996 The anchorage mechanics of deep-rooted larch, *Larix europea* × *L. japonica*. J. Exp. Bot 47, 1509–1517.

Crook M J and Ennos A R 1997 The increase in anchorage with tree size of the tropical tap rooted tree *Mallotus wrayi*, King (Euphorbiaceae). *In* Plant Biomechanics. Eds. G Jeronimidis and J F V Vincent. pp 31–36. Centre for Biomimetics, Reading, UK.

Crook M J, Ennos A R and Banks J R 1997 The function of buttress roots: a comparative study of the anchorage systems of buttressed (*Aglaia* and *Nephelium ramboutan* species) and non-buttressed (*Mallotus wrayi*) tropical trees. J. Exp. Bot. 48, 1703–1716.

Danjon F, Bert D, Godin C and Trichet P 1999 Structural root architecture of 5-year-old *Pinus pinaster* measured by 3D digitising and analysed with AMAPmod. Plant Soil 217, 49–63.

Ennos A R 1989 The mechanics of anchorage in seedlings of sunflowers *Helianthus anuus* L. New Phytologist 113, 185–92.

Ennos A R 1993 The scaling of root anchorage. J. Theor. Biol. 161, 61–75.

Ennos A R 1995 Development of buttresses in rain forest trees: the influence of mechanical stress. *In* Wind and Trees. Eds M P Coutts and J Grace. pp 293–301. Cambridge University Press, UK.

Fayle D C F 1968 Radial growth in tree roots. Technical report no. 9. Faculty of Forestry, University of Toronto, Canada.

Fournier M, Bailleres H and Chanson B 1994 Tree biomechanics: growth, cumulative prestresses and reorientations. Biomimetics 2, 229-51.

Fraser A I and Gardiner J B H 1967 Rooting and stability in Sitka spruce. Forestry Commission Bulletin 40. HMSO, London.

Gardiner B A, Stacey G R, Belcher R E and Wood C J 1997 Field and wind tunnel assessments of the implications of respacing and thinning for tree stability. Forestry 70, 233–252.

Gartner B L 1997 Trees have higher longitudinal growth strains in their stems than in their roots. Int. J. Pl. Sci. 1584, 418–423.

Guitard D G 1987 Mécanique du Materiau Bois et Composites. Cepadues Editions, France.

Hintikka V 1972 Wind-induced movements in forest trees. Metsäntutkimuslitoksen Julkaisuja 76, 1–56.

Köstler J N, Bruckner E and Bibelriether H 1968 Die Wurzeln der Waldbäume. Verlag Paul Parey, Hamburg & Berlin.

Mattheck C and Breloer H 1995 The body language of trees. A handbook for failure analysis. HMSO, London.

Nicoll B C and Ray D 1996 Adaptive growth of tree root systems in response to wind action and site conditions. Tree Physiology 16, 899–904.

Nykänen M-L, Peltola H, Quine C P, Kellomäki S and Broadgate M 1997 Factors affecting snow damage of trees with particular reference to European conditions. Silva Fennica 31, 193–213.

Peltola H and Kellomäki S 1993 A mechanistic model for calculating windthrow and stem breakage of Scots pines at stand edge. Silva Fennica 27, 99–111.

Quine C P 1995 Assessing the risk of wind damage to forests: practice and pitfalls. *In* Wind and Trees. Eds M P Coutts and J Grace. pp 379–403. Cambridge University Press, UK.

Stokes A, Berthier S, Sacriste S and Martin F 1998 Variations in maturation strains and root shape in root systems of Maritime pine (*Pinus pinaster* Ait.). Trees 12, 334–839.

Stokes A, Drexhage M and Guitard D G 1999 A method for predicting the site of failure in trees under mechanical loading. *In* The Supporting Roots of Trees and Woody Plants: Form, Function and Physiology. Ed. A Stokes. Developments in Plant and Soil Sciences, Kluwer Academic Publishers. (*In press*).

Stokes A and Guitard D G 1997 Tree root response to mechanical stress. *In* The Biology of Root Formation and Development. Eds A Altman and Y Waisel. pp 227–236. Plenum Publishing, NY.

Stokes A, Martin F, Sacriste S and Fourcaud T 1997 Adaptation of tree roots to wind loading: the relationship between mechanical behaviour and wood formation. *In* Plant Biomechanics. Eds G Jeronimidis and J F V Vincent. pp 339–346. Centre for Biomimetics, Reading, UK.

Stokes A and Mattheck C 1996 Variation of wood strength in roots of forest trees. J. Exp. Bot. 47, 691–699.

Timell T E 1986 Compression Wood in Gymnosperms. Springer-Verlag, Berlin.

Watson A J 1995 Measurement of wind-induced tree root stresses in New Zealand. *In* Wind and Trees. Eds M P Coutts and J Grace. pp 220–226. Cambridge University Press, UK.

Watson A J 1999 Wind-induced forces in the near-surface lateral roots of radiata pine. For. Ecol. Mgmnt. (*In press*).

A. Stokes (ed.), *The Supporting Roots of Trees and Woody Plants: Form, Function and Physiology*, 31–39.

Root deformation in plantations of container-grown Scots pine trees: effects on root growth, tree stability and stem straightness

A. Lindström* and G. Rune
Dalarna University, S-776 98 Garpenberg, Sweden

Key words: Pinus sylvestris, potted plants, root distribution, spiralled roots, stem base crookedness, winching test

Abstract

Root system deformation was studied in 23 Scots pine (*Pinus sylvestris* L.) stands in central Sweden. The study comprised both plantations created with container-grown plants (Paperpot) and natural stands including young (7–9 year old) and older (19–24 year old) trees. Trees were measured with regards to distribution of roots, root deformation, stability, stem straightness and wood properties in stumps. Root distribution was most uniform for naturally regenerated trees. Older trees generally showed a better root distribution than young trees. The young planted trees displayed a high frequency of severely spiralled root systems, while only a few of the older trees had spiralled root systems. No severe root deformations were observed on naturally regenerated trees. Naturally regenerated trees were more stable than those which had been planted. Differences in bending moment, when trees were pulled to an angle of 10°, were considerable between young planted and naturally regenerated trees, but less pronounced for the older trees. Young planted trees had the highest frequency of severely crooked stem bases, while naturally regenerated trees had the straightest mode of growth. Tensile strength in peripheral wood samples of the stumps was substantially lower for planted than for naturally regenerated trees. Strain values to breakage of wood samples, taken from the root collar and the central- and peripheral part of the stump were lower for planted trees. The conclusions from this study are that root distribution, tree stability and stem straightness of planted Paperpot-grown trees will improve after a certain time and approach the state of naturally regenerated trees. As trees grow older, early established crooked stem bases will be compensated by radial growth and the tree will appear straighter. Inside the stem, however, problems may still remain with abnormal fibre direction and compression wood together with inferior root strength due to fibre disturbances as a result of spiralled roots.

Introduction

During the 1970s, many nurseries in Sweden changed from bare-root to containerised plant production. During the first ten years of containerised plant production, the Paperpot container was the most frequently used. Throughout the years, many billions of Paperpot seedlings have been planted in Sweden (Nyström 1989). Studies have now shown that containers of the Paperpot type, with smooth inside walls, cause different types of root deformation of which spiralling is the best known defect (Balisky et al. 1995; Burdett et al. 1986; Lindström 1998). If containers have impenetrable walls, lateral roots cannot extend horizontally and therefore they either circle within the container or they will grow vertically to the bottom. The importance of an evenly spread and undisturbed root system for the juvenile anchorage of a tree is well known (e.g. Balisky et al. 1995; Lindgren and Örlander 1978; Lindström and Håkansson 1995).

A serious effect of a weakly anchored root system is that mechanical instability may occur in plantations. Previous studies of young planted container-grown Scots pine (*Pinus sylvestris* L.) (Hultén and Jansson 1978; Håkansson and Lindström 1989), lodgepole pine (*Pinus Contorta* Dougl.) and Douglas fir (*Pseudotsuga menziesii* var. *glauca* (Beissn.) Franco) (Halter and Chanway 1993) indicated that they had

* FAX No: +46 225 261 00. E-mail: ali@du.se
Plant and Soil is the original source of publication of this article. It is recommended that this article is cited as: *Plant and Soil* **217:** 29–37, 1999.

a root form that could cause instability. The state of juvenile stability in a tree is known to have a major impact on the future development of stem form (Moss 1971). Instability may result in crooked stem bases (see Rosvall 1994), the extent of which depends on the bending forces exerted on the tree, e.g. wind and snow. Other reported negative effects from root deformation are reduced growth (Halter et al. 1993; Lindström and Persson 1996) and deteriorated vitality due to root disease caused by fungal infections (Livingston 1990).

The ultimate consequences of root deformation are uprooting due to a weak root anchorage and root breakage caused by fibre disturbances within the stump. Uprooting and root breakage have been observed in young plantations of container-grown Scots pine (Lindström and Håkansson 1995) and lodgepole pine (see Rosvall 1994). Earlier investigations established a relationship between visible root deformations such as root spiralling and instability (Håkansson and Lindström 1989). However, studies indicate that root deformations will be hidden by secondary growth when trees grow older (Gillgren 1971; Stefansson 1978). Earlier established and visible fibre disturbances with bark remains inside the stump may contribute to inferior stability and increased risk for future root breakage in mature trees.

To our knowledge, there is little information on tree stability and stem deformation in Scots pine plantations grown in Paperpot. Neither is it known how root deformation, stability and stem deformation develop, as the stand ages. The objectives of this study were to investigate young and older plantations of Paperpot-grown plants and naturally regenerated trees with a special emphasis on root development and deformations, stability and stem straightness. Since fibre disturbances within the root system may cause a weak connection between stem and root, we also measured the fibre tensile strength and strain during mechanical loading in stumps of older Scots pine trees.

Materials and methods

Root system deformation was studied in 23 Scots pine stands in central Sweden. The study was based on non-experimental surveys performed in 1990 and 1993 on randomly selected sites (59°36′N-60°26′N, 12°21′E-16°41′E; 30-230 m elevation). The study comprised young (7–9 year old) and older (19–24 year old) trees in naturally regenerated stands and plantations of Paperpot-grown plants (Table 1). Seeds from seed orchards were sown in commercial nurseries in 75 ml containers (Paperpot FH 408, Lännen Tehtat Oy, Finland, 1002 seedlings/m^2) filled with peat. Seedlings were grown for 3 to 7 months before planting.

Most sites were similar with regards to soil type (sandy loam till) and moisture (dry and mesic soil). Mechanical soil treatment by a scarifier of harrow type was carried out only on the planted sites. In each of the young and older stands, 18 and 10 trees respectively were randomly selected for stability tests.

Stem base crookedness was estimated by using a digital protractor (Lucas Anglestar, model DP 45, USA). The angle was registered at the point where greatest angle was found. Basal crookedness of stems was assigned to four classes; straight (0–5°), slightly crooked (6–30°), crooked (31–45°) and very crooked (>46°). In order to measure stability, trees were pulled by a wire attached to a winch. Pulling was carried on until the tree leaned 10° from its initial position. The pulling force was then registered by a dynamometer (Figure 1). Bending moment was calculated as: $B=N\times h$, where B=bending moment (Nm), N=horizontal pulling force (N) and h=distance (m) from ground level to attachment point of the pulling wire on the stem.

After measurements of stability were completed, the root systems of six and five trees were manually excavated on each site with young and older trees respectively. The investigation of young trees was done three years before the investigation of older trees. For young trees the ends of roots were cut off so that the stump and root bases fitted into a cylinder with a 10 cm radius. Measurements were made on roots that crossed the cylinder's casing and roots that crossed the hemisphere's casing and roots in four quadrants (Figure 2A, B). The measurements included the number and cross calipered diameter of roots $\geq$3 mm at a distance of 10 cm from the central root axis. Later, when roots of older trees was collected, the measurement cylinder for young roots were too small. Therefore the distance at which roots of older trees were measured was extended from 10 to 30 cm. The diameter of measured roots was also increased, from $\geq$3 mm to $\geq$5 mm for older trees. Consequently, the root area measure is not used when comparing root development of young and older trees.

An index of evenness of root distribution in the four quadrants, a root area index (RAI), was calculated as the ratio between root area in the quadrant with the largest root area and the total root area. For an evenly distributed root system RAI is 0.25, and higher RAI

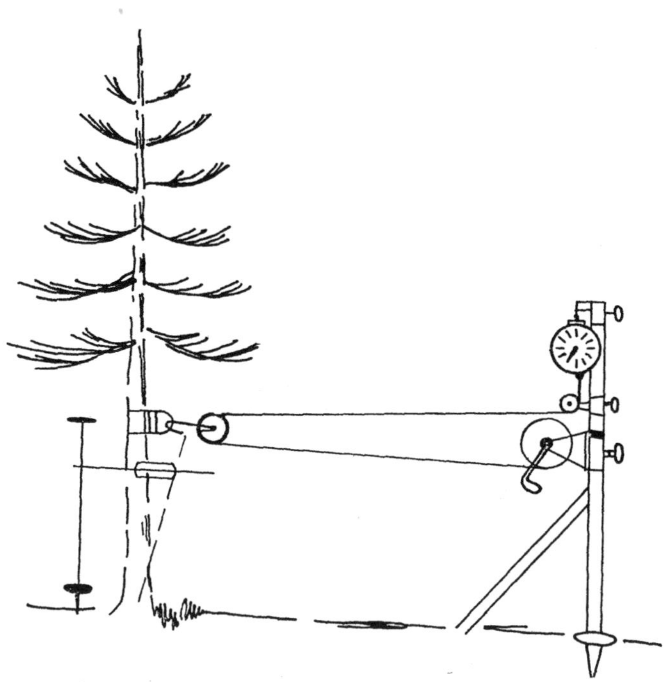

Figure 1. Equipment for measuring tree stability (from Hultén and Jansson 1978).

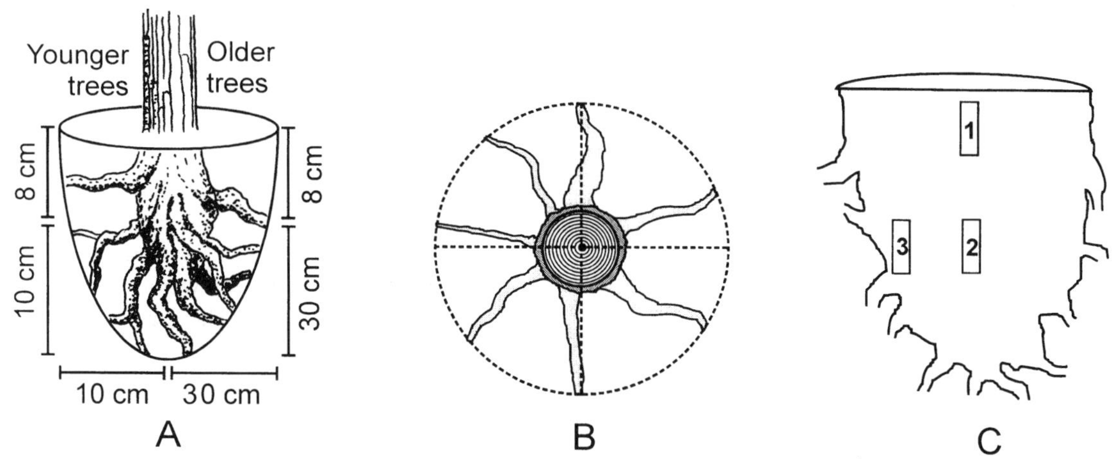

Figure 2. Separation of root systems of young and older trees into roots that crossed the cylinder's casing, roots that crossed the surface of the hemisphere's casing (A) and roots in four quadrants (B). Position of wood samples selected for tensile strength test. Samples were taken in 1) root collar, 2) central and 3) peripheral part of the stump (C).

Table 1. Age, average diameter (20 cm above ground surface), average height, number of sites and tested Scots pine trees. Standard error within brackets

Type of tree	Age	Dia., mm	Height, cm	Number of sites	Number of tested trees
Paperpot	7–9	50 (0.9)	187 (3.2)	5	90
Paperpot	19–21	130 (3.2)	702 (13.5)	10	100
Natural regen.	7–9	54 (1.0)	251 (5.5)	6	108
Natural regen.	22–24	124 (5.7)	872 (37.9)	2	20

indices indicate a more asymmetrical root system. In addition, the number of spiralled roots with a diameter ≥3 mm (young trees) and ≥5 mm (older trees) were counted on each root system. The spiralling of a root system was classified as *negligible* if the number of added turns for spiralling roots was ≤1 turn, *moderate* if spiralling was >1 but ≤3 turns, *severe* if spiralling was >3 turns or if a single root or more had spiralled >1 turn.

Bark remains inside the stumps may indicate that the roots are affected by deformations hidden by secondary growth. Five stumps from each of older Paperpot-grown plants and naturally regenerated plants were randomly selected and were vertically split through the pith. The sectioned surfaces were visually evaluated with regards to remains of bark (% of total cross section area) using pictures from digital image analyses.

Tensile strength and strain of wood samples during loading could be measures of the effects of fibre disturbances in roots. In order to determine tensile strength and strain in the above mentioned stumps, three vertically oriented 10 cm^3 ($10 \times 1 \times 1$ cm) wood samples were notched in the root collar, the central and the peripheral part of the stump (Figure 2C). The wood samples had a waist with a thickness of 5 mm according to the description of Bodig and Jayne (1982). The wood samples were then conditioned at 50% relative humidity in a climate chamber for seven days. After that, the moisture content (% of dry weight) of wood samples ranged from 8.2 to 9.7%. The samples were tested in tension in a universal testing machine (Hounsfield 5000, England) at a crosshead speed of 40 mm/min. When breakage of the wood fibres occurred, force and tensile strain ((tensile mm/original sample length)×100) were registered.

Statistical analyses of differences in morphological characters between trees were made by pairwise comparisons (df=1) using the χ^2 test (Zar 1974). Differences in bark remains and tensile strength and strain of wood samples in stumps were evaluated using ANOVA and root distribution in the soil using a Student's *t*-test. To describe the relationship between stem diameter and bending moment when pulling trees, and stem diameter and root area for excavated trees the function was used: $Y = e^a \times D_{0.20}{}^b$ where Y=bending moment (Nm) or root area (mm^2), D=cross calipered stem dia. at 20 cm height (mm) and a and b=constants. For comparisons between naturally regenerated and planted trees we have used a common model where the dependent variables bending moment and root area depend not only on diameter but also on regeneration method.

Results

Naturally regenerated trees were straighter than Paperpot-grown trees in both the young (p=0.008, n=198) and older (p=0.002, n=120) stands. The young Paperpot-grown trees had a higher frequency of crooked to very crooked stem bases than young naturally regenerated trees (p=<0.001, n=198). The frequency of planted trees with crooked to very crooked stem bases was less for older than for young trees (p=0.040, n=190) (Table 2).

Naturally regenerated trees were more stable than planted ones. Differences in bending moment, when trees were pulled to an angle of 10°, were considerable between young planted and naturally regenerated trees (p=<0.001) but less pronounced for the older trees (p=0.035) (Figure 3).

The mean numbers of roots per root system were similar between planted and naturally regenerated trees for both the young (14 roots, SD±4.2) and older trees (22 roots, SD±7.9). However, the roots

Table 2. Proportion of Scots pine trees with straight (0–5°), slightly crooked (6–30°), crooked (31–45°) and very crooked (>46°) stem bases

Type of tree	Age	Proportion of stem bases, %				Number of tested trees
		Straight stem bases	Slightly crooked	Crooked	Very crooked	
Paperpot	7–9	37	47	13	3	90
Paperpot	19–21	38	55	5	2	100
Natural regen.	7–9	56	42	0	2	108
Natural regen.	22–24	75	25	0	0	20

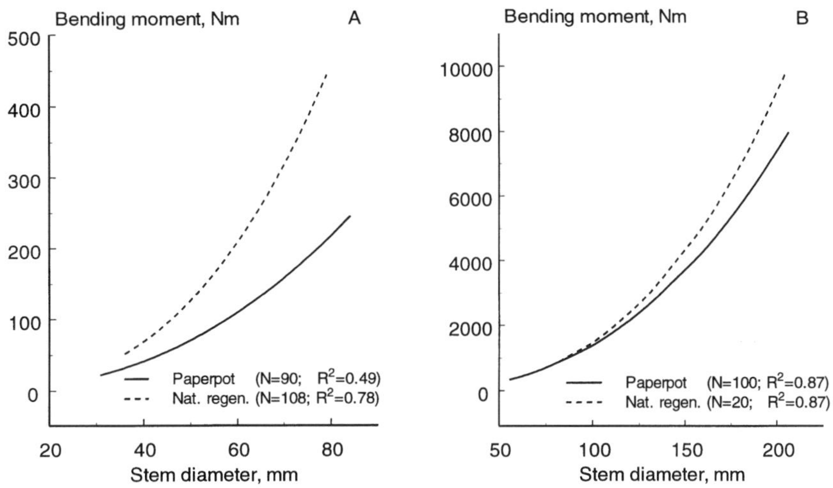

Figure 3. Relationship between stem diameter and bending moment when pulling to a 10° angle according to the function $B = e^{a} \times D_{0.20}{}^{b}$ (B=bending moment (Nm), D=cross calipered stem diameter at 20 cm height (mm), a and b=constants) for A) 7–9 year and B) 19–24 year old naturally regenerated and planted Scots pine trees.

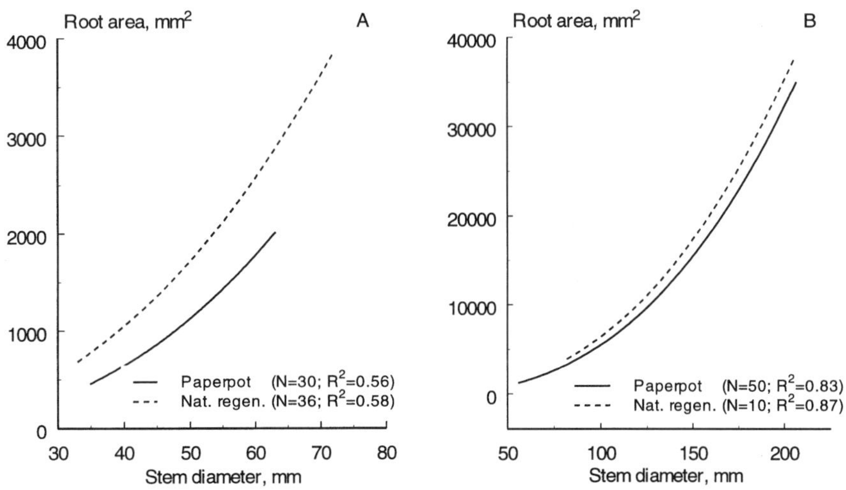

Figure 4. Relationship between stem diameter and total root area for young (A) and older (B) planted and naturally regenerated trees according to the function $R = e^{a} \times D_{0.20}{}^{b}$ (R=root area (mm^2), D=cross calipered stem diameter at 20 cm height (mm), a and b=constants).

Table 3. Root area index (RAI)* of young and older Paperpot-grown and naturally regenerated trees and $P(t)$-values

	7–9 year old	19–24 year old	P(t)
Paperpot	0.51 (n=30)	0.40 (n=50)	<0.001
Natural regen.	0.45 (n=36)	0.38 (n=10)	0.028
P(t)	0.026	0.131	

*The ratio between the root area in the quadrant with the largest root area and total root area. Lower values imply a more uniform root distribution.

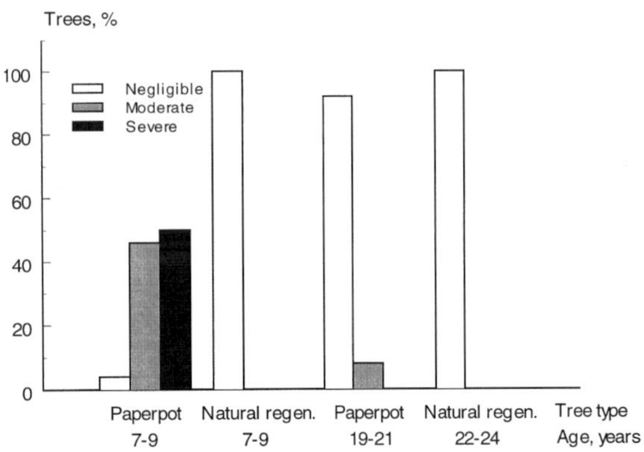

Figure 5. Proportion (%) of trees with different degrees of spiralled roots divided among types of trees. The spiralling of a root system was classified as *negligible* if the number of added turns for spiralling roots was ≤ 1 turn, *moderate* if spiralling was >1 but ≤3 turns, *severe* if spiralling was >3 turns or if a single root or more had spiralled >1 turn (n = 10–50).

of naturally regenerated young trees had substantially larger areas, i.e. sum of root cross sectional areas at a fixed distance, than those of the young planted trees (p=<0.001). For older trees this difference was less pronounced (p=0.032) (Figure 4). Young naturally regenerated trees had a more uniform root distribution (lower RAI) than young planted trees, while this difference was less obvious for older trees. Older trees generally showed a better root distribution than young trees (Table 3). Very few root systems of young planted trees were classified with negligible spiralling compared to naturally regenerated trees (p=<0.001, n=66). For the older trees, differences in spiralling between roots from planted and naturally regenerated trees were small (p=0.355, n=60) (Figure 5).

Bark remains in the cross sectional area of split stumps were lower for naturally regenerated (0.5%) than planted trees (5.8%) (p=0.009, df=1, n=5). Highest tensile strength values were obtained in the root collar of split stumps. Significant differences between different site of sample in root stumps were obtained both for tensile strength (p<0.001, df=2, n=10) and strain (p=0.010, df=2, n=10). The tensile strength in peripheral wood samples was significantly lower for planted than for naturally regenerated trees. For all sites of the stump, strain values to breakage were lower for planted trees (p=0.006, df=1, n=15) (Figure 6).

Discussion

In this survey, naturally regenerated trees showed the straightest mode of growth while young Paperpot-grown trees had the highest frequency of crooked stem bases. These results are supported by previous studies showing that stem sweeps are rare in naturally regenerated stands (Burdett et al. 1986) and sown stands (Mattsson and Rune 1998) of lodgepole pine. The high frequency of crooked stem bases of planted trees in this study indicates that they have been unstable during a period of their development. In the winching tests, we found that young Paperpot-grown trees were more unstable than naturally regenerated trees. One reason for this instability can be that planted trees had a lower root area in relation to the stem diameter compared with naturally regenerated trees. A large root area is usually associated with good stability (Lindgren and Örlander 1978). Another explanation is that the naturally regenerated trees had the most uniform root distribution (see Langerud et al. 1988). It is well established that the form of the root system is a principal determinant of tree stability (Coutts 1983).

Only the sites planted with Paperpot-grown plants were scarified. This could have affected the stability comparison with the unscarified naturally regenerated sites. Lindström and Håkansson (1995) showed that tree stability in young lodgepole stands was improved by scarification. These results indicate that root deformation of Paperpot-grown trees in this study was likely to be the main reason for inferior mechanical stability compared to naturally regenerated trees.

Differences in bending moment, when trees were pulled to an angle of 10°, were considerable between young planted and naturally regenerated trees, but less pronounced for older trees. The young Paperpot-grown trees also displayed a higher frequency of crooked to very crooked stem bases (16%) than older ones (7%). This observation supports Burdett et al. (1986) who suggested that, as trees grow older, they

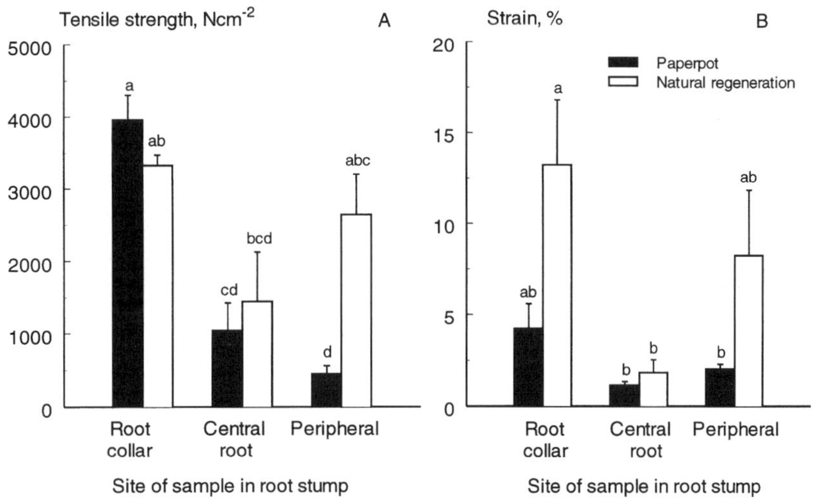

Figure 6. Tensile strength (A) and strain values to breakage (B) in wood samples taken from the root collar, central and peripheral part of the stump. Tensile strength (N/cm^2) and strain (%) were registered at breakage of wood samples. Vertical bars show standard error of the treatment mean, total number of sample in a treatment=5. Treatments with a common letter above bars do not differ significantly at the 5% level according to Tukey's test.

become straighter and the bow of the stem will be compensated by radial growth.

We found that roots of naturally regenerated trees had a larger cross sectional area than those of planted trees. Studies of Nichols and Alm (1983) with naturally regenerated plants and plants reared in Paperpots support this result. Root distribution was better for older planted trees (RAI=0.40) than for the younger planted trees (RAI=0.51). The fact that the root distribution is improved as trees age has also been reported earlier by Nielsen (1998). It was also found that 50% of the root systems of young Paperpot-grown trees had severe spiralling, whereas older Paperpot-grown trees displayed a low frequency of spiralled root systems. Gillgren (1971) showed that initially entangled pine roots successively fuse and in the course of time, a root system develops with normal exterior appearance and function. Van Eerden (1978) suggested that, as time elapses, the initial effects of containerisation on the root system will be surmounted and he also (Van Eerden 1982) predicted that such aberrations would be overcome six years after planting. However, in this study structural abnormalities were observed within root system of Paperpot-grown trees, and these abnormalities were clearly apparent 19 to 21 years after planting.

When Scots pine trees reach about 20 years of age in central Sweden, radial growth makes it difficult to see root deformations on the outside of root stumps. Most of the spiralling roots are now occluded and the surface of the root is covered with a more or less continuous cambium layer. However, inside the root, problems still remain with reduced tensile strength and strain. When section surfaces of divided roots were examined, it was found that occluded root deformations of planted trees were expressed as abnormal fibre disturbances and remains of bark. These abnormalities could have a negative effect on the strength of the root system, which was shown in the results from the tensile strength and strain tests of wood samples from the stumps. The difference in wood tensile strength between the planted and naturally regenerated trees was particularly large in the occluding zone. These results indicate that these trees, in spite of an enclosed root spiralling 20 years after planting, still have a weakness in the upper part of the root system. The low tensile strength and strain values in stumps indicate an increased risk for root breakage in older container-grown trees.

In summary, our studies show that root distribution, tree stability and stem straightness of planted Paperpot-grown trees will improve after a certain time and approach the state of naturally regenerated trees. As trees grow older, early established root deformations and stem base crookedness will not be visible and the tree will appear straighter.

However, the high incidence of stem base crookedness in container-grown plantations is a cause for concern, because crooked stems may have an adverse effect on timber quality and wood capture (Pfeifer

1982). A number of studies concerning the formation of compression wood (Burdon 1975; Fayle 1976; Low 1964; Timell 1986), show that the formation of compression wood mainly depends on stem sway and leaning of trees including stem base crookedness. Warensjö and Lundgren (1998) showed that compression wood has an impact on bow and crook in butt logs of Norway spruce (*Picea abies* (L.) Karst).

The use of container types that cause spiralling roots would generally result in similar effects as we have seen in Paperpot-grown plantations. Root spiralling is foremost associated with early container designs, but may also occur in hard wall containers frequently used today (Lindström and Håkansson 1995). Future problems with instable cultures may be reduced by the development of new container types. Promising results have been obtained with containers that permit lateral root growth through slots in the wall, i.e. systems that permit pruning of roots, which inhibit spiralling (Lindström 1998; Lindström and Persson 1996).

In the past, most studies have tended to focus on the stability and root deformations of young container-grown trees, and not on the effects of instability on the timber quality in mature forests. The degree to which stem base crookedness affects the value of a tree would depend on the final use of the timber and conversion technology used. In Sweden, the first planted stands established with container-grown seedlings are now mature for thinning. Today it is not possible to predict the economical consequences of containerisation as the extent of stem base crookedness and its impact on wood quality is not sufficiently investigated. This study indicates that we may face future economical losses in container-grown stands due to deteriorated wood properties from instability. Besides further studies of stem straightness and wood quality in these stands, more studies are needed to develop container systems and planting techniques that guarantee juvenile stability of planted trees.

Acknowledgement

This study was financed by the Swedish National Board of Forestry, the Swedish Forestry Research Foundation and the Foundation for Knowledge and Competence Development. Our special thanks to Claes Hellqvist and Lars Håkansson (Dalarna University), for help with field work and data processing. We are indebted to Dr Mats Nylinder (Swedish University of Agricultural Sciences), Dr Anders Mattsson, Dr Erik Ståhl and Dr Bengt Gyldberg (Dalarna University) for valuable comments on the manuscript.

References

Balisky A C, Salonius P, Walli C and Brinkman D 1995 Seedling roots and forest floor: Misplaced and neglected aspects of British Columbia's reforestation effort? For. Chron. 71, 59–65.

Bodig J and Jayne B A 1982 Mechanics of wood and wood composites. Van Nostrand Reinhold Co., New York. 712 p.

Burdett A N, Coates R, Eremko R and Martin P A F 1986 Toppling in British Colombia's lodgepole pine plantations: significance, cause and prevention. For. Chron. 62, 433–439.

Burdon R D 1975 Compression wood in *Pinus radiata* clones on four different sites. N.Z. J. For. Sci. 5, 152–164.

Coutts M P 1983 Root architecture and tree stability. Plant Soil 71, 171–188.

Fayle D C F 1976 Stem sway affects ring width and compression wood formation in exposed root bases. For. Sci. 22, 193–194.

Gillgren I 1971 Om rotsnurr, rotdeformation och rotstrangulering. Examensarbete i ämnet skogsskötsel utfört hos SCA. Skogshögskolan, 34 p.

Halter M R and Chanway C P 1993 Growth and root morphology of planted and naturally regenerated Douglas-fir and lodgepole pine. Ann. For. Sci. 50, 71–77.

Halter M R, Chanway C P and Harper G J 1993 Growth reduction and root deformation of containerized lodgepole pine saplings 11 years after planting. For. Ecol. Manage. 56, 131–146.

Hultén H and Jansson K Å 1978 Stability and root deformation of pine plants (*Pinus sylvestris*). [In Swedish with English summary.] Royal College of For., Dep. of Reforest. Rep. 93, 84 p.

Håkansson L and Lindström A 1989 Försök med olika behållartyper – Resultat av stabilitets- och rotundersökningar 9 år efter plantering. SLU, Inst. för Skogsprod., Stencil 52, 40 p.

Langerud B R, Sandvik M and Sjovold A 1988 Root-symmetry of conifer plants in field. [In Norwegian with English summary.] Comm. of the Norw. For. Res. Inst. 40, 1–20.

Lindgren O and Örlander G 1978 A study on root development and stability of 6 to 7-year old container plants. *In* Proc. of the Root Form of Planted Trees Symp., Victoria, B. C., Can., May 16–19, 1978. Eds E Van Eerden and J M Kinghorn. pp 142–144. Can. For. Serv., B. C., Joint Rep. No. 8.

Lindström A 1998 Root deformation and its implications for containerized seedling establishment and future quality development [In Swedish with English summary.] *In* Root Development and Stability. Conf. held at Garpenberg, Swed., 30 Sept.–1 Oct. 1997. Ed. C Almqvist. pp 51–60. The For. Res. Inst. of Swed. Rep. No. 7.

Lindström A and Håkansson L 1995 Going to the root of the evil. Canadian silviculture magazine 3, 14–17.

Lindström A and Persson B 1996 Tallföryngringens och trädens kvalitet påverkas av odlingskrukans utformning. Plantnytt nr 4, 1–6.

Livingston W H 1990 *Armillaria ostoyae* in young spruce plantations. Can. J. For. Res. 20, 1773–1778.

Low A J 1964 A study of compression wood in Scots pine (*Pinus silvestris* L.). Forestry 37, 179–201.

Mattsson M and Rune G 1998 Root development in lodgepole pine from direct seeding and planting. [In Swedish with English summary.] *In* Root Development and Stability. Conf. held at

Garpenberg, Swed., 30 Sep.–1 Oct. 1997. Ed. C Almqvist. pp 82–87. The For. Res. Inst. of Swed. Rep. No. 7.

Moss A 1971 An investigation of basal sweep of lodgepole and shore pines in Great Britain. Forestry 44, 43–65.

Nichols T J and Alm A A 1983 Root development of container-reared, nursery-grown, and naturally regenerated pine seedlings. Can. J. For. Res. 13, 239–245.

Nielsen C N 1998 Root development of bare rooted seedlings after field establishment and consequences for wind stability. *In* Root Development and Stability. Conf. held at Garpenberg, Swed., 30 Sept.–1 Oct. 1997. Ed. C Almqvist. pp 66–71. The For. Res. Inst. of Swed. Rep. No. 7.

Nyström C 1989 Svensk plantproduktion 1989. Plantnytt nr 6, 1–4.

Pfeifer A R 1982 Factors that contribute to basal sweep in lodgepole pine. Irish For. 59, 7–16.

Rosvall O 1994 Stability in lodgepole pine and resistance to wind and snow loads. [In Swedish with English summary.] The For. Res. Inst. of Swed. Rep. No. 2, 47 p.

Stefansson E 1978 Root quality of pine plantations established with seedlings grown in multipots. *In* Proc. of the Root Form of Planted Trees Symp., Victoria, B. C., Can., May 16–19, 1978. Eds E Van Eerden and J M Kinghorn. pp 114–118. Can. For. Serv., B. C., Joint Rep. No. 8.

Timell T E 1986 Compression wood in Gymnosperms. Vols 1–3, Springer-Verlag, Berlin.

Warensjö M and Lundgren C 1998 Impact of compression wood on deformations of sawn wood of spruce (*Picea abies* (L.) Karst.). [In Swedish with English summary.] Swed. Univ. of Agric. Sci. Dep. of For. Products. Rep. 255, 38 p.

Van Eerden E 1978 Roots of planted trees in central British Colombia. *In* Proc. of the Root Form of Planted Trees Symp., Victoria, B. C., Can., May 16–19, 1978. Eds E Van Eerden and J M Kinghorn. pp 201–209. Can. For. Serv., B. C., Joint Rep. No. 8.

Van Eerden E 1982 Root form of planted trees in central British Columbia. *In* Proc. of the Can. Contain. Tree seedl. Symp., Toronto, Ont., Can., Aug. 1981. Eds J B Scarratt, C Glerum and C A Plexman. pp 401–409. Great Lakes For. Res. Cent., Sault Ste. Marie, Ont., Can.

Zar J 1974 Biostatistical analyses. Prentice-Hall, Inc., Englewood Cliffs, N. J., 718 p.

A. Stokes (ed.), The Supporting Roots of Trees and Woody Plants: Form, Function and Physiology, 41–49.

Root strength, growth, and rates of decay: root reinforcement changes of two tree species and their contribution to slope stability

Alex Watson[1,*], Chris Phillips[1] and Michael Marden[2]
[1]*Landcare Research, P.O. Box 69, Lincoln, Canterbury, New Zealand and* [2]*Landcare Research, P.O. Box 445, Gisborne, New Zealand*

Key words: root decay, root growth, root reinforcement, root site-occupancy, root strength, slope stability

Abstract

Information on live root-wood strength, rates of root decay and root growth of both radiata pine (*Pinus radiata* D. Don) and kanuka (*Kunzea ericoides* (A. Rich.) Joy Thomps. var. *ericoides*) are combined to form a generalized conceptual model of changes in nett root reinforcement. The model provides an initial opportunity to rank the plant species having specific below-ground rooting habits that can be used to control erosion, and when linked with extreme flood probability can be used to indicate the risk of a storm likely to cause slope instability in the period between clear-felling and regrowth. Erosion-susceptible slopes planted 1 year after clearfelling in radiata pine at 1250 stems ha^{-1} regain root site-occupancy in 4.7 years, an interval during which there is an 80% chance of experiencing an extreme flood. Similarly for radiata planted at 800 and 400 stems ha^{-1}, root site-occupancy is regained in 5.6 and 7.5 years, and the probability of occurrence of an extreme event within these periods is 85 and 90%, respectively. For erosion-susceptible slopes on which kanuka has become established, the probability of a significant event within the 2.8 years prior to root site-occupancy is 60%. Slopes felled of radiata pine are potentially more vulnerable to the stresses promoting slope instability, at least in the earlier years.

Introduction

Information on rates of root growth, live root wood strength, rates of decay, and site-occupancy has been obtained from ongoing studies over the last decade or so, which have been aimed at improving our understanding of the role vegetation plays in contributing to slope stability through root reinforcement (Ekanayake et al., 1997; Marden et al., 1991; O'Loughlin and Watson, 1979; Phillips and Watson, 1994; Watson and O'Loughlin, 1990; Watson et al., 1995, 1997). Studies involving root strength loss after clear-felling and rates of root growth of replacement vegetation indicate inadequate root–soil reinforcement for a period of up to 6 years after logging. This roughly corresponds to the observed period of increased frequency of slope instability.

The rate at which a tree planting site is occupied by roots depends on the lateral root growth rate and the planting density. Root site-occupancy is based on the use of these parameters to estimate the length of time required to obtain root system overlap within the upper soil layers, and is taken as a theoretical starting point for plantation forests to begin contributing to slope stability. The concept projects the tree root system onto the ground as a circle. Trees, i.e. circles, are grown at the selected planting density, increasing circle diameter over time until root site-occupancy is attained. At site-occupancy the nett root reinforcement of the soil by roots, i.e. the sum of the reinforcement by the residual decaying roots of the cut trees and that by the new roots of the regenerating trees, is considered to be 100%. If the nett root reinforcement is less than 100%, areas devoid of roots will remain between adjacent trees.

Much of the hill country of the East Cape region of the North Island of New Zealand is prone to serious erosion. The original podocarp-hardwood forest was cleared to create farmland during the latter half

* FAX No: + 64 3 325 2418. E-mail: watsona@landcare.cri.nz
Plant and Soil is the original source of publication of this article. It is recommended that this article is cited as: *Plant and Soil* **217**: 39–47, 1999.

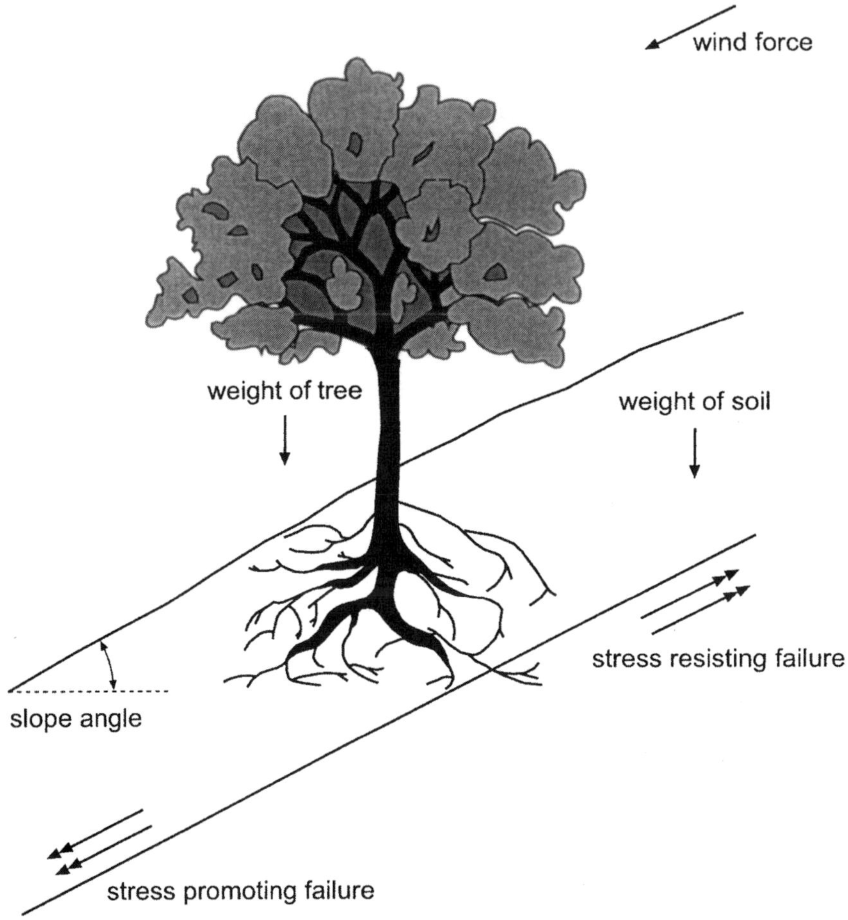

Figure 1. A generalized representation of a part-forested slope. The forces resisting failure depend on the soil strength plus the additional soil reinforcement supplied by root soil cohesion. The forces promoting failure depend on the weight of soil and trees, slope angle, and wind force. Diagram modified from Wu and Swanston (1980).

of the nineteenth century. Increased erosion, coupled with productivity decline following 70–100 years of intensive grazing, has resulted in several cycles of abandonment and natural reversion over much of the area.

Over recent decades, large areas of farmland have reverted to indigenous scrub, in which Kanuka (*Kunzea ericoides* (A. Rich.) Joy Thomps. var. *ericoides*) forms the major component. These areas have either been re-cleared for continued pastoral development or are now being considered for forestry conversion. The former situation has often been associated with an increase in the frequency of translational landslides (Figure 1).

This region also has a history of extreme floods which, when combined with the inherently unstable geology and the original forest removal, have been major contributors to erosion of the East Cape hill country. In response to a particularly severe event in 1988, the New Zealand Government established a scheme to convert marginal pastoral farmland in the more erosion-susceptible areas to radiata pine (*Pinus radiata* D. Don) dominated plantation forestry (Ministry of Forestry, 1994).

On the erosion-prone hill country in the East Cape region, the recommended minimum initial planting density for radiata pine is 1250 stems ha^{-1}. Over recent years there has been some concern at an increasing trend towards lower planting densities, due in part to genetically improved and therefore more expensive planting material, and increased labour costs. The stand density and the size of the individual root systems are major factors that influence the degree to which trees enhance slope stability. The rate at which trees contribute to stability is largely dependent on the time taken to reach site-occupancy, the magnitude of the root–soil reinforcement, and their modification of the soil-water regime by evapotranspiration. At lower planting densities trees provide a much reduced level of stability for several years after planting, as their ability to promote the stronger, drier soils needed for slope stabilization is seriously diminished. If the trees are removed by clear-felling, the beneficial effects of evapotranspiration are lost immediately, and thereafter slope stability is likely to decline at a rate largely determined by the magnitude of the live root-wood strength and the rate at which this strength is lost after the death of the parent trees.

In this paper we present data on live root-wood strength, rates of root decay and root growth of both radiata pine and kanuka. A comparative model of rates of root reinforcement changes that occur between clear-felling and regrowth of vegetation is also presented. The implications for slope stability in relation to the probability of an extreme storm event occurring on slopes covered by an undisturbed naturally regenerating tree species (kanuka) and three planting regimes of a plantation forestry species (radiata pine) are evaluated and discussed.

Site characteristics and sources of data

Radiata pine root samples used to generate live root/root decay information (O'Loughlin and Watson, 1979) were collected from trees growing on dissected hill country, approximately 30 km north of Christchurch (43°32′S, 172°35′E), New Zealand. Here the annual rainfall is 774 mm (Ashley Forest, Station No. H32252, altitude 107 m) (Tomlinson and Sansom, 1994).

The balance of the data were collected from two study sites in the Gisborne region, east of the Raukumara Range in the East Cape hill country. The radiata pine stands were at Mangatu Forest in the headwaters of the Waipaoa and Mangatu Rivers. Mean annual rainfall at the forest headquarters is 1350 mm (NZ Met. Service, 1984). The kanuka stands were growing north of Gisborne (38°30′S, 178°04′E) on poorly utilized farmland that, in part, had been allowed to revert to indigenous scrub. Mean annual rainfall at this site is 1409 mm. Descriptions of the topography and soils for the radiata pine and kanuka sites are available in Watson and O'Loughlin (1990) and Watson et al. (1997), respectively.

The climate of the East Cape is temperate marine, with warm dry summers and cool wet winters. Extreme floods, generally resulting from high-intensity rainfall during tropical cyclones have been a predominant factor contributing to the unstable nature of the hill country east of the Raukumara Range. Although such storms are relatively infrequent, there were four major events during the 1980s, the largest of which occurred in March 1988, when up to 900 mm fell with in 5 days. In all there have been 29 extreme floods of the Waipaoa River (catchment area 1580 km^2) during this century (Kelliher et al., 1995).

Methods

Root strength and decay rates

Tensile strength of individual tree roots, collected from live trees or stumps of known cutting date, is the most common parameter measured to determine live root strength and rates of root strength loss with decay (O'Loughlin and Watson, 1979). Root samples from selected trees were sealed in plastic bags and, when necessary, kept in cool storage to preserve root-moisture content prior to testing. Detailed description of sampling techniques and laboratory testing are described in Watson et al. (1997). Briefly, the root is clamped at both ends in the testing device and the resistance to an applied force necessary to break it under tension is taken as the measure of the tensile strength.

Rates of root growth

Root growth information comes from two larger studies designed to investigate temporal changes in root morphology and biomass of radiata pine (Watson and O'Loughlin, 1990) and kanuka (Watson et al., 1995). A total of 28 root systems, 13 radiata pine and 15 kanuka, were excavated using high-pressure water. The technique required selection of trees growing on slopes steep enough to allow an easy flow of sluiced material away from the excavation site. Sluicing began far enough downhill from the stump to ensure that as the excavation proceeded up slope it was of sufficient depth to expose the deeper vertical roots. As the root systems were exposed, they were propped in position and their architecture mapped and measured (Table 1).

The excavated radiata pine were taken from three even-aged planted stands. Age classes for kanuka were determined by taking the mean of the annual ring counts of the excavated trees within each stand.

Table 1. Site parameters, age, number and size of sampled trees and dimensions of their root systems. Figures in parentheses represent the standard error of means

	Radiata pine			Kanuka		
Mean age (years)	8 (0)	16 (0)	25 (0)	6 (0.7)	16 (1.4)	32 (2.7)
Stand density[(a)] (stems/ha^{-1})	240	250	270	16 000	12 800	3900
No. of trees	5	5	3	5	5	5
Mean tree height (m)	9.5 (–)	21.1 (–)	30.4 (–)	6.1 (0.5)	6.7 (0.3)	13.6 (0.3)
Mean DBH[(b)] (mm)	170 (10.7)	400 (25.9)	550 (27.7)	45 (4.4)	66 (7.2)	127 (15.3)
Mean max. root length (m)	3.5 (0.3)	4.2 (0.5)	9.1 (1.0)	1.5 (0.2)	3.0 (0.4)	3.6 (0.7)
Max. root length (m)	4.7	6.4	10.4	1.9	4.5	6.1
Mean max root depth (m)	1.8 (0.1)	2.4 (0.1)	2.9 (0.1)	1.3 (0.3)	1.6 (0.2)	1.3 (0.1)
Max. root depth (m)	2.1	2.6	3.1	2.2	2.1	1.5

[a]Data for radiata pine are final stand densities and are from different stands.
[b]DBH=Diameter at Breast Height (height measured to 1.4 m on up hill side of tree).

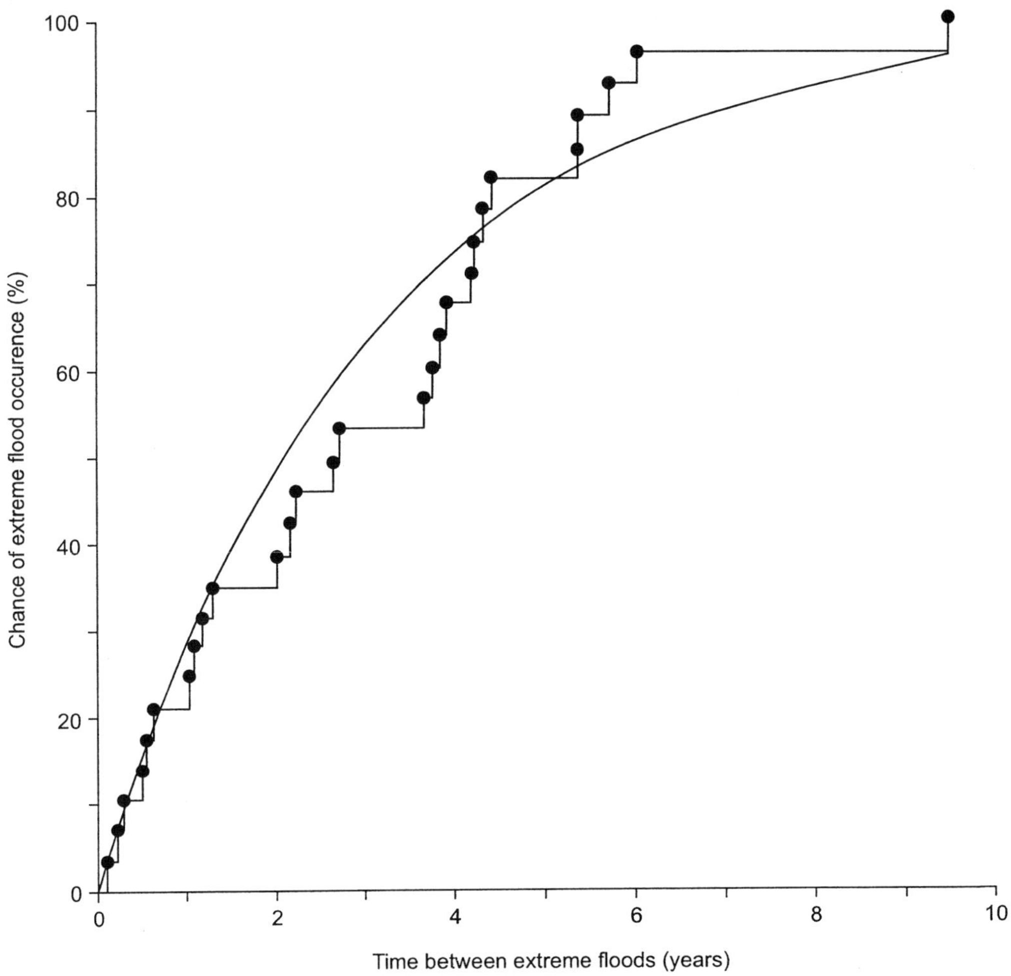

Figure 2. The relationship of time (years) between extreme floods of the Waipaoa River at Kanakanaia (discharge >1500 m^3 s^{-1}) and the chance of their occurrence. Analysis is based on time interval between 29 extreme floods this century. The step function represents field observations, the curve gives their expected chance of occurrence as represented by the cumulative exponential probability function described in Kelliher et al. (1995).

Landslide risk estimation

The analysis of landslide risk is based on the proposal that extreme river floods provide a surrogate for severe rain storms associated with slope instability. The proposal is supported by historical landslide data from the Waipaoa catchment. In this approach, the catchment area of the Waipaoa (1580 km^2), a major regional river flowing southward from the Raukumara Range, was treated as a large areal rain gauge (Kelliher et al., 1995). Extreme flood events were defined as having discharge rates exceeding the average annual flow (15 m^3 s^{-1}) by 2 orders of magnitude. In all, there have been 29 events since 1900. A standard probability analysis was used to produce a relationship of time between extreme floods and the risk of their occurrence (Figure 2). As all data were used to define the probability distribution function, independent error estimates could not be computed.

Results and discussion

Root strength and rates of decay

The mean maximum live tensile strength of 188 small-diameter (1.3–13.9 mm) radiata pine roots was 17.6±0.9 MPa at a mean diameter of 5.3±0.2 mm (Table 2). The range in tensile strength was 7.7–37.5 MPa (O'Loughlin and Watson, 1979). After felling the parent trees, the root-wood tensile strength declined on average 5.9 MPa per year (Figure 3).

The mean maximum live tensile strength of 64 small-diameter (1.1–13.5 mm) kanuka roots was 32.5±3.2 MPa at a mean diameter of 4.7±0.3 mm (Table 2). The range in tensile strength was 15.5–80.0 MPa. Mean tensile root-wood strength increased by 33% to reach a maximum of 43.1±2.8 MPa, 12 months after the cutting of the parent trees, and did not fall below the mean live root-wood strength until about 24 months after tree felling (Watson et al., 1997). The mean rate of loss of tensile strength of kanuka root-wood over the 3-year period after attaining maximum tensile strength, was 9.2 MPa per year (Figure 3).

The initial increase in kanuka root tensile strength and the delay in the onset of root decay is discussed in Watson et al. (1997). To summarize, loss of root-wood moisture after felling is assumed to be accompanied by a decrease in root diameter. As tensile strength is a function of the inverse square of the radius, a small decrease in root diameter can give an appreciable increase in the calculated root tensile strength. The length of time between the death of the parent tree and the onset of root decay is species dependent. In radiata pine, a soft-wood species, the time will be short, possibly a few weeks (O'Loughlin and Watson, 1979), but in more durable timbers it will be much greater. For kanuka, the time between felling and the onset of root decay is about 12 months (Figure 3).

Root growth and site-occupancy

In the 8 years following planting, the mean lateral growth rate of radiata pine roots was 0.44 m $year^{-1}$ (Table 1). For trees planted at 1250 stems ha^{-1} (a recommended slope stabilization regime), 800 stems ha^{-1} (a commercial forestry regime) and 400 stems ha^{-1} (an agro-forestry regime), root site-occupancy at this average root growth rate would occur in 3.7, 4.6, and 6.5 years, respectively.

Uniform canopy heights are a feature of kanuka stands and indicate a relatively short colonization or establishment period. The average lateral growth rate of kanuka roots during the first 6 years was 0.25 m $year^{-1}$ (Table 1), and at a natural establishment density of 16 000 stems ha^{-1} site-occupancy by the root systems would be achieved in 1.8 years. Although kanuka root systems are considerably smaller than those of radiata pine, the difference in individual biomass is more than compensated for by the higher kanuka stand density.

Information on root strength and rates of decay when combined with root growth data enable the construction of a generalized conceptual model of the changes in nett root reinforcement provided by kanuka and radiata pine following felling through the early phase of the subsequent planting cycle (Figure 4). Nett root reinforcement is assumed to attain a maximum value at root site-occupancy. If tree replanting is delayed, say 1 year, nett root reinforcement of both tree species is at a minimum between 2 and 2.5 years after felling. The root reinforcement by radiata pine, established at 1250, 800, and 400 stems ha^{-1}, will decline by between 70 and 80% before the roots of the replacement trees start to contribute to soil strength. As the onset of decay of kanuka roots is delayed 1 year (Figure 3), the maximum decline in nett root reinforcement after felling is about 25%. Slopes felled of radiata pine are potentially more vulnerable to the stresses promoting slope instability, as their lower root contribution to soil strength indicates slope failure could be triggered by storm events of lower

Table 2. Root-wood strength and root diameter of radiata pine and kanuka. Figures in parentheses represent the standard error of means

Months since cutting	Radiata pine		Kanuka	
	Mean tensile strength (MPa)	Mean root diameter (mm) [*n*]	Mean tensile strength (MPa)	Mean root diameter (mm) [*n*]
0	17.62 (0.46)	5.3 [188], (0.2)	32.45 (1.59)	4.7 [64], (0.3)
3	14.37 (0.54)	5.6 [105], (0.2)	–	–
6	–	–	36.36 (1.36)	5.6 [74], (0.4)
9	12.31 (0.56)	6.2 [134], (0.2)	–	–
12	–	–	43.13 (1.42)	5.9 [80], (0.3)
14	11.04 (0.39)	6.8 [140], (0.2)	–	–
18	–	–	34.94 (1.56)	5.2 [89], (0.3)
24	–	–	31.64 (1.38)	6.1 [100], (0.3)
29	3.33 (0.47)	8.3 [51], (0.5)	–	–
30	–	–	28.30 (1.41)	6.1 [79], (0.3)
36	–	–	24.20 (1.67)	7.0 [67], (0.4)
42	–	–	19.15 (1.04)	5.3 [68], (0.2)
48	–	–	15.53 (0.94)	6.3 [87], (0.2)

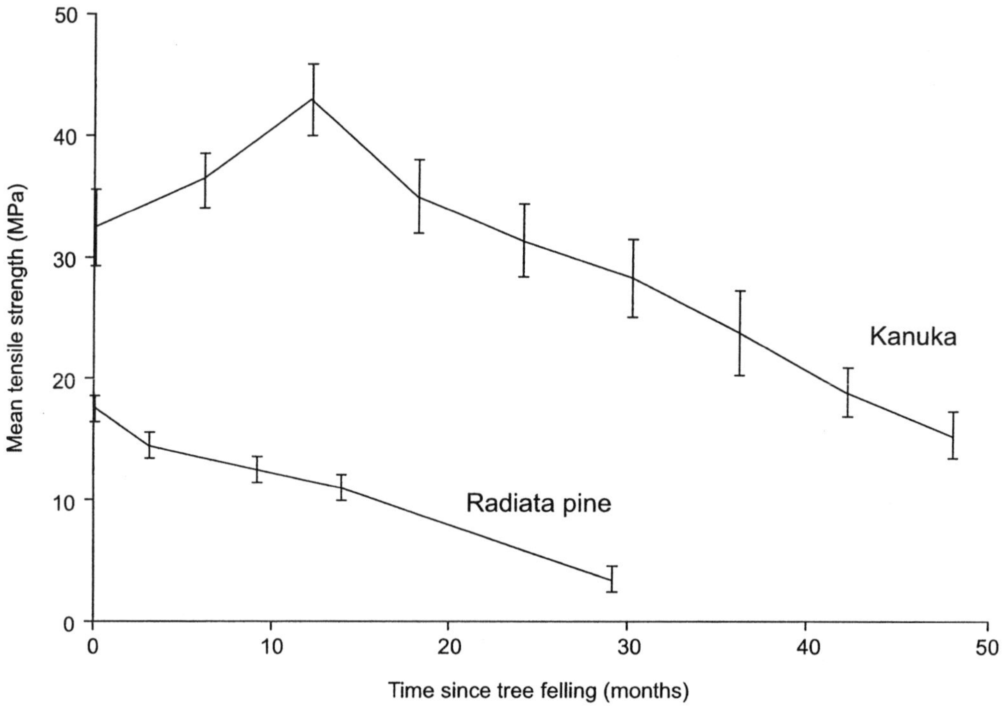

Figure 3. Plot of mean maximum tensile strength against time elapsed since tree felling for kanuka roots <20 mm diameter and radiata pine roots <18 mm diameter.

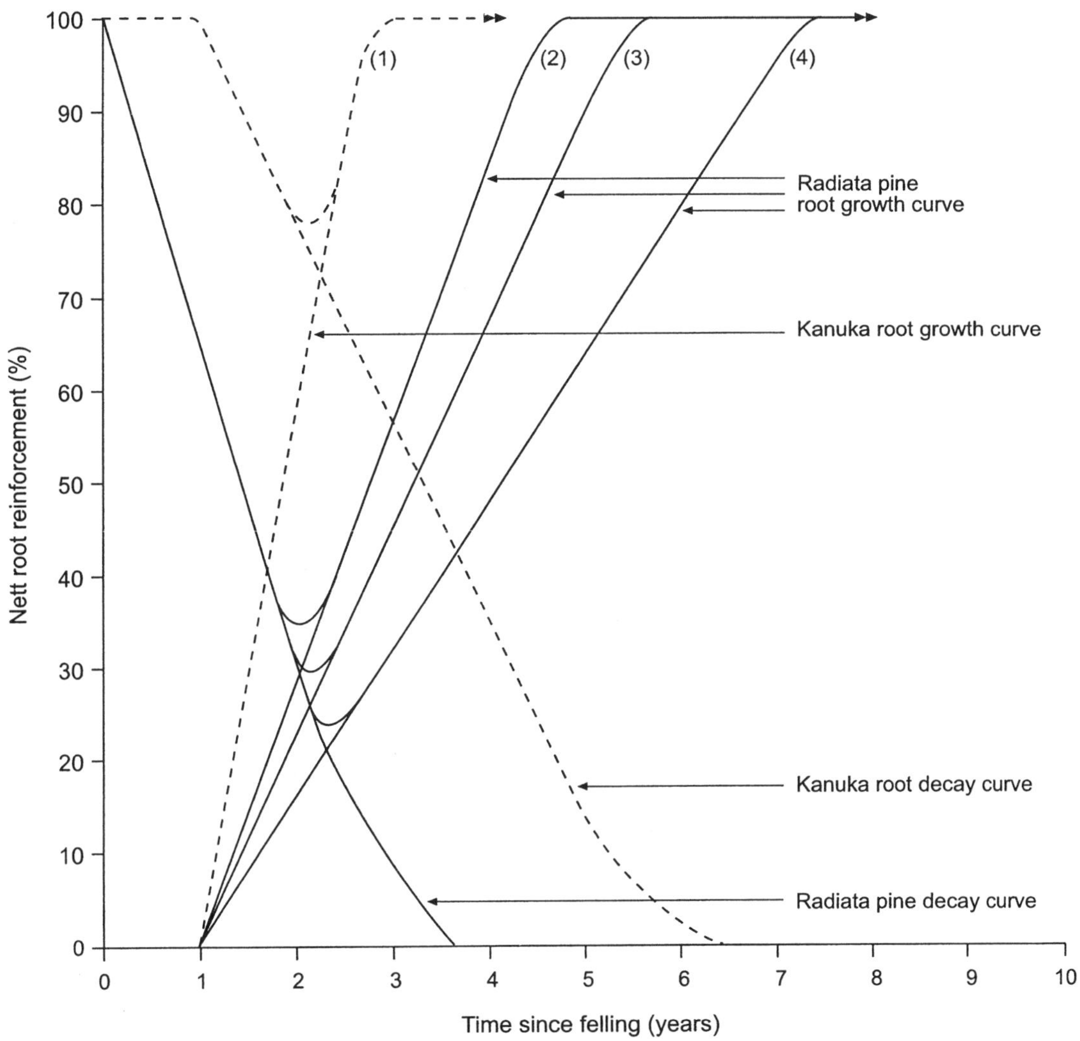

Figure 4. Comparative model of relative root reinforcement changes after clear felling (at year 0) and planting (at year 1) of kanuka and radiata pine. Growth curve (1) represents an initial natural kanuka establishment density of 16 000 stems ha^{-1}. Growth curves (2), (3) and (4) represent initial radiata pine planting densities of 1250 stem ha^{-1} (a recommended slope stabilisation regime), 800 stems ha^{-1} (a commercial forestry regime), and 400 stems ha^{-1} (an agro-forestry regime), respectively.

rainfall duration and intensity than those required to fail a slope cleared of kanuka.

If we now assume that the probability of an extreme flood on the Waipaoa River can be used to indicate the likelihood of a storm to cause slope instability, the curve in Figure 2 suggests that there is about a 30% chance of at least one extreme event every year (observed data suggests 25%) and a greater than 99% chance every 10 years. Erosion-susceptible slopes replanted 1 year after felling in radiata pine at 1250 stems ha^{-1} regain root site-occupancy in 4.7 years, an interval during which there is an 80% chance of experiencing an extreme flood. Similarly for radiata replanted at 800 and 400 stems ha^{-1}, root site-occupancy is regained in 5.6 and 7.5 years during which the probability of occurrence of an extreme event is 85 and 90%, respectively. For erosion-susceptible slopes on which kanuka has become established, the probability of a significant event within the 2.8 years prior to root site-occupancy is 60%. Although the difference in time that either species would leave a slope vulnerable to a major storm is between 2 and 5 years, for slopes growing radiata pine at the planting densities suggested, the increase in probability of such an event happening is in the order of 20–30%. This increase in risk probability approximates the 12-month delay in the onset of kanuka root decay – a delay which may in part account for the greater level of stability provided by kanuka on clearcut slopes.

Conclusions

The general understanding of how vegetation contributes to slope stability is relatively well advanced. However, a detailed appreciation of the parameters that could be used in selecting species that have specific below-ground qualities that can be used to control shallow mass movement is not as well developed. In New Zealand, the choice of plant species for erosion control are often chosen by default. Species selection is frequently decided on features such as availability, drought tolerance, altitudinal range, or perceived below-ground characteristics based on above-ground growth, rather than on their performance attributes in terms of stabilization. Our conceptual model provides an initial opportunity to rank plant species that have specific below-ground rooting habits that can be used to control erosion. In particular, as the model allows the use of more detailed information on below-ground plant characteristics coupled with an opportunity to manipulate planting density, it opens the possibility for a more cost-effective targetting of plant materials for landslide control and prevention.

The underlying objective of this paper was to explore the concept of root site-occupancy in relation to stand density and by using a simple model and a flood frequency probability curve, give forest managers an appreciation of the risk involved in leaving erosion-prone slopes without the benefit of adequate root reinforcement; in another words, supply a workable tool that tended more to the practical than academic. It was not our intention to provide a rigorous risk prediction analysis. To be able to estimate the risk to $\pm 10\%$ was considered a reasonable goal, given the diversity of input parameters.

Many studies have shown that tree roots improve soil strength and slope stability (O'Loughlin and Ziemer, 1982; Phillips et al., 1990; Sidle 1991; Wu and Erb, 1988), and many models have also been developed to assess vegetation effects on slope stability. A limited number of models allow comparisons between vegetation types or management regimes. Of these, the most recent group tend to be distributed physically based models that can be coupled with GIS (Sidle and Wu, in press). However, several non-distributed and time-oriented models have been developed to evaluate effects of timber harvesting or vegetation management on landslide occurrence (Benda and Zhang, 1990; Sidle, 1992; Ziemer et al., 1991). Our approach has been similar to the latter group. This study demonstrates that our conceptual model when coupled with data on root strength, root decay, and root growth is useful for evaluating vegetation conversion or clearance issues on unstable hill country such as found in the East Cape region of New Zealand.

Acknowledgements

We thank our colleagues Dr Jagath Ekanayake and Donna Rowan for assistance during the data collection phases of the various projects upon which this paper is based; and the Wood Technology Dept., School of Forestry, Canterbury University for the generous use of their Instron Universal Testing Machine. This study was funded by the Foundation for Research, Science and Technology (Contracts C09406, C09606, C09807).

References

Benda L and Zhang W 1990 Accounting for the stochastic occurrence of landslides when predicting sediment yields. IAHS-AISH Publ. 192, 115–127.

Ekanayake J C, Marden M, Watson A J and Rowan D 1997 Tree roots and slope stability: a comparison between pinus radiata and kanuka. NZ J For. Sci. 27(2), 216–233.

Kelliher F M, Marden M, Watson A J and Arulchelvam I M 1995 Estimating the risk of landsliding using historical extreme river flood data. J. Hydrol. (NZ) 33(2): 123–129.

Marden M, Phillips C J and Rowan D 1991. Declining soil loss with increasing age of plantation forests in the Uawa Catchment, East Coast Region, North Island, New Zealand. *In* Proceedings of International Conference on Sustainable Land Management, Nov. 17-23, Napier, New Zealand. pp. 358–361.

Ministry of Forestry 1994 A Guide to the East Coast Forestry Project. Ministry of Forestry, Wellington, New Zealand.

New Zealand Meteorological Service 1984 Rainfall normals for New Zealand 1951–1980. New Zealand Meteorological Service Misc. Publ. No. 185.

O'Loughlin C L and Watson A J 1979 Root-wood strength deterioration in radiata pine after clearfelling. NZ J. For. Sci. 9(3), 284–293.

O'Loughlin C L and Ziemer R R 1982 The importance of root strength and deterioration rates upon edaphic stability in steepland forests. *In* Carbon Uptake and Allocation in Subalpine Ecosystems as a Key to Management. Proceedings of an International Union of Forest Research Organizations workshop P.I. 107-00 Ecology of Subalpine Zones, August 2–3 Oregon State University, Corvallis, OR, USA. Ed. R H Warring. pp.70–78.

Phillips C J and Watson A J 1994 Structural tree root research in New Zealand. Manaaki Whenua Press, Landcare Research Science Series No. 7., Lincoln, New Zealand. 71 pp.

Phillips C J, Marden M and Pearce A J 1990 Effectiveness of reforestation in prevention and control of landsliding during large cyclonic storms. *In* Proceedings of International Union of Forest Research Organisations XIX World Congress, Volume 1, pp 340–350.

Sidle R 1991 A conceptual model of changes in root cohesion in response to vegetation management. J. Envir. Qual. 20, 43–52.

Sidle R 1992 A theoretical model of the effects of timber harvesting on slope stability. Wat. Resour. Res. 28, 1897–1910.

Sidle R and Wu W (In press) Simulating effects of timber harvesting on the temporal and spatial distribution of shallow landslides. Z. Geomorph.

Tomlinson A I and Sansom J 1994 Rainfall normals for New Zealand for the period 1961 to 1990. NIWA Science and Technology Series No. 3.

Watson A J and O'Loughlin C L 1990 Structural root morphology and biomass of three age classes of *Pinus radiata*. NZ J. For. Sci. 20(3) 97–110.

Watson A J, Marden M and Rowan D 1995 Tree species performance and slope stability. *In* Vegetation and Slopes, Ed. D H Barker. pp 161–171. Thomas Telford Press, London.

Watson A J, Marden M and Rowan D 1997 Root-wood strength deterioration in kanuka after clearfelling. NZ J. For Sci. 27(2), 205–215.

Wu T H and Erb R 1988 Study of soil-root interaction. J. Geotech. Eng. 114(12), 1351–1375.

Wu TH and Swanston 1980 risks of landslides in shallow soils and its relation to clearcutting in Southeastern Alaska. For. Sci. 26(3) 495–510.

Ziemer R R, Lewis J, Rice R M and Lisle T E 1991 Modelling the cumulative watershed effects of forest management strategies. J. Envir. Qual. 20, 36–42.

A. Stokes (ed.), The Supporting Roots of Trees and Woody Plants: Form, Function and Physiology, 51–57.

A study on the adaptability mechanism of tree roots on steep slopes

Madhusudan B. Shrestha*, M. Horiuchi, Y. Yamadera and T. Miyazaki
*Shinshu University, Faculty of Agriculture, Department of Forestry, Kami Ina Gun, Minami Minowamura 8304, Nagano-Ken, 399-4511, Japan**

Key words: centrifugal direction, centripetal direction, down-slope, geotroprism, lateral roots, liminal angles, tap-root, up-slope

Abstract

Woody vegetation affects the stability of steep mountain slopes mainly by increasing the shear strength of soil *via* root reinforcement. Plant roots on steep terrain tend to elongate uphill. As the slope angle increases, the root growth uphill is greater. It is assumed that gravitational force may be one of the factors responsible for this orientation of roots. Therefore, we carried out a study of root growth under applied centrifugal force to determine the effects of root gravitropism on steep slopes. *Pinus densiflora* was sown in plastic pots and placed at different distances from the same center (at different radii) on a continuous rotary circular platform for five months. Two other stationary plots, one in the horizontal plane and the other with an inclined slope of 45°, were set up as control plots. This experiment showed that the direction of root elongation changes extensively with respect to the centrifugal force applied on the plant. The tap-root elongates towards the resultant direction of the sum of gravity and centrifugal force, whereas the lateral roots elongated in the direction opposite that of centrifugal force. The control plots also showed the same tendency i.e. lateral roots elongated in the uphill direction in the inclined plot, and the tap-root developed towards the gravity in the horizontal plot.

Introduction

Roots of a mature plant may grow in many directions, including upwards. The direction adopted may be under genetic control since root systems of different species have characteristic morphologies owing, in part, to the degree by which the individual root deviates from the vertical (Kutschera, 1960). Rufelt (1962) proposed that the liminal angle is the resultant of the opposing positive (downward) and negative (upward) geotropic growth, each being differently affected by external and internal conditions.

It is widely believed that the root cap is geotropic, and acts both as a sensor of the direction of gravity and as a source of root growth regulators. It has been suggested that the regulator accumulates within the lower half of the root and causes the retardation of growth, thus leading to the bending of roots. The root cap is not the only portion of the root tip that is important for geotropism. The tissue between the cap and growing zone may also play a role in gravity. A study carried out by Jackson and Barlow (1980) suggested that the tip is not the site of gravity perception but is essential for geotropism, as it is the principal site for gravity perception. Studies have shown that the movement of dense, starch-containing amyloplasts under the force of gravity, is the initiating physical event of geotropism (Audus 1962, 1972; Iversen and Larsen, 1973; Volkman and Sievers, 1979). In another study, geotropism was regained some hours after the roots were removed from hormone solutions, in association with the reformation of starch in the proplastids of the cap. Since the removal of the root cap immediately removes the ability to perceive gravity, there can be no lateral movement of the inhibitor in the elongating zone (Juniper, 1976).

Little information is available on the adaptable mechanism of trees on steep slopes. The elongation of roots towards the vector of gravity is due to the production of amyloplast in the root cap, which changes

* Fax: +81 0265 725259.
E-mail: sabo8my@gipmc.shishu-u.ac.jp

the permeability of Ca^{2+} ions in cells, resulting in the deviation in growth (Takahashi, 1993). As well as gravitropism, light, aeration, water, mechanical stimuli and other environmental factors play significant roles in regulating the direction of elongation (Coutts and Nicoll, 1991; Darwin, 1880; Sachs 1887). Tap-roots are commonly supposed to exhibit positive geotropism, and lateral roots are sometimes plagiogeotropic although the evidence for this is conflicting (Lake and Slack, 1961).

On steep slopes, tree roots are generally oriented uphill and stabilize the soil (Yamadera, 1992). It has been found that the formation of tree roots differs according to the steepness of the slope and is species specific. Studies have shown that lateral roots of *Picea sitchensis* also grow uphill (Coutts and Nicoll, 1991); however, the mechanism for such phenomena has not been explained. As roots are highly effective in strengthening loose soil particles (Karizumi, 1979), it can be presumed that roots will be useful for stabilizing mountain slopes.

In the present study we investigated the change in root orientation on slopes, by applying centrifugal force on plants. We assumed that the reason for the change in root orientation on slopes may be due to gravitational force acting on roots. We also considered that the different reactions of root growth on slopes may be due to the responses that roots experience due to gravity. Many studies have been carried out on gravitropism in crop plants, but in this study a woody plant, *Pinus densiflora* was selected. *Pinus densiflora* is comparatively drought tolerant, grows in barren land and has a large tap-root with horizontal lateral roots.

Materials and methods

This experiment was carried out under greenhouse conditions at Shinshu University research farm in June, 1997. In the experiment, hard plastic pots were filled with dry sand and sown with *Pinus densiflora*. The pots were then placed on a rotary circular platform with different distances from the center (at different radii in two circumferences, Figure 1). The pots were 170 mm dia., 200 mm tall and the depth of the sand was 170 mm. The plants were watered twice daily and supplied with a 1000x diluted solution (N.P.K. = 5.10.5) once or twice a month. The experiment was carried out for a period of five months.

The non-stop continuous rotating circular platform was set up at 20 revolutions per/min. for five months. The applied centrifugal force is calculated from the formula:

$$F = M^{*}V^{2}/r,$$

Where M is the weight of the plant, r is the radius of the circular platform (m) and V is the applied velocity ($0.67^{*}\pi^{*}$r m/s). Therefore, the applied centrifugal force was: $F = 1.85$ N (for small radius, $r_1 = 0.41$ m) and $F = 2.57$ N (for large radius, $r_2 = 0.585$ m). The centrifugal force for r_1 is denoted by S.C.P (small centrifugal plot) and that of r_2 is L.C.P. (large centrifugal plot).

In addition to the rotational plot, two other stationary plots were set up as controls: one in the horizontal plane and the second with an inclination of 45° to the horizontal surface (Figure 2). After germination, the plants were thinned to leave 4–5 seedlings per pot in all three plots. The roots of seeded *Pinus densiflora* were excavated and growth height and liminal angles were measured. The excavation and measurement was done during October and November of the same year. The mean height of the shoot was 4.5 cm. T-tests were used to determine the difference between liminal angles of roots in different treatments.

Results

The influence of centrifugal force on the tap-root

The liminal angles of the tap-roots were found to be 1.67° ± 1.73° in the horizontal plot, 2.00° ± 2.26° in the inclined plot and 9.50° ± 0.98° in both L.C.P.s and S.C.P.s. The liminal angles of tap-roots in the horizontal plots and the centrifugal plots were significantly different ($p < 0.01$). Similarly, the differences between the inclined plots and the centrifugal plots were also highly significant ($p < 0.01$). These results suggest that the liminal angle of the tap-root changes significantly as centrifugal force is applied. However, the difference between the S.C.P. and L.C.P.s was found to be insignificant.

Influence of centrifugal force on first-order lateral roots

Changes in the liminal angles in the horizontal and inclined plots

In the horizontal plots, the mean liminal angle of lateral roots was found to be 76.7° ± 2.7°. However, in the inclined plots, the liminal angle up-slope was greater and that down-slope was smaller (Table 1).

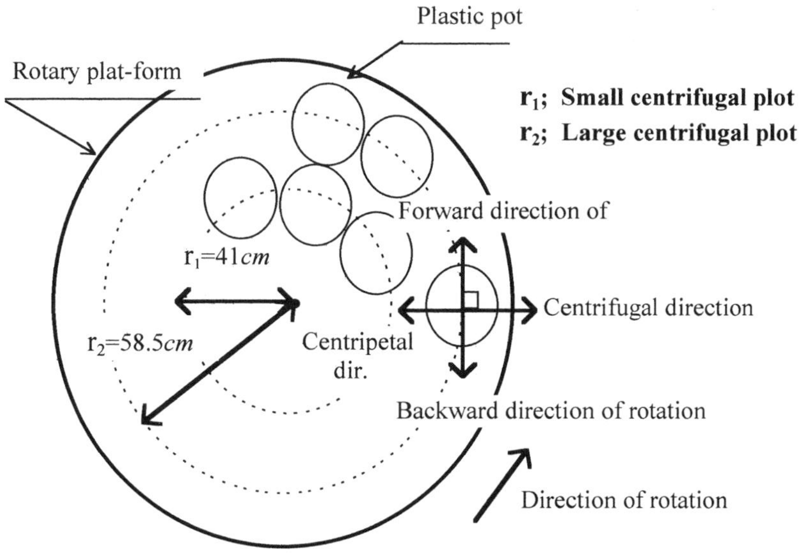

Figure 1. A model diagram of the centrifugal plot.

Table 1. The liminal angles of the lateral roots in the inclined plot

Plots	Direction	Mean (n)	Standard error
Horizontal	All	76.68° (55)	± 2.69°
Inclined	Up-slope	95.15° (13)	± 11.27°
	Down-slope	63.18° (11)	± 9.09°
	Horizontal	83.24° (20)	± 7.70°

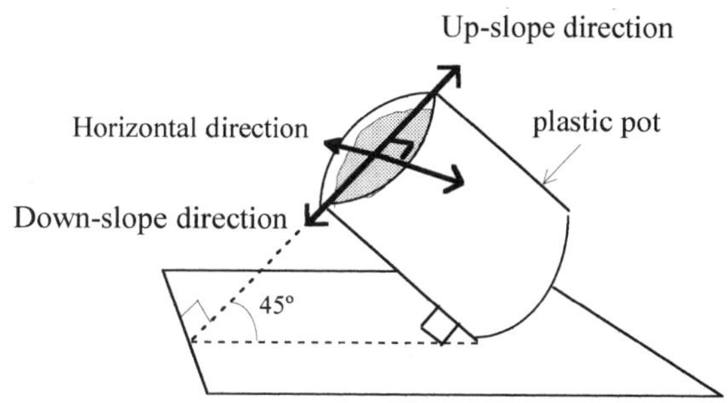

Figure 2. A model diagram of the inclined plot.

The mean liminal angle up-hill increased by 18.5° whereas that down-hill decreased by 13.5°. However, there was no significant difference in the average liminal angle between the horizontal plot and that in the horizontal direction in the inclined plot. These results indicate that the lateral roots elongate more up-hill compared to those down-hill and are analogous to those observed on steep mountain slopes.

Changes in the mean liminal angles of lateral roots in the centrifugal environment

There was no significant difference between mean liminal angles in the centripetal direction of the S.C.P.s and the L.C.P.s (Table 2). However, the difference between these mean liminal angles and those growing in the horizontal plot was found to be significantly different (Table 2, $p < 0.01$). Differences between the mean liminal angles in the centripetal direction and those in the centrifugal, forward and backward directions of rotation, were found to be highly significant (Table 2, $p < 0.01$).

There were no significant differences in mean liminal angle in the centrifugal direction compared to that in the horizontal plot, nor in the anti-clockwise and clockwise directions in the S.C.P.s and L.C.P.s compared to those in the horizontal plot (Table 2).

Response of the tap- and lateral roots to centrifugal force

The degree of root displacement was obtained by subtracting the liminal angle (the angle of deviation from a plumb line) in the applied centrifugal plot by that in the horizontal plot. The degree of displacement of the tap-root was found to be 7.83° ± 0.98° (Figures 3–7), which suggests that it grows more towards the centrifugal direction (by approx. 8°) when force is ap-

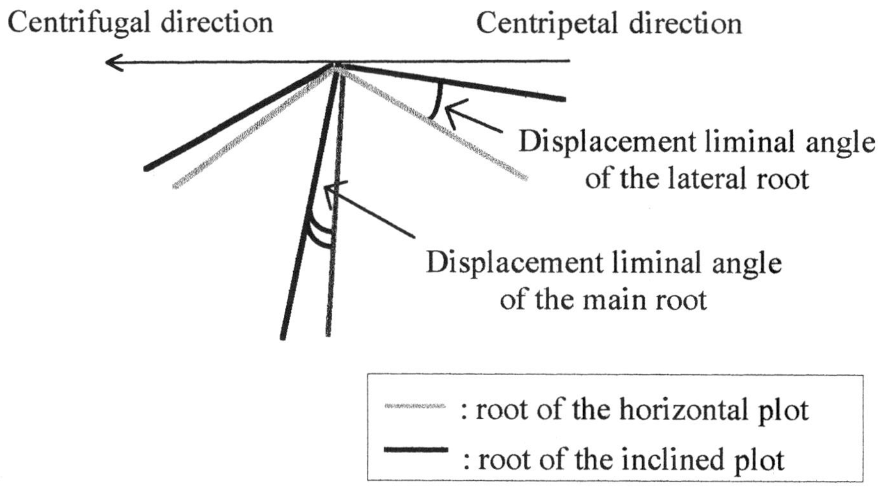

Figure 3. Diagrammatic representation of the displaced liminal angle in the centrifugal plot (cross-section).

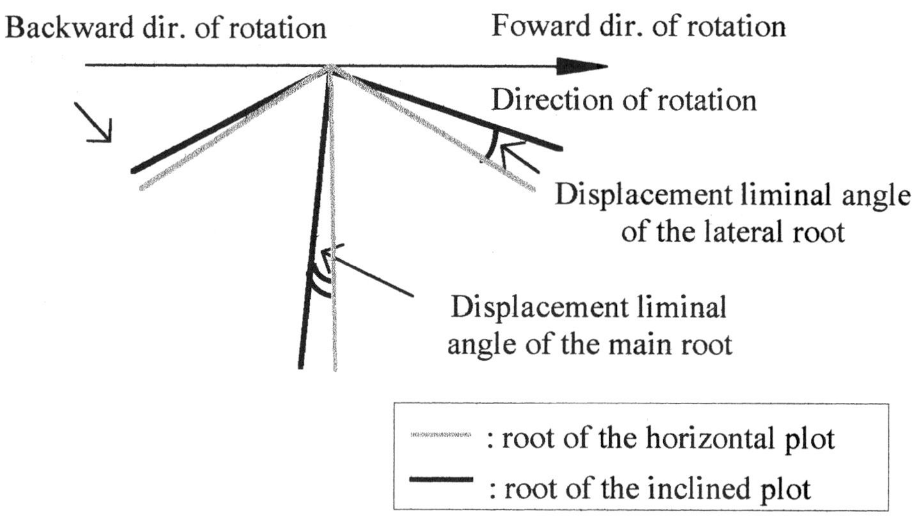

Figure 4. Diagrammatic representation of the displaced liminal angle in the centrifugal plot (front-view).

plied. The degree of displacement of lateral roots was found to be 4.43° ± 5.52° in the centripetal direction, 2.73° ± 4.83° in the centrifugal direction, 0.85° ± 3.87° in the forward rotation direction and −0.09° ± 3.39° in the backward-rotation direction (Figures 3–7). The obtained values of displacement in the lateral roots were smaller compared to that of the tap-root. The difference between the amount of displacement of the tap-root compared to that of the lateral roots in the centrifugal direction was significant ($p < 0.05$). Besides this, amount of displacement of the tap-root in the other two directions (forward and backward) was highly significant ($p < 0.01$). These results imply that the tap-roots are more sensitive to centrifugal force than lateral roots.

The degree of displacement of lateral roots in the inclined plots was 19.63° ± 11.94° up-slope, 15.57° ± 9.62° down-slope and 6.48° ± 7.70° in the horizontal direction (Figure 8). In the tap root, the displacement was very slight (0.33° ± 2.26°). Therefore, the tap-root growing in the inclined plane elongated towards the direction of gravity, as in the horizontal plane, since the obtained value is nearly zero (0.33). How-

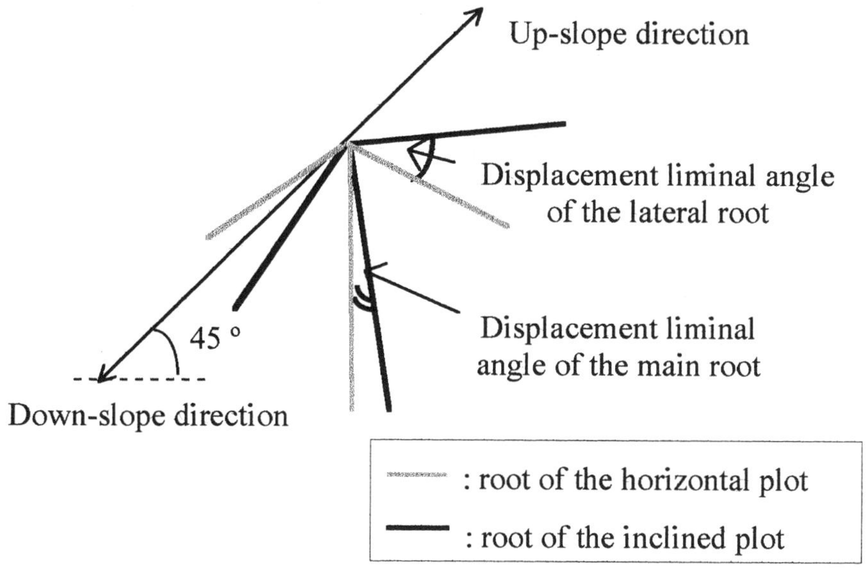

Figure 5. Diagrammatic representation of the displaced liminal angle in the inclined plot (cross-section).

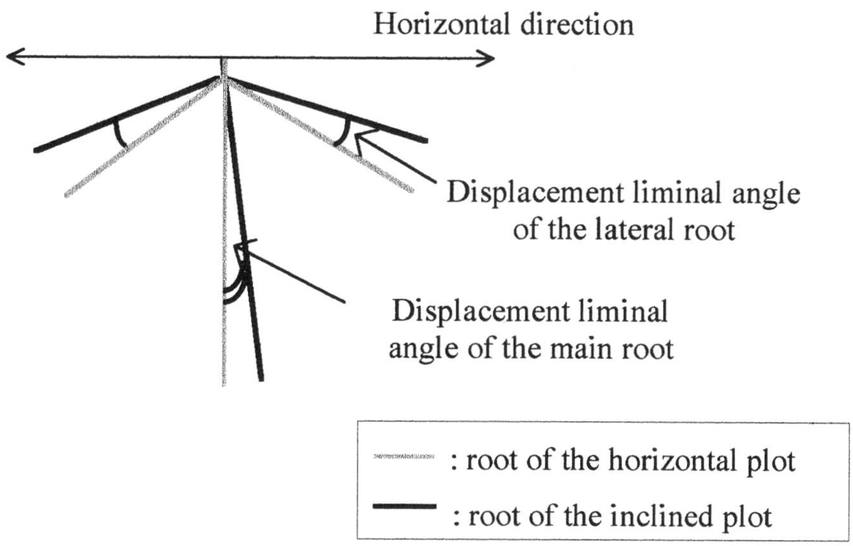

Figure 6. Diagrammatic representation of the displaced liminal angle in the inclined plot (front-view).

ever, the lateral roots significantly elongated up-slope by 20° and decreased by 16° down-slope.

Discussion

It can be concluded that tap-roots and lateral roots behave differently when grown under centrifugal force. The tap-root responds strongly and develops in the direction of the centrifugal force whereas lateral roots elongate in the opposite direction to that of the main root. The differences in the liminal angles of lateral roots were thought to be due to the strength of the applied centrifugal force. Lateral roots elongate towards the centripetal direction, i.e. not in the direction of the force. The influence of the applied centrifugal force had little effect on the clockwise and anti-clockwise. Responses of lateral roots in the centrifugal plot were similar to those in the inclined plot. The liminal angles of lateral roots in the centripetal direction and the up-slope direction were also found to be analogous. As the difference in liminal angles between the clockwise

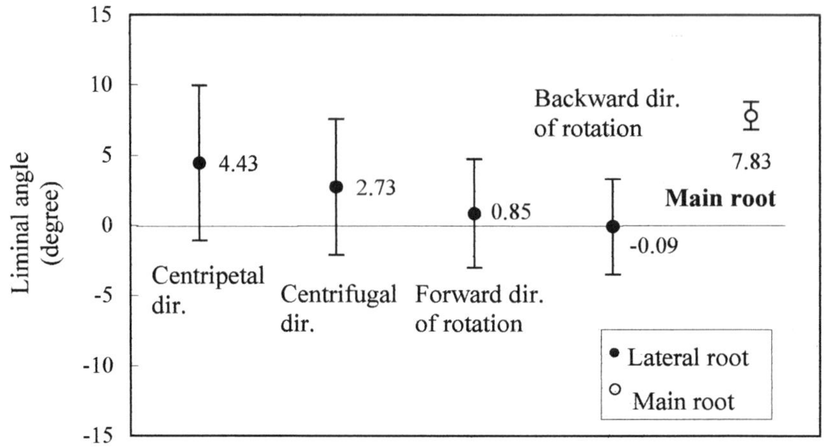

Figure 7. Displacement of mean liminal angles of the tap- and lateral roots in the centrifugal plot.

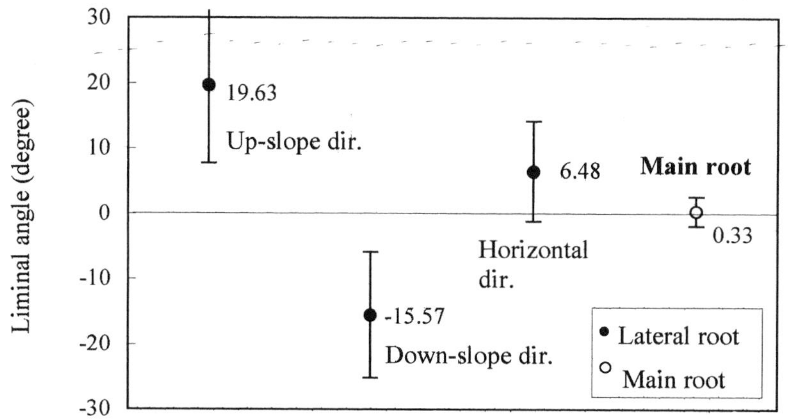

Figure 8. Displacement of mean liminal angles of tap- and lateral roots in the inclined plot.

Table 2. The liminal angles of the lateral roots in the centrifugal plots

Plots	Direction	Mean (n)	Standard error
Horizontal	All	76.68° (55)	± 2.69°
Small centrifugal	Centripetal	92.67° (8)	± 8.63°
	Centrifugal	76.17° (12)	± 6.59°
	Forward	78.82° (6)	± 4.47°
	Backward	78.67° (8)	± 2.85°
Large centrifugal	Centripetal	91.10° (11)	± 9.53°
	Centrifugal	70.67° (17)	± 7.76°
	Forward	74.25° (14)	± 7.73°
	Backward	75.67° (4)	± 4.88°

and anti-clockwise directions of rotation was found not to be significant in the S.C.P.s and L.C.P.s, centrifugal force had no effect. The reason for this might be because of the small centrifugal force used in the experiment.

The upward elongation of lateral roots in the inclined plot might be the result of the weight (soil or water in the pot) that the roots experienced. As the weight increases or as the slope angle increases, the liminal angle may deviate extensively from the vector of the force implied. When the root enters a region where the surface is elevated, as in the Coutts and Nicoll's (1991) experiment, the internal tendency to upward growth is manifested because conditions causing downward deflection are no longer present. To investigate the outcome more specifically, experiments on moderate or more steep slopes (less or more than 45°) are desirable. We set up only one centrifugal experimental plot in this study, other plots are necessary with different centrifugal forces to clarify the outcome more clearly.

The uphill elongation characteristics of lateral roots in the inclined plot may be influenced by light, soil-moisture and nutrients etc. Lake and Slack (1961) have shown that exposure to light is a prerequisite for geotropism in some plants. This is because light is necessary for the formation of a growth inhibitor of roots, which afterwards becomes asymmetrically disposed under the influence of gravity. The experimental plastic pots that we used were all opaque, thus influence of light can be neglected. Even though we watered or provided solution very cautiously, in order to provide an equal distribution of water inside the pot, measurements and considerations of these factors are necessary to find out their effects on root elongation.

The decrease in liminal angles of lateral roots is consistent with observations by Coutts and Nicoll (1991) on *Picea sitchensis*. These authors found that there was also a tendency for upward growth in long lateral roots which developed from the upper part of the root system. From the work carried out, they suggested that lateral roots which grew towards the soil surface, bent downwards again, seeking moisture and nutrients. The mechanisms for this downwards deflection of upward growing lateral roots are still unknown. We did not observe these deflections in lateral root development on either the inclined plot or the centrifugal plot.

On investigating the relationship between root elongation and slope steepness, it was observed that the elongation direction of roots changes significantly as the slope becomes steeper. It was found that the growth of roots of arboreal plants and herbaceous plants differs in the direction of elongation corresponding to the steepness of the slope (Yamadera, 1993). The tap-root of the woody plant elongates towards the vector of gravity regardless of the slope angle, however, numbers of lateral uphill roots increase and downhill roots decrease as the slope becomes steeper. Furthermore, the angle between the downhill lateral roots and the main root is smaller whereas the angle between uphill lateral roots and the tap-root becomes greater. These results are similar to our observations in the inclined plot for *Pinus densiflora*. The upwardly growing roots do not only play an effective and vital role in fixing or stabilizing weathered soil strata on steep slopes, but have important implications for plant growth and survival as they can explore the nutrient-rich layers of the soil without being damaged by exposure.

Investigating the adaptability of different species of woody plants on slopes is important for stabilizing steep, mountain or cut slopes. If we know the mechanism of root formation of these plants, it would be helpful in selecting the most appropriate species for particular areas.

References

Audus L J 1962 The mechanism of the perception of gravity by plants. *In* Biological Receptor Mechanisms. Ed. J W L Beament Symposia of the Society for Experimental Biology. Vol. 16, pp 197–226. Cambridge University Press, Cambridge.

Coutts M P and Nicoll B C 1991 Orientation of the lateral roots of trees 1. Upward growth of surface roots and deflection near the soil surface. New Phytol. 119, 227–234.

Darwin C 1881 The Power Movement in Plants D. Appleton and Company.

Iversen T-H and Larsen P 1973 Movement of amyloplast in the statocytes of geotropically stimulated roots. The pre-inversion effect. Physiol. Plant. 28, 172–181.

Jackson M B and Barlow P W 1980 Root geotropism and the role of growth regulators from the cap: a re-examination. Plant Cell Environ. 4, 107–123.

Juniper B E 1976 Geotropism. Ann. Rev. Plant Physiol. 27, 385–406.

Karizumi N 1979 Zyumoku konkei zu setsu. seibundo Shinkosha Publishing Co. Ltd., Nishikicho, Kanda, Chiyoda ku, Tokyo.

Kutschera L 1960 Wurzelatlas mitteleuropaischer Ackerunkrauter und Kulturepflanzen. DLG- Verlags-GmbH, Frankfurt-am-Main.

Lake J V and Slack G 1961 Dependence on light of geotropism in plant roots. Nature 191, 300–302.

Morita S and Abe J 1994 Root Hand Book. Root Research Committee, Japan.

Rufelt H 1962 Plagiotropism in roots. *In* Encyclopedia of Plant Physiology Vol. 17: Physiology of Movements, Part 2. Ed. W Ruhland. pp 322–343. Springer-Verlag, Berlin, Gottingen, Heidelberg.

Sachs J 1887 Lectures on the Physiology of Plants. The Clarendon Press. Oxford, UK.

Takahashi H and Kan H 1987 Utsusen no syokubutsu gaku. Kabusiki gaisya gakkai syuppan senta, Tokyo, Japan. pp 53–95.

Volkmann D and Seivers A 1979 Graviperception in multicellular organs. *In* Encyclopedia of Plant Physiology, New Series. Vol. 3: Physiology of Movements. Eds W Haupt and M E Feinleib. pp 573–600. Springer-Verlag, Berlin, Heidelberg, New York.

Yamadera Y, Anpo A and Yoshida H 1993 Sizenkankyou wo saisei suru tame no midorino sekkei. JAGREE, Tokyo, Japan. 169 p.

STRUCTURAL ROOT SYSTEM ARCHITECTURE IN FOREST TREES

A. Stokes (ed.), The Supporting Roots of Trees and Woody Plants: Form, Function and Physiology, 61–75.

Structural root architecture of 5-year-old *Pinus pinaster* measured by 3D digitising and analysed with AMAPmod

F. Danjon[1,*], D. Bert[1], C. Godin[2] and P. Trichet[3]

[1] *INRA Croissance et Production, Recherches Forestières B.P. 45, 33611 Gazinet Cedex, France;*
[2] *CIRAD - GERDAT, Unité de Modélisation des Plantes B.P. 5035, 34032 Montpellier Cedex, France and*
[3] *INRA Ecophysiologie et Nutrition, Recherches Forestières B.P. 45, 33611 Gazinet Cedex, France*

Key words: biomass, forest tree, *Pinus pinaster*, root system, stem straightness, three-dimensional analysis

Abstract

Pinus pinaster (Ait.) is a high yielding forest tree, producing nearly a fourth of French marketed timber essentially from intensively managed stands located in southwestern France, in the Landes Forest. This species has generally a poor stem straightness, especially when it grows in poor sandy podzol of the Landes Forest, affected by summer droughts and winter floods. Above- and below-ground architecture and biomass as well as stem straightness were measured on twenty-nine 5-year-old planted trees uprooted by pulling with a lumbering crane. A very precise numeric representation of the geometry and topology of structural root architecture was gained using a low-magnetic-field digitising device (Danjon et al., 1998; Sinoquet and Rivet, 1997). Data were analysed with AMAPmod, a database software designed to analyse plant topological structures (Godin et al., 1997). Several characteristics of root architecture were extracted by queries including root number, length, diameter, volume, spatial position, ramification order, branching angle and inter-laterals length. Differences between root systems originated from their dimensions, but also from the proportion of deep roots and the taproot size, which represented 8% of the total root volume. The proportion of root volume in the zone of rapid taper was negatively correlated with the proportion of root volume in the taproot indicating a compensation between taproot and main lateral root volume. Among all studied root characteristics the maximal rooting depth, the proportion of deep roots and the root partitioning coefficient were correlated with stem straightness.

Abbreviations: 3D – three-dimensional; SAL – stump angle of lean; ZRT – zone of rapid taper; RPC – root partitioning coefficient

Introduction

Root architecture is described by the geometric deployment of roots in the 3-dimensional space and by the topology, i.e. how individual roots are connected to each other through branching (Lynch, 1995). Root system architecture is one fundamental aspect of forest tree productivity, determining water and nutrient uptake efficiency, below-ground competition and interactions between roots, soil and micro- organisms (Lynch, 1995). It determines also resistance to uprooting and stem straightness (Balneaves and De La Mare, 1989; Coutts, 1983; Mason, 1985; Stokes et al., 1996).

Root growth models provide a flexible tool for examining the effects of several factors whose importance is difficult to assess in the field (Brown and Kulasiri, 1994). They should be included partly in process-based models, but the number of functions required to describe a root system is considerable (Brown et al., 1997). Architecture and three-dimensional (3D) spatial distribution data of whole root systems is needed to establish these functions.

Tree root systems are often highly variable (Henderson et al., 1983b). A reliable characterisation

* FAX No: +33 55 668 0223. E-mail: fred@pierroton.inra.fr
Plant and Soil is the original source of publication of this article. It is recommended that this article is cited as: *Plant and Soil* **217:**, 49–63, 1999.

of their structure require excavation or uprooting of the entire root system. Only few joint topology and geometry measurements of tree root systems have been made so far. They consisted in a manual assessment of 3D coordinates of a small sample of forest trees (Henderson et al., 1983a; Puhe, 1994). A new method for the joint measurement and analysis of tree root architecture was adapted by Danjon et al. (1999) from a technique developed by Sinoquet and Rivet (1997) to measure aerial parts of trees. Coordinates of root segments were measured jointly with the topology using a 3D low-magnetic-field digitising device. Output data were analysed with the AMAPmod software specially developed to handle plant architectures (Godin et al., 1997; Reffye et al., 1995).

The Landes Forest covers one million hectares in southwestern France, row seeded or planted with maritime pine (*Pinus pinaster* Ait.), producing nearly a fourth of French marketed timber. Maritime pine is a rapid growing tree, taking typically 40–50 years to reach clear-cut size in intensively managed stands. The soils of the Landes Forest are poor (Saur, 1989) and water is a predominant limiting factor to growth in summer when the root zone is limited by a hard pan or by high winter water table (Berbigier et al., 1991). Maritime pine has a poor stem straightness already as sapling when growing on humid or semi-humid sandy podzol (Arbez, 1979). Damage by windthrow causes regularly important economic losses in the Landes Forest (Berthier et al., 1997). A breeding program for maritime pine started in 1960. A low juvenile - mature genetic correlation in stem height growth was found which may originate from differences in resource allocations to root development (Danjon, 1994).

Only few papers have been published concerning the root systems architecture of *Pinus pinaster* (Auberlinder, 1982; Chaperon et al., 1991; Maugé, 1987) and none of them presented a comprehensive study of the root system architecture. A better knowledge of the rooting in young maritime pine was needed, especially to investigate the relations to stem straightness (Colin et al., 1996). It was also required in a study of biomass allocation including the above- and below-ground architecture.

Twenty-nine 5-year-old maritime pines were mechanically uprooted and measured using a 3D digitising device. This paper presents results from analyses of root architecture at the individual tree level made with the AMAPmod software.

Material and methods

Tree material

The 'L' plot is a 5.3 ha stand located 20 km southwest from Bordeaux (58 m altitude; latitude 44° 44′ N, longitude 0° 46′ W). The soil is a medium humid sandy podzol on a flat location, and the plot is surrounded by a 0.5 to 1.20 m deep ditch. Stumps from the previous rotation were removed from the soil. The field has been ploughed in the row direction, then perpendicularly, and finally smoothed with a single action two gang disk harrow without fertilisation. A 4×2 m row maritime pine plantation was established in spring 1992 with seedlings from local provenance grown one year (1 + 0) in 400 cm^3 rectangular peat planting bricks. Rows were oriented in the 148° direction. A mechanical weeding has been done during the first three years with a disk harrow. The mean yearly rainfall is 900 mm and the mean annual temperature is 13.5 °C (Granier and Loustau, 1994). The water table may fluctuate close to the ground surface in rainy winters and sink to approximately 1.5 m deep in midsummer. A discontinuous and deep (>50 cm) hard pan was found in some points but it was not mapped. The mean wind direction is north-west. A 6 months monitoring of wind near the site showed that winds blow rarely from South or East, and blow frequently and with the highest speed either from North (10°) or from West (275°) (Berbigier, unpublished data). The mortality at age 5 averaged fifteen trees per ha.

Measurements

In February 1997, twenty-nine classes with 218 pines have been defined from the height at the age of 4 years. The 29 maritime pines with the mean height in each class were sampled. Height, diameter at breast height (dbh) and stem straightness were measured on standing trees. Two whole-stem straightness scores were subjectively assessed at right angles to each other on a scale of 0 (worst) to 20 (best) (William and Lambeth, 1989). Stem straightness was also assessed by the stump angle of lean (SAL) on the first meter (Danjon, 1994). After forking manually the forest floor within one meter from the stump, trees were carefully pulled upwards by a logging crane with a chain attached around the base of the trunk. Large broken roots were manually excavated afterwards and fastened to the corresponding place in the root system. All entire root systems were weighed after 6 months air drying. The dry biomass was calculated from the drying of a sample of seven root systems at 60 °C. The stem biomass was calculated from the whole stem fresh weight and from a sample of five oven-dried stem disks per tree (Porté, 1999).

The X, Y, Z coordinates of roots were measured with a 3D digitiser (3 SPACE Fastrak, Polbemus, Long Ranger option) using low frequency electromagnetic fields sensing (Polhemus, 1993). The measurement was driven by a specifically written software called 'Diplami' running under Windows 95 on PC computers (Sinoquet and Rivet, 1997; Sinoquet et al., 1997). One diameter was recorded at each position measured. A plastic dial calliper was used because large metallic items are likely to perturb the measurement.

According to Atger and Edelin (1995), the taproot is the first order root, roots originating on the taproot are second-order roots, roots initiated on second-order are third-order roots and so on. According to Henderson et al. (1983a), 'central axis' referred to the vertical line passing through the center of the growth rings at soil level. The distance to the central axis is the 'radial distance'. 'Angle' was used for the direction relative to the horizontal plane, measured downwards. 'Forks' were recorded when a root split into two equally sized roots subtending equal angles to the parent root. The position and diameter of all branching points were recorded. Between branching points, measurements were made all 20 centimetres when the root was straight and all 2 cm when the root was highly bent or tapered. The root part between two measurement points was termed a 'segment'. All structural roots with a proximal diameter larger than 2 mm were measured.

As long horizontal roots were dug out up to an extension of 5 m, the spatial coordinates of such roots could not be recorded and only diameters at consecutive distances of 20 cm were measured. The lacking geometrical information was inferred by AMAPmod. They were considered as straight roots with azimuth and angle of the last digitised segment.

Calculations

The output data file was analysed using the AMAPmod software (Godin et al., 1997) handling branched structures at several scales: characteristics of root segments, roots or whole root systems were obtained. AMAPmod provides also 3D graphical reconstructions for checking the data. Additional computations and figures were done with Splus (Statistical Sciences, 1993).

The distance between the two corresponding measurement points was used as the segment length. The segment volume was calculated as a cone from its length and proximal and distal diameters. The root length or volume was computed as the sum of length or volume of all its segments. Vectors representing each segment of a root were summed; the angle of this sum toward the horizontal plane was used as the 'mean root angle' (Batschelets, 1981). The 'branching angle' was defined as the angle between the root of higher order and the parent root segment. A root 'bend' where one of its segments subtend an angle greater than 60° to the parent segment. The apical unbranched zone was excluded in calculation of inter-laterals length because it was broken.

The distribution of root length according to their diameter was made following Böhm (1979). *Very fine* and *fine* roots (<0.2 cm) were pooled into the small roots. In spatial root distribution computations, the length or volume of all segments which center is located in a given elementary surface were simply pooled. It generated the spotted pattern found in low root density zones in Figures 1 and 2. According to Nicoll et al. (1997), the stump was not included in root volume computations. It represents usually a major part of the total coarse root volume and has a very specific function regarding root anchorage or water and nutrient uptake and transport. The stump was arbitrarily defined as the taproot from soil level to 15 cm deep.

The 'zone of rapid tapel' ('ZRT', Wilson, 1975) is of major importance for anchorage (Coutts, 1987). Six large trees from this study exhibited a significant decrease in longitudinal residual strains of maturation at a distance of 20 cm from the central axis (Stokes et al., 1998). Thus, root parts located within 25 cm radial distance and above 15 cm depth were considered to be located in the ZRT. Root length and volume per diameter class or per position were divided by the total root length or volume so as to be analysed as size-independent proportions. Biomass allocation to the roots was studied as the root partitioning coefficient or RPC (Canham et al., 1996), i.e. the proportion of roots in the root + stump + stem volume.

As a stratified sampling was used covering the whole stand area, several characteristics at the stand level were computed. 3 sets of 9, 10 and 10 trees were selected from their stem biomass, defining increasing 'vigour classes', CI, CII and CIII. In order to test if a character is tree size dependant, the correlation

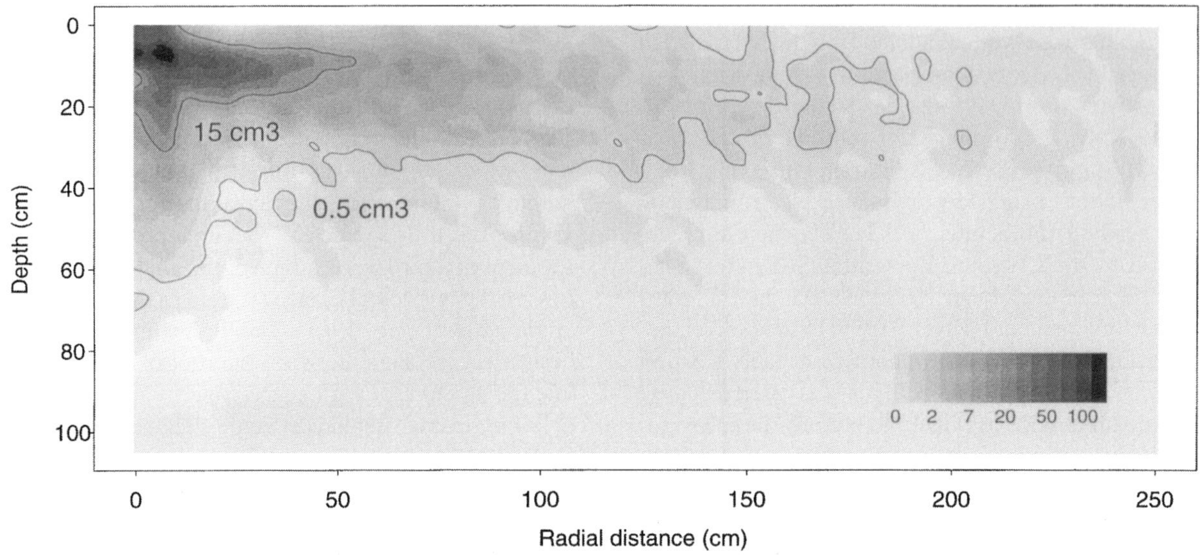

Figure 1. Cumulative 2D spatial distribution of root volume (cm^3 within a surface area of 25 cm^2) as a function of depth and radial distance: tree mean. Shades of grey-scale define evenly spaced isocontours of log (root volume +1).

with stem biomass was computed. Relations between variables and distribution of trees into groups can be assessed using multivariable analysis: a principal component analysis (PCA) has been made on scaled values of variables (Statistical Science, 1993). According to Cailliez et Gueneau (1972), a sub-sample of variables was selected, retaining only structurally independent variables. For instance, from the five variables describing the aerial part dimensions, i.e. stem height, diameter at breast height, diameter at soil level, stem volume, stem biomass, only the stem biomass was included. Only results concerning the 4 first principal components are reported.

Results

Aerial parts and root sizes

The volume or biomass of root and stump together equalled approximately half of the stem volume or biomass (Table 1). Compared to the small trees (CI), the larger ones (CIII) had approximately twice the root volume, root biomass, root length, stem volume, stump volume and number of roots. All variables corresponding to the root system size were closely correlated to the stem biomass (Tables 1 and 2). Trees had developed a coarse root length density larger than 5 m/m^2 (50 km/ha) during the four years after planting. In the three vigour classes, the RPC averaged 25% (Table 3), it was not correlated to the tree size.

Stem straightness and roots

The correlation between stem straightness and SAL was not significant ($p = -0.23$). The stem straightness was not significantly related to the stem biomass nor to any of the characters listed in Table 3 and 4, apart a positive correlation with maximal rooting depth ($p = 0.47^{***}$) and with the proportion of deep root volume and a negative correlation with the proportion of root volume in the zone of rapid taper and with the proportion of oblique roots. The corresponding correlations with SAL were not significant. The correlation between stem straightness and RPC was only 0.31, but reached 0.68^{****} when only three trees with a short taproot and low stem straightness were removed. These correlations suggest that the stem is straighter if the root system is more developed in depth and if the RPC is larger.

Root branching

The mean total proximal root area of all roots (68 cm^2) was close to the mean stem diameter at ground level (52 cm^2) (Table 3). The mean number of forks was low. The mean maximum branching order was close to 4. The proportions of 2nd, 3rd and 4th order roots averaged respectively 30, 60 and 10% of the total root

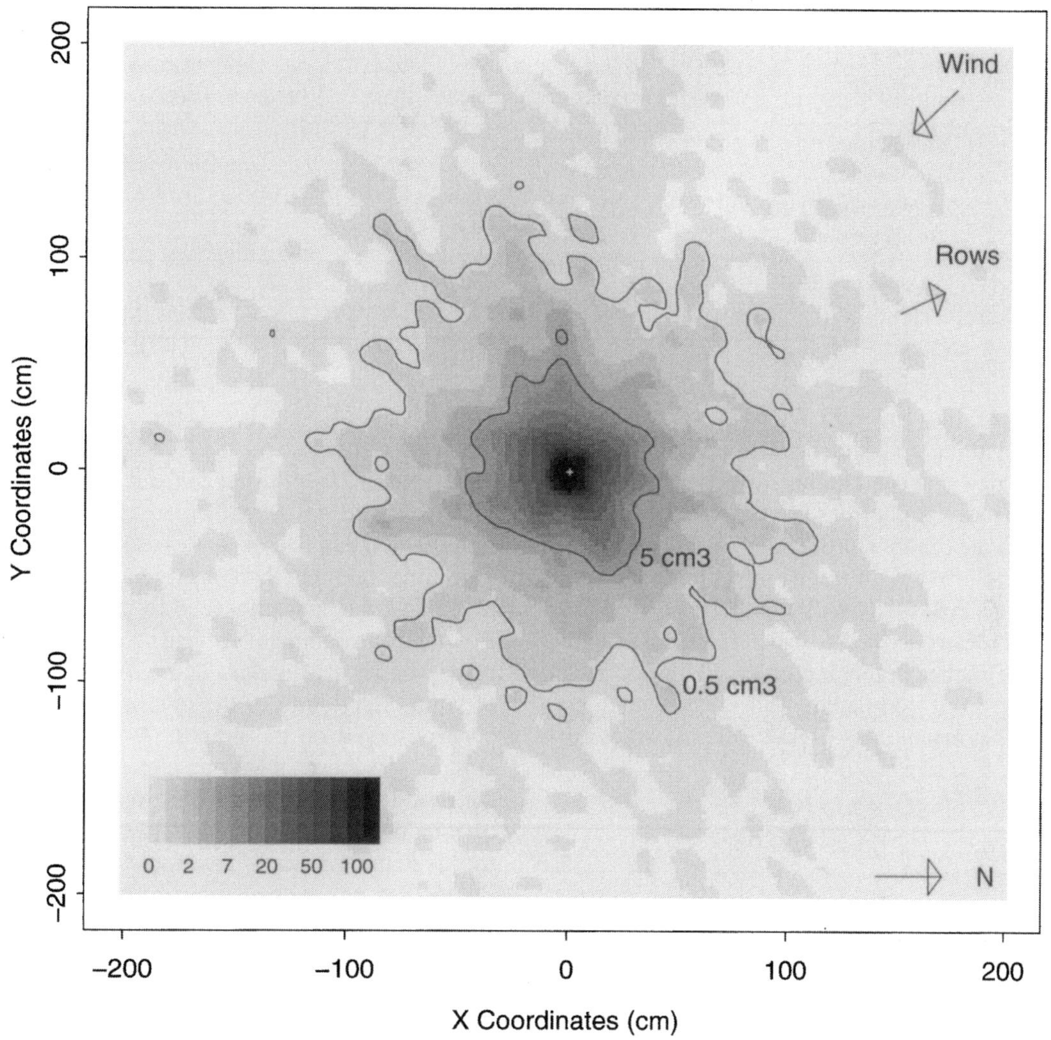

Figure 2. Cumulative 2D spatial distribution of surface (depth < 30 cm) root volume (cm^3 within a surface area of 100 cm^2) in the horizontal plane (i.e. integrated over depth): tree mean. Shades of grey-scale define evenly spaced isocontours of log (root volume +1).

Table 1. Above- and below-ground tree dimensions. S-D is standard deviation, C.V. is coefficient of variation, CI, CII and CIII are vigour classes ($1m^3 = 10^6$ cm^3, $1\,t = 10^6 g$)

	Stand /ha	Tree			Vigour class			Correlation with stem biomass
		Mean	S-D	C.V.	CI	CII	CIII	
Stem volume (cm^3)	$7.6.10^6$	6150	2500	0.407	3890	5690	8640	0.94****
Stump volume (cm^3)	$0.89.10^6$	671	366	0.545	438	532	1020	0.92****
Root volume(cm^3)	$2.7.10^6$	2150	1020	0.472	1270	2020	3090	0.81****
Stem biomass (g)	$2.8.10^6$	2260	964	0.426	1420	2080	3200	
Root + stump biomass (g)	$1.4.10^6$	1140	463	0.405	704	1130	1560	0.87****
Root length (cm)	$4.9.10^6$	4000	1260	0.314	3150	3700	5060	0.76****
Number of roots	12.10^4	98.3	31.2	0.317	74	94.2	124	0.71****

Table 2. Correlations between general dimensions of the trees. Levels of significance = *: <5%, **: <1%,***: <0.1%,****: <0.01%

	DBH	Stump volume	Root length	Root volume	Number of roots	Maximal root depth
Stump volune	0.75****					
Root length	0.7****	0.78****				
Root volume	0.75****	0.79****	0.84****			
Number of roots	0.7****	0.71****	0.81****	0.76****		
Maximal root depth	−0.04	0.28	0.22	0.16	0.27	
RPC	−0.03	0.14	0.38*	0.58***	0.3	0.16

number. A medium positive correlation was observed between the tree size and number of second, third and fourth order roots. Eight trees had a distinctly shorter taproot, i.e. a maximum taproot depth smaller than 13 cm. These trees had a significantly lower mean number of second order roots (22 vs. 29 Pr.= 1.1%).

Second order roots had a wider branching angle than roots of higher branching order (Table 3), both variables were not correlated. The mean branching angle and the mean inter-laterals length were not related to the tree size.

Spatial distribution of roots

The maximal rooting depth averaged only 20% of the mean height (Table 3), it showed low correlations with above- and below-ground dimension variables (Table 2). The maximum taproot depth averaged only 45 cm. Half of the root volume was located in the 10 upper centimetre of soil and nearly 50% of the root volume was located in the zone of rapid taper (Table 4), but only 6% of the root volume was located in the deep soil layers. The taproot averaged 39% of the deep root volume and 51% when the taproot depth is larger than 13 cm. The proportion of deep roots showed a high coefficient of variation (Table 4). The large trees had a lower proportion of shallow roots.

The taproot represented 8% of the root volume, with a large coefficient of variation (Table 4). The proportion of taproot volume was not related to the tree size, it reached 10% when the taproot depth was larger than 13 cm. Correlations between the proportion of root volume within the ZRT on one hand and the proportion of taproot volume and deep root volume on the other hand reached −0.63*** and −0.77****, respectively.

Below 20 cm deep, the root volume was concentrated within 20 cm distance from the taproot (Figure 1). There were no secondary sinker nor long oblique roots. The mean lateral root extension was more than four times greater than the mean maximal rooting depth and was not influenced by the tree size (Table 3). The mean spread of surface roots in the horizontal plane was mostly regular (Figure 2). An obvious effect of wind direction or of the row orientation could not been observed on the graphs. A slightly lower root volume was found from one meter radial distance on the row. This was probably due to a more difficult extraction of the roots growing under the neighbour tree root systems.

Most of the second order roots originated in the 10 first centimetre depth, on the stump, in particular because 8 trees had a short taproot (Figure 3). Vertical roots and oblique roots represented each only 6% of root volume excluding the taproot, both proportions had a high variability, but they were not related to the tree size (Table 4). The five trees having more than 10% oblique root volume had less than 10% vertical root volume and the five trees having more than 10% vertical root volume had less than 10% oblique root volume. The mean length between bends reached 100 cm (Table 4), and was smaller for larger trees, it reached 1.15 meter in shallow roots.

Root diameter distribution

The very large and large roots averaged nearly 70% of root volume, but only 15% of root length. The largest trees had a greater proportion of very large root volume and length and a smaller proportion of medium diameter roots, simply by a size effect, it was also shown in the correlation matrix (Table 5). Small roots represented only 10% of root volume but 40%

Table 3. Mean values of above- and below-ground characters. ρ_{sb} is the correlation coefficient with stem biomass

Character	Mean	S-D	C.V.	CI	CII	CIII	ρ_{sb}
Stem height (cm)	323	50	0.16	276	322	366	0.86****
Stem diameter at ground level (cm)	8.1	1.2	0.15	7.1	8	9.3	0.71****
Stem diameter at breast height (cm)	4.7	1.1	0.23	3.7	4.6	5.6	0.88****
Stem straightness (40 = very straight)	25	7	0.28	26	2	25	−0.06
Stump angle of lean (degree)	4	2.3	0.57	3.2	4.6	4	0.15
RPC (%)	24	4.8	0.20	23	24	24	0.08
Root length/root volume (cm/cm^3)	2	0.63	0.31	2.6	1.9	1.7	−0.5**
Total proximal root area of all roots (cm^2)	68	29	0.42	45	64	94	0.75****
Maximum branching order	4.2	0.47	0.11	4	4.1	4.4	0.3
Number of second order roots	27	6.7	0.25	23	27	31	0.6***
Number of third order roots	59	21	0.35	44	55	77	0.61***
Number of fourth order roots	11	9.2	0.83	5.7	11	16	0.58***
Number of forks	3.2	1.6	0.50	3.1	3.1	3.4	0.13
Mean inter-laterals length (cm)	14	2.9	0.21	14	14	15	−0.04
Mean branching angle of 2nd order roots (degree)	70	8.9	0.13	73	69	68	−0.1
Mean branching angle of 3rd, 4th and 5th order roots (degree)	66	7	0.11	63	68	68	0.19
Mean length between bends (cm)	101	42	0.42	130	94	81	−0.43**
Maximum root depth (cm)	65	15	0.23	65	59	71	0.14
Maximum taproot depth (cm)	46	27	0.58	50	49	38	−0.13
Maximum root radial extension (cm)	280	106	0.38	278	274	288	0.11

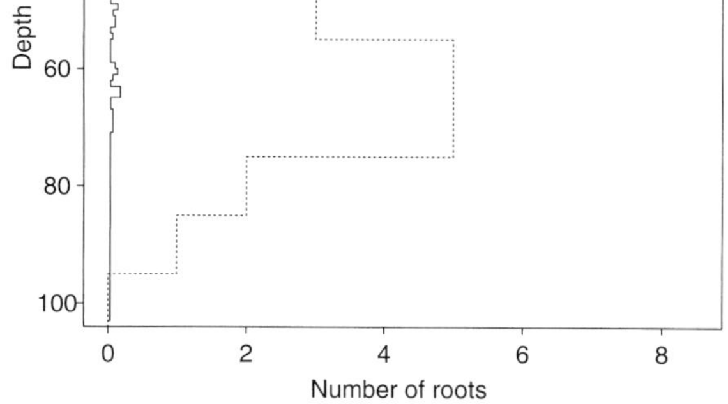

Figure 3. Mean vertical distribution of 2nd order root origin points by 1 cm classes (solid line) and vertical distribution of taproot tips by 10 cm classes (dotted line).

of root length (Table 4). It should be noted that the proportion of lost roots was greater for medium and small roots.

Relationships between root system descriptors

The two first components of the principal component analysis on individual tree characteristics explained nearly half of the variance (Figure 4a).

Dimensional variables including root volume, number of roots and stem biomass had high positive loadings for the first component (PC1): PC1 was a combination of characters representing the above- and below-ground tree size.

A high positive value for PC2 represented a high proportion of root volume in the taproot (PV.Taproot), in the stump (PV.Stump) and in the deep roots (PV.Deep root) and a high maximum depth, but a low proportion of root volume in the ZRT (PV.ZRT). PC2 was a combination of variables accounting for the rooting depth and for the stump and taproot size.

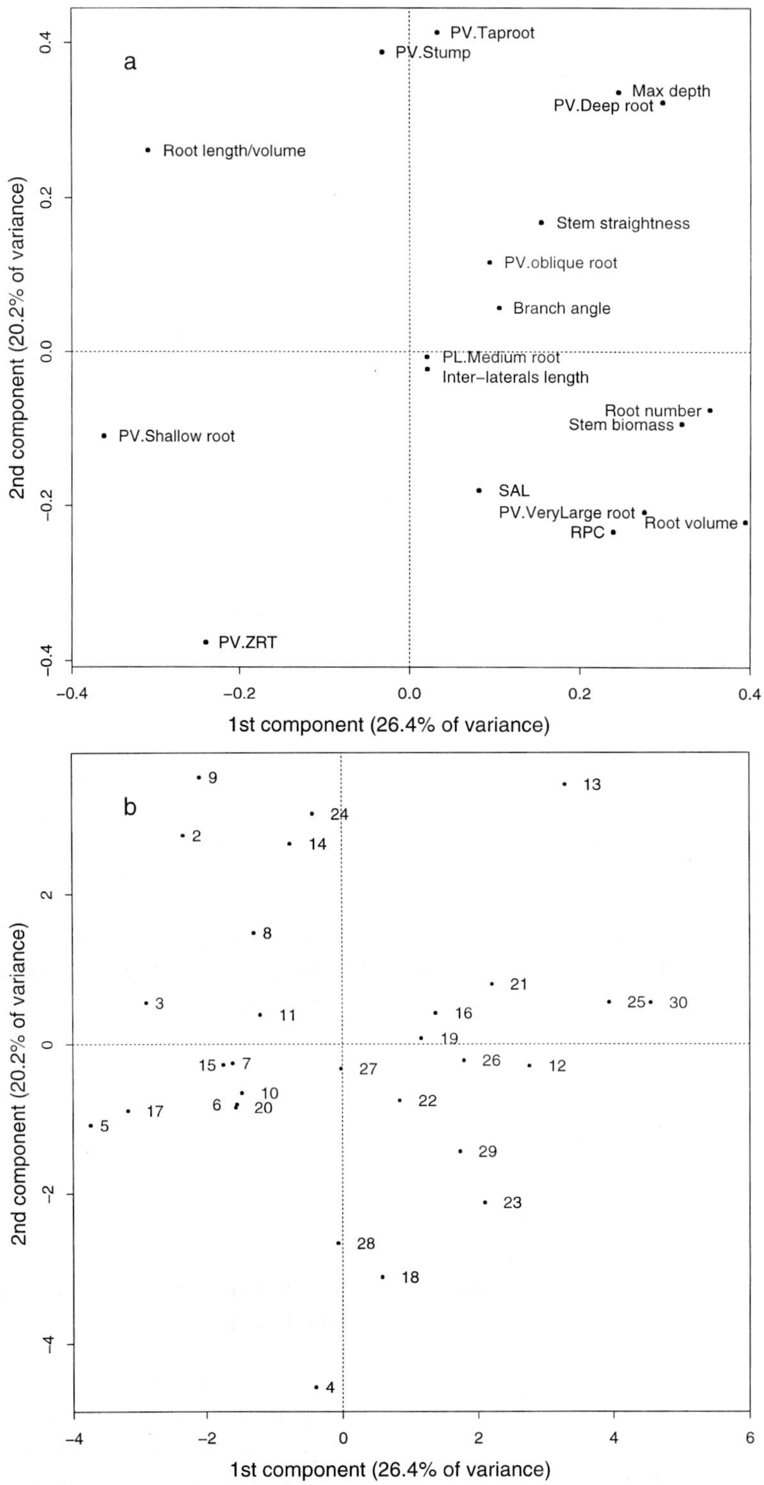

Figure 4. Principal component analysis of tree characteristics. Scores for PC1 and PC2: (a) loadings for the 18 original variables; (b) loadings for the 29 trees of the sample. PV refers to 'proportion of root volume' and PL refers to 'proportion of root length'.

Table 4. Proportion of root volume and length by position and by diameter class in %. Stump excluded from volume. Root diameter classes: very large (>2 cm), large (2–1 cm), medium (1–0.5 cm), small (≤ 0.5 cm)

Proportion (%)	Mean	S-D	C.V.	CI	CII	CIII	ρ_{sb}
Shallow root volume (above depth of 10 cm)	53	14	0.27	58	57	43	−0.45**
Deep root volume (below depth of 30 cm)	5.8	4	0.7	5.8	4.2	7.4	0.21
Root volume within ZRT	48	7.4	0.15	48	51	46	−0.18
Taproot volume	7.7	7.3	0.96	9.4	6.9	6.6	−0.05
Horizontal roots, volume (angle <30°)[1]	88	11	0.12	88	90	86	−0.02
Oblique root volume (angle between 39° and 50°)[1]	6	7.9	1.3	4.1	5.9	7.9	0.06
Vertical root volume (angle >50°)[1]	5.9	8.2	1.4	8.4	3.9	5.7	−0.04
Very large root volume	38	10	0.27	34	37	43	0.48**
Large roots volume	30	6.3	0.21	27	33	30	−0.06
Medium root volume	22	6.2	0.28	26	21	20	−0.42*
Small root volume	9.5	4.6	0.49	13	8.8	7.1	−0.4*
Very large root length	4.3	1.6	0.38	3.4	4.2	5.1	0.39*
Large root length	12	3.5	0.29	9.5	14	13	0.27
Medium root length	33	6.5	0.2	32	33	33	−0.03
Small root length	51	8.3	0.16	55	50	49	−0.17

[1]Computed from the 'mean root angle', taproot excluded.

Maximal rooting depth and proportion of deep roots had also medium and positive loadings for PC1 indicating that larger trees had generally a deeper rooting. In the same way, the loadings for shallow root volume (PV.Shallow root) and for the root length/volume ratio indicated that larger trees had thicker roots and a lower proportion of shallow roots. The RPC was positively related to the tree size and negatively related to the rooting depth. Finally, a better stem straightness was moderately associated to deep rooting.

All trees with a large taproot were located in the upper part of Figure 4b. One of them (tree 13) associated a large size and a deep rooting. The three trees with a shallow rooting (n° 4, 18 and 28) had also a medium size.

The 3rd and the 4th components accounted for another 20% of the total variation (Figure 5a). A high negative value for PC3 represented a low proportion of medium root length and a short mean inter-laterals length, the correlation between both variables reached 0.5. PC4 was mainly a contrast between the proportion of oblique roots and stem straightness: none of the six trees with a higher proportion of oblique roots had a large stem straightness score.

Tree 14 had long sparsely branched second order roots with a medium diameter (Figure 5b). Trees in the upper part of the plot had a straight stem. Tree 24 had many oblique roots, a poor stem straightness and a low RPC. It should be noted that PC6 (5.9% of variance) accounted mainly for the mean branching angle of 3rd, 4th and 5th order roots, tree 7 having a very large loading for PC6, i.e. a wide mean branching angle.

Discussion

Biomass allocation

The biomass allocation to coarse roots was high compared to allocation to the stem. The extracted root biomass, including stump, averaged 50% of the stem biomass. About 0.35 tons of biomass per hectare per year were allocated to the coarse roots and to the stump, during the four growing seasons in the field. In a nearby 16-year-old maritime pine stand, the extracted root biomass reached 11.3 t/ha and averaged 30% of the stem biomass and only 18% of the stem + branch biomass (Lemoine et al., 1986).

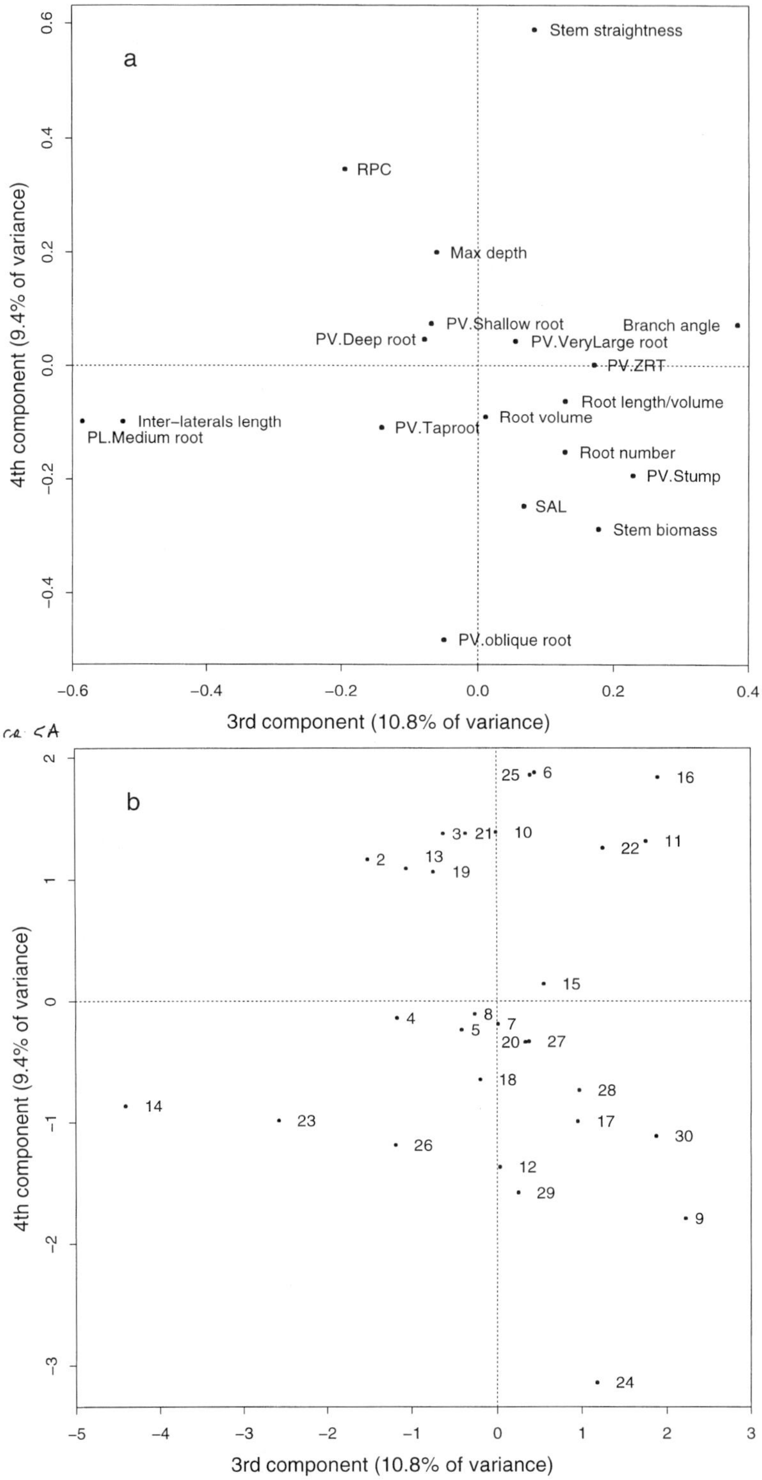

Figure 5. PCA of tree characteristics. Scores for PC3 and PC4: (a) loadings for the 18 original variables; (b) loadings for the 29 trees of the sample.

Table 5. Correlation matrix between root volume and length proportions and above- and below-ground parameters

Proportion (%)	Stem straightness	DBH	Root volume	Root length	Number of roots	Maximal root depth	RPC
Shallow root volume	−0.19	−0.37*	−0.56***	−0.48**	−0.55***	−0.5**	−0.33*
Deep root volume	0.43**	0.15	0.24	0.17	0.31	0.81****	0.06
Root volume within ZRT	−0.37*	0.02	−0.18	−0.14	−0.2	−0.72****	−0.05
Taproot volume	0.14	−0.16	−0.28	−0.33*	−0.11	0.43*	−0.39*
Oblique root volume	−0.31*	0.17	0.06	0.06	0.29	0.15	−0.08
Vertical root volume	0.04	0.09	0.02	0.13	−0.03	0.13	0.02
Very large root volume	0.1	0.53**	0.63***	0.28	0.39*	−0.04	0.35*
Large root volume	0.02	−0.06	−0.03	0	−0.08	0.08	0.11
Medium root volume	−0.18	−0.49**	−0.55***	−0.31	−0.34*	−0.02	−0.35*
Medium root length	−0.17	−0.12	0.07	−0.05	−0.09	−0.06	0.14
Small root length	0.15	−0.12	−0.37*	−0.06	−0.02	0.05	−0.37*

Consistent relationships between stem biomass or volume and root volume indicated that the stem volume is a good predictor of root system size. The RPC had the same coefficient of variation than height, it ranged from 14% to 33%, indicating a substantial variability in allocation strategies in this sample. It may be assessed by only weighting the roots with a diameter larger than 1 cm, which represented 70% of the root volume. The above-mentioned hypothesis concerning a high genetic variability in biomass allocation to the roots is not invalidated by these results. Height and dbh was also found to be highly correlated to the root system size in 30 to 40-year-old *Pseudotsuga menziesii* (Mirb.) Franco trees (Bloomberg and Hall, 1986), in 8, 16 and 25-year-old *Pinus radiata* D. Don trees (Watson and O'Loughlin, 1990) and in 20- to 28-year-old *Quercus petraea* (Matt.) Liebl. (Drexhage et al., 1999). The genetic variability of RPC was assessed in 11-year-old *Picea sitchensis* (Bong.) Carr. where the mean RPC of five clones ranged between 12% and 19% (Nicoll et al., 1997).

The stump partitioning coefficient, i.e. the proportion of stump in the root + stump + stem volume, varied from 4% to 14% and was not related to the RPC nor to the stem size. Small stumps consisted in 7 cm deep disks, large stumps had a low tapered cone shape and reached 15 cm depth. In 8-year-old planted *Pinus radiata*, the stump contained 28% to 53% of the coarse root biomass and in 16- and 25-year-old *Pinus radiata* it reached 43 to 59% of the coarse root biomass (Watson and O'Loughlin, 1990).

Spatial distribution

The mean length between bends was large, indicating that the roots were generally straight, in particular the long horizontal surface roots. In 16-year-old plantation grown *Picea sitchensis*, the mean length between bends was four times shorter (Henderson et al., 1983a).

The maximal root extension was underestimated because all lateral roots were broken. However, maximal root extension reached far beyond the mid points (1 m and 2 m) to the neighbouring trees. In the studied stand, the onset of intra-specific competition for upper soil layer resources occurred before the fifth year. In 8-year-old *Pinus radiata*, Watson and O'Loughlin (1990) also observed that the roots that extended furthest from the tree stump were generally those growing above a soil depth of 10 cm. In poor soils, trees generally develop less tapered and less branched roots (Atger, 1991). In young *Pseudotsuga menziesii* trees, sparsely branched 'pioneer' roots can extend considerable distances, quickly establishing the main structural elements of the root system over a wide area (McMinn, 1963). On the other hand, such a rooting strategy is unsuitable in stands after crown closure and the onset of an intense competition (McMinn, 1963).

Several early successional forest tree species had a significantly greater proportion of deep roots than late successional or tolerant species (Gale and Grigal, 1987). The latter are better adapted to sites where resources are concentrated near the soil surface as the result of biocycling and soil development. The low nutrient content of sandy podzol may explain why young maritime pine develop a shallow root system, ensuring an availability to nutrient turn-over although it is an early successional species. Traditional weeding operations consist in dragging a sledge with three heavy rolling cylinders fitted with perpendicular blades (termed in French 'rouleau landais'). It is likely to cut many surface roots, reducing wind stability and growth. The shallow rooting pattern originate in particular in the concentration of emission of second order roots in the 10 upper centimetre of soil. *In Quercus robur* L. seedlings, the vertical distribution of laterals along the taproot is uniform from soil level to 1 m depth (Belgrand et al., 1987).

Whereas the total root biomass and the surface lateral root system generally remain unaffected by the site conditions, by the soil preparation and by the establishment techniques, several authors reported a high variability in the vertical rooting of forest trees (Drexhage and Gruber, 1998; Somerville, 1979; Stone and Kalisz, 1991). Although deep penetrating roots were important for water and nutrient supply, a small number of deep roots with small diameter may be sufficient to maintain transpiration during drought (Breda et al., 1993). In *Quercus petraea* saplings from natural regeneration, variability in amount of vertical roots was also high, their proportion tended to decrease with increasing stem diameter (Drexhage et al., 1999).

The high variability of proportion of root volume in the taproot and the low maximal rooting depth may be explained by the establishment technique: development of the seedling taproot was hampered by the first year growing conditions in planting bricks. Cremière (1994) observed that *Pinus pinaster* seedlings planted in peat planting bricks had a better vigour, survival, stability and root development than seedlings grown in paper-pots. Directly seeded pines are more likely to develop a strong taproot, and to have a straight stem (Auberlinder, 1982). It was also reported for *Pinus radiata* (Somerville, 1979). In young trees, a well developed taproot is likely to be a desirable survival characteristic (McMinn, 1963). The variability in deep rooting may originate also from soil conditions. However, the typical root deformations caused by a hard pan or by a high water table level (Maugé, 1987) were never observed in the sample studied. The low number of bends observed indicate a low level of root tangle.

For exploitation of soil resources beyond 40 cm depth, coarse roots were essentially located in 20 cm radial distance from the central axis. Access to deep soil layers was size-independent. During the summer, the main deep soil layer resource is water, for which competition occurs mainly at the stand level. In *Pinus radiata* (Somerville, 1979), *Pseudotsuga menziesii* (McMinn, 1963) and boreal conifers (Eis, 1974), the soil volume was well exploited throughout its depth by roots emanating from the taproot and oblique roots. In older maritime pines, secondary sinkers develop in the vicinity of the stump enlarging the area colonised by the deep roots (Maugé, 1987).

The correlation between both characters used for describing the stem straightness was low. Though it is a subjective measurement, the stem straightness visual scoring is more likely than the SAL to take into account all straightness defects along the whole stem. It is generally not exactly known which root characteristics best promote tree stability (Stokes et al., 1997). Both PCA and correlations results showed that the stem straightness is influenced by several root characteristics including the proportion of deep roots, the taproot development, the proportion of roots in the ZRT and the RPC. This pattern may be explained by the following hypothesis. A larger RPC provide generally a better stability, which is also dependant from the root system shape. Several trees have failed to regenerate vertical roots after planting, the resulting anchorage deficiency was partly compensated by regeneration of oblique roots or by enlarging the roots in the ZRT. In 8 year-old *Pinus radiata*, early stand stability was determined by a good taproot regeneration after planting (Balneaves and De La Mare, 1989). Auberlinder (1982) had also shown that in maritime pine saplings, unstable trees had a very high branch to root ratio, an irregular distribution of roots and a higher aerial parts early growth in the first 3 years: in stable trees vertical roots averaged 40% of total root cross-sectional area vs. 25% in unstable trees. Stem form defects may also be due to aerial competition.

For several forest tree species, the number of 2nd order roots is rather low, ranging from 3 to 15 and it is related to the diameter of the tree (Henderson et al., 1983a; Wu et al., 1988). In 3- to 9-year-old *Pinus taeda* and *Pinus echinata* Mill., the number of 2nd order roots per root system ranged from 14 to 58, with a mean of 28 roots per tree (Carlson and Harrington, 1987), close to the present value of 31.2. The large number of second order roots in *Pinus pinaster* may originate from the large taproot size, it is one of a few tree species that conserve a tap root at maturity (Stokes et al., 1998). In our sample, the number of roots in each order was partly determined by the tree size. Infrequency of branching is also characteristic of low vigour in *Pseudotsuga menziesii* (McMinn, 1963).

Forks occurred rarely in our young *Pinus pinaster* root systems grown on sandy podzol and weeded using a disk harrow. In 16-year-old *Picea sitchensis* more than half of the ramifications were forks and proliferations (Henderson et al., 1983a). Forks originate generally in damage to apex, but damage to root apex may have often resulted in a replacement root which had the same size as the parent root, therefore the number of damages could not be estimated from the number of forks.

The mean branching angle of second order roots was wider than the branching angle of higher order roots. Most of the second order roots originated on the stump, they first grew in an oblique downward direction (with a 15° angle relative to the horizontal plane). In 10 to 30 centimetre radial distance they ran parallel to the soil surface. The optimum branching angle to resist vertical pull-out is 90° for second order roots and between 0° and 20° for third order roots (Stokes et al., 1996). On the other hand, lateral branches need to subtend large angles to their parents in order to exploit separate soil regions (Henderson, 1983b). In *Larix decidua* Mill. seedlings the mean second order root branching angle was larger than 80° and in 20-year-old *Picea sitchensis* it averaged 58° (Stokes et al., 1996).

Conclusion

An efficient and rapid measurement and analysis method allowed the measuring of a fairly large sample of 29 maritime pine root systems, 2850 roots and 22 500 points were measured. It may be used for agronomic testing (i.e. comparing several treatments) involving several dozens of root systems. A general description of the root system architecture in young planted maritime pines could be obtained. Besides the variability due to the dimension of the tree, the main source of variability was found in the proportion of root volume in the deep roots and in the taproot, this may be partly due to the plantation technique. It has clouded relationships between several studied parameters. The stem straightness is influenced both by the biomass partitioning to the roots and by the regeneration of a taproot or vertical roots after planting. In our sample most of the characters describing root architecture such as branch angle, number of forks or mean inter-laterals length were not tree size dependant.

Parameter distributions estimated for the root systems may be used in root growth simulation model (Colin-Belgrand et al., 1989).

Acknowledgements

We thank F. Courdier, B. Médard, B. Issenhuth, P. Rivet, L. Wehrlen and J.-P. Chambon for technical support and J. Dauzat for the device. Financial support was provided by the European Community (contract no. AIR3-CT93-1269). We also thank two anonymous reviewers for their helpful comments.

References

Arbez M 1979 Contrôle génétique et mécanismes d'acquisition des défauts de rectitude du tronc chez le pin maritime landais. *In* proc. of 104e Congrès national des sociétés savantes, Bordeaux, Sciences, n^{o} fasc. II.

Atger C 1991 L'architecture est-elle influencée par le milieu? *In* proc. conf. 'L'ARBRE, Biologie et Développement', Montpellier, France. Ed. C Edelin. pp 72–84. Naturalia Monspeliensia n^{o} h. s.

Atger C and Edelin C 1995 A case of sympodial branching based on endogenous determinism in root system: *Platanus hybrida* Brot. Acta Bot. Gallica 142, 23–30.

Auberlinder V 1982 De l'instabilité du pin maritime. Annales de Recherche Sylvicoles, A. Fo. Cel, Paris. pp 139–178.

Balneaves J M and de la Mare P J 1989 Root patterns of *Pinus radiata* on five ripping treatments in a Canterbury forest. N.Z. J. For. Sci. 19, 29–40.

Batschelet E 1981 Circular statistics in biology. Academic Press, London. 371 p.

Berbigier P, Diawara A and Loustau D 1991 Étude microclimatique de l'effet de sécheresse sur l'évaporation d'une plantation de pins maritimes et du sous-bois. Ann. Sci. For. 22, 157–177.

Berthier S, Stokes A and Guitard D 1997 Effects of mechanical stress on wood formation in maritime pine (*Pinus pinaster*) roots. *In* proc. of the 2nd workshop of IUFRO working party S5.01.04 'Connection between sylviculture and wood quality through modelling approaches and simulation software', Kruger National Park, South Africa, 26–31 August 1996. INRA, Versailles, France.

Bloomberg W J and Hall A A 1986 Effects of laminated root rot on relationships between stem growth and root-system size, morphology, and spatial distribution in Douglas-fir. Forest Sci. 32, 202–219.

Böhm W 1979 Methods of studying root systems. Ecological studies n° 33. Eds W D Billing, F Golley, O L Lange and J S Olson. Springer Verlag, Berlin. 188 p.

Belgrand M, Dreyer E, Joannes H, Velter C and Sciuller I 1987 A semi-automated data processing system for root growth analysis: application to a growing oak seedling. Tree Physiol. 3, 393–404.

Breda N, Cochard H, Dreyer E and Granier A 1993 Field comparison of transpiration, stomatal conductance and vulnerability to cavitation of *Quercus petraea* and *Quercus robur* under water stress. Ann. Sci. For. 50, 571–582.

Brown T N and Kulasiri D 1994 Simulation of *Pinus radiata* root system structure for ecosystem management applications. Simulation 62, 42–57.

Brown T N, Kulasiri D and Gaunt R E 1997 A root-morphology based simulation for plant soil microbial ecosystem modelling. Ecol. Model. 99, 275–287.

Cailliez F and Gueneau P 1972 Analyse en composantes principales des propriétés technologiques des bois malgaches. Ann. Sci. For. 30, 215–266.

Canham C D, Berkowitz A R, Kelly V R, Lovett G M, Ollinger S V and Schnurr J 1996 Biomass allocation and multiple resource limitation in tree seedlings. Can. J. For. Res. 26, 1521–1530.

Carlson W C and Harrington C A 1987 Cross-sectional area relationships in root systems of loblolly and shortleaf pine. Can. J. For. Res. 17, 556–558.

Chaperon H, Hinschberger F, Haury P and Alazard P. 1991 Étude comparative du développement de plants de pin maritime issus de boutures et de semis. Ann. de Rech. Sylvicoles, AFOCEL, Paris. 1991, 15–34.

Colin F, Danjon F and Wehrlen L 1996 Études racinaires au sein du programme 'Croissance' de l'INRA (*Quercus petraea* et *Pinus pinaster*). Rev. For. Fr. 46, 165–172.

Colin-Belgrand M, Joannes H, Dreyer E and Pagès L 1989 A new data processing system for root growth and ramification analysis: Description of methods. Ann. Sci. For. 46, 305–309.

Coutts M P 1983 Root architecture and tree stability. Plant Soil 71, 171–188.

Coutts M P 1987 Developmental process in tree root systems. Can. J. For. Res. 17, 761–767.

Crémière L 1994 Conteneurs, quelle conséquences pour le pin maritime. Information – Forêt, AFOCEL-ARMEF, Paris, 1, 69–88.

Danjon F 1994 Heritabilities and genetic correlations for estimated growth curve parameters in maritime pine. Theor. Appl. Genet. 89, 911–921.

Danjon F 1995 Observed selection effects on height growth, diameter and stem form in maritime pine. Silvae Genet. 44, 10–19.

Danjon F, Sinoquet H, Godin C, Colin F and Drexhage M 1999 Characterisation of structural tree root architecture using 3D digitising and AMAPmod software. Plant Soil 211, 241–258.

Drexhage M and Gruber F 1998 Architecture of the skeletal root system of 40-year-old *Picea abies* on strongly acidified soils in the Harz Mountains (Germany). Can. J. For. Res. 28, 13–22.

Drexhage M, Chauvière M, Colin F and Nielsen C N N 1999 Development of structural root architecture and allometry of *Quercus petraea*. Can. J. For. Res. 29, 600–608.

Eis S 1974 Root system morphology of western hemlock, western red cedar and Douglas-fir. Can. J. For. Res. 4, 28–38.

Gale M R and Grigal D F 1987 Vertical root distributions of northern tree species in relation to successional status. Can. J. For. Res. 17, 829–834.

Godin C, Costes E and Caraglio Y 1997 Exploring plant topological structure with the AMAPmod software: an outline. Silva Fennica 31, 355–366.

Granier A and Loustau D, 1994. Measuring and modelling the transpiration of a maritime pine canopy from sap-flow data. Agric. For. Meteorol. 71, 61–82.

Henderson R, Ford E D, Renshaw E and Deans J D 1983a Morphology of the structural root system of sitka spruce: 1. Analysis and quantitative description. Forestry 56, 121–135.

Henderson R, Ford E D and Renshaw E 1983b Morphology of the structural root system of sitka spruce: 2. Computer simulation of rooting pattern. Forestry 56, 137–153.

Lemoine B, Gelpe J, Ranger J and Nys C 1986 Biomasses et croissance du pin maritime. Étude de la variabilité dans un peuplement de 16 ans. Ann. Sci. Forest. 43, 67–84.

Lynch J 1995 Root architecture and plant productivity. Plant Physiol. 109, 7–13.

Mason E G 1985 Causes of juvenile instability of *Pins radiata* in New Zealand. N.Z. J. For. Sci. 15, 263–280.

Maugé J P 1987 Le pin maritime, premier résineux de France. Institut pour le Développement Forestier, Paris, France. 192 p.

McMinn R G 1963 Characteristics of Douglas-fir root systems. Can. J. Bot. 41, 105–122.

Nicoll B C, Easton E P, Milner A D, Walker C and Coutts M P 1997 Wind stability factors in tree selection: distribution of biomass within root systems of sitka spruce clones. *In* Wind and wind related damage to trees, Eds M Coutts and J Grace. pp 276–301. Cambridge University Press.

Polhemus 1993 3SPACE user's manual, Kaiser Aerospace & Electronics Company, Colchester, Vermont, USA. 158 p.

Porté, A. 1999 Modélisation des effets du bilan hydrique sur la production primaire et la croissance d'un couvert de pin maritime (*Pinus pinaster* Ait.) en Lande Humide. Ph.D. thesis, Université de Paris-Sud Orsay. 172 p.

Puhe J 1994 Die Wurzelentwicklung der Fichte (*Picea abies* [L.] Karst.) bei unterschiedlichen chemischen Bodenbedingungen. Berichte des Forschungszentrum Waldökosysteme der Universität Göttingen, Reihe A, Band 108, 129 p.

Reffye P de, Houllier F, Blaise F, Barthélémy D, Dauzat J and Auclair D 1995 A model simulating above- and below-ground tree architecture with agroforestry applications. Agrofor. Syst. 30, 175–197.

Saur E 1989 Alimentation oligo-minérale du pin maritime (*Pinus pinaster* Soland in Ait.) en relation avec quelques caractéristiques physico-chimiques des sols sableux des Landes de Gascogne. Ann. Sci. For. 46, 119–129.

Sinoquet H and Rivet P 1997 Measurement and visualization of the architecture of an adult tree based on a three-dimensional digitising device. Trees-Struct. Funct. 11, 265–270.

Sinoquet H, Rivet P and Godin C 1997 Assessment of the three-dimensional architecture of walnut trees using digitising. Silva Fennica 31, 265–273.

Somerville A 1979 Root anchorage and root morphology of *Pinus radiata* on a range of ripping treatments. N.Z. J. For. Sci. 9, 294–315.

Statistical Sciences 1993 S-PLUS User's Manual, Version 3.2, Seattle: StatSci, a division of MathSoft, Inc., 450 p.

Stokes A, Ball J, Fitter A H, Brain P and Coutts M P 1996 An experimental investigation of the resistance of model root systems to uprooting. Ann. Bot. 78, 415–421.

Stokes A, Berthier S, Sacriste S and Martin F 1998 Variations in root shape and maturation strains in the root systems of maritime pine. Trees 12, 334–339.

Stone E L and Kalisz P J 1991 On the maximum extent of tree roots. Forest Ecol. Manage. 46, 59–102.

Watson A and O'Loughlin C 1990 Structural root morphology and biomass of three age-classes of *Pinus radiata*. N.Z. J. For. Sci. 20, 97–110.

Williams C G and Lambeth C C 1989 Bole straightness measurement for advanced-generation loblolly pine genetic tests. Silvae Genet. 38, 212–217.

Wilson B F 1975 Distribution of secondary thickening in tree root systems. *In* The development and function of roots. Eds J G Torrey and D T Clarkson. pp 197–219. Academic Press Inc., London.

Wu T H, Bettadapura D P and Beal P E 1988 A statistical model of root geometry. For. Sci. 34, 980–997.

A. Stokes (ed.), *The Supporting Roots of Trees and Woody Plants: Form, Function and Physiology*, 77–81.

Genetics of root architecture in 1-year-old *Pinus pinaster* measured with the WinRHIZO image analysis system: preliminary results

F. Danjon*, D. Pot, A. Raffin and F. Courdier
INRA, Recherches Forestières B.P. 45, 33611 Gazinet Cedex, France.

Introduction

In south-west France, one million hectares of Maritime pine (*Pinus pinaster* Aït.) produce nearly a quarter of French marketed timber, but the production is hampered by heavy stem straightness defects (Arbez, 1979). Damage by windthrow regularly causes important economic losses to stands growing in the poor sandy podzol of the Landes Forest, affected by summer droughts and winter floods. An advanced breeding program was started in 1960, aiming at providing improved seed material. Trees have been selected for height growth and stem straightness by plus tree selection followed by two cycles of recurrent selection (Danjon, 1995). Such an improved population has been available commercially since 1985, and its performances need to be tested. The breeding level 'G1' is the result of one cycle plus trees selection plus one cycle recurrent selection at age 10 years (Durel, 1992).

The local Maritime pine provenance shows a fairly low juvenile mature correlation for vigour, which may originate in various biomass allocation strategies to roots and shoots (Danjon, 1994b, 1995; Kremer, 1992): some families may favour the development of their aerial parts (the 'sprinters') and other families may develop larger root systems (the 'stayers'). After the onset of high inter-tree competition, the stayers outrun the sprinters, since water and nutriments are the main limiting factor. Much work has been carried out on the genetics of Maritime pine especially with regards to height growth (Kremer, 1981, 1992; Lascoux and Kremer, 1994). However the investigation of full tree architecture and biomass, including the root systems, has only recently started (Colin et al., 1996).

In order to investigate the juvenile genetics of rooting and to assess the correlation with mature traits, a 89 half-sib families retrospective progeny test was installed in spring 1995. After one growing season, the root and aerial parts of 1300 trees were measured by classical techniques including the measurement of biomass, and by scanning followed by image analysis of all root systems: new computer image analysis systems have been developed recently (Arsenault et al., 1995) providing rapid two-dimensional measurement techniques of several features of root architecture. Older manual methods (Tennant, 1975) were tedious and measured mainly the root length. Preliminary results concerning the variables measured using image analysis were reported in this paper.

Materials and methods

A juvenile progeny test was sown in seed beds in three complete blocks in May 1995 in one to three adjacent rows per block and per family, at a 7 × 7 cm spacing. This test was comprised of 89 half-sib polycross progenies of plus trees 'G0 breeding level' sampled from eight districts in the Landes Forest. As control, five provenances from Corsica, Morocco and France, and three additional levels of genetic breeding (wild population, 'G1' and Corsica × Landes hybrid) were also included. Height, mortality and occurrence of forks were measured in September 1995 on the 16 000 sown trees. In January 1996, 6000 seedlings were rapidly measured (height, depth of the taproot, depth of the deepest root, presence of a fork, diameter of the collar, diameter of the longest root, number of main roots) and planted in an open field as bare-root plants.

In May 1996, 1295 additional trees were lifted with forks, and their roots and aerial parts were measured with more details, by classical techniques including biomass weighting and 15 characters such as height and diameter of roots and shoots, number of branches. All root systems were also measured by scanning followed by image analysis.

* Tel: 335-5797-9013. E-mail: fred@pierroton.inra.fr

For image analysis, the WinRHIZO software (Régent Instruments Inc., 1996) was used. The accuracy of measurements was directly checked on screen. The scanner was used in transparency mode, without immersion in water, providing contrasted pictures without shadings. All images were saved for further analysis. The software has a specific method for overlapping correction (Arsenault et al., 1995). Seven characters were assessed by the system: total root surface, number of forks (ramifications), number of root tips and root overlap, root length per diameter class. Three diameter classes were used, i.e. 'fine' (0–0.5 mm), 'medium' (0.5–3 mm) and 'large' (3–12 mm).

The taproot has a major importance in Maritime pine root systems, it is one of a few tree species that conserve a tap root at maturity, therefore, the presence of the tap root will alter the pattern of stress within the root system (Stokes, 1999). The taproot could not directly be measured with the WinRHIZO software, however, it was measured manually.

The variance analyses were made with the following models:

$$(1) Y_{ijk} = \mu + \text{FAMILY}_i + \text{block}_j + (\text{FAMILY} \cdot \text{block})_{ij} + \epsilon_{ijk},$$

$$(2) Y_{ijkl} = \mu + \text{district}_i + \text{FAMILY}_j(\text{district}_i) + \text{block}_k + (\text{district} \cdot \text{block})_{ik} + \varepsilon_{ijkl}$$

$$(3) Y_{ijk} = \mu + \text{provenance}_i + \text{block}_j + (\text{provenance} \cdot \text{block})_{ij} + \epsilon_{ijk}$$

where: FAMILY_i is the ith family, random effect; block_j is the jth block, fixed effect; district_i is the ith district, fixed effect; provenance_i is the ith provenance or breeding level, fixed effect; and ϵ_{ijk} is the random residual error

Restricted maximum likelihood (REML) estimations of heritabilities (with their standard errors) and genetic correlations were computed with the Select-INRA software (INRA, 1991; Mangin and Vincourt, 1992). Narrow sense individual heritabilities were estimated as:

$$h^2_{\text{ns}} = \frac{4\sigma_F^2}{\sigma_F^2 + \sigma_E^2}$$

Results and discussion

The image analysis system provided a rapid (a few minutes per root system) and efficient analysis of root images: almost all roots were taken into account by the system, only the very fine ones were not detected. Hence, the number of root forks was greater than the number of tips, indicating that many root crosses were counted as forks (Table 1). The seedlings had a large amount of fine roots, which were strongly entangled after uprooting. The number of tips, forks, crosses and the root length were highly correlated (Table 1). These four variables all account for the amount of fine roots.

Table 1. Overall means and phenotypic correlations between collar diameter and variables measured with WinRHIZO (Mean and SD for 'Crosses': 104 and 80)

	Collar diameter	Root length	Root projected area	Number of: tips	forks
Overall mean	0.44 cm	235 cm	20 cm^2	162	480
Standard deviation	0.16	136	13	96	351
Root length	0.62				
Root project area	0.78	0.94			
Tips	0.58	0.86	0.78		
Forks	0.46	0.93	0.80	0.86	
Crosses	0.41	0.92	0.76	0.80	0.97

Despite the problems specific to root studies, average heritabilities (0.4 – 0.5, Table 2) were yielded for root systems characteristics indicating that the bias from cultivation and lifting methods was not too high. Heritability of 1-year height was 0.9. Heritability estimations might be biased upwards by the layout: all individuals from one given family originated only from three locations in the seed beds. A very low heritability was yielded for the subjective manual measure of number of large roots. Dimensional characters appeared more heritable than architectural characters: heritabilities for the total root length or root projected area were higher than heritabilities for the fine root or large root length.

District effects were high for all variables except for the large root number (Table 3). The seed-sources differed by their overall size, but not by the relative amount of root length in each root class. The northern district had the best vigour, and the largest amount of roots, and the most southern district had the worst vigour and the lowest amount of roots. However no north/south or east/west gradient or major differences could be shown for vigour and root characteristics between the six other Landes Forest seed sources, despite climatic gradients.

Table 2. Analyses of variance results for the (family + block + family · block) model. Estimated narrow-sense heritabilities and standard errors in the plus tree progenies

Factors	Collar diameter	Coarse root number	Root area	Root length: Total	Root length: 'Fine'	Root length: 'Medium'	Root length: 'Large'	Root biomass
Family	****		****	****	****	****	****	****
Block		****	****	***	****	****	***	***
Interaction								
Heritability	0.56	0.06	0.49	0.48	0.37	0.46	0.40	0.48
Standard error	0.14	NA	0.13	0.12	0.08	0.09	0.10	0.12

Significance level: *: < 5%, ** < 1%; *** < 0.1%; **** < 0.01%.

Table 3. Analyses of variance results for the (district + [family in district] + block + district · block) model

Factors	Collar diameter	Large root number	Root area	Root length: Total	Root length: 'Fine'	Root length: 'Medium'	Root length: 'Large'	Root biomass
District	****		****	***	***	****	***	****
Family in district	****		****	****	****	****	****	****
Block		****	****	***	****	****	****	****
District · block								

Differences between provenances and between breeding generations were non significant or poorly significant (Table 4). Large differences between provenances and breeding generations were observed in mature Maritime pine (Danjon, 1994a).

Fairly high genetic and residual correlations were yielded between collar diameter, root biomass and large root length (Table 5). It is due to a part-whole correlation, i.e., the large roots and especially the taproot constitute the largest part of the root biomass, and the taproot weight is greatly influenced by its upper diameter, the collar diameter. Thus Nicoll et al. (1995) do not include the stump in root volume computations, as it usually represents a major part of the total below-ground volume and has a very specific function regarding root anchorage or water and nutrient uptake and transport. Medium genetic and residual correlation were found between collar diameter and fine or medium root biomass. The genetic correlation between fine root length and large root length was higher ($\rho_g = 0.83$) than the corresponding residual correlation ($\rho_r = 0.57$): an indirect selection for a larger root biomass using the collar diameter as a selection criteria will also provide a genetic gain on fine root length.

Conclusion

WinRHIZO provides a very powerful tool for estimating rapidly parameters of root architecture in small plants, the scanning surface reaching 29.7 × 36 cm. Larger root systems can also be analysed by this way, but only from photographs or numeric pictures. Advanced analysis features implemented in the WinRHIZO pro version (topology, fractal dimension, amount of root per depth) were not used, but the images have been conserved for further analysis. A better measurement of finer roots could have been made by immersion in water, but the time spent for the measurement would have increased. Only features of the two-dimensional geometry have been measured. The geometry can hardly be measured by such a technique. All the more so, the three-dimensional geometry could not be assessed, and it was lost during the uprooting.

Heritabilities of root characteristics were fairly high, showing a good correlation with collar diameter. However effects of natural selection by climatic gradients could not been detected.

Acknowledgement

We thank E. Michelozzi, O. Lagardère, B. Montoussé and L. Saléra for technical assistance. Financial

Table 4. Analyses of variance results for the (provenance + block + provenance · block) model

Factors	Collar diameter	Large root number	Root area	Root length				Root biomass
				Total	'Fine'	'Medium'	'Large'	
Provenance	*				*			*
Block			*	*	*	***		
Interaction	**				*			

Table 5. Estimated genetic correlations (above) and residual correlations (below) in the plus tree progenies

Variables	Collar diameter	Large root number	Root area	Root length			
				Total	'Fine'	'Medium'	'Large'
Coarse root	(0)						
number	*0.29*						
Root	0.88	(0)					
area	*0.79*	*0.44*					
Root	0.78	(0)	0.97				
length	*0.66*	*0.47*	*0.52*				
'Fine' root	0.73	(0)	0.88	0.97			
lenght	*0.5*	*0.33*	*0.76*	*0.9*			
'Medium' root	0.73	(0)	0.93	0.9	0.95		
length	*0.62*	*0.44*	*0.9*	*0.98*	*0.82*		
'Large' root	0.94	(0)	0.98	0.9	0.83	0.87	
length	*0.84*	*0.3*	*0.87*	*0.77*	*0.57*	*0.72*	
Root	0.94	(0)	0.94	0.81	0.78	0.79	0.97
biomass	*0.89*	*0.35*	*0.84*	*0.72*	*0.54*	*0.68*	*0.89*

support was provided by the European Community (contract no. AIR3-CT93-1269 "Rootarch" project).

References

Arbez M 1979 Contrôle génétique et mécanismes d'acquisition des défauts de rectitude du tronc chez le pin maritime landais. Proc. of 104e Congrès National des Sociétés Savantes, Bordeaux, sciences no fasc. II.

Arsenault J L, Pouleur S and Messier C 1995 WinRHIZO, a root-measuring system with a unique overlap correction method. Hortscience 30, 306.

Colin F, Danjon F and Wehrlen L 1996 Etudes racinaires au sein du programme 'Croissance' de l'INRA (*Quercus petraea* et *Pinus pinaster*). Rev. For. Fr. 46, 165–172.

Danjon F 1994 Stand Features and Height Growth in a 36-Year-Old Maritime Pine (*Pinus pinaster* Ait.) Provenance test. Silvae Genet. 43, 52–62.

Danjon F 1994 Heritabilities and genetic correlations for estimated growth curve parameters in maritime pine. Theor. Appl. Genet. 89, 911–921.

Danjon F 1995 Observed selection effects on height growth, diameter and stem form in maritime pine. Silvae Genet. 44, 10–19.

Durel C-E 1992 Gains génétiques attendus après sélection sur index en seconde génération d'amélioration du pin maritime. Rev. For. Fr. 44, 341–355.

INRA 1991 Sélect - manuel d'utilisation. 104 pp.

Kremer A 1981 Déterminisme de la croissance en hauteur du pin maritime (*Pinus pinaster* Ait.) III. Évolution des composantes de la variance phénotypique et génotypique Ann. Sci. Forest. 38, 355–375.

Kremer A 1992 Predictions of age–age correlations of total height based on serial correlations between height increments in maritime pine (*Pinus pinaster* Ait.) Theor. Appl. Genet. 85, 152–158.

Lascoux M and Kremer A 1994 Effect of the purposive choice of families on the estimates of the juvenile mature correlation derived from retrospective tests. Can. J. For. Res. 24, 756–761.

Mangin B and Vincourt P 1992 Schémas de sélection: de la représentation généalogique au modèle statistique. Élaboration du modèle. Genet. Sel. Evol. 24, 71–84.

Nicoll B C, Easton E P, Milner A D, Walker C and Coutts M P 1995 Wind stability factors in tree selection: distribution of biomass within root systems of Sitka spruce clones. *In* Wind and Wind Related Damage to Trees. Eds M Coutts and J Grace. pp 276–301. Cambridge University Press, New York.

Régent Instruments Inc. 1996 Mac/WinRHIZO introduction manual and reference manual. 36 & 40 p.

Stokes A 1999 Strain distribution during anchorage failure of *Pinus pinaster* Ait. at different ages and tree growth response to wind-induced root movement. Plant Soil 217, 17–27.

Tennant, D. 1975 A test of modified line intersect method for estimating root length J. Ecol. 63, 995-1001.

A. Stokes (ed.), The Supporting Roots of Trees and Woody Plants: Form, Function and Physiology, 83–96.

Root architecture, early development and basal sweep in containerized and bare-rooted Douglas fir (*Pseudotsuga menziesii*)

Erik Sundström[1] and Michael Keane[2,*]
[1]*FoReCo, Plangatan 5, 90336 Umeå, Sweden (Tel/Fax: 46 90 139768; E-mail: eriksund@algonet.se) and*
[2]*Coillte Teoranta, Research & Development, Newtownmountkennedy, Co. Wicklow, Ireland*

Key words: architecture, bare root, basal sweep, container, dominant root, Douglas fir, root, symmetry

Abstract

Seedlings grown in containers often have deformed, spiralling, kinking root systems (reported especially in pine growing in colder climates like Fenno-Scandinavia). The current study examines the effect of containers on the root systems of Douglas fir planted in Ireland. A sample of sixteen 10-year-old Douglas fir trees, planted as either bare-rooted transplants (2/1) or containerized seedlings (Paperpot 610 (2/0)) on an acid brown earth site in Ireland, were excavated for root achitectural studies. In addition to thorough above-ground measurements, an assessment of basal sweep was also carried out. Root systems were systematically described and horizontal angles, cross-sectional areas (CSA), and maximum depths of all roots > 5 mm in diameter were recorded. Various biomass ratios and estimates for dominant roots were also calculated. Symmetry of the root systems, max. and min. numbers, and CSA of roots for sectors of various sizes were compared for the two plant types. The initially (when planted) bigger bare-rooted transplants were still larger than the container-planted trees after ten years. Of the total above-ground fresh weight, the container stock had allocated more to the crown and less to the stem compared to bare-roots. The difference in root weights between stock types was small. Basal sweep occurred on average in 50% of the bare-roots and 35% of the container trees. The direction of the basal sweep leaning was concentrated to the NE, which coincided with the direction of the slope and the prevailing wind direction. No effect of planting position in relation to the direction of site preparation was found for basal sweep. The size of the root system, fresh weight and total cross-sectional area was on average for all trees correlated to both DBH and shoot fresh weights. For containers separately, however, only root area was correlated to DBH and stem fresh weight. In spite of the difference in the above-ground size, there were no significant differences in root numbers and root area (CSA) between the two plant types. No difference in rooting depth between plant types was found. When splitting the root system into 120° horizontal sectors (1/3 of the root system) the highest numbers of roots were concentrated in the NW direction. The highest amount of root area tended to be concentrated along a NW-SE diagonal, with a dominance for the latter (SE). Sectors without dominant roots (expected to be the future main structural, stabilizing roots), varied in size (94–178°) but were on average wider in bare-roots and on average concentrated in the NE direction and the downhill side of the slope.

Introduction

Ever since containerised tree seedlings came in use in the late 1960s, there has been an ongoing debate regarding how containerised seedlings perform, e.g. survival, growth, root symmetry and stability, stem deformation, etc., in comparison with bare-rooted and naturally established seedlings of various species.

Comparisons of survival and growth between bare-rooted and containerised Douglas fir (*Pseudotsuga menziesii*) have given contradicting results, but it seems that container stock more often has better survival (Hobbs and Wearstler, 1983; McDonald, 1991; Michaud, 1983; Muhle et al., 1979; Owston and Stein,

* FAX No: +353 1 2011199. E-mail: mgkeane@indigo.ie
Plant and Soil is the original source of publication of this article. It is recommended that this article is cited as: *Plant and Soil* **217:** 65–78, 1999.

1978; Schmaltz, 1990; Strittmatter, 1977) but suffers slightly in growth (Helgerson et al., 1992; Newton et al., 1993; Van den Driessche, 1992) compared to bare-rooted stock. The better growth of bare-rooted plants is often explained by the difference in seedling size at the time of planting; i.e. bare-root plants are often larger than container plants.

Several studies have also examined differences in root growth of the two stock types. Containerised seedlings of species that can not develop adventitious roots, especially pine (*Pinus* sp.), planted in countries with cold climates have been found to have restricted and deformed root systems as a result of the shape of the container (Halter et al., 1993; Huuri, 1978; Lindström, 1978; Van Erden and Arnott, 1974). Improved container designs have reduced these problems to some extent (Lindström and Håkansson, 1994). Bare-rooted seedlings, however, have also been found to have deformed root systems, often said to be a result of plant handling or the planting method, e.g. J-shaped and L-shaped root systems (Van Erden, 1978; Wass and Smith, 1994).

Root studies of planted Douglas fir, however, have not shown the same negative effects on root morphology due to the container as have the pine species (Arnott, 1978; Iske, 1990; Long, 1978; Preisig et al., 1979).

The less developed root systems in containerised seedlings have caused stability problems later in the stand development in both Scots and lodgepole pines in the Nordic countries (Håkansson and Lindström, 1994; Hulten and Jansson, 1978, Lindgren and Örlander, 1990; Lindström, 1990; Rosvall, 1994). Poor stability at an early stage may cause toppling of seedlings (Burdett et al., 1986), which later on can give stem deformities, like basal sweep. The basal sweep problem, however, was seen as a problem mainly related to the use of bare-rooted seedlings. It was suggested that the use of containers could reduce the frequency of severe basal sweep in bare-root planted lodgepole pine in Ireland (Carey and Hendrick, 1986; Pfeifer, 1982). The problem with toppling of seedlings, which later causes basal sweep stem deformities, has also been observed in young, 2-5 years old Douglas fir plantations.

This study is a comparison of the field performance of bare-rooted (2/1) and containerized (Paperpot 610, 2/0) Douglas fir ten years after planting on an acid brown earth site in Ireland. The objectives were to study the effects of the two stock types on the above-ground performance: survival, growth and stem form; as well as on the root system morphology: size, depth and symmetry of the root systems.

Materials and methods

Site

The experiment was established on a reforestation site in Woodstock property, Knocktopher forest in County Kilkenny. The previous crop of Norway spruce (Yield class 16) was clearfelled in 1983. Elevation is 115 m with an easterly aspect and slopes varying between 7 and 12°. The soil is an acid brown earth with a thin peat layer. The site was prepared after clearfelling by ripping (breaking the soil layer with a vertical steel blade) at two meters spacing.

Planting of bare-rooted Douglas fir transplants (2/1) was carried out in February 1985. In July the same year, three sizes (608, 610, 615, where the first digit refers to the diameter, and the following two digits to the height in cm of the container) of containerized Paperpot seedlings (2/0) were planted. A total of 15 plots (22 $\times$ 22 m) were planted at 2$\times$2 m spacing with the following distribution of plant types: 5 plots of transplants, 2 plots of 608s, 7 plots of 610s and 1 plot of 615s.

Sample for root excavation

Due to the unavailability of a complete range of container sizes, sample trees for root excavation were selected only from Pp 610 containers and from bare-rooted transplants. A total of 16 trees, eight transplants and eight containers, from four plots was selected for intense measurements. Sample trees were selected within the constraint of $\pm$ 10% deviation from the plot mean breast height diameter (DBH).

Measurements

At plot level, an assessment of DBH, height and stem quality (basal sweep) was carried out in the central 6$\times$6 planting positions in all 15 plots. Basal sweep was assessed as the deviation of stem centre at breast height from the vertical compared to the centre at stem base. Distance and direction of the deviation as well as planting position in relation to the ripped line were recorded for the basal sweep assessment.

At sample tree level, above- and below-ground assessments were carried out in the following ways.

Above-ground

All live branches were cut off and weighed at the site. The trees were then cut, leaving a 2 m high stem as a lever for pulling out the root systems. The root systems were carefully cleaned (while the tree was still standing) using an air-knife (a high-pressure air-blasting-tool), loosened and pulled completely out of the ground. The stems were cut at ground level and the whole stem was weighed. Stem diameter was measured at 0.25, 0.75, 1.5 m and thereafter at every meter. Any type of crown or stem damage was recorded.

Below-ground

All 16 root systems were, after thorough cleaning, weighed and then transported to an indoor facility for detailed measurements. For measuring, each root system was turned upside down, centered and mounted on a root measuring table. For each single root, horizontal angle and diameter (>5 mm) were recorded at three radii from stem centre (15, 30 and 60 cm). The diameters were taken in two directions on every occasion to get the cross-sectional area of the root. Maximum depth was recorded only once for each root and that in the section where it occurred. Notes of tap roots and breakage were taken.

Calculations and analyses

Arithmetic means for height, diameter breast height (DBH), basal area, stem volume, and arcsin-transformed survival-rates were calculated for each plot and plant type. Basal area calculation was based on DBH. Stem volume per tree was for the sample trees calculated as the sum of the volume of each stem section as:

$$Volume = \Sigma(A_t + A_b)/2^*L \qquad (1)$$

where A_t= area of each section's top, A_b= area of each section's bottom, L = the length of each section. This 'true' volume of the sample trees was used for estimating volume of the height and diameter measured trees of the plot assessment, using secondary regression equations.

Below ground

The Numbers (Nos) of all roots > 5 mm in diameter and their Cross Sectional root Area (CSA) were totalled for each radius separately and for the entire root system. Nos and CSA were summed up for sectors of various widths (45, 90, 120 and 180°) of the entire root system as well as radii-wise. Sectors were rotated in order to localise the horizontal position of maximum and minimum Nos and CSAs. Dominant roots were calculated in three different ways: a) The 90%-rule; where all roots, starting with the biggest, up to 90% of total CSA were defined as dominant roots (Coutts, 1983); b) The 8%-rule; where all roots >8% of total CSA were defined as dominant roots (Nielsen, 1997); c) Average-rule; where all roots > the average CSA-root were defined as dominant roots. The widest sector without dominant roots was determined by rotating the root system horizontally. The average maximum depth as well as the proportion maximum depths in each of the three sections were calculated for each root system and means for plant types were generated.

Differences between plant types in both above and below ground parameters were compared in an analysis of variance (SPSS, Anon. 1990), based on plot level means as well as on means on single tree level (the latter only for excavated trees). Correlations between above and below ground parameters were calculated and tested with Spearman's bivariate correlation coefficient (SPSS, Anon. 1990).

Results

Plot data

Survival and tree size

On average, bare-rooted transplants of Douglas fir were bigger than containerized Paperpot 610s ten years after planting (Figure 1). Height, diameter (DBH) and stem volume were all significantly greater in bare-roots, as well as volume per hectare (even with the higher survival rate for the containers). The average height difference between the two stock types was 0.47 m in 1995, which in absolute values is almost the same as it was in 1988, three years after planting (1.74 for bare-roots and 1.32 for containers). The relative height difference has, however, decreased from 32% to 8% in the last eight years.

Additional information regarding the smaller (Pp 608) and the larger (Pp 615) container types, are presented separately (Table 1). Statistical comparisons of the three different container sizes were not applicable, however, due to the low numbers of replicates.

Basal sweep

On average, every second bare-rooted transplant and every third containerized tree ($p = 0.023$) had a basal

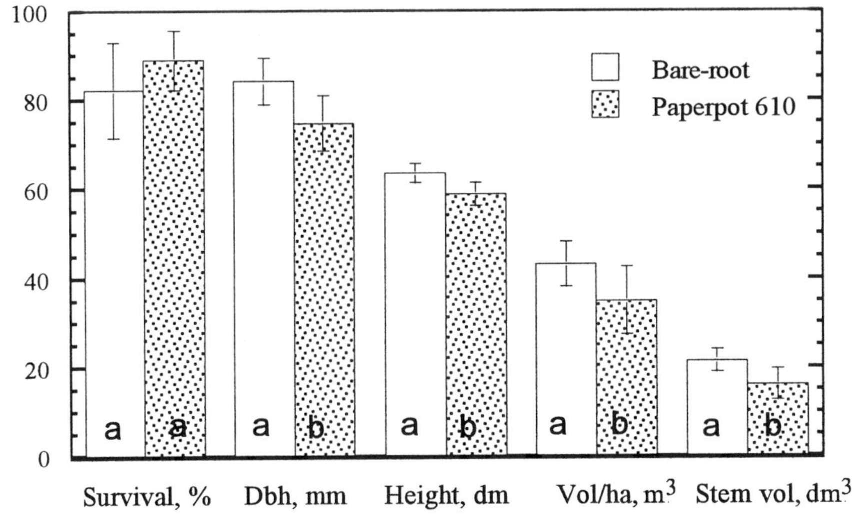

Figure 1. Basic stand information in 1995 for area KTR 6/85 planted with bare-rooted transplants and containerised (Paperpot 610) Douglas fir in 1985. Bars with different letters are significantly different at 0.05 level. Error bars show variation as standard deviation.

Table 1. Survival, basal sweep, DBH, height and volume ha^{-1} of 10-yr-old Douglas fir, planted in 1985 as bare-rooted transplants and 3 sizes of containers (Pp 608, 610, 615)

Stock type	No. of plots	Survival %	Basal sweep (> 10 cm) %	DBH cm	Height m	Volume/ha m^3
Bare-root	2	84.7	54.0	8.82	6.48	50.2
Paperpot 608	2	84.7	39.0	7.86	6.08	40.0
Paperpot 610	2	88.9	35.0	7.81	6.06	41.4
Bare-root	1	86.1	45.2	8.17	6.21	42.1
Paperpot 610	1	83.3	26.7	6.49	5.51	21.5
Paperpot 615	1	83.3	50.0	8.15	6.20	40.0

Table 2. The proportion of trees deviating from stem centre at breast height compared to stem centre at ground level (basal sweep) in 10-yr-old Douglas fir, planted in 1985 as bare-rooted transplants and as containers (Pp 610)

Stock type	No. of plots	Basal sweep (> 0 cm) %	Std. error	Basal sweep (> 10 cm) %	Std. error	Basal sweep (> 15 cm) %	Std. error	Basal sweep (> 25 cm) %	Std. error
Bare-root	5	80.8 A*	4.6	51.7 A	6.8	26.1 A	5.3	7.9 A	2.7
Paperpot 610	7	71.7 B	1.2	36.9 B	2.3	13.0 A	4.2	3.8 A	1.7

*Different letters indicate significant differences at $p < 0.05$.

sweep deviating more than 10 cm from stem centre at ground level (Table 2). The average distance of deviation from stem centre of trees leaning more than 10 cm was 22 cm (≈10°) for transplants and 18 cm (≈8°) for containers. Planting position was not related to the occurrence of basal sweep (Table 3).

The direction of the leaning of trees with basal sweep was unevenly distributed. Among trees with basal sweep a higher proportion (64% compared to 37.5% in an even distribution) was leaning towards east to north-east, a direction coinciding with the downwards direction of the slope as well as with the

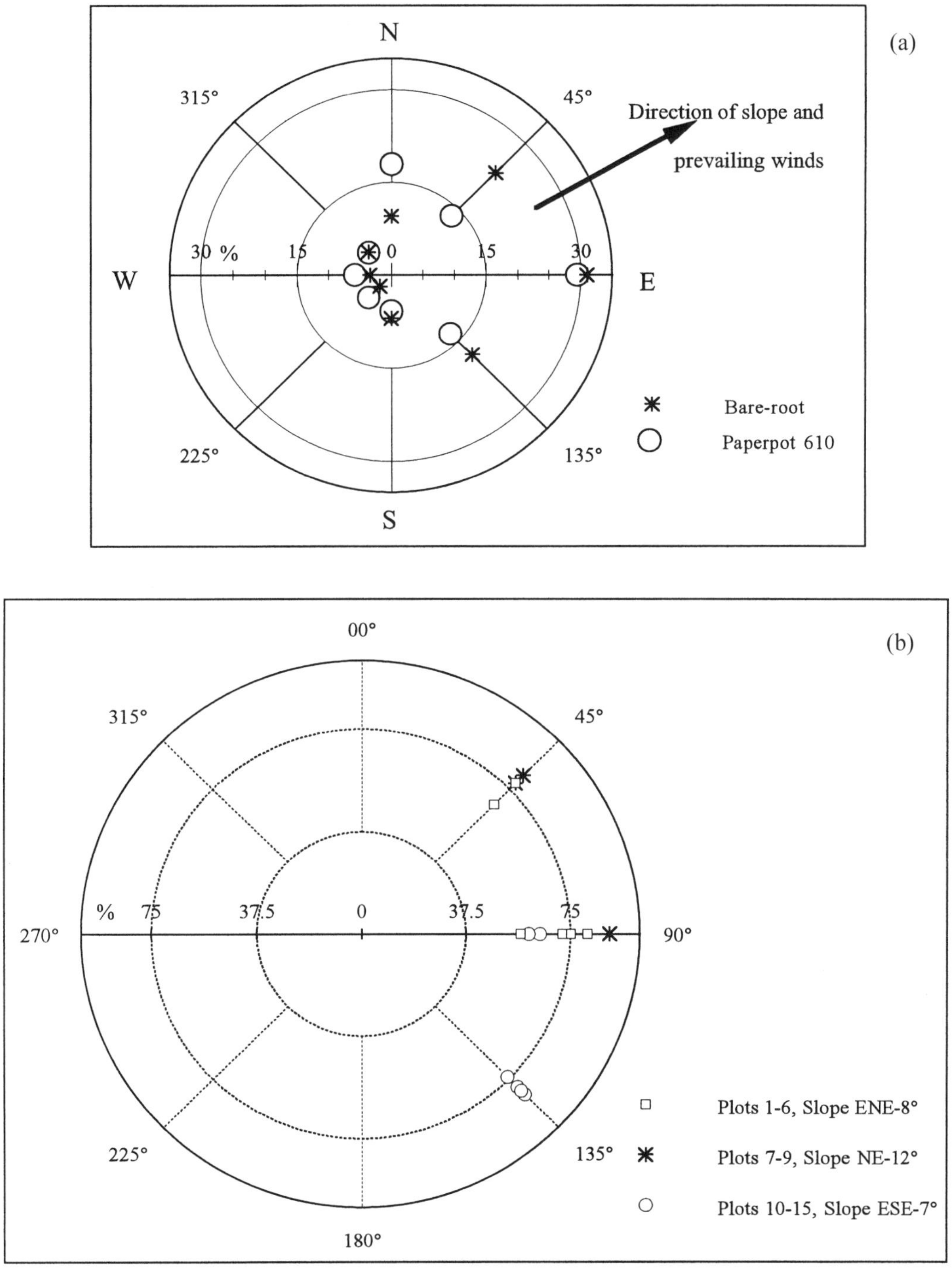

Figure 2. The proportion of trees (%) with basal sweep (> 10cm from stem centre) in each of eight 45° sectors for Douglas fir planted as either bare-rooted transplants or containerized Paperpot 610. (b) Proportion of trees with basal sweep in the 135° wide sectors with the maximum proportion of basal sweep. Symbols indicate the dominant lean direction in each of the 15 plots, which vary in three different slope directions.

Table 3. Number of planted trees (N), and number and frequency (%) of basal sweep (>10 cm) for 10-yr-old bare-rooted transplants and containerised (Pp 610) Douglas fir trees, planted in three planting positions in relation to the ripped line; north, south and centre

Planting position	Planted, N Bare-root	Planted, N Pp 610	Basal sweep, N Bare-root	Basal sweep, N Pp 610	Basal sweep, % Bare-root	Basal sweep, % Pp 610
North	53	10	40	8	75 A*	80 A
Cente	56	190	46	133	82 A	70 A
South	35	19	30	15	86 A	79 A

*Different letters indicate significant differences at $p < 0.05$.

prevailing wind direction (Figure 2a). This was more pronounced for bare-rooted transplants (72%) than for containerized (60%) trees. The direction of the slope varied between plots, and the lean direction of trees with basal sweep tended to follow this direction (Figure 2b).

Sample tree data

Tree size

The 16 trees excavated for root studies were considered as representative trees, compared to the plot average data. Among the excavated trees, bare-rooted transplants were 20% taller than containerized trees (8% for the plot average) and 5% larger in diameter (DBH) (13% plot average). Stem volume per tree was for transplants on average 0.024 m^3 compared to 0.019 m^3 for containerized seedlings. The difference in height was the only significant difference ($p < 0.05$).

Weights and root/shoot ratios

There were no significant differences in fresh weights between the two stock types, either above or below ground. When separating the shoot into stem and crown, bare-roots had allocated more into stems whereas containers had relatively heavier crowns (Figure 3).

This was also found in ratios of fresh weights, where the crown/above-ratio was significantly higher in containerised trees compared to bare-rooted trees (Figure 4). Accordingly, the average root/crown ratio was higher for bare-rooted trees but only statistically significant at the single tree level. There were no significant differences between plant types in ratios such as the total cross sectional area of roots > 5 mm (CSA), although the mean ratio between basal area at breast height (BaBH) and CSA in the first section (0.15 m radius) was slightly higher in bare-roots (Figure 4).

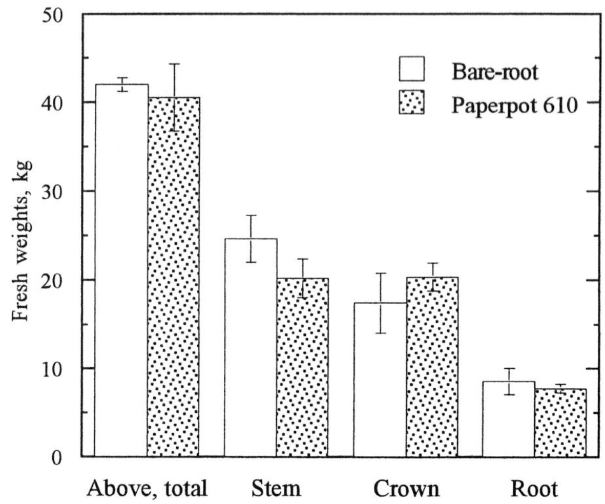

Figure 3. Fresh weights in different sections of the tree for 10-yr-old Douglas fir, planted as either bare-rooted transplants or as containerized Paperpot 610. No differences between bare-root and paperpot 610 were significant at $p < 0.05$.

On average, DBH and above-ground fresh weights were both significantly correlated to both root fresh weight and root CSA (Table 4). When testing these correlations separately for each plant type, however, root fresh weight was only correlated with stem fresh weights of bare-roots. Root CSA, on the other hand, showed strong correlation with stem fresh weights but only in containerized trees.

Root architecture

Number and ramification of roots

Both plant types had, on average, 21 roots ($\geq$ 5 mm) at 15 cm from the centre of the root system. Nearly the same numbers of roots were found at 30 cm, whereas at 60 cm the numbers of roots had decreased. There were no significant differences in root numbers between the two plant types. Numbers of roots were, within the present size-range of trees, not correlated to tree size.

The ramification patterns of the roots did not differ much between the two stock types (Figure 5). On average, 70% of the roots measured in the first section (0–15 cm) had branched 1.5 times before reaching the second radius at 30 cm. In the third section (30–60 cm), 40% of the roots of bare-roots and 33% of the containers continued to the third radius, branching in the third section 1.5 and 1.9 times, respectively.

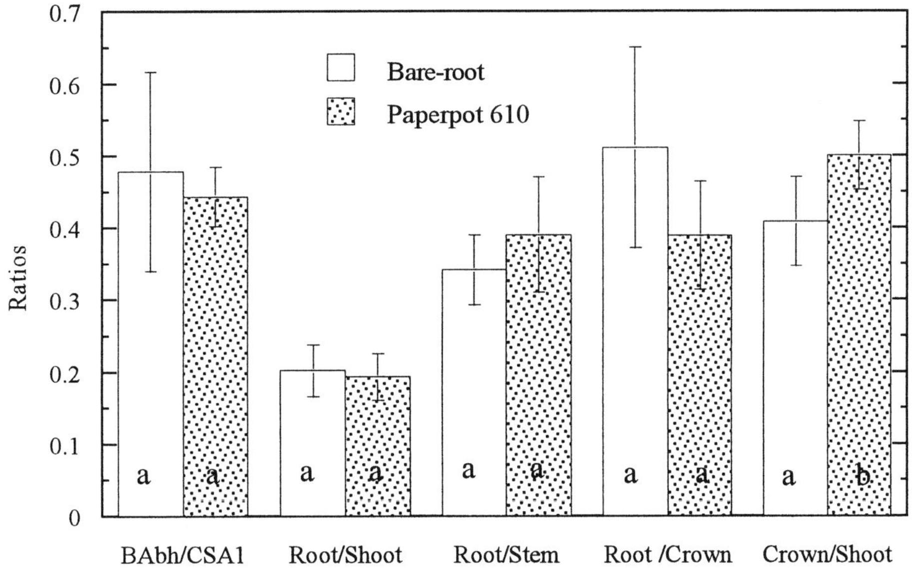

Figure 4. Fresh weight biomass ratios between different sections of the tree plus the ratio between the basal area at breast height (BaBH) and the total cross sectional area of all roots at 15 cm from stump centre (CSA1) for Douglas fir planted as bareroots or containers. Different letters indicate significant difference at $p < 0.05$.

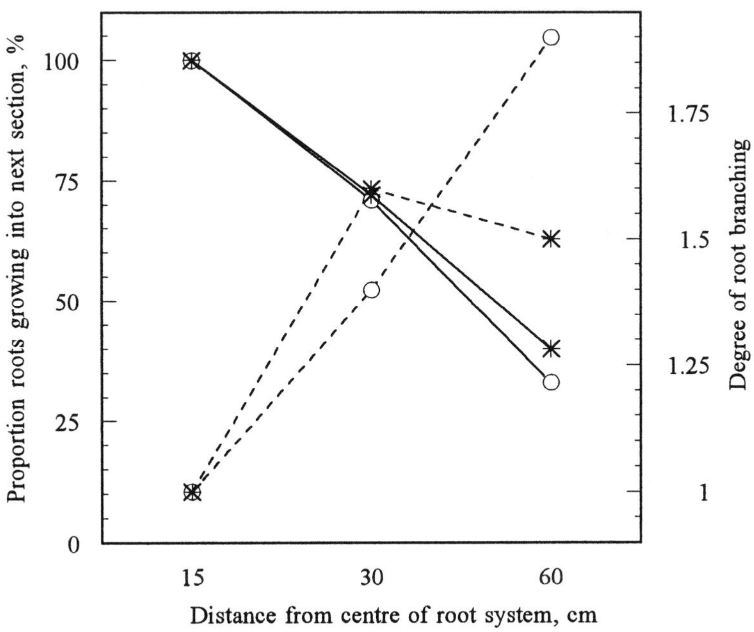

Figure 5. Root ramification for bare-rooted transplants (stars) and containerised Pp 610 (circles) of 10-yr-old Douglas fir measured at three radii (15, 30 and 60 cm). Proportion (%) roots growing into the following section (solid lines, left y-axis) and degree of branching (broken lines, right y-axis).

Table 4. Spearman's correlation coefficients between diameter (DBH) and fresh weights (stem, tot. above ground) on the one hand, and fresh root weight and cross-sectional area of roots (CSA1 = 1st radius, CSAtot = whole root system) on the other, for bare-rooted transplants and containerised (Pp 610) 10-yr-old Douglas fir

	DBH			Stem fresh weight			Above ground fresh weight		
	Bare-root	Pp 610	All	Bare-root	Pp 610	All	Bare-root	Pp 610	All
CSA1	0.65	0.83*	0.65**	0.55	0.97**	0.56*	0.62	0.50	0.57*
CSAtot	0.65	0.76*	0.61*	0.57	0.85**	0.52*	0.50	0.55	0.55*
Root weight	0.53	0.58	0.65**	0.71*	0.64	0.70**	0.55	0.55	0.54*

*Significant correlations at $p < 0.05$, **Significant correlations at $p < 0.01$.

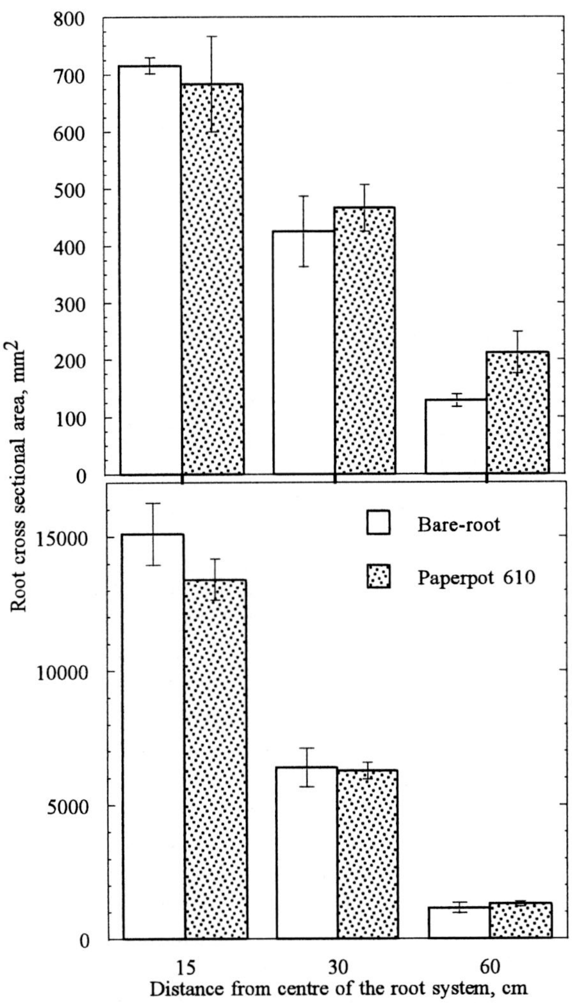

Figure 6. Mean cross-sectional area (CSA) of all (bottom) and the average (top) roots at three radii from the centre of the root system for bare-rooted transplants and containerised 10-yr-old Douglas fir. Error bars show standard deviation. No differences between stock types were significant at $p < 0.05$.

Root depth

Average rooting depth was around 20 cm for all roots in the root system irrespective of stock type and distance from the stem centre. For dominant roots, average root depth was slightly deeper and around 25 cm for both stock types. Maximum depth for each root system was on average between 60 and 70 cm and a little less away from the centre. Rooting depth was registered for each individual root only once, and that in the section where it occurred. For roots measured at both first and second radii, a higher proportion had the maximum depth in the second section, indicating that most roots were growing downwards.

Root area

Root cross sectional area (CSA) of all roots at the first radius (15 cm) made up 65% of the total root CSA and was, on average, slightly but not significantly higher for bare-rooted compared to the containerized trees (Figure 6). At 30 cm (30% of total CSA) and 60 cm (5% of total CSA) from the centre of the root system there was no difference in root CSA between the two stock types.

The root CSA of the average root was around 700 mm^2 at the first radius (15 cm), which corresponds to a root diameter of 26 mm (Figure 6). At first radius, the average root CSA of bare-roots was slightly larger than the average container root, but further away container roots were larger and significantly so at the 60 cm radius. Root area was correlated to tree size only for containerised trees (Table 4).

Root symmetry

By splitting the root system into three 120° sectors, it was found that bare-rooted transplants had 60% of their roots and 58% of the CSA in those sectors with the highest number of roots and the highest proportion

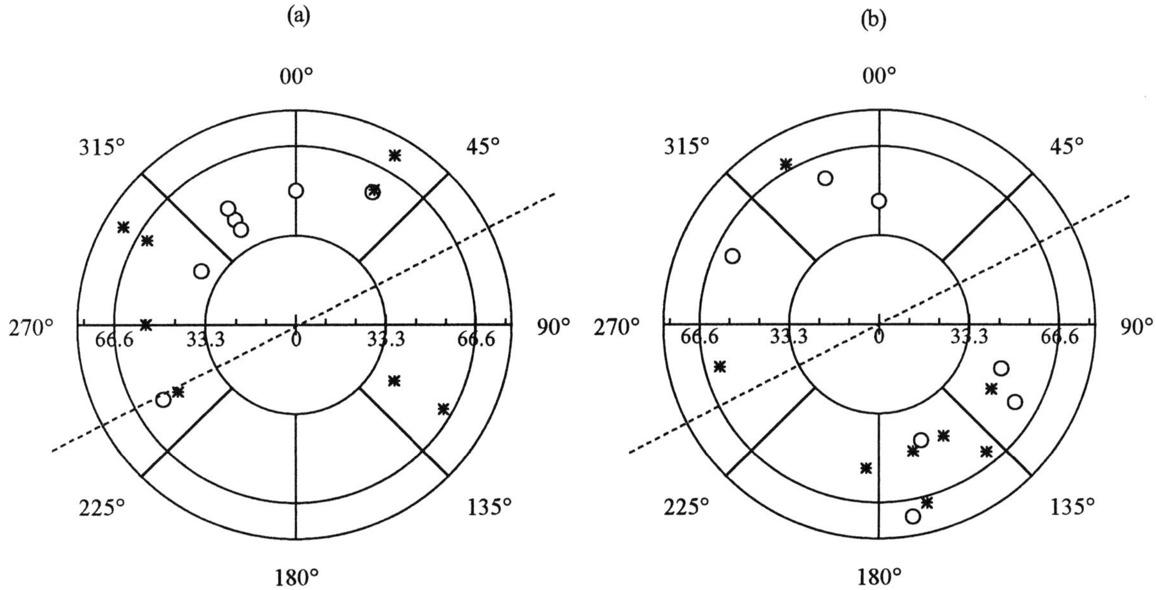

Figure 7. Centre (angle,°) of 120° sectors with the highest proportions (radius,%) of roots (ideal = 33.3%) for individual Douglas fir trees planted as either bare-rooted transplants (stars) or as containerized Pp 610 (circles). Numbers of roots (a) and root area (CSA) (b). Direction of site preparation (broken line) is marked.

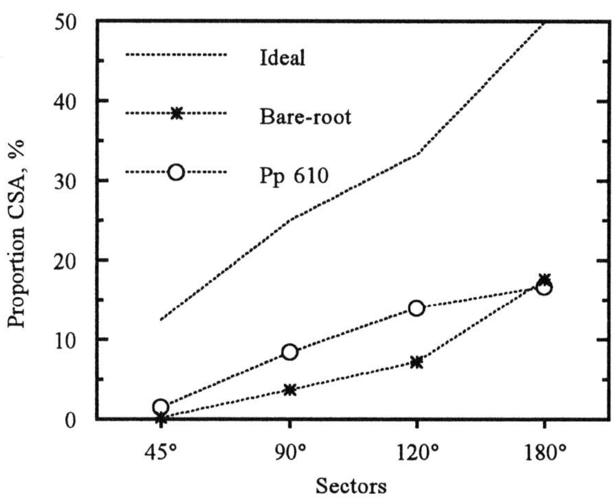

Figure 8. Proportion root area (CSA) in the sector with the lowest proportion root area for four different sector sizes; 45°(12.5%), 90°(25%), 120°(33.3%) and 180°(50%) for 10-yr-old Douglas fir planted as bare-rooted transplants or as containers (Pp 610). Average for stock type.

of root area (CSA) (ideal symmetry = 33.3%). Corresponding figures for containers were 49% and 59%, respectively. Sectors with the maximum numbers of roots had their centres in the NW direction, whereas maximum CSAs were found either in SE or NW (Figure 7). The direction of ripping of the site preparation stretched from NE to SW.

In sectors of varying sizes (45°, 90°, 120° and 180°) containing the minimum proportion of root area CSA, bare-rooted transplants had less than containerised in two of the sector sizes (90° and 120°) (Figure 8).

Dominant roots

Dominant roots were defined according to three different rules: (a) all roots ≥ the average CSA-root, (b) all roots with > 8% of total CSA, and (c) all roots making up to 90% of total CSA. According to the 8% -rule, bare-rooted trees had on average 4.4 dominant roots compared with 3.6 for containers (Figure 9a). The corresponding figures according to the 'average' -rule were 6 and 5, and for the 90% -rule, 7.5 and 7.6 dominant roots, respectively. The number of dominant lateral roots was positively correlated to total number of roots according to the 90% -rule, negatively according to the 8% -rule, and least correlated according to the 'average' - rule (Figure 9b).

The maximum width of a sector without dominant roots was around 175° according to the 8% -rule, less according to the other two definitions but significantly wider for bare-rooted trees compared to containerised for the 'average' -rule (Table 5). Dominant roots were most frequently missing downhill in the NE direction of the root system (Figure 10).

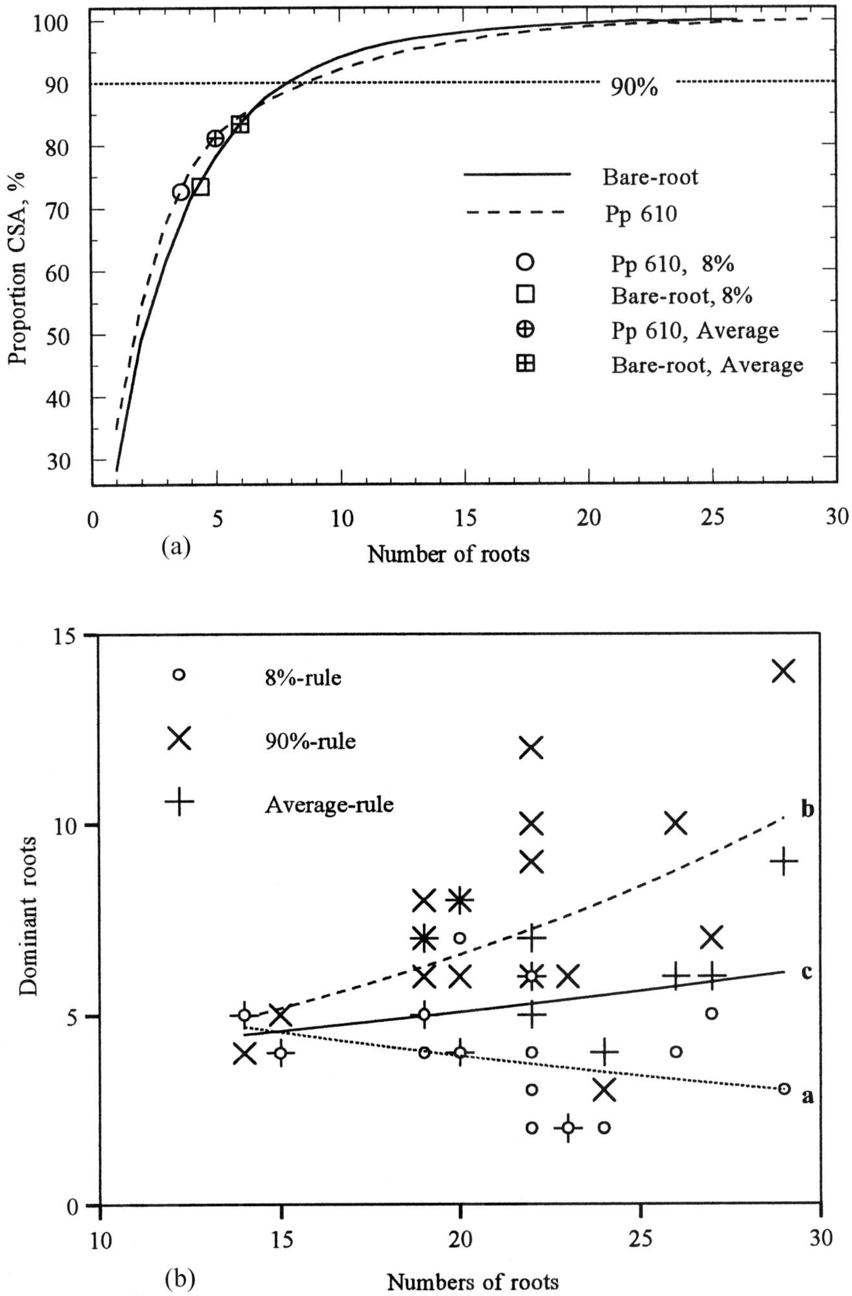

Figure 9. (a) Mean cumulatative root CSA for 10-yr-old Douglas fir planted as either bare-rooted transplants (solid line, squares) or as containerized Paperpot 610 (broken line, circles). Three ways of defining 'Dominant' roots are marked in the graph; a; roots $\geq$ average CSA root (crossed), b; roots > 8% of total CSA (empty), c; all roots making up 90% of total CSA (dotted line). (b) Dominant roots in relation to total number of roots for three different definitions of dominant roots: a; roots > 8% of total CSA (dotted line), b; all roots making up 90% of total CSA (broken line) c; roots $\geq$ average CSA root (solid line).

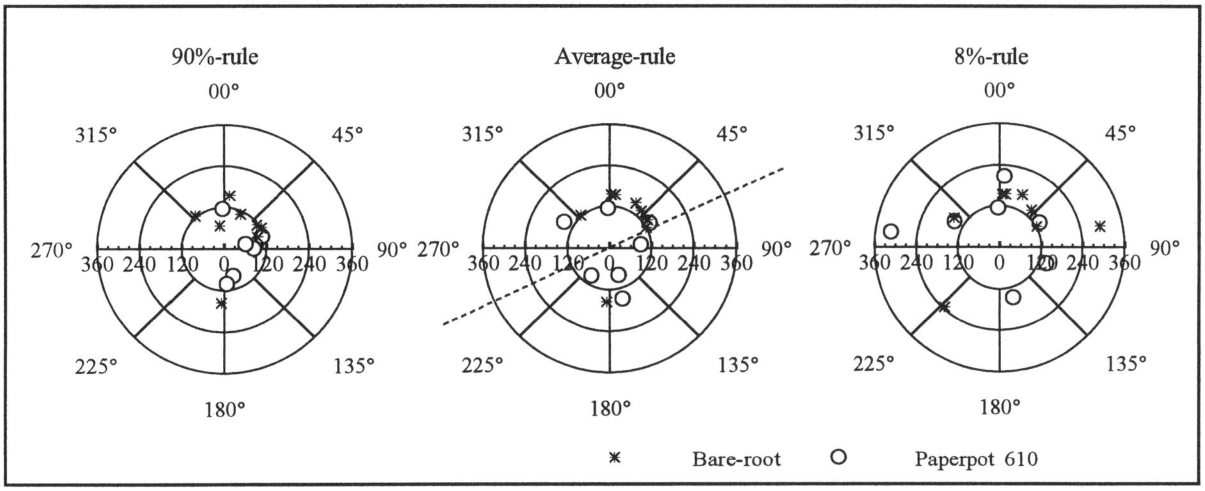

Figure 10. Location and width of the widest section without dominant roots for 16 10-yr-old Douglas fir trees planted as either bare-rooted transplants (stars) or as containerized Paperpot 610 (circles). Dominant roots are defined according to three different rules; All roots up to 90% of total CSA (left), roots ≥ average CSA-root (centre), and roots ≥ 8% of total CSA (right). Horizontal angles (°) show the centre and radius the size (width in °) of the widest sector without dominant roots.

Table 5. Average maximum width (°) of sectors without dominant roots for 10-yr-old Douglas fir planted as either bare-rooted transplants or as containers, according to three different ways of defining dominant roots: a) roots > the average CSA root, b) all roots making up to 90% of total CSA, c) roots with > 8% of total CSA

Stock type	Max. sector width, 90% -rule		Max. sector width, average -rule		Max. sector width, 8% -rule	
	Average	St. dev.	Average	St. dev.	Average	St. dev.
Bare-root	119 a*	28	142a	14	178a	58
Paperpot 610	94 a	19	116b	29	174a	69

*Different letters indicate significant differences at $p < 0.05$.

Discussion

Survival, growth and tree size

On average ten years after planting, survival was 80-90% and slightly but not significantly better for containerized Douglas fir compared to the bare-rooted transplants. The initially 1.5–2 years older and taller bare-rooted seedlings were still bigger after ten years than the containers. The average height difference between the stock types (around 0.5 m in absolute values) was, however, almost the same as it was three years after planting. Thus the initial advantage in height had not increased in seven years, but rather decreased in relative terms, supporting the findings of Van den Driessche (1992) who stated that the time advantage of larger stock decreased with time. The difference in height together with the 13% larger diameter at breast height in bare-roots resulted in 30% more volume per stem in the transplants than in the containers. Even when taking survival into consideration, volume per hectare was still significantly higher for transplants. This confirms the findings of Newton et al. (1993) who found a strong positive correlation in Douglas fir between initial height and long-term growth, but contrasting McDonald (1991) who stated that container stock outperformed bare-root stock, without however presenting data on the initial seedling height. An interesting observation was that fresh weights of shoots and roots were very similar for the two stock types in spite of the difference in stem volume. This was due to the fact that containerized trees had both absolutely and relatively larger crowns than bare-rooted transplants. The reason for container trees allocating more biomass into crowns than bare-roots is unknown but could either be due to the difference in age and development or an issue of genetics/provenances, since different seed sources were used for the two stock types. It is, however, reasonable to assume that the bigger crowns of the container trees will in the coming years lead to a relatively greater stem growth increment than will those of the bare-rooted trees. The overall fresh weight root/shoot ratio (≈ 0.2) however, did not differ between stock types and thus root/shoot ratio would not be a factor of relevance to tree stability in this study.

Stem form and root architecture

Basal sweep is a stem deformity often found in Douglas fir plantations. It is mainly a man-made problem, due to an imbalance between a large shoot and a small root system, partly caused by the continuous drive to improve growth through genetic improvement and provenance selection. This was reported already in 1957 by Wright and Baldwin, who showed a positive correlation between growth rate and stem deformation rate in a Scots pine provenance test. Other studies with lodgepole pine have also confirmed this (OïDriscoll, 1980; Rosvall, 1994). What we tend to forget is that a vigorous fast-growing shoot has to be supported and balanced by a vigorous root system (Nielsen, 1992). Other measures said to affect the frequency of basal sweep are site preparation, planting technique and stock type. To reduce the problem with basal sweep in fast-growing lodgepole pine in Ireland, it was recommended that the use of containerized seedlings would give more symmetrical root systems than bare-rooted transplants and thus improve the stability (Carey and Hendrick, 1986; Pfeifer, 1982).

In our study with containerised and bare-rooted Douglas fir, a slight basal sweep (> 10 cm deviation of stem centre at breast height from stem centre at ground level) was found in half of the bare-rooted and a third of the containerized trees. This would, however, still leave us with between 1000 (bare-roots) and 1400 (containers) trees with acceptable stem form, for the selection of the final stand. Trees with slight basal sweep were on average leaning eight (containers) to ten degrees (bare-roots), which is less than the 12°-limit that Fitzsimons (1988) stated as the maximum allowable average at stand level for Irish-grown lodgepole pine being accepted as saw-logs. The proportion of trees with severe basal sweep (> 25 cm) was below 10%, which will not cause any reduction of the stand value since those trees will be removed at the first thinning.

There are reasons to believe that the higher frequency of basal sweep found in bare-root stock in this study was related to seedling size rather than to root variables, since the bare-rooted stock was planted with bigger seedlings, and there were no noticeable differences between the two stock types at root level. Moreover, tree size was still positively correlated with occurrence of basal sweep after ten years.

The direction of the basal sweep lean was mainly eastwards, which coincided with the downhill side as well as the exposure to the prevailing south-westerly winds. However, the site was located on the leeward side of the hill and moreover sheltered from westerly winds by mature stands to the west, and therefore it is less likely that wind exposure had any major impact on the young seedling's stability. The direction of slopes varied slightly (ENE-ESE) between plots and since there was a tendency for the direction of basal sweep to follow the slope variation, one may assume that slope plays a role in basal sweep.

McMinn (1963) reported very asymmetrical root systems of Douglas fir growing on slopes, with fewer and shorter roots on the downhill side. Eis (1974) stated that root growth on the downhill side was often more developed, however, not universally so and that many trees had the greatest root development uphill or along the contour. In two separate root studies of Sitka spruce trees, Nicoll et al. (1995) found most roots on the down-slope side of trees, whereas Nicoll and Ray (1996) found most root mass up-slope. In both studies, however, root mass was concentrated on the side of the tree away from the prevailing wind direction. Although trees were not selected for basal sweep in the sample of the 16 root-excavated trees in this study, proportionally fewer roots, less root area and almost no dominant lateral roots were found on the downhill side of the slope.

Root architecture

The extensive measurements of the below-ground compartment in this study revealed only a few differences between the two stock types, which is in accordance with other root studies of containerised and bare-rooted Douglas fir (Arnott, 1978; Newton and Cole, 1991). There were no signs of any root deformities, neither J-shaped root systems as a result of poor planting of bare-roots nor spiralling or constraints caused by the shape of the container. It seems as if Douglas fir has a remarkable capacity to overcome deformities quickly, as Stein (1984) found after wrenching root systems of Douglas fir. Negative effects of wrenching had diminished by one year and had disappeared completely by five years.

In our study fresh weights, numbers of roots, root area, rooting depth and root ramification were all similar for the two stock types in spite of the difference in shoot size. There was quite a large variation in root symmetry between individual trees in both stock types which was related to physical barriers in the soil, like stones or old stumps, similar to findings reported by Quine et al. (1991). The more general effect of slope, as discussed above, was of more relevance to the root symmetry than to the stock type. Both numbers of roots and root area seemed to concentrate along the contour lines instead of uphill or downhill. When considering the width of sectors containing no dominant lateral roots (average -rule), bare-roots had significantly larger sectors without dominant roots than containers. The average sector in bare-roots where dominant roots were absent was 140° wide, which is more than the *critical* value of 120° that Nielsen (1977) stated as being *critical* for future stability. For both stock types, dominant roots were most frequently missing at the downhill side of the slope.

The ripped line of the site preparation, however, was also up and down the slope and across the contour line and it is therefore difficult to distinguish if the root symmetry is merely affected by slope or if the site preparation has had some contribution to it.

Conclusions

When planting seedlings of Douglas fir, the choice between bare-rooted or containerised stock types seemed to be of minor importance for factors such as survival and growth as well as for stem form and root development. This holds even for a comparison with the nowadays abandoned paper-pot container system, which is known to cause negative effects on root development and tree stability, at least in pine species and in areas with colder climatic conditions. The slightly better growth as well as the moderately higher rates of basal sweep found in the bare-rooted trees, were most likely effects of difference in initial seedling size rather than effects of the stock type used. Planting seedlings with balanced root/shoot ratios are probably the single most important factor for improved stem form and stability in Douglas fir. Since planting in sloping terrain seemed to have negative effects on the stem form and root development of Douglas fir, extra notice should be given under such conditions.

Acknowledgement

The authors thank Joe Kilbride, Donal OïHare and Niall Farrelly for helpful assistance in fieldwork and in the tedious work involved with measuring root systems. Finally we thank the Irish Forest Service for assistance in making this study possible.

References

Anon 1990 SPSS Reference Guide. 950 pp. SPSS Inc., 444 N. Michigan Av., Chicago, Ill. ISBN 0-13-177858-7.

Arnott J T 1978 Root development of container-grown and bareroot stock: coastal British Columbia. In The Root Form of Planted Trees. Proc. Symp., 16–19 May 1978. Eds E van Eerden and J M Kinghorn. pp. 257–267. Victoria, B.C., Canada. British Columbia Ministry of Forests, Victoria.

Burdett A N, Coates H, Eremenko R and Martin P A F 1986 Toppling in British Columbia's lodgepole plantations: Significance, Cause and Prevention. For. Chron. 62, 433–439.

Carey M L and Hendrick E 1986 Lodgepole pine in the Republic of Ireland. I. Site types, ground preparation and nutrition. For. Ecol. Man. 15: 4, 301–317.

Coutts M P 1983 Development of the structural root system of sitka spruce. Forestry 56, 1–16.

Eis S 1974 Root system of western hemlock, western red cedar, and Douglas fir. Can. J. For. Res., 4, 28–38.

Fitzsimons B 1988 An assessment of the extent of basal sweep in south coastal lodgepole pine. Irish For. 45: 1, 21–34.

Håkansson L and Lindström A 1994 Stability in 20-years old containerised Scots pine plantations. Dep. For. Yield Res., Swed. Univ. Agric. Sci. Inf. Rep. 87, 22 p.

Halter M R, Chanway C P and Harper G J 1993 Growth reduction and root deformation of containerised lodgepole pine saplings 11 years after planting. For. Ecol. Man., 56, 131–146.

Helgerson O T, Tesch S D, Hobbs S D and McNabb D H 1992 Effects of stocktype, shading, and species on reforestation of a droughty site in southwest Oregon. Northwest Science, 66: 2, 57–61.

Hobbs S D and Wearstler K A 1983 Performance of three Douglas-fir stocktypes on a skeletal soil. Tree Planters' Notes, 34: 3, 11–14.

Hulten H and Jansson K-Å 1978 Stability and root deformation of pine plants (*Pinus silvestris*). In The Root Form of Planted Trees. Proc. Symp., 16–19 May 1978. Victoria, B.C., Canada. Eds E van Eerden and J M Kinghorn. British Columbia Ministry of Forests, Victoria. pp. 145–150.

Huuri O 1978 Effects of various treatments at planting and of soft wall containers on the development of Scots pine. In The Root Form of Planted Trees. Proc. Symp., 16–19 May 1978. Victoria, B.C., Canada. Eds E van Eerden and J M Kinghorn. British Columbia Ministry of Forests, Victoria. pp. 101–108.

Iske K W 1990 Comparison of different plant assortments in Douglas fir. Allgem. Forstzeitschrift. No. 9–10, 235–237.

Lindgren O and Örlander G 1978 A study on root development and stability of 6-to-7-year old container plants. In The Root Form of Planted Trees. Proc. Symp., 16–19 May 1978. Victoria, B.C., Canada. Eds E van Eerden and J M Kinghorn. British Columbia Ministry of Forests, Victoria. pp. 142–145.

Lindström A 1978 Root deformation in different types of plant growing systems and possibilities of reducing root deformation. Dep. Forest Prod., Swed. Univ. Agric. Sci., Garpenberg, Sweden. Rep. 91. Environm. Can. Translation, OOENV TR-1710, 93 p.

Lindström A 1990 Stability in young stands of containerised Scots pine (*Pinus sylvestris*). Dep. For. Yield Res., Swed. Univ. Agric. Sci. Inf. Rep. 57, 28 p.

Lindström A and Håkansson L 1994 Going to the root of all evil. About root deformities and stability. Small Scale Forestry, Dep. of For. Yield, Swed. Univ. Agric. Sci., Newsletter 2, pp. 19–23.

Long J N 1978 Root system form and its relationship to growth in young planted conifers. In The Root Form of Planted Trees. Proc. Symp., 16–19 May 1978. Victoria, B.C., Canada. Eds E van

Eerden and J M Kinghorn. British Columbia Ministry of Forests, Victoria. pp. 222–234.

McDonald P M 1991 Container seedlings outperform bareroot stock: survival and growth after 10 years. New-Forests 5: 2, 147–156.

McMinn R G 1963 Characteristics of Douglas fir root systems. Can. J. Bot., 41, 105–122.

Michaud D 1983 Effects of cultivation conditions and sorting of plants on the growth of Douglas fir. Annales de Recherches Sylvicoles, AFOCEL, 156–189.

Muhle O, Rohrig E, Rossa G, Block W and Gissmann B 1979 Studies on Douglas fir plantings in 1974 and 1975 in northwest Germany. Aus dem Walde, Mitteilungen aus der Niedersachsischen Landesforstverwaltung. No. 29, 7–42.

Newton M and Cole E C 1991 Root development in planted Douglas fir under varying competitive stress. Can. J. For. Res., 2: 1, 25–31.

Newton M, Cole E C and White D E 1993 Tall planting stock for enhanced growth and domination of brush in the Douglas-fir region. New Forests, 7: 2, 107–121.

Nicoll B C and Ray D 1996 Adaptive growth of tree root systems in response to wind action and site conditions. Tree Physiol. 16: 891–898.

Nicoll B C, Easton E P, Milner A D, Walker C and Coutts M P 1995 Wind stability factors in tree selection: distribution of biomass within root systems of Sitka spruce clones. *In* Wind and Trees. Eds M P Coutts and J Grace. pp 276–292. Cambridge University Press, Cambridge, UK.

Nielsen C N 1992 Will traditional conifer tree breeding for enhanced stem production reduce wind stability? Silvae Genetica, 41: 307–318.

Nielsen C N 1997 The calculation module for structural roots in the TREEARCH-database. Internal report, 8. The Arboretum, Royal Vet. Agric. Univ., Copenhagen.

O'Driscoll J 1980 The importance of lodgepole pine in Irish Forestry. Irish For., 37: 1, 7–22.

Owston P W and Stein W I 1978 Survival, growth, and root form of containerized and bare-root Douglas-firs and noble firs seven years after planting. *In* The Root Form of Planted Trees. Proc. Symp., 16–19 May 1978. Victoria, B.C., Canada. Eds E van Eerden and J M Kinghorn. British Columbia Ministry of Forests, Victoria. pp. 216–221.

Pfeifer A L 1982 Factors that contribute to basal sweep in lodgepole pine. Irish For. 39: 1, 7–16.

Preisig C L, Carlson W C and Promnitz L C 1979 Comparative root system morphologies of seeded-in-place, bareroot, and containerized Douglas-fir seedlings after outplanting. Can. J. For. Res. 9: 3, 399–405.

Quine C P, Bernard A C, Coutts M P and Reynard B R 1991 Effects of mounds and stumps on the root architecture of Sitka spruce on a peaty gley restocking site. Forestry 64, 385–401.

Rosvall O 1994 Stability in lodgepole pine and resistance to wind and snow loads. For. Res. Inst., Sweden. Rep. 2. 45 p.

Schmaltz J 1990 Investigations on Douglas fir plants in soft-wall containers. Forst und Holz, 45: 17, 502–504.

Stein W I 1984 Wrenching Douglas fir seedlings in August: immediate but no lasting effects. Research Paper, Pacific Northwest, USDA, Forest Service, PNW-317, 12 pp.

Strittmatter W 1977 Assessing the reasons for success or failure in the use of containerized plants. Allgemeine Forstzeitschrift. 32: 39, 971–973.

Van den Driessche R 1992 Absolute and relative growth of Douglas-fir seedlings of different sizes. Tree Physiology, 10: 2, 141–152.

Van Erden E 1978 Roots of planted trees in central British Columbia. In The Root Form of Planted Trees. Proc. Symp., 16–19 May 1978, Victoria, B.C., Canada. Eds E van Eerden and J M Kinghorn. British Columbia Ministry of Forests, Victoria. pp. 201–207.

Van Erden E and Arnott J T 1974 Root growth of container-grown stock after planting. In North American containerised tree seedling Symp., Aug. 1974, Denver, Colorado, USA. Eds R W Tinus, W I Stein and W E Balmer. Great Plains Agric. Counc. Publ. 68, pp. 393–397.

Wass E F and Smith R B 1994 Impacts of soil disturbance on root systems of Douglas fir and lodgepole pine seedlings. Pacific and Yukon Reg. Can. For. Service. Inform. Rep. BC-X-348. 22 p.

Wright J W and Baldwin H I 1957 The 1938 International Union Scots pine provenance test in New Hampshire. Silv. Gen., 6, 2–14.

A. Stokes (ed.), The Supporting Roots of Trees and Woody Plants: Form, Function and Physiology, 97–108.

The effect of cultivation technique on root and shoot biomass production by young Sitka spruce (*Picea sitchensis* (Bong.) Carr.) trees on surface water gley soils

J. M. Wills[1], E. Sundström[2], J. J. Gardiner[3] and M. Keane[4]
[1]*Department of Crop Science, Horticulture and Forestry, University College Dublin, Belfield, Dublin 4, Ireland. Tel/FAX No: +353 1 2762585; E-mail: jliu@esatclear.ie.*
[2]*FoReCo, Plangatan 5, 903 36 Umeå, Sweden*
Tel/FAX No: +46 90 139 768; E-mail: Erik.Sundstrom@ssko.slu.se
[3]*Department of Crop Science, Horticulture and Forestry, University College Dublin, Belfield, Dublin 4, Ireland*
Tel: +353 1 706 7767; FAX No: +353 1 7061104; E-mail: valerie.guilfoyle@agriculture.ucd.ie
[4]*Research and Technology, Coillte, Newtownmountkennedy, Co. Wicklow, Ireland*
Tel: +353 1 201 1111; FAX No: +353 1 201 1199; E-mail: mgkeane@indigo.ie

Key words: root biomass, shoot biomass, Sitka spruce (*Picea sitchensis*), cultivation, mole drainage, mounding, surface water gley

Abstract

In Ireland much of the land available for plantation establishment requires some degree of soil cultivation and drainage to improve its suitability for tree growth. The method of cultivation and drainage normally varies depending upon the soil type and its drainage characteristics. Little research has been carried out on the impact of practices such as mounding and mole drainage upon rooting and biomass production in young crops. The research reported in this paper was carried out on four sites where the general soil type was surface water gley. Three of the sites studied were afforestations on old farmland, while the fourth site was a reforestation of a windblown stand. The cultivation methods investigated included mounding, mole drainage with mounds, mole drainage only and ripping. Sitka spruce trees ranging in age from five to thirty years were excavated from these experimental sites to describe the impact of different soil cultivation techniques upon root architecture and above and below ground biomass production. The results showed that while the effect of mound drainage upon water table level varied from site to site, the establishment of trees on mounds led to an increase in above and below ground biomass production. Mole drainage of these sites without mounds, did not lead to any increase in root or shoot development when compared with trees growing on uncultivated/undrained ground.

Abbreviations: SED – standard error of the difference between two means

Introduction

The area becoming available for afforestation and reforestation in Ireland has been steadily increasing over the last number of years. Following successful establishment a major problem facing many of these plantations is the threat of wind-throw before they reach the end of their rotation. Wet mineral soils often have high water tables that contribute to the development of very shallow root systems (Ray and Nicoll, 1998). These soils are dense and heavy with a high clay content. Sitka spruce grows exceptionally well on such sites producing yield classes in the region of 20–24 m^3/ha/yr.

Site preparation to encourage the development of deep symmetrical root systems is essential on these sites not only for successful establishment but also to prevent wind-throw later in the life of the plantation (Quine et al., 1991). Good site preparation helps to

* *Plant and Soil* is the original source of publication of this article. It is recommended that this article is cited as: *Plant and Soil* **217:** 79–90, 1999.

quickly establish a vigorous healthy crop by increasing soil drainage and soil aeration, suppressing vegetation and by providing a raised planting position and increasing soil temperature (Armstrong et al., 1976). In recent years mechanical mounding has become a particularly popular site preparation technique. There has been much evidence of the benefits of mounding on early tree establishment and growth (Sutton, 1993). Mounding increases soil aeration and temperature, it provides a weed free planting area and increases the amount of rootable soil volume (Sutton, 1993). However, little research has been carried out concerning the impact of such techniques on root development.

Experimentation dealing with soil cultivation techniques for the establishment of forest trees has been ongoing in Ireland for at least 40 years (O'Carroll, 1972). Most of this work was aimed at improving the establishment and early growth of crops planted on difficult sites. The assessments carried out generally involved the measurement of crop growth parameters and little attention was given to rooting and long term crop stability. During the course of the work described in this paper a number of cultivation experiments were located. Trees were excavated from the experiments to describe the impact of different cultivation techniques upon the root growth of Sitka spruce. The general objective was to investigate if soil cultivation increases tree growth and root development and which cultivation treatments may be the most effective on the site types investigated.

Materials and methods

Site description

Three of the sites described in this paper were experimental areas while the fourth site was an area of commercial forestry. Prior to afforestation in 1990, site 1 at Ballygar, Co. Galway was neglected farmland of small fields, some overgrown with furze (*Ulex* spp.) and white-thom (*Crataegus monogyna*). Site 2 at Strokestown, Co. Roscommon was used for agriculture and was previously cultivated. It was afforested in 1990. Site 3 at Rossmore, Co. Laois was a surface water gley overlain by a shallow peat layer. The site was formally used for rough grazing and may have been cultivated in earlier times. Site 4 at Derrybrien, Co. Galway was a commercial site which was originally planted in 1931 with Sitka spruce. The stand was prematurely felled in 1967 due to wind-throw and reforested in 1968 with Sitka spruce, yield class 24. Rainfall at all sites is generally in excess of 1000 mm per annum. Further details of the location and physical characteristics of each of the sites are presented in Table 1 and Table 2.

Site cultivation

The cultivation experiments at Ballygar, Strokestown, and Rossmore were established to investigate the impact of various cultivation and establishment techniques on the survival, growth and stability of Sitka spruce. At Ballygar and Strokestown experiments were established in 1990. The cultivation treatments included mole drainage, mechanical mounding and mole drainage with mounds (Table 3). Mechanical mounding involves the excavation of deep open drains (= 1 m in depth) at regular intervals (8–16 m). The spoil from these drains is used to form 2500 raised planting positions (mounds) per hectare at 2 m × 2 m intervals. According to the classification by Sutton (1993) the mounds would be described as mesoid, minmounds. Mole drainage involves creating an underground drain, using a bullet shaped tine at a depth between 30 cm and 45 cm. Wings can be attached to the shaft of the tine to create maximum soil shattering. The latter is called ripping. Moling and ripping are usually carried out with a drain spacing of 2 m. Mole drainage is occasionally combined with hinged mounds or inversod (Sutton, 1993).

At Rossmore, ripping and ploughing treatments were included amongst other cultivation treatments in an experiment which was established in 1977 (Table 3). An area that had been mechanically mounded in 1968 at Derrybrien forest was selected for investigation. This site was selected in order to investigate the impact of mound planting upon the growth of older crops. The site is one of the oldest known mechanically mounded sites in the country.

Experimental design

The experimental design employed at all sites (except Derrybrien) was a randomised block design, with three blocks (Table 3). Each treatment appeared once within each block. At Ballygar and Strokestown there were four treatments: mechanical mounding; mole drains with mounds; mole drainage only, and no cultivation. The mole and mound treatment was carried out us-

Table 1. The location and physical characteristics on each of the experimental sites (Wills, 1999)

Forest	Location	Elevation	Aspect	Exposure
Ballygar	52° 03′ N; 08° 02′ W	65 m	Flat, with slope generally of 1° or less	Sheltered
Strokestown	58° 45′ N; 08° 05′ W	100 m	South-west, with a slope of approx. 2°	Exposed
Rossmore	52° 04′ N; 07° 01′ W	305 m	North Slight slope	Very exposed
Derrybrien	53° 04′ N; 08° 35′ W	65 m	North. Slope of approx. 1°–2°	Sheltered

ing a Bogmiser mounding machine.[1] At Rossmore the experiment included double mouldboard and single mould board ploughing and ripping treatments. For the purposes of this study trees were taken only from the ripped treatment. The planting position was 30 cm to one side of the rip line. However, due to restrictions in the budget and time for the study, trees were only selected from one block. The site at Derrybrien was an area of commercial forestry and was not a designed experiment. All trees were slit planted at the side of the rip or mole drain and on top of mounds. Bare-root planting stock was used at all of the sites.

Tree selection

The number of trees selected at each site was based on a statistical estimate of the variance in tree height or DBH (diameter at breast height). To estimate the variance, tree height was measured on the sites where the crop had not yet closed canopy and DBH was measured on the older crops. These estimates indicated that the number of trees which should be excavated from each plot in order to estimate above and below ground biomass to within ± 5% of the true means at the 95% confidence level ranged from 10–74 for the experimental sites in question. Clearly it was not possible to excavate the large sample produced by the statistical estimate. The maximum number of trees permitted by the resources available, were chosen for detailed above and below ground measurement. All treatments in the experiments at Ballygar and Strokestown were sampled. Due to the young age of the experiments at these sites a smaller study of older trees was undertaken at Rossmore and Derrybrien. These sites were selected as they supported older trees on ripped and mounded soil cultivation treatments. Although it was not possible to sample trees from these sites rigorously, it was hoped that a smaller sample would give some indication of how older trees respond to ripping and mounding.

At the experimental sites in Ballygar and Strokestown three trees of average height were randomly selected from each plot (i.e. nine trees per treatment) (Table 4). Detailed above and below ground measurements were carried out on these trees in June 1995 at Strokestown (after 4.5 growing seasons) and in September 1995 at Ballygar (after 4.75 growing seasons). When additional resources were made available, a second sample was taken in January and February 1997 (after 6 growing seasons). Average height trees were again randomly selected from each plot based on the assessment taken in 1995. Four trees were selected from the control, mole and mole and mound treatments. In an attempt to establish if distance from the mound drain had an effect on root development and tree growth in the machine mound treatments, trees were randomly selected according to their distance from the mound drain. At Ballygar 12 trees were selected from the machine mound plots. Three trees were selected at distances of 1, 3, 5 and 7 m from the mound drain. Trees at Strokestown were sampled in a similar way, but due the closer drain spacing (mound drain centres at 12 m apart) there were no trees at 7 m from the drain and therefore only nine trees were selected from the machine mound treatment.

At Rossmore and Derrybrien tree selection was based on mean DBH. The DBH of each tree in the selected plots was measured. Based on this DBH assessment, 10 trees of mean DBH at Rossmore were randomly selected from the appropriate plots. A plot of approximately 50 m × 50 m was laid out in the mechanically mounded area at Derrybrien. Nine

---

* A machine which creates a mole drain and a planting mound at the same time, developed by Ballintubber Enterprises, Ireland.

Table 2. Site descriptions including geology, soil type and vegetation

Site	Geology	Soil type	Vegetation	Crop
1. Ballygar	Old red sandstone and some glacial drift of mixed origin	Uppermost layer of shallow medium loam changing quickly to an impervious gley of a sandy/stony type	*Poa* spp., *Juncus* spp.	Sitka spruce. Planted 1990 at 2 m × 2 m spacing
2. Strokestown	Mixed sandstone – limestone drift and glacial lake deposits	Upper layer of medium loam changing to an impervious gley of a sandy/stony type	*Poa* spp., *Juncus* spp. and some *rubus* spp.	Sitka spruce. Planted 1990 at 2 m × 2 m spacing
3. Rossmore	Climatic peat on coal measure and Yoredale shale formations with mixed glacial drift	Surface water gley overlain by a shallow peat layer	*Poa* spp., *Juncus* spp. with some *Celluna vulgaris*	Sitka spruce. Planted 1977 at 2 m × 2 m spacing
4. Derrybrien	Old red sandstone	Grey brown podzolic overlying a layer of impermeable gley	No vegetation	Sitka spruce. Planted 1968 at 2 m × 2 m spacing

Table 3. Experimental design and drainage at each of the sites

Site	Experimental design	Treatment	Distance between drains	Depth of drains
1. Ballygar	Randomised block design (3 blocks)	Machine mounding	18.0 m	0.45 m
		Mole and mound	2.0 m	0.35–0.40 m
		Mole	2.0 m	0.35–0.40 m
		No cultivation	–	–
2. Strokestown	Randomised block design (3 blocks)	Machine mounding	16.0 m	0.45 m
		Mole and mound	2.0 m	0.35–0.40 m
		Mole	2.0 m	0.35–0.40 m
		No cultivation	–	–
3. Rossmore	Randomised block design (3 blocks)	Ripping	2.0 m	0.80 m
4. Derrybrien	Not experimental	Machine mounding	18.0 m	0.50 m

Table 4. The number of trees excavated from each site

Site	Treatment	Year	No. of trees excavated
1. Ballygar	Machine mounding	1995	9
		1997	3 trees @ 1 m from drain
			3 trees @ 3 m from drain
			3 trees @ 5 m from drain
			3 trees @ 7 m from drain
			36 trees in total
	Mole and mound	1995	9
		1997	12
	Mole	1995	9
		1997	12
	No cultivation	1995	9
		1997	12
2. Strokestown	Machine mounding	1995	9
		1997	3 trees @ 1 m from drain
			3 trees @ 3 m from drain
			3 trees @ 5 m from drain
			27 trees in total
	Mole and mound	1995	9
		1997	12
	Mole	1995	9
		1997	12
	No cultivation	1995	9
		1997	12
3. Rossmore	Ripping	1997	10
4. Derrybrien	Machine mounding	1997	9

trees of mean DBH were randomly selected and their distance to the nearest drain was noted. All trees were excavated for detailed above and below ground measurements in 1997.

A total of 226 were sampled from all sites. The average DBH of the sample trees at Ballygar was 1.5 cm in 1995, and 2.6 cm in 1997. At Strokestown the average DBH of the sample trees was 4.4 cm in 1995 and 6.6 cm in 1997. The average DBH of the sample trees at Rossmore and Derrybrien was 18 and 22 cm respectively.

Tree measurements

Following their location in the field, the sample trees were harvested and root and shoot biomass were determined. Stem biomass was measured by weighing the fresh weight of the stem in the field. Two stem disks were then taken, one from the base of the stem and another 1 m from the top of the stem. These disks were taken to the laboratory, where their fresh weight was noted. They were dried in an oven at 105 °C for 48 hours, at which time the dry weight was taken. From this data the moisture content of the sample was calculated. The percentage of moisture in the sample was used to calculate the dry weight of the complete stem. Crown (branches and foliage) biomass was measured by weighing all the branches and foliage in the field. Representative samples were dried as above and used to calculated the dry weight of the entire crown.

Following the harvesting of the above ground part of the trees the root systems were excavated. The method used was similar to that of Nicoll et al. (1995). Trees were rocked by hand (for the smaller trees) and with a winch (larger trees) to loosen the root system and peripheral roots were cut at approximately 1.0–1.5

m circumference from the stem while the root systems were still in the soil. The root systems were cleaned using an air knife. Using this method of extraction very little root breakage occurred and almost all roots greater than 5 mm in diameter as well as some of the finer roots were preserved. The dry biomass of the structural roots (roots > 5 mm in diameter) was calculated in a similar way to the above ground biomass. Following the procedures developed by Nielsen (1995). Roots systems of the trees at Strokestown and Ballygar were cut at a radius of 60 cm. The larger root systems of the trees at Rossmore and Derrybrien were cut at a radius of 100 and 120 cm respectively.

Statistical analysis

Analysis of the data was carried out using the SAS software system for Windows version 6.11 (Anon., 1990). Each data-set was analysed as a randomised block design using the General Linear Models procedure (GLM). Six separate models were analysed for each site at each sampling period (1995 and 1997). Branch and foliage biomass, stem biomass, total shoot biomass, root biomass and total tree biomass were the dependant variables with treatment (machine mound, mole and mound, mole and no cultivation) and block as the independent variables. Thus the models run for each dependant variable were: dependant variable = treat + block. The GLM procedure computes an ANOVA and least square means for each model, standard errors of the least square means (SEM) and a pairwise, two-way *T*-test on each pair of means.

In the case of the data for Ballygar and Strokestown in 1997 the means for each plot in the analysis were weighted due to the fact that mean values for plots in the machine mound treatment were made from a larger number of trees than the other treatments. Root and shoot biomass were logarithmically transformed (natural log) to permit the analysis of the root:shoot ratio (a ratio does not have a normal distribution and so data in this form must be transformed to permit an ANOVA). Thus the root:shoot ratio was analysed as natural log of the root biomass minus the natural log of the shoot biomass.

Results

General observations at Ballygar and Strokestown

In general, trees that were planted on mounds had a higher total biomass production than trees which were not planted in a raised planting position. At both Ballygar and Strokestown distance from the mound drain had no effect on root or shoot biomass (Tables 5 and 6). The exception to this were the trees planted at Ballygar, where the tree root biomass was significantly larger at 3 m from the drain compared to 1 m from the drain. The block effect was not significant at $\alpha = 0.05$ at either Strokestown or Ballygar. Between the growing seasons of 1995 and 1997 shoot biomass increased by on average 142 and 222% at Strokestown and Ballygar respectively. Root biomass increased on average by 77 and 122% respectively. DBH was generally larger for the trees planted on mounds compared to trees planted on the flat. However, the difference between mound planted trees and flat planted trees was most significant at both sites after 6 growing seasons (Table 7).

Ballygar

In the Ballygar experiment trees planted on mounds had significantly ($\alpha = 0.05$) higher branch and foliage biomass, stem biomass and root biomass after 4.75 and 6 growing seasons (Tables 8 and 9). However, this increased biomass production was more or less equally distributed between the above ground parts of the trees and the root systems, so that no significant change was observed in the root:shoot ratio.

Strokestown

At Strokestown differences in biomass production were not as pronounced as at Ballygar (Tables 10 and 11). In 1995 total biomass production was significantly higher for trees in the machine mound treatment and mole and mound treatment compared to trees in the mole and uncultivated plots at probabilities of $\alpha = 0.1$ and $\alpha = 0.05$ respectively. Branch and foliage biomass was not significantly ($\alpha = 0.05$) higher for trees that had been planted on mounds compared to trees planted on the flat. Stem biomass, however, was significantly higher for the trees planted on mounds, although it was only significant at $\alpha = 0.06$ when the machine mound and control treatment were compared. Trees after 6 growing seasons in the machine mound treatment had a significantly ($\alpha = 0.05$) higher total shoot biomass and stem biomass than trees in the mole or uncultivated treatment. Trees in the mole and mound treatment also had a higher stem and total shoot biomass than trees in the mole or uncultivated treated plots but this was only significant at $\alpha = 0.09$. Branch and foliage biomass although higher for trees planted

Table 5. Summary of ANOVA and pairwise *T*-test of mean shoot and root biomass (kg dry weight) and mean root:shoot ratio with increasing distance from the mound drain for trees harvested in 1997 at Ballygar after 6 growing seasons. (Treatments with different letters indicate a significant difference at $\alpha = 0.05$)

Summary of ANOVA

	Branch + foliage biomass		Stem biomass		Total shoot biomass		Root biomass		Total biomass		Root:shoot ratio	
	df	Pr>F	df	Pr>F	df	Pr>F	df	Pr>F	df	Pr>F	df	Pr>F
Treatment	3	0.61	3	0.92	3	0.73	3	0.09	3	0.57	3	0.27
Block	2	0.59	2	0.94	2	0.72	2	0.24	2	0.57	2	0.91
Error	6	–	6	–	6	–	6	–	6	–	6	–

Mean biomass values and results of pairwise *T*-test

Treatment	Branch + foliage biomass		Stem biomass		Total shoot biomass		Root biomass		Total biomass		Root:shoot ratio	
1 m	1.43	A	1.16	A	2.58	A	0.62	A	3.21	A	0.25	A
3 m	1.61	A	1.22	A	2.83	A	0.90	B	3.73	A	0.35	A
5 m	1.77	A	1.27	A	3.05	A	0.77	A,B	3.81	A	0.29	A
7 m	1.67	A	1.19	A	2.86	A	0.73	A,B	3.59	A	0.25	A
SED	0.255		0.176		0.401		0.085		0.444		0.129	

on mounds compared to trees not planted on mounds was not significantly different between treatments.

In general, mound planted trees at Strokestown, excavated after 4.5 and 6 growing seasons had significantly higher ($\alpha = 0.05$) total biomass production and root biomass production than trees planted on the flat. The exception to this was the trees harvested after 6 growing seasons from the mole and mound plots. The root biomass and total biomass produced by those trees was higher than that of trees from the uncultivated and moled plots. However this result was only significant at the 92% level. The root: shoot ratio was significantly higher ($\alpha = 0.05$) for trees planted on mounds in 1995 but by 1997 there was no difference in root:shoot ratios between treatments.

Rossmore and Derrybrien

Trees planted on machine mounds at Derrybrien had a higher root:shoot ratio than the trees planted in the ripped site at Rossmore (Table 12).

Shoot root relationships

The relationship between shoot and root biomass of trees on ripped plots at Rossmore and those on machine mounds at Derrybrien indicated that trees planted on machine mounds may have a higher proportion of root compared to shoot biomass than trees planted on ripped or moled sites (Figure 1).

Discussion

The method of soil cultivation can be of fundamental importance for the survival and early growth of trees planted on wet mineral soils. Previous research has shown that methods which provide a raised planting position can lead to superior early growth (Armstrong et al., 1976). The results presented in this paper confirm this for Sitka spruce planted on mounds in the western part of Ireland. Total biomass production by trees planted on mounds was significantly greater at both the Ballygar and Strokestown sites by comparison with trees planted beside mole lines or without soil cultivation. This effect was apparent after 4.5 growing seasons, and was still evident after 6 years. In absolute terms this difference was greater at Strokestown (5.5 kg) than at Ballygar (2.2 kg) (Tables 8 and 10), although, the relative increase in total biomass was greater a Ballygar (i.e. trees in the machine mound treatment at Strokestown were 1.5 times larger than those in the uncultivated plots while at Ballygar they were 2.6 times larger). The early growth of trees, at both sites was greater on mounded plots than in the mole drained plots. In fact, biomass production by trees established on the moled plots was not significantly different from that of trees established without soil cultivation and drainage, although Hendrick (1989) had reported that mole drainage res-

Table 6. Summary of ANOVA and pairwise *T*-test of mean shoot and root biomass (kg dry weight) and mean root:shoot ratio with increasing distance from the mound drain for trees harvested in 1997 at Strokestown after 6 growing seasons. (Treatments with different letters indicate a significant difference at $\alpha = 0.05$)

	Summary of ANOVA											
	Branch + foliage biomass		Stem biomass		Total shoot biomass		Root biomass		Total biomass		Root:shoot ratio	
	df	Pr>F	df	Pr>F	df	Pr>F	df	Pr>F	df	Pr>F	df	Pr>F
Treatment	2	0.57	2	0.60	2	0.82	2	0.42	2	0.66	2	0.47
Block	2	0.01	2	0.71	2	0.04	2	0.72	2	0.04	2	0.49
Error	4	–	4	–	4	–	4	–	4	–	4	–

Treatment	Mean biomass values and results of pairwise *T*-test											
	Branch + foliage biomass		Stem biomass		Total shoot biomass		Root biomass		Total biomass		Root:shoot ratio	
1 m	7.59	A	5.68	A	13.27	A	3.19	A	16.81	A	0.27	A
3 m	7.29	A	6.28	A	13.57	A	3.54	A	16.64	A	0.23	A
5 m	7.85	A	6.00	A	14.84	A	3.30	A	17.57	A	0.27	A
SED	0.488		0.558		0.891		0.466		1.013		0.143	

Table 7. Summary of ANOVA and pairwise *T*-test of mean DBH (cm) for trees harvested at Ballygar (after 4.75 and 6 growing seasons) and at Strokestown (after 4.5 and 6 growing seasons). (Treatments with different letters indicate a significant difference at $\alpha = 0.05$)

	Summary of ANOVA							
	Ballygar				Strokestown			
	After 4.75 growing seasons		After 6 growing seasons		After 4.5 growing seasons		After 6 growing seasons	
	df	Pr>F	df	Pr>F	df	Pr>F	df	Pr>F
Treatment	3	0.01	3	0.01	3	0.29	3	0.01
Block	2	0.29	2	0.43	2	0.51	2	0.11
Error	6	–	6	–	6	–	6	–

Treatment	Mean DBH (cm) and results of pairwise *T*-test							
	Ballygar				Strokestown			
	After 4.75 growing seasons		After 6 growing seasons		After 4.5 growing seasons		After 6 growing seasons	
Machine mound	1.84	A, B	2.97	A	4.50	A	7.22	A
Mole and mound	1.89	A	3.06	A	4.66	A	6.62	A
Mole	1.28	B	1.99	B	4.07	A	5.76	B
No Cultivation	0.80	C	1.51	B	4.27	A	5.80	B
SED	0.246		0.320		0.291		0.319	

Table 8. Summary of ANOVA and pairwise *T*-test of mean shoot and root biomass (kg dry weight) and mean root:shoot ratio for trees harvested in 1995 at Ballygar after 4.75 growing seasons. (Treatments with different letters indicate a significant difference at $\alpha = 0.05$)

	Summary of ANOVA											
	Branch + foliage biomass		Stem biomass		Total shoot biomass		Root biomass		Total biomass		Root:shoot ratio	
	df	Pr>F	df	Pr>F	df	Pr>F	df	Pr>F	df	Pr>F	df	Pr>F
Treatment	3	0.0002	3	0.001	3	0.0002	3	0.001	3	0.0003	3	0.59
Block	2	0.88	2	0.14	2	0.44	2	0.66	2	0.50	2	0.73
Error	6	–	6	–	6	–	6	–				
	Mean biomass values and results of pairwise *T*-test											
	Branch + foliage biomass		Stem biomass		Total shoot biomass		Root biomass		Total biomass		Root:shoot ratio	
Machine mound	0.60	A	0.42	A	1.01	A	0.38	A	1.39	A	0.37	A
Mole and mound	0.55	A	0.50	A	1.05	A	0.39	A	1.45	A	0.38	A
Mole	0.28	B	0.18	B	0.46	B	0.16	B	0.62	B	0.34	A
No cultivation	0.17	B	0.12	B	0.29	B	0.10	B	0.40	B	0.36	A
SED	0.045		0.056		0.086		0.045		0.126		0.080	

Table 9. Summary of ANOVA and pairwise *T*-test of mean shoot and root biomass (kg dry weight) and mean root:shoot ratio for trees harvested in 1997 at Ballygar after 6 growing seasons. (Treatments with different letters indicate a significant difference at $\alpha = 0.05$)

	Summary of ANOVA											
	Branch + foliage biomass		Stem biomass		Total shoot biomass		Root biomass		Total biomass		Root:shoot ratio	
	df	Pr>F	df	Pr>F	df	Pr>F	df	Pr>F	df	Pr>F	df	Pr>F
Treatment	3	0.03	3	0.003	3	0.01	3	0.01	3	0.01	3	0.22
Block	2	0.37	2	0.43	2	0.38	2	0.35	2	0.37	2	0.73
Error	6	–	6	–	6	–	6	–	6	–	6	–
	Mean biomass values and results of pairwise *T*-test											
	Branch + foliage biomass		Stem biomass		Total shoot biomass		Root biomass		Total biomass		Root:shoot ratio	
Machine mound	1.63	A	1.23	A	2.86	A	0.75	A	3.61	A	0.29	A
Mole and mound	1.67	A	1.20	A	2.87	A	0.79	A	3.66	A	0.28	A
Mole	0.87	B	0.69	B	1.56	B	0.36	B	1.92	B	0.22	A
No cultivation	0.64	B	0.49	B	1.13	B	0.26	B	1.40	B	0.23	A
SED	0.306		0.134		0.433		0.120		0.550		0.111	

Table 10. Summary of ANOVA and pairwise *T*-test of mean shoot and root biomass (kg dry weight) and mean root: shoot ratio for trees harvested in 1995 at Strokestown after 4.5 growing seasons. (Treatments with different letters indicate a significant difference at $\alpha = 0.05$)

Summary of ANOVA

	Branch + foliage biomass		Stem biomass		Total shoot biomass		Root biomass		Total biomass		Root:shoot ratio	
	df	Pr>F	df	Pr>F	df	Pr>F	df	Pr>F	df	Pr>F	df	Pr>F
Treatment	3	0.23	3	0.04	3	0.11	3	0.01	3	0.05	3	0.01
Block	2	0.59	2	0.40	2	0.51	2	0.82	2	0.60	2	0.07
Error	6	–	6	–	6	–	6	–	6	–	6	–

Mean biomass values and results of pairwise *T*-test

	Branch + foliage biomass		Stem biomass		Total shoot biomass		Root biomass		Total biomass		Root:shoot ratio	
Machine mound	2.74	A	2.43	A,B	5.17	A,B	1.75	A	6.92	A	0.27	A
Mole and mound	2.74	A	2.54	A	5.28	A	1.86	A	7.14	A	0.29	A
Mole	2.07	A	1.82	C	3.89	B	1.10	B	4.99	B	0.23	B
No cultivation	2.13	A	1.89	B,C	4.01	A,B	1.15	B	5.16	B	0.23	B
SED	0.382		0.233		0.585		0.156		0.731		0.051	

Table 11. Summary of ANOVA and pairwise *T*-test of mean shoot and root biomass (kg dry weight) and mean root: shoot ratio for trees harvested in 1997 at Strokestown after 6 growing seasons. (Treatments with different letters indicate a significant difference at $\alpha = 0.05$)

Summary of ANOVA

	Branch + foliage biomass		Stem biomass		Total shoot biomass		Root biomass		Total biomass		Root:shoot ratio	
	df	Pr>F	df	Pr>F	df	Pr>F	df	Pr>F	df	Pr>F	df	Pr>F
Treatment	3	0.15	3	0.02	3	0.04	3	0.01	3	0.03	3	0.37
Block	2	0.43	2	0.27	2	0.30	2	0.29	2	0.29	2	0.79
Error	6	–	6	–	6	–	6	–	6	–	6	–

Mean biomass values and results of pairwise *T*-test

	Branch + foliage biomass		Stem biomass		Total shoot biomass		Root biomass		Total biomass		Root:shoot ratio	
Machine mound	7.57	A	5.99	A	13.56	A	3.29	A	16.85	A	0.24	A
Mole and mound	7.20	A	5.28	A,B	12.48	A,B	2.81	A,B	15.30	A,B	0.23	A
Mole	5.13	A	4.08	B	9.21	B	2.06	B	11.26	B	0.23	A
No cultivation	5.28	A	3.98	B	9.26	B	2.07	B	11.38	B	0.23	A
SED	1.187		0.549		1.490		0.319		1.779		0.062	

Table 12. Mean shoot and root biomass (kg dry weight) and mean root:shoot ratio for trees at Derrybrien (29 yrs. old) and Rossmore (20 yrs)

Site	Branch + foliage biomass	Stem biomass	Total shoot biomass	Root biomass	Total biomass	Root:shoot ratio
Rossmore	29.92	73.05	102.96	18.67	121.63	0.18
Derrybrien	57.42	104.79	162.21	45.72	207.93	0.28

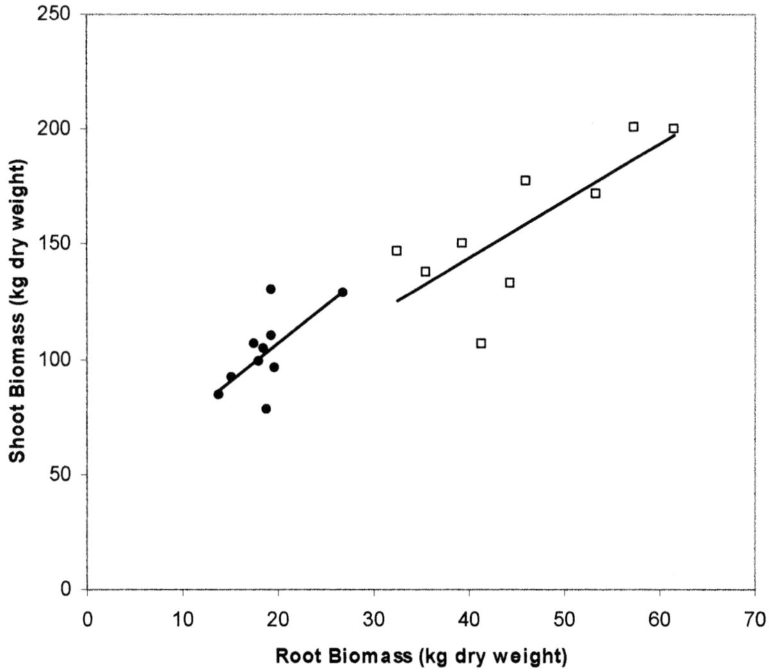

Figure 1. Relationship between shoot and root biomas for trees planted on machine mounds (□, $y = 2.48x + 44.86$, $R^2 = 0.60$, $P = 0.01$) and in moled/ripped plots (●, $y = 3.28x + 41.67$, $R^2 = 0.44$, $P = 0.04$).

ulted in an increase in diameter growth in 14–16 year old Sitka spruce trees growing on surface water gley soils.

The large differences in tree size found between the two sites at Strokestown and Ballygar was probably due to better natural drainage and a higher level of fertility in the surface soil horizons at Strokestown compared to Ballygar. In most cases, however, mound planted trees had greater total biomass production, greater stem biomass and DBH, greater foliage biomass and greater root biomass than trees planted on the flat at both times of sampling and at both sites. While this may not be due to improved drainage, there arise the possibilities that it is due to improved weed control, more favourable soil temperatures and moisture conditions within the planting mound. However, root:shoot ratios did not seem to be affected by soil cultivation. Trees which were planted on mounds appeared to have slightly higher root:shoot ratios. However, this result was only significant for the trees excavated after 4.5 growing seasons from the Strokestown site. Hence, a repeat sample was taken, two growing seasons later and the effect was not found. These results indicate that although trees planted on mounds have a higher root biomass than trees planted on the flat, the proportion of root biomass to shoot biomass was not increased and therefore does not appear to be a proportional increase in the allocation of biomass to the roots as proposed by Nicoll and Ray (1995). While trees planted on mounds grew faster than trees planted on the flat, the root:shoot ratios remain the same after the first six years after growth. However, the indication from the older trees, excavated at Derrybrien and Rossmore, is that this position may not be maintained over the entire rotation.

The relationship between shoot and root biomass indicated that more mature trees planted on machine mounded sites may have a higher proportion of root biomass compared to shoot biomass, than trees planted on the flat in moled or ripped ground (Figure 1). However, these trees were not sampled from replicated experiments and, therefore, no definite conclusions can be drawn from the results of the tree biomass production collected from these sites.

Acknowledgements

This project was part funded by COFORD, (National Council for Forest Research and Development in Ireland) and Coillte Teoranta, the Irish Forestry Board. The authors would like to thank Niall Farrelly, Donal O'Hare and the Coilite research foresters, Joe Freeman and Frank Collins for their help and perseverance in the field.

References

Anon., 1990. SAS/STAT Users Guide, Version 6. Fourth Edition. Volume 1 and 2. SAS Institute Inc. Cary, NC, USA. 1686 p.

Armstrong W., Booth T. C., Priestly P. and Read D. J., 1976. The relationship between soil aeration, stability and growth of Sitka spruce (*Picea sitchensis* (Bong.) Carr.) on upland peaty gleys. J. Appl. Ecol., 13: 585–591.

Hendrick E., 1989. The effect of cultivation method on the growth and root anchorage of Sitka spruce. Irish Forestry, 46: 19–28.

Nicoll B. C., Easton E. P., Milner A. D., Walker C. and Coutts M. P., 1995. Wind stability factors in tree selection: distribution of biomass within root systems of Sitka spruce clones. In 'Wind and Trees' M. P. Coutts and J. Grace, (Eds), Cambridge Univ. Press.

Nielsen C. C. N., 1995. "Structural root measurement in the TREEARCH-database (ROOTARCH-method)." Project TREEARCH, Internal report NR. 2, March 1995. Arboretum, R. Vet. a. Agric University, Denmark.

O'Carroll R. N., 1972. Studies on fertilization, soil cultivation and planting techniques in their effects on growth and yield of forest crops. Unpublished thesis. University College Dublin, Ireland.

Quine C. P., Burnand A. C., Coutts M. P. and Reynard B. R., 1991. Effects of mounds and stumps on the root architecture of Sitka spruce on a peaty gley restocking site. Forestry, Vol. 64, No. 4.

Ray D. and Nicoll C., 1998. The effect of soil water table depth on root-plate development and stability of Sitka spruce. Forestry, Vol. 71, No. 2.

Sutton R. F., 1993. Mounding site preparation: A review of European and North American experience. New Forests, 7: 151–192.

Wills J. M., 1999. The development of Sitka spruce crops (*Picea sitchensis* (Bong.) Carr.) on wet mineral soils following site cultivation. Unpublished thesis. University College Dublin, Ireland.

A. Stokes (ed.), *The Supporting Roots of Trees and Woody Plants: Form, Function and Physiology*, 109–118.

Influence of cultivation method on root grafting in Norway spruce (*Picea abies* (L.) Karst.)

Tiit Külla[1] and Krista Lõhmus[2,*]
[1]*Faculty of Forestry, Estonian Agricultural University, 5 Kreutzwaldi St., 51014 Tartu, Estonia and*
[2]*University of Tartu, Institute of Geography, 46 Vanemuise St., 51014 Tartu, Estonia*

Key words: Norway spruce (*Picea abies* (L.) Karst.), root characteristics, rooting density, root grafting, site characteristics

Abstract

Root grafting is the process by which a functional union of two or more roots subsequent to their formation is formed. The above- and below-ground parts of three Norway spruce stands (natural stand, Umbric Luvisol; row-culture and group-culture, Planosol; stand ages 40, 43 and 43, respectively) of high site quality (I) were investigated. Stand densities were 1550, 1783 and 1722 stems ha^{-1}, respectively. In all investigated stands, root grafting was most sensitive to tree spacing. Grafts were observed in case the distance between the trees was 0.7–1.2 m. Grafts occurred always in areas of higher rooting density, in a row of the row-culture and within a tree group in the group-culture. Root grafting was enhanced in case of a narrower humus horizon in the group culture compared with the row-culture, 16.5 and 30 cm, respectively; the humus horizon contained 99% and 95% of conducting roots with $d \geq 5$ mm, respectively. Root graftings occurred in 75% of excavated trees in the group-culture, in 37.5% of excavated trees in the row-culture and in 33.3% of excavated trees in the natural stand. Stand age was 24 years in the row-culture and 22 years in the group-culture at the beginning of root grafting. No grafts occurred between two suppressed trees, whereas in 86–100% of all cases, at least one tree was dominant or codominant. In row- and group-cultivated Norway spruce stands, the initial minimum diameter of the grafted root without bark was from 1 to 3 cm in 63% of cases. Grafting of roots with $d < 1$ cm or $d > 10$ cm was rare or absent. Root grafting had usually begun at the root age of 10–20 years (46% of cases).

Introduction

Root graft is a functional union of two or more roots subsequent to their formation (Graham and Bormann, 1966). Grafts are commonly established between roots of the same tree or between roots of neighbouring trees of the same species. Formation of a root graft begins due to the mechanical pressure between roots, which is caused by their secondary growth; hence, grafts are formed only from intensively thickening roots. Secondary growth is most rapid in the basal parts of conducting roots. High mechanical pressure causes wounding and as the cambium is very susceptible to it, wounded cambial cells die almost without exception (Romberger et al., 1993). Thinning of the bark occurs at the contact surface of roots and compressed ray cells proliferate initially to form a layer of callus (Savel'eva, 1969; Zimmermann et al., 1980). This proliferation continues until the free space between the roots is filled with cells and continuity of the cambial ring is restored. Regenerating cambium always develops internally simultaneously with callus mass and its development begins from existing cambium (Romberger et al., 1993). Initially, different roots form then joint annual rings. Dense contact between roots, thinning of bark at contact surface and thickening of roots beneath their contact surface are the early stages of root grafting, but the process does not always result in formation of a root graft with joint annual rings.

Usually, root grafting is more frequent in deciduous trees than in conifers (Savel'eva, 1969), but it depends on the tree species biotic and abiotic con-

* FAX No: +372-7-375-825. E-mail: krista@ut.ee
Plant and Soil is the original source of publication of this article. It is recommended that this article is cited as: *Plant and Soil* **217:** 91–100, 1999.

ditions of the site. Root grafting is common in Norway spruce (Köstler et al., 1968; Külla and Lõhmus, 1996; Save'l'eva, 1969; Sutton, 1969) and can occur in dense stands in young age. Graham and Bormann (1966) have found grafts already established in 3-year-old Norway spruce. In a 15-year-old group-cultivated stand of Norway spruce (5000 stems per ha), 31.3% of trees had root grafts between neighbouring trees (Savel'eva, 1969). In the present paper the terminology concerning root and root system is given after Sutton and Tinus (1983). By the age of 15–20 years, the tap root system of a Norway spruce has developed into an adventitious system in well-drained soils, where most of the first and higher order roots spread horizontally in upper soil horizons (Serebryakov, 1962). At that age, the tap root has usually decayed and new sinker roots emerge from large lateral roots near the stem (Köstler et al., 1968; Külla and Lõhmus, 1996; Savel'eva, 1969; Serebryakov, 1962; Sutton, 1969). In extremely wet sites with a high ground water level, no sinker roots are formed, while grafts occur between the roots of the same tree. Thus, the grafting of vertically spreading roots, i.e. tap roots and sinker roots, is very rare in Norway spruce (Savel'eva, 1969).

Root grafting has some advantages, for example, stability of trees with joint root systems can be higher. Survival of suppressed trees can be enhanced by assimilate and nutrient support from grafted neighbour trees (Graham and Bormann, 1966). The most essential disadvantage of root grafting in Norway spruce stands is promotion of the spreading of root and butt rot, inflicted especially by *Heterobasidion annosum* (Stenlid, 1986; Tamminen, 1985). Butt rot is the rot which spreads from roots up the tree stem. The root graft represents the most rapid pathway for infection to spread from a diseased tree to an uninfected tree (Stenlid, 1986). Thus, formation of root grafts should be prevented by forest management. The important factors affecting root grafting in Norway spruce should be elucidated.

The above- and below-ground parts of three dense Norway spruce stands (40-year-old natural stand, Umbric Luvisol; 43-year-old row-culture and group-culture, Planosol; stand densities were 1550, 1783 and 1722 stems ha^{-1}, respectively) at the *Oxalis* site of high site quality (I) were investigated with the aim:

1. to analyse the effect of cultivation method (different spacing) on root grafting and rooting density,
2. to estimate the influence of tree and root size on the grafting extent and frequency,
3. to analyse the distribution of root age at the beginning of root grafting and to estimate stand age at the beginning of grafting; and
4. to estimate the response of root grafting and the vertical distribution of roots to humus horizon thickness.

Material and methods

The studied Norway spruce stands are described in (Külla and Lõhmus, 1996; Lõhmus and Oja, 1983; Lõhmus et al., 1991). The characteristics of investigated stands are given in Table 1. The natural stand was situated in Estonia, 58° 42′ N and 26° 44′ E; the row- and group-cultivated spruce stands were 50 km SW of it. The thickness of the forest floor (O) horizon was 3–4 cm in all stands, the mean thickness of the humus (A) horizon being 30 cm in the row-culture, 16.5 cm in the group culture and 25 cm in the natural spruce stand. Investigation methods for the plantations and for the natural stand of Norway spruce were different. In all cases, the tree group involved in the study was chosen in accordance with the distribution of diameter at breast height (DBH) in the corresponding stand and the results were assumed to be representative for the whole stand.

Group- and row-cultivated Norway spruce stands

The row culture with 2×2 m planting density (2500 plants per ha) was established in 1951. In the group culture, nine plants were planted regularly on 2.3×3.0 m plots in the same year, with a distance of 3.5 m between the plots (2300 plants per ha). Both stands were harvested in 1994 due to damage caused by elk; no harvesting was done earlier. The response of the above-ground parts of the stands to the cultivation method was analysed by comparing the following stand characteristics: mean height, mean DBH, basal area, standing volume and number of stems per ha. The DBH distributions of living and dead trees in row- and group-cultivated stands were compared. Tree diameter variability was significantly higher in the group-culture, than in the row-culture ($P < 0.01$, F-test). The number of suppressed trees that survived and the number of both dominant and codominant trees were larger in the group-culture compared to the row culture.

Trench wall (Böhm, 1979; Bengough et al., 1992; Kolesnikov, 1971) method was used to study the vertical distribution of conducting roots. Trench depth

Table 1. Stand characteristics in natural, row- and group-cultivated stands of 40- and 43-year old Norway spruce in Estonia

Cultivation method	Canopy composition	Age, yr	Stems ha^{-1}	Mean height, m	Mean DBH, cm	Basal area, $m^2\ ha^{-1}$	Stem overbark volume $m^3\ ha^{-1}$
Natural	10 S, P, B	40	1550	18.6±0.9	19.4±0.5	48.9	442
Row	10 S	43	1783	19.1±0.7	17.4±0.3	42.5	406
Group	10 S	43	1722	17.9±0.8	18.1±0.7	44.3	410

S – *Picea abies*, P – *Pinus sylvestris*, B – *Betula pendula*

of was 1 m in both stands, the length of the trench was 6.0 m in the group-cultivated spruce stand and 2.7 m in the row-cultivated stand. The trench in the group-culture was longer due to a longer distance (3.5 m) between the tree groups. In the row-culture, the trench was taken at an equal distance from the rows (1 m), while in the group-culture the trench was taken at equal distances from the groups (1.75 m). The location of roots on trench wall was mapped. All root diameters ≥ 5 mm were measured and thinner conducting roots were counted. The vertical distribution of the d ≥ 5 mm root cross-section area (mm^2m^{-1}) and d < 5 mm root counts by soil horizons were estimated for both stands.

Skeleton method was applied to study the horizontal distribution of roots and root grafting (Böhm, 1979; Kolesnikov, 1971). A 3.9×4.2 m^2 area in the row-cultivated stand and an area of 6.0×4.5 m^2 in the group-cultivated stand were excavated to a depth of 20 cm of mineral soil. In the group-culture, nine root systems (one of them from a dead tree) were excavated. In the row-culture, four root systems were excavated to the full extent and another four were excavated to half the extent. Neither root nor butt rot was found in the excavated stumps of living trees, despite the occurrence of rot in 24.3% of stumps in row-culture (Külla and Lõhmus, 1996). A grid net of crossing strings (grid size 30×30 cm) was used for the mapping of root systems. The minimal diameter of the roots to be mapped, was 5 mm. Roots with a diameter smaller than 30 mm were mapped having an equal diameter. Only root systems of living trees were taken into account. The length of all mapped roots was measured in order to calculate rooting densities of roots with d ≥ 5 mm, and was expressed as the length of roots per 1 m^2 of stand area ($m\ m^{-2}$). All root grafts were collected and the growth rings, maximum and minimum diameters (bark excluded) and the cross-sectional area were measured on air-dry transverse discs. The age at which root grafting began was determined, and root maximum and minimum diameters at that age were measured. Graft age was defined as the maximum age of all graft forming roots, i.e. the annual rings of the oldest root at the time of collecting were counted.

Table 2. Distribution of biomass and production in 40-year-old natural Norway spruce stand

Fraction	Biomass ($Mg\ ha^{-1}$)	Increment ($Mg\ ha^{-1}\ hr^{-1}$)
Stem	192.0	9.0
Branches	18.6	2.6
Needles	21.6	3.1
Roots	48.3	[a] 4.0
Total	280.5	[a] 18.7

[a] Without roots of diameter less than 1.0 mm.

Natural Norway spruce stand

Investigation methods for the naturally regenerated spruce stand were different and applied with the aim to estimate the biomass and production (Table 2), as well as the spatial distribution of the above-ground and below-ground parts. Preliminary results were obtained on the basis of seven felled model trees and six excavated root systems as described in Lõhmus and Oja (1983), Oja and Lõhmus (1985, 1995) and Lõhmus et al. (1991). In this study, they are given to characterize the stand and the spatial distribution of roots. Data on root grafting in the stand have not been used earlier. Neither root nor butt rot was found on model trees.

Horizontal and vertical distribution of roots in the natural Norway spruce stand

The position of all trees in the study area was mapped. Model trees were neighbours and represented the diameter distribution of the stand (Lõhmus et al., 1991). In excavated root systems, 130 lateral and sinker roots emerging from stump were investigated: the maximum and minimum overbark diameters at root collar and root length were measured. The direction of the first order lateral roots proceeded in all cases almost straight from the stump (Oja and Lõhmus, 1995). All 130 lateral and sinker roots were divided, according to root length, into ten zones of equal length for analysis of the spatial configuration of the biomass of conducting roots (Lõhmus et al., 1991). Total biomass of horizontal roots was distributed around the trunk according to hyperbolic Equation (1) (Lõhmus and Oja, 1983):

$$R = 6.3 + 28.93L^{-1}, r = 0.99, P < 0.0001, \quad (1)$$

where R is the portion of roots in the ring layer with a width of 0.5 m (%), and L (m) is the external radius of the layer. A step of 0.5 m was used for the values of L, the maximal value of L being 7.5 m; r is the correlation coefficient and P is level of probability. More than 75% of root biomass was located near the trunk in the ring layer with a width 1 m.

Root mass (%) and geometrical surface area (%) in different diameter classes as well as root area index (m^2 m^{-2}) were estimated (Lõhmus et al., 1991). Root area index (RAI) was 3.5 m^2 m^{-2}; the ratio of conducting to mycorrhizal short roots in root surface area was 46:54. Compared with leaf area index (LAI), 26 m^2 m^{-2} , RAI was small (Lõhmus et al., 1991).

The vertical and horizontal distributions of fine roots ($d < 1$ mm) were estimated from 45 and 71 soil cores (cutting area of a core 16.4 cm^2), respectively as described in (Lõhmus et al., 1991; Oja and Lõhmus, 1985). More than 80% of roots were located in humus (A) horizon and only 5% of fine roots were growing in the forest floor. The horizontal distribution of fine roots was random and uniform; their vertical distribution is described by Equation (2):

$$FR = 26.2e^{-0.0474LD}, r = 0.97, P < 0.001, \quad (2)$$

where FR is the percentage of fine root mass in the 5 cm soil layer (taking the total mass of fine roots in the forest floor and in the upper 40 cm layer as 100%; LD is the mean depth of the layer (2.5, 7.5, ..., 37.5 cm).

Root length from the tree up to the root graft, and the diameters at root base and before entrance to the root graft were measured.

Results

Root grafting in the natural spruce stand

Only two root grafts were found between two different model trees. One of the trees was the codominant, the other being suppressed. Their breast height diameters were 23.0 cm and 16 cm, respectively. The distance between the trees was 0.7 m; distances between the other model trees without grafting were longer than 2 m. One root grafting occurred at the site where the first order lateral roots of the suppressed tree grew into the root system of the codominant tree. Elongation of the root stopped at a root length of 72 cm. The diameters at root collar were 9.0 and 4.7 cm and near the root graft 3.3 and 2.9 cm, respectively. The distance from the tree up to the second root graft was 55 cm for the codominant tree and 54 cm for the suppressed tree. Basal root diameters for the codominant tree were 3.0 and 1.8 and for the suppressed tree 2.6 and 2.1 cm, respectively. The proximal diameters of the root near the root graft were 0.8 and 0.9 cm for codominant tree and 0.6 and 0.7 cm for suppressed tree. Root graftings occurred in 33.3% of excavated trees, the number of grafts per tree being 0.33.

Horizontal and vertical distribution of roots in group- and row-cultivated Norway spruce stands

Horizontal distribution

Horizontal projections of excavated root systems in the forest floor and in the uppermost 20 cm horizon for the row- and group-cultures are presented in Figures 1 and 2. As seen in Figure 1, the distance between the rows was not always 2 m, while the spacing of the trees in a row varied even more depending on planting method and on stand differentiation. In the row-culture in the area of 3.9×4.2 m^2, the total length of mapped roots d ≥ 5 mm was 108.6 m. In the following calculation, the number of root systems in the mapped area was taken equal to six (four whole and four half-root systems). The calculated length of d ≥ 5 mm roots per tree was 18.1 m and per stand area 3.2 m m^{-2}.

In the group culture (Figure 2) nine stumps were situated on the 3.9×4.2 m^2 area, one of them originating from a dead tree. Only the root systems of living trees were taken into account in root length measurement. The total length of all mapped roots was 170.5 m; the mean length ($\pm$ standard error) per tree was 21.3 $\pm$ 8.7 m and rooting density per stand area was 3.7 m m^{-2}.

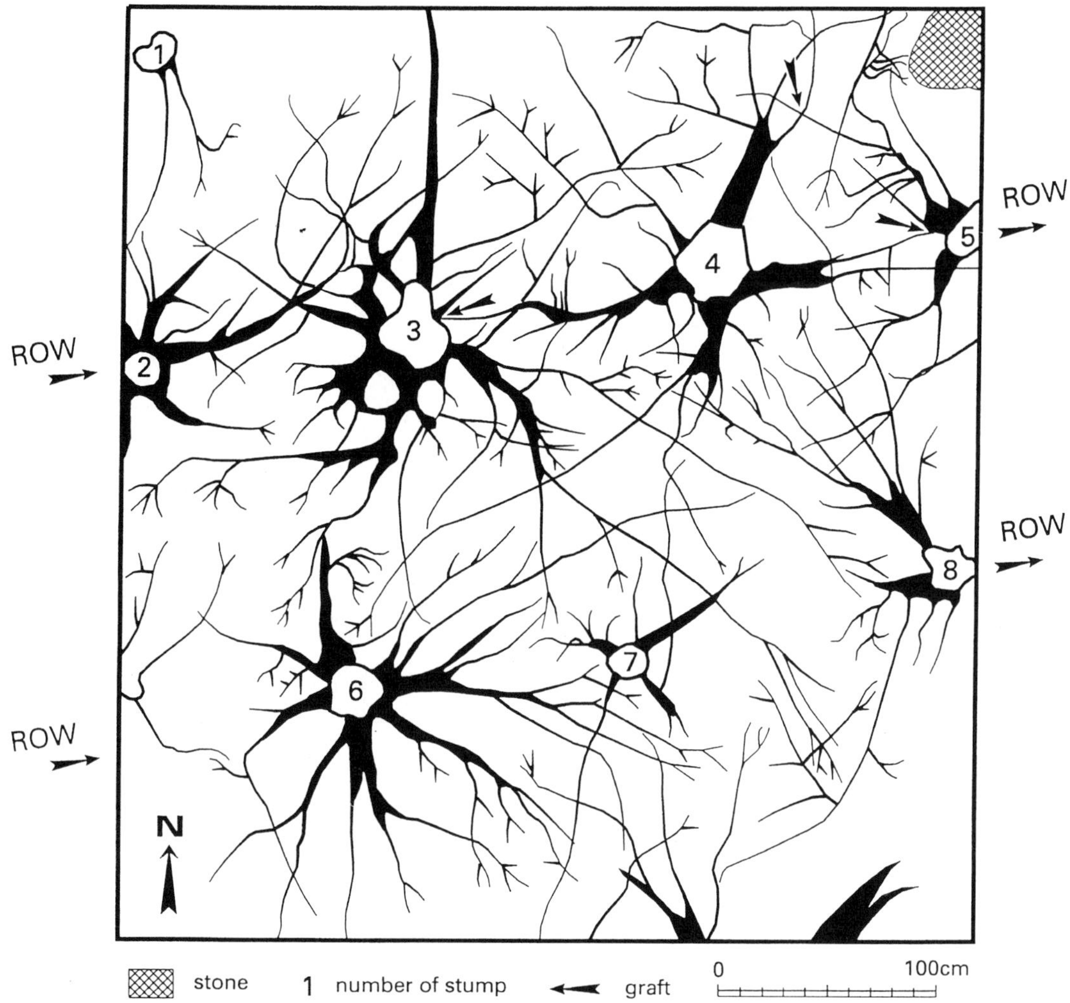

Figure 1. The horizontal projection of excavated root systems in row-cultivated Norway spruce stand in forest floor and uppermost 20 cm soil layer. The minimum diameter of mapped roots was 5 mm and roots less than 30 mm in diameter were taken as equal in diameter. Root grafts are indicated by arrows.

Vertical distribution

The root cross-section area of d $\geq$ 5 roots per unit area of vertical trench wall was 456 mm^2 m^{-2} in the row-culture and 525 mm^2 m^{-2} in the group-culture. As the trenches in the row-and group-cultures were placed at equal distances for both from the rows and tree groups, the given vertical rooting densities represent the minimal values. The diameter of roots with d $\geq$ 5 mm on trench wall ranged from 9.0 to 49.0 mm in the row-culture and from 5.0 to 22.5 mm in the group-culture. The number of roots d $\geq$ 5 mm per 1 m trench in the row- and group-cultures were 4.4 and 4.5, respectively. In the row-culture, 94.8% of d $\geq$ 5 mm root cross-section area was located in the humus (A) horizon, while in the group-culture the respective figure was 99.4%. As the thickness of the humus horizon in the group-culture was almost two times as small as in the row-culture, roots were more superficial in the group-culture than in the row-culture. Thus, although stand mean diameters and heights were not different (t-test, level of probability $P \geq 0.05$) and other stand characteristics were similar (Table 1), the horizontal and vertical distributions of conducting roots d $\geq$ mm were different.

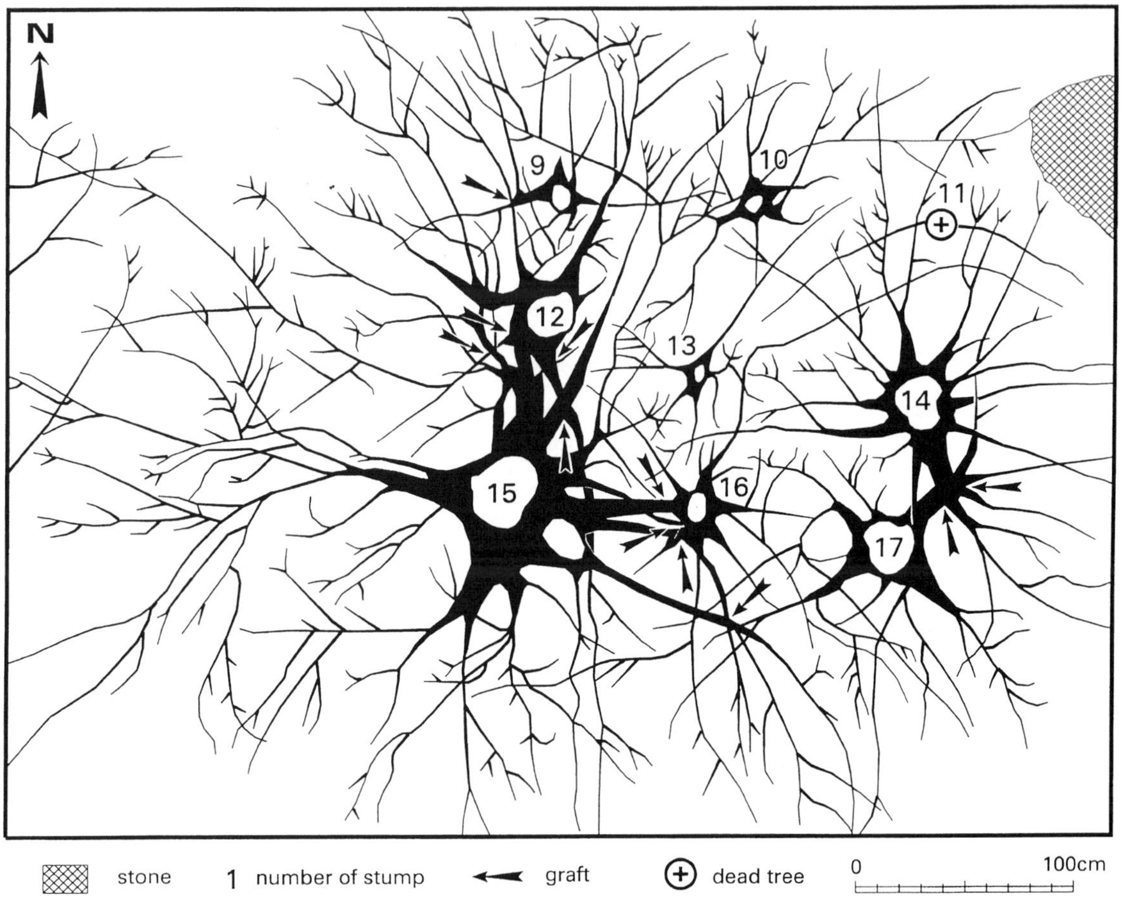

Figure 2. The horizontal projection of excavated root systems in group-cultivated Norway spruce stand in forest floor and uppermost 20 cm soil layer. The minimum diameter of mapped roots was 5 mm and roots less than 30 mm in diameter were taken as equal in diameter. Root grafts are indicated by arrows.

Root grafting in group- and row-cultivated Norway spruce stands

The positions of root grafts in excavated areas in row- and group-cultivated Norway spruce stands are indicated with arrows in Figure 1 and Figure 2, respectively. Data on measured root grafts, distances between trees with grafting roots, root ages and diameters are presented in Table 3. Grafts occurred always in areas of higher rooting density, in rows in the row-culture and inside a tree group in the group-culture, when the distance between the trees was as small as 0.8–1.2 m. No grafts occurred between two suppressed trees; in 86% of cases at least one tree was a dominant or a codominant. Grafting density in the row- and group-cultures is presented schematically in Figure 3A and Figure 3B, respectively, where bold solid lines between the trees indicate root grafts. Broken lines between the trees (Figure 3A, Figure 3B) indicate close contact between the roots as well as morphological changes which may lead to root grafting: thinning of bark at contact surface and thickening of roots beneath their contact surface. Usually, the graft section was not perpendicular to the axis of a grafted root due to different angles between grafted roots. Thus, the maximal root diameter being usually overestimated, the smaller diameter was used to characterize root diameter at the beginning of grafting. Minimal root diameter without bark in the air dry stage varied from 0.4 to 8.3 cm. Root age at the beginning of grafting varied from 7 to 35 years, graft age from 16 to 39 years and the maximal number of joint annual rings from 2 to 21. The relative distributions of root age and diameter groups are presented in Tables 4 and 5. Grafting process resulted in formation of the

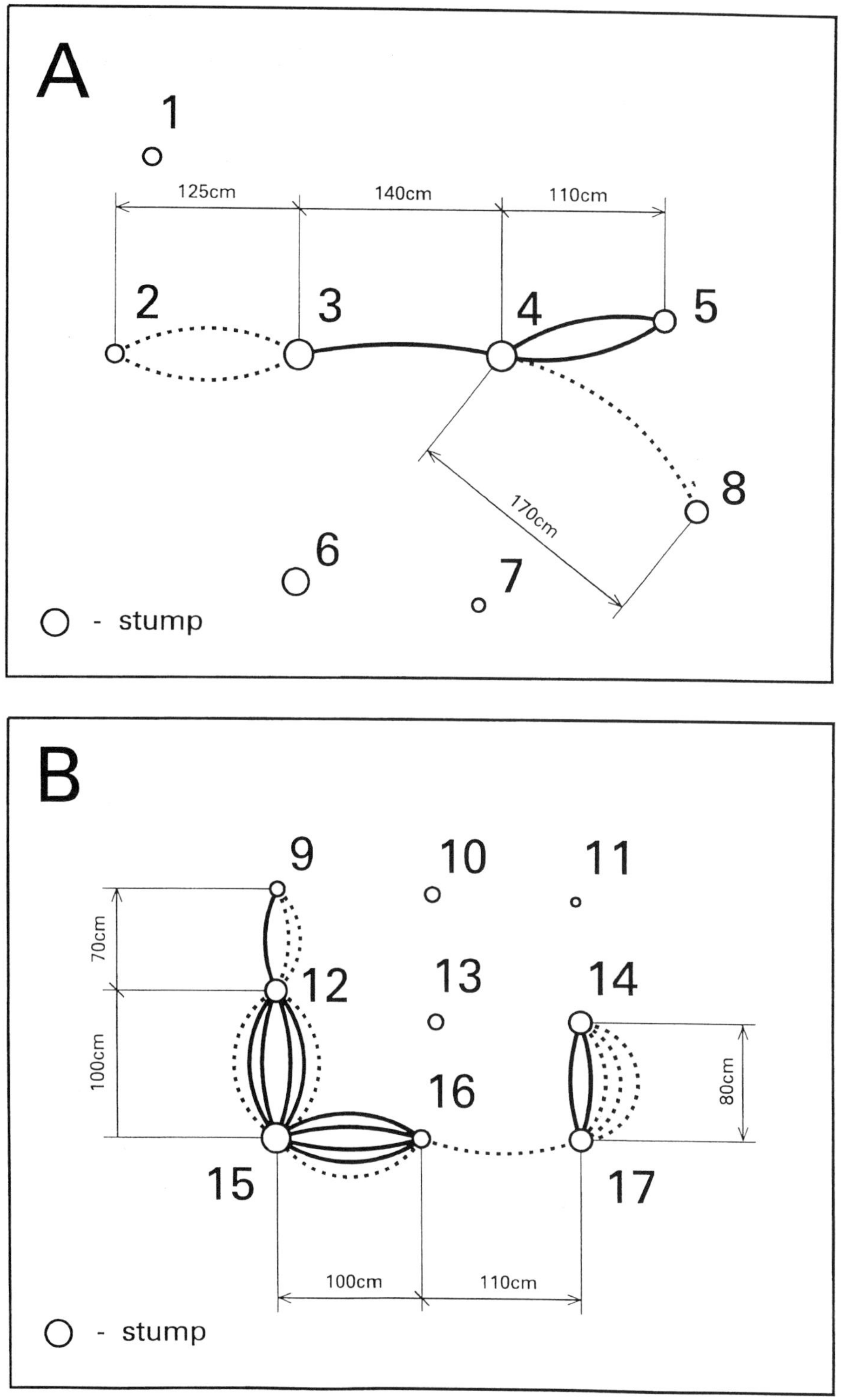

Figure 3. Grafting density in 43-year-old (A) row- and (B) group -cultures of Norway spruce. Bold solid lines between the trees indicate tree grafts, broken lines indicate the close contact between the roots and the morphological changes which may lead to grafting of roots (thinning of bark at contact surface and thickening of roots beneath their contact surface). No. are trees and diameter of a circle corresponds to stump diameter.

Table 3. Characteristics or root grafts in row(R)- and group(G)-cultivated Norway spruce stands. Root RD1 and RD2 are the maximum and minimum diameters of a root, tree No 1 is the bigger and tree No 2 the smaller tree with grafted roots (see Figures 1–3); GD1 and GD2 are the maximum and minimum diameters of graft cross-section

Tree No 1	Tree No 2	Distance between trees, m	Root No	[a]Root age, years	Root diameter RD1, cm	Root diameter RD2, cm	Graft age, years	Graft diameter GD1 cm	Graft diameter GD2 cm	Graft cross-section area, cm^2
4	5	1.1	1 R	20	4.5	3.0	39	9.0	7.0	48.0
4	5	1.1	2 R	18	4.0	2.3	39	9.0	7.0	48.0
4	5	1.1	3 R	7	0.5	0.4	39	9.0	7.0	48.0
4	5	1.2	4 R	11	2.0	1.2	25	5.1	2.6	12.0
4	5	1.2	5 R	13	1.4	1.2	25	5.1	2.6	12.0
14	17	0.8	1 G	16	3.7	3.0	29	6.5	3.6	20.3
14	17	0.8	2 G	18	1.4	1.4	29	6.5	3.6	20.3
14	17	0.8	3 G	35	16.4	8.3	35	19.0	10.0	167.3
14	17	0.8	4 G	19	5.7	3.2	35	19.0	10.0	167.3
14	17	0.8	5 G	10	2.2	2.0	35	19.0	10.0	167.3
15	16	1.0	6 G	12	2.8	1.3	35	15.2	11.3	126.8
15	16	1.0	7 G	24	3.4	3.0	35	15.2	11.3	126.8
15	16	1.0	8 G	26	7.4	5.2	36	12.0	6.5	65.8
15	16	1.0	9 G	12	3.5	3.0	36	12.0	6.5	65.8
15	16	1.0	10 G	26	7.0	3.6	39	11.0	5.5	65.0
15	16	1.0	11 G	14	5.6	2.4	39	11.0	5.5	65.0
15	16	1.0	12 G	19	1.5	1.3	39	11.0	5.5	65.0
15	12	1.0	14 G	15	3.7	1.3	22	5.4	2.9	11.5
15	12	1.0	15 G	20	1.7	1.3	22	5.4	2.9	11.5
15	12	1.0	16 G	10	2.4	2.2	16	5.6	3.4	13.5
15	12	1.0	17 G	9	2.2	1.7	16	5.6	3.4	13.5
15	16	1.0	17 G	31	5.1	3.2	37	8.4	4.0	29.3
15	16	1.0	18 G	29	2.5	2.2	37	8.4	4.0	29.3
15	12	1.0	19 G	22	6.5	5.7	27	12.8	5.0	65.3
15	12	1.0	20 G	24	3.7	3.6	27	12.8	5.0	65.3
15	12	1.0	21 G	22	6.5	6.2	27	11.9	6.6	70.3
15	12	1.0	22 G	24	4.0	3.0	27	11.9	6.6	70.3
15	12	1.0	23 G	8	2.3	1.8	27	11.9	6.6	70.3

[a]Root age at the beginning of root grafting.

first joint annual rings in the row-culture and in the group-culture when the stands were 24 and 22 years old, respectively.

Root graftings occurred in 75% of trees in the group-culture and for 37.5% of trees in the row-culture. The number of grafts per tree in the group- and row-cultivated stand was 1.375 and 0.375, respectively (see Figure 3A, B).

Discussion

In all investigated stands, root grafting was most sensitive to the dense spacing of the trees. This can be explained by the larger amount of root mass near the stem. Grafts always occurred in areas of higher rooting density, in rows in the row-culture and inside a tree group in the group-culture. As it was observed in this study, stand density appeared irrelevant. Both induction of grafting and the number of grafts per tree in Norway spruce depended on cultivation method. Root grafting was enhanced by the narrower humus horizon; according to Savel'eva (1969) and Sutton and Tinus (1983), the narrower rooting zone enhances root grafting.

Use of trench wall method leads to underestimation of the proportion of fine roots, because the method

Table 4. Relative distribution of root age at the beginning of root grafting

Root age (yr)	< 10	10–20	20–30	30–40	≤40
Percentage, %	17	46	30	7	100

Table 5. Relative distribution of root minimal diameter at the beginning of root grafting

Root diameter (cm)	< 1	1–2	2–3	3–5	5–10	10 <
Percentage, %	3	33	30	17	17	100

yields only the relative vertical distribution of thicker lateral roots (Bengough et al., 1992). Comparison of root excavation and trench wall method on our Norway spruce stands allowed to conclude that the latter method is suitable for estimating the relative vertical distribution of roots d $\geq$ 5 mm. Also, due to root overlapping and vertical spreading of sinker roots, actual rooting density is higher than estimated by the skeleton method. The estimated values were used to compare rooting density in investigated plantations. To establish whether a graft is formed or not, it is nessessary to examine its cross section. However, we observed that in case of highly intertwined roots, which were impossible to separate, there existed no joint annual rings or other joint tissues. Tree social status influences root grafting in conifers. Yli-Vakkuri (1953) found in Scots pine that most grafts were established between a suppressed tree and a vigorous tree. In our stands, no grafts occurred between two suppressed trees, in 86–100% of cases, at least one tree was a dominant or a codominant. Our results did not confirm the ability of Norway spruce to form grafts in very young age (Graham and Bormann, 1966; Sutton, 1969). Stands were 22 and 24 years old when the first root grafts were formed. Reconstruction of stand age at the beginning of root grafting and the results of temporal variability of grafting frequency are to be regarded with care. The number of annual rings in thinner laterals emerging from stumps could be smaller than actual root age (Coutts, 1983). Missing rings are often apparent in roots (Fayle, 1968), but also in trunks and branches. Krause and Eckstein (1992) found that for several tree species including Norway spruce, that minimum and maximum increments are mostly related throughout the entire tree; missing rings at one point of a tree correspond to narrow rings in the other parts of the tree. Root grafting frequency depended on root age and diameter at the beginning of the process (Tables 4 and 5). Our data did not provide evidence that young root age enhances grafting (Savel'eva, 1969; Sutton and Tinus, 1983).

Tree diameter variability was significantly higher in the group-culture than in the row-culture. More suppressed trees survived, and more dominant and codominant trees were observed in the group-culture compared with the row-culture. This is in accordance with Graham and Bormann (1966) that survival of suppressed trees can increase due to assimilate and nutrient support from grafted neighbour trees.

As the root graft represents the most rapid pathway for secondary infection to spread from a butt-rot diseased tree to an uninfected tree (Stenlid, 1986; Tamminen, 1985), its formation should be avoided. On the basis of our and earlier studies (Köstler et al., 1968; Külla and Lõhmus, 1996; Savel'eva, 1969; Serebryakov, 1962; Stenlid, 1986; Sutton, 1969; Tamminen, 1985), we suggest the following stand management to avoid infection by root grafting: planting density should be lower than 2500 stems ha^{-1}; distances shorter than 1.5–2 m between trees should be avoided. When there exists danger of *Heterobasidion annosum* infection, stand thinning should be completed before the stand age of 15–20 years; after that time no thinning is recommended until the final harvesting. In younger stands, root grafting is rare or absent as shown in our study. Recently exposed stump surfaces and wounds in stems and roots brought about by thinning represent the primary (spore) infection routes (Stenlid, 1986).

Acknowledgements

This study was partly supported by the Estonian Science Foundation Grant No 2487. We thank Dr Kalle Remm for the help in preparing the figures and Mrs Ester Jaigma for revising the English text.

References

Bengough A G, MacKenzie C J and Diggle A J 1992 Core-Break and profile wall methods: how many roots do you really count? *In* Root Ecology and its Practical Application. Eds L Kutschera, E Hübl, E Lichtenegger, H Persson and M Sobotik. pp 743–746. Proceedings of 3. ISRR Symposium, September 2nd–6th 1991, Wien.

Böhm W 1979 Methods of Studying Root Systems. Ecological Studies 33. Springer Verlag, Berlin-Heidelberg. 188 p.

Coutts M P 1983 Development of the structural root system of Sitka spruce. Forestry 56, 1: 1–16

Fayle D C F 1968 Radial growth in tree roots. University of Toronto, Faculty of Forestry, Tech. Rep. No. 9, Toronto, Ontario, 183 p.

Graham B F and Bormann F H 1966 Natural root grafts. Bot. Rev. 32, 255–292.

Kolesnikov V A 1971 The Root System of Fruit Plants. Mir Publishers, Moscow, USSR. 269 p.

Krause K and Eckstein D 1992 Holzzuwachs an Ästen, Stamm und Wurzeln bei normaler und extremer Witterung (Annual growth throughout the whole tree during normal and extreme climatic conditions) *In* Luftverunreinigungen und Waldschäden am Standort 'Postturm', Forstamt Farchau/Ratzeburg. Eds W Michaelis and J Bauch. pp 215–242. GKSS-Forschungszentrum Geesthacht GmbH, Geesthacht.

Köstler J N, Brückner E and Bibelriether H 1968 Die Wurzeln der Waldbäume (The Root Systems of Forest Trees). Paul Parey Publishers, Hamburg-Berlin. 284 p.

Külla T and Lõhmus K 1996 Assessment of cultivation method on spatial distribution of roots in middle-aged spruce stands. Forestry Research No 27, 36–46 (in Estonian).

Lõhmus K and Oja T 1983 On methods of investigating the underground part of forest stands. Lesovedenie (Soviet Forestry) 4, 56–62 (in Russian).

Lõhmus K, Lasn R and Oja T 1991 The influence of climatic and soil physical conditions on growth and morphology of Norway spruce roots. *In* Plant Roots and their Environment. Developments in Agricultural and Managed-Forest Ecology 24. Eds B L McMichael and H Persson. pp 233–239. Elsevier, Amsterdam.

Oja T and Lõhmus K 1985 Horizontal distribution of Norway spruce roots in a middle-aged stand. Lesovedenie (Soviet Forestry) 1, 44–47 (in Russian).

Oja T and Lõhmus K 1995 Norway spruce roots: structure and layout. In Workshop on Root Ecology, Abstracts, Tartu, June 12th 1995. Ed. K Lõhmus. pp 41–45. University of Tartu, Tartu, Estonia (in Estonian).

Romberger J A Hejnowcz Z and Hill J F 1993 Plant Structure: Function and Development. Springer-Verlag, Berlin Heidelberg New York. 524 p.

Savel'eva L S 1969 Grafting of Tree Root Systems. Lesnaja Promychlennostj Publishers, Moscow, USSR. 72 p.

Serebryakov 1962 Ecological Morphology of Plants. Vyschaja chkola Publishers, Moscow, USSR. 378 p.

Stenlid J 1986 Biochemical and Ecological Aspects of the Infection Biology of *Heterobasidion annosum*. PhD Thesis. Swedish University of Agricultural Sciences, Uppsala, Sweden. 37 p.

Sutton R F 1969 Form and Development of Conifer Root Systems. Commonwealth Agricultural Bureaux, Farnham Royal, Bucks., England. 131 p.

Sutton R F and Tinus R W 1983 Root and Root System Terminology. Forest Science Monograph 24, 1983. 137 p.

Tamminen P 1985 Butt-rot in Norway spruce in southern Finland. Commun. Inst For. Fenn. 127, 1–52.

Yli-Vakkuri P 1953 Studies of organic root grafts between trees in *Pinus sylvestris* stands. *Acta for. fenn.* 60, No 3. 117 p. Helsinki (in Finnish).

Zimmermann M H, Brown C L and Tyree M T 1980 Trees Structure and Function. Springer-Verlag, New York Inc., USA. 336 p.

GROWTH AND DEVELOPMENT

A. Stokes (ed.), The Supporting Roots of Trees and Woody Plants: Form, Function and Physiology, 121–130.

Comparison of radial increment and volume growth in stems and roots of *Quercus petraea*

Michael Drexhage *, Françoise Huber and Francis Colin
INRA, Centre de Nancy, Unité Croissance, Production et Qualité des Bois, 54280 Champenoux, France

Key words: biomass partitioning, dendrochronology, radial growth, radial increment, root growth, *Quercus petraea*

Abstract

Dendrochronological studies dealing with roots, stems and branches are very rare or often take the form of short notes. The difficulties of detecting rings and of quantifying the radial growth in roots have already been described for various species. In oak the anatomical root structure differs from stemwood. The roots are radial-porous or diffuse-porous, and there is often no clear distinction between individual rings. In our study visual and radiographic techniques were used to examine radial increment in roots of sessile oak (*Quercus petraea* L.) which was compared with radial growth in branches and along the stem. Coarse roots were cut from four 30- to 34-year-old trees that had been uprooted mechanically and disks were taken at different distances from the stem-root base. Ring widths were measured in the stem at height of 0.3 m, at breast height (1.3 m), beneath the crown, in branches of the crown, and in roots every 20 cm. The ring widths were cross-dated, and the heterogeneity of growth within a root and within the root system were analysed. Asymmetric growth frequently occurred in roots so that ovals, I-beam and T-beam shapes were developed. With the method used in our study the annual growth layers close to the central cylinder could be distinguished as well as beneath the bark. Pointer years were detected in all sections of the tree and permitted correction of ring widths in roots. Root system, stem and branch showed a basic similarity in their radial sequence of ring width. The annual biomass increment was weaker and more variable with several consecutive changes in the roots than in the stems. The root/shoot ratio reached a minimum rather early, beginning at the cambial age of 20 years.

Introduction

Sessile oak (*Quercus petraea* (M.) Liebl.) is one of the dominant species in nemoral forests of central Europe and is an economically important forest tree in France. Although silvicultural management, forest productivity and wood quality of oak trees has been intensively investigated for many years (Bary-Lenger and Nebout, 1993; Becker et al., 1995; Zhang et al., 1994), dendrochronological studies which include the growth of shoots and roots are very rare or often take the form of short notes (Schweingruber, 1996). Little work has been carried out on root growth, because root wood is more difficult to obtain and its economic importance is negligible.

* FAX No: +333 8339 4034. E-mail: drexhage@nancy.inra.fr
Plant and Soil is the original source of publication of this article. It is recommended that this article is cited as: *Plant and Soil* **217:** 101–110, 1999.

Thus, there are innumerable studies dealing with current and past radial growth in stems to analyse the oak dieback in France (Landmann et al., 1993), the long-term growth trends of trees (Badeau et al., 1996) and oak forests (Lebourgeois, 1997). One current and non-destructive technique is to take one wood-core per tree at breast height to analyse the year rings for a high number of individuals on stand level (Lebourgeois, 1997). There are some papers dealing with radial growth in branches and stems to analyse the hydraulic architecture of red oak (White, 1993) and sessile oak (Hatsch, 1997). Only two studies dealing with roots, stems and branches of oak are known to the authors. Gasson (1985, 1987) has investigated several aspects of the secondary xylem anatomy in several oak species, and Krause and Eckstein (1993) have compared annual growth with contemporary climatic conditions in 125-year-old pedunculate and sessile oak.

As is the case for many hardwood species, the anatomical root structure in oak differs from stemwood (Cutler, 1976; Gasson, 1985). The oak roots are radial-porous or diffuse-porous, and there is often no clear distinction between individual rings. In addition, root wood anatomy can vary in the similar way to wood from the aerial part of the tree, depending on environmental conditions. Although tree ring boundaries were difficult to recognise in the roots and annual rings were frequently missed, considerable similarities were reported between different roots of the same tree, and the mean ring-width series of the root system fitted the mean ring series of the trunk (Krause and Eckstein, 1993).

Our objective was to compare radial increment and volume growth in above- and below-stump parts of the tree in order to calculate the wood production and to find more information about the biomass distribution between root-stem-crown concerning a mechanical and physiological balance. In a preliminary study, two recording techniques were used, a visual method to measure ring widths in stems and branches and a combination of visual and radiographic methods to determine and verify tree ring measurements in roots of different diameter and at different distances from the stem-root base. In our case, the tree ring boundaries were difficult to recognise in the roots. Thus, we adopted the X-ray technique that Bedeneau and Pagès (1984) used to examine tree rings in birch and sweet chesnut roots. The radial and growth increment of branches, stems and roots for four 30- to 34-year-old sessile oaks from a naturally regenerated area in north-eastern France were determined to analyse their response to climate and to estimate the biomass partitioning within the tree.

Material and methods

Site details and extraction of the trees

The naturally regenerated 30- to 34-year-old sessile oak (*Quercus petraea* (M.) Liebl.) is located in the forest of Champenoux in north-eastern France (237 m altitude, latitude 48°44′ N, longitude 6°14′ E). Stand density was 8400 trees per hectare. The codominant tree was *Carpinus betulus* L. with a contribution of 15 to 40%. The site was relatively flat, and the soil was a weakly leached brown earth with a silty clay texture developed over a pseudogley horizon. The soil depths ranged from 0.7 to 1.0 m. They were characterised by low base saturations, and pH(H_2O) of 4.8–5.0 at the forest floor and 6.0–6.5 at a depth of 1 m. Mean annual rainfall was 769 mm (mean of 1975–1996) and the mean temperature was 9.2 °C. In 1997 four dominant trees were felled leaving a 1-m-high stump for the extraction of the root system (cf. Colin et al., 1995; Drexhage et al., 1999). Stem and root system dimensions are given in Table 1.

Ring width measurements

On the felled tree, disks were collected (1) at the stem height of 0.3 m above the forest floor (= stem base), (2) at breast height (1.3 m), (3) beneath the beginning of the crown, (4) in the centre of the crown and (5) from two branches in the centre of the crown. On the root system, the position of each major root (main first-order root and tap root) was mapped.

Two methods were used for the ring width measurements. In the first, the disks collected from the aerial part of the tree were sanded and prepared with a scalpel, if necessary. Stem and branch ring widths were measured along the four cardinal directions, whereas the ring widths on the stem base were measured every 45 ° due to the more eccentric pattern of annual rings.

In the second, disks were collected from major roots and the tap root at 20-cm distances from the stem-root base. The larger disks from the tap root 20 cm beneath the stump and from horizontal coarse roots near the stump were sanded. The radius with the largest number of tree rings was selected (Krause and Morin, 1995). Tree rings are usually measured on only one trace due to the irregular pattern of annual rings. Such a trace could not always follow a straight line as the rings were often eccentric. In addition, the traces were prepared using a scalpel to recognise the ring boundaries. To verify the visual measurements and to obtain clearly-defined tree ring boundaries, radiographs were prepared. The root disks were prepared as followed: the root disk was sawn into a rectangular soloid; the soloid was placed underneath a twin-bladed circular saw and a lath of 1.8 mm was cut out. The sample had to be cut perpendicular to the direction of the fibres, otherwise the X-ray picture would not be clear (Schweingruber, 1988). The procedure was repeated until a high quality strip could be obtained. The strips were cleaned using compressed air. X-ray preparation in principle follows the procedure described by Polge and Nicholls (1972) and Schweingruber (1988). The radiographs were analysed on a light table which was installed between the stereo-

Table 1. Tree parameters of four sessile oaks

Tree	DBH (cm)	Age (y)		Heights (m)		Volume (10^{-3} m^2)		Biomass (kg)		R/S
		$H_{0.3}$	$H_{1.3}$	total	to Crown	Stem	Roots	Stem	Roots	
12	11	34	30	12.3	7.2	61.2	7.2	43.5	5.1	0.1
13	12	33	30	12.0	5.9	59.9	6.6	44.4	4.9	0.1
14	11	32	28	11.8	6.1	46.1	8.7	29.0	5.5	0.2
16	11	33	29	11.4	6.6	43.4	5.2	28.2	3.4	0.1

Note: Age at the height of 0.3 m ($H_{0.3}$) and 1.3m ($H_{1.3}$); age, DBH, height, root biomass, and stem volume were measured whereas stem biomass and root volume were calculated; the mean moisture content of the wood was 12%; roots are coarse roots with a diameter > 10 mm were oven dried at 104 ° for 24 h.

microscope and the travelling stage. For both methods, visual and X-ray, all ring widths were measured with a precision of 0.1 mm, using a micrometer and semi-automatic ring width measuring machine connected to a Macintosh computer. Three different cut locations for the strip preparation were sometimes necessary for the same disk. The relationship between ring widths measured visually and X-ray pictures was analysed by using correlation analysis.

The coarse roots were oven dried to determine the biomass. The stem biomass (kilogram) was estimated by using *StemVolume* under bark of fresh wood (cubic metre) and wood density (kilogram/cubic metre) equations:

$$StemVolume = \frac{CSA_b + CSA_c}{2} Length \tag{1}$$

where CSA_b is the surface of the stem base, CSA_c is the surface at the beginning of the crown and *Length* is the length between them.

The disks were oven dried to determine their *DryWeight* and the *Volume* under bark of fresh wood was measured to calculate the *WoodDensity*:

$$WoodDensity = \frac{DryWeight}{Volume} \tag{2}$$

The mean moisture content of the disks taken from different parts of root systems, stems and branches was 12%.

Dendrochronological analysis

The tree-ring widths were plotted as a time-series, and the data were visually checked on a light table. All analyses started at the stem-root base and progressed toward its apex to ensure correct dating of year rings (Krause and Eckstein, 1993). Step by step each subsequent disk was compared with the preceding one. The cross-dating was done within the four or eight different radii in stem and branch disks, within different tree heights and within roots and between roots and stem. The exact position of year rings in roots was determined by analysing characteristic pointer years (abrupt growth change).

The annual wood increment was calculated throughout the entire tree. The circumference of roots, branches and the bole are continually expanding, so that we developed following procedure: In the stem and branches the annual growth increment n was calculated by determining the cross-sectional area (CSA) with the geometric mean of four radii and eight radii for the stem base, using the following equations (Biging and Wensel, 1988):

$$R_{in} = \Sigma_{j=1}^{n} r_{ij} \tag{3}$$

$$CSA_n = \Sigma_{i=1}^{4} R_{in}^2 \frac{\pi}{400} \tag{4}$$

$$CSA_n = \Sigma_{i=1}^{8} R_{in}^2 \frac{\pi}{800} \tag{5}$$

where r_i is the ring width of ith direction and R_n is the radius of nth year measured on the disk from the centre to the bark. The radius (millimetres) of year n was calculated by summing the annual ring width from the first n years beginning at the centre.

In roots the cambial growth is frequently asymmetric, especially near the stem-root base. Initially roots became elliptical in the first ten years and then developed a slab shape or T-beam shape. The annual growth increment was calculated by determining the root CSA increment (square centimetres) with the geometric mean of radii by using following equation (Figure 1):

$$CSA_n = \left[r_n^2 + \left(r_n \frac{\tilde{R}_2}{\tilde{R}_1} \right)^2 + \left(r_n \frac{\tilde{R}_3}{\tilde{R}_1} \right)^2 + \left(r_n \frac{\tilde{R}_4}{\tilde{R}_1} \right)^2 \right] \frac{\pi}{400} \tag{6}$$

where r_n is the ring width of nth year and $\tilde{R}$ is the radius measured with a calliper in four directions on

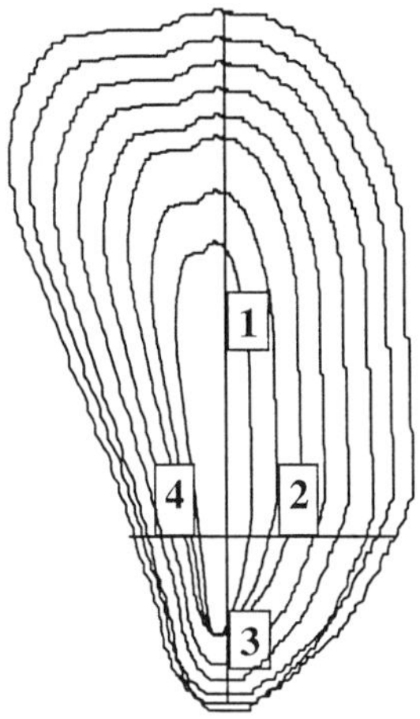

$$CSA_n = \left[r_n + \left(r_n \frac{\tilde{R}_2}{\tilde{R}_1} \right)^2 + \left(r_n \frac{\tilde{R}_3}{\tilde{R}_1} \right)^2 + \left(r_n \frac{\tilde{R}_4}{\tilde{R}_1} \right)^2 \right] \frac{\pi}{400}$$

Figure 1. Procedure for calculating the cross-sectional area in asymmetrically shaped roots. The index r_n is the ring width of nth year and $\tilde{R}$ is the radius measured with a calliper in four directions on the disk from the centre to the bark. $\tilde{R}_1$ was measured on the same trace as that used for ring width measurements. The ratios between $\tilde{R}_1$ and R_2, $\tilde{R}_3$, $\tilde{R}_4$ were calculated to estimate the growth increment.

the disk from the centre to the bark. $\tilde{R}_1$ was measured on the same trace as that used for ring width measurements. The ratios between $\tilde{R}_1$ and $\tilde{R}_2$, $\tilde{R}_3$, $\tilde{R}_4$ were calculated to estimate the growth increment. The procedures were carried out separately for the first ten years (ellipse form) and for the consecutive years (slap and T-beam form), where it was possible. The radius (millimetres) of year n was calculated by summing the annual ring width from the first n years beginning at the centre.

The annual growth layer GL was calculated by using following equation:

$$GL = CSA_n - CSA_{n-1} \qquad (7)$$

In the root, ring widths and growth layer profiles were compared at equal distances from the stem-root base. The same variables were compared between roots and stem.

The annual biomass increments of the root systems were calculated using the annual CSA increment and the linear regressions for the relationship between root CSA (rCSA) and root biomass (rbiomass in grams) as the dependent variable (Drexhage et al., 1999):

$$rbiomass = a + brCSA \qquad (8)$$

where a and b are constants. The linear regression models for the relationship between the root biomass and root CSA were compared by using the analysis of covariance strategy in the PROC MIXED procedure of the *SAS* ® System for Mixed Models (Litell et al., 1996).

Results

Comparison between visual and X-ray measurements in roots

Tree-ring borders are difficult to recognise on oak root sections. For most of the samples, the X-ray pictures were only partially clear. The clarity could change within one strip from the centre to the bark (Figure 2). Ring-porosity was apparent at the periphery of root sections near the stem-root base. It was not always possible to find the position perpendicular to the direction of the fibres, especially for the large root disks near the stem-root base. On the other hand, 80% of the X-ray pictures for the root disks with a diameter of 2.5 cm and smaller were clear. The tree rings in those roots were small in relation to the wide rings of the large root disks, so that X-ray pictures were useful for measuring ring widths. The ring boundaries could be localised, and thus the ring widths could be measured by identifying the large vessels on the radiographs. These values were used to justify the ring data that were determined visually. The values of ring widths differed between the visual and the X-ray method (Figure 3). Nevertheless, corrections for the data after 1976 were possible by cross-dating the ring widths of the roots with those of the stem-base (Figure 4) and by comparing the ring widths measured by the visual and the X-ray method, as shown for one root of tree 13 (Figure 5). Similar characteristic years reductions and pronounced increases in radial growth were found for all four trees after 1976. The ring widths varied more before 1976 when the trees were young.

Comparison between characteristic years and ring width

Reductions in radial growth in the years 1971, 1976, 1979 and 1983 and five consecutive changes beginning in the years of 1986, 1987 every second year were detected in all tree components as shown for tree 12 (Figure 6). Some of them could be correlated with climate events. The growth reductions in 1971, 1976, 1991 could be explained by summer drought (mean

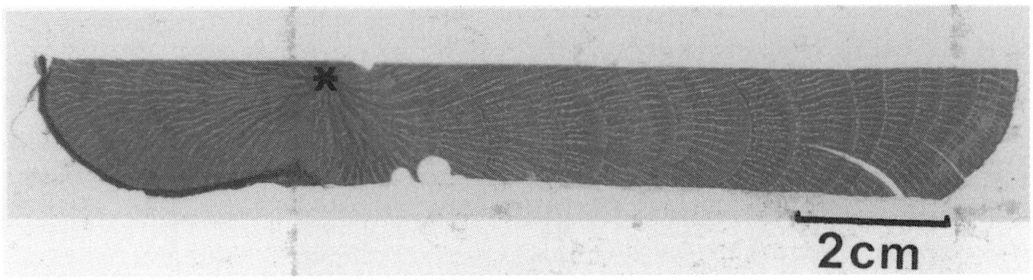

Figure 2. Positive of an X-ray picture of oak root wood cut at the distance of 20 cm from the stem-root base. *: anatomical centre.

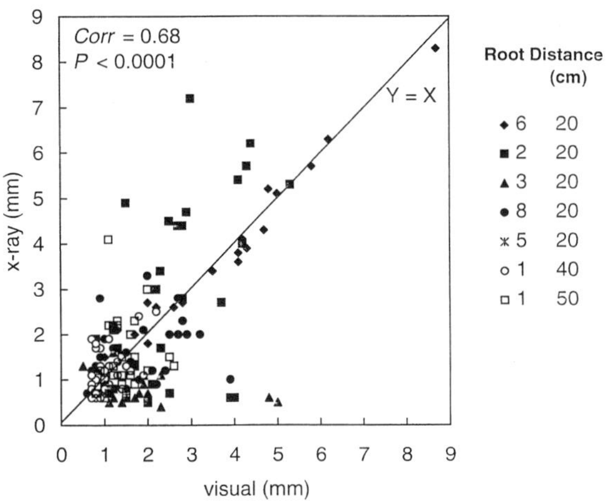

Figure 3. Relationship between ring widths measured visually (*X*) and on X-ray pictures (*Y*) for 7 root disks of oak tree 13 at distances of 20, 40, and 50 cm from the stem-root base.

annual precipitation: 601, 480, 564 mm) and in 1979, 1986, 1987 by strong frost in January and February (monthly means of −1 to −5°C).

A thinning effect could not be directly detected. Nevertheless, after the relatively dry year of 1989 (648 mm), an over-proportionally high increase in radial growth was found in the roots of tree 12 in the following year. This high release in 1990 might be explained by the heavy thinning in the winter of 1989.

Growth increment

The growth rings were concentric in the first 5 to 6 years. After this period, asymmetric growth occurred frequently; initially roots became elliptical, then developed an I-beam shape or a slab shape. Close to the stem, the root developed a buttress T-beam shape. Growth rings were then widest on the upper side of the root. The taproot close to the stem developed a more rosette form with bulges. As only one trace with the largest distance between centre and the bark was selected to measure the ring widths in the roots, Equation (6) was developed to calculate the annual growth increment.

Consistent linear regressions were found for root CSA and root biomass as the dependent variable (Figure 7, Table 2). Although the analysis of covariance indicated that the regression relationships were the same for the four trees ($P > 0.15$), the root biomass was calculated by using the regression coefficients for each tree.

The annual increment of stem biomass can be described with a sigmoid curve for all four trees (Figure 8). The secondary growth slightly increased in the first 10 to 15 years, passed a period of a more rapid increment between 15 to 20 years, and slowed down again after 30 years. As compared with stem growth, the root biomass increment was much weaker, and varied several times between the cambial age of 10 to 20 years.

The long-term relationship between stem and root biomass development was analysed as root/shoot ratio in relation to the cambial age (Figure 9). The proportion of root biomass was high in the initial tree stage and rapidly decreased down to a ratio of 0.1. This minimum was reached at ages of 16 to 29 years for the trees 12, 13 and 16. The minimum for the tree 14 was reached at the age of 11 years, but the ratios increased again with the cambial age, varied and maintained the high level of 0.2 up to the present time.

Discussion

Tree-ring analysis and root shape

The X-ray technique was used to count tree rings in birch and sweet chesnut roots (Bedeneau and Pagès, 1984). The preparations for making radiographs are time consuming, because more than one cut was necessary to get a clear X-ray picture. The preparation and sawing required 15 min for a large strip. Clear radiographs were difficult to obtain from large root

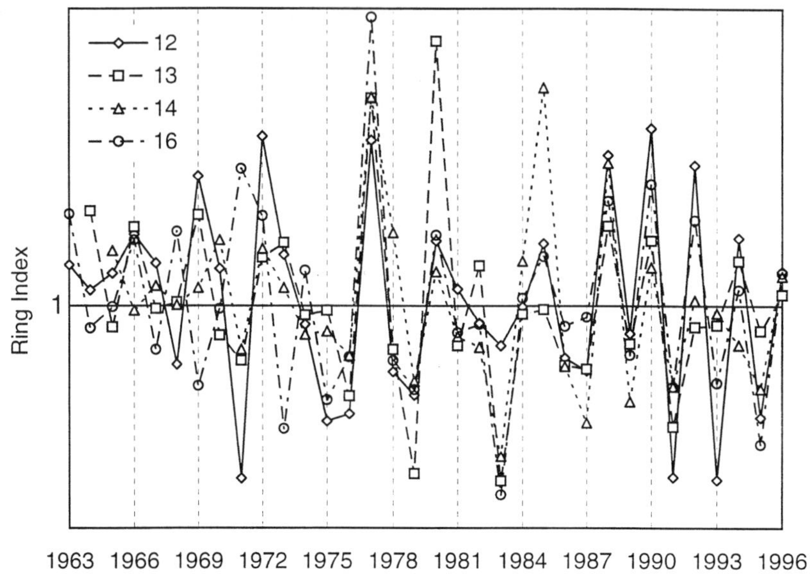

Figure 4. Mean growth layer indices for the stem bases of four 30- to 34-year-old oak trees.

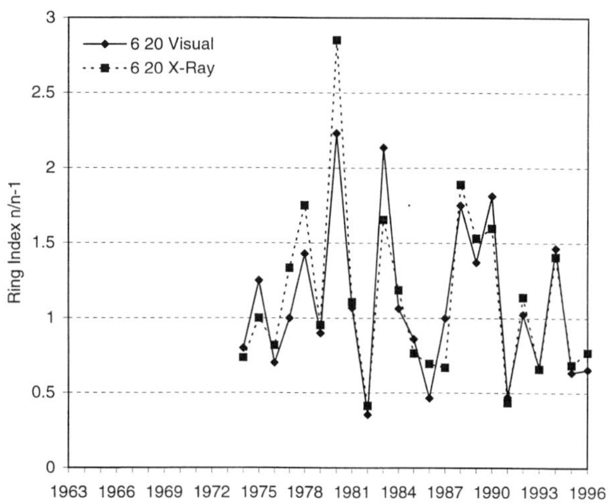

Figure 5. Tree-ring series measured visually and on an X-ray picture for one root disk of an oak tree 13 at 20 cm distance from the stem-root base.

Table 2. Regression equations for the relationship between root cross-sectional area (cm^2) and root biomass (g) calculated for the four 30- to 34-year-old oak trees (refer to Figure 7)

Tree	n	a	b	R^2	SEx	SEy
12	11	n.a.	13.69	0.92	0.79	103.68
13	12	n.a.	11.79	0.80	0.95	134.42
14	9	n.a.	14.15	0.97	0.65	92.86
15	9	n.a.	11.09	0.73	0.90	98.15

Note: Model form is $y = a + b\,x$, where y is root biomass, x is root cross-sectional area, and a and b are constants; n is number of the coarse roots per tree; n.a., equation is forced through the origin because the intercept was statistically non significant ($P > 0.05$).

strips. The wood had to be cut perpendicular to the direction of the fibres and the lath had to reach hygroscopic equilibrium (Schweingruber, 1988). However, the radiographic technique was used in our study for two reasons: firstly, to verify the ring widths visually measured on sanded root disks, and secondly, to get clear pictures of tree rings in roots where it was not visually possible from the disk. The oak roots are radial-porous or diffuse-porous, and there is often no clear distinction between individual rings. On the other hand, ring-porosity was frequently visible at the root periphery as described for roots of 125-year old oaks (Krause and Eckstein, 1993). Nevertheless, the ring borders could be localised on the cross section of the X-ray pictures by identifying the large vessels, which are produced early in the growing season (Gasson, 1987; Woodcock, 1989). Furthermore, our trials suggest that this method could be interesting for roots with small diameters, because the cuts in a perpendicular direction were more successful. The X-ray pictures were clear and the narrow tree rings which were not visible on the root disks, were easy to recognise.

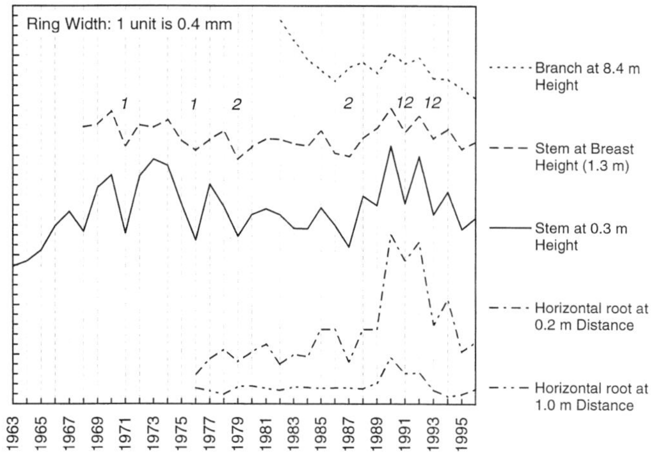

Figure 6. Tree-ring series for one branch at 8.4 m height, the stem at breast height (1.3 m) and at 0.3 m height and for one horizontal root at distances of 0.2 and 1.0 m from the stem-root base. *1*: water deficit, *2*: frost periods in January and February, *12*: frost periods and water deficit in the same year.

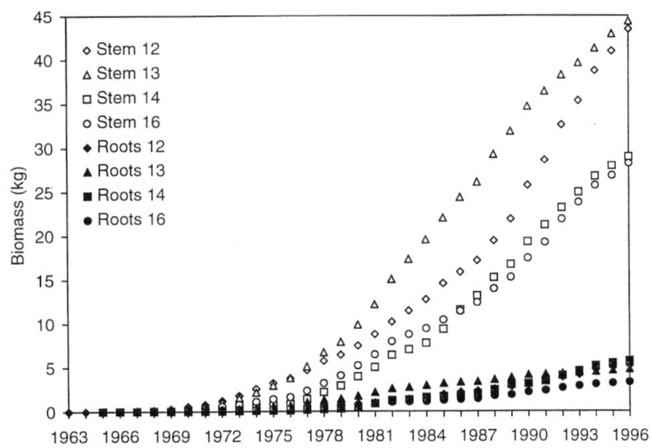

Figure 8. The annual biomass increment for the stems and for the root systems of four 30- to 34-year-old oak trees.

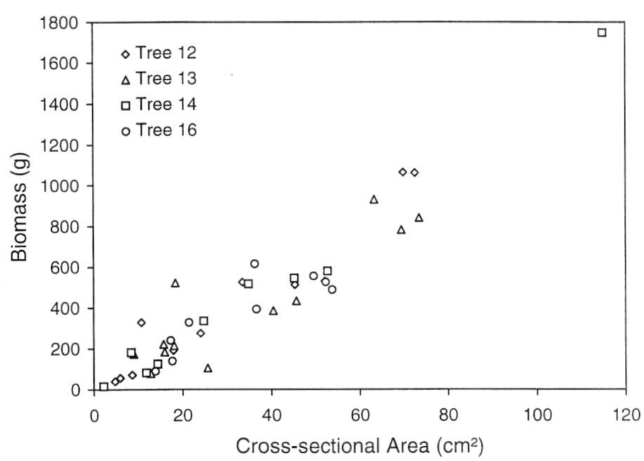

Figure 7. Coarse root biomass (*Y*) against root cross-sectional area (*X*) for four 30- to 34-year-old oak trees.

It was difficult to verify the exact calendar dates in taproot disks collected at 20 and 30 cm beneath the forest floor. Thus, the stem-root base was analysed first and was used for cross-dating and for further calculations in the roots. Nevertheless, the ring widths and the number of rings were highly variable within the roots due to the technical problems. Certainly a number of rings were not found in oak roots, but we have not attached much importance to their quantification, as Krause and Eckstein (1993) have done for 125-year-old oaks. They reported a considerable number of missing rings in the roots which could be attributed to insufficient rainfall. However, the exact position of the missing rings was often difficult to find (Krause and Morin, 1995). Another problem of verifying the number of years that occurred near the

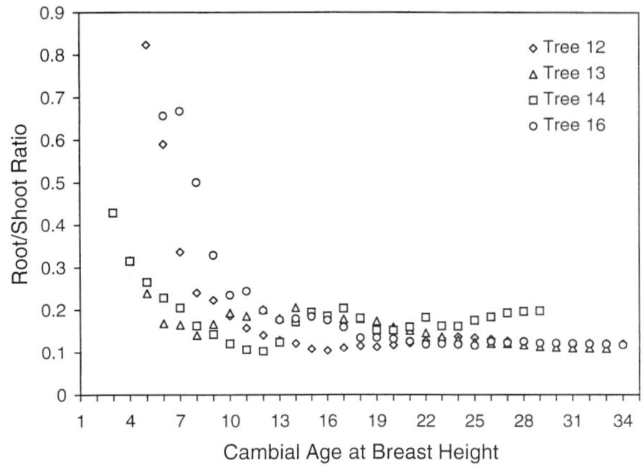

Figure 9. Root/shoot ratio of biomass depending on the cambial age determined at breast height for four 30- to 34-year-old oak trees.

stem-root base was reported by Telewski (1993), who proposed an analysis of the radial increment at the root/shoot interface for determining the exact age of a tree.

Nicoll and Ray (1996) have shown different types of cross-sectional shape of roots in Sitka spruce and have explained these by the spatial location within the root system and by adaptive growth in response to tree growth and wind movement. The adaptive growth in structural roots due to external factors can certainly be generalised for all angiosperms and conifers. The root system of oak was often described as a mixed model of a heart- and sinker-root system with horizontal superficial roots, a pronounced tap root, from which lateral and obliquy roots descend, and sinker roots (Drexhage et al., 1999; Köstler et al., 1968; Lucot and Bruckert, 1992), so that similar root shapes as described for spruce were to be expected. The shapes changed with tree growth as a result of mechanical stress (Mattheck and Breloer, 1992; Nicoll and Ray, 1996). The asymmetric root growth had consequences on estimating the annual growth layer increment and made all further calculations more difficult.

Characteristic years and ring width

The trees investigated were young and the time-series were very short. Thus, the relationships between event years and abrupt growth changes were analysed qualitatively for characteristic climate observations. In our study, branches, stem and roots respond to the climate variability, although the ring widths were also variable within and between trees. Pointer years were useful for the cross-dating the ring widths of roots with the stem-base. Anyway, the number of rings and the ring widths in roots have to be corrected due to the technical problem of correct registration.

Growth reductions were detected in dry years and in years with heavy frost periods in January and February. The same pointer years of 1990, 1992, 1994, that were detected in all components of the oak trees, strongly influenced the annual growth in branches of Norway spruce trees (Deleuze et al., 1998) and the radial increment in stems of 40-year-old sessile oak (Barbaroux, personal communication) both from stands near our investigation plot. In contrast to our study, Krause and Eckstein (1993) used statistical methods to analyse the nature of the influence of climate on root growth. They found a very weak response to the climate variability in 125-year-old oak trees from a stand in northern Germany.

Root and stem increment

Allometric relationships which include coarse roots are rare. Furthermore, there is little information about the timing of structural root response to release and on the distribution of biomass to roots, trunks and branches (Krause and Morin, 1995; Urban et al., 1994). The models for estimating root biomass increment are usually based on relationships between tree diameter and coarse root biomass (Bartelink, 1998). In our study, we present some empirical data based on regressions of root CSA and biomass and on calculations of annual increment in the stem and the coarse roots.

As already reported for other species by Kozlowski (1971), the cambial growth varied among individual trees of the same stand, and in different parts of the root system. Due to mechanical constraints the cambial growth in roots is much more irregular than in stems or branches (Krause and Eckstein, 1993). In our study, the biomass in the stem increased gradually during three consecutive periods, beginning with an initial slow increase followed by a more rapid increment that decreased after 30 years. The root system biomass tended to be much weaker than the stem. The current increments were more variable in the roots. The long-term relationship between stem and root development was as to be expected, and started with a high proportion of root biomass in the initial young stage and decreased rapidly down to 0.1 with increasing cambial age. A similar tendency was reported for 32- to 36-year-old red pine (Fayle, 1975). The current annual volume increment in red pine roots as a percentage of that of stem and roots reached a minimum at the age of 26 to 30 years. For sessile oak, several changes in the root/shoot ratio were detected between cambial age of 10 to 19 years due to the variability in root growth increment. The increment variability might be explained by external factors such as climate or thinning, but these factors could not really be discriminated in our study.

For Douglas-fir and beech a simulation of thinning effect revealed a slow decrease in distribution to the stem, to fine roots and foliage and an increased investment in branches and roots (Bartelink, 1998). The temporary increase in the root/shoot ratio might be explained by the preferential allocation of resources to the roots that stabilise the tree (Coutts, 1987; Stokes and Guitard, 1997). A thinning effect on root growth might be analysed by a control population without any treatments as shown recently for 120-year-old white spruce (Urban et al., 1994).

Our calculations must be handled with caution, because none of the growth layers was exactly round, the central cylinder was eccentric and the root shapes were irregular. Subtle approximation procedures based on ratios of radii were carried out to calculate the annual biomass increment in the roots. These calculations will be scrutinised by measuring ring widths on root disks in several directions to analyse the running form of growth layers.

For young 1 to 1.4 m high oaks from a natural regeneration, the stem biomass ranged from 30 to 40 g and the root biomass ranged from 12 to 32 g (Drexhage, unpublished). Compared to these values our calculations slightly over-estimated the stem biomass, whereas the estimated root biomass were in the same range from 14 to 30 g.

Conclusion

In principle, ring width measurements in roots are possible and feasible for increment studies by combining several preparation techniques. For obtaining clear X-ray pictures, the saw technique has to be improved. The root data could be corrected by comparison of X-ray pictures with tree rings measured in the stem base. Annual increment and biomass partitioning are variable and differed between individuals, but were broadly similar in direction.

We suggest that dendrochronology techniques will help to increase our understanding of root system development and of the relative gain and loss in thickening of structural roots compared to the stem which might be changed with time in periods of physical stresses such as drought and frost. Furthermore, these techniques can be applied to analyse the magnitude and timing of response of branches, trunks and structural roots to release and to calculate the biomass partitioning following thinning.

Acknowledgements

We are grateful to the following people for their help: L Wehrlen, M Ravart, F Bernier and the BECKER company who assisted with the field work. The forester Mr Lefort from the forest office Champenoux put the research site at our disposal. We are deeply indebted to M Loubère for the idea using the X-ray method, C Collet for providing helpful comments on the manuscript, and A Nys for correcting the English. We are indebted as well to the referees for a number of valuable suggestions that helped us to improve the initial manuscript. The work was funded by the European Commission (FAIR-BM-96-2554).

References

Badeau V, Becker M, Bert D, Dupouey J L, Lebourgeois F and Picard, J F 1996 Long-term growth trends of trees: ten years of dendrochronological studies in France. *In* Growth trends in European Forests. Eds H Spiecker, K Mielikäinen, M. Köhl and J P Skovsgaard. pp 167–181. European Forest Institute Research Report No. 5. Springer, Berlin.

Bartelink H H 1998 A model of dry matter partitioning in trees. Tree Physiol. 18, 91–101.

Bary-Lenger A and Nebout J P 1993 Les chênes pédonculé et sessile en France et en Belgique. Editions du Perron, Alleur-Liège, Belgium. 604 p.

Becker M, Bert G D, Bouchon J, Dupouey J L, Picard J F and Ulrich E 1995 Long-term changes in forest productivity in northeastern France: the dendroecological approach. *In* Forest decline and atmospheric deposition effects in the French mountains. Eds G Landmann and G Bonneau. pp 143–156. Springer, Berlin, Germany.

Bedeneau M and Pagès L 1984 Etude des cernes d'accroissement ligneux du système racinaire d'arbres traités en taillis. Ann. Sci. For. 41, 59–68.

Biging G S and Wensel L C 1988 The effect of eccentricity on the estimation of the basal area and basal area increment of coniferous trees. For. Sci. 34, 621–633.

Colin F, Danjon F and Wehrlen L 1995 Etudes racinaires au sein du programme 'Croissance' de l'INRA (*Quercus petraea* et *Pinus pinaster*). Rev. For. Fr. 157, 165–172.

Coutts M P 1987 Developmental processes in tree root systems. Can. J. For. Res. 17, 761–767.

Cutler D F 1976 Variation in root wood anatomy. *In* Wood Structure in Biological and Technological Research. Eds P Baas, A J

Bolton, D M Catling. pp 143–156. Leiden Botanical series No. 3. Leiden University Press, The Netherlands.

Deleuze C, Loubère M, Pouderoux S and Rittié D 1998 Premiers résultats des liaisons branches-tiges – base de houppier fonctionnel – données du clinal d'épicéas d'Amance 97 – effet de la densité. Unpublished Report ENGREF/INRA Nancy, France. 36 p.

Drexhage M, Chauviere M, Colin F and Nielsen C N N 1999. Development of structural root architecture and allometry of *Quercus petraea*. Can. J. For. Res. 29, 600–608.

Fayle D C F 1975 Distribution of radial growth during the development of red pine root systems. Can. J. For. Res. 5, 608–625.

Gasson P 1985 Automatic measurement of vessel lumen area and diameter with particular reference to pedunculate oak and common beech. IAWA Bulletin 6, 219–237.

Gasson P 1987 Some implications of anatomical variations in the wood of pedunculate oak (*Quercus robur* L.), including comparisons with common beech (*Fagus sylvatica* L.). IAWA Bulletin 8, 149–166.

Hatsch E 1997 Répartition de l'aubier et acquisition de la forme de la tige chez le chêne sessile (*Quercus petraea* (Matt) Liebl): analyse, modélisation et relation avec le développement du houppier. PhD thesis ENGREF/INRA Nancy, France. 183 p.

Köstler J N, Brückner E and Bibelriether E 1968 Die Wurzeln der Waldbäume. Verlag Paul Parey, Hamburg, Germany. 284 p.

Kozlowski T T 1971 Growth and development of trees. Volume II. Cambial growth, root growth, and reproductive growth. Academic Press Inc., New York, USA. 514 p.

Krause C and Eckstein D 1993 Dendrochronology of roots. Dendrochronologia 11, 9–23.

Krause C and Morin H 1995 Changes in radial increment in stems and roots of balsam fir (*Abies balsamea* (L.) Mill.) after defoliation by spruce budworm. The Forestry Chronicle 71, 747–754.

Landmann G, Becker M, Delatour C, Dreyer E and Dupouey J L 1993 Oak dieback in France: historical and recent records, possible causes, current investigations. *In* Zustand und Gefährdung der Laubwälder. Rundgespräche der Kommission für Ökologie, Vol. 5. pp 97–114. Verlag Dr. Friedrich Pfeil, München, Germany.

Lebourgeois F 1997 RENECOFOR – Etude dendrochronologique des 102 peuplements. du réseau. Ed. Office National des Forêts, Département des Recherches Techniques, Fontainebleau, France. 307 p.

Littell R C, Milliken G A, Stroup W W and Wolfinger R D 1996 SAS ®System for Mixed Models, SAS Institutue Inc., Cary, USA. 633 p.

Lucot E and Bruckert S 1992 Organisation du système racinaire du chêne pédonculé (*Quercus robur*) développé en conditions édaphiques non contraignantes (sol brun lessivé colluvial). Ann. Sci. For. 49, 465–479.

Mattheck C and Breloer H 1992 Der Wurzelquerschnitt als Protokoll der Lastgeschichte. Allgemeine Forst-und Jagd-Zeitung 163, 142–145.

Nicoll B C and Ray D 1996 Adaptive growth of tree root systems in response to wind and site conditions. Tree Physiol. 16, 891–898.

Polge H and Nicholls JWP 1972 Quantitative radiography and the densitometric analysis of wood. Wood Sci. 5, 51–59.

Schweingruber F H 1988 Tree rings: basics and applications of dendrochronology. Kluwer Academic Publishers, Dordrecht, The Netherlands. 276 p.

Schweingruber F H 1996 Tree rings and environment. Dendroecology. Swiss Federal Institute for Forest, Snow and Landscape Research, Birmensdorf. Paul Haupt Publishers, Berne, Suisse. 609 p.

Stokes A and Guitard D 1997 Tree root response to mechanical stress. *In* The Biology of Root Formation and Development. Eds A Altman and Y Waisel. pp 227–236. Basic Life Sciences, Vol. 65. Plenum Publishing Corporation, New York, USA.

Telewski F W 1993 Determining the germination date of woody plants: a proposed method for locating the root/shoot interface. Tree-Ring Bulletin 53, 13–16.

Urban S T, Lieffers V J and MacDonald S E 1994 Release in radial growth in the trunk and structural roots of white spruce as measured by dendrochronology. Can. J. For. Res. 24, 1550–1556.

White A D 1993 Relationships between foliar number and cross sectional areas of sapwood and annual rings in red oak (*Quercus rubra*) crowns. Can. J. For. Res. 23, 1245–1251.

Woodcock D W 1989 Climate sensitivity of wood-anatomical features in a ring-porous oak (*Quercus macrocarpa*). Can. J. For. Res. 19, 639–644.

Zhang S Y, Nepveu G and Owoundi R 1994 Intratree and intertree variation in selected wood quality characteristics of European oak (*Quercus petraea* and *Quercus robur*). Can. J. For. Res. 24, 1818–1823.

A. Stokes (ed.), The Supporting Roots of Trees and Woody Plants: Form, Function and Physiology, 131–135.

Tree root system enhancement with paclobutrazol

Gary W. Watson
The Morton Arboretum, Lisle, IL 60532, USA

Key words: paclobutrazol, tree growth regulator, root growth, transplanting

Abstract

Paclobutrazol (PBZ) was applied as a basal soil injection or drench at 0.8 g a.i. per cm trunk diameter. PBZ treatment was effective in stimulating fine root development of established healthy and chlorotic pin oaks (*Quercus palustris*) and declining white oaks (*Quercus alba*). PBZ treatment may be effective in stabilizing declining trees with insufficient fine root development. Elongation and dry weight of first year regenerated roots of newly planted black maple trees (*Acer nigrum*) were increased by PBZ treatment. This could result in faster establishment of transplanted trees.

Introduction

Poor quality soils and small planting spaces on urban sites often limit root growth. Root development of landscape trees can be inadequate to support the crown (Watson, 1991), especially when root space is more constricted than crown space, and following root loss. An underdeveloped root system can be responsible for low vigor, stress, dieback and decline. Root system enhancement and crown size control are two approaches to restoring the balance between the crown and root system.

The growth regulator paclobutrazol (PBZ), a gibberellin biosynthesis inhibitor, has been shown to reduce the shoot growth of many species (Davis et al., 1985), and is commonly used to control the regrowth of trees under utility lines. This growth regulator has promise in other landscape situations, wherever control of plant size is needed. In some situations, PBZ is also known to increase certain aspects of root growth (Ashokan et al., 1995; Atkinson and Crisp, 1983; Bausher and Yelenosky, 1989; Davis et al., 1985; Wang and Faust, 1986). As a result of reduced shoot growth and/or root growth stimulation, the effect of PBZ can be to increase root/shoot ratio (Atkinson and Crisp, 1983; Marquard, 1985; Ruter, 1994; Swietlik and Miller, 1983).

When applied as a basal soil injection or drench, only a few roots at the base of the tree are in direct contact with the PBZ. PBZ is absorbed by roots and translocated in the xylem only, towards the branch tips (Couture, 1982). The stimulation of root development must be an indirect effect, since the chemical is never in contact with roots more than 20–30 cm from the base of the tree. Photosynthesis is not reduced by PBZ treatments (Dejong and Doyle, 1984; Rieger and Scalabrelli, 1990; Warren et al., 1991) and increases in root growth may be due partially to increased carbohydrate supply to roots (Davis et al., 1985; Warren et al., 1991). Higher levels of ABA often associated with PBZ treatment have been shown to maintain growth of roots under drought stress (Kozlowski and Pallardy, 1997). An increase in fine root development implies a more favorable root/crown balance and less stress in treated trees. Reduced water use and improved water status has been reported after treatment with PBZ (Ruter and Martin, 1994; Swietlik and Miller, 1983; Wample and Culver, 1983; Warren et al., 1991). Apparent improvements in vigor, color and drought resistance may be related to a greater capacity of the root system to absorb moisture and mineral nutrients from the soil (Atkinson and Crisp, 1983).

Preventing decline of established trees

Cultural methods that improve the soil environment can be used successfully to improve the growth of landscape tree root systems. Surface mulching can increase fine root density up to 10 times just below

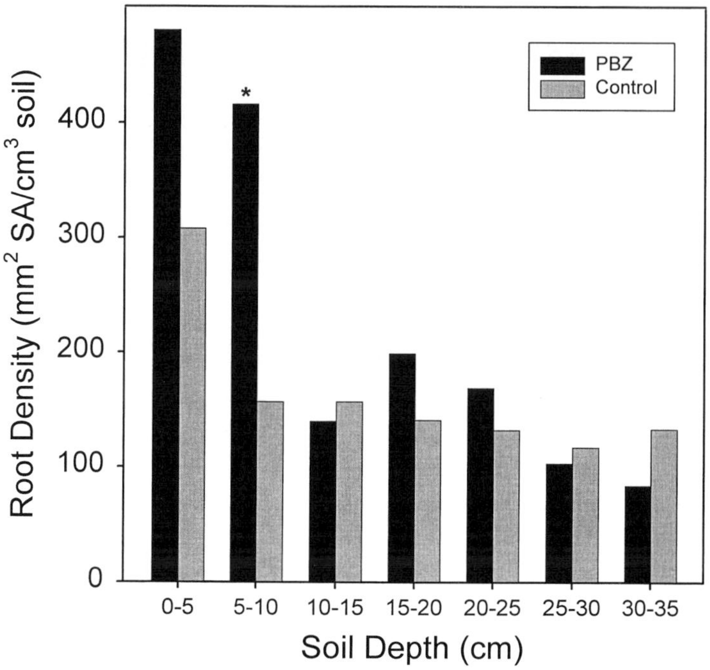

Figure 1. Fine root development of pin oaks after PBZ basal soil injection treatment (PBZ). Roots were sampled 1 m from soil injection site. The greatest root density increases were in the 10 cm thick topsoil layer. An * indicates significant difference from the control

the soil surface (Green and Watson, 1989) and the humus from the decomposing mulch can provide effective additional root space volume (Watson, 1988). This root system enhancement can lead to improved crown appearance and growth. (Green and Watson, 1989; Himelick and Watson, 1990). When such cultural methods are impractical, growth regulators such as paclobutrazol may be an effective alternative.

Two rows of pin oak (*Quercus palustris*) trees, between 21 and 27 cm dbh were selected from a large, evenly spaced block of trees. There were seven trees in each row, in excellent condition and with full crowns. One row of trees was treated with paclobutrazol (0.8 g active ingredient per cm trunk caliper) as a basal soil injection. The trees in the other row served as untreated controls. Three years later, fine root development was measured using root 7.5 cm diameter density cores. Soil was washed from the roots, oak roots were separated from other roots and debris by hand, and root surface area was measured with a digital image analysis system.

PBZ increased fine root density of pin oaks by 92 percent in the upper 10 cm of soil (Fig. 1). At approximately 10 cm, there was an abrupt change to more dense and clayey soil horizon below which increases in root density were inconsistent and not significant. This implies that poor quality soils may have to be improved before PBZ can be effective in improving root development. Combining mulching with paclobutrazol treatment may be even more effective. At this 0.8 g a.i./cm caliper rate on trees with full crowns, the visual difference in appearance between treated and control pin oaks was negligible, and limited to a slightly greener leaf color without reduction in leaf size or shoot length.

Because of an observed relationship between poor root development and foliar chlorosis of pin oaks in urban landscapes, 10 chlorotic pin oaks were treated with paclobutrazol at a similar rate in hopes that increased root development would improve color. After 3 years, root density of chlorotic trees increased 3-fold compared to untreated controls. Color of treated trees also improved while color of untreated trees continued to get more yellow (Fig. 2). The differences were not significant.

Mature white oaks (*Quercus alba*) were treated with PBZ (0.8 g a.i. per cm trunk caliper). The trees were 50–70 cm dbh with spreading open-grown crowns. All were in a state of decline as judged by evidence of dieback, tufted growth, and chlorosis in

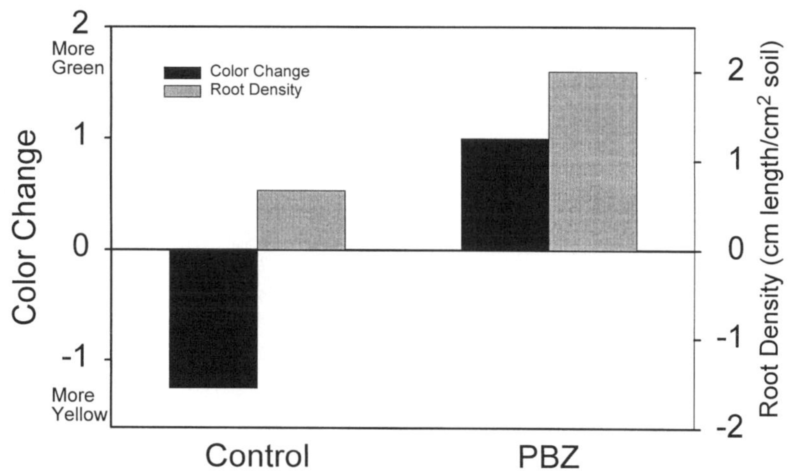

Figure 2. Fine root density of chlorotic pin oaks increased 3-fold by paclolbutrazol treatment compared to untreated controls. Color of treated trees also improved while color of untreated trees continued to get more yellow. The differences were not significant.

Figure 3. White oak treated with a PBZ basal soil injection. Photos were taken prior to treatment (left), 7 years after treatment (right).

the crown. There were four treated trees and four untreated controls.

Analysis of soil core samples revealed that PBZ treatment increased root density of the white oaks from 280 to 460 mm^2 root surface area/cm^3 soil. The first sign of crown improvement (slightly greener color) of the white oaks was observed the second year after treatment. At the time of the root sampling, three full seasons after treatment, the crowns of the declining white oaks were showing signs of improvement. Color and vigor was improving. Unfortunately, at that time, the majority of the trees included in the study were cut down for reasons unrelated to the study, and too few remained for meaningful crown growth data to be collected once improvements started to become obvious in the remaining trees. The only surviving declining tree that was treated with PBZ continued to show crown improvement through 1996, 7 years after a single treatment (Fig. 3). New growth was deep green and vigorous and the leaves were not noticeably smaller. The leaves were much less scorched by midsummer compared to leaves of a nearby untreated tree. Such long lasting effects of a single PBZ treatment has been reported previously (Burch et al., 1996).

Growth regulator treatment is not a quick cure for mature declining trees. It will take several years for any visible crown response to develop. Rapidly declining trees could die before the treatment would have a chance to stabilize them. PBZ treatments may prove to be more useful on trees in the very early stages of

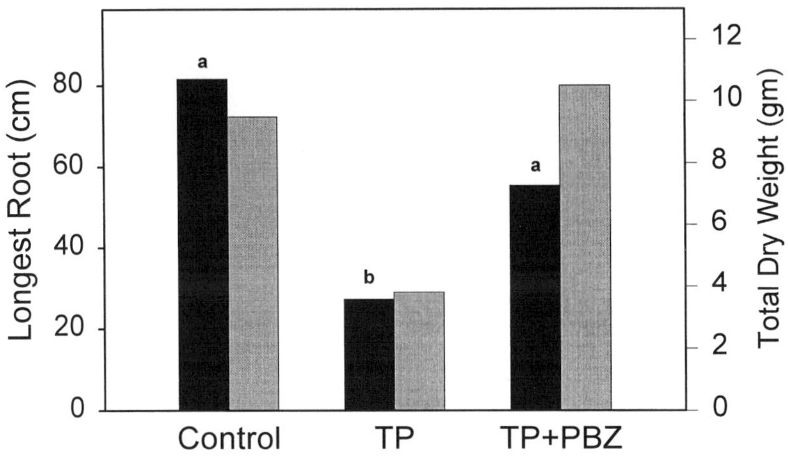

Figure 4. The length of regenerated maple roots was increased significantly by PBZ treatment after transplanting. This could speed the regrowth of the natural root spread. Dry weight of regenerated maple roots was increased 275% by PBZ treatment after transplanting, though not significantly.

decline, or as preventive maintenance on mature, but still healthy trees.

Faster establishment of transplanted trees

Use of growth regulators to increase root development after transplanting is not new. Auxins have been shown to increase the number of regenerated roots on bare root and ball-and-burlap stock (Lumis, 1982; Magley and Struve, 1983; Prager and Lumis, 1983; Struve and Moser, 1984; Watson, 1987). While auxins promote root initiation, they can inhibit root elongation (Kramer and Kozlowski, 1960). More regenerated roots can result in an excessively high root density in a very small soil volume. The moisture held in this small volume of soil is constant, so the additional roots may not lead to faster establishment and better growth (Watson, 1987).

In situations such as transplanting, where roots have been severed and must grow long distances to reestablish the original root spread and occupy the original soil volume, rapid elongation rate is very important, perhaps more important than increased root initiation. Root density data (Atkinson and Crisp, 1983) and visual observation of roots in the studies with mature trees treated with paclobutrazol suggested that the fine roots were more profusely branched, but its effect on root elongation has not been reported.

In a study designed to investigate the effects of paclobutrazol on root elongation rate, ten 7-cm caliper black maples (*Acer nigrum*) were transplanted in late fall of 1996 and five non-transplanted were selected as controls. Half of the transplanted trees were treated with PBZ as a basal soil drench at 0.8 g a.i. per inch caliper prior to budbreak the following spring.

At the end of the first season after transplanting, root regeneration was measured by harvesting all of the roots growing from a 30-cm wide, 20-cm deep section of the root ball perimeter. Roots of non-transplanted trees had also been severed in this same area to observe root regrowth. The longest root from each tree was measured, and all roots were oven dried and weighed after harvesting.

PBZ significantly increased the length of regenerated roots of transplanted trees. The longest roots of transplanted trees treated with PBZ averaged 55 cm. Without PBZ the longest root of transplanted trees was 27 cm and significantly shorter than both PBZ treated and non-transplanted trees (Fig. 4). PBZ increased root dry weight 275% in the transplanted trees (Fig. 4).

This data shows that both root elongation and branching appear to be increased by PBZ treatment. It usually takes several years for a transplanted tree to regrow the normal spreading root system that was lost during harvest. Increased root elongation would result in faster replacement of the normal spread of the root system. Increased branching would mean faster proliferation of fine roots into the enlarged soil volume occupied by the root system of PBZ treated trees. Based on the first year data from these maples, it is possible that transplanted trees could become established in half the normal time after PBZ treatment.

Paclobutrazol treatments can increase root growth and show promise for use as a long-term treatment for trees under stress resulting from underdeveloped root systems. Trees showing early signs of stress an decline may benefit the most. The response is probably not rapid enough in large trees to be useful as a treatment for rapidly declining trees. Trees that have restricted root environments may benefit, especially when combined with mulching or other root environment improving procedures. Transplanted trees or trees that have lost large portions of their root system from construction activities may regenerate the lost roots faster, and be stressed for a shorter period if treated with paclobutrazol. Treatment as part of a plant health care program may also prove beneficial in maintaining a high root/shoot ratio and high vigor. Much more research is needed to understand the potential benefits of paclobutrazol treatments on landscape trees.

References

Ashokan PK, Chaney WR and Premachandra GS 1995 Soil applied paclobutrazol affects leaf water relations and growth of American elm (*Ulmus americana* L.) seedlings. Plant Growth Reg. Soc. Am. Quar. 23, 1–12.

Atkinson D and Crisp CM 1983 The effect of some plant growth regulators and herbicides on root system morphology and activity. Acta Hortic. 136, 21–28.

Bausher MC and Yelenosky G 1986 Sensitivity of potted citrus plants to top sprays and soil applications of paclobutrazol. HortSci. 21, 141–143.

Burch PL, Wells RH and Kline WN 1996 Red maple and silver maple growth evaluated 10 years after application of paclobutrazol tree growth regulator. J. Arboricult. 22, 61–66.

Couture R 1982 A new experimental growth regulator from ICI. Proc. Growth Regulat. Soc. Am. 9, 59.

Davis TD, Sankhla N, Walser RH and Upadhyaya A 1985 Promotion of adventitious root formation on cuttings by paclobutrazol. HortSci. 20, 883–884.

Davis TD, Steffens GL and Sankhla N 1988 Triazole plant growth regulators. Hort. Rev. 10, 63–105.

DeJong TM and Doyle JF 1984 Leaf gas exchange and growth responses of mature 'Fantasia' nectarine trees to paclobutrazol. J. Am. Soc. Hort. Sci. 109, 878–882.

Green TL and Watson GW 1989 Effects of turfgrass and mulch on the establishment and growth of bare-root sugar maples. J. Arboricult. 15, 268–272.

Himelick EB and Watson GW 1990 Reduction of oak chlorosis with wood chip mulch treatments. J. Arboricult. 16, 275–278.

Kozlowski TT and Pallardy SG 1997 Growth Control in Woody Plants. Academic Press, New York, 641 pp.

Kramer PJ and Kozlowski TT 1960 Physiology of Trees/Woody Plants. Mcgraw-Hill, New York.

Lumis GP 1982 Stimulating root regeneration of landscape-size red oak with auxin root sprays. J. Arboricult. 8, 325–326.

Magley SB and Struve DK 1983 Effects of three transplant methods on survival, growth and root regeneration of caliper pin oaks. J. Environ. Hortic. 1, 59–62.

Marquard RD 1985 Chemical growth regulation of pecan seedlings. HortSci. 20, 919–921.

Prager CM and Lumis GP 1983 IBA and some IBA-synergists increases of root regeneration of landscape-sized and seedling trees. J. Arboricult. 9, 117–123.

Rieger M and Scalabrelli G 1990 Paclobutrazol, root growth, hydraulic conductivity and nutrient uptake of 'Nemaguard' peach. HortSci. 25, 95–98.

Ruter JM 1994 Growth and landscape establishment of Pyracantha and Juniperus after application of paclobutrazol. HortSci. 29, 1318–1320.

Ruter JM and Martin CA 1994 Effects of contrasting climate and paclobutrazol on the growth and water use of two container-grown landscape plants. J. Environ. Hort. 12, 27–32.

Struve DK and Mozer BC 1984 Auxin effects on root regeneration of scarlet oak seedlings. J. Am. Soc. Hortic. Sci. 109, 91–95.

Swietlik D and Miller SS 1983 The effect of paclobutrazol on growth and response to water stress of apple seedlings. J. Am. Soc. Hortic. Sci. 108, 1076–1080.

Wample RL and Culver EB 1983 The influence of paclobutrazol, a new growth regulator, on sunflowers. J. Am. Soc. Hortic. Sci. 108, 122–125.

Wang SY and Faust M 1986 Effect of growth retardants on root formation and polyamine content in apple seedlings. J. Am. Soc. Hortic. Sci. 111, 912–917.

Warren SL, Blazich FA and Thetford M 1991 Whole-plant response of selected woody landscape species to uniconazole. J. Environ. Hort. 9, 163–167.

Watson GW 1987 Are auxins practical for B&B trees? Am. Nurseryman 166, 183–184.

Watson GW 1991 Attaining root:crown balance in landscape trees. J. Arboricult. 17, 211–216.

Watson GW 1996 Tree root system enhancement with paclobutrazol. J. Arboricult. 22, 211–217.

A. Stokes (ed.), The Supporting Roots of Trees and Woody Plants: Form, Function and Physiology, 137–143.

Clonal strategies of a woody weed: *Melia azedarach*

G.M. Tourn[1,*], M.F. Menvielle[2], A.L. Scopel[3] and B. Pidal[1]
[1]*Cátedra de Botánica Agrícola, Facultad de Agronomía, UBA. Avda. San Martín 4453, 1417-BA, Argentina;* [2]*Administración de Parques Nacionales, Santa Fe 690, 1059-BA, Argentina and* [3]*IFEVA-Departamento de Ecología, Facultad de Agronomía, UBA. Avda. San Martín 4453, 1417-BA, Argentina*

Key words: clonal growth, disturbance, invasions, *Melia azedarach*, root buds, root sprouts

Abstract

Melia azedarach L., a weedy tree that typically reproduces by seeds, may exhibit clonal growth following disturbance (e.g. fire, herbivory, animal injury). A dynamic and holistic study (*sensu* Hallé et al., 1978) was carried out in order to determine the architecture and the vegetative growth strategies, in plants collected from undisturbed and disturbed areas in the 'El Palmar' National Park. The architectural and morphological observations can be summarized as follows: (a) Tree architecture is determined by a monopodial trunk that produces tiers of orthotropic monopodial branches. (b) The root system consists of a tap root with woody primary lateral roots that exhibit plagiotropic growth and are typically found within the first few centimeters (ca. 4 cm) of the soil profile. (c) Root buds were produced in all (fire treated and control) plagiotropic root cuts when incubated under controlled conditions. Root suckers developed from the differentiation of parenchymatous cells produced by meristematic activity in the cambial zone. The ecological implications of vegetative reproduction on the invasive capacity of this species are discussed.

Introduction

Chinaberry (*Melia azedarach* L.) is a fleshy-fruited tree from Asia (Mabberley, 1984; Miller, 1990; Pennington, 1981) that has been introduced to Argentina to be used as a shade-tree, timber or locust barrier (Cozzo, 1972; Ragonese and García, 1980). In the El Palmar National Park (Entre Ríos, Argentina), this species has become an agressive invader, and is currently present in ca. 60% of the park with different degrees of invasion, from low- (isolated individuals) to severe-forming dense thickets (Movia and Menvielle, 1990). Results from ongoing research have shown that these trees are able to reproduce vegetatively from both stumps and roots when the trees undergo physical damage (i.e. herbivory, fire) (Menvielle and Scopel, 1999). Previous studies on this species have focused on taxonomic features (reviewed by Mabberley, 1984), while no description of those plant traits that may be relevant to understand its invasive capacity exists.

The ability to recognize a potential invader depends on the identification of traits that are positively correlated with invasion success. Lists of attributes of weedy species have been compiled by many authors (Baker, 1974; Roy, 1990; Rejmánek, 1995), but it is only in the present decade that several attempts have been made to develop predictive models based on both species characterisitcs and ecological information. The juvenile period, seed/fruit production, seed dispersal, seed bank type and vegetative reproduction are plant traits shown to have predictive value (Panetta, 1993; Reichard, 1997; Richardson et al., 1990). Once an individual has the opportunity to invade a specific ecosystem and becomes established, the possibility of enduring in that system is higher if it exhibits some sort of vegetative reproduction (clonal growth), in which perenniality will result from the dispersal of ramets through time, and mobility from their dispersal through space (Mogie and Hutchings, 1990).

Abrahamson (1980) and Jeník (1994) each provide a comprehensive description of the different organs

* FAX No: 514-8730/8739. E-mail: gmtourn@mail.retina.ar
Plant and Soil is the original source of publication of this article. It is recommended that this article is cited as: *Plant and Soil* **217:** 111–117, 1999.

involved in clonal growth and reiteration of woody plants (Jeník, 1994); the most common types include: stolons, epicormic buds, lignotubers (mainly shrubs), stems and roots. A limited number of species produce long horizontal roots that give rise to shoots (Jeník, 1994). Poplars (*Populus* spp) – Reim, 1929; Schier, 1973 – *Prunus spinosa* (Rauh, 1937), *Fagus grandiflora* (Jones and Raynal, 1988), *Quercus virginiana* (Tomlinson, 1980), *Psidium guianense* (Hallé et al., 1978) and *Liquidambar styraciflua* (Kormanik and Brown, 1967), share the ability to produce shoots from adventitious root buds. In natural conditions, vigorous sprouting, specially from stems and root buds, typically occur in these species when subjected to some sort of injury like that caused by herbivores. Physical agents, like fire, that remove top growth may also trigger extensive and rapid vegetative growth, allowing total canopy coverage to recover to the predisturbance level after a few years. Despite the wide range of genera sharing this feature, detailed studies on the anatomy and structure of these bud-sprouting roots and the mechanisms activating their development are scant (Jones and Raynal, 1988; Rinne et al., 1987).

In this work, we carried out a dynamic and holistic study (*sensu* Hallé et al., 1978), from the seedling to the adult stage, in order to determine tree morphology and architecture in *M. azedarach*, and the sprouting capacity of roots growing in disturbed and undisturbed conditions was tested in the laboratory. The ecological implications of this type of clonal growth on the invasive capacity of the species are discussed.

Materials and methods

Architectural and morphological observations were made on plants growing at the 'El Palmar' National Park (31°55′ S, 58°16′ W), during 1997–1998 (one growing season) by comparing the successive stages of growth (from juvenile to adult) identifying all categories of axes present (at least ten individuals per stage). More detailed morphological observations were also made on plants obtained from seeds and grown in the field at the Lucien Hauman Botanical Garden, Faculty of Agronomy, Buenos Aires City (34°35′ S, 58°29′ W).

Architectural analysis was based on observations of the following morphological characters (Hallé et al., 1978): type of growth (rhythmic or continuous); branching pattern (presence or absence of vegetative branching, rhythmic, continuous or diffuse branching); morphological differentiation of axes (orthotropy or plagiotropy) and the position of flowering branches (terminal or lateral). To determine the morphology of the root system, observations were made on individual plants, from seedlings to 12-yr old trees (n=10).

The sprouting capacity of plagiotropic roots belonging to fire-treated and control trees (ca. 12 yr old) was tested under controlled laboratory conditions. In January 1996, 1 month after an experimental fire, root sections were severed from plagiotropic roots located between 2 and 4 cm of depth. Several root samples were taken from at least five trees per treatment. Roots were classified in three categories according to their diameter: small, medium and large; wich corresponded to 6 (±0.7), 13 (±0.9) and 23.5 (±1.4) mm respectively (roots between one to five yr old). Three to four 10-cm long root sections per diameter class and treatment were placed on moistened filter paper in plastic boxes, and incubated at alternating temperature (20–30 °C) for 30–45 d (n=3–6). Water was added as needed and the plastic boxes were wrapped with clear polyethylene to prevent desiccation. The number of roots producing sprouts were counted at regular intervals (7 d) throughout the incubation period.

Results

Architecture

Our observations show that tree architecture (Figure 1 A–E) is determined by a monopodial trunk (Figure 1

Figure 1. Morphological and architectural aspects of *Melia azedarach* adult trees during autumn and winter. (A) General view of the crown architecture showing the acrotonic vegetative branches and basitonic reproductive branches; (B) A coppice of adult trees (7–8 years of growth) originated from a root bud, showing the sympodial branching pattern (Arrows indicate several trunks); (C) Diagrammatic representation and photograph of 2 shoot units. The shoot unit of the precedent year consists of a shoot with determinate growth arising from the preformed apical bud (see the bud scale scars) it includes the peduncle scars at the basal nodes, a proleptic branch near the dead apex (∗) and the dormant buds. The current year shoot unit shows the basitonic fruit clusters and the acrotonic dormant buds; (D) Distal part of the axis showing leaves with axillary buds and the apical bud before winter; (E) An adult tree developing from seedling, note just one orthotropic sympodial trunk. Abbreviations: ab – axillary buds; apb – apical bud; db – dormant buds; fc – fruit clusters; l – leaves; ps – peduncle scars; su – shoot unit.

Figure 1.

Table 1. Sprouting capacity of plagiotropic root segments from fire-treated and control *M. azedarach* trees under laboratory conditions

	Root diameter		
	Small	Medium	Large
Control trees	1.2 (0.3)[a]	4.6 (1.3)	9.5 (1.2)
Fire-treated trees	2.0 (0.5)	3.2 (1.2)	10.0 (3.1)

[a]Mean number of sprouts/root section±SE (n=3–6).

E) which grows rhythmically and produces tiers of orthotropic monopodial branches resembling Rauh's model (Hallé et al., 1978). In certain cases, the distal part of the crown exhibits sympodial growth by substitution due to the death of apical meristems (Figure 1 A–B).

The architectural unit is represented by a trunk and the orthotropic branches with radial symmetry (leaves with spiral arrangement) bearing deciduous normal foliage leaves with a single bud in each leaf axil. The inflorescence is an axillary panicle – many flowered, ultimate branchlets often cymose, that develop from the basal nodes (preformed buds) of the current-year shoot unit (Figure 1 C). The proleptic branches (Figure 1 B-C) grow out from the distal nodes (neoformed buds) of the shoot unit (acrotony). Traumatic reiterations grow out from horizontal root buds as a response to root damage (e.g. fire, animal injury). Profuse coppicing (traumatic reiterations) is also a common feature in damaged trees or stumps from fallen trees; no vegetative regeneration has been observed in trees that had not suffered some degree of disturbance.

Root system

The root system consists of a tap root and lateral (first order, second order, etc.) roots. The longest first order lateral roots, exhibit plagiotropic growth and are found close to the soil surface (ca. 4 cm).

The incubation of excised root segments in a growth chamber led to the development of callus tissue, the initiation and differentiation of adventitious buds, and the growth of adventitious sprouts. Approximately 20 d after the start of the incubation period, the development of a compact yellowish mass of cells (callus) was observed at both ends of the severed root segments, in the lenticels and in the cracks present in the bark of fire-treated trees (Figure 2 A). At a later stage, 30–45 d, adventitious shoots began to differentiate from the callus cells produced by the activity of lateral meristems (Figure 2 B). While callus proliferation occured massively, shoots developed only at the proximal end of the root segments. All root segments produced adventitious shoots regardless of previous treatment. The mean number of sprouts produced was related to root diameter; the larger root sections (23.5 mm) produced 5–8 times more shoots than the smaller diameter class (6 mm) (Table 1). However, results from field experiments in which prescribed fires were applied in different times of the year consistently show that fire-treated trees produce significantly more adventitious shoots from roots than their untreated counterparts (4.5 vs. 1 shoots under the crown of the tree, for fire-treated and control trees, respectively) (Menvielle and Scopel, 1999).

Discussion

In an earlier description, Pennington (1981) classified *M. azedarach* as a monopodial tree. As several species of the Meliaceae family, which exhibit rhytmic growth and produce tiers of orthotropic monopodial branches, this species fits into Rauh's model. Regulation of the growth pattern depends on the apical meristem, when this meristems dies, it is replaced by the uppermost lateral meristems and the tree reverts to a sympodial growth habit (Hallé et al., 1978). This dying back of branch apices has been cited by Miller (1990) in a revision of the *Meliaceae* genera. This too, may be frequently observed in *M. azedarach* trees growing at the El Palmar National Park. The timing and intensity of frosts during late autumn and winter, which are common in this temperate region, may be responsible for this phenomenon.

The morphology and histological origins of newly-initiated sprouts in trees is a topic that has been poorly investigated (Jeník, 1994; Schier, 1973), and the factors involved in the induction of root sprouting have been studied only for a few native or commercially valuable species (Jones & Raynal, 1987, 1988; Maini and Horton, 1966). The origins of root shoots may be different in different species. From histological observations of released buds and the later formation of suckers in sweetgum *(L. styraciflua)*, Kormanik and Brown (1967) have shown that root suckers actually developed from buds that had formed during the early life of the root and remained suppressed. In the case of *M. azedarach*, no preformed root buds could be found in the material excised from either fire-treated or control plants. In turn, our growth chamber experi-

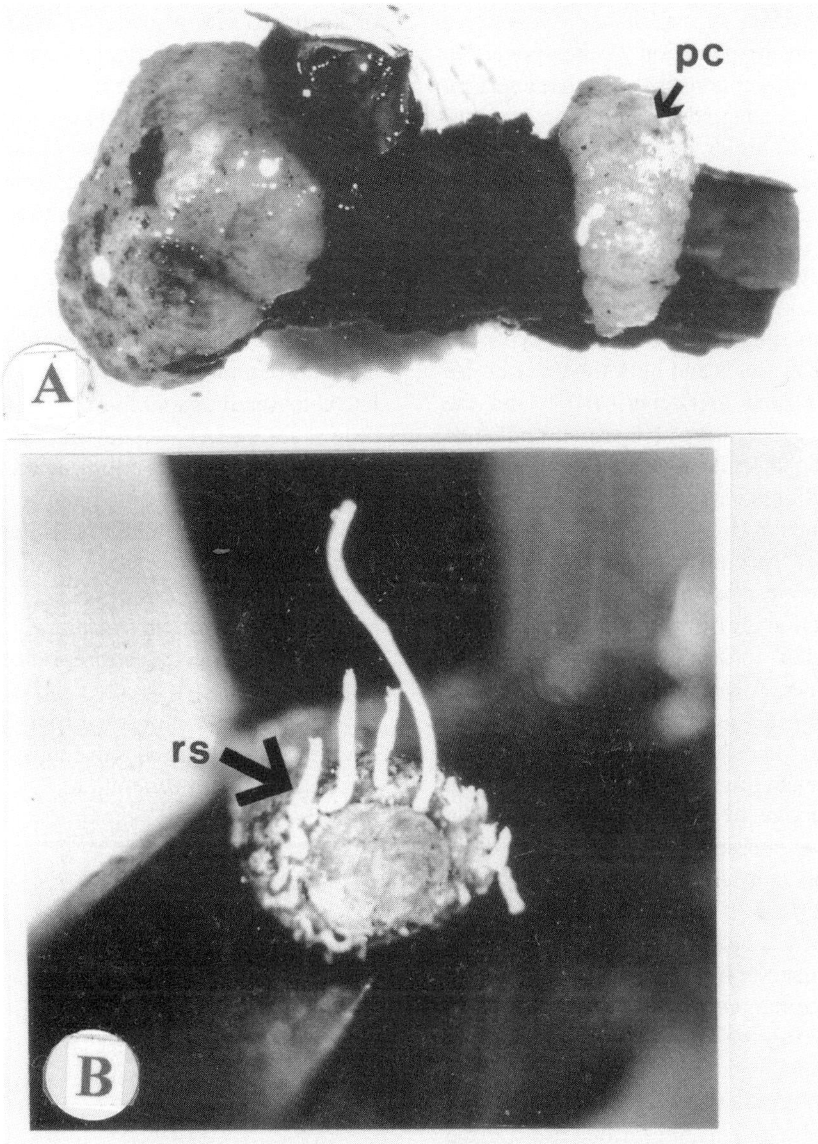

Figure 2. Two different stages of callus development. (A) Root segment of a fire-treated tree with parenchymatous cells produced by the lateral meristems at the transverse root surface (proximal end) and at the cracked bark; (B) Cross section of a root segment showing the parenchymatous cells produced by cambium activity and differentiation of root suckers. Abbreviations: pc – parenchymatous cells; rs – root suckers.

ments show that root severing was sufficient to induce the initiation of callus and differentiation of root buds *de novo*. These results are supported by field observations in which root suckers develop as a response to animal injury or a major disturbance such as a fire or tree felling. This same pattern of response has been observed in the case of aspens (Barnes, 1966; Maini and Horton, 1966) and american beeches (Jones and Raynal, 1986, 1987, 1988), two species in which clonality has been extensively studied. In an early study, Brown (1935) described the origin and ontogeny of root suckers arising from the phellogen of aspen roots. Removal of apical dominance, exposure to elevated temperatures and injury on superficial roots (Jones and Raynal, 1986) are factors that have been identified to stimulate root suckering under natural conditions.

Most woody species that initially colonize an area following disturbance by fire or logging do so from long-term seed reserves in the soil (Whelan, 1997).

However, once the plant is established, vegetative growth allows for the perpetuation of the species in the system. Sprouting is a conspicuous characteristic among woody species adapted to fire-prone environments (Menges and Kohfeldt, 1995). Greenleaf manzanita (*Arctostaphylos patula*), bigleaf maple (*Acer macrophyllum*) and tanbark (*Lithocarpus densiflora*) in western North America, sprout readily from root crowns or stumps if a disturbance eliminates aboveground organs (Radosevich et al., 1997). Mallee eucalyptus (*Eucalyptus* spp.) – Jacobs, 1955 – typically sprouts from epicormic buds and lignotubers after fire (Cunningham 1960 cited in Harper, 1977) and the Autralian *Eucalyptus tetrodonta* also produce adventitious shoots from roots (Jeník, 1994). In a study of the dynamics of vegetation after fire, Trabaud (1987) and Trabaud and Lepart (1980, 1981) found that around 80% of the recorded species at the study site could immediately regenerate after fire through sprouting. Vegetative reproduction is also wide spread among invasive woody species. In a comprehensive study in which 349 woody species were compared for its invasiveness, Reichard (1997) found that 44% of the woody invaders reproduced vegetatively, while only 25% of noninvasive species had this trait. Root suckering species are likely to become troublesome weeds as is the case of the invasive 'marabú' (*Dichrostachys glomerata*) of the human-induced savanna in Cuba (cited in Jeník, 1994).

Melia azedarach has behaved as an aggressive invader in the El Palmar National Park. From a few foci identified at the time the park was created in 1965, it has expanded to 60% of the area corresponding to the park. Several attempts have been made in the past to control the spread of this alien species, which involved the use of chemical treatments, prescribed fires or tree fellings. These treatments, far from halting its advance, induced prolific resprouting which has almost certainly contributed to the increase of the population not only in terms of area occupied but also in density, transforming a few individuals into a dense stand. Our study provides the first description of the initiation and development of adventitious buds and suckers from excised root segments of *M. azedarach*. The growth chamber experiments with excised roots show that the formation of a callus occurs whenever a root is damaged or severed. Thus, fire or animal-mediated injury, such as that caused by wild pigs or other feral animals, may be equally effective in triggering proliferation of callus tissue and formation of adventitious buds and shoots. The main difference between the two types of disturbance is a matter of scale: whereas animals have only localized effects, fires can affect hundreds of hectares.

Results from ongoing experiments indicate that the chances of a root sprout to survive into a new individual are much higher than those of a seedling (Menvielle and Scopel, 1999). Although seedling emergence during one growing season under the crown of a tree may be 40× the number of adventitious shoots (160 m^{-2} and 4 m^{-2} for seedlings and shoots, respectively), after 2 years, these numbers level off due to the differential survivorship of shoots (ca. 40%) relative to seedlings (0.5–3%). In the same period of time, root suckers accumulate more than 200 times more biomass than seedlings (300 g vs. <1 g, DW), and can effectively outcompete sub-canopy vegetation. Furthermore, within a 2-year time frame, root suckers can attain reproductive stage. The recognition of plagiotropic roots as key structures in shoot regeneration is critical to understand the process of invasion and in selecting appropriate methods of control. Experimental work in the field combining fire, tree felling and herbicides is currently under way to find out the best strategy to reduce adult populations while minimizing tree regeneration.

Acknowledgements

This work was financially supported by grants from Antorchas Foundation (#A-13532/1-85), FONCyT (PICT 00344), Consejo Nacional de Investigaciones Científicas y Tecnológicas (PID 226) and the University of Buenos Aires (UBA TG 05).

References

Abrahamson W G 1980 Demography and vegetative reproduction. *In* Demography and Evolution in Plant Populations. Ed. O T Solbrig. University of California Press, Berkeley, USA.

Baker H G 1974 The evolution of weeds. Ann. Rev. Ecol. Syst. 5, 1–24.

Barnes B V 1966 The clonal growth habit of American aspens. Ecol. 47, 439–447.

Brown A B 1935 Cambial activity, root habit and sucker development in two species of poplar. New Phytol. 34, 163–179.

Cozzo D 1972 Informaciones sobre el cultivo del Paraíso gigante (*Melia azedarach*) en Misiones, Argentina. Rev. For. Arg. III 4, 127–133.

Hallé F, Oldeman R A A and Tomlinson P B 1978 Tropical trees and forests. Springer, Berlin. 441 p.

Harper J L 1977 Population biology of Plants. Academic Press Inc., San Diego, CA, USA. 892 p.

Jacobs M R 1955 Growth habits of the eucalypts. Commonwelth Goverment Printer, Canberra: Commonwealth of Australia, Forest and Timber Bureau.

Jeník J 1994 Clonal growth in woody plants: a review. Folia Geobot. Phytotax. Praha 29, 291–306.

Jones R H and Raynal D J 1986. Spatial distribution and development of root sprouts in *Fagus grandifolia* (*Fagaceae*). Am. J. Bot. 73, 1723–1731.

Jones R H and Raynal D J 1987 Root sprouting in American beech (*Fagus grandifolia*): production, survival and the effect of parent tree vigor. Can. J. For. Res. 17, 539–544.

Jones R H and Raynal D J 1988 Root sprouting in American beech (*Fagus grandifolia*): effects of root injury, root exposure and season. For. Ecol. Manag. 25, 79–90.

Kormanik P P and Brown C L 1967 Root buds and the development of the root suckers in sweet gum. For. Sci. 13, 338–345.

Mabberley D J 1984 A monograph of *Melia* in Asia and the Pacific. The history of white cedar and persian lilac. Gard. Bull. Sing. 37, 49–64.

Maini J S and Horton K W 1966 Vegetative propagation of *Populus* spp. I. Influence of temperature on formation and initial growth of aspen suckers. Can. J. Bot. 44, 1183–1189.

Menges E S and Kohfeldt N 1995 Life history strategies of Florida scrub plants in relation to fire. Bull. Torr. Bot. Club 122, 282–297.

Menvielle M F and Scopel A L 1999 Life history traits of an alien tree invading a national park in temperate South America. 84th Annual Meeting, Ecological Society of America. Book of Abstracts, p 282

Miller N G 1990 The Genera of *Meliaceae* in the Southeastern United States. J. Arnol. Arb. 71, 453–486.

Mogie M and Hutchings M J 1990 Phylogeny, ontogeny and clonal growth in vascular plants. *In* Clonal Growth in Plants: Regulation and Function. Eds Van Groenendael and H de Kroon. pp 3–22. Academic Press, The Hague, The Netherlands.

Movia C and Menvielle F 1990 Mapa de vegetación del Parque Nacional El Palmar. APN, Bs.As., Argentina.

Panetta F D 1993 A system of assessing proposed plant introductions for weed potential. Pl. Protect. Quart. 8, 10–14.

Pennington T D 1981 A Monograph of Neotropical *Meliaceae*. *In* Flora Neotropica. pp 1–27. The New York Botanical Garden.

Pennington T D and Styles B T 1975 A generic Monograph of the *Meliaceae*. Blumea 22, 419–540.

Radosevich S, Holt J and Ghersa C 1997 Weed Ecology. Implications for Management. John Wiley & Sons Inc., New York. 589 p.

Ragonese A and García A 1980 Identidad botánica de los paraísos gigantes (*Melia azedarach* var. gigantea cv "Gotz" y cv "Garrasino"). INTA-Castelar, IDiA III, 110–113.

Rauh W 1937 Die Bildung von Hypokotyl- und Wurzelsprossen unh ihre Bedeutung für die Wuchsformen der Pflanzen. Nova Acta Leopoldiana, N.F. 4, 393–553.

Reichard S E 1997 Prevention of invasive plants Introductions on national and local levels. *In* Assessment and Management of Plant Invasions. Eds J O Luken and J W Thieret. pp 215–227. Springer, New York, USA.

Reim P 1929 Die Vermehrungsbiologie der Aspe auf Grundlage des im Estland und Finnland gesammelten Untershuchungsmaterial. Univ. Tartu, Tartu.

Rejmánek M 1995 What makes a species invasive? *In* Plant Invasions. Eds P Pysûk, K Prach, M Rejmánek, P M Wade. SPB Academic Publishing, The Hague, Netherlands.

Richardson D M, Cowling R M and LeMaitre D C 1990 Assesing the risk of invasive success in Pinus and Banksia in South African mountain fynbos. J. Veg. Sci. 1, 629–642.

Rinne P, Kauppi A and Fern A 1987 Induction of adventitious buds and sprouts on birch seedlings (*Betula pendula* Roth and *B. Pubescens* Ehrh.). Can. J. For. Res. 17, 545–555.

Roy J 1990 In search of the characteristics of plant invaders. *In* Biological Invasions in Europe and the Mediterranean Basin. Eds F di Castri, J Hansen and M Debussche. pp 335–352. Kluwer, Dordrecht, Netherlands.

Schier G A 1973 Origin and development of aspen suckers. Can. J. Bot. Res. 3/1, 45–53.

Tomlinson P B 1980 The biology of trees native to tropical Florida. Harvard Forest, Mass. Allston, Mass.: Harvard University Printing Office. 480 p.

Trabaud L 1987 Fire and the survivals traits of plants. *In* The Role of Fire in Ecological Systems. pp 65–89. The Hague: SPB Academic.

Trabaud L and Lepart J 1980 Diversity and stability in Garrigue ecosystems after fire. Vegetatio 43, 49–57.

Trabaud L and Lepart J 1981 Changes in floristic composition of a *Quercus coccifera* L. Garrigue in relation to different fire regimes. Vegetatio 46, 105–116.

Whelan R J 1997 The Ecology of Fire. Cambridge University Press. New York. 346 p.

A. Stokes (ed.), The Supporting Roots of Trees and Woody Plants: Form, Function and Physiology, 145–156.

Clonal variation in above- and below-ground growth responses of *Populus tremuloides* Michaux: Influence of soil warming and nutrient availability

John S. King[1,*], Kurt S. Pregitzer[1] and Donald R. Zak[2]
[1]*School of Forestry and Wood Products, Michigan Technological University, Houghton, MI 49931, USA and* [2]*School of Natural Resources and Environment, University of Michigan, Ann Arbor, MI 48109, USA*

Key words: biomass partitioning, clonal variation, N-availability, *Populus tremuloides*, root longevity, soil warming

Abstract

Trembling aspen (*Populus tremuloides* Michx.) is the most widely distributed tree species in North America making it important to terrestrial carbon and nutrient cycles. Due to anthropogenic climate change high latitude temperatures are expected to increase, making it necessary to assess the feedback between above- and below-ground carbon pools to increased temperature at sites of both high and low N-availability. We grew four clones of aspen at two levels of soil temperature and two levels of soil N-availability for 98 days and quantified photosynthesis, growth, biomass allocation, and root length production and mortality. High soil temperature increased rates of photosynthesis (65%), resulting in greater whole-plant growth (37%) through increases in roots, stems, and foliage; however these increases generally occurred only in soil of high N-availability. Root:shoot biomass allocation varied between clones but was unaffected by the soil temperature or N-availability treatments. Root length production and mortality increased at elevated soil temperature, but this response was modified by soil N-availability. At high soil temperature, soil N-availability had little effect on root dynamics, while at low soil temperature, high soil N-availability increased both the production and mortality (turnover) of roots. We conclude that trembling aspen has the potential for substantially greater growth and root turnover under conditions of warmer soil at sites of both high and low N-availability, but that allometric patterns of growth are under strong genetic, rather than environmental control.

Introduction

The expected increase in average global temperature of 1 to 3.5 °C within the next 100 years will likely be more pronounced at high northern latitudes in autumn and winter (Houghton et al., 1996). High latitudes are thought to be especially vulnerable to global climate change because of the complex feedbacks between climate, thermohaline circulation of the oceans, the influence of ice and snow cover on surface albedo, and responses of terrestrial plant communities (Houghton et al., 1996; Ganopolski et al., 1998). Shifts in net primary production, biomass partitioning, and carbon cycling in forest ecosystems will play a major role in the biosphere's response to the rising levels of atmospheric CO_2 (Amthor, 1995; Melillo et al., 1996; Norby et al., 1996) and increasing nitrogen deposition (Galloway et al., 1995; Vitousek et al., 1997). Of particular concern is whether warming of soils at high latitudes will result in these areas becoming net sources, rather than sinks, for atmospheric carbon in the future (Billings et al., 1982; Hobbie, 1996; Oechel and Vourlitis, 1996).

A major process delivering carbon (C) to northern forest soils is the production and turnover of fine roots (Fogel, 1983; Hendrick and Pregitzer, 1993). Whole-plant growth, photosynthesis, and fine root production and turnover have all been shown to be positively correlated to increases in air temperature, nutrient availability, and atmospheric CO_2 (Pregitzer et al., 1995; Berntson and Bazzaz, 1996; Tissue et al., 1996). Increased soil temperatures will occur during

* E-mail: jsking@mtu.edu
Plant and Soil is the original source of publication of this article. It is recommended that this article is cited as: *Plant and Soil* **217**: 119–130, 1999.

any general warming of the planet (Peterjohn et al., 1994), yet the effects of warmer soils on the dynamics of C cycling in northern forests has received relatively little attention. Hendrick and Pregitzer (1993) attributed greater rates of fine root mortality in a 'southern' sugar maple forest to higher soil temperature, and Lawrence and Oechel (1983) and Marshall and Waring (1985) have presented evidence that root respiration and turnover increase with soil temperature. Using minirhizotrons, Forbes et al. (1997) have provided some of the first evidence that elevated soil temperature directly decreases root longevity in *Lolium perenne* L. As both temperature and the availability of certain nutrients are increasing concurrently with atmospheric CO_2, it is important to observe the interaction of these other factors on relationships governing carbon storage in northern forest ecosystems.

Trembling aspen (*Populus tremuloides* Michaux) is the most widely distributed tree species in North America, occurring over a range that spans 111° of longitude and 48° of latitude (Perala, 1990). It is a major commercial species, providing fiber for pulp and flakeboard production. Ecologically, it is an early-successional species, occurring on soils of a wide range of pH and N-availability (Barnes and Wagner, 1981; Perala, 1990). It is very shade intolerant and is characterized by high photosynthetic rates and rapid growth. Regeneration is primarily by vegetative propagation, especially after disturbance, and clonal variation in physiology and phenology may confer an adaptive advantage to some genetic lines in response to changing environmental conditions (Kubiske et al., 1998). Finally, aspen productivity in the Great Lakes region of the midwestern United States has been estimated at 2.9 mg C ha^{-1} $year^{-1}$ (live biomass) on good sites managed with a 40-year rotation (Alban and Perala, 1992), illustrating the potential C sink strength of aspen ecosystems across its range.

In order to investigate the effects of increased soil temperature and nutrient availability on carbon gain and loss of an important north-temperate forest species, we grew four clones of aspen for 98 days at two levels of soil temperature and two levels of nutrient availability at the University of Michigan Biological Station, Pellston, Michigan. We hypothesized that: (1) increased levels of soil temperature and N-availability would result in higher photosynthetic rates and greater whole plant growth; (2) growth at elevated soil temperature would be accompanied by increased rates of root production and turnover; and (3) increased nutrient availability would decrease biomass allocation to roots regardless of soil temperature.

Methods

The study site was located at the University of Michigan Biological Station near Pellston, Michigan (45°34′ N, 84°40′ W). The experimental plant material consisted of four previously described clones of *Populus tremuloides* which differ in leaf phenology and physiological characteristics (Barnes, 1959). Two of the clones (42 and 51) are characterized by early-senescing foliage (mid-October) and high photosynthetic rates and two (1 and 8) by late-senescing foliage (mid-November) and relatively lower photosynthetic rates (Kubiske et al., 1998). Roots of each clone were collected at the Biological Station in October 1993, and freshly propagated for this study at the University of Michigan.

On June 17, 1997, cuttings of the clonal material were planted in 3 L pots, each ramet being randomly assigned to high or low soil N-availability, and high or low soil temperature. The soil media has been described previously (Pregitzer et al., 1995; Randlett et al., 1996) and resulted in levels of N-availability characteristic of the range found across the northern Great Lakes region (Zak and Pregitzer, 1990). Briefly, high N-availability soil consisted of a homogenized A-horizon of a Kalkaska series soil (sandy, mixed, frigid, Entic Haplorthod) with a net N-mineralization rate of 348 μg N g^{-1} d^{-1}. The low N-availability soil was a 4:1 mix of Rubicon series C-horizon soil (sandy, mixed, frigid, Entic Haplorthod) with A-horizon of the Kalkaska series soil; it had a net N-mineralization rate of 45 μg N g^{-1} d^{-1}.

Soil temperature was controlled by exposing pots to ambient air temperature (high temperature treatment) or by cooling them with a constant-flow water bath. The water bath consisted of a groundwater supply pumped into sealed wooden boxes, each with an inlet and outlet. Pots were sealed in the boxes with silicone sealant and a drain hole installed beneath each. As the experimental treatments were replicated three times, three cooling boxes were constructed containing 24 pots each. The cooling boxes were randomly interspersed with three boxes of 24 pots built with the same dimensions using 2.54 cm thick foam insulation. Foam insulation was used to buffer rapid temperature changes over the course of the day. Ten-minute soil temperature averages were measured in a pot located at the center and edge of two of the high temperature and low temperature groups of pots with chromel-constantan theromcouples connected to a Li-Cor 1000 datalogger. Thermocouples were buried at mid-pot depth and moved to different pots in the same relative locations after each harvest. Plants were exposed to natural light, air temperature and precipitation throughout the course of the study. Volumetric soil moisture in the pots was monitored using a Trane TDR (time domain reflectometry) system and water added during extended dry periods.

Three destructive harvests were conducted at approximately one-month intervals. On the day preceding each harvest, photosynthesis was measured on the plants to be harvested using a LICOR 6400 portable photosynthesis system in open mode equipped with a chamber with internal light source (670 nm red LED). Photosynthetic photon flux density (PPFD) was maintained at 1800 μmol m^{-2} s^{-1} and the system was calibrated for zero and span at each sample date using a gas standard of known CO_2 concentration. Measurements were taken on the most recent fully expanded leaf between 10:00 AM and 14:00 PM Eastern Standard Time. During the harvests, plants were separated into roots (live plus dead), stems and leaves. Roots were extracted from soil by removing pots and spraying low-pressure water through the root-soil mass suspended over a 2-mm mesh screen. Root systems were carefully inspected for any adhering soil, which was removed by hand. Plant leaf area was measured with a Licor LI-3000 leaf area meter (Licor, Inc., Lincoln, Nebraska). Stem height was recorded and all biomass components were dried to constant weight in a forced air oven at 65 °C.

To monitor root length production and mortality, six glass aquaria were set up adjacent to the pot study with identical plant material and treatments. The aquaria were partitioned into eight cells with a volume equal to that of the pots. Soil temperature, N-availability and clone were randomized within the aquaria, and all treatment combinations were replicated three times. As the aquaria did not drain freely, a clear polyvinyl chloride plastic rain-out shelter was installed to control water inputs and volumetric soil moisture was monitored using the TDR system. If soil moisture greatly exceeded that of the pots, water was withheld until evapotranspiration reduced it to the level of the pots. Similarly, water was added during extended dry periods. Roots growing within a predetermined area against one glass wall of each cell were traced onto acetate sheets approximately every 10 days, and cumulative root length production and mortality were determined over the course of the study following the method of Berntson and Bazzaz (1996). Between tracing dates the aquaria walls were covered with aluminum foil to shield the soil from light.

Statistical analyses

Because individual ramets varied in size at the beginning of the experiment, all statistical tests included an analysis of covariance (ANCOVA) to adjust for initial plant size. Data were log-transformed to normalize variances across treatments, meeting the assumptions of analysis of variance (ANOVA). This was confirmed by inspection of residuals and normal probability plots. Treatment main effects and interactions were analyzed as a completely random design (CRD) with three replications (Sokal and Rholf, 1995); the experiment was not blocked since there was no indication of environmental gradients within the experimental area. Shifts in biomass allocation were evaluated using allometric analysis (Troughton, 1955; Ledig et al., 1970), a statistically robust method of separating treatment effects from ontogenetic effects on growth. Allometric analysis consists of comparing the slopes of the log-log relationship between plant parts over time as affected by the experimental treatments. Statistically significant treatment effects are indicated by non-parallel slopes and tested using an ANCOVA between the respective plant parts (Niklas, 1994). We performed allometric analyses on the distribution of biomass between roots and shoots, and root and stems to remove possible effects of varying leaf phenology. Root production and mortality data were analyzed using repeated measures analysis following Pregitzer et al. (1995). To account for serial correlation of the root tracings, orthogonal quadratic polynomials were fitted to the data as calculated for unequally spaced sample dates (Robson, 1959; Meredith and Stehman, 1991). Coefficients from the polynomials were then used as primary data to test for significant treatment effects using standard ANOVA (Steel and Torrie, 1980).

Results

Temperature control

Over the course of the experiment, the groundwater cooling system effectively maintained temperature separation between soils in the low and high temperature treatments, although the separation was highly

Table 1. Monthly mean soil temperature (standard deviation) of four different clones of aspen ramets grown at two levels of soil N-availability and temperature for 98 days at the University of Michigan Biological Station, Pellston, Michigan. Pot codes are defined as: H-Ctr – high temperature, center pot; L-Ctr – low temperature, center pot; H-Edge – high temperature, edge pot; L-Edge – low temperature, edge pot

Month	H-Ctr	L-Ctr	H-Edge	L-Edge
June (N=1431)	19.83 (5.10)	13.14 (2.37)	21.61 (5.67)	14.55 (1.78)
July (N=3516)	19.27 (4.21)	12.24 (0.69)	19.41 (4.64)	12.42 (0.86)
August (N=5226)	17.15 (3.90)	13.20 (0.87)	17.25 (3.86)	13.68 (1.40)
September (N=2640)	15.60 (3.41)	13.22 (0.74)	15.81 (3.17)	13.59 (1.22)

variable (Figure 1A). Low temperature pots averaged 13 °C over the course of the experiment, while the high temperature pots varied between 10 and 29 °C. Comparison of thermocouples at the center of the boxes and at the edge for both temperature treatments indicated that there was not much of a gradient within the boxes (Table 1; Figure 1B, C). Mean soil temperature differences between the low and high treatments (24 h averages) for June, July, August and September were 6.88 °C, 7.01 °C, 3.76 °C, and 2.30 °C, respectively. Although our control over temperature differences was limited, the water-cooled system accurately tracked the daily and seasonal fluctuations in temperature, which we feel is more realistic than a square-wave treatment (Figure 1). The spike in the low temperature treatment in early July (Figure 1B) was due to a 4-hour electrical failure, giving an indication of the ambient heat load on the pots.

Photosynthesis

Clear treatment effects were apparent on photosynthetic rates that persisted for the duration of the experiment (Figure 2). Clone was not a significant factor at any time, but high soil N-availability resulted in 65% greater photosynthetic rates, averaged over all three harvests. This stimulation of photosynthesis was augmented by high soil temperature at the first and last harvests (53% and 26%, respectively), although the statistical significance of soil temperature was only marginal at the last harvest (Figure 2). Rates of photosynthesis increased throughout the experiment, even though leaves had begun to change color and senesce by the last harvest.

Biomass accumulation

By the end of the experiment, total plant dry mass ranged from approximately 2 g to over 9 g (Figure 3A) and soil temperature and soil N availability were significant as main effects (Table 2). High soil temperature resulted in larger plants (37%), but this generally occurred only under high N availability, although the temperature by N-availability interaction was marginally significant (P = 0.08). High soil N-availability usually resulted in larger plants (62%) regardless of soil temperature. Root systems comprised the largest fraction of total plant mass, 63% on average (Figure 3B), and therefore strongly influenced the whole-plant response to the treatments. Soil temperature, however, was only marginally significant for root biomass (Table 2). Averaged over other factors, the increase in root mass due to high soil N-availability was 48%. Stems were the next largest fraction, comprising 20% of total mass on average, and were significantly affected by a temperature by soil N-availability interaction (Table 2). Stem mass increased significantly under high soil N at high temperature (135%), but there was little or no response at low temperature (Figure 3C). Finally, leaf mass was significantly affected by clone and a temperature × soil N-availability interaction (Table 2). Averaged over other factors, clone 42 had the smallest leaf mass (0.55 g), followed by clone 1 (0.64 g), clone 8 (0.74 g), and then clone 51 (0.81g). The leaf response to the temperature × soil N interaction was similar to that for stem mass (Figure 3D); leaf mass increased under high N-availability at high temperature (222%), but at low temperature there was little or no response to soil N-availability.

Root length production and turnover

Net root length production (production minus mortality) increased at a rapid rate early in the experiment, leveled off and finally declined by the end of the experiment (Figure 4A). It increased at high temper-

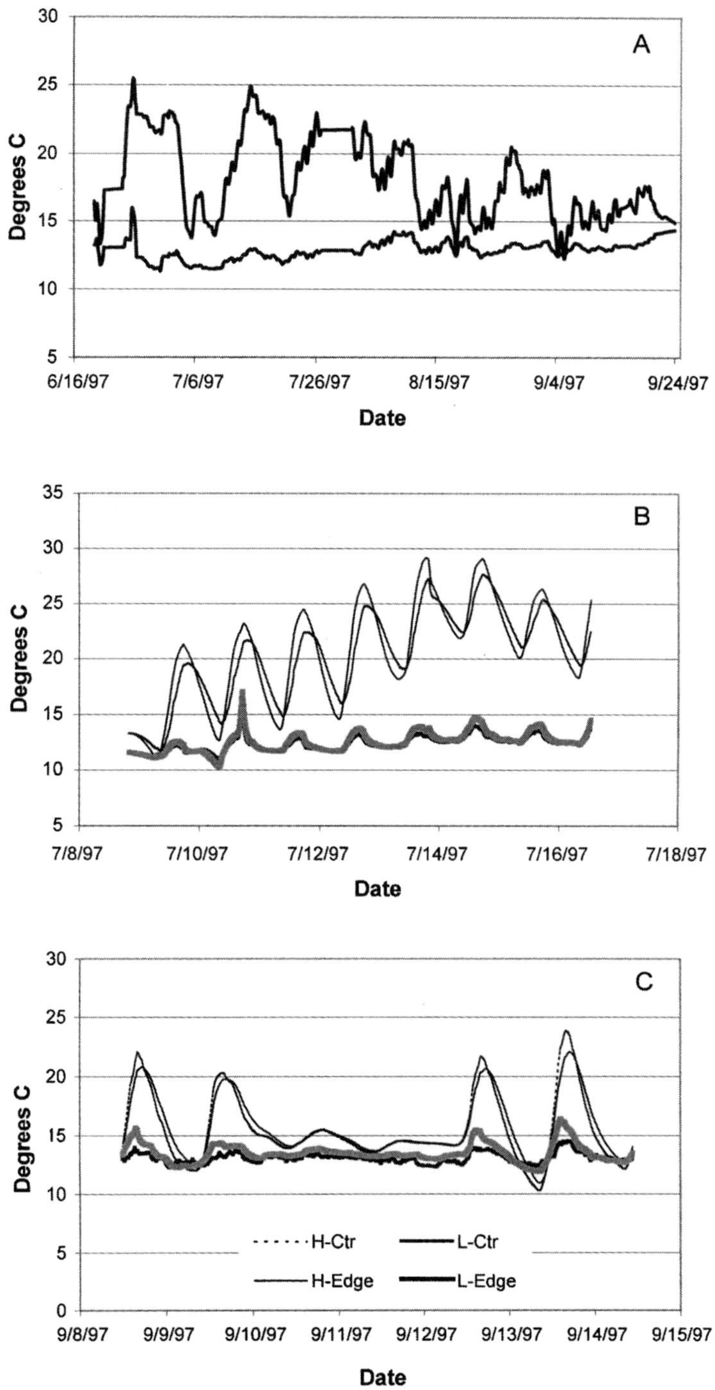

Figure 1. Pot temperature measured at mid-pot depth over the course of the growing season of four clones of aspen grown at two levels of soil N-availability and temperature for 98 days at the University of Michigan Biological Station, Pellston, Michigan. Panel A – moving average (3.5 days) soil temperature in the high and low temperature treatments for the entire experiment; Panel B – soil temperature for a week early in the experiment; Panel C – soil temperature for a week late in the experiment. The legend for panels B and C is as defined in Table 1.

ature, but the rate of increase was modified by soil N-availability. This was indicated by a significant temperature effect in the mean and quadratric terms of the orthogonal polynomials used to analyze root produc-

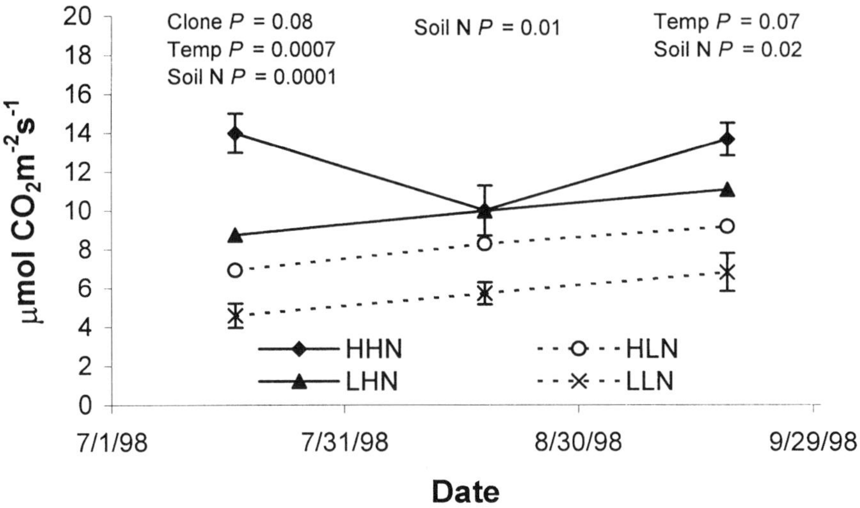

Figure 2. Mean photosynthetic rates of four aspen clones grown at two levels of soil N-availability and temperature for 98 days at the University of Michigan Biological Station, Pellston, Michigan. Bars are one standard error of the mean. Treatment codes are: HHN = high temperature, high N-availability; HLN = high temperature, low N-availability; LHN = low temperature, high N-availability; LLN = low temperature, low N-availability.

tion responses (P = 0.0009 and P = 0.0008, respectively). It was also affected by soil N-availability in the linear term (P = 0.04) and marginally by a temperature × soil N-availability interaction in the quadratic term (P = 0.06). High soil N-availability resulted in greater rates of net root production, but this only occurred at low temperature. Total root production (cumulative production plus cumulative mortality) responded similarly (Figure 4B). Temperature was significant for the mean (P = 0.0004) and quadratric (P = 0.01) terms, while soil N-availability was significant in the linear term (P = 0.004). Finally, the rate of root mortality (both mean and linear terms) was significantly increased at high temperature (P = 0.0005), and the rate of increase (quadratic term) was influenced by soil N-availability (P = 0.0032) (Figure 4C).

Root:shoot biomass partitioning

The pattern of response of all biomass fractions to the treatments was unique and consistent for each clone; that is, roots responded similarly to the stems, which responded similarly to the leaves (Figure 3, panels B through D). This suggests there was no change in biomass allocation among plant parts induced by the treatments, which was tested with the allometric analyses (Table 3; Figure 5). Of the experimental treatments, only clone had a significant effect on the partitioning of biomass between the covariate (root mass) and response variables (shoot mass and stem mass) (Table 3). The clones exhibited a range of biomass allocation to shoots relative to roots, with Clone 8 having the highest and Clone 1 the lowest; the other two clones being intermediate (Figure 5A). This pattern persisted for the partitioning of biomass between roots and stems (Figure 5B). The R-square values for the ANCOVA models used in this analysis were 0.57 for roots:shoots and 0.77 for roots:stems, indicating a reasonable degree of confidence in fitting linear models.

Discussion

The importance of soil temperature

The quantity of C contained in terrestrial soils is a balance between rates of production and decomposition of organic litter (Olson, 1963). Modern society is altering environmental factors that directly affect forest physiology, growth and foodweb interactions. Therefore, it is likely that the accumulation of carbon in biomass and soils of future forests will differ from that of the past. Temperature exerts major control over global primary production (Lieth, 1975) and decomposition (Waring and Schlesinger, 1985), and therefore should be foremost in our investigations of forest responses to global change. An unanswered yet critical question is

Table 2. P-values of ANOVA on biomass accumulation in four clones of aspen grown at two levels of soil N-availability and temperature for 98 days at the University of Michigan Biological Station, Pellston, Michigan

Source	Total[1]	Root	Stem	Leaf
Initial Mass	*ns*	*ns*	*ns*	*ns*
Clone	*ns*	*ns*	*ns*	0.0089
Temperature	0.0471	0.0783	*ns*	0.0004
Temp*Clone	*ns*	*ns*	*ns*	*ns*
Soil N	0.0009	0.0059	0.0015	0.0001
Soil N*Clone	*ns*	*ns*	*ns*	0.0279
Temp*Soil N	0.0846	*ns*	0.0308	0.0289
Temp*Soil N*Clone	*ns*	*ns*	*ns*	*ns*

[1] *ns* indicates non-significant values.

Table 3. P-values of ANCOVA on biomass allocation in four clones of aspen grown at two levels of soil N-availability and temperature for 98 days at the University of Michigan Biological Station, Pellston, Michigan

Source[1]	Shoot[2]	Stem
Root*Temp	*ns*	*ns*
Root*Soil N	*ns*	*ns*
Root*Temp*Soil N	*ns*	*ns*
Root*Clone	0.0330	0.0043
Root*Temp*Clone	*ns*	*ns*
Root*Soil N*Clone	*ns*	*ns*
Root*Temp*Soil N*Clone	*ns*	*ns*

[1] Denotes interaction term in the model between the covariate (root mass) and other experimental factors. Significant values indicate non-parallel slopes in the response between different levels of a factor.
[2] *ns* indicates non-significant values.

how warmer soils of the future will affect the production and turnover of fine roots, a major pathway for delivery of energy to carbon-limited soil microorganisms (Zak et al., 1993). Furthermore, soil warming has been shown to increase N-mineralization in forest soils (Peterjohn et al., 1994), making it necessary to assess the effects of higher temperatures in soils of varying N-availability.

We have attempted to address these questions using a simple, inexpensive technology that can be applied almost anywhere. The continuous flow of groundwater around the sealed pots maintained the low temperature pots at approximately 13 °C, which is the average growing season soil temperature for forests of northern Lower Michigan (Hendrick and Pregitzer, 1997). We achieved good separation between the high and low soil temperature treatments, and pot temperatures precisely tracked daily and seasonal fluctuations in ambient conditions (Figure 1). The system provided minimal control over the magnitude of separation between high and low temperature pots, however we feel that the precise tracking of ambient temperature fluctuations more than compensated for this by exposing the plants to realistic temperature profiles as opposed to a square-wave treatment. One shortcoming of the technique, however, is that the difference in amplitude of daily temperature fluctuations between the high and low temperature pots (Figure 1B and C) may have unrecognized affects on physiology.

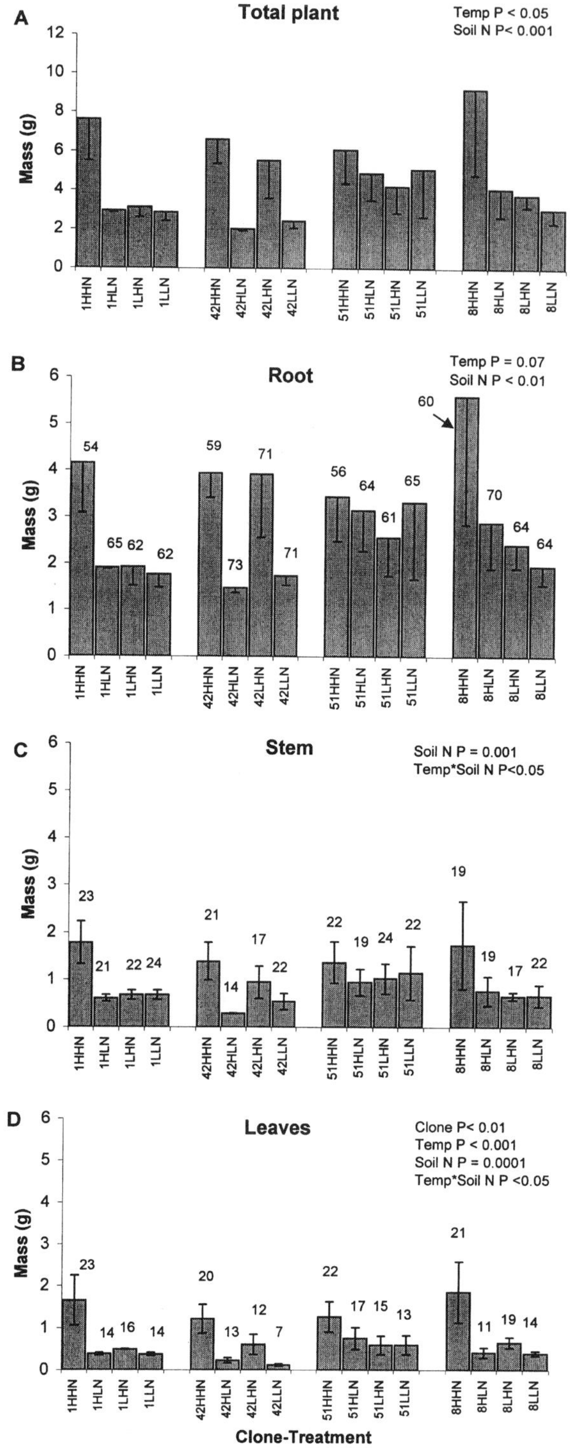

Figure 3. Mean mass of plant parts of four aspen clones grown at two levels of soil N-availability and temperature for 98 days at the University of Michigan Biological Station, Pellston, Michigan. Bars are one standard error of the mean. Treatment codes are: clone number; H, L = high and low temperature, respectively; HN, LN = high and low nitrogen availability, respectively. Numbers above bars are percentage total plant biomass comprised by respective plant parts. Note change of scale for Total plant.

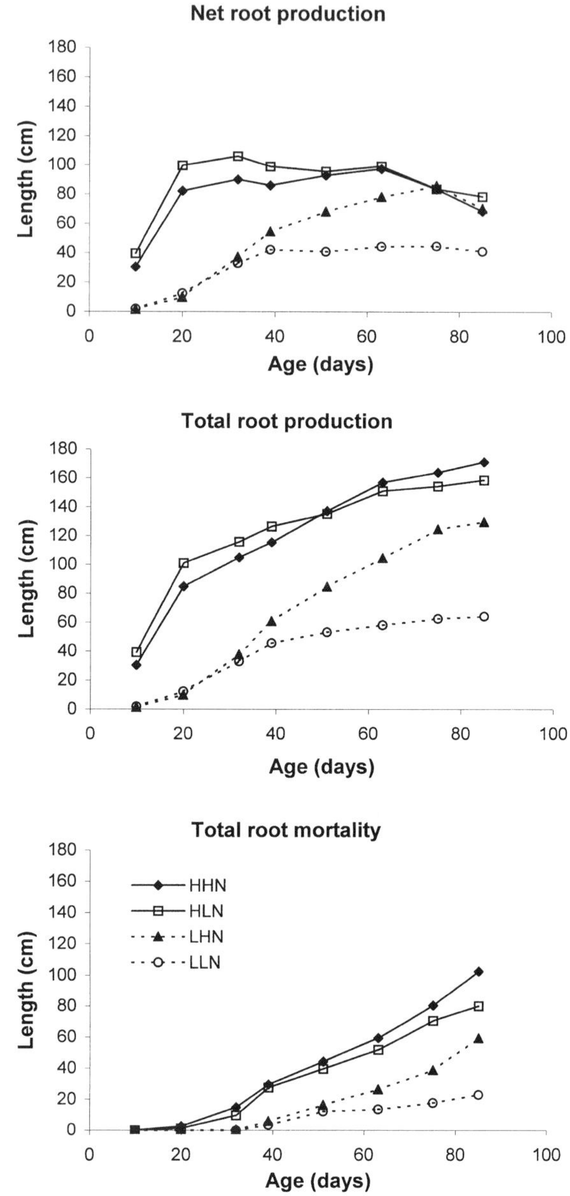

Figure 4. Cumulative root length production and mortality of four clones of aspen grown at two levels of soil N-availability and temperature for 98 days at the University of Michigan Biological Station, Pellston, Michigan.

Photosynthesis and growth

Our first hypothesis was that trembling aspen would exhibit increased rates of photosynthesis and growth at elevated soil temperatures. Numerous investigators have demonstrated a strong, positive relationship between photosynthesis in forest tree species and soil temperature (Day et al., 1991; Lippu and Puttonen,

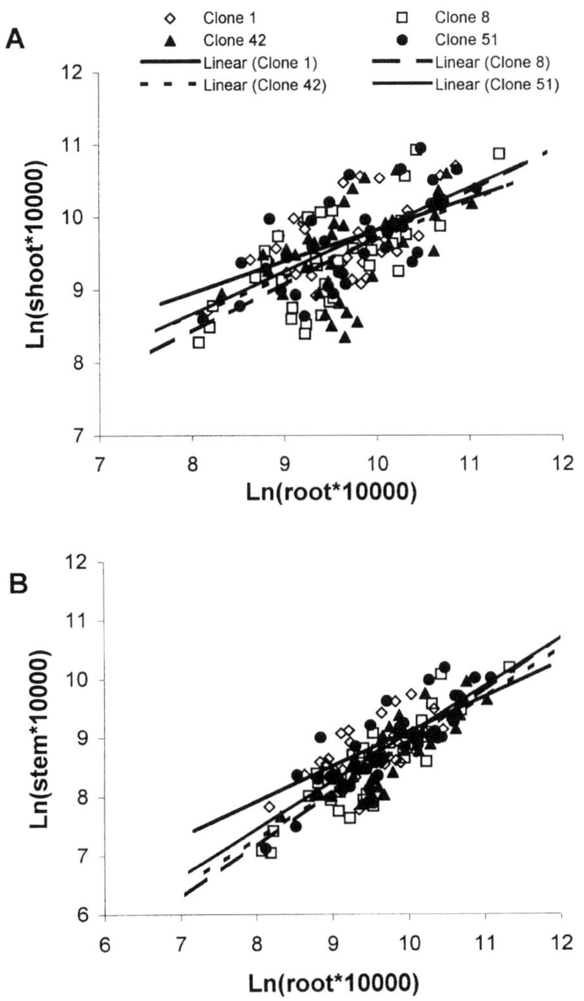

Figure 5. Allometric relationships between plant parts of four clones of aspen grown at two levels of soil N-availability and temperature for 98 days at the University of Michigan Biological Station, Pellston, Michigan. Panel A – ln shoot vs. ln root; Panel B – ln stem vs. ln root.

1991; Landhäusser et al., 1996); moreover, soil temperature may be a better predictor of carbon assimilation than air temperature (Delucia, 1986; Delucia and Smith, 1987; Jurik et al., 1988; Schwarz et al., 1997). Our results with trembling aspen are in agreement with these findings and support our first hypothesis. Photosynthesis increased 65% in response to higher soil N-availability and this response was augmented by the additive effect of elevated soil temperature for the duration of the experiment (Figure 2). The enhanced carbon assimilation translated into greater total plant biomass (Figure 3) due to increases in all organ systems (roots, leaves, stems), again primarily driven by high soil N-availability, but augmented by elevated soil temperature (Table 2). This synergistic effect of various soil properties on growth has been reported by others (Parsons et al., 1994) and will probably be an important controller of forest responses to changing environmental conditions. Increased biomass production at elevated soil temperature has also been reported by other investigators (Lawrence and Oechel, 1984; Kaspar and Bland, 1992; Landhäusser et al., 1996), and in conjunction with 'fertilization' from elevated atmospheric CO_2 and nitrogen deposition, may contribute to enhanced carbon sequestration in future forest ecosystems.

Although clonal variation in trembling aspen can be important for leaf phenology and carbon gain (Barnes, 1959; Kubiske et al., 1998), it did not appear to greatly affect photosynthesis or growth in this study (Table 2; Figure 2). There was a marginally significant effect of clone on photosynthesis ($P = 0.08$) early in the experiment, but this did not persist over the course of the growing season. In addition, certain clones experienced earlier leaf senescence resulting in lower final leaf biomass (Figure 3D). These clonal responses were of insufficient magnitude to have a statistically significant effect on final biomass accumulation (Figure 3A), which was dominated by the response of roots and stems.

Root length production and turnover

Although root longevity in poplar has been shown to decline due to infection by arbuscular mycorrhizal fungi (Hooker et al., 1995), very little work has been done on the effects of soil warming on the production and turnover of roots in forest tree species (Kaspar and Bland, 1992). Redmond (1955) demonstrated that rootlet mortality in yellow birch (*Betula alleghaniensis* Britton) was correlated to soil temperature, and Tabbush (1986) reported higher numbers of new roots in Sitka spruce (*Picea sitchensis* (Bong.) Carr.) and Douglas-fir (*Pseudotsuga menziesii* (Mirbel) Franco) grown at higher soil temperatures. Tryon and Chapin (1983) found increased root elongation rates in a variety of taiga tree species as soils warmed over the course of the growing season. It is also known that soil temperature affects root growth, water relations and survival of out-planted seedlings (Lyr and Hoffman, 1967; Lopushinsky and Kaufman, 1984; Sands, 1984; Grossnickle and Blake, 1985).

We hypothesized that the rate of root production and mortality (turnover) would increase at elevated soil temperature, resulting in greater rates of carbon input to soils. Our study supports this hypothesis, showing that elevated soil temperature resulted in increased cumulative root production and mortality (Figure 4). In agreement with Pregitzer et al. (1995), we found soil N-availability to exert strong control over root dynamics and it modified the temperature response. At low soil temperature, root production and mortality increased much more in high N-availability soil compared to low N-availability soil, resulting in greater net root production. In contrast, at high soil temperature, N-availability had little effect on root dynamics. If similar responses to soil warming occur in aspen-dominated ecosystems, our data suggest that substantial increases in root turnover are possible at sites of both high and low soil N-availability. A possible mechanism for this might be increased N-mineralization in low N-availability soils, resulting in greater root production and decreased root longevity.

Although elevated soil temperature may lead to greater litter inputs to forest soils through higher rates of root turnover (Figure 4C), it may also contribute to greater ecosystem losses of C through its effects on autotrophic respiration. It is well established that root respiration increases with temperature (Billings et al., 1977; Linder and Troeng, 1981; Lawrence and Oechel, 1983; Burton et al., 1996; Boone et al., 1998), with a $Q_{10} \geq 2$ (see Burton et al., 1996). Higher maintenance respiration rates may be associated with the shorter root life spans, as has been shown for leaves (Eissenstat, 1992). In addition, the higher production of roots at elevated soil temperature (Figure 4B) should increase root growth respiration (Waring and Schlesinger, 1985), and combined with greater maintenance respiration, could significantly increase fluxes of C from forest soils to the atmosphere. It should be noted, however, that not all investigators have found simple, direct relationships between root respiration and soil temperature (Fitter et al., 1997, 1998).

Midway through the study, the rate of increase in cumulative root production declined while cumulative mortality continued to increase at an increasing rate (Figure 4B, C). This response was especially pronounced at high soil temperature and resulted in a declining rate of net root production (Figure 4A). Greater root mortality relative to production, combined with the generally greater root biomass at high temperature, would indicate substantially greater inputs of carbon to warmer soils late in the growing season. A key uncertainty regarding C storage in soils is the degree to which increasing rates of decomposition will consume the additional root carbon.

Biomass allocation

As mentioned, experimental warming of forest soil has been shown to increase N-mineralization (Peterjohn et al., 1994) and increased N-availability has been shown to reduce biomass allocation to roots in tree seedlings (Griffin et al., 1995; Gebauer et al., 1996). Therefore, if soils warm appreciably in response to global change it is possible this will be accompanied by reduced allocation of biomass to roots, representing reduced carbon inputs belowground. We hypothesized that high soil N-availability would reduce biomass allocation to roots irrespective of soil temperature. Contrary to our hypothesis, we found that the only factor controlling biomass allocation in this study was clone (Table 3). The allometric analysis indicated that Clone 1 had the greatest relative allocation of biomass to roots and Clone 8 the least (Figure 5). As both of these clones are of the late leaf-fall phenology type, it is clear that biomass allocation patterns may vary independent of leaf phenology. In addition, these results suggest there is probably a range of biomass allocation patterns in the field, depending on the clonal composition of the populations, and that biomass allocation is under strong genetic rather than environmental control (Gedroc et al., 1996; King et al., 1999). From a global change perspective, the relevant finding is that trembling aspen will not alter biomass allocation and therefore proportional C inputs to soils in response to soil warming. As this experiment was done with very small plants grown in pots, however, further studies using field-grown trees are necessary to determine the generality of these findings.

Soil temperature, northern forests and global change

If results from this study of small plants grown in pots extends to more natural ecosystems, trembling aspen may have the potential for enhanced photosynthesis and growth in response to soil warming at sites of both high and low N-availability. An immediate consequence of this will be greater inputs of C and nutrients to forest soils due to greater rates of biomass production above- and belowground. In addition, higher rates of root production and mortality should substantially augment C delivery to forest soils and possibly result in greater nutrient availability in the future. The net effect of these responses on C storage in forest ecosystems will hinge on the degree to which decomposition and C transformations in the soil are stimulated by a warmer environment.

Acknowledgements

We would like to acknowledge the continued support of our global change research by the University of Michigan Biological Station. Special thanks go to Bob Van de Kopple and Chris Vogel for technical advice and helpful discussions. Mark Kubiske provided invaluable assistance with some of the statistical analyses. Jared DeForest helped with the harvests, and Mike Hyslop and Dave Karnosky provided computer equipment for the root turnover analysis. This research was supported by the U.S. Department of Energy PER and U.S. Forest Service Global Change Research programs.

References

Alban D H and Perala D A 1992 Carbon storage in Lake States aspen ecosystems. Can. J. For. Res. 22, 1107–1110.

Amthor J S 1995 Terrestrial higher-plant response to increasing atmospheric [CO_2] in relation to the global carbon cycle. Global Change Biol. 1, 254–274.

Barnes B V 1959 Natural variation and clonal development of *Populus tremuloides* and *P. grandidentata* in Northern Lower Michigan. Ph.D. Thesis, The University of Michigan, Ann Arbor. 344 p.

Barnes B V and Wagner Jr W H 1981 Michigan trees. The University of Michigan Press, Ann Arbor. 383 p.

Berntson G M and Bazzaz F A 1996 The allometry of root production and loss inseedlings of *Acer rubrum* (Aceraceae) and *Betula papyrifera* (Betulaceae): Implications for root dynamics in elevated CO_2. Am. J. Bot. 83, 608–616.

Billings W D, Peterson K M, Shaver G R and Trent A W 1977 Root growth, respiration, and carbon dioxide evolution in an arctic tundra soil. Art. Alp. Res. 9, 129–137.

Billings W D, Luken J O, Mortensen D A and Peterson K M 1982 Arctic tundra: a source or sink for atmospheric carbon dioxide in a changing environment? Oecologia 53, 7–11.

Boone, R D, Nadelhoffer K J, Canary J D and Kaye J P 1998. Roots exert a strong influence on the temperature sensitivity of soil respiration. Nature 396: 570–572.

Burton A J, Pregitzer K S, Zogg G P and Zak D R 1996 Latitudinal variation in sugar maple fine root respiration. Can. J. For. Res. 26, 1761–1768.

Day T A, Heckathorn S A and Delucia E H 1991 Limitations of photosynthesis in *Pinus taeda* L. (loblolly pine) at low soil temperatures. Plant Physiol. 96, 1246–1254.

Delucia E H 1986 Effect of low root temperature on net photosynthesis, stomatal conductance and carbohydrate concentration in Engelman spruce (*Picea engelmannii* Parry ex Engelm.) seedlings. Tree Physiol. 2, 143–154.

Delucia E H and Smith W K 1987 Air and soil temperature limitation on photosynthesis in Engelman spruce during summer. Can. J. For. Res. 17, 527–533.

Eissenstat D M 1992 Costs and benefits of constructing roots of small diameter. J. Plant Nutr. 15, 763–782.

Fitter A H, Graves J D, Wolfenden J, Self G K, Brown T K, Bogie D and Mansfield T A 1997 Root production and turnover and carbon budgets of two contrasting grasslands under ambient and elevated atmospheric carbon dioxide concentrations. New Phytol. 137, 247–255.

Fitter A H, Graves J D, Self G K, Brown T K, Bogie D S and Taylor K 1998 Root production, turnover and respiration under two grassland types along an altitudinal gradient: influence of temperature and solar radiation. Oecologia 114, 20–30.

Fogel R 1983 Root turnover and productivity of coniferous forests. Plant Soil 71, 75–85.

Galloway J N, Schlesinger W H, Levy II H, Michaels A and Schnoor J L 1995 Nitrogen fixation: Anthropogenic enhancement-environmental response. Global Biogeochem. Cycles 9, 235–252.

Ganopolski A, Kubatzki C, Claussen M, Brovkin V and Petoukhov V 1998 The influence of vegetation-atmosphere-ocean interaction on climate during the Mid-Holocene. Science 280, 1916–1919.

Gebauer R L E, Reynolds J F and Strain B R 1996 Allometric relations and growth in *Pinus taeda*: the effect of elevated CO_2 and changing N availability. New Phytol. 134, 85–93.

Gedroc J J, McConnaghay K D M and Coleman J S 1996 Plasticity in root/shoot partitioning: optimal, ontogenetic, or both? Func. Ecol. 10, 44–50.

Griffin K L, Winner W E and Strain B R 1995 Growth and dry matter partitioning in loblolly and ponderosa pine seedlings in response to carbon and nitrogen availability. New Phytol. 129, 547–556.

Grossnickle S C and Blake T J 1985 Acclimation of cold-stored jack pine and white spruce seedlings: effect of soil temperature on water relation patterns. Can. J. For. Res. 15, 544–550.

Hendrick R L and Pregitzer K S 1993 Patterns of fine root mortality in two sugar maple forests. Nature 361, 59–61.

Hendrick R L and Pregitzer K S 1997 The relationship between fine root demography and the soil environment in northern hardwood forests. Ecoscience 4, 99–105.

Hobbie S A 1996 Temperature and plant species control over litter decomposition in Alaskan tundra. Ecol. Mono. 66, 503–522.

Hooker, J E, Black K E, Perry R L and Atkinson D 1995 Arbuscular mycorrhizal fungi induced alteration to root longevity in poplar. Plant Soil 172, 327–329.

Houghton J T, Meira Filho L G, Callander B A, Harris N, Kattenberg A and Maskell K 1996 Technical summary. *In* Climate change 1995: The science of climate change. Contribution of Working Group 1 to the Second Assessment Report of the Intergovernmental Panel on Climate Change. Eds J T Houghton, L G Meira Filho, B A Callander, N Harris, A Kattenberg and K Maskell. pp 13–49. Cambridge University Press, Cambridge.

Jurik T W, Briggs G M and Gates D M 1988 Springtime recovery of photosynthetic activity of white pine in Michigan. Can. J. Bot. 66, 138–141.

Kaspar T C and Bland W L 1992 Soil temperature and root growth. Soil Sci. 154, 290–299.

King J S, Albaugh T J, Allen H L and Kress L W 1999 Stand-level allometry in *Pinus taeda* as affected by irrigation and fertilization. Tree Physiol. (*In review*).

Kubiske M, Pregitzer K S, Zak D R and Mikan C J 1998 Growth and carbon allocation of *Populus tremuloides* clones in response to atmospheric CO_2 and soil nitrogen availability. New Phytol. 140, 251–260.

Landhäusser S M, Wein R W and Lange P 1996 Gas exchange and growth of three arctic tree-line tree species under different soil temperature and drought preconditioning regimes. Can. J. Bot. 74, 686–693.

Lawrence W T and Oechel W C 1983 Effect of soil temperature on the carbon exchange of seedlings. I. Root respiration. Can. J. For. Res. 3, 840–849.

Lawrence W T and Oechel W C 1984 Soil temperature effects on carbon exchange in taiga trees. Tasks Veg. Sci. 13, 127–136.

Ledig F T, Bormann F H and Wenger K F 1970 The distribution of dry matter growth between shoot and roots in loblolly pine. Bot. Gaz. 131, 349–359.

Lieth H 1975 Modeling the primary productivity of the world. *In* Primary Productivity of the Biosphere. Eds H Lieth and R H Whittaker. pp 237–263. Springer-Verlag, New York.

Linder S and Troeng E 1981 The seasonal variation in stem and coarse root respiration of a 20-year old Scots pine (*Pinus sylvestris* L.). Mitt. Fors. Bund. Wien 142, 125–139.

Lippu J and Puttonen P 1991 Soil temperature limitations on gas exchange in 1-year-old *Pinus sylvestris* (L.) seedlings. Scand. J. For. Res. 6, 73–78.

Lopushinsky W and Kaufman M R 1984 Effects of cold soil on water relations and spring growth of Douglas-fir seedlings. For. Sci. 30, 628–634.

Lyr H and Hoffman G 1967 Growth rates and growth periodicity of tree roots. Int. Rev. For. Res. 2, 181–236.

Marshall J D and Waring R H 1985 Predicting fine root production and turnover by monitoring root starch and soil temperature. Can. J. For. Res. 15, 791–800.

Melillo J M, Prentice I C, Farquhar G D, Schulze E.-D and Sala O E 1996 Terrestrial biotic responses to environmental change and feedbacks to climate. *In* Climate change 1995: The science of climate change. Contribution of Working Group 1 to the Second Assessment Report of the Intergovernmental Panel on Climate-Change. Eds J T Houghton, L G Meira Filho, B A Callander, N Harris, AKattenberg and K Maskell. pp 449–481. Cambridge University Press, Cambridge.

Meredith M P and Stehman, S V 1991 Repeated measures experiments in forestry: focus on analysis of response curves. Can. J. For. Res. 21, 957–965.

Niklas K J 1994 Plant allometry: The scaling of forms and process. University of Chicago Press, Chicago, IL. 395 p.

Norby R J, Wullschleger S D and Gunderson C A 1996 Tree responses to elevated CO_2 and implications for forests. *In* Carbon dioxide and terrestrial ecosystems. Eds G W Koch and H A Mooney. pp 1–21. Academic Press, Inc., San Diego.

Oechel W C and Vourlitis G L 1996 Direct effects of elevated CO_2 on arctic plant and ecosystem function. *In* Carbon dioxide and terrestrial ecosystems. Eds G W Koch and H A Mooney. pp 163–176. Academic Press, Inc., San Diego.

Olson J S 1963 Energy storage and the balance of producers and decomposers in ecological systems. Ecology 44, 322–331.

Parsons A N, Welker J M, Wookey P A, Press M C, Callaghan T V and Lee J A 1994 Growth responses of four sub-Arctic dwarf shrubs to simulated environmental change. J. Ecol. 82, 307–318.

Perala D A 1990 *Populus tremuloides* Michx., trembling aspen. *In* Silvics of North America, Volume 2, Hardwoods. Eds R M B Burns and B H Honkala. pp 555–569. USDA Forest Service Handbook Number 654, Washington DC.

Peterjohn W T, Mellilo J M, Steudler P A, Newkirk K M, Bowles F P and Aber J D 1994 Responses of trace gas fluxes and N availability to experimentally elevated soil temperature. Ecol. App. 4, 617–625.

Pregitzer K S, Hendrick R L and Fogel 1993 The demography of fine roots in response to patches of water and nitrogen. New Phytol. 125, 575–580.

Pregitzer K S, Zak D R, Curtis P S, Kubiske M E, Teeri J A and Vogel C S 1995 Atmospheric CO_2, soil nitrogen and turnover of fine roots. New Phytol. 129, 579–585.

Randlett D L, Zak D R, Pregitzer K S and Curtis P S 1996 Elevated atmospheric carbon dioxide and leaf litter chemistry: Influences on microbial respiration and net nitrogen mineralization. Soil Sci. Soc. Am. J. 60, 1571–1577.

Redmond D R 1955 Studies in forest pathology. XV. Rootlets, mycorrhiza, and soil temperatures in relation to birch dieback. Can. J. Bot. 33, 595–627.

Robson D S 1959 A simple method for constructing orthogonal polynomials when the independent variable is unequally spaced. Biometrics 15, 187–191.

Sands R 1984 Transplanting stress in Radiata pine. Aust. For. Res. 14, 67–72.

Schwarz P A, Fahey T J and Dawson T E 1997 Seasonal air and soil temperature effects on photosynthesis in red spruce (*Picea rubens*) saplings. Tree Physiol. 17, 187–194.

Sokal R R and Rohlf F J 1995 Biometry. 3rd edition. W.H. Freeman and Company, NewYork. 887 p.

Steel R G D and Torrie J H 1980 Principles and procedures of statistics. 2nd edition. McGraw-Hill, New York. 633 p.

Tabbush P M 1986 Rough handling, soil temperature, and root development in outplanted Sitka spruce and Douglas fir. Can. J. For. Res. 16, 1385–1388.

Tissue D T, Thomas R B and Strain B R 1996 Growth and photosynthesis of loblolly pine (*Pinus taeda*) after exposure to elevated CO_2 for 19 months in the field. Tree Physiol. 16, 49–59.

Troughton A 1955 The application of the allometric formula to the study of the relationship between roots and shoots of young grass plants. Ag. Prog. 30, 59–65.

Tryon P R and Chapin III F S 1983 Temperature control over root growth and root biomass in taiga forest trees. Can. J. For. Res. 13, 827–833.

Vitousek P M, Aber J D, Howarth R W, Likens G E, Matson P A, Schindler D W, Schlesinger W H and Tilman D G 1997 Human alterations of the global nitrogen cycle: sources and consequences. Ecol. Appl. 7, 737–750.

Waring R H and Schlesinger W H 1985 Forest ecosystems: Concepts and management. Academic Press, San Diego. 340 p.

Zak D R and Pregitzer K S 1990 Spatial and temporal variability of nitrogen cycling innorthern Lower Michigan. For. Sci. 36, 367–380.

Zak D R, Pregitzer K S, Curtis P S, Teeri J A, Fogel R and Randlett D L 1993 Elevated atmospheric CO_2 and feedback between carbon and nitrogen cycles. Plant Soil 151, 105–117.

A. Stokes (ed.), The Supporting Roots of Trees and Woody Plants: Form, Function and Physiology, 157–169.

Above- and below-ground growth of white spruce seedlings with roots divided into different substrates with or without controlled-release fertilizer

M.J. Krasowski[1,*], J.N. Owens[2], L.E. Tackaberry[3] and H.B. Massicotte[3]
[1]*British Columbia Ministry of Forests, Red Rock Research Station, 18 435 Forest Nursey Rd., Prince George, B.C. V2N 5Y7, Canada;* [2]*School of Forest Biology, Department of Biology, University of Victoria, P.O. Box 1700, Victoria, B.C., V8W 2Y2, Canada and* [3]*College of Science and Management, Forestry Program, University of Northern British Columbia, Prince George, B.C. V2N 4Z9, Canada*

Key words: conifer, fertilizing at planting, mineral, nutrients, organic

Abstract

Thirty-two one-year-old white spruce (*Picea glauca* (Moench) Voss) seedlings were grown outdoors for one season in 35 L pots buried in the soil. The pots were vertically split in half. One compartment (mineral) was filled with loamy sand. The bottom of the other compartment (organic) was filled with 10 cm sand topped with 15 cm of organic substrates. Two seedling types (16 seedlings each), (i) polystyroblock-grown and (ii) peat-board grown with mechanical root pruning had their root systems split approximately in half into each of the vertical compartments. Controlled-release (26-12-6 N-P-K) fertilizer was added to one or to none (control) of the compartments. Above-ground growth was positively affected by fertilizer placed in either soil compartmnent. Nutrient content of the foliage was greater in fertilized than in unfertilized seedlings; N and P concentrations were significantly increased. Bud reflushing occurred frequently in fertilized seedlings. Unfertilized container-grown seedlings had the fewest roots in either soil compartment. Unfertilized mechanically-pruned seedlings had significantly greater root length, root surface area, and more root tips in mineral than in organic compartments. They also had more P in current-year leaves than did unfertilized container-grown seedlings. Fertilizer added to mineral compartments significantly affected root growth in these compartments only, whereas when added to organic compartments it affected root growth in both compartments. Root systems of the two seedling types were differently affected by fertilizer: in mechanically-pruned seedlings, the number of roots was reduced but their length and diameter increased; in container-grown seedlings, root proliferation was stimulated and this increased total root length and root surface area. Five ectomycorrhizal morphotypes were identified. E-strain was the most abundant. Except for *Cenococcum*, all morphotypes were present in nursery stock prior to planting. Changes in distribution of morphotypes after planting appeared related to root health condition rather than to applied fertilizer.

Introduction

Slow growth (growth check) during the first post-planting years has been reported for various forest species including *P. glauca* and its hybrids with *P. engelmannii* Parry Engelm. (Burdett et al. 1984; Mullin, 1964; Sutton, 1972; Vyse, 1981). Growth check occurs even on sites where resource supply does not appear to be growth limiting (Hallsby, 1995). A positive effect of the presence of surface organic matter on growth of planted seedlings has been recognized (McMinn, 1982) and site preparation treatments that mix organic soil into the planting microsite have been successfully tried (Hallsby, 1995; McMinn, 1982; Sutton, 1993). Methods of organic matter treatment that would optimize seedling establishment are yet to be developed (Hallsby, 1995). However, Balisky et al.

* E-mail: marek.krasowski@gems1.gov.bc.ca
Manufacturers are mentioned for reference purposes only. This mention does not constitute endorsement and the study should not be viewed as an evaluation of any brand-name products.
Plant and Soil is the original source of publication of this article. It is recommended that this article is cited as: *Plant and Soil* **217:** 131–143, 1999.

(1995) considered the forest floor as an appropriate rooting medium for planted tree seedlings. They proposed that root morphologies and planting strategies would have to be site-specifically modified to promote the development of roots into the organic/mineral soil interface. These authors suggested that planting into microsites consisting of rotten wood or duff might be advantageous to seedling establishment while planting into mineral soil could be detrimental. This still remains controversial.

Organic horizons of forest soils are an important reservoir of mineral nutrients, especially of nitrogen Jurgensen et al. (1997). Adding fertilizer to organic matter may increase its decomposition rates (Salonius, 1972). Several types of controlled-release fertilizers are now available. In colder climatic zones and at high elevations, tree planting can be limited by long-lasting snow cover and short growing season. It is uncertain whether conifer seedlings can substantially benefit from controlled-release fertilizer when planted late in the season. Growth performance of seedlings treated with fertilizer at planting was inconsistent in British Columbia (Anonymous, 1995) but some trials showed promising results (van den Driessche, 1988). The identification of causes for inconsistent growth performance of fertilizer-treated seedlings from various types of planting stock was recommended (Anonymous, 1995).

Mycorrhizal associations between fungi and roots of conifer species are common and are believed to be important and beneficial to the growth of forest trees. Organic soil substrates well support organisms favorable to trees, icluding mycorrhizal fungi (Jurgensen et al., 1997) but it is not known whether roots located in mineral and organic soils differ in their ability to form mycorrhizae with different species of fungi. Decreased colonization of conifer roots by mycorrhizal fungi with increasing rates of fertilizer release have been reported (e.g. Crowley et al., 1986; Gagnon et al., 1995). However, little is known about effects of fertilizer applications on the abundance and composition of mycorrhizae on conifer roots located in different soil substrates.

The objectives of this study were to evaluate (i) above-ground growth responses to controlled-release fertilizer under short (late planting) growing season, (ii) differences in above and below-ground growth responses to the the placement of the fertilizer package in either the mineral or organic soil compartments, compared to unfertilized controls, (iii) the development of root system portions divided into different types of soil and as influenced by the aplied fertilizer, (iv) the effect of fertilizer treatments on seedlings nutritional status, (v) the influence of soil types and fertilizer treatments on the composition and abundance of ectomycorhizal morphotopes on roots of studied seedlings, and (vi) compare the above listed responses between two different types of white spruce planting stock.

Materials and methods

Plant material

A single white spruce seed source of central British Columbia origin was sown in mid-March 1996 into (1) PSB 415B styroblocks®[1] and (2) Vapo[2] peat trays. The PSB container cavities are 149 mm deep, 35 mm in diameter, the plants are spaced approximately 45×45 mm. The Vapo peat boards are 80 mm thick and are held in plastic trays designed for crosswise cutting of the peat boards. The trays have perforated bottoms but no lateral restrictions to root growth. Plants are spaced 50×50 mm (Parviainen and Tervo, 1989). The peat boards were heat-sterilized and calcinated at manufacture. Both stock types were grown at Red Rock Research Station in Prince George, British Columbia in the same greenhouse and the same watering and fertilizing regime until late September, 1996. Vapo seedlings were mechanically pruned during nursery growth three times (last time one week before lifting to separate individual seedlings) using an electric knife run through slots in the sides of the plastic tray containing the peat boards. About four hundred seedlings of each stock type were lifted in October 1996, and stored frozen (−2 °C) until June, 1997.

Divided-root system experiment

On June 16, 1997, sixteen seedlings of heights in the range of 110–140 mm and root collar diameters between 2.2–2.8 mm) were selected from each stock type. Roots were gently washed and each seedling was individually planted into a 33 L plastic pot (with bottom drainage holes) divided vertically in half (Figure 1) by a sheet of aluminum sealed to the pot with silicone caulk. A small notch was cut in the middle of the top edge of the aluminum sheet to allow placing

[1] Beaver Plastic Limited, Edmonton, Alberta.
[2] Kekkilä OY, Vapo Group, Tuusula, Finland.

Table 1. Soil nutrient elements in the three types of root growth media used in the divided-root experiment with white spruce seedlings

Element	Units	Nursery-bed soil (mineral compartment)	Soil organic mix	Masonry sand (bottom of organic compartment)
Total C	%	0.86	9.94	0.19
Total N	%	0.05	0.19	0.01
$N(NH_4+)$	ppm	0.32	1.21	0.25
$N(NO_3-)$	ppm	6.90	0.44	2.31
C/N ratio		18.97	52.99	15.88
Available P	ppm	162.2	111.4	59.7
Exchangeable K	cmol +/kg	0.19	0.45	0.21
Exchangeable Al	cmol+/kg	0.03	0.01	0
Exchangeable Na	cmol+/kg	0.02	0.01	0.06
Exchangeable Fe	cmol+/kg	0	0	0
Exchangeable Mg	cmol+/kg	1.05	1.74	1.39
Exchangeable Ca	cmol+/kg	2.76	10.47	3.25
Exchangeable Mn	cmol+/kg	0.01	0	0
S	%	0.0055	0.0199	0.0036
Effective CEC*	cmol+/kg	4.10	12.8	4.90
pH		6.10	6.93	7.38

*Effective exchangeable cations (CEC) are expressed as the sum of seven cations (Ca, Mg, K, Mn, Fe, Al, and Na).
% = percent of dry weight.

a seedling with its root collar below the soil surface. Each seedling had its roots split approximately in half into each of the two compartments separated by the metal sheet. Tap roots in both stock types had been air-pruned during nursery growth and were not treated any different than other roots while dividing the root systems. In each pot, one compartment (the mineral compartment) contained non-sterilized loamy sand taken from the nursery bed. The second compartment (organic compartment) had the bottom 10 cm filled with non-sterilized masonry sand (to improve drainage) topped with 15 cm of commercially available organic mix containing compost, peat, and decomposing ground bark.[3] Samples of the soil media were taken for chemical analysis (Table 1).

Seedlings of each stock type were then randomly assigned to one of three fertilizer treatments: 6 pots to the control treatment (no fertilizer added), 5 pots to fertilizer into the organic compartment, and 5 to the fertilizer into the mineral compartment. One fertilizer–containing package[4] (9 g; 26.33% total N, 12.00% phosphate, and 6.02% soluble potash) in polyurethane-coated granules providing 24.9% coated slow-release N, 5.27% coated slow-release P, 5.13% slowly available soluble K, and 6.01% S (combined)) was placed 2 cm away from the outermost roots at the depth of about 12 cm in each fertilized compartment. The rate of fertilizer release is temperature dependent and at soil temperatures typical for the central interior of British Columbia the fertilizer would be slowly released over up to 3 growing seasons (N.C. Anderson, Reforestation Technologies International, person. comm., 1999). The 32 pots were buried to their rims into the soil of the nursery bed, watered only once after planting, then left at ambient outdoor conditions until the end of the growing season. The pots were frequently checked and germinating weeds removed.

Seedling measurements and ectomycorrhizae assessment

Seedlings were removed from pots on October 2, 1997 and their shoots and roots were separated at the root collar. Shoots were individually packaged and labeled for identification. Subsamples of foliage from the nursery year and current year shoots were taken for scanning and determination of projected leaf sur-

[3] The Answer Garden Products, Ltd., Abbotsford, B.C., Canada.

[4] Reforestation Technologies International, Salinas, California, USA.

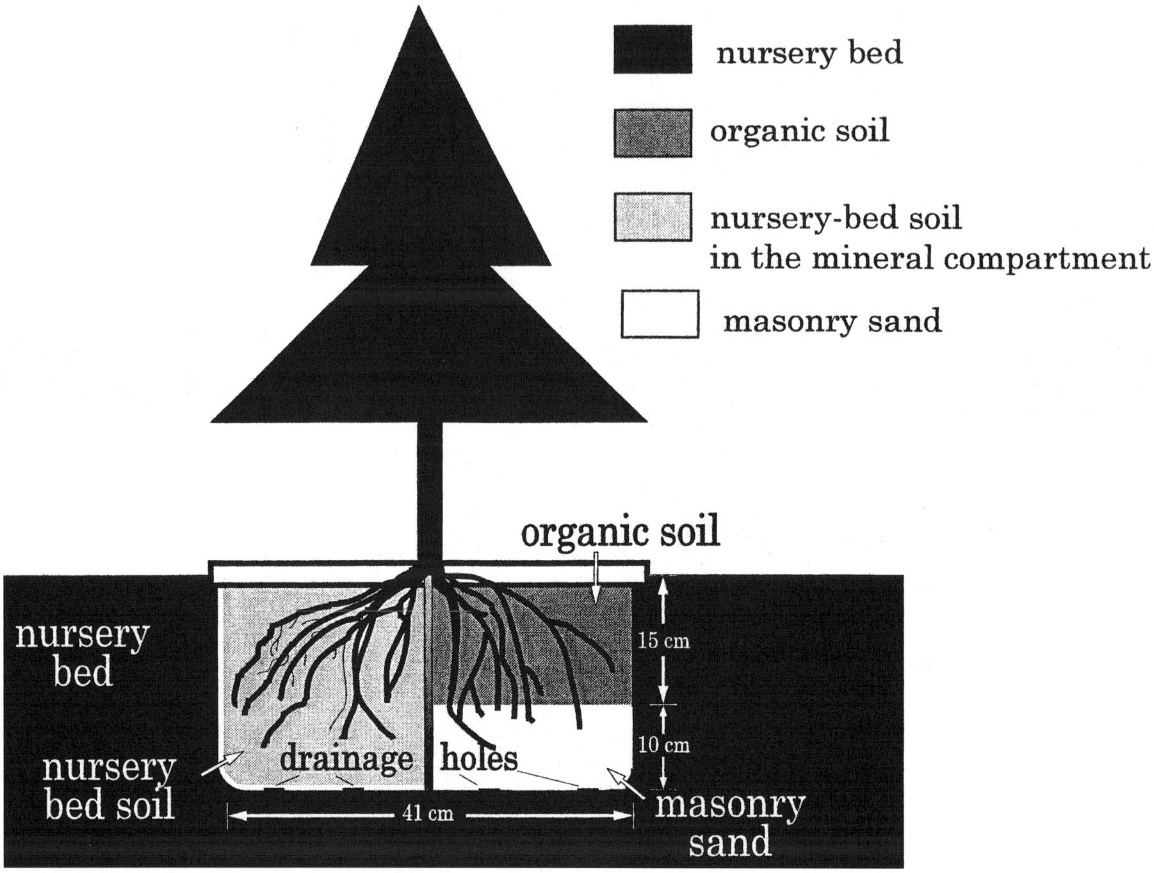

Figure 1. Schematic drawing (not to scale) of the arrangement of soil substrates in pots used in the divided-root experiment.

face area,[5] then the subsamples were dried to constant weight and the relationship between projected leaf area and dry weight was determined for each year-portion of each seedling. The remaining foliage and stems were also dried to constant weight and determined separately so the total leaf projected area of each year of growth of each seedling could be calculated from the ratios derived from the subsampled foliage. The subsamples and the rest of the foliage were then combined and sent for nutrient analysis to the analytical laboratory of the British Columbia Ministry of Forests (BCMoF). Roots from different pot compartments were separated and individually labeled. The roots were gently washed to remove adhering soil. Roots from each pot compartment were subdivided to smaller portions to reduce overlapping, scanned on the Hewlett-Packard C2521B scanner and digitally analyzed.[6] Immediately after scanning, the roots were packed in plastic bags and sent to the University of Northern British Columbia (Prince George Campus) for ectomycorrhizae assessment. Entire root-system portions from each split compartment were examined to determine mycorrhizal status. Characterization and confirmation of each type of ectomycorrhiza was done using bright field optics. Macroscopic characteristics and abundance counts were done using a dissection microscope; to confirm mantle, emanating hyphae, Hartig net and rhizomorph features, frequent root squash mounts were prepared and examined under a compound microscope (Massicotte, 1994). An abundance estimate of each ectomycorrhiza was made for all laterals and for each entire split root system (64 in total). Split root system abundance measures were statistcally analyzed for E-strain fungi only as the distribution and/or low abundance of other fungi did not allow for statistical comparisons.

[5] WinNeedle software, Regent Instruments, Quebec City, Qc, Canada.

[6] WinRhizo software, Regent Instruments.

Samples of the three soil types filling the pot compartments as well as foliage collected from the trees grown with split roots were analyzed at the BCMoF analytical laboratory. The procedures were based on methods described by Carter (1993) and Kalra and Maynard (1991). For plant tissue elements, ground tissue samples were strong acid/microwave digested and analyzed with ARL 3560 ICP spectrometer. Plant tissue nitrogen and carbon were determined by combustion analysis on the Fisons NA-1500 NCS elemental analyzer. Soil exchangeable cations were determined by a 0.1 *M* barium chloride extraction. Available ammonium-N and nitrate-N were extracted from soil samples with 2 *N* KCl and measured with the Technicon colorimetric auto-analyzer. Available P was determined by the Bray–P1 extraction method followed by colorimetric analysis for ortho-phosphate on the UV-visible spectrophotometer. Finely-ground soil subsamples were used to determine total N and C with Fisons NA-1500 NCS elemental analyzer and sulfur with the Leco SC–132 sulfur elemental analyzer. Soil pH was determined on a 1:1 water slurry except for the organic substrate where 1:2 water slurry was used.

Experimental design and statistical analysis

The experiment used a completely randomized design and all variables not separated by pot compartments or years of growth (foliage only) were analyzed by one-way ANOVA using SAS 6.9[7] statistical package. The mean square (MS) of a tree nested in fertilizer treatment and stock type were used as an error term for testing effects of all sources. For variables determined by pot compartment (e.g. root length in each compartment) the design was treated as a split-plot with fertilizer treatment and stock type being the crossed main plot factors and the compartment the split-plot factor. The same was done for foliage nutrient content analysis, with foliage year of origin being the split-plot factor. Effects of the stock type, fertilizer treatment and the interaction between these two were tested with MS of a tree nested in stock type and fertilizer treatment. All other effects were tested with MS of the compartment (or the year for foliage) nested in all higher factors. The same model was used when analyzing root length and root surface area distribution by multivariate analysis of variance (MANOVA). Diagnostic statistics were performed using the Univariate Procedure of SAS. No data transformations were necessary.

[7] The SAS Institute, Cary, N.C., USA.

Results

Ambient air temperature and rainfall during the period of the experiment

Mean air temperatures (1.2 m above ground) at the Environment Canada meteorological station (Prince George airport) located in the proximity of the research station were: 13.6 °C, 15.7 °C, 15.9 °C, and 11.5 °C, for June, July, August, and September, 1997, respectively. Total precipiation and the number of days with measurable precipitation (in brackets) were: 55.2 mm (8), 73.0 mm (19), 38.2 mm (11), and 49.0 mm (14) for June 16–30, July, August, and September, respectively.

Shoot growth

At planting, Vapo seedlings were taller (p=0.02) (mean height 133 mm) and thicker (p=0.02) at the stem base (mean diameter 2.7 mm) than PSB seedlings (113 mm and 2.4 mm, respectively). At the end of the 1997 growing season, planting stock had no significant effect on seedling height, stem diameter, or any other morphological variable listed in Table 2. Only the fertilizer treatment significantly affected the final 1997 seedling height (p=0.0001), the height increment (p=0.001), the final 1997 stem–base diameter (p=0.0001) and its increment (p=0.0001), total dry weight of stems (p=0.04), dry weight of the 1997 leaves (p=0.02), projected area of the 1997 leaves (p=0.04) and dry weight of the root system (both pot compartments combined) (p=0.0001). Means of all impacted variables were always greater in fertilizer-treated seedlings than in untreated controls. However, pot compartment choice for placement of fertilizer did not affect shoot growth (Table 2). In summary, fertilized seedlings had more shoot growth than unfertilized seedlings. Almost half of the seedlings in pots reflushed buds in mid-summer but only two were from unfertilized plants – both of PSB stock type. Among the dozen fertilized plants that reflushed, eight received fertilizer into the mineral compartments and both stock types were equally represented.

Root growth

Unfertilized PSB seedlings developed comparatively few root tips in either soil (Figure 2). Unfertilized Vapo seedlings had copious root tips in the mineral soil and few in the organic soil (Figure 2). Fertilizing PSB seedlings in mineral compartments stimulated

Table 2. Morphological characteristics (least square means) of seedlings grown with their root systems split into organic and mineral compartments with or without slow release fertilizer

	Fertilizer placement		
	None	In mineral	In organic
Variable			
Height (mm)	182b	230a	228a
1997 height increment (mm)	65b	109a	97a
Stem-base diameter (mm)	4.09b	5.96a	5.81a
1997 stem diameter increment (mm)	1.43b	3.42a	3.27a
Dry weight (g) of:			
all leaves*	2.08a	3.04a	2.84a
all stems	1.57b	2.36a	2.57a
new (1997) leaves	1.06b	2.08a	1.77a
root system (compartments combined)	1.02b	1.96a	2.01a
New (1997) leaves projected surface area (mm^2)	6136b	11132a	9334a

*1996 (nursery-grown) + 1997 foliage.
Note: Means of the same variable followed by the same letters are not significantly different from each other at $\alpha = 0.05$.

root proliferation in these compartments only. Fertilizer placed into organic compartments significantly stimulated root proliferation in this compartment in both types of planting stock. Fertilizing Vapo seedlings in either compartment significantly reduced root tip numbers in the mineral soil. In both stock types, fertilizing mineral compartments had no marked effect on root numbers in organic compartments (Figure 2). The differences in root tip numbers between the stock types, pot compartments, and fertilizer treatments resulted in a significant three-way interaction of these three factors (Table 3).

Root length and root surface area were analyzed using the same analytical model as for root tip numbers. Surprisingly, results of these ANOVAs did not show a significant three-way interaction effect on root length (p=0.17) and root surface area (p=0.22). However, pair-wise comparisons of least square means of different three-way combinations indicated that such interaction effect should be expected for both these variables. Because of this discrepancy, least square means of all three-way combinations are shown for root length (Figure 3) and root surface area (Figure 4). Significantly different means for both variables were separated on the basis of calculated least significant differences.

Root length and root surface area of unfertilized PSB seedlings were comparatively low in both compartments (Figures 3–4). Fertilizing PSB seedlings in mineral compartments increased root tip number, root length, and root surface area in these compartments only (Figures 2–4). Fertilizing organic compartments increased the length and surface area of PSB roots in both compartments, compared to unfertilized seedlings (Figures 3–4). Root length and root surface area of unfertilized Vapo seedlings were much greater in mineral than in organic compartments; although root number of fertilized Vapo seedlings declined in fertilized mineral compartments (Figure 2), root length was not reduced and root surface area significantly increased (Figures 3–4). Vapo seedlings responded to fertilizing mineral compartments by elongating and thickening their roots and reducing root proliferation. In spite of this reduction, there were still more roots than in unfertilized organic compartments of Vapo or PSB seedlings. Fertilizing organic compartments of Vapo seedlings stimulated root proliferation in these compartments and increased the length and surface area of the roots (Figures 2–4). Responses of roots in mineral compartments to fertilizing organic compartments were different in the two stock types. In PSB, root tip numbers rose correspondingly in both compartments (Figure 2). Root length and root surface area in mineral compartments also increased (Figures 3–4). In Vapo, root numbers in mineral compartments declined the same as in response to their direct fertilizing but root length and surface area did not change, compared to unfertilized seedlings (Figures 2–4).

Table 3. Results of analysis of variance on root tip numbers of white spruce seedlings grown with root systems divided into organic and mineral compartments with or without controlled-release fertilizer

Source	DF	MS	F	p>F	Error term
STK	1	3662099	0.7	0.4255	MS Tree (STK Fert)
Fert	2	23128947	4.1	0.0275	MS Tree (STK Fert)
STK*Fert	2	62611690	11.2	0.0003	MS Tree (STK Fert)
Tree (STK Fert)	26	5585536	1.6	0.1033	MS Error
Comp	1	70123100	20.8	0.0001	MS Error
Comp * STK	2	29650960	8.8	0.0064	MS Error
Comp $*$ Fert	2	20380424	6.0	0.0070	MS Error
Comp*Fert*STK	2	14244234	4.2	0.0259	MS Error
Error	26	3378566			

STK = planting stock, Fert = fertilizer treatment, Comp = divided-pot compartment, DF = degrees of freedom, MS = mean square.

Experimental effects on radial growth of roots were further examined by MANOVA using root length and root surface area distribution divided into three root diameter classes: (1) less than 0.4 mm, (2) 0.4 to 0.8 mm, inclusive, and (3) more than 0.8 mm. Outcomes were similar for both root length and root surface area and differed little whether absolute values or percents were analyzed. Roots and root parts thinner than 0.4 mm constituted more than half of the total length of root systems, regardless of the stock type, compartment, and fertilizer treatment. Roots thicker than 0.4 mm were significant contributors to the root system surface area (not shown). Differences to root surface area distribution due to fertilizer treatments were more conspicuous than the differences in the distribution of root length. Both stock types, when unfertilized, had the greatest percent of their root length and root surface area in the smallest root diameter class, especially so in the Vapo seedlings. Fertilizer application shifted more root length and root surface area to the two greater root diameter classes in both stock types, but more so in Vapo than in PSB seedlings and more in mineral than in organic compartments.

Soil nutrients, foliage nutrient concentration, and nutrient content

The organic soil had a higher C/N ratio than did either of the two mineral substrates (Table 1). It had more N in the form of ammonia whereas the two sands had more N in nitrate. The total exchangeable cations was greater in the organic mix than in the mineral substrates. The nursery-bed soil (mineral compartment) had more available phosphorus than either the organic mix or the sand used for filling the bottom of the organic compartment, possibly due to prior use of the nursery bed for seedling production and related fertilizer applications. All three soil substrates had close to neutral pH (Table 1).

Regardless of foliage age and fertilizer treatment, concentrations of macronutrients and micronutrients in leaves of seedlings in the divided-root experiment were adequate according to Chapin and Van Cleve (1989). Concentrations of macroelements contained in the fertilizer were highest in leaves of fertilized plants, regardless of which compartment was fertilized. However, differences existed among N, P, and K concentrations in foliage of different age, and/or between the stock types. N concentration following fertilizing was increased by approximately two-fold in nursery foliage and by 2.5 times in new, current-year foliage, compared to unfertilized seedlings that had 1.1% (percent of dry weight) N concentration, regardless of leaf age. This was reflected in foliage N content (concentration $\times$ dry weight) as shown in Table 4. Significant interaction effects between fertilizer treatment and foliage age were detected for N concentration (p=0.0001) and N content (p=0.0076). No significant differences were found between stock types.

Unfertilized PSB seedlings had low P concentration (0.12%) compared to unfertilized Vapo seedlings (0.28%) (p=0.02). Vapo seedlings had P concentrations similar to that of fertilized seedlings (Table 4). Fertilized and unfertilized Vapo seedlings had significantly greater concentration of K (0.71% and 0.82%, respectively) than unfertilized PSB seedlings (0.52%) but not fertilized PSB seedlings (0.75%). Different age

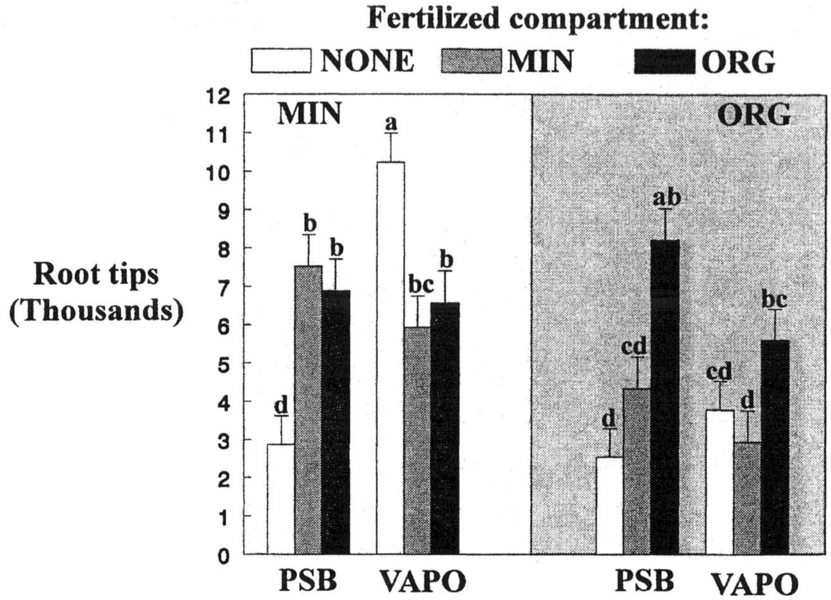

Figure 2. Mean numbers of root tips (Figure 2), aggregate root lengths (Figure 3), and root surface areas (Figure 4) of white spruce seedling root system portions present in compartments of different soil types and with or without controlled-release fertilizer. The clear panels on the left show responses of root system portions located in mineral compartments whereas the shaded panels on the right show responses of roots in organic compartments. Bars marked with the same letters show means not significantly different from each other (across both panels) at α = 0.05.

leaves of Vapo seedlings had similar K concentrations while new leaves of PSB seedlings had 0.13% higher K concentration than nursery foliage. These differences caused interaction effects of stock type with fertilizer treatment (p=0.0005) and stock type with foliage age (p=0.02) on the concentration of K. Contents of K in foliage are shown in Table 4.

Concentrations of other elements were lower in new leaves than in nursery leaves but their contents were greater in new foliage due to its greater biomass (Table 4). Iron, whose content was also much lower in new than in nursery leaves (not shown), was an exception.

Ectomycorrhizae and root condition assessment

Except for *Cenococcum,* all morphotypes found on roots from the divided-root experiment were present in nursery stock prior to planting. *Thelephora* was the most abundant morphotype that colonized PSB seedlings during nursery growth. *Amphinema* and *Mycelium radicis atrovirens* (MRA) were common on PSB seedlings but E-strain occurred at low levels. In contrast, E-strain was the most abundant nursery morphotype on Vapo seedlings which also had moderately abundant *Thelephora. Amphinema* and MRA did not occur on Vapo roots prior to outplanting.

Five ectomycorrhizal morphotypes were characterized on roots and a small number of roots on two seedlings were colonized by unidentified fungi. The most abundant ectomycorrhiza was E-strain. It was absent on the entire root system of only one seedling. Generally, E-strain was less abundant on seedlings with poorer root systems (those systems with many roots that were shrivelled, darkened, or with damaged rhizodermal layers) and these occurred more often in PSB, especially in unfertilized seedlings. E-strain abundance was analyzed by ANOVA and significant effects of the pot compartment (p=0.01) and of stock type (p=0.0001) were found. More roots were colonized by E-strain in mineral (75%) than in organic (61%) compartments and in Vapo (85%) than in PSB (51%) seedlings.

Amphinema was the second most common ectomycorrhiza, although much less abundant than E-strain. It occurred in 42 of 64 pot compartments, more frequently in PSB (29 compartments) than in Vapo seedlings, and more abundantly in PSB than in Vapo (17.8% versus 1.9% of colonized root tips, respectively). *Thelephora* and MRA occurred at similar lower levels, rarely occupying more than 5% (*Thelephora)* or 1% (MRA) of roots. Both appeared to associate more frequently with older roots, such as those in the root

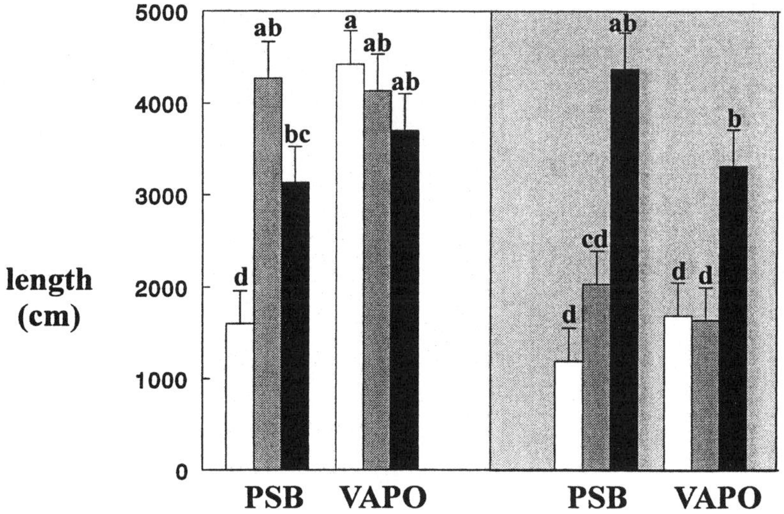

Figure 3.

collar area. MRA also seemed to associate with root systems in poorer condition, often those on which E-strain levels were low. *Thelephora* occurred in 56 of 64 compartments. MRA occurred in both soil types but in 29 PSB compartments compared to only 14 Vapo. *Cenococcum* occurred at very low levels, often on less than 1% of the root system. It occurred on roots of six PSB and seven Vapo seedlings and in all but two seedlings only in the mineral compartment.

Percent of non-colonized or lightly colonized root tips was greater in organic (22%) than in mineral (6%) compartments. This difference occurred in PSB (32% and 7% for organic and mineral compartments, respectively) but not in Vapo seedlings (approximately 10% for each compartment.

Discussion

Controlled-release fertilizer stimulated growth of white spruce seedlings of both stock types in spite of late planting. Austin and Strand (1960), Rothacher and Franklin (1964), and Carlson and Preisig (1981) obtained similar results with different fertilizer formulas and with different conifers. Our study covered only the first growing season and may not indicate long-term trends in growth performance of fertilized seedlings. One reason for inconsistency of planting stock performance treated with slow release fertilizer (Anonymous, 1995) may be the variation among different trials in fertilizer formulas, stock types, and site, especially soil characteristics.

Frequent bud reflushing in potted seedlings might have resulted from the confinement of the soil in pots causing concentration of fertilizer around the roots. Most of the reflushing occurred in seedlings with fertilized mineral compartments. Differences between the compartments in moisture fluctuation would not be surprising and these could modify the timing or rates of fertilizer release. The contribution of lammas growth to the current-year shoot extension was so minimal that it was not separately measured. Coutts and Philipson (1976) showed a negligible effect of high N concentration on shoot elongation from the first flush of growth of Sitka spruce seedlings, but a very significant effect on shoot growth from the second flush. However, their fertilizer application treatments were different (not a controlled-release fertilizer) than in our study and a long, constant photoperiod they used apparently sustained shoot elongation after the second flush. The controlled-release fertilizer used in our study stimulated shoot elongation enough to cause numerous bud flushes in mid-summer but not enough to sustain the resultant shoot elongation under ambient, decreasing day length. Bud reflushing may result in delayed maturation of shoot tips causing their succulence and concerns have been expressed about a potential predisposition of shoots in fertilizer-treated seedlings to frost and desiccation injury (Brockley, 1988). Yet, *Picea glauca* treated with the same controlled-release fertilizer as the one used in this study suffered significantly less winter damage

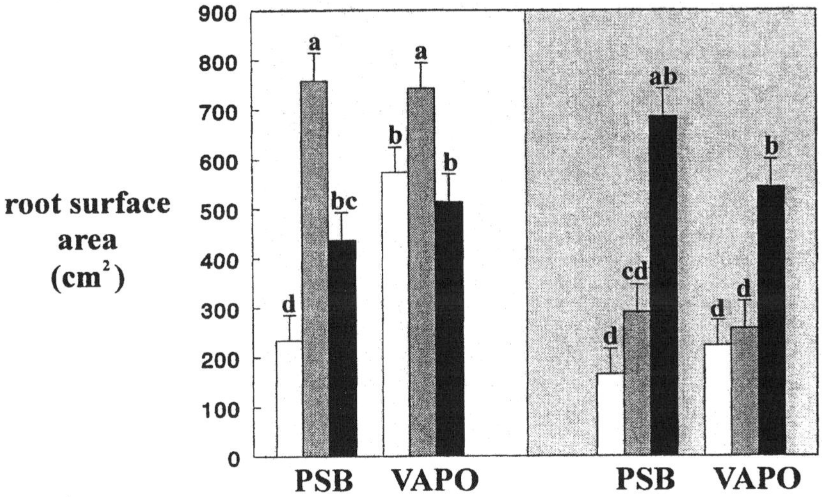

Figure 4.

attributed to desiccation than did unfertilized seedlings (Krasowski, 1998).

The study reported here accounted only for nutrients in the foliage at the end of the experiment and not for nutrient uptake, translocation, and losses due to leaching. Nevertheless, total nutrient pools of all elements other than apparently unavailable Fe were greater in fertilized than in unfertilized seedlings. Nutrient pools reflect the plant's nutrient uptake (Chapin and Van Cleve, 1989) which must have been greater in fertilized than in unfertilized plants. Only N accumulated in fertilized plants at a level that significantly increased its concentration in leaves. However, fertilizer N was supplied in much greater proportion than P and K.

The most striking difference between root growth of unfertilized PSB and Vapo seedlings was the great proliferation of roots of the latter in mineral compartments. Vapo seedlings had significantly higher percent of their roots colonized by mycorrhizal fungi than did PSB seedlings They also had fewer roots in poor condition. A reduced root colonization by E-strain fungus was particularly common in unfertilized PSB seedlings. These factors could affect nutrient uptake and be the cause of the observed differences in P and K contents between the two stock types. While the growth of both stock types benefited from the applied controlled-release fertilizer, it appears that the fertilizer was much more critical to the improved growth of PSB than Vapo roots.

Studies on Sitka spruce (Coutts and Philipson, 1976; Philipson and Coutts, 1977) showed that nutrients, particularly N, stimulated radial growth of roots in direct contact with fertilizer. There was little growth stimulation of unfertilized portions of Sitka spruce roots even though the supplied nutrients were apparently translocated to these roots. In white spruce seedlings examined in this study, fertilizer had a localized effect on root proliferation, elongation, and radial growth mainly when placed into mineral compartments. When applied to organic compartments, the fertilizer effect was no longer localized but the growth of roots was differently affected in the two stock types. Localized proliferation of roots could be viewed as an adaptive mechanism of root systems allowing them to exploit local concentrations of nutrients in the soil (Drew, 1975; Drew and Saker, 1975; Robinson, 1994; van Vuuren et al., 1996). This would explain responses of PSB seedlings to fertilizer addition and the response of Vapo seedlings to fertilizing the organic compartments. In the absence of localized nutritional stimulus in the soil, root proliferation and root growth of white spruce seedlings depended on the soil type and on properties of root systems resulting from different nursery cultures. Based on results of this trial and on results obtained by Coutts and Philipson (1976) and Philipson and Coutts (1977), it might be expected that the symmetry of root systems treated with localized supply of fertilizer could be altered. However, Carlson and Preisig (1981) reported no such alterations in Douglas-fir supplied with slow-release fertilizer at planting.

Differences in root growth between the two unfertilized stock types in mineral soil were not reflected

Table 4. Nutrient content of white spruce seedling foliage of plants from the divided root experiment

Fertilizer treatment		None		In mineral		In organic	
Foliage produced in		1996	1997	1996	1997	1996	1997
Element	Units						
C	$g \times 10^{-1}$	0.5244b	0.5193b	0.4581b	1.064a	0.540b	0.941a
Ca	$g \times 10^{-1}$	0.0090ab	0.0051b	0.0083ab	0.0122a	0.0108a	0.0130a
K	$g \times 10^{-1}$	0.0072b	0.0071b	0.0060b	0.0162a	0.0081b	0.0161a
Mg	$g \times 10^{-1}$	0.0020b	0.0016b	0.0019b	0.0030a	0.0020b	0.0022ab
Mn	$g \times 10^{-6}$	438.8cd	245.4d	494.6bc	843.7a	563.3bc	800.6ab
N	$g \times 10^{-1}$	0.0076c	0.0076c	0.0176b	0.0512a	0.0121bc	0.0532a
P	$g \times 10^{-1}$	0.0021b	0.0023b	0.0022b	0.0058a	0.0030b	0.0060a
S	$g \times 10^{-1}$	0.0013b	0.0011b	0.0017b	0.0031a	0.0021ab	0.0029a
Foliage dry weight	$g \times 10^{-1}$	1.02c	1.06c	0.96c	2.07a	1.10c	1.77b

Note: Means of the same variable followed by the same letters are not significantly different from each other at $\alpha = 0.05$.

in corresponding differences to the shoot growth of these stock types. However, in another portion of these studies (Krasowski, 1998) shoots and roots of Vapo seedlings grew significantly more than in PSB seedlings, without ferilizer but over a month-longer growing season. The poor root growth in organic soil was reversed by adding fertilizer. Stimulation of all aspects of root growth in the organic soil must have been mediated by the addition of at least one of the elements contained in the fertilizer. We did not use natural forest soils in this experiment but its results suggest that additional studies on different types of forest soils are needed before the recommendations for planting conifer seedlings with roots located in organic layers of the forest floor (Balisky et al., 1995) could be broadly accepted. At least in some organic soils, root proliferation during the seedling-establishment phase could be enhanced by applying slow release fertilizer, especially when container-grown seedlings are planted. Seedlings planted in organic soil and not supplied with fertilizer may have underdeveloped root systems.

Coutts (1982) reported stimulation of root growth in wet soil but poor growth and dying of some roots in dry soils. These responses were independent of the soil type (sandy loam or peat were used). Our study was intentionally conducted under ambient conditons without supplemental irrigation and differences in moisture fluctuation between the mineral and organic soils were likely, especially in August which was the driest month during the study period. Surface organic soil may dry excessively on warm days (Sutton, 1991) and this may have occurred in our study. However, the stimulation of root growth in organic compartments by fertilizer suggests that nutrient immobilization rather than low soil moisture limited root growth in these compartments.

The differences in the composition and abundance of ectomycorrhizae appeared to be related more to root health status than to the different root-growth media and fertilizer treatments. Since root assessment revealed more poor and dying roots in unfertilized than in fertilized seedlings and in PSB than Vapo seedlings, interaction effects involving stock type and fertilizer treatment on ectomycorrhizae composition could be indirect. It is not known why PSB seedlings had more shriveled and dying roots than Vapo. In unfertilized PSB seedlings, the proliferation of lateral roots appeared suppressed. This was consistent with observations made in other studies. Excavated root systems of white spruce PSB seedlings often resembled the expansion of their nursery configuration (Krasowski et al., 1996). These root systems had numerous long, vertically oriented laterals, and fewer branches than 'fibrous' root systems of Vapo seedlings.

Dahlberg (1990) found no evidence that the presence of organic layer in forest soils could affect the establishment and development of ectomycorrhizae in planted seedlings. In our study, the only mycorrhizal fungus that seemed to prefer one type of soil was *Cenococcum*. It was more common in roots located in mineral than in organic soil, even though its abundance was extremely low. Since organic compartments had sand in their bottom portions, it is possible that the two occurrences of *Cenococcum* in organic compartments were in sand rather than in organic soil. Inoculation of Sitka spruce seedlings with *Cenococcum* did not improve growth of seedlings planted on clearcuts in Alaska (Loopstra et al. 1988; Shaw et al., 1987). Resident fungi provided better nutritional benefits to the host plants than inoculated species of three fungi, including *Cenococcum* (Sidle and Shaw, 1987).

Controlled-release fertilizer used in this study did not reduce colonization of roots by mycorrhizal fungi. Reductions in the degree of root colonization by fungi could be related to the suppression of fungal development by high, toxic concentrations of fertilizer salts in the soil (Maronek et al., 1981, 1982). This was not evident in the current study indicating that there was no toxic effect of the applied fertilizer on mycorrhizal fungi. *Thelephora,* a common morphotype during nursery growth, was partly replaced by other fungi after planting. Low levels of *Amphinema* and MRA on roots of Vapo may reflect the absence of these fungi on nursery roots of this stock type. The fact that seedlings came from greenhouse culture and were growing outdoors for only one season may account for the absence of other mycorrhizal fungi. The apparent reverse relationship between the abundance of E-strain on healthy root systems and MRA on poorer root systems is interesting. Numerous reports have proposed that MRA may display semi-pathogenic behaviour on weak or senescent roots (De la Bastide and Kendrick, 1990; Haug et al., 1988; Livingston and Blaschke 1984).

Successful planting of white spruce seedlings and its future growth may depend on the attributes of seedling's root system and on the soil substrate in which these roots are placed. Results of this study are not readily transferable to natural field situatuations with real forest floors. The organic materials of forest floors vary in their physical and chemical properites but ususally have lower pH and C:N ratio than the organic mix used in this study (P. Sanborn, soil scientist, Prince George Forest Region, person. comm., 1998). Also, the organic material used in this study was homogenized and lacked the natural vertical and lateral variability of forest floors. Nevertheless, this study indicated that slow release fertilizer can improve the growth of shoots and roots of white spruce seedlings in spite of the short growing season resulting from late planting. It also showed that root growth of different types of white spruce planting stock differed in relation to the type of the growing medium and in response to the application of controlled-release fertilizer. Testing the effectiveness of controlled-release fertilizer on growth of shoots and roots of trees planted into natural and varied forest soils would be a logical expansion of the reported study.

Acknowledgement

We thank Adam (Oli) Caputa (Red Rock Research Station) for technical assistance and Peter Ott (biometrician, Research Branch BCMoF, Victoria) for biometric consultation. Appreciation is extended to Clive Dawson (BCMoF analytical laboratory, Victoria) for soil and plant tissue analyses. This study was supported by the NSERC Industry/University Partnership Grant to J N Owens and by the British Columbia Ministry of Forests. This work is a portion of a dissertation by the first author submitted to the University of Victoria in partial fulfillment for the Ph.D.

References

Anonymous 1995 Gromax^{TM} and fertilization at time of planting: a provincial summary of operational and research experience. Regeneration Note #7, Forest Renewal Section, Silviculture Practices Branch, British Columbia Ministry of Forests.

Austin R C and Strand R F 1960 The use of slowly soluble fertilizers in forest planting in the Pacific Northwest. J. For. 58, 619–627.

Balisky A C, Salonius P, Walli C and Brinkman D 1995 Seedling roots and forest floor: misplaced and neglected aspects of British Columbia's reforestation effort? For. Chron. 71, 59–65.

Burdett A N, Herring L J and Thompson C J 1984 Early growth of planted spruce. Can. J. For. Res. 14, 644–651.

Brockley R P 1988 The effects of fertilization on the early growth of planted seedlings: a problem analysis. Forest Resource Development Agreement (FRDA). Report #011, Forestry Canada, British Columbia Ministry of Forests joint publication, Victoria B.C. pp. 16.

Carlson W C and Preisig L 1981 Effects of controlled-release fertilizers on the shoot and root development of Douglas-fir seedlings. Can. J. For. Res. 11, 230–242.

Carter M. 1993. Soil sampling and methods of analysis. Canadian Society of Soil Science. 1993.

Chapin F S III and Van Cleve K 1989 Approaches to studying nutrient uptake, use and loss in plants. *In* Plant Physiological Ecology, Eds R W Pearcy, J Ehleringer, HA Mooney and P W Rundel. pp. 185–207. Chapman and Hall, New York.

Coutts M P 1982 Growth of Sitka spruce seedlings with roots divided between soils of unequal matric potential. New Phytol 92, 49–61.

Coutts M P and Philipson J J 1976 The influence of mineral nutrition on the root development of trees I. The growth of Sitka spruce with divided root systems J. Exp. Bot. 27, 1102–1111.

Crowley D E, Maronek D M and Hendrix J W 1986. Effect of slow release fertilizers on formation of mycorrhizae and growth of container grown pine seedlings. J. Environ. Hort. 4, 97–101.

Dahlberg A 1990 Effect of soil humus cover on the establishment and development of mycorrhiza on containerized *Pinus sylvestris* L. and *Pinus contorta* ssp. *latifolia* Engelm. after outplanting. Scand. J. For. Res. 5, 103–112.

De la Bastide P Y and Kendrick B 1990 The *in vitro* effects of benomyl on disease tolerance, ectomycorrhiza formation, and growth of white pine *Pinus strobus* seedlings. Can. J. Bot. 68, 444–448.

Drew M C 1975 Comparison of the effects of a localized supply of phosphate, nitrate, ammonium and potassium on the growth of the seminal root system, and the shoot, in barley. New Phytol. 75, 479–90.

Drew M C and Saker L R 1975 Nutrient supply and the growth of the seminal root system in barley. J. Exp. Bot. 26(90), 79–90.

Gagnon J, Langlois C G, Bouchard D and Le Tacon F 1995 Growth and ectomycorrhizal formation of container-grown Douglas-fir seedlings inoculated with *Laccaria bicolor* under four levels of nitrogen fertilization. Can. J. For. Res. 25, 1953–1961.

Hallsby G 1995 Field performance of outplanted Norway spruce: effects of organic matter amendments and site preparation. Can. J. For. Res. 25, 1356–1367.

Haug I, Weber G and Oberwinkler F 1988 Intracellular infection by fungi in mycorrhizae of damaged spruce trees. Eur. J. For. Path. 18, 112–120.

Jurgensen M F, Harvey A E, Graham R T, Page-Dumroese D S, Tonn J R, Larsen M J and Jain T B 1997 Impacts of timber harvesting on soil organic matter, nitrogen, productivity, and health of inland nortwest forests. For. Sci. 43, 234–251.

Kalra Y P and Maynard D G 1991 Methods Manual for Forest Soil and Plant Analysis. Forestry Canada Information Report NOR-X-319.

Krasowski M J 1998. The development of roots and root systems in white spruce (*Picea glauca* ([Moench] Voss) seedlings and the influence of cultural treatments on root morphology, anatomy and the capacity to conduct water. Ph.D. dissertation, University of Victoria, Victoria, B.C. 273 p.

Krasowski M J, Letchford T, Caputa A, Bergerud W A, Ott P K, 1996 The susceptibility of white spruce seedlings to overwinter injury and their post-injury field responses. New Forests 12, 261–278.

Livingston W H and Blaschke H 1984 Deterioration of mycorrhizal short roots and occurrence of *Mycelium radicis atrovirens* on declining Norway spruce in Bavaria. Eur. J. For. Path. 14, 340–348.

Loopstra E M, Shaw C D III and Sidle R C 1988 Ectomycorrhizal inoculation fails to improve performance of Sitka Spruce seedlings on clearcuts in southeastern Alaska. West. J. Appl. For. 43, 110–112.

Maronek D M, Hendrix J W and Stevens C D 1981 Fertility-mycorrhizal interactions in production of containerized pine oak seedlings. Scientia Horticulturae 15, 283–289.

Maronek D M, Hendrix J W and Cornelius P L 1982 Slow-release fertilizers for optimizing mycorrhizal development in container-grown pine seedlings inoculated with *Pisolithus tinctorius*. HortScience 107, 1104–1110.

Massicotte H B 1994 Characterization of ectomycorrhizal morphotypes. *In* Practical Methods in Mycorrhiza Research. Eds M Brundrett, L Melville and L Peterson, Chap 11, pp. 88–94. Mycologue Publications.

McMinn R G 1982 Ecology of site preparation to improve performance of planted white spruce in northern latitudes. *In* Forest regeneration at high latitudes: experiences from northern British Columbia. USDA For. Serv. Proc. Northwest For. Range Exp. Stn. Misc. Rep. 82–1 pp. 25–32.

Mullin R E 1964 Reduction in growth of white spruce after outplanting. For. Chron. 40, 488–494.

Parviainen J and Tervo L 1989 A new approach for production of containerised coniferous seedlings using peat sheets coupled with root pruning. Forestry 62 (suppl.), 87–94.

Philipson J J and Coutts M P 1977 The influence of mineral nutrition on the root development of trees. II. The effect of specific nutrient elements on the growth of individual roots of Sitka spruce. J. Exp. Bot. 28, 864–871

Robinson D 1994 The responses of plants to non-uniform supplies of nutrients. New Phytol. 127, 635–674.

Rothacher J S and Franklin J F 1964 Fertilizer pellets improve growth of planted Douglas-fir on an unfavorable site. U.S. For. Serv. Tree Plant. Notes 67, 9–11.

Salonius P O 1972 Microbial response to fertilizer treatments in organic forest soils. Soil Sci. 114, 12–19.

Shaw C G III, Sidle R C and Harris A S 1987 Evaluation of planting sites common to a southeast Alaska clear-cut. III. Effects of microsite type and ectomycorrhizal inoculation on growth and survival of Sitka spruce seedlings. Can. J. For. Res. 17, 334–339.

Sidle R C and Shaw C G III 1987 Evaluation of planting sites common to a southeast Alaska clear-cut. IV. Nutrient levels in Ectomycorrhizal Sitka spruce seedlings. Can. J. For. Res. 17, 340–345.

Sutton R F 1972 Constraints on the growth of young white spruce. *In* White spruce the ecology of northern resource. Ed. R G McMinn, Can. For. Serv. North. For. Res. Cent. Inf. Rep. Nor-X-40. pp. 24–39.

Sutton, R F 1991 Soil properties and root development in forest trees: A review Forestry Canada, Ont. Region, Sault Ste. Marie, ON. Inf. Rep. O-X-413. 42 p.

Sutton R F 1993 Mounding site preparation: a review of European and North American experience. New For. 7, 151–192

van den Driessche R 1988 Response of Douglas-fir (*Pseudotsuga menziesii* [Mirb.] Franco) to some different fertilizers applied at planting. New For. 2, 89–100.

van Vuuren M M I, Robinson D and Griffiths B S 1996 Nutrient inflow and root proliferation during the exploitation of a temporally and spatially discrete sources of nitrogen in soil. Plant Soil 178, 185–192.

Vyse A 1981 Growth of young spruce plantations in interior British Columbia. For. Chron. 57, 174–180.

A. Stokes (ed.), The Supporting Roots of Trees and Woody Plants: Form, Function and Physiology, 171–178.

Heritability of first-order lateral root number in *Quercus*: implication for artificial regeneration of stands

Paul P. Kormanik[1], Shi-Jean S. Sung[1], Taryn L. Kormanik[2], Stanley J. Zarnoch[3] and Scott Schlarbaum[4]
[1]*Institute of Tree/Root Biology, Southern Research Station, USDA-Forest Service, 320 Green Street, Athens, GA 30602, USA;* [2]*Department of Agricultural and Environmental Sciences, University of Georgia, Athens, GA 30602, USA;* [3]*Biometrics, Southern Research Station, USDA-Forest Service, Asheville, NC 28802, USA and* [4]*Forest Genetics, Department of Forestry, Wildlife and Fisheries, University of Tennessee, Knoxville, TN 37901, USA*

Key words: first-order lateral roots, heritability estimates, *Quercus*, shade tolerance

Abstract

Natural regeneration of oak (*Quercus*) species in the USA has been easy to obtain on the lower quality xeric sites (site index ≤20 m at age 50) by developing advanced oak reproduction before stands are harvested. This approach has not been successful with *Q. rubra, Q. pagoda*, or *Q. alba* growing on highly productive river bottom, cove or other mesic sites (site index ≥23 m at age 50) because developing seedlings are overtopped by initially faster-growing and more shade-tolerant competitor species common on these sites. Artificial regeneration to increase percentages of these three valuable multiple-use species in specific stands has not been satisfactory because of limitations imposed by planting stock quality and by the traditional technology used to manage newly established stands. After the number of first-order lateral roots (FOLR) produced by an individual seedling was shown to be a highly heritable trait, artificial regeneration trials were started by the USDA Forest Service's Institute of Tree/Root Biology. A nursery fertility protocol was developed that could reliably and consistently produce seedlings of specific sizes. Thereafter, a seedling evaluation system was developed utilizing the parameters of FOLR, root collar diameter (RCD), and height (HT) for selecting seedling for outplanting. On mesic sites, survival and early growth for *Q. rubra* has been very good, with height growth of 4–5 m in 3–8 years. Trials are now being conducted with *Q. pagoda* and *Q. alba*. Methods for developing evaluation systems for oak species grown in nurseries under conditions different than the ones used in this study are discussed.

Abbreviations: FOLR – first-order lateral roots; RCD – root collar diameter; HT – height; NRO – northern red oak; WO – white oak; ITRB – Institute Tree/Root Biology; CBO – cherrybark oak; PAR – photosynthetic active radiation

Introduction

Members of the oak genus (*Quercus* spp) represent an important component of many forests in the eastern half of the United States as well as in many European countries. Oak species provide important forest lumber products and are of major importance for other uses such as mast production for numerous birds and wildlife species. Oaks also have other recreational values, and are of aesthetic importance in both urban and rural forests. They occupy a range of sites in the USA from the very fertile and productive river bottoms and mesic sites to the more xeric and less productive upland sites. Maintaining a significant oak component in new stands depends on having advanced oak regeneration when the current stand is harvested. Advanced oak regeneration on the lower quality upland sites (site index ≤ 20 m at age 50) is not difficult to obtain because competitive species are absent or techniques for obtaining advance regeneration have been developed (Sander, 1972; Sander et al., 1976; Sander and Clark, 1971). Obtaining oak reproduction on the higher quality sites has been difficult because of the competition from faster growing shade tolerant species that are released or that develop when the stand is thinned to encourage advanced oak regeneration (Lorimer, 1993). Clark (1993) reports that many foresters fear that the oak forests of Eastern United States may be experiencing radical changes in structure and composition and may be unstable on all but the more xeric sites because of competition from more shade tolerant species.

Some oak species like northern red oak (NRO) and white oak (WO) (*Quercus rubra* and *Q. alba*, respectively) have been more difficult to regenerate than others. Unless technology is developed for their retention in future stands, many believe that NRO may eventually disappear from the most productive mesic sites in the eastern hardwood forests (Kellision, 1993). Many have advocated the selection or shelterwood regeneration systems for obtaining advanced regeneration on these mesic sites such as been so effective on the lower quality upland sites (Loftis, 1983a,b). However, after almost 50 years of various modifications of these regeneration systems, very few new stands have been obtained on the higher quality sites (site index $\geq$ 23 m at age 50) even though some early partial regeneration successes have been reported. These poor regeneration results can usually be attributed to overestimating the shade tolerance of oak and their slow response to release compared to the rapid growth of competing shade tolerant species that quickly occupy the site (Loftis, 1990a; Nowacki et al., 1990). Unfortunately, even if one is successful in obtaining advanced oak regeneration, success is still not assured because when the final canopy is removed, the well-established, more shade tolerant species will generally occupy the site (Hodges and Gardiner, 1993).

Ten to 15 years ago the USDA Forest Services' Institute of Tree/Root Biology (ITRB) began a series of investigations to determine the feasibility of developing advanced regeneration of NRO in forest tree nurseries as 1-0 seedlings. These seedlings were to be used for modifying traditional artificial regeneration technology, or to enrich existing stands, or to develop entirely new stands with highly competitive individual seedlings. This extensive research effort now encompasses several broad scientific disciplines which include determining heritability values for various seedling root characteristics, basic seedling physiology, development of nursery protocols to produce seedlings, and developing systems for evaluating seedling competitive potential as well as soil modification procedures to maximize early oak seedling development. While 15 oak species have been studied in the nursery, NRO, WO and cherrybark oak (CBO) (*Q. pagoda*) have received the greatest effort in evaluating nursery seedlings in conjunction with field testing.

The discovery of the importance of first-order lateral root (FOLR) number to the competitive ability exhibited by loblolly pine (*Pinus taeda*) and sweetgum (*Liquidambar styraciflua*) seedlings was the catalyst for continued research with oak species (Kormanik, 1986; Kormanik et al., 1990). Kormanik and Muse (1986) reported on early nursery and field trials with these species and they concluded that "regardless of phenotypic characteristics of mother trees, their progeny can be stratified by number of first-order-lateral roots (FOLR), and that seedlings with fewest lateral roots will be least competitive in the nursery and after outplanting in a forest environment". This report stimulated interest in potential widespread application of artificial regeneration for oak species in various sections of the United States through more thorough scrutiny of oak seedlings' root systems (Kormanik, 1989; Schultz, 1988; Schultz and Thompson, 1990). However, before valid heritability estimates could be obtained or a biologically sound seedling evaluation system developed, a nursery protocol was needed to permit consistency in seedling production so valid statistical comparisons could be made among families and different nursery locations for different years.

Materials and methods

Nursery protocol

Fifteen oak species have been used during development of a nursery protocol that would reliably produce seedlings of specific sizes in southeastern US nurseries (Kormanik et al., 1994a). Although many oak species are being evaluated for this nursery research protocol, only open pollinated half-sib progeny from specific NRO, CBO, and WO mother trees have been tested in sufficient quantities to obtain reliable, but albeit tentative heritability estimates for FOLR. These species will be addressed here. It is quite likely that modification of the protocol will be required when oak, or even other hardwoods, are produced in regions with significantly different environmental and edaphic conditions than exist in the southeastern United States (Kormanik et al., 1994a,b, 1995).

Briefly, the nursery fertility protocol that we developed appears to be effective, if not optimal, for most hardwood species (Kormanik et al., 1994a). The soil fertility levels for this nursery protocol are adjusted annually before acorns are fall sown. The levels of Ca, K, P, Mg, Cu, Zn and B are set at 500, 130, 100, 50, 0.3–3, 3–8, and 0.4–1.2 ppm, respectively, based on soil test results from standard Mehlich extraction. The desired seedling bed density is between 54–57/m^2. Eight to 12 N applications (as NH_4NO_3) are required during the growing season depending upon species and seedling size desired. The first two N applications are 17 kg/ha, the third at 56 kg/ha and subsequent ones at 168 kg/ha. Nitrogen applications start in mid-May and continue at 10-day intervals until mid-September. Stopping N top dressing at this time permits 60–90 days for seedling to harden off before first frost is anticipated at our nursery location. Irrigation is provided as needed until seedlings are well established in mid-June and thereafter when moisture tensiometers register 30–40 centibars at 15–20 cm depth. When seedlings approach the desired height, i.e., 1.0–1.2 m for NRO and CBO or 0.60–0.70 m for WO, either top dressing of N, irrigation frequency, or both can be altered to reduce stem elongation in favor of RCD and root expansion. It is important to maintain consistency in nursery practices if valid annual comparisons among progeny from various mother trees are to be undertaken. For example, a significant change in seedling bed density affects development of lateral roots and can significantly affect FOLR numbers and RCD without materially affecting seedling height. This results in a spindly seedling that may not survive or compete satisfactorily for several years.

FOLR: heritability

All heritability trials were conducted at the ITRB Experimental Nursery located approximately 7 km from the US Forest Service Laboratory in Athens, Georgia. The soil fertility protocol described earlier was followed and nursery layout was uniform for all heritability trials regardless of oak species being evaluated. The experimental nursery beds have been described in detail elsewhere (Kormanik, 1986), are 1.2×18.2 m, and have 120 rows with approximately 0.15 m between rows. There are 10–12 contiguous rows per mother tree per replication depending on whether 10 or 12 progeny groups are being evaluated. Thirteen sowing spots are included in each row and there are two replications per trial.

Family mean heritability (h^2) estimates, on a plot mean basis, are computed for FOLR number according to Kormanik et al. (1990) as:

Table 1. First-order lateral roots (FOLR) family means broad sense heritability estimates for three *Quercus* species in several studies[a]

Study	Heritability[2]	Standard Error[3]
Q. rubra (NRO)		
1	0.898	0.153
2	0.870	0.114
3	0.843	0.235
Q. pagoda (CBO)		
1	0.904	0.104
2	0.783	0.271
3	0.558	0.428
Q. alba (WO)		
1	0.843	0.138
2	0.918	0.073

[a] 12 families were used in each test except for *Q. pagoda*, test 1, which had nine families. There were always two replications in each test.

$$h^2 = \frac{MS_F - MS_{RF}}{MS_F}$$

where MS_F is mean square for family, and MS_{RF} is mean square for the replication×family interaction, which are obtained from analysis of variance for the randomized block design. An estimate of the standard error of h^2 is computed as:

$$S.E.(h^2) = \left[\frac{2d^2(d+n-2)}{n(d-2)^2(d-4)} \cdot \frac{MS_{RF}^2}{MS_F^2}\right]^{0.5}$$

where d is degrees of freedom for family, and n is degrees of freedom for the replication×family interaction.

Table 1 contains heritability estimates for northern red, white and cherrybark oaks from different experiments conducted over the past 8 years. Where multiple studies values are shown, the mother trees and years are different. Except for cherrybark oak, which represents acorn collections primarily from coastal plain of South Carolina, the original mother trees of NRO and WO were from scattered areas in the states of Georgia, North Carolina, Virginia, Tennessee and Kentucky. Heritability estimates for NRO families in this test number 3 (Table 1) were also obtained from seedlings produced in Tennessee by a different nursery protocol. The seedlings grown in Tennessee were significantly

Table 2. Morphological and growth characteristics of 1-0 northern red oak (*Quercus rubra*) seedling from which seedling evaluations are made[a]

Family	FOLR# x,	FOLR# Range	% Seedling with FOLR# $\leq x$,	RCD (mm) x,	RCD (mm) Range	HT (cm) x,	HT (cm) Range
1-14-915	6.4	0–26	58	9.8	3.9-18.7	125	22-233
2-10-540	5.6	0–23	54	8.4	3.7-15.8	135	40-262
2-19-630	4.6	0–25	57	9.0	2.9-17.1	131	43-238
4-14-2459	5.9	0–20	52	10.5	3.4-18.9	125	22-253
2-29-565	5.2	0–24	58	9.8	3.2-19.2	134	18-260
2-6-735	4.0	0–16	63	9.1	3.4-17.6	117	28-211
3-3-526	4.9	0–19	54	9.1	3.3-15.7	114	33-243
4-2-902	3.6	0–15	62	8.4	3.6-15.	124	46-223
4-14-200	5.0	0–20	58	8.6	3.0-15.9	120	41-232
4-27-100	4.2	0–19	56	9.3	2.9-17.9	141	33-246
4-4-882	4.0	0–19	57	8.8	3.9-16.4	126	47-243
2-23-850	2.3	0–25	72	6.2	1.0-16.9	73	8-207

[a]Within a half-sibling group, individuals whose development exceeds the mean values for FOLR number, RCD and HT should be acceptable for artificial regeneration.

smaller in root collar diameter (RCD) and height (HT) and the FOLR were smaller in diameter and not as well developed as those grown at our experimental nursery, but the h^2 estimates were comparable.[1] The h^2 estimates for FOLR reported here for oak (Table 1) are comparable to those reported for loblolly pine (*Pinus taeda*) which were obtained from seedlings produced under different nursery protocols (Kormanik and Ruehle, 1989; Kormanik et al., 1990). This suggests that genetic control of root morphological traits may be quite significant and need to be considered when evaluating seedlings of other species.

Seedling evaluation

Major emphasis for evaluating competitive potential of oak seedlings in the United States has emphasized seedlings' RCD and heights, but both of these parameters have varied considerably depending on nursery protocol and geographical location of nurseries. Established quality standards for oak seedlings have not been developed in the United States or elsewhere in the world and may be reflected in the absence of wide acceptance of artificial regeneration of oak species. Tentatively we have begun to set standards for outplanting stock of the three oak species reported here which involves a combination of FOLR numbers, RCD, and HT. We select NRO, CBO and WO seedlings with FOLR numbers $\geq$5 and RCD $\geq$7 mm. Heights for NRO and CBO should be $\geq$1 m, but WO are acceptable at heights of $\geq$0.60 m. However, final determination of these standards must await both more nursery and field trials under different environmental and edaphic conditions to verify their utility. A single standard for evaluating all oak species may be unlikely but a procedure to develop a standard for any species appears feasible using these three variables in conjunction with a consistently applied nursery protocol adapted to individual locations.

Our oak seedling evaluation concept began by periodically excavating seedlings in the nursery and under forest conditions. In the forest stands, oak seedling decline became obvious after they were overtopped by faster growing competitors or when canopy closure occurred. WO seedlings survived better than NRO or CBO in the forest due to their early fall germination because it prevents the acorn desiccation that is so injurious to the recalcitrant oak acorns. NRO and CBO however, may remain on the forest floor for several months without germinating and viability can be significantly reduced through desiccation as has been reported by others (Bonner and Vozzo, 1987). Normally, however, the advantage of WO fall germination is shortlived, and if not released they pass out of the stand in 2–3 years. All surviving seedlings beginning to decline in vigor in forest stands were found to have few, if any, FOLR. Observing root morphology of individual seedlings in the forest involved

[1] Data on file at Scott Schlarbaum, Department of Forestry, Knoxville, Tennessee.

Table 3. Morphological and growth characteristics of 1-0 cherrybark oak (*Quercus pagoda*) seedling from which seedling evaluations are made[a]

Family	FOLR#		% Seedling with FOLR#	RCD (mm)		HT (cm)	
	$x_,$	Range	$\leq x_,$	$x_,$	Range	$x_,$	Range
Combic 1	4.9	0–18	52	7.3	2.1-12.1	102	33-160
Combic Int	4.8	0–22	56	7.6	3.2-12.5	103	23-161
Combic 2	4.7	0–15	59	7.7	2.0-13.1	105	23-174
Gillin 20	3.1	0–14	64	6.1	2.6-10.8	94	24-160
Kirby 17	4.4	0–15	56	6.8	2.9-10.8	104	37-156
Mahan 18	3.4	0–16	56	6.2	2.4-11.1	103	40-172
Moraux 19	4.4	0–17	56	6.4	1.9-10.6	95	26-165
Museum	4.9	0–17	47	6.6	2.1-12.4	105	31-189
Orchard	4.4	0–16	57	6.8	2.4-13.6	102	24-172

[a]Within a half-sibling group, individuals whose development exceeds the mean values for FOLR number, RCD and HT should be acceptable for artificial regeneration.

destructive sampling, and, thus, repeated observations on the same seedling were not possible. However, measurement of photosynthetic active radiation (PAR) indicated less than 10% of full sunlight was reaching most newly developing seedlings that were in declining vigor. This low light intensity at the forest floor has been reported by others and is not sufficient for long term development of oak seedlings (Hodges and Gardiner, 1993).

We began evaluating nursery seedlings as soon as most seedlings had completed their initial flush, and by mid-June a comparable absence of FOLR development was evident among the smaller and less competitive individuals. This was comparable to what was observed with low vigor seedlings in the forest. If oak seedlings were overtopped in the nursery by siblings by mid-July, they seldom were ever free to grow again. When seedlings bed density was reduced to alter competition among seedlings, the least competitive seedlings still produced few FOLR even though they had larger stem calipers. Oak seedlings that produced few FOLR represented at least 50% of the population from given half-sibling lots. These seedlings that were not competitive in the nursery also were not able to compete successfully after outplanting (Kormanik et al., 1998).

Seedling data collected earlier for obtaining heritability estimates were addressed again in light of the field responses that were being observed from the initial NRO plantings. This nursery data indicated for NRO, WO and CBO that the mean number of FOLR for open pollinated half-sib progeny generally averaged between 4 and 6 while the range in FOLR numbers varied widely among these three species from 0 to 40. It was evident that the FOLR mean values of seedlings were significantly affected by the skewed frequency distribution because seedlings with the fewest FOLR were occurring with the greatest frequency even under ideal nursery conditions. Examples of the skewness of seedlings' morphological characteristics are presented in Tables 2, 3 and 4 for NRO, CBO and WO respectively.[2] These species examples are for the first heritability values for each species shown in Table 1. All acorns used in these specific trials were from single tree selections. The NRO were from the USDA Forest Service's Watuga Seed Orchard located in western Tennessee, the CBO from the Coastal Plain of South Carolina and the WO from various locations on the Piedmont Plateau region of Georgia.

Each table contains the family means and ranges for FOLR, percentage of seedlings with FOLR numbers≤the mean FOLR, RCD and HT. These values are all important parameters for evaluating oak seedlings regardless of nursery production protocols used at a particular location. In many nursery trials involving these three species, as well as 12 other oak species, we have found that the individuals with FOLR numbers≥than the crops' mean FOLR values will represent the more competitive individuals of that species or sibling lot. The least competitive individuals usually have a single taproot with only fine, poorly developed, hair-like lateral roots that atrophy rapidly when seedlings are lifted from the nursery bed or after

[2] Data for other tests on file at ITRB, USDA Forest Service, Athens, GA 30602.

Table 4. Morphological and growth characteristics of 1-0 white oak (*Quercus alba*) seedling from which seedling evaluations are made[1]

	FOLR#		% Seedling with FOLR#	RCD (mm)		HT (cm)	
Family	$\bar{x}$	Range	$\leq\bar{x}$	$\bar{x}$	Range	$\bar{x}$	Range
KYWO 11	3.4	0–17	63	7.5	2.0-13.8	46	12-128
KYWO 31	3.0	0–19	66	8.4	3.0-15.3	35	8-104
NAWO-01	2.5	0–16	67	7.2	3.2-13.6	28	8-87
NAWO 23	4.1	0–22	62	8.2	2.9-15.2	48	8-142
SAWO 28	6.2	0–36	64	8.4	2.2-19.3	45	12-171
NAWO-24	3.8	0–21	60	7.9	2.8-15.0	39	14-147
NAWO-29	2.3	0–18	67	6.5	1.0-15.3	25	8-100
SAWO 12	2.8	0–16	58	7.4	2.4-13.0	36	11-112
NAWO-28	2.1	0–18	72	6.5	2.5-11.6	24	6-68
SAWO 14	4.6	0–32	64	8.6	2.7-18.4	46	12-151
SAWO 3	3.9	0–27	61	8.7	3.6-18.0	39	12-110
SAWO 7	3.6	0–23	61	9.0	2.7-17.4	48	12-134

[a]Within a half-sibling group, individuals whose development exceeds the mean values for FOLR number, RCD and HT should be acceptable for artificial regeneration.

seedlings are outplanted. When these least desirable seedlings are outplanted, they will become overtopped and die within 2–3 years. The poor performance often reported from artificial regeneration attempts with NRO on high quality mesic sites can often be related back to seedling size and poor vigor of the planting stock (McGee and Loftis, 1986). This combination of inferior seedling traits resulted in rapid seedling suppression by competing vegetation and such seedlings are eliminated from the forest soon after outplanting. This may simply be a function of the absence of lateral root development and the competitive potential of individuals with few FOLR. It is unlikely, however, that any single morphological trait will be able to realistically quantify a seedling's competitive potential.

These three oak species, like most oak species, are shade intolerant and during the first several years after outplanting allocate significant amounts of carbon production to root development. Unfortunately, without adequate sunlight, the root system appears to be a poor carbon sink, and lateral root development is significantly reduced and over a relatively short period begins to atrophy (Sung et al., 1998). Within a few years the seedlings lose their tenuous competitive position and die. This is also the sequence observed in the understory for naturally regenerated oak by others (Hodges and Gardiner, 1993; Lorimer, 1993).

Regeneration application

The oak seedlings selected for general outplanting should be chosen from the top 50% of the crop and should have FOLR numbers ≥ the mean FOLR number from the crop being evaluated. This selection criteria will assure the forester of obtaining the taller stems with the larger RCD. If, however, outplantings are to serve primarily as mast producing areas and be potential seed orchards, we recommend selecting only those seedlings in the top 25–30% of the crop. These seedlings will have the largest number of FOLR and be the tallest with the largest RCD. These are easily identified even before they are lifted. The largest NRO seedlings produced under these exacting nursery conditions typically develop quickly and may begin to produce acorns in 5–8 years under the proper field conditions. Under present seed orchards management conditions, NRO and CBO may attain RCD of from 10–15 cm and heights of 4–6 m during this 5–8 year period. In regenerated forest stands, without supplemental irrigation and fertilization, heights may be comparable but their RCD may be reduced by about 50 percent and the seedlings may need to be released from competition between year three and five (Kormanik et al., 1998). White oak have, in our limited tests, grown at about half the rate of the two red oaks but we have not observed early acorn production. Since acorn production is under considerable genetic control, it is risky to cull individuals from potential seed orchards until more definitive research on acorn production is obtained (Schlarbaum et al., 1994).

Seedlings produced in different environments due to geographic location may not develop as quickly as those in the southeastern United States. However, seedling evaluation for potential competition ability may still be readily accomplished. The parameters for FOLR, RCD and HT can be obtained by random sampling of 100–200 individual seedlings to obtain data for these mean values. In environmental conditions where seedlings may attain 1 year heights of only 25–35 cm, it is unrealistic to expect many FOLR to develop and seedlings with 2–3 FOLR may represent competitive seedlings under those conditions. However, we currently have serious reservations for growing the seedlings as 2-0 or 1-1 planting stock to improve seedlings' root and stem characteristics in eastern United States. With proper nursery management, oak seedlings of outstanding competitive potential can easily be produced in one growing season provided the acorns are fall sown. We find that much second year growth is directed to root development that is removed when the 2-year-old seedlings are lifted. The second year does not result in much improvement in root characteristics of seedlings if these seedlings were not competitive as 1-0 nursery stock.

Acknowledgment

This research was funded by US Department of Energy Grant DE-AI09-76SR00870 and by Georgia Forest Commission.

References

Bonner F T and Vozzo J R 1987 Seed biology and technology of *Quercus*. Gen. Tech. Rep. SO-66. New Orleans, LA: US Department of Agriculture, Forest Service, Southern Forest Experiment Station. 21 pp.

Clark F B 1993 A historical perspective of oak regeneration. In Proceedings: Oak regeneration: Serious Problems Practical Recommendations USDA For. Serv. Gen. Tech Rep. SE-84. pp 3–13. Knoxville, TN. Sept. 8–10, 1992.

Hodges J D and Gardiner E S 1993 Ecology and physiology of oak regeneration. In Proceeding: Oak Regeneration: Serious Problems Practical Recommendations USDA For. Serv. Gen. Tech Rep. SE-84. pp 54–65. Knoxville, TN. Sept. 8–10, 1992.

Kellison R C 1993 Oak regeneration – Where do we go from here? In Proceedings: Oak Regeneration: Serious Problems Practical Recommendations USDA For. Serv. Gen. Tech Rep. SE-84. pp 308–315. Knoxville, TN. Sept. 8–10, 1992.

Kormanik P P 1986 Lateral root morphology as an expression of sweetgum seedling quality. For. Sci. 32, 595–604.

Kormanik P P 1989 Frequency distribution of first-order lateral roots in forest tree seedlings: silvicultural implications. *In* Proceeding: 5th Biennial Southern Silvicultural Conf. pp 101–105. Memphis, TN. Nov 1988. USDA For. Serv. Gen. Tech. Report SO-74.

Kormanik P P and Muse H D 1986 Lateral roots a potential indicator of nursery seedling quality. *In* TAPPI Proceedings: 1986 Research and Development Conference. pp 187–190. Raleigh, NC. Sept 28–Oct 1.

Kormanik P P and Ruehle J L 1989 First-order lateral root development: Something to consider in mother tree and progeny assessment. *In* Proceedings: 20th Southern Forest Tree Improvement Conference. pp 220–227. 26–30 June 1989, Charleston, SC. Sponsored Publ #42 of the SFTIC.

Kormanik P P, Ruehle J L and Muse H D 1990 Frequency distribution and heritability of first-order lateral roots in loblolly pine seedlings. For. Sci. 36, 802–814.

Kormanik P P, Sung SS and Kormanik TL 1994a Toward a single nursery protocol for oak seedlings. pp 89–98. *In* Proceedings: 22nd Southern Forest Tree Improvement Conference. 14–17 June 1993, Atlanta, GA.

Kormanik P P, Sung S S and Kormanik T L 1994b Irrigating and fertilizing to grow better nursery seedlings. *In* Proceedings: Northeastern and Intermountain Forest and Conservation Nursery Association. pp 115–121. 2–5 Aug 1993, St. Louis, MO. Gen. Tech. Report RM-243.

Kormanik P P, Sung S S, Kormanik T L and Zarnoch S J 1995 Oak regeneration – Why big is better. *In* National Proceedings: Forest and Conservation Nursery Association. pp 117–123, 1995. Thomas D. Landis and Bert Cregg, Tech. Coordinators.

Kormanik P P, Sung S S, Kass D L and Schlarbaum S 1998 Effect of seedling size and first-order-lateral roots on early development of northern red oak on mesic sites. *In* Proceedings: Ninth Biennial Southern Silvicultural Research Conference. pp 247–252. Clemson, SC. February 1997.

Loftis D L 1983a Regenerating red oak on productive sites in the southern Appalachians: a research approach. *In* Proceedings: 2nd Bienn. South. Silvic. Res. Conf. Gen. Tech. Rep. SE-24. pp 144–150.

Loftis D L 1983b Regenerating Southern Appalachian mixed hardwoods with the shelterwood method. South. J. Appl. For. 7(4), 212–217.

Loftis D L 1990a Predicting Post-harvest performance of advanced red oak production in the southern Appalachians. For. Sci. 36, 908–916.

Lorimer C G 1993 Causes of the oak regeneration problems. *In* Proceedings: Oak Regeneration: Serious Problems Practical Recommendations USDA For. Serv. Gen. Tech Rep. SE-84. pp 14–39. Knoxville, TN. Sept. 8–10, 1992.

McGee C E and Loftis D L 1986 Planted oaks perform poorly in North Carolina and Tennessee. North. J. Appl. For. 3, 114–116.

Nowacki G J, Abrams M D and Lorimer G G 1990 Composition, structure, and historical development of northern red oak stands along an edaphic gradient in north-central Wisconsin. For. Sci. 32, 276–292.

Sander I L 1972 Size of advance reproduction: Key to growth following harvest cutting. USDA For. Serv. Res. Pap. NC-79. 6 pp.

Sander I L and Clark F B 1971 Reproduction of upland hardwood forests in the central states. Agric. Handb. 405 USDA For. Serv., Washington, DC, 25 pp.

Sander I L, Johnson P S and Watt R F 1976 A guide for evaluating the adequacy of oak advance reproduction. USDA For. Serv. Gen. Tech. Rep. NC-23. 7 pp.

Schlarbaum S E, McConnell J L, Barber L R, Cox R A, Grant J F, Kormanik P P, LaFarge T, Lambdin P L, Oak S W, Proffitt J R, C K and Tibbs R T 1994 Research and management in a young northern red oak seedling seed orchard. *In* Proceedings: 22nd Southern Forest Tree Improvement Conference. pp 52–58. 14–17 June 1993, Atlanta, GA.

Schultz R C 1988 The nature of hardwood seedling root development: second year insights from the Hardwood Nursery cooperative. *In* Proceedings: Northeast Area Nursery Men's Conf. pp 26–39. July 11–14, Saratoga Springs, NY.

Schultz R C and Thompson J R 1990 Nursery practices that improve hardwood seedling root morphology. Tree Planters Notes 41, 21–32.

Sung S S, Kormanik P P and Zarnoch S J 1998 Photosynthesis and biomass allocation in oak seedlings grown under shade. *In* Proceedings: Ninth Biennial Southern Silvicultural Research Conference. pp 227–233. Clemson, SC. February 1997.

A. Stokes (ed.), *The Supporting Roots of Trees and Woody Plants: Form, Function and Physiology*, 179–184.

Root-infecting fungi associated with a decline of longleaf pine in the southeastern United States

William J. Otrosina[1,*], Diane Bannwart[2] and Ronald W. Roncadori[2]
[1]*USDA Forest Service, Southern Research Station, Tree Root Biology, 320 Green Street, Athens, Georgia 30602, USA and* [2]*University of Georgia, Department of Plant Pathology, Athens, Georgia 30602, USA*

Key words: decline, *Heterobasidion annosum*, *Leptographium procerum*, *Leptographium terebrantis*, longleaf pine, symptoms

Abstract

A 35-year-old longleaf pine stand exhibited trees in various stages of decline. A study was conducted to determine root-infecting fungi and other abnormalities associated with varying degrees of crown symptoms. A four-class crown symptom rating system was devised according to ascending symptom severity. *Leptographium procerum* and *L. terebrantis* were significantly associated with increasing crown symptom severity. *Heterobasidion annosum* was also isolated in higher frequency as crown symptoms increased. Also, evidence of insects on roots increased as did amount of resinosis observed. Edaphic and silvicultural factors may interact with these pathogens and insects to pose a pathological limitation on longer-term management objectives. Further research is needed to determine relationships among various edaphic, silvicultural, and biological factors associated with the decline syndrome on this site.

Introduction

Longleaf pine, *Pinus palustris* Mill., is the most important component of a plant community that supports a major proportion of the biological diversity in the coastal plain of the southern United States (Martin et al., 1993). The natural range of longleaf pine includes the Atlantic and Gulf Coastal Plains from southeastern Virginia to eastern Texas and south through the northern two-thirds of Florida at elevations ranging from 0 to 200 m. There is also a mountain province of this species in central Alabama and northwestern Georgia that extends to 600 m elevation. Ultisols are the predominant soil order within the natural range of this tree species, with Typic Paleudults and Plinthic Paleudults being typically associated with natural longleaf pine stands. Deep sandy Entisols such as Quartzipsamments also support longleaf pine extending in elevation from near sea level in Florida through 185 m in Georgia and South Carolina (Boyer, 1990).

Due to factors such as conversion from forest to agriculture or other forest tree species, and past failures in artificial regeneration of this species, the longleaf pine ecosystem has declined from its original (historical) geographic range of 35 million hectares to less than two million hectares (Clark, 1984). At present, there is considerable interest and effort spent in restoration of this species to more of its historical range.

Increased mortality is being observed in certain longleaf pine stands in the 30–45-year-old age class and this mortality often occurs within two years after the administration of prescribed fire (Otrosina, 1998). One puzzling aspect of these observations is that this tree species evolved with periodic fires and depends upon fire for serving natural functions such as seed bed preparation and control of brown spot needle disease caused by *Mycosphaerella dearnessii* Barr (= *Scirrhia acicola* (Dearn.) Siggers). In general, longleaf pine has been characterized as resistant or tolerant to most diseases that affect other southern pine species (Boyer, 1990). These diseases include fusiform rust,

* FAX No: (706) 559-4291. E-mail: otrosina@negia.net or wotrosin/srs_athens@fs.fed.us
Plant and Soil is the original source of publication of this article. It is recommended that this article is cited as: *Plant and Soil* **217:** 145–150, 1999.

caused by *Cronartium quercuum* (Berk.) Miyabe ex Shirai f. sp. *fusiforme* (Hedgc. & N. Hunt) Burdsall & G. Snow, annosum root disease, caused by *Heterobasidion annosum* (Fr.) Bref., and other root diseases. The purported root disease tolerance notwithstanding, root pathogenic fungi such as *Leptographium* species and *Heterobasidion annosum* have been isolated in high frequencies from declining longleaf pine trees on many physiographically appropriate sites; however, their role in causing mortality is unclear (Otrosina et al., 1995; Otrosina, 1998). Coinciding with this decline is a serious lack of information on pathological factors and their relationships to symptom development.

Knowledge of which root-infecting fungi are associated with longleaf pine decline would provide important information relevant to understanding their roles in stand deterioration. Root diseases can thwart management objectives by adversely impacting on stand structure many decades after establishment. Seed production losses and losses in habitat for endangered species such as the red-cockaded woodpecker (*Picoides borealis*) are some potential impacts of root diseases in longleaf pine. Thus, achievement of restoration goals in this species is consequential and at least partly depends upon understanding belowground pathological dynamics (Otrosina and Garbelotto, 1998). To this end, the objectives of this study are to (1) determine what relationships exist between root pathogenic fungi and above-ground symptoms, (2) determine the root infecting fungi associated with declining longleaf pine, and (3) advance hypotheses regarding the role of these fungi in longleaf pine ecosystems.

Materials and methods

Study site description

A thirty-five-year-old longleaf pine stand with trees in various stages of decline was selected at the Savannah River Natural Resource Management and Research Institute near New Ellenton, South Carolina. The stand had been thinned to approximately 4 m^2/ha basal area five years prior to this study to develop an advanced natural seed orchard. The average height of the dominant and co-dominant trees was 18–22 m tall with diameters at breast height ranging from 30 to 38 cm. Canopy gaps from mortality ranged from 0.1 to 0.3 ha, although symptomatic trees not identified with established gaps occurred randomly throughout the stand. The soil series existing on the study site is Dothan (Typic Kandiudult), characterized by a sandy upper soil layer, a heavier clay layer occurring within 20–30 cm of the surface, and a hardpan at 30–45 cm depth.

Sampling method

After a preliminary survey of the stand, a classification system of crown symptoms was devised based upon progressive severity. Four crown symptom classes were created: (1) healthy, full crown, predominantly dark green foliage, with less than 10% of foliage being off-color (yellow or brown); (2) early decline, visibly reduced foliar density, with 10% to 25% of the foliage greenish-yellow to yellow, possibly with some drooping needles; (3) moderate decline, foliar density 10% to 50% of class 1 trees, 25% to 50% of crown either yellow, yellow-green, brown, or a combination of these, staghorns in the crown evident; and (4) late decline, foliar density less than 50% of class 1 trees, and the crown having greater than 50% yellow or brown foliage along with defoliated branches scattered throughout the crown. Symptom progression on sampled trees over two growing seasons was followed by noting the visual appearance of crowns during periodic site visits.

Every four to six weeks, over two growing seasons, representative trees within each crown symptom class were selected and sampled by means of root excavation. Fifty-two trees were sampled, representing the four crown symptom classes. There were 10, 19, 14, and 9 trees in symptom class 1, 2, 3, and 4, respectively. Two to four lateral roots that were within 25 cm of the soil surface from selected trees of each symptom class were exposed to approximately two meters distal from the root collar. The average diameter of roots sampled ranged from 6.8 to 11.3 cm. After excavation, resinosis, blue-stain, or evidence of insect galleries was recorded. Resinosis was detected on root bark as discolored, resin-soaked patches, often with encrusted soil, or as resin-soaked xylem in sample cores. Evidence of stain was defined as the presence of blue discoloration in the sample core xylem. No attempt to identify insects was made.

Roots were then sampled for presence of root-infecting fungi with an increment hammer designed to remove 5.0 mm diameter cores of woody tissue to a depth of approximately 3 cm. Symptomatic areas along the length of the excavated roots characterized by resinosis, necrosis, or insect attack were sampled preferentially. At least six cores were obtained from individual symptomatic areas along the exposed root length. When no obvious symptomatic areas were encountered on an excavated root, at least six core samples were taken at the root collar, the midpoint of the excavated length, and at the distal (approximately 2 m from the root collar) end of the root length sampled. Individual excavated root lengths sampled ranged from 1.5 to 2.5 m. Samples were placed in polyethylene bags and kept on ice in a cooler until arrival at the laboratory. All samples were stored at 5 °C and processed within 3 days of collection.

Twelve declining saplings ranging in size from the grass stage to 1.0 m in height, and growing near the drip-lines of selected symptomatic trees, were excavated. The tap root and lateral root systems extending 0.5 m from the root collar were collected. Representative samples from these roots were obtained at several locations along their length and stored as described above.

Root cores and root samples from saplings were rinsed in tap water and dipped in 95% ethanol. Segments of tissue approximately 0.5 to 1.0 cm long were plated on to 2% Difco malt extract agar (MEA) or 1.25% MEA amended with 0.2 g/L cycloheximide and 0.2 g/L streptomycin sulfate (Hicks et al., 1980). Plates were incubated at approximately 25 °C on a laboratory bench and evaluated after one week and every two days thereafter for a period of 21 days. Any ophiostomoid fungi observed on cycloheximide amended MEA were transferred onto 2% MEA for purification and subsequent identification. Presence of *Heterobasidion annosum* on unamended MEA was also recorded by observations of its anamorph, *Spiniger meineckellus* (A.J. Olson) Stalpers.

Data analysis

Data on root pathogenic fungal species, presence of root staining, resinosis, and evidence of insect galleries were analyzed on a tree basis (Otrosina et al., 1997), and expressed as a proportion of trees within a symptom class having the particular variable present in at least one root. Since the symptom class scores were discreet and ordinal in scale, the Mantel-Haenszel Chi-Square was used to test for increasing or decreasing trends in the proportions of each variable with respect to symptom class.

Results

Based upon observations of individual trees over two growing seasons, foliar symptoms usually began with thinning crowns, followed by discoloration, defoliation, and death. Discoloration generally began as a slight paling or yellowing of the needles, which became progressively reddish then changed to brown. Trees in early stages of decline (class 2), tended to develop dieback in the terminal ends of branches. In some cases, trees became discolored before noticeable needle thinning had occurred. Drooping foliage and heavy cone crops also characterized latter-stage symptoms. Symptoms of some trees progressed from class 1 to class 2 or from class 2 to class 3 within one month and remained as class 2 or 3 for the two growing seasons. During the course of this study, only 9 symptom class 4 trees were observed on the entire study site.

Ophiostoma piceae (Münch) Syd. & P. Syd., *Leptographium procerum* (W.B. Kendr.) M.J. Wingf., and *Leptographium terebrantis* Barras & Perry were the predominant blue-stain fungi isolated from sampled trees. In general, there was a significant trend in the proportion of trees from which at least one of these fungi was isolated with respect to symptom severity ($p = 0.001$). No relationship was observed between a particular blue-stain fungus and a specific symptom class. Of the symptom class 4 trees, 88.9% yielded one or more of these fungi. The proportion of trees from which *L. procerum* or *L. terebrantis* was isolated revealed a significant increasing trend as symptom severity increased ($p = 0.001$ and $p = 0.008$, respectively) (Table 1). No significant trends were found in proportion of trees with *O. piceae* with respect to increasing symptom severity ($p = 0.115$). *Heterobasidion annosum* tended to be isolated more frequently as crown severity increased ($p = 0.075$) (Table 1).

The proportion of trees with resinosis in roots increased with symptom severity ($p = 0.004$). Resinosis in roots was observed in 30% of healthy (symptom class 1) trees. Only one blue-stain fungal species, *O. piceae,* was obtained from a tree with root resinosis in this symptom class. Presence of blue-staining in roots was not associated with symptom severity ($p = 0.495$) (Table 2).

Proportions of tree roots with insect galleries for symptom classes 1, 2, 3, and 4 ($n = 10, 19, 14$, and 9) were 0.10, 0.11, 0.21, and 0.67, respectively, and were significantly associated with progressive crown symptom severity (Chi-Square = 8.69, $p = 0.003$). All insect galleries were associated with resinous lesions and at

Table 1. The proportion of longleaf pine trees within four progressive crown symptom severity classes having root pathogenic fungi

Symptom class	N	*Leptographium procerum*	*Leptographium terebrantis*	*Ophiostoma piceae*	*Heterobasidion annosum*
		Fungal Species			
		Proportion			
1	10	0.0	0.0	0.10	0.0
2	19	0.26	0.26	0.32	0.16
3	14	0.29	0.36	0.36	0.14
4	9	0.78	0.56	0.44	0.33
Mantel-Haenszel Chi-square		11.46	7.06	2.48	3.16
	p	$p = 0.001$	$p = 0.008$	$p = 0.115$	$p = 0.075$

Table 2. Proportion of longleaf pine trees in four progressive crown symptom severity classes exhibiting blue-stain and resinosis in roots

Root Condition	Symptom Severity Class 1	2	3	4	Chi-Square	p
Resinosis	0.30	0.63	0.64	1.00	8.408	0.004
Stain	0.10	0.05	0.0	0.22	0.466	0.495

least one of the three species of blue-stain fungi, although no relationship was evident between particular fungal species and insect galleries.

Ten of the twelve symptomatic saplings had resinosis associated with some portion of the root system. Of the 12 sapling roots sampled, 11 yielded *L. procerum*, three yielded *L. terebrantis*, and two yielded *O. piceae*. Two of the 12 saplings were infected with *H. annosum* and *L. procerum*. Blue-stain was evident in roots of only six of the saplings. In one case, excavation revealed that a sapling root infected with *L. procerum* was in contact with but not grafted to a mature sampled tree root also infected with this fungus. None of the saplings sampled had evidence of insect attack.

Discussion

Presence of pathogenic blue-stain fungi in roots of longleaf pine has not been widely reported. This tree species has been considered resistant or highly tolerant to many diseases, including root diseases, that severely affect other pine species in the southern United States. Recent evidence suggests that certain root infecting blue-stain fungi may play a role in the decline of longleaf pine following prescribed burning (Otrosina et al., 1995; Otrosina and Garbelotto, 1998). It is hypothesized that on certain sites, prescribed burning of stands within the 30–45-year-old age class causes stress that, in some as yet unknown way, may predispose tree roots to more rapid colonization by these fungi (Otrosina, 1998). In our study, three species of root infecting blue-stain fungi, *L. terebrantis*, *L. procerum*, and *O. piceae* are widespread in this longleaf pine seed production area.

The presence of *L. procerum* and *L. terebrantis* in roots increased with increasing above-ground symptom severity while presence of *O. piceae* was not related to above-ground symptoms (Table 1). *O. piceae* is regarded as a weak pathogen (Nevill and Alexander, 1992) and may have opportunistically colonized root segments that were debilitated by unknown factors. The pathogenicity of *L. terebrantis* and *L. procerum* on other conifer species has been demonstrated (Klepzig et al., 1991; Nevill et al., 1995; Nevill and Alexander 1992; Rane and Tattar, 1987). This is the first report of these root-infecting fungi associated with above-ground symptoms in longleaf pine.

The fungus *H. annosum*, unlike the blue-stain fungi, is a root-rot fungus that can cause structural instability in conifers in addition to death of root tissue. Its presence on the site is not surprising and it may be a factor in the observed decline. Thinning stumps that are capable of being colonized by the fungus exist on this site, although we did not excavate stump roots to determine the extent of *H. annosum* colonization. No basidiocarps were observed at the root collars of stumps in proximity to sampled trees. Colonized stumps are the primary means of residual tree infection by this fungus (Otrosina and Cobb, 1989). The *H. annosum* in roots of longleaf pine that we found in our study is interesting, since this tree species is reportedly tolerant of annosum root disease (Froelich et al., 1977). Mortality-induced gaps in this seed production area are reminiscent of disease centers characteristic of *H. annosum* (Otrosina and Cobb, 1989). Spreading from distal portions of residual tree roots, where colonized stump root tissues are initially contacted, the fungus moves toward the root collar (Otrosina and Cobb, 1989). Some evidence based on the presence of basidiocarps in longleaf pine thinning stumps seems to indicate the widespread presence of *H. annosum* throughout the range of this tree species (Otrosina, unpublished observations). On the other hand, a high frequency of stumps producing basidiocarps does not necessarily indicate the presence of annosum root disease within a given longleaf pine stand. These observations indicate the need for further research on *H. annosum* infection dynamics in this tree species. Based on the reported tolerance of longleaf pine to annosum root disease (Froelich et al., 1977), we speculate that circumstances such as eroded soils that prevent development of deep root systems, long intervals between prescribed fire, or mechanical damage to residual tree root systems by harvesting equipment may be required for this fungus to cause significant amounts of root disease in this tree species. These conditions seem to be present in our study.

It is possible that *L. terebrantis* and *L. procerum* may be secondary root colonizers of trees with more advanced symptoms and thus play no role in the decline. However, 94% of the root isolations that recovered these two pathogens were associated with resinous lesions. The proportion of trees with resinous roots also increased with increasing symptom severity (Table 2). Presence of resinosis associated with isolation of these fungi implies a host defensive response to infection (Parmeter et al., 1989). Because production of resinous compounds is energetically costly to the tree, these fungi may play an active role in the observed decline syndrome. Resin was observed to infiltrate deep into the root xylem and may also interfere with water conduction (Parmeter et al., 1992).

Also, as symptom severity increased, the proportion of trees with root-feeding insect galleries also increased. Presence of galleries in roots was always associated with resinosis and with the isolation of one or more of the three blue-stain fungal species. Certain root colonizing insects may be vectors of ophiostomoid fungi involved in root diseases of other conifer species (Harrington and Cobb; 1988, Klepzig et al., 1991; Nevill and Alexander, 1992), although little is known about the insects that attack roots of longleaf pine.

L. procerum and the presence of resinosis was associated with the majority of symptomatic seedlings we excavated adjacent to sampled trees. This fungus may pose a threat to subsequent regeneration, as we have observed contacts between a sapling root infected with*L. procerum* and a mature tree root also infected with this fungus.

Root infection by the blue-stain fungi may be indicators of stress from various agents. For example, basal areas greater than approximately 30 m^2/ha have been shown to increase incidence of *L. procerum* and *L. terebrantis* in southern pine beetle (*Dendroctonus frontalis* Zimm.) control plots adjacent to southern pine beetle-attacked plots (Otrosina et al., 1997). A recent study in a 35-year-old longleaf pine stand showed isolation frequencies of blue-stain fungi and *H. annosum* to be higher in prescribed burned plots than in unburned control plots (Otrosina and Garbelotto, 1998). In our study, shallow soils that have a hardpan layer at 30 to 45 cm depth tend to concentrate lateral roots near the surface where damage can occur. Deep, well-developed tap roots characteristic of longleaf pine do not develop on this site, as evidenced by our observations of nearby blown down, 25 m tall trees with an abrupt 30–45 cm root zone below which no tap root or lateral root exists. On this study site, root damage from equipment such as brush cutters, cone harvesting machinery, equipment associated with prescribed burning, as well as prescribed burning itself, may provide susceptible portions of root tissue for insect attraction and fungal colonization. Thirty percent of class 1 trees had resinosis (Table 2) which may indicate abiotic root damage. Due to these factors, a pathological rotation age may be reached that thwarts various longer-term management objectives on these sites.

Considerable circumstantial evidence exists indicating that the decline of longleaf pine on this seed production area may be due to a complex of factors involving root pathogenic fungi, insects, edaphic conditions, and silvicultural practices. Inasmuch as seed production is critical for restoration or reforestation of longleaf pine throughout its range, and because decline problems we have addressed in this study are manifested later in the life of a stand, caution should be exercised when selecting sites for this management goal. Young and apparently healthy stands growing on historically and physiographically correct sites may be selected and managed for seed production only to become unproductive during their peak seed-bearing period. Certainly, eroded, shallow soils ought to be avoided in planning longleaf pine seed production areas. Also, care should be taken to minimize residual tree damage by mechanized equipment. Further research must be conducted that addresses specific causal roles played by various silvicultural, pathological, and edaphic factors on this site as well as other sites where similar decline and mortality in longleaf pine stands are observed.

Acknowledgements

The authors wish to thank the Savannah River Natural Resources Management and Research Institute, P.O. Box 700, New Ellenton, South Carolina 29809 for providing the research area, funding, and facilities; and the United States Department of Energy, Savannah River Site, New Ellenton, South Carolina 29809 for financial support of this research. We also wish to thank Dr. Stanley J. Zarnoch, USDA Forest Service, Southern Research Station, Asheville, NC 28802 for statistical advice and analyses.

References

Boyer W D 1990 Longleaf pine. In Silvics of North America, Vol. 1, Conifers. R M Burns and B H Honkala, Technical Coordinators. pp 405–412. Forest Service, United States Department of Agriculture Handbook 654. Washington, D.C. 675 p.

Clark T D 1984 The greening of the South. University Press of Kentucky. Lexington, KY. 168 p.

Froelich R C, Kuhlman E G, Hodges C S, Weiss M J and Nichols J D 1977 *Fomes annosus* in the south-guidelines for prevention. USDA Forest Service, State and Private Forestry, Southeastern Area, Atlanta, Georgia, USA. 17 p.

Harrington T C and Cobb F W Jr. 1988 Leptographium root diseases on conifers. APS Press, St. Paul, Minnesota, USA. 149 p.

Hicks B R, Cobb F W Jr. and Gersper P L 1980 Isolation of *Ceratocystis wageneri* from forest soil with a selective medium. Phytopathology 70: 880–883.

Klepzig K D, Raffa K F and Smalley E B 1991 Association of an insect fungal complex with red pine decline in Wisconsin. For. Sci. 37: 1119–1139.

Martin W H, Boyce S G and Echternacht H C 1993 Biodiversity of the southeastern United States: Upland terrestrial communities. Wiley & Sons, New York, NY, USA. 200 p.

Nevill R J and Alexander S A 1992 Pathogenicity of three fungal associates of *Hylobius pales* and *Pissodes nemorensis* (Coleoptera : Curculionidae) to eastern white pine. Can. J. For. Res. 22: 1438–1440.

Nevill R J, Kelley W D, Hess N J and Perry T J 1995 Pathogenicity to loblolly pines of fungi recovered from trees attacked by southern pine beetles. Southern J. Appl. For. 19: 78–83.

Otrosina W J and Cobb F W Jr 1989 Biology, ecology, and epidemiology of *Heterobasidion annosum*. In Proceedings of the Symposium on Research and Management of Annosus Root Disease (*Heterobasidion annosum*) in Western North America. Technical Coordinators W J Otrosina and R F Scharpf. pp 26–33. Forest Service, United States Department of Agriculture General Technical Report # 116, Berkeley, California, USA. 177 p.

Otrosina W J, White L W and Walkinshaw C H 1995 *Heterobasidion annosum* and blue-stain fungi in roots of longleaf pine are associated with increased mortality following prescribed burning. Phytopathology 85: 1197 (abstr.)

Otrosina W J, Hess N J, Zarnoch S J, Perry T J and Jones J P 1997 Blue-stain fungi associated with roots of southern pine trees attacked by the southern pine beetle, *Dendroctonus frontalis*. Plant Disease 81: 942–945.

Otrosina W J 1998 Diseases of forest trees: consequences of "exotic ecosystems". *In* Proceedings of the Ninth Biennial Southern Silvicultural Research Conference. Ed. T Waldrop. pp 103–106. Forest Service, United States Department of Agriculture, Southern Research Station. Gen. Tech. Rep., SRS-20. Asheville, North Carolina, USA. 628 p.

Otrosina W J and Garbelotto M 1998 Root diseases and exotic ecosystems: implications for long-term site productivity. *In* Root and Butt Rots of Forest Trees. Eds C Delatours, B Marcais, J J Guillaumin and B Lung-Escarmant. pp 275–283. INRA, Paris. 459 p.

Parmeter J R, Slaughter G W, Chen M M Wood and D L Stubbs H A 1989 Single and mixed inoculations of ponderosa pines with fungal associates of *Dendroctonus* sp. Phytopathology 79: 768–772.

Parmeter J R, Slaughter G W, Chen M M and Wood D L 1992 Rate and depth of sapwood occlusion following inoculation of pines with blue-stain fungi. For. Sci. 38: 34–44.

Rane K A and Tattar T A 1987 Pathogenicity of blue-stain fungi associated with *Dendroctonus terebrans*. Plant Dis. 71: 879–883.

MODELLING OF GROWTH, DEVELOPMENT AND ARCHITECTURE

A. Stokes (ed.), The Supporting Roots of Trees and Woody Plants: Form, Function and Physiology, 187–194.

Why model root system architecture?

Loïc Pagès

INRA, Centre d'Avignon, Ecophysiologie et Horticulture, Domaine St-Paul, Site Agroparc, 84914, Avignon Cedex 9, France

Key words: architecture, mathematical model, method, root system, simulation, three-dimensional

Abstract

In this short discussion, I shall attempt to classify some of the various models of root system architecture according to the different purposes they can serve. A major objective, which is recurrent in many modelling approaches, was to provide better representations of the root system as an input for uptake models or crop models. I show how the required representations followed the concepts of the uptake modelling approaches. A second objective was to study the morphology and morphogenesis of the root system per se. Some examples show how the models of root system architecture can provide new insights in studying root development, comparing genotypes, and quantifying the effects of the environment on the root system. Another promising domain in which such models were shown to be helpful is the integration at the root system level of physiological processes, especially on carbon allocation. This integration allows new studies of the functional relationships between the root system and the shoot system. Methodological problems remain a crucial point in the observation of root system. Therefore, several authors have tried to show how theoretical root systems, as they are simulated by architectural models, may help in the design and optimisation of new experiments. The possible integration of local and global processes, or processes at different space and time scales into a consistent frame (the model), is also very valuable in root research since observations are generally scattered in both time and space. Thus, this review shows the very large potential of such models in the general scope of root system and plant research. These models could probably serve additional purposes that have not been considered up to now, like the study of anchorage, in which the representation of the architecture of the proximal root system is probably a key step. For the next future, models will appear for integrating more physiological processes, relating to both the structure (morphogenesis, architecture) of the root system, and to its function (uptake, and resource allocation). The difficulty will be to find good compromises between the need for integrating more processes, and the complexity of such models.

Introduction

I would like to ask this question – why model root system architecture? – because this modelling looks like a difficult task, since the root system is a complex and adaptive object. One might wonder whether it is madness, or even masochism, to keep such an ambitious goal. In this paper, I would like to show that it is really necessary to have detailed representations of the root system in order to get a better understanding of its functioning. In addition, modelling will be shown to be a tool for optimising, and hence reducing, the hard experimental work which is associated to the study of root systems.

Asking this question – why model root system architecture? – is also a means to visit some of the different objectives that have been devoted to the modelling approaches, and therefore to present different models that have been built with various view points. By the way, such a short review cannot be exhaustive.

I will focus on those models that describe the complete root system, and not only individual roots, since the root system, within the plant, is really 'a system' in the sense that its components (the roots) are connected in an organised network, and interact by physiological relationships. As every complex system, it requires models specifically built at the relevant organisation level, to drive its study. Describing the root system architecture means describing both its shape (i.e., spatial configuration of the system) and its structure (i.e., description of its components and of their topological relationships).

Representing the uptake system within the soil

Most of the models that simulate the development and production of crops in relation to climatic conditions (radiation, temperature, water, nutrients) require a representation of the root system in order to predict its water or/and nutrient uptake. These processes of water and nutrient uptake, when considered at the soil–plant level, involve several distinct sub-processes. In a first step, the resource is transferred from a given soil location towards the vicinity of the root: it is a soil transfer. In a second step, the resource must cross the external tissues of the root until the xylem vessels (uptake *sensu stricto*) and must be transported along the root, through the sap conducting system. This second step is a within-plant transfer.

The soil to root transfer

During this step, the distance from the soil resources to the roots is obviously of prime importance, especially for immobile resources. Therefore, it is especially important to describe the spatial distribution of roots. which determines the path lengths during the soil to root transfer. That is why the root system is often described in terms of root length density, by a one-dimensional root profile (i.e. root length density versus depth, Fig. 1). From a given root length density, it is possible to calculate a mean distance between the roots, and so to estimate the length of the soil transfer. However, this type of description assumes that root distribution in a soil layer is well represented by a given density, i.e., that roots are uniformly distributed within this layer. Several authors have shown that actual root distributions cannot be assumed to be regular in the soil as a consequence of soil heterogeneity and morphological properties of the root system (e.g., Pellerin and Pagès, 1996; Tardieu et al., 1992). It can have great consequences concerning uptake, as shown by Tardieu et al. (1992). In their study on a maize crop in a loamy soil, Pellerin and Pagès (1996) showed on observed horizontal root maps that the root distribution was highly structured in space (Fig. 2). Firstly, roots tended to be more dense under the row that under the inter-row. Secondly, roots tended to be clustered, and so competed for water and nutrients, whereas large

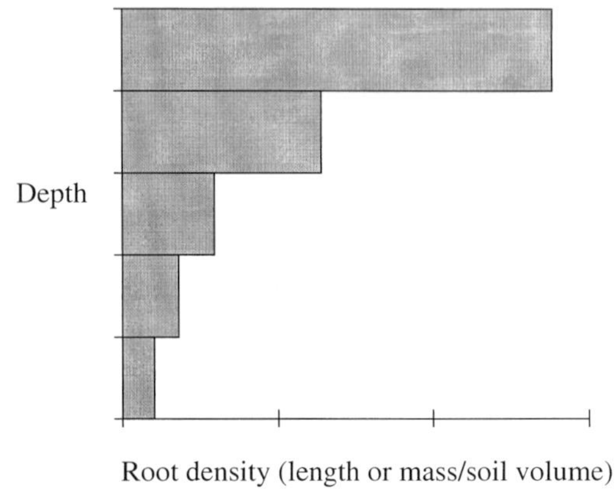

Figure 1. Root profile, typical representation of the root distribution.

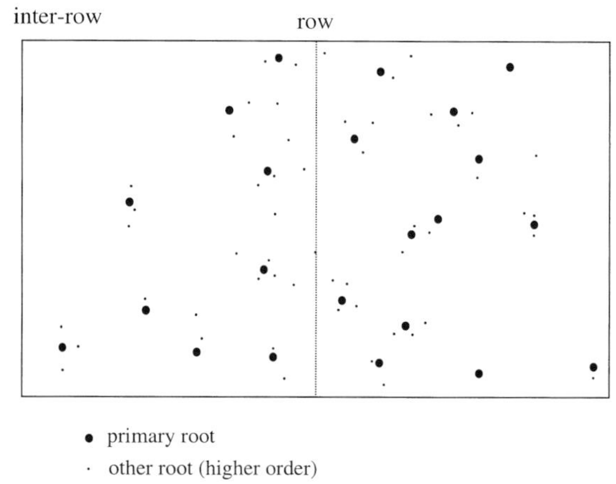

Figure 2. Horizontal root map (at a 30-cm depth), in a maize crop. Each dot represents a root (after Pellerin and Pagès, 1996).

spaces existed between the clusters, in which the soil resources could hardly be used by the plant. Thirdly, the different types of roots (primary, secondary, etc.) were not randomly distributed in this pattern, but they followed some structural rules. Lateral roots (secondary and tertiary) are generally around, and at a short distance, from primary roots.

Modelling the root system architecture, and simulating root maps by calculating cross sections in three-dimensional simulated structures, has been shown (Pagès and Aries, 1988) to be a natural way for representing this type of complex pattern. The peculiarities of the two-dimensional root patterns are direct consequences of the root system morphology. In this case (Fig. 2, after Pellerin and Pagès, 1996), the root density gradients from the row to the inter-row must be interpreted in relation to the crop structure (effect of the row in this case). Moreover, the roots are ag-

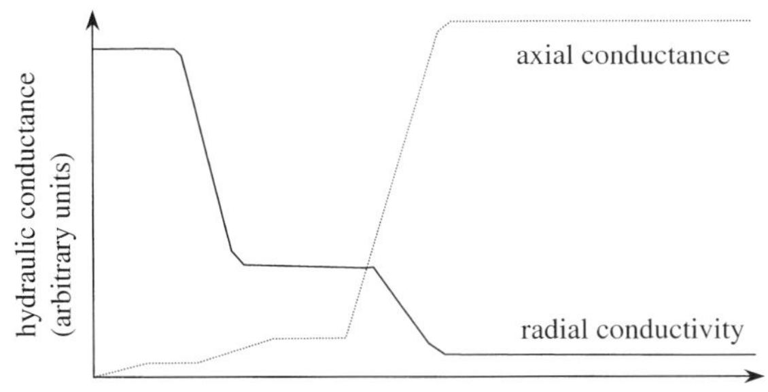

Figure 3. Variation of axial conductance and radial conductivity along the maize primary roots (after Doussan et al., 1998b).

gregated in space, because they are connected to each other. Such evidences have in fact large consequences on the global root distribution which were often neglected. The clustering intensity, and the distances from lateral roots to their mother roots, depend on the length of each individual root. It is common, if not universal, that fine lateral roots are also very short roots.

Uptake and transport

During this second step of the global uptake process, involving a radial and longitudinal transport of the soil resources within the root system, it seems important to take into account the structure of the root system, i.e., the different types of roots, their ontogenetic evolution with age, and the way they are connected.

With the progress in measurement techniques, more and more work is being done that show the important differences in water conductance that exist from one type of root to the other (e.g., Varney and Canny, 1993), and along the roots (e.g., Doussan et al., 1998b; Peterson et al., 1993). Fig. 3 (after Doussan et al., 1998b) shows an example of such variations along the primary roots of maize. The radial conductance tends to decrease with distance to the apex, while the axial conductivity tends to increase. It is worth noting the large variations (at least one order of magnitude) that were found in conductance values.

The branching points may also be places of high resistance (Peterson et al., 1993).

This information, about the heterogeneity in the root system functioning, can be associated rather naturally to architectural descriptions. In architectural models, the root system is generally coded into a set of root segments (Pagès and Aries, 1988). Each root segment is the formal representation of a root piece, and contains attributes: spatial coordinates, age, root type, connection to other segments, etc. Functional attributes may be given or calculated in addition to these developmental attributes: radial conductance, axial conductance, connection resistance, etc. This type of association between structural and functional information opens the way for many interesting applications (see last chapter).

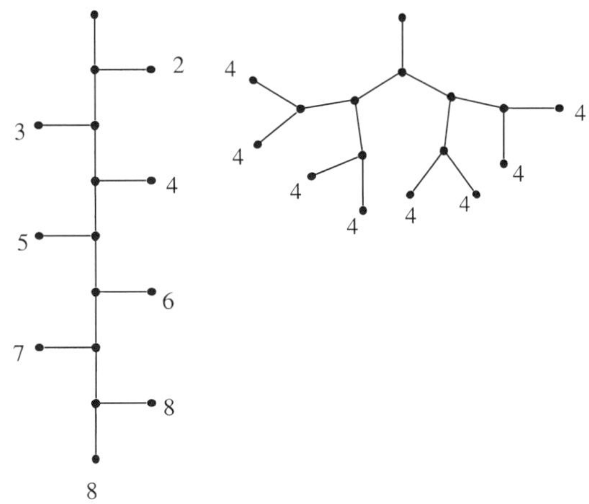

Figure 4. Two extreme types of topological structures: herringbone (left) and dichotomous (right). The topology can be characterised by two parameters: magnitude, and altitude. In this case, both structures have the same magnitude (number of links), but different altitudes (maximal path length).

Studying effects of the genotype and the environment on the root system

For studying the morphogenesis of the root system per se, at the root system level, it is important to get tools in order to make objective comparisons between different root systems. These objective comparisons are a prerequisite step for studying either genetic effects, or environmental effects, on the root system. The general principle of the method is to substitute to the real root system (complex object) a model with a set of parameters (simpler object). The comparison relies on the set of parameters, so that it may be quantitative, and use mathematical tools like statistics. Therefore, the problem is to choose a model (a formal representation of the root system) that is relevant for the current investigation, and simple enough for allowing a comparison relying on a limited number of parameters.

Different approaches have been followed to achieve this task. Fitter (1986, 1996) suggested to use the formalism that was developed for the description of hydraulic networks. Branching patterns were considered as binary trees, made up of links and nodes, on which parameters like magnitude and altitude can be observed (Fig. 4). These parameters allow a quantification of the structure, and particularly to distinguish various intermediate structures between the extremes: the herringbone pattern on one hand, and the dichotomous pattern on the other. This model description has been used to show how irrigation and fertilisation can affect the structure on different plants (Fitter, 1986). Later on, different applications have been done based on this method, in order to quantify different effects, related to the genotype, and to biotic or abiotic external factors.

Unlike this approach, which is essentially static, other authors (Diggle, 1988; Fitter et al., 1991; Pagès and Aries, 1988) have proposed to use developmental models, that try to mimic the dynamics of the root system development. These models simulate the developmental processes that concur to the extension of the root system, namely growth and branching. Thus, the parameters of these models are associated to these developmental processes, and allow the calibration of growth and branching functions. This type of model looks relevant for analysing how the developmental processes (e.g., growth, branching) are affected by the different factors under study. It helps during the different stages of the study: analysis (identification of the processes involved, quantification of these processes) and synthesis (estimating global consequences of local variations).

Methodological tools

It is very common to state that root research has been hampered by methodological problems. Among these methodological problems, some rely on the particular techniques which are used for the observations. These techniques are often imperfect: they give incomplete and scattered pieces of information, they are often destructive, and they may give artefacts.

Therefore, tools are required as complement to these techniques for gathering various data obtained at different space or time locations. It is generally difficult, especially in the field, to observe and describe the entire root system, and it is still more difficult to monitor continuously its development. Conversely, methods exist for excavating limited parts of the root system at a given stage, concerning either the structural root system (e.g., excavation by water or air pressure), or the fine root system (e.g., soil monoliths). Van Noordwijk et al. (1994) and Spek and Van Noordwijk (1994) showed how a fractal model could be valuable to gather observations on extensive tree root systems, allowing to build entire root systems (theoretical ones) from local and limited field observations. Using this approach, in which root systems are approximated to fractal objects (i.e., with internal auto-similarities), it becomes possible to estimate global characteristics, such as total root length.

Architecture models were also used in the scope of virtual experiments. Pagès and Bengough (1997) simulated mini-rhizotron observations in maize crop, and compared different experimental designs (Fig. 5). For that, they simulated the root system of a crop at different stages, and calculated virtual pictures obtained from tubes (cylinders) crossing these crop root systems (Fig. 6). They quantified for instance the possible bias on root depth dynamics when it is estimated using this mini-rhizotron method. The strength of the modelling approach is its ability to make nearly perfect experiments, with many replicates, but on theoretical objects which are only approximations of real root systems. Since the simulated root systems are perfectly known (as sets of segments with three-dimensional coordinates, resulting from a well-defined model), the intersections between the virtual mini-rhizotron tubes and the root systems are also totally explicit. The alternative way in this scope is to make real experiments, with many practical difficulties, leading to various uncertainties, and limited numbers of replicates, but which rely on real objects. No doubt that both approaches (theoretical versus empirical) should interact in order to optimise data acquisition and analysis.

Integrated plant physiology

The study of the root system development and functioning, when achieved at the root system and whole plant level, requires specific tools that allow merging quantitative hypotheses. Merging and quantifying interactive processes is often the only means to test hypotheses, and to discriminate alternatives. For example, the between-root competition for carbohydrates has often been considered as a major factor which determines various features of the root system architecture (see the review by Coutts, 1987). Thaler and Pagès (1996) confirmed this high sensitivity of the root system to carbohydrate competition in Rubber tree (*Hevea brasiliensis*) seedlings. On this species with a typical rythmic growth (Fig. 7), they observed

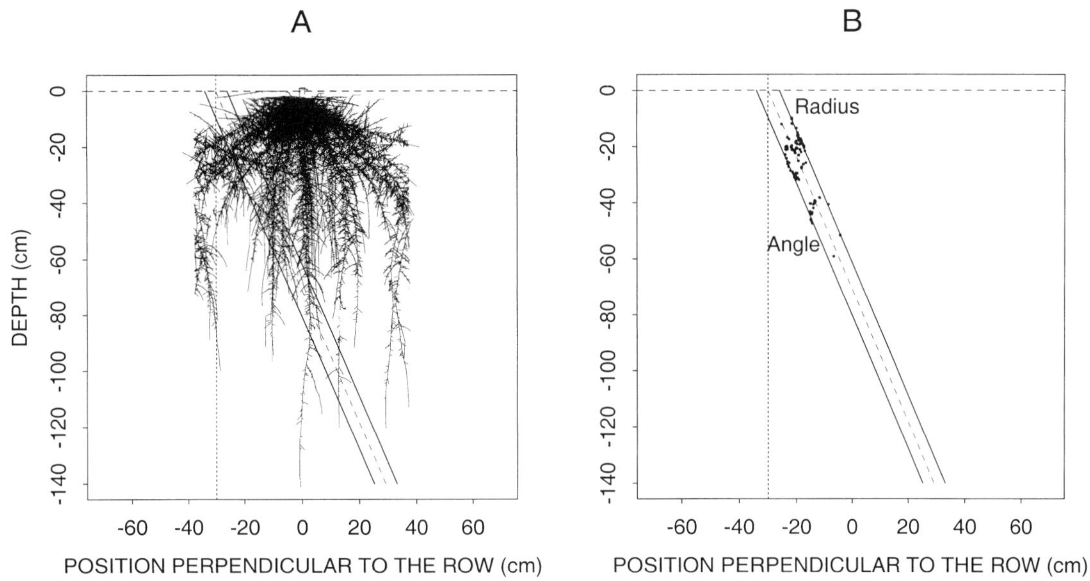

Figure 5. Projection of the simulated root system of a maize crop (A), and its intersection with a minirhizotron tube (B) (after Pagès and Bengough, 1997).

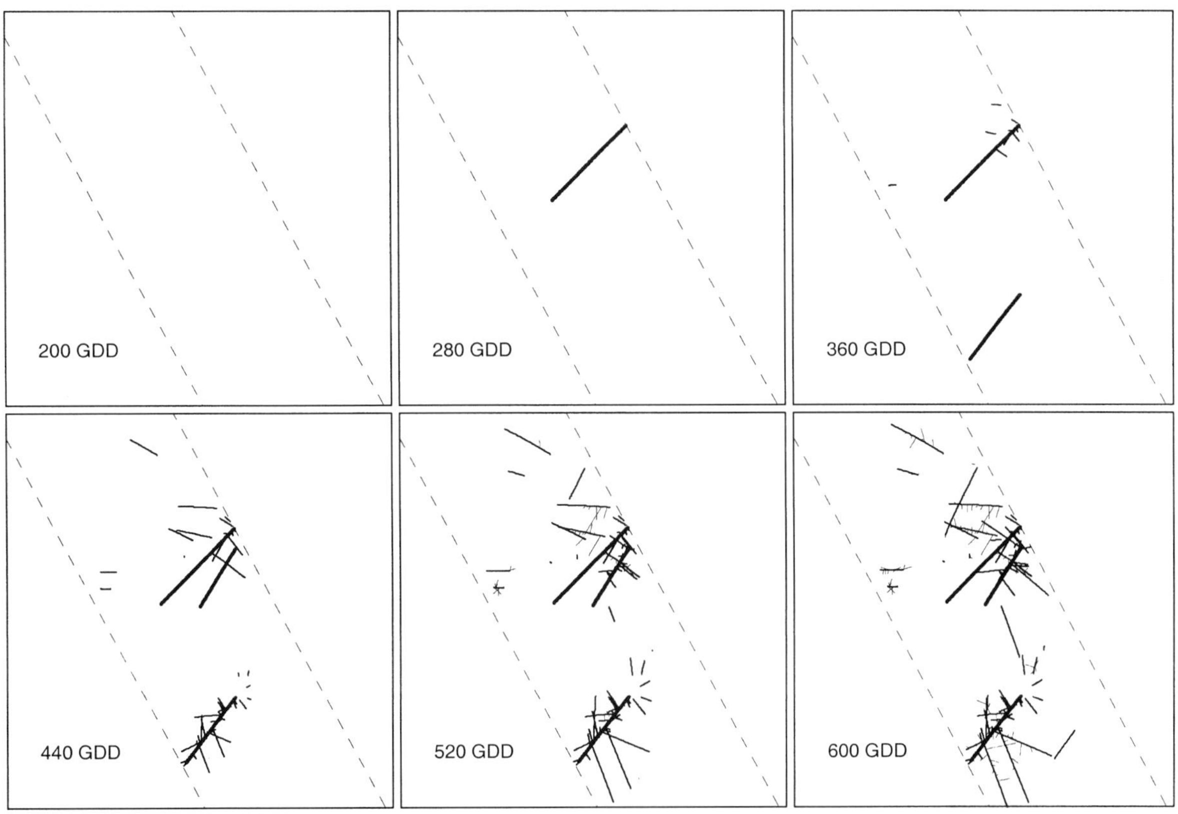

Figure 6. Virtual pictures calculated for simulated crops at different stages (from 200 to 600 growing degree days). The tube (dashed line) is gradually colonised by different types of roots (solid lines of different widths) (after Pagès and Bengough, 1997).

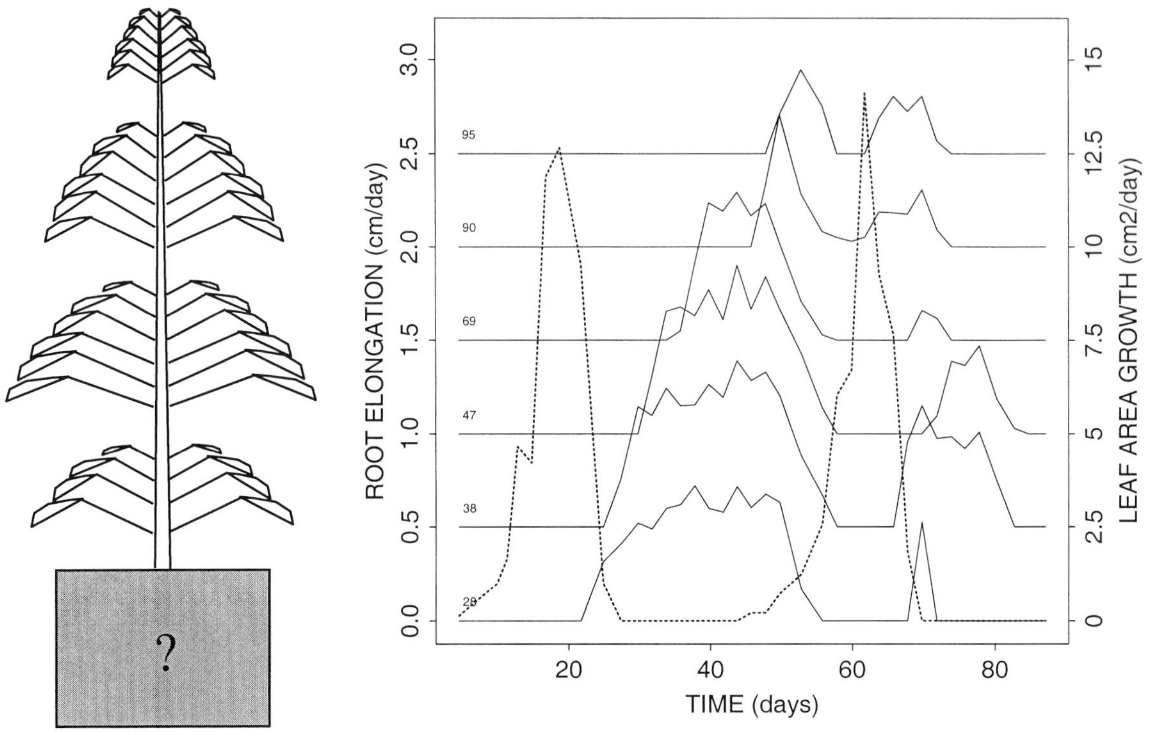

Figure 7. Young rubber tree seedling with a typical rhythmic growth habit (left), and growth patterns for leaf area (dashed line, on the right), and for various roots (solid lines, on the right) (after Thaler and Pagès, 1996).

periodic root growth patterns with a decrease (or stop) in root growth rate when the leaves grow, and a resumption in root growth when the leaves stop growing. The different roots were more or less sensitive according to their position within the architecture, the finer being the more sensitive. A source-sink model describing carbohydrate allocation, associated with a morphogenetic model describing the organ appearance in the plant topology, allowed the quantitative representation of these organ interactions (Thaler and Pagès, 1998). By simulating root growth patterns similar to the observed ones, the model validated the hypothesis that root growth was mainly determined by the carbohydrate availability within the whole plant.

Some trends for the future

These modelling approaches of the root system architecture have just started during the last 10 years, and they are still reaching their maturity level. So, many of the presented works will be developed further in the next years.

Studying further the morphogenetic processes which concur to the root system development is an important topic, and it is probable that the simulation models of root system architecture will still play an important role in it. One of their major role will be to set a straightforward link between the global architecture of the root system on one hand, and the basic edification processes on the other. It is worth noting that much effort have been done in the last years to formalise the processes of growth and branching, which are the most natural to deal with. In the future, other important developmental processes, like reiteration (local duplication of a more or less complete architecture), or root decay, should be given more emphasis, like in the recent works from Vercambre et al. (1998), Jourdan et al. (1995) and Jourdan and Rey (1997).

From a methodological point of view, these models of the root system architecture should become a common help to make different observations and different methods more complementary. They can be a common frame in which data originating from different sources can be gathered. When the simulated structures are considered as realistic enough, they can become the

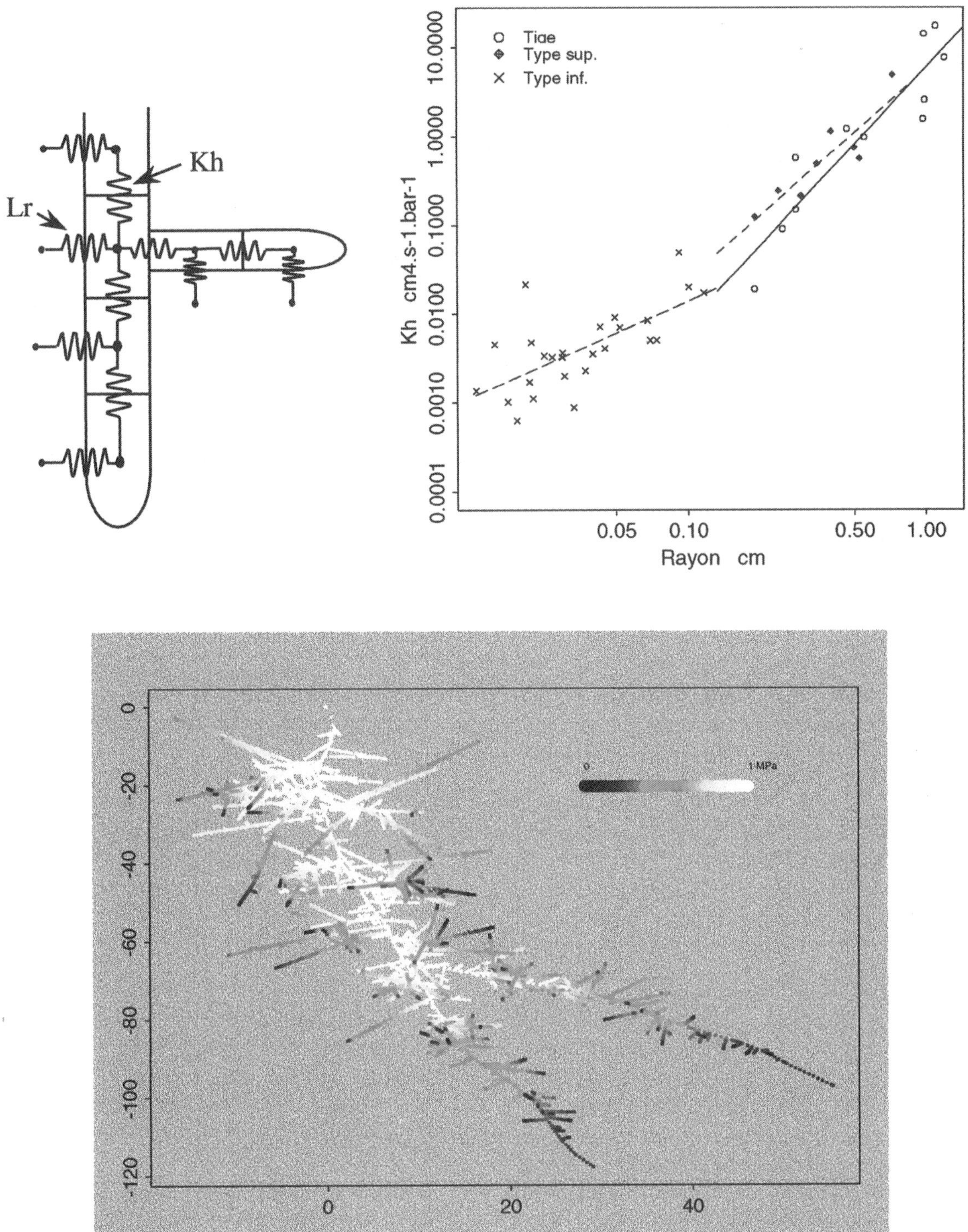

Figure 8. Modelling the root system hydraulic architecture, using Ohm law (top left: Lr, radial conductance; Kh, axial conductance), and measurements of axial conductances in relation to the root radius (top right). The diagram at the bottom presents the map of water potentials as it is calculated by the model when the external water potential is uniform and equal to 0, and the root water potential is fixed at −1 MPa at the root base (after Vercambre, Pagès and Doussan, 1998, unpublished results).

support for virtual experiments, allowing the conception of optimized sampling schemes (see Grabarnik et al., 1998).

In order to reach a further step of realism, models of root system architecture will need to deal more and more explicitly with both the external constraints (the soil particularly), and with the resources availability (e.g., water and carbohydrate). Thus, there is a clear need for merging different models in order to get a more integrated view upon the complex interacting system.

A first example has been cited in the previous chapter concerning the association between a model of root system development and a source-sink model for carbohydrates (Thaler and Pagès, 1998).

A second example is given by the work from Clausnitzer and Hopmans (1994). Since the root system modifies deeply the soil medium by its own functioning (water uptake especially), their objective was to merge a model of root system development with a model of water uptake. In this attempt, the root system development is simulated simultaneously to water uptake and soil transport. Therefore, at each time step, the root tips experience a new set a soil properties, related to the new set of water content values in the soil, and root growth is calculated for each root tip according to these new characteristics.

A third example of promising work has been done by Doussan et al. (1998a,b), who showed the importance, for predicting water uptake, of considering simultaneously the architecture of the root system and the hydraulic properties of the roots (i.e., their radial and axial water conductance). They merged a model of the root system architecture with a model of water transfer within the roots. Using literature data on the water conductance of maize roots, they could estimate the parameters of the model and they studied the global consequences of the root resistances to water transport (Fig. 8). For example, the results show that the water flux into the root depends on the position along the root, and the type of root. Thus, the model was shown to be an efficient means to gather data from different origins regarding water conductance within the root system, in order to obtain integrated results, at the root system level. On the maize root system (Doussan et al., 1998b), these global results were shown to account for the experimental data on the distribution of water influx obtained by Varney and Canny (1993).

References

Clausnitzer V and Hopmans JW 1994 Simultaneous modeling of transient three-dimensional root growth and soil water flow. Plant Soil 164, 299–314.

Coutts MP 1987 Developmental processes in tree root systems. Can. J. For. Res. 17, 761–767.

Diggle AJ 1988 ROOTMAP – a model in three-dimensional coordinates of the growth and structure of fibrous root systems. Plant Soil 105, 169–178.

Doussan C, Pagès L and Vercambre G 1998a Modelling the hydraulic architecture of root systems: an integrated approach of water absorption. I. Model description. Ann. Bot. 81, 213–223.

Doussan C, Vercambre G and Pagès L 1998b Modelling the hydraulic architecture of root systems: an integrated approach of water absorption. II. Distribution of conductances and consequences. Ann. Bot. 81, 225–232.

Fitter AH 1986 The topology and geometry of plant root systems: influence of watering rate on root system topology in *Trifolium pratense*. Ann. Bot. 58, 91–101.

Fitter AH 1996 Characteristics and functions of root systems. *In* Eds Y Waisel, A Eshel and U Kafkafi. pp 1–20. Plant Roots, the Hidden Half. Marcel Dekker, New York, 2nd edition.

Fitter AH, Stickland TR, Harvey ML and Wilson GW 1991 Architectural analysis of plant root systems. I. Architectural correlates of exploitation efficiency. New Phytol. 118, 375–382.

Grabarnik P, Pagès L and Bengough AG 1998 Geometrical properties of simulated maize root systems: consequences for length density and intersection density. Plant Soil 200, 157–167.

Jourdan C and Rey H 1997 Modelling and simulation of the architecture and development of the oil palm root system. I. The model. Plant Soil 190, 217–233.

Jourdan C, Rey H and Guédon Y 1995 Architectural analysis and modelling of the branching process of the young oil-palm root system. Plant Soil 177, 63–72.

Pagès L and Aries F 1988. SARAH: modèle de simulation de la croissance, du développement, et de l'architecture des systèmes racinaires. Agronomie 8, 889–896.

Pagès L and Bengough AG 1997 Modelling minirhizotron observations to test experimental procedures. Plant Soil 189, 81–89.

Pellerin S and Pagès L 1996 Evaluation in field conditions of a three-dimensional architectural model of the maize root system: comparison of simulated and observed horizontal root maps. Plant Soil 178, 101–112.

Peterson CA, Murrmann M and Steudle E 1993 Location of the major barriers to water and ion movement in young roots of *Zea mays* L. Planta 189, 288–297.

Spek LY and Van Noordwijk M 1994 Proximal root diameters as predictors of total root system size for fractal branching models. II. Numerical model. Plant Soil 164, 119–128.

Tardieu F, Bruckler F and Lafolie F 1992 Root clumping may affect the root water potential and the resistance to soil-root water transport. Plant Soil 140, 291–301.

Thaler P and Pagès L 1996 Periodicity in the development of the root system of young rubber trees. Relationship with shoot development. Plant Cell Environ. 19, 56–64.

Thaler P and Pagès L 1998 Modelling the influence of assimilate availability on root growth and architecture. Plant Soil 201, 307–320.

Van Noordwijk M, Spek LY and De Willigen P 1994 Proximal root diameters as predictors of total root system size for fractal branching models. I. Theory. Plant Soil 164, 107–118.

Varney GT and Canny MJ 1993 Rates of water uptake into the mature root system of maize plants. New Phytol. 123, 775–786.

A. Stokes (ed.), The Supporting Roots of Trees and Woody Plants: Form, Function and Physiology, 195–207.

A model simulating interactions between plant shoot and root architecture in a non-homogeneous environment

F. Blaise[1,*], T. Fourcaud[1,2], A. Stokes[2] and P. de Reffye[1]
[1]*Programme Modélisation des Plantes, CIRAD-AMIS, BP 5035, 34032 Montpellier Cedex 1, France and* [2]*LRBB, Domaine de l'Hermitage, BP 10, 33610 Cestas Gazinet, France**

Key words: AMAPpara, competition, ecophysiology, root system, simulation, 3D models

Abstract

Modelling of plant structure and growth has undergone major changes in the last few decades. Two major lines of research have been pursued: the integration of ecophysiological knowledge in process-based models which usually lack a description of plant topology and geometry, and the generation of 3-D virtual plants using morphogenetic models, which simulate the architectural development in a stable and homogeneous environment. There is now a trend to merge these two approaches i.e. to link plant architecture and function. This trend is based on the recognition that plant structure: (i) is the joint output of physiological processes and the morphogenetic programme of the plant, (ii) determines its external environment and regulates its functioning accordingly (iii) directly conditions the physiological processes within the plant e.g. allocation of photosynthates. Therefore, double regulation of the ecophysiological processes by the plant and its neighbours, and of the morphogenetic program by the availability of the internal and external resources produced or consumed by these physiological processes, is explicitly considered in the same model. Such an approach is being implemented in the software, AMAPpara, which simulates the parallel functioning and growth of plants which interact with each other. AMAPpara also includes an architectural description of aerial and root systems. The plant morphogenetic program is a priori parameterised according to the selected species i.e. topology, geometry and allometric relationships are considered for botanical elements such as leaves, as well as the maximum life-span of leaves and roots etc. The actual architectural development of the plant is then regulated by physical constraints e.g. competition for space and light, and physiological processes e.g. water transpiration and carbon assimilation. Such processes themselves depend on the structure and biophysical environment of the plant e.g. internal hydraulic architecture, competition for space among neighbouring trees and light interception. This paper presents a new model which investigates the influence of the environment on root system development and the resulting consequences on the growth of the whole plant.

Introduction

The 3D simulation of plant growth and architecture usually concerns the aerial parts, which are easier to observe and the data less difficult to collect than for the below-ground organs. The current approach is mainly morphogenetic and uses algorithms as computer grammar (Prusinkiewicz et al. 1997) or stochastic processes (Reffye et al., 1995, Barczi et al., 1997) which simulate the growth of the topological structure, according to standard, botanical, empirical laws. It is well known that plant architecture is governed by a genetic process. The 'architectural models' of trees (Hallé et al., 1978) or flowering structures are typical examples. Local conditions in the plant environment can modify plant development but are considered a secondary effect. Such models provide no information about the biomass produced, which depends on available light, nor the growth rate, which depends on temperature. Thus, the same plant can be obtained with different climatic conditions. Therefore, in the case of a pure morphogenetic simulation, the sizes of the plant organs fixed in the 3D plant model, are derived from direct observations and not from a functional computation of the photosynthesis. Thus, the plant is 'clothed' only after the topological skeleton has been computed.

* Fax: +33 4 67 59 38 58. E-mail: frederic.blaise@cirad.fr

Recent applications of this morphogenetic approach have been carried out on root systems. Young root systems show differentiated axes (orthotropic and plagiotropic) and a strong organization of architecture (Atger et al., 1994). Although there are no real internodes as in the shoot axis, it has been shown that one can define a virtual inter-node, which is an optimal discretization of root length and allows architecture to be simulated. Using this method, the first realistic 3D root systems have been simulated with reasonable validations for *Elais guinensis*, (Oil palm) (Jourdan et al., 1995a, b, c) and *Hevea brasiliensis*, (rubber tree) (Le Roux et al., 1994; Pages et al., 1995).

Numerous models also exist which deal with carbon allocation in plants divided into simple compartments (Cannell et al. 1994, Makkela 1986). Such models do not belong to the architectural model family and so are not discussed further.

A new approach involving modelling of the fabrication of vegetal matter by estimating photosynthesis and assimilate transfer inside the plant, has been made using the following software: LIGNUM (Perttunen et al., 1996), GROGRA (Kurth 1996), and AMAPpara (Blaise, 1991, Blaise et al. 1997, 1998, Reffye et al. 1996, 1997, Fourcaud et al. 1997). LIGNUM (Pertunen et al., 1996) is concerned only with the shoot growth of individual *Pinus sylvestris* trees. GROGA (Kurth, 1996) also takes into consideration the interaction between trees within in a forest stand. AMAPpara (Reffye et al. 1996) uses information about water use efficiency, and integrates the interactions between shoots and roots during plant growth in a given physical environment.

In a model of root growth and sap flow (Clausnitzer and Hopmans, 1994), the functioning of the whole plant was considered and the biomass fabricated according to transpiration and water use efficiency. However, the aerial structure of the plant was not described and was simplified into classical compartments. A 3D simulation of root system growth was carried out using purely opportunistic behaviour, which depended on local hydric conditions encountered in the soil (Clausnitzer and Hopmans, 1994). Such an assumption does not comply well with the laws of morphogenesis. In the creation of a new organ it is considered that the 'proposal' is instigated by the internal organisation of the plant, and that the 'production' is governed by local environmental conditions. Thus the development of plant organs has important consequences for growth and architecture. The present paper uses the same assumptions as Clausnitzer and Hopmans (1994) for plant functioning, i.e. with regards to transpiration and water use efficiency, but more emphasis is placed on the interaction between both shoot and root architecture during growth. The effect of the environment on plant development is then taken into account using similar methods to those of Clausnitzer and Hopmans (1994).

Biological and morphogenetic assumptions for 3D simulation of root systems

A morphogenetic approach to plant architecture (above-ground parts only) has been carried out based on the stochastic functioning of buds according to physiological age (Reffye, 1995). Time is divided into cycles of growth, and creation or differentiation of the organs is controlled by an automaton named the 'reference axis.' Measurements made on architecture are then analysed and converted into probabilities which control the behaviour of the bud for growth, death, and branching at each step of development. AMAPmod software (Godin et al., 1997) enables the biologist to fit the empirical data with theoretical laws. It is then possible to simulate architecture if plant geometry is combined with the topological structure defined by the organs produced during the cycles of growth, according to the stochastic laws and the differentiation steps. Jourdan and Rey (1995c) have adapted this method and applied it to root systems of Oil palms (Figure 1). The below-ground plant organs were found to parallel those above-ground with regards to architecture. Three types of vegetative units were distinguished: leaves (or root hairs), terminal growth units and sapwood units. Although there are no inter-nodes in roots, it is possible to define virtual inter-nodes using a given length, found by choosing the optimum fit of root distribution determined by Markov chain theory (Jourdan, 1995a). This newly defined length thus corresponds to a shoot inter-node, and enables the simulation to monitor the production of apical meristems. It is also possible to recognize different steps of differentiation for the root axis, thus the theory of the 'reference axis' can be applied.

Description of the model AMAPpara

AMAPpara uses an elementary hypothesis for the physiological functioning of plants (Reffye, et al. 1996). A step by step procedure is used to run

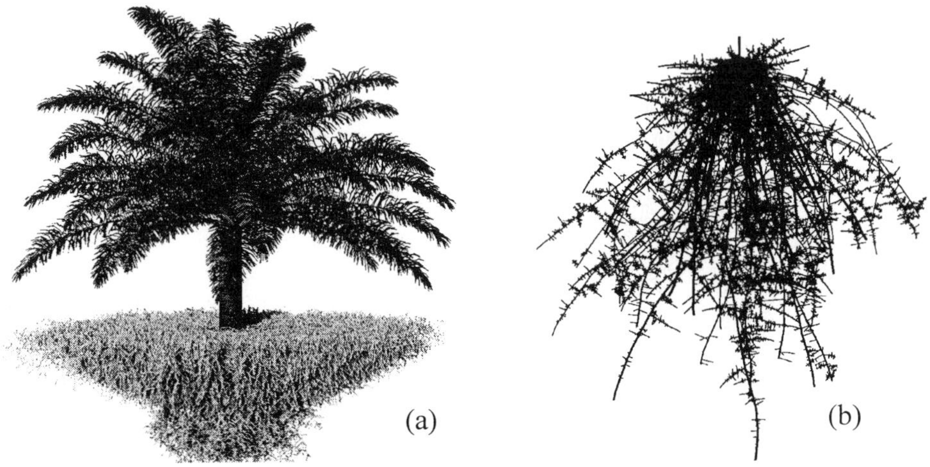

Figure 1. Simulation of the root-system of the palm-tree by the AMAPsim software (Cirad): (a) entire tree simulation with both aerial and root system, (b) young root system (Jourdan and Rey, 1997).

the model, each step corresponding to one cycle of growth. Therefore all the parameters of the model are integrated into each period of growth (one year for a tree). The tree is considered as a hydraulic structure, which is composed of two ramified systems linked together: the roots and the shoots. This structure is able to draw up soil water and mineral elements by the roots and transport them via the sapwood to the leaves. It is assumed that the plant is placed in an ideal environment and that there is a proportional relationship between accumulated fresh vegetal matter production and transpiration. Until now, the root system played a neutral role in the AMAPpara models, and was presumed to have a proportional biomass to that of the above-ground parts. It was also assumed that the root system was without significant hydraulic resistance due to lack of hydric stress. Competition for space between root systems of neighbouring plants has not yet been calculated. The main features of the model AMAPpara are summarized below and are described in more detail in Reffye et al. (1996, 1997).

The functioning of the growth cycle

One step of calculation corresponding to a cycle of growth is composed of three main parts:

Biomass allocation and primary growth

The allocation of fresh matter reserves fabricated at the previous cycle is shared between the different compartments according to their own sinks (Figure 2). Compartments can either be a set of organs e.g. an entire root system, or individual structures e.g. internodes or leaves. The way the organs are assembled in a set defines a model of allocation. The fresh matter is used for primary growth and for the expansion of the organs made at the previous cycle. The construction of the organ depends on empirical and allometric rules, which define the shape according to the volume of matter. The organs are then stacked onto the geometrical structure of the previous cycle according to various botanical laws e.g. phyllotaxy and branching angle. A 3D simulation of growth can then be ensured. We assume that the architecture set in place at the beginning of the cycle does not change during the cycle duration.

Hydraulic structure and water transpiration

Secondly, the external architecture is transformed into a hydraulic structure, which is split into a set of elements corresponding to vegetative units. For the shoot and root organs, three types of vegetative units are distinguished: the leaves (or root hairs), the terminal growth units and the sapwood units. Each of these units has a hydraulic resistivity and regulates water transport. Water flow is calculated in each part of the plant using a global assembly method for these hydraulic elements. The boundary conditions are given by the water potentials. We assume that the physical parameters e.g. temperature, light and water potential do not change during the cycle duration and that the environment of the plant is homogeneous. The amount

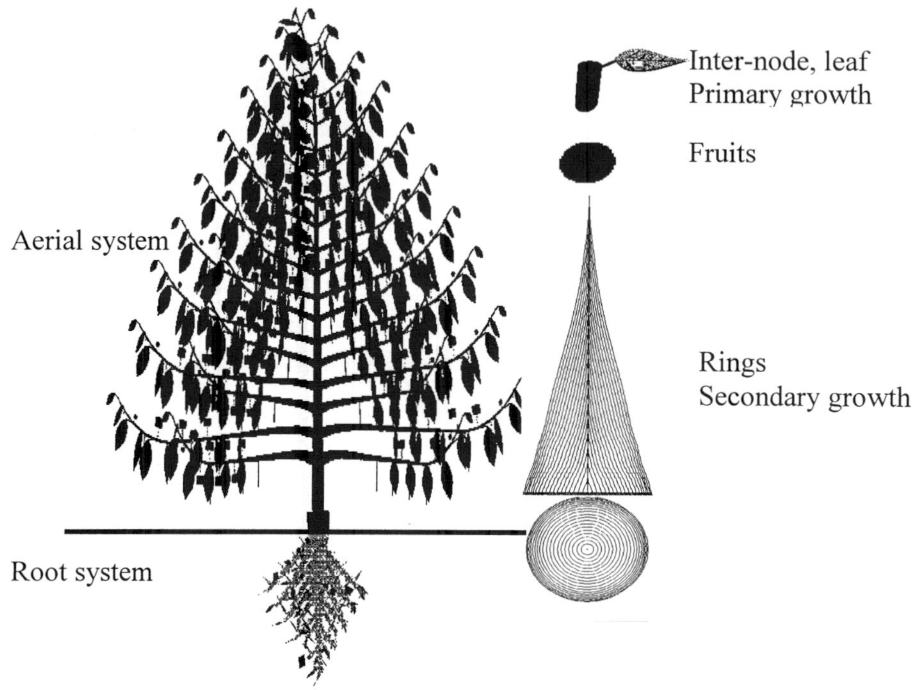

Figure 2. Sinks for the biomass in the above and below-ground tree architecture. Sinks can be primary compartments i.e. aerial and root systems or single organs i.e. inter-nodes, leaves.

of transpiration is calculated using:

$$Q(t) = \frac{\Delta\Psi(t)}{R(t)}\Delta t \qquad (1)$$

where: *Q(t)* is quantity of water at the cycle t, $\Delta\Psi$(t) is water potential, *R(t)* is hydraulic resistance of both shoots and roots in series, Δ t is duration of cycle.

Matter fabrication and secondary growth

The hypothetical relationship between transpiration and assimilate production allows the quantity of fresh matter made by the leaves to be calculated i.e. water use efficiency (WUE),. We assume that WUE is constant during a cycle. The ratio between dry matter and fresh matter is set to 10%. It is assumed that the respiration factor is deducted in the WUE notion as we deal with plant architecture, i.e. fixed carbon, which is not subject to respiration. Therefore, Equation (1) becomes:

$$Qm(t) = Wue(t)Q(t) \qquad (2)$$

where: *Wue(t)* is Water use efficiency at cycle t, *Qm(t)* is quantity of fabricated fresh matter.

Part of the matter fabricated is used by the cambium to build the new layer along the vegetative axis, which corresponds to the new growth ring. The allocation of the matter is assumed to be uniform along the layer, which is considered as an organ and has its own sink. The remaining material is kept and constitutes the reserves for the next cycle of growth.

The growth of the plant as a recurrence formulation

The growth of the plant during the current cycle is ensured by the reserves stocked in the previous cycles and also according to its inherent morphogenetic behaviour. The shape of each organ can be computed from its volume of matter and it is possible to link the reserves between cycles using the architectural data and mathematical equations based on recurrence formulation (Reffye, 1996). Therefore:

$$Q_N = F(E, f_N, A, B, C, ..., Q_{N-1}, ..., Q_K, , , Q_0 \qquad (3)$$

with Q_N the volume of fresh matter fabricated at cycle *n*, Q_0 the seed reserve at the starting cycle of the growth, *E* the potential of matter production depending on WUE (hydric potential assumed constant), f_N the number of assimilator organs at cycle n of growth,

A, B, C synthetic parameters, resulting in the combination of allometric rules, resistivities and sinks of the organs.

For example, consider the following simple case:

- Immediate and complete expansion of organs at their cycle of creation and a time life duration of one cycle for the leaves (in trees where the cycle lasts one year).
- No hydraulic resistance in the aerial axis nor in the root system. Only the leaves offer resistance.
- Allocation of biomass in a constant ratio between above- and below-ground architecture during growth.

Therefore, for biomass fabricated at each cycle:

$$Q_N = \frac{E Q_{N-1}}{A + B\left(\frac{Q_{N-1}}{f_N}\right)^C} \tag{4}$$

General results on the biomass production per growth cycle, independent of the architectural model, are obtained using eq. (4). These results are determined from the hydraulic resistance of the leaf blade and petiole, and depend also on the following allometric laws:

- Growth is exponential during youth and organs increase in size.
- Growth is stabilized after several cycles because the hydraulic constraints are mainly located in the leaves. The organs reach a size limit. Growth in height becomes linear and increase in weight depends on the architectural model: the greater the number of branches, the larger the matter production, but the smaller the organs.
- Growth can slow down if the hydraulic resistance of the axis becomes significant e.g. in old trees, or if cavitation occurs).

Simulation of the tree/environment interactions

The tree and its environment

Besides the action of the light, other physically impeding phenomena exist within the crown of a tree, or between the crowns of neighbouring trees. These phenomena are known collectively as 'architechural shyness'. This shyness expresses the capacity in certain species to avoid the crowns of neighbouring trees by limiting development of their own branches before contact. As a result, open spaces form between neighbouring crowns (Figure 3). However, reconstituting architectural shyness is not just a matter of positioning within a given space, trees which have been simulated independently from each other. Competition between trees must be taken into consideration (Blaise, 1991, Blaise et al., 1996). The simulation process allows for the growth simulation of several trees at a time, by using the topology of the hierarchical structure at the plant level. Competition occurs at different levels. Many physically impeding phenomena within the crown of a tree or between the crowns of neighbouring trees can cause a 'pruning' of the plant axes. The root system can also be subject to this kind of interaction.

Discretization of space and trees

The major problem to overcome with regards to discretization of objects in their environment, is the resolution of collisions between entities with a known geometry. The most direct solution is to use analytic geometry that, with the help of classic tools, is usually satisfactory in most cases. This approach can only be used for scenes composed of a limited number of volumes and surfaces with simple geometry. It is not, however, conceivable in treating plant structures with a great number of elements. Therefore, we use also the voxel space technique which has been created to simulate the interactions between a tree and its environment during growth (Green, 1989)

Voxel space is defined as a region in 3-D space which becomes discrete by following a regular mesh of parallelograms or voxels. The dimension of each voxel is called the discretization unit. Due to this discrete space, it is easier and more rapid to determine some relationships, such as proximity or intersection between geometrical objects, that with the use of analytical geometry. Thus plant organs are not considered as geometrical entities, but as groups of voxels. A voxel is therefore considered occupied, once it has been traversed by an organ.

Detection and treatment of the impediment

An impediment occurs each time an axis penetrates into a voxel which is already occupied. Depending on the case, different treatments can be carried out. For mechanical reasons, The probability of collision between two axes depends on their 'topological spacing'. For example, two axes on the same growth unit are not likely to impede each other, despite their proximity. However, Two geometrically similar axes belonging to different main branches, have a strong chance of collision. Once the eventual collision is detected, and the impediment demonstrated, an adequate

Figure 3. Simulation of crown impediment during the growth of two *Plumeria*.

treatment must be carried out. It is not important to determine the exact intersections between the internodes. It is only the absence or existence of a collision which is of interest. This would mean inhibition or continuation of growth i.e. in the case of an impediment, the bud dies.

The dimension of the voxel discreting space can greatly modify the resulting architecture, (Figure 4). However, the principle of spatial discretization should not be considered as a method with random results. Indeed, the different architectures obtained with voxels of different dimensions gradually evolve and converge towards a 'limit' topology that corresponds to an asymptotic value for the unit of discretization. In other words, a minimal discretization unit exists under which the obtained architectures are similar (Blaise, 1991).

Extension of the AMAPpara model for the root-system

Botanical features

The root system is built in the same way as that of the shoot. The axis is generated with virtual inter-nodes as previously defined. Each axis bears root-hairs, which are modelled as single flattened squares whose surface absorbs water. A life-time is defined for these organs. Primary growth results in the establishment of new virtual inter-nodes and root-hairs, and ensures an increase in axis length. At the end of each cycle, a new ring is added along the axis and secondary growth is carried out. The distribution of the biomass along the axis is assumed uniform from the tip to the base of the root-system. These under-ground organs also need to have sinks, and allometric rules need to be applied. Thus the functioning of the above and below parts of plant architecture is treated as symmetrical.

Under-ground modelling of the spatial environment

The spatial environment where the root-system develops, can also be discretized into voxels. Each voxel can contain not only roots but also stones, thus rendering it heterogeneous. This in turn limits the available space for the development of the axes. These two factors allow the simulation procedure to handle root growth.

Simulation results of the model

Influence of root system architecture on plant development

The allocation of assimilates to the various compartments and organs in a tree can be explicitly modelled (Figure 2). The sinks can be elementary organs e.g. leaves, internodes, fruits, growth rings or larger compartments e.g. the root or aerial system. Note that the matter repartition can be carried out theoretically and independently of the architectural model.

Competition between the aerial and root system has major consequences on whole plant development. The AMAPpara software can obtain very different kinds of plants according to the biomass repartition between the two systems. Therefore, combination of various possibilities for biomass allocation may yield different results in terms of plant size and morphology for the same architectural model. For a given species, it is therefore necessary to test which assimilate partitioning model is the most relevant. The software AMAPpara can currently simulate three alternative models:

In model M1, resources are shared initially between both the shoot and root systems, then secondly among compartments defined by branching

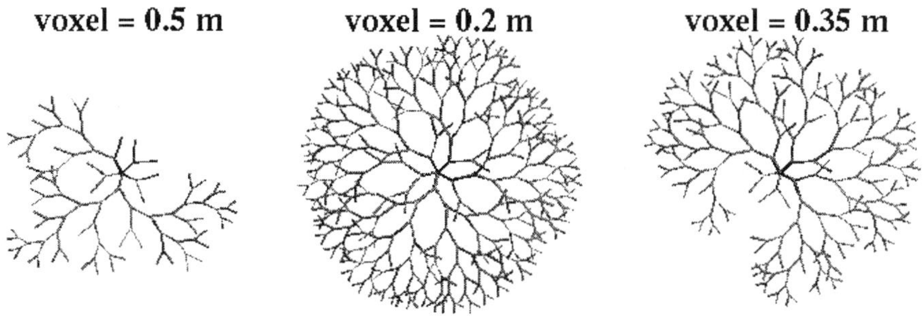

Figure 4. Simulation of the impediment inside a tier of branches of *Terminalia* with different sizes of voxels.

orders within each system. Assimilates are then partitioned among the organs within each compartment. This model is similar to traditional eco-physiological models and does not really take the whole architecture into account.

In model M2, assimilates are also partitioned initially between the shoot and root systems in a constant proportion during growth, but are then directly shared amongst each organ according to the sinks within each system. Thus the partitioning of the biomass between these two structures does not depend on the numbers of organs in each structure. Therefore there is no competition for carbon between the shoot and root system, and no imbalance of above and below-ground resource allocation can occur during growth.

In model M3, the organs of shoot and root systems compete for the same carbon reserves, depending on their sinks. Therefore the architectural model has a strong influence. The more organs a system possesses, the more attractive it is for resources. Thus, a strong imbalance can occur between shoot and root systems during growth, if they have different architectural models.

Simulations of growth according to models of biomass repartition

Only models M2 and M3 will be discussed, as M1 does not consider plant architecture. The matter production depends on the hydraulic resistance of the plant and the biomass repartition model is used. The hydraulic resistance of the assimilating organs i.e. leaves and root-hairs depends on the surface resistance available for water flow i.e. leaf blades and root-hairs, and also on the axis resistance i.e. petioles and root-hairs. The hydraulic architecture of the plant is very simple if the woody axes offer a weak resistance. The assimilating organs are all set in parallel and the total resistance of the two structures can be computed by the following formula:

$$\begin{array}{l} \frac{1}{R_1} = \sum_{i=1}^{N1} \frac{1}{r_i^1} \\ \{ \frac{1}{R_2} = \sum^{N2} i = 1 \frac{1}{r_i^2} \\ R_P = R_1 + R_2 \end{array}$$

where: r_i^1 is the hydraulic resistance of the leaf number i for above-ground architecture, R_1 is the hydraulic resistance of above-ground architecture, r_i^2 is the hydraulic resistance of the root-hair number i for the below-ground architecture, R_2 is the hydraulic resistance of below-ground architecture, Rp is the hydraulic resistance of the both structures in series.

If one system offers an infinite resistance the total resistance of the plant is infinite and growth stops because there is no more transpiration. If one system offers a negligible resistance towards the other, growth will be controlled by the other system.

Three typical architectural types will be chosen for the simulation of plant growth using models M2 and M3. These architectural types, ‘Corner’, ‘Leeuwenberg’ and ‘Rauh,’ are taken from those described by Hallé et al. (1978). The ‘Corner’ type consists of a single stem bearing simple fronds e.g. palm trees. The ‘Leeuwenberg’ type is entirely sympodial e.g. *Plumeria*. The ‘Rauh’ type is pure monopodial, with different branching orders as e.g. *Pinus* sp. These different architectural forms have previously been simulated for the shoots only (Reffye et al. 1996). In order to simulate the interaction between shoot and root growth, it is necessary to choose a typical shoot type e.g. Rauh (limited to second order branching), and combine it with different architectural types of root system i.e. Corner, Leeuwenberg and Rauh. Although these model root systems are not necessarily realistic, the results provide us with an indication of how the architecture of the below-ground parts may influence that of the shoot system.

The more the plant has organs of a particular size, the less will be its hydraulic resistance. Therefore, the most successful architectural type with regards to this factor, would be the Leeuwenberg model, followed by Rauh and then Corner. The biomass repartition model monitors the size of these assimilating organs and also the water exchange occurring at leaf blade and root-hair surfaces. If either the shoot or root system out-competes the other in obtaining assimilates, the latter will have very small organs, and this will inhibit its growth. Results of simulations using models M2 and M3, for plants the same age are presented below.

Results for model M2

The biomass repartition ratio between the shoot and root systems is the same for the three different cases presented i.e. Rauh shoot architecture with different root architectural types.

In the Corner type root system, only one large tap-root exists, and in the Rauh model, a few large roots are present (Figure 5). In these systems, biomass is allocated and shared between few organs, whereas in the Leeuwenberg model, the same biomass is allocated to many different roots. In model M2, the hydraulic resistance is lowest in the most highly branched root system i.e. with Leeuwenberg type architecture. Therefore, the above-ground organs will be the largest in the plant with a Leeuwenberg type root system. The influence of biomass allocation is thus less than that of the hydraulic network.

If the number of cycles increases indefinitely, we observe that (Figure 6):

- The growth of plant I is controlled mainly by the Corner root system type.
- In plant II, there is an equilibrium in growth between the two systems.
- The growth of plant III is controlled by the above-ground Rauh type architecture.

Results for model M3

The conditions for the simulation are the same as those used for Model M2, except for the rules concerning biomass allocation. In this model, the repartition of matter between the above- and below-ground systems depends on the number of organs and the sinks which supply them. The plant with the smallest aerial organs possesses the most highly branched root system (Figure 7), unlike the same plant (plant III) simulated using model M2 (Figure 6). Therefore, the Leeuwenberg type root system is more efficient at obtaining assimilates than either the Corner or Rauh type systems. However, this efficiency of the root system impedes the growth of the aerial parts. The best shoot growth is ensured if the ratio between the number of organs of the two systems remains proportional during growth. The only plant with a balanced equilibrium between the two systems is that built with Rauh type architecture for both above- and below-ground architecture.

Root system growth in an non homogeneous environment

The computing structure and tools which must be defined in order to take into account the physically impeding phenomena in crowns have already been defined. The same tools are used to simulate the effect of the environment on root growth i.e. the voxel space technique.

Physical impediment can occur in the soil in two ways:

- Impediment caused by a collision inside a voxel between roots coming from the same or from different root systems.
- Impediment coming from the heterogeneity of the soil structure which involves stones of different sizes scattered into the below–ground space.

Impediment caused by root system architecture

For the simulation of impeded and free-growing root systems in a non-homogenous soil, a single plant was chosen possessing a Corner type shoot architecture and a Leeuwenberg type root system, which in this case represents a rhizome e.g. ginger (Bell, 1976). The diameter of the rhizome was kept constant, therefore the length of an axis was proportional to its volume. The model M2 was chosen to calculate biomass allocation between shoots and roots. The results show that the free-growing plant was taller than the plant whose root system was impeded, because the biomass fabricated was greater (Figure 8). The root system was also better developed and had a lower hydraulic resistance. However the branches were relatively small because the biomass was shared between many organs. The root system of the impeded plant was spread over a larger surface compared to that of the control plant (Figure 8). As the hydraulic resistance is bigger, there is therefore less matter to share between

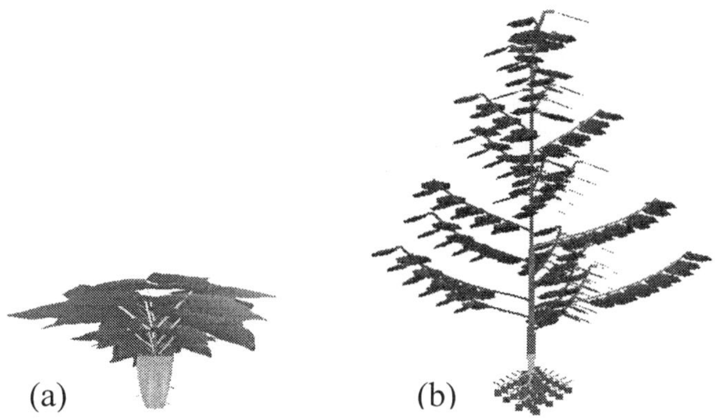

Figure 5. Simulation of the biomass repartition between shoots and roots: (a) Corner model - plant with tap-root (b) Rauh model - plant with a few large roots.

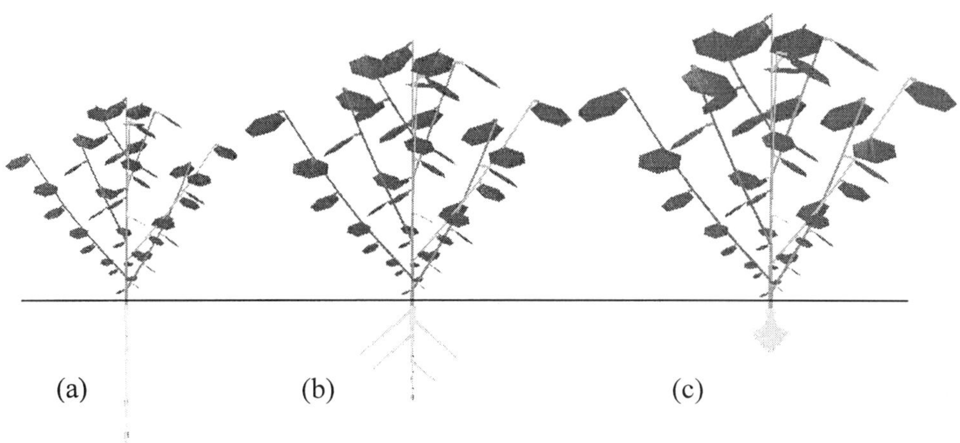

Figure 6. Model M2: influence of the architectural model of the root system on the development of the aerial system. The greater the branching, the less the hydraulic resistance: (a) Corner, (b) Rauh, (c) Leeuwenberg.

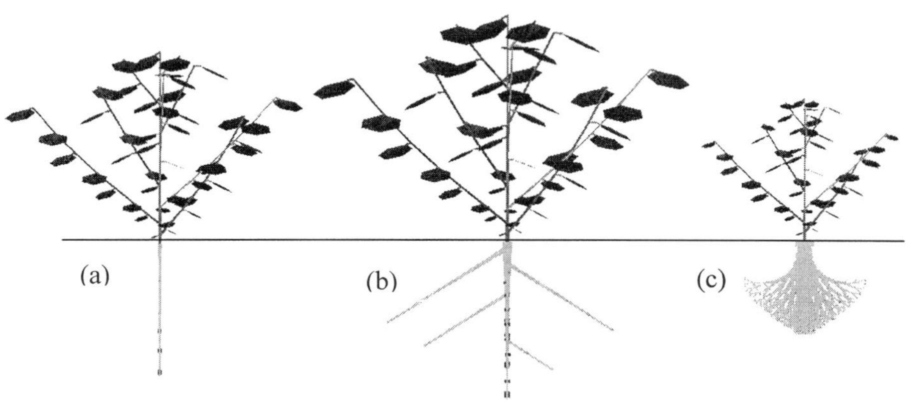

Figure 7. Model M3: influence of the architectural type of root system on development of the shoot system. The highly branched root system needs more biomass and out-competes the shoot system: (a) Corner, (b), (c) Leeuwenberg.

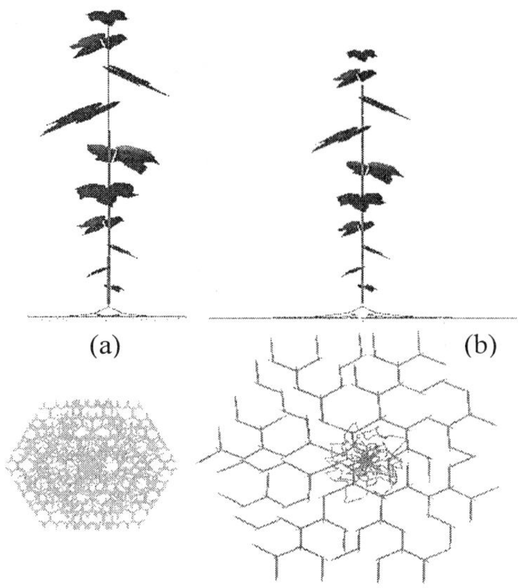

Figure 8. Interaction between shoots and roots in a simple virtual plant with a rhizome: (a) plant without impediment, (b) plant with impediment.

different organs of the plant. Nevertheless, the proportion of biomass per organ was greater because the impediment reduces the number of branches. Therefore the plant becomes more sensitive to the spatial environment.

These results can be extended to interactions between many plants, where root competition is a major factor to consider when modelling growth (Figure 9). The roots develop once they reach an area with less physical constraints. These longer root branches can then benefit from available assimilates.

Impediment caused by the soil heterogeneity

In order to simulate a complex soil, the underground space is scattered with small, numerous cubes, or impeding voxels (Figure 10). Inside each of these cubes, the space is considered as 'toxic' for root growth, so that growth stops definitively if a root apex grows inside it. These voxels are not to be confused with those used to discretize the soil, and which are not 'toxic'. A strong impediment to plant growth can result if the density of harmful voxels is high. One of the results of the impediment will be to decrease the number of new organs and to increase the size of the remaining ones.

The use of voxels is thus especially important when simulating plant growth in a plantation. Using simplistic Rauh type root systems, a simulation of shoot and root competition in a stand of *Abies* and *Pinus* was carried out. *Abies* was given typical laterally extending root systems and *Pinus*, tap-root type systems (Köstler et al. 1968). The results of the simulation show that some roots did not enter into voxels and thus continued growing around them, whilst other roots did grow into the voxel space and 'died' (Figure 11). The living roots became greater in length as more biomass was allocated to the remaining roots not 'killed' by the voxels. Above-ground, the tallest trees were those with the biggest root systems and with some physical impediment between crowns. In those trees closely surrounded by others, there is less room for branch development, therefore, more assimilates can be allocated to stem growth. However, if root system growth is strongly impeded, that of the aerial system will also be reduced.

Conclusion

The present paper aims at extending the previous research carried out by the authors, mainly concerned with shoot growth, to the root system. The interactions between the shoots and roots can be modelled and placed into two categories. The first category concerns the *internal* competition for biomass between shoot and root systems, whereas the second is concerned with the *external* interaction between organs created during growth and the space available.

In the first category we can consider independently various models for biomass allocation depending on the sinks available, as well as various architectural models. As in previous work by Clauznitser and

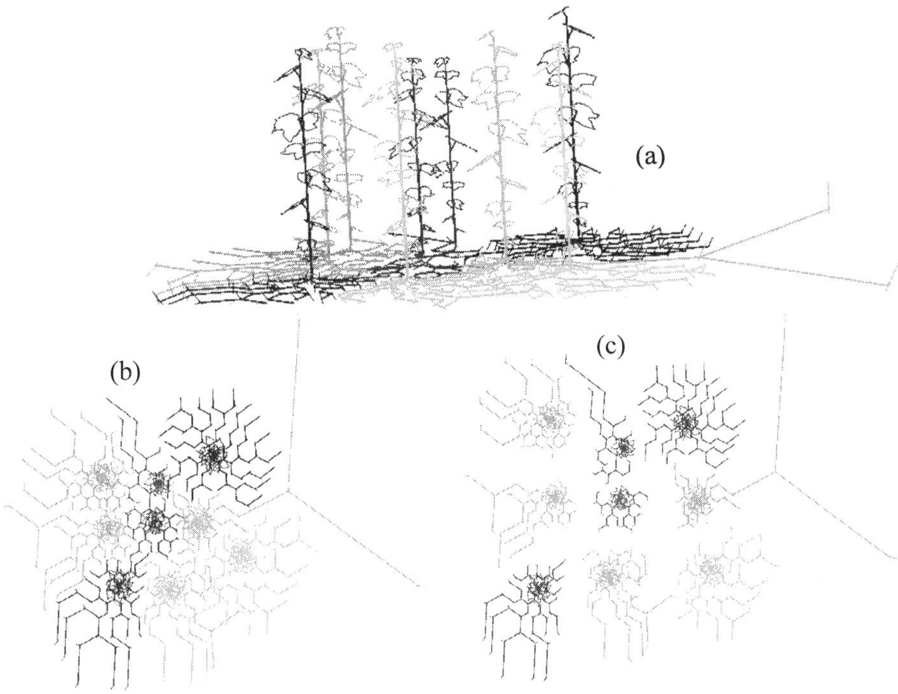

Figure 9. Effect of impediment on root extension inside a group of plants: (a) profile of the set, (b) rhizome organisation, (c) separated rhizomes of (b).

Hopmans (1994), carbon fabrication depends on water transpiration and WUE. For example, a simple case of this type of model would be one where a constant ratio of biomass allocation is kept between shoot and root systems. Therefore, shoots and roots would be the main compartments and would not compete for carbon. However if the architectural model of the shoot system is fixed, but the architectural type of root system can be modified, interaction between above- and below-ground growth can be observed: the greater the branching of the root system, the less the hydraulic resistance. Therefore, the Leeuwenberg type root system can improve plant transpiration compared to that possessing a Corner type root system, hence, the whole plant is bigger. Nevertheless, the branches of the highly ramified Leeuwenberg root system are smaller than the less branched Corner root system, as there are more sinks which need supplying. The space available for the underground resources is thus reduced.

A stronger interaction between root and shoot growth was found in plants where where all the organs of the whole plant compete for the same quantity of biomass. The most highly branched root systems will out-compete the shoot system, thus reducing the total surface of the assimilating aerial organs. Therefore, shoot growth will decrease, with leaves becoming smaller and smaller. In this system, only small plants can be built, unless a constant ratio can be kept during growth, in the numbers of organs of above- and below-ground systems.

The second category of interactions concerns the interaction of the plant structure with the local environment. The spatial environment is discretized with voxels and it is possible to change the direction of a root axis depending on local conditions (Blaise, 1991; Pagés, 1995). A physical impediment can be applied to a root system growing in a non-homogeneous soil, e.g. one which contains randomly distributed stones. The results of the simulation show that the impediment can cause a reduction in plant growth. However, roots increase in length as more carbon is allocated to those remaining roots which have not been 'killed' by the obstacles (voxels).

To summarize, the AMAPpara model can now consider both shoots and roots at the same level for processing plant growth. It is possible to contemplate complex simulations involving above- and below-ground interactions on the growth of virtual plants. However, it will be necessary in the future to use more

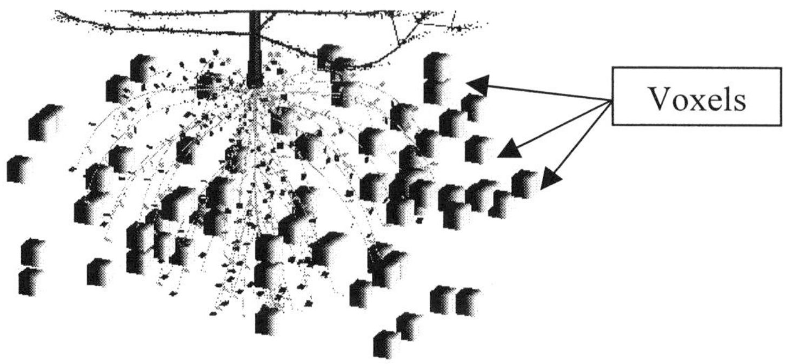

Figure 10. Representation of soil heterogeneity with voxels.

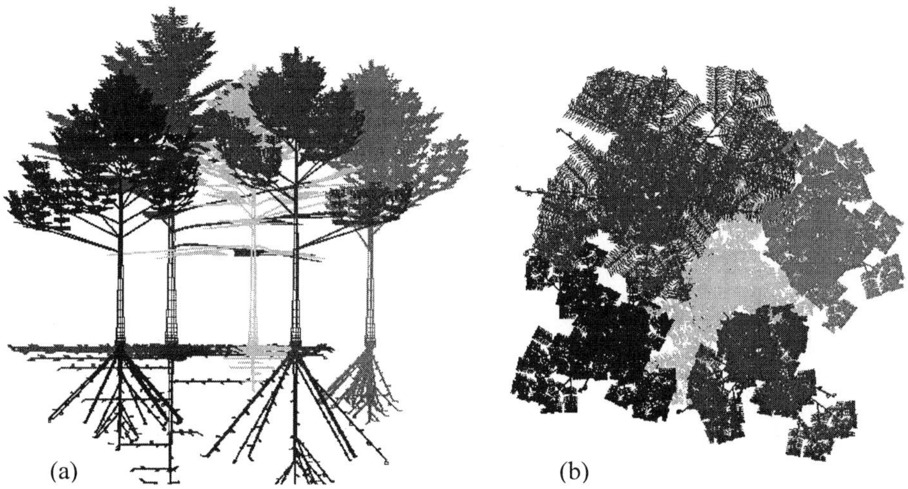

Figure 11. Above and below ground competition for space in a grove of conifers (Fir and pine trees). The height and biomass of each tree depend on the local spatial conditions: (a) profile, (b) aerial view.

realistic types of root system architecture, and to correlate the type of architecture with plant age, soil type and nutrient availability (Fitter et al., 1991, Fitter and Stickland, 1991).

References

Atger C and Edelin C 1994 Premières données sur l'architecture comparée des systèmes racinaires et caulinaires des arbres. Can. J. Bot., 72, 963–975.

Bell A 1976 Computerized vegetative mobility in rhizomatous plants. Lindenmayer, Rozenberg. North Holland Company.

Barczi J.-F, Reffye P and Caraglio Y 1997 Essai sur l'identification et la mise en oeuvre des paramètres nécessaires à la simulation d'une architecture végétale: le logiciel AMAPsim. *In* Eds J Bouchon, P (de) Reffye and D Barthélémy. Modelisation et simulation de l'architecture des végétaux. pp 255–423. Science Update. I.N.R.A. Edition.

Blaise F 1991 Simulation du parallélisme dans la croissance des plantes et applications.Thèse de doctorat, Université Louis Pasteur, Strasbourg, France.

Blaise F, Houllier F and Reffye P 1996 Simulation of tree architecture and growth in a forest stand: AMAPpara software. *In* Connection between silviculture and wood quality through modelling approaches and simulation softwares IUFRO WP S5.01.04 Workshop, Hook, Sweden. Ed. G. Nepveu INRA. p 46–55.

Blaise F, Barczi J F, Jaeger M, Dinoouard P and Reffye P 1998 Simulation of the growth of plants – modelling of metamorphosis and spatial interactions in the architecture and development of plants. *In* Cyberworlds 6. Eds T L Kuni and A Luciani. pp. 81–109, Springer-Verlag Tokyo.

Cannell M G R and Dewar R C 1994 Carbon allocation in trees: a review of concepts for modelling. Adv. in Ecol. Res. 25, 59–104.

Clausnitzer V and Hopmans J W 1994 Simultaneous modelling of transient three dimensional root growth and soil water flow. Plant Soil 164, 299–314.

Fitter A H, Stickland T R, Harvey M L and Wilson G W 1991 Architectural analysis of plant root systems. I Architectural correlates of exploitation efficiency. New Phytol. 118, 375–382.

Fitter A H and Stickland T R 1991 Architectural analysis of plant root systems 2. Influence of nutrient supply on architecture in contrasting plant species. New Phytol. 118, 383–389.

Fourcaud T, Blaise F, Reffye P, Houllier F and Barthélémy D 1997 Plant growth simulation based on ecophysiological processes – Influence of architectural modifications on tree growth. *In* Plant Biomechanics conference proceedings I. Ed. G Jeronimidis and J F V Vincent. pp. 331–336. Centre for Biomimetics, University of Reading.

Godin C, Guédon Y, Costes E and Caraglio Y 1997 Measuring and analysing plants with the AMAPmod software. *In* Advances in computational life sciences: plants to ecosystems, Ed. M T Michalewicz. pp 53–84. CSIRO, Australia.

Green N 1989 Voxel space automata: modelling with stochastic growth processes in voxel space. Computer Graphics, 233, 175–184.

Hallé F, Oldeman R A A and Tomlinson P B 1978 Tropical Trees and Forest. Springer-Verlag, Berlin, Heidelberg, New York.

Jourdan C, Rey H and Guedon Y 1995a Modelling of the branching process of the root system of young oil palm seedlings. A Markovian approach. *In* Structure and Function of Roots. Ed. F. Baluska. pp 113–119. Kluwer Acadmic Publishers.

Jourdan C, and Rey H 1995b Architectural analysis and modelling of the branching process of the young oil-palm (*Elais Guinensis* Jacq.) root system. I. The model. Plant Soil. 190, 217–233.

Jourdan C, and Rey H 1995c Architectural analysis and modelling of the branching process of the young oil-palm (*Elais Guinensis* Jacq.) root system. II. Estimation of root paramaters using the RACINES postprocessor. Plant Soil. 190, 235–246.

Köstler J N, Brückner E and Bibelriether H 1968 Die Wurzeln der Waldbäume. Verlag Paul Parey, Hamburg and Berlin.

Kellomäki S and Strandman H 1995 A model for the structural growth of young Scots pine crowns based on light interception by shoots, Ecological Modelling, 80, 237–250.

Kurth W 1996 Some new formalisms for modelling the interactions between plant architecture, competition and carbon allocation. Proc. 4th Workshop on Individual-based Structural and Functional Models in Ecology, Wallenfels, Bayreuther Forum für Ökologie.

Le Roux Y and Pagès L 1994 Développement et polymorphisme racinaire chez de jeunes semi d'hévéa (*Hevea brasiliensis*). Can. J. Bot. 72, 924–932.

Mäkelä A 1986 Implications of the pipe model theory on dry matter partitioning and height growth in trees. J. Theor. Biol. 123, 103–120.

Pagès L, Le Roux Y and Thaler P 1995 Modélisation de l'architecture racinaire. Plant. Rech. Dév. 2, 19–30.

Perttunen J, Sievänen R, Nikinmaa E, Salminen H, Saarenmaa H and Väkevä J 1996 LIGNUM: A tree model based on simple structural units. Ann. Bot. 77, 87–98.

Prusinkiewicz P, Hanan J, Hammel M, and Mech R 1997 L-systems: from the theory to visual models of plants. *In* Advances in computational life sciences: plants to ecosystems. Ed. M T. Michalewicz. pp 1–27. CSIRO, Australia.

Reffye P, Barthélémy D, Blaise F, Fourcaud T and Houllier F 1996 Essai sur la simulation des relations fonctionnelles entre architecture racinaire, architecture caulinaire et croissance en épaisseur des axes chez les végétaux. Proc. L'Arbre. Biologie et Développement, 3ème colloque, Montpellier France. Naturalia Monspeliensa, Ed. C. Edelin.

Reffye P, Fourcaud T, Blaise F, Barthélémy D and Houllier F 1996 A functional model of tree growth and tree architecture, Silva Fennica 31, 297–311.

Reffye P, Houllier F, Blaise F, Barthélémy D, Dauzat J and Auclair D 1995 A model simulating above-and below- ground tree architecture with agroforestry applications, Agroforestry Systems 30, 175–197.

Reffye P, Houllier F, Blaise F and Fourcaud T 1997 Essai sur les relations entre l'architecture d'un arbre et la grosseur de ses axes végétatifs. *In* Modélisation et simulation de l'architecture des arbres. Eds J Bouchon, P Reffye and D Barthélémy, pp 255–423. INRA, Science Update.

Shinozaki K, Yoda K, Hozumi K and Kira T 1964 A quantitative analysis of plant form - the pipe model theory. I. Basic analysis. Jap. J. Ecol., 14, 97–105.

Valentine H T 1985 Tree-growth models: derivations employing the pipe-model theory. J. Theor. Biol. 117, 579–585.

A. Stokes (ed.), The Supporting Roots of Trees and Woody Plants: Form, Function and Physiology, 209–215.

Why are laterals less affected than main axes by homogeneous unfavourable physical conditions? A model-based hypothesis

Philippe Thaler[1] and Loïc Pagès[2]
[1]*Cirad-CP, Programme Hévéa, BP 5035, F34032 Montpellier, Cedex 1, France and* [1]*INRA Centre d'Avignon, Ecophysiologie et Horticulture, F84914 Avignon, Cedex 9, France*

Key words: carbon allocation, *Hevea brasiliensis*, root growth, root system architecture, simulation model, soil impedance

Abstract

When plants develop in strong soils, growth of the root system is generally depressed. However, branching and elongation of branches are often less affected than growth of the main axes, whenever the whole root system encounters even-impeded conditions. On the basis of a model simulating root growth and architecture as related to assimilate availability, we propose a simple hypothesis to explain such behaviour. In the model, growth of each root depends on its own elongation potential, which is estimated by its apical diameter. The potential elongation rate–apical diameter relationship is the same for all the roots of the system and is described by a monomolecular function. Our hypothesis is that the effect of soil strength can be simulated by introducing an impedance factor in the definition of root maximum potential elongation rate, common to the whole root system. When such impedance factor is applied, it affects more the potential of larger roots (main axes) than that of thinner roots (secondary and tertiary branches). Simulations provided in high impedance conditions led to root systems characterised by short taproots, whereas growth of secondary roots was unaffected and growth of tertiary roots was enhanced. Actual branching density was also higher, although branching rules have been unchanged. Such simulated systems where similar to that observed in strong soils. Friction laws or pore size can be involved in the larger reduction of the potential growth of main axes. Moreover, when growth of main axes is restricted, assimilate availability becomes higher for branches and that could explain that their growth could be increased in a homogeneous strong soil.

Introduction

When plants develop in unfavourable physical conditions, characterised by a high soil strength or impedance, growth of the root system is depressed, (Bengough and Mullins, 1990; Goss, 1977). However, the different root types and the different developmental processes seem to be differentially affected by soil strength. Branching (initiation and/or development of lateral meristems) is often enhanced (Goss, 1977) and elongation of branches is less affected than that of main axes (Misra and Gibbons, 1996) and could be sometimes enhanced (Goss and Scott Russel, 1980). Mechanisms relying mechanical impedance to the elongation of main axes have been studied by many authors (Wilson et al., 1977; review by Bengough and Mullins, 1990): it is considered that physical constraints directly reduces cell elongation rate and/or division rate. On the other hand, explanations of the apparent increase in branch development in such conditions remain scarce (see, for instance, the review by Bennie, 1996). Competition for resources and active correlations (decrease of the inhibition exerted by main axes on the development of laterals) can explain the observed patterns when main roots encounter unfavourable conditions whereas laterals remain in better conditions (Riedacker et al., 1982) but such compensatory mechanisms are not likely to explain the relative, or even absolute, positive effects observed on laterals placed in the same environment than main axes. Some authors have proposed that growth of primary

* *Plant and Soil* is the original source of publication of this article. It is recommended that this article is cited as: *Plant and Soil* **217:** 151–157, 1999.

axes can induce horizontal 'cracks' in the surrounding soil hence providing a local reduction in impedance that could be favourable to the growth of laterals (Barley et al., 1965; Misra and Gibbons, 1996). But such explanation seems limited to very particular conditions. A more suitable hypothesis is that pore size in the considered soils is such that main and larger roots are impeded, but not lateral, thinner roots (Bengough and Mullins, 1990). However, the penetration ability of roots is known to increase with root diameter (Materechera et al., 1992) and this should rather be favourable to the elongation of main axes.

The work presented herein is an analysis of the different responses of taproot and laterals facing a homogeneous constraining soil, on the basis of carbohydrate allocation within the root system. To this end, we used a specific model that describes rubber tree root system architecture and development as related to assimilate availability (Thaler and Pagès, 1998).

The model

Briefly, the model (Thaler and Pagès, 1998) developed for rubber tree (*Hevea brasiliensis*), considers each root to be a component of a system of axes – characterised by its spatial position and connections within the system – and as an individual sink competing for carbohydrates. Architectural bases of the model were developed by Pagès et al. (1989) and adapted to rubber tree by Pagès et al. (1995). At each time step, the root system, represented as a set of segments, is extended by the application of developmental processes: emission of new roots, growth and branching. The emission process generates the beginning of seminal (taproot and early secondary roots) and adventitious (late secondary roots) axes. Branching is assumed to occur accropetaly and is described by the inter-branch distance, which is specific to each root type. Root typology was established by Le Roux and Pagès (1994). The model developed by Thaler and Pagès (1998) differs with the previous architectural models by its way of simulating the growing process. In the present model, elongation and radial growth are no more simulated by typical growth functions of time, which parameters are fixed at the beginning of each root's growth. In order to rely root growth to assimilate availability, each root is assumed to be an individual sink competing for carbohydrates. The carbon requirements for growth are satisfied on the basis of the demand of the growing organs, which include the carbon involved in structural growth and the carbon used by growth respiration. Daily carbohydrate availability after maintenance respiration is calculated together with carbohydrate demand for the growth of leaf, stem and root. Growth occurs at potential rate for all sinks when sufficient carbon is available. If this is not possible, the fraction of potential growth that can be supported is calculated as the ratio of the carbon available after maintenance respiration to the sum of carbon requirements for potential growth of all the sinks. Since this fraction is the same for all sinks, variations in relative growth rates between organs depend on variations in the daily potential growth of each sink. Carbohydrate input is provided by seed reserves, photosynthesis and plant reserves. In the present version of the model, daily photosynthesis is proportional to leaf surface, specified by a specific rate which variations reflect light changes. The reserves are released according to constant rates. Empirical logistic relations fitted to observed growth describe stem potential elongation and leaf area potential growth. Stem potential radial growth and leaf specific mass potential growths are assumed to be constant.

The key point of the model is the determination of each root's growth potential. In most of allocation models, this potential is estimated by growth curves in conditions of non-limiting assimilate supply (Marcelis et al., 1989). But such a method is hardly suitable in our case since extensive variations in both growth rate and duration were recorded within root types (Le Roux and Pagès, 1994), so that growth curves for faster and longer-term growing roots are not likely to represent the growth potential for a given root type. Observed relationship between root elongation rate and apical diameter (Thaler and Pagès, 1996b) led us to consider the apical diameter of any root to be a convenient morphologic indicator of its elongation potential (Thaler and Pagès, 1998). The upper limit of the plot representing elongation rate versus apical diameter is considered to represent the potential elongation rate corresponding to each apical diameter. This relation is described by a monomolecular function which reflects the strong increase in potential elongation rate with diameter for low diameters (< 0.7 mm), and the lower increase for larger diameters (Figure 1).

$$Ep = E\max * (1 - \exp^{-b(D-D0)/E\max})$$

Ep: potential elongation rate (cm d^{-1})
Emax: maximum potential elongation rate (cm d^{-1})
D: apical diameter (mm)
$D0$: threshold diameter under which the root does not elongate (mm)

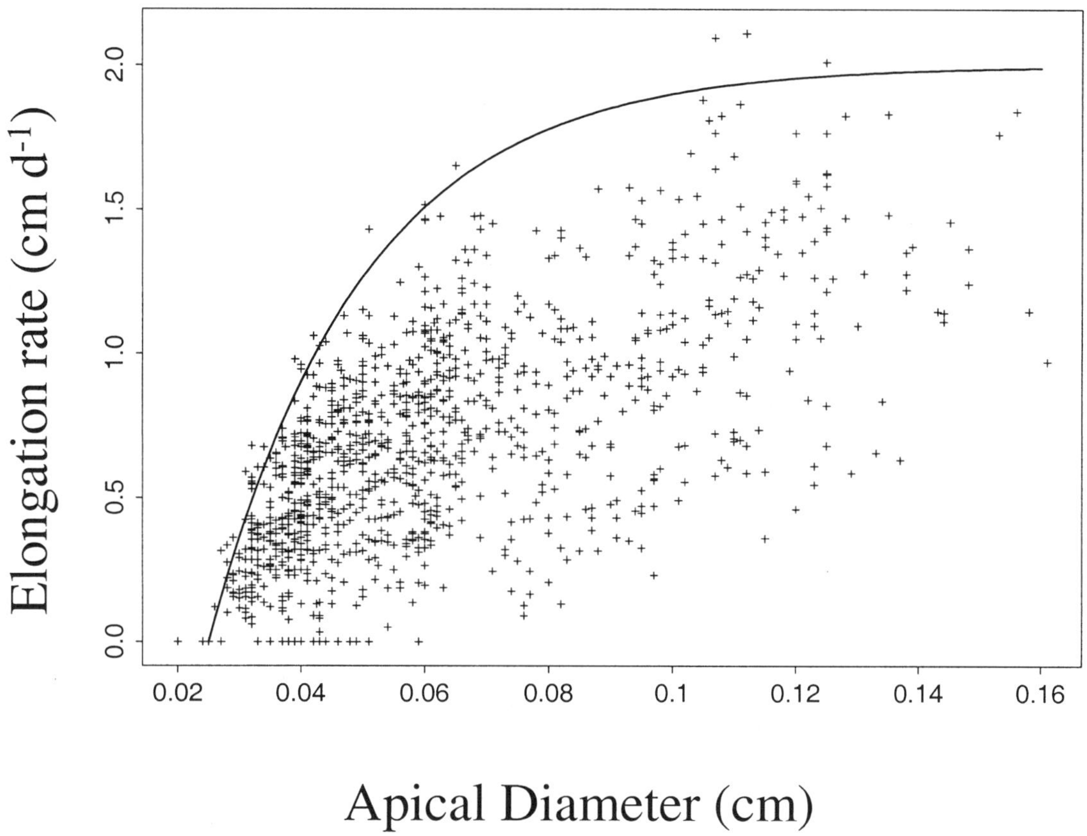

$$Ep = Emax \times (1\text{-}exp^{(-b(D-D0)/Emax)})$$

Figure 1. Apical diameter (cm)–elongation rate relationship for taproots and secondary roots. Each point represents data for one root at a given date. At each date 6 roots were measured on each plant (10) of the experiment. The envelope curve (—) is a monomolecular function adjusted to 95% quantiles.

b: initial increase in elongation rate with diameter (slope at threshold diameter, cm d^{-2}).

Moreover, the potential growth itself is controlled by the supply/demand ratio. For a given root, the apical diameter is not constant, but increases when this ratio is high and decreases when it is low (Thaler and Pagès, 1996b). Variations in apical diameter are simulated by successive functions:

During primordium development: for each root type, a potential initial (at emergence date) diameter and a daily potential diameter increment are defined. When assimilate supply is higher than demand, the primordium grows at its potential rate. When demand is higher than supply, a proportion of the potential increment is achieved. If the diameter at the emergence date is lower than a fixed threshold, the root fails to emerge.

After emergence: from its value at emergence, the diameter for each root will increase when the supply/demand ratio is higher than a given threshold, or decrease when this ratio is lower than another threshold. These thresholds and the (constant) increasing and decreasing rates are input parameters of the model. The elongation rate–apical diameter relationship and the rule describing the evolution of the apical diameter are common to all the root types described by Le Roux and Pagès (1994) on rubber tree: taproot, early secondary roots (ESR), acropetal secondary roots (ASR), late secondary roots (LSR), tertiary roots for all secondary root type, and quaternary roots.

However, each root type has a specific initial (at emergence) potential diameter: 2.0 mm for the taproot, 0.8 mm for the ESR, 0.5 mm for the ASR and 0.37 mm for tertiary roots. Thereby, starting from an initial value depending on root type, the potential elongation rate for all roots within the system will vary following a unique rule. Growth dynamics of any root is so controlled daily by a two step variation according to assimilate availability:

– variation in the proportion of potential growth actually achieved
– variation in the potential growth itself.

(Variation in apical diameter for the next time step.) We used this model to study the effects of uniform mechanical constraints on the development of a root system. We have considered that, from an intrinsic potential, external factors such as soil strength determine a conditional potential, as proposed to model the influence of temperature in classical growth models (Grossman and Dejong, 1994). We made the hypothesis that the variation induced by soil impedance affects the maximum potential elongation rate (Emax) defined for all the roots of the system. Introducing an impedance factor (I) to the definition of the potential elongation rate simulates such variation.

$$E\max_{\text{impeded}} = I * E\max$$

$$Ep = I * E\max * (1 - \exp^{-b(D-D0)/I*E\max})$$

We made no additional hypotheses on possible active correlations or specific effects on branching or any different behaviour between the different root types in response to the level of mechanical impedance.

Methods

Materials and methods used to establish the model and its parameters were described in Thaler and Pages (1996a,b, 1997).

We have simulated root systems developing with different impedance factors (I), using a set of default values for the other parameters of the model (Thaler and Pagès, 1998), except for Emax which was chosen as 2.0 cm d^{-1} (default value: 1.7 cm d^{-1}). Three impedance factors were considered: 1.0 (low impedance), 0.75 and 0.5 (high impedance). Duration of the simulation was 77 days. The model displays graphs representing the kinetics of mass, length, diameter for individual organs or per root type, but also the structure of modelled systems, i.e. the position of the different axes and the connections between them. Geometrical parameters (growth and branching angles) were not taken into account in the present work. Analyses rely on the examination of these graphs.

Results

When I varied between 1.0 (low impedance) and 0.5 (high impedance), curves representing the elongation rate-apical diameter relationships diverged more markedly for diameters larger than 0.5 mm. Therefore, the elongation potential for roots with large diameter, like taproot, was more sensitive to impedance variations than the elongation potential for thinner roots like ASR and tertiary roots (Figure 2).

The overall growth of the root system, characterised by its dry weight, was slightly affected by variations in I (data not shown). On the other hand, Figure 3 shows that the different root types had various reactions to changes in this parameter: elongation of taproot was strongly depressed when impedance increased, secondary roots were slightly affected, whereas elongation of tertiary roots was enhanced (temporal fluctuations in root growth rates were due to variations in assimilate availability consecutive to the rhythmic development of the shoot) Thaler and Pagès, 1996a. Conditions that lead to a decrease of the maximum potential elongation of roots thus resulted in root systems with a relative (secondary roots) or absolute (tertiary roots) enhancement of the growth of branches.

Figure 4 shows the structure of the root systems resulting from simulations with impedance factor of 1.0 and 0.5. Tertiary roots were not shown. Differential responses of secondary roots and taproots resulted in clearly different root structures. When impedance was low (I = 1.0), the root system was characterised by a long taproot bearing short branches, whereas higher impedance (I = 0.5) resulted in a shorter taproot bearing a larger proportion of longer laterals. Length of the longest lateral roots, the ESR, was similar in both cases, so that the horizontal expansion/vertical expansion ratio was largely higher with higher impedance. Moreover, when impedance was low, some segments of the taproot were unbranched (the branches failed to emerge since assimilate availability was too low during the development of their primordium, as a result of high demand exerted by taproot and by the rhythmic growth of shoot). While such bare segments were very short with higher impedance as a result of lower de-

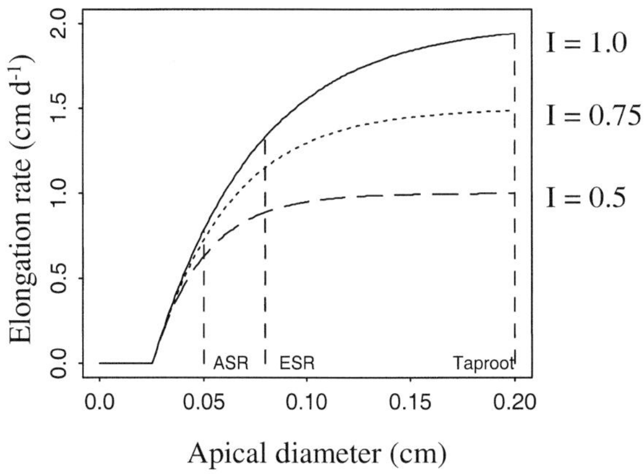

$$Ep = I \cdot Emax \cdot (1\text{-}exp^{(-b(D-D0)/I.Emax)})$$

Figure 2. Evolution of simulated potential elongation rate (cm d^{-1}) as a function of apical diameter with three impedance factors: 1.0 (—, low impedance); 0.75 (- - -, moderate impedance); 0.5 (– –, high impedance).

mand exerted by taproot. Higher impedance can so result in root systems with a higher overall branching density, although the process of branching has been unchanged.

Discussion

Simulations produced with a model merging allocation and architectural features, showed that root development patterns encountered in uniform constraining soils can be reproduced by introducing an impedance factor which reduces the maximum potential elongation rate common to all the roots of the system.

Such factor induces larger variations in the potential and actual growth for larger, fast growing roots – as taproots – than for thinner, slow growing roots as ASR and tertiary roots. Such behaviour of the model resulted from the observed differences between the initial apical diameter of the different root types (Le Roux and Pagès, 1994).

By this way, we can reproduce root development patterns encountered in impeded soil: decrease in the elongation of main axes, increase in the branching density and increase in elongation of branches, leading to less hierarchized root systems (Misra and Gibbons, 1996). It was not necessary to introduce hormonal correlations or to suppose differences in the impedance faced by main axes versus laterals.

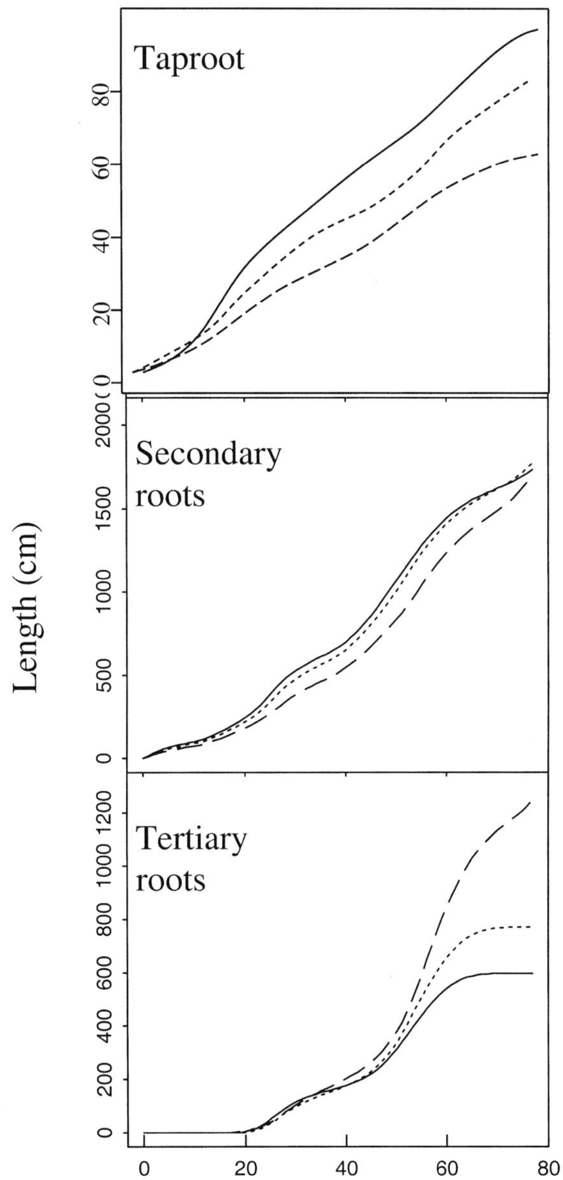

Figure 3. Evolution of cumulated length for the taproot, secondary roots and tertiary roots of root systems simulated with three impedance factors: 1.0 (—, low impedance); 0.75 (- - -, moderate impedance); 0.5 (– –, high impedance).

An explanation of the higher decrease recorded for main axes could be that the friction encountered by a given root is proportional to its elongation rate, so that fast growing roots could be more affected by soil strength. In the same way, pore size in the con-

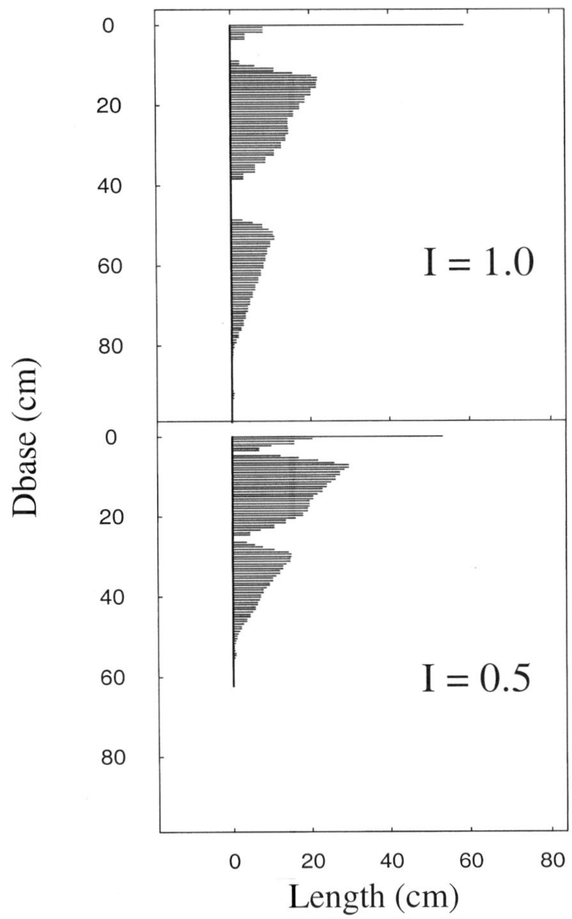

Figure 4. Simulation of root development with two impedance factors: 1.0 (low impedance) and 0.5 (high impedance). Horizontal lines represent the final length (cm) of each secondary roots as a function of its insertion position (Dbase, cm) along the taproot, shown by the vertical line.

sidered soils could be such that main and larger roots are impeded, but not thinner roots.

Moreover, in our model, consequences of such limitation of taproot growth potential will result in a decrease of its carbohydrate demand also. Therefrom, the availability of carbohydrates is likely to become higher for the growth of secondary and tertiary roots. This results in the absolute increase in the length of laterals that can be observed in a homogeneous unfavourable environment (Goss and Scott Russel, 1980). Such behaviour is reinforced since we consider sink strength of the different roots to be controlled by assimilate availability: taproot lower demand induced by high impedance will result in an increase in the potential growth for its branches, without explicit involvement of hormonal correlations. Since the growth of lateral primordia also depends on assimilate availability, a decrease in taproot demand allows a larger number of lateral primordia to reach the threshold diameter which is required to the emergence of the root, thus increasing the actual branching density, as observed in media with high mechanical impedance (Goss and Scott Russel, 1980; Misra and Gibbons, 1996).

Generally speaking it can be supposed that axes with high growth potential are likely to be dependent on external conditions, which could limit the achievement of such potential, while thinner roots are not or few dependent on soil conditions, but are limited by their own growth potential, dependent on meristem size, estimated by the apical diameter (Thaler and Pagès, 1997).

References

Barley K P, Farell D A and Greacen E L 1965 The influence of soil strength on the penetration of a loam by plant roots. Aust. J. Soil Res. 3, 69–79.

Bennie A T P 1996 Growth and mechanical impedance. In Plant roots: The Hidden Half. Eds Y Waisel, A Eshel and U Kafkafi. pp 453–470. Marcel Dekker Inc., New York, USA.

Bengough A G and Mullins C E 1990 Mechanical impedance to root growth: a review of experimental techniques and root growth responses. J. Soil Sci. 41, 341–358.

Goss M J 1977 Effects of mechanical impedance on root growth in barley (*Hordeum vulgare* L.). I. Effects on the elongation and branching of seminal root axes. J. Exp. Bot. 28, 96–111.

Goss M J and Scott Russel R 1980 Effects of mechanical impedance on root growth in barley (*Hordeum vulgare* L.). III. Observations on the mechanism of response. J. Exp. Bot. 31, 577–588.

Grossman Y F and Dejong T M 1994 PEACH: a simulation model of reproductive and vegetative growth in peach trees. Tree Physiol. 14, 329–345.

Le Roux Y and (Pagès L 1994 Développement et polymorphisme racinaires chez de jeunes semis d'hévéa (*Hevea brasiliensis*). Can. J. Bot. 72, 924–932.

Marcelis L F M 1993 Simulation of biomass allocation in greenhouse crops – a review. Acta Hort. 328, 49–67.

Materechera S A, Alston A M, Kirby J L and Dexter A R 1992 Influence of root diameter on the penetration of seminal roots into a compacted subsoil. Plant Soil 144, 297–303.

Misra R K and Gibbons A K 1996 Growth and morphology of eucalypt seedling-roots, in relation to soil strength arising from compaction. Plant Soil 182, 1–11.

Pagès L, Jordan M O and Picard D 1989 A simulation model of the three-dimensional architecture of the maize root system. Plant Soil 119, 147–154.

Pagès L, Le Roux Y and Thaler P 1995 Modélisation de l'architecture racinaire. Plant. Rech. Dév. 2, 19–34.

Riedacker A, Dexheimer J, Takavol R and Aiaoui H 1982 Modifications expérimentales de la morphogenèse et des tropismes dans le système racinaire de jeunes chênes. Can. J. Bot. 60, 765–778.

Thaler P and Pagès L 1996a Periodicity in the development of the root system of young rubber trees (*Hevea brasiliensis* Müll-Arg.). Relationship with shoot development. Plant Cell Environ. 19, 56–64.

Thaler P and Pagès L 1996b Root apical diameter and root elongation rate of rubber seedlings (*Hevea brasiliensis*) show parallel responses to photoassimilate availability. Physiol. Plant. 97, 365–371.

Thaler P and Pagès L 1997 Competition within the root system of rubber seedlings (*Hevea brasiliensis*) studied by root pruning and blockage. J. Exp. Bot. 48, 1451–1459.

Thaler P and Pagès L 1998 Modelling the influence of assimilate availability on root growth and architecture. Plant Soil 201, 307–320.

Wilson A J, Robards A W and Goss M J 1977 Effects of mechanical impedance on root growth in barley (*Hordeum vulgare* L.). II. Effects on cell development in seminal roots. J. Exp. Bot. 28, 1216–1227.

RESPONSES TO ABIOTIC STRESS

A. Stokes (ed.), *The Supporting Roots of Trees and Woody Plants: Form, Function and Physiology*, 219–225.

The effects of wind speed and direction on radial growth of structural roots

Bruce C. Nicoll and Andrew J. Dunn
*Forestry Commission Research Agency, Northern Research Station, Roslin, Midlothian EH25 9SY, Scotland, UK**

Abstract

We used dendrochronological techniques to explore the relationships between tree wind loading and structural root annual radial growth. Study trees were shallow rooted 46-year-old Sitka spruce that had grown in an exposed upland plantation experiencing strong prevailing wind from the south west. We constructed growth ring chronologies using measurements of annual growth ring widths on stem sections and from the largest structural roots, and related them to wind climate data for the life of the trees. Separate root chronologies were constructed from samples from each quadrant around the trees relative to the prevailing wind direction, at 0.5 m, 0.75 m, 1.0 m and 1.25 m from the tree centre, i.e. 16 separate root chronologies. In this preliminary analysis, we correlated yearly growth data from the chronologies with mean hourly and maximum hourly wind speed in each month of the corresponding year. There were few significant correlations between wind speed and stem chronologies, but many positive correlations with the root chronologies. The analysis indicated an overall increase in allocation of assimilates to the structural roots in response to wind. Allocation to structural roots on the lee side was most strongly correlated with maximum wind speeds and this allocation appeared to be at the expense of roots further from the tree, that would have had less of a structural role. Development of windward roots was correlated well with maximum gusts in the corresponding years. There were, however, few significant correlations between wind maximum gusts and thickening of roots in the two sectors perpendicular to the wind direction. These results indicate that these adaptive changes in allocation within root systems resulted from flexing of structural roots in the vertical plane, rather than from twisting. We propose that the amount of secondary thickening in any part of the root system in any one year is proportional to the amount of bending stress experienced at that point. The time of year that the roots experience stress appears to be important: significant correlations between wind speed and root growth ring chronologies were predominantly from wind movement during January and February, and July to October. The January and February correlations could result from storage of assimilates in rays adjacent to the dormant cambium, in regions that experience the most stress. These stored assimilates would enhance radial growth when the cambium becomes active in the spring. There was little influence of wind on root development during spring and early summer when most assimilates are used in shoot growth and stem thickening.

Introduction

As lateral roots of trees thicken to become structural roots, the development of their size and shape is influenced by root movement induced by wind acting on the crown. As early as 1803, Thomas Knight concluded from experiments where he controlled the movement of trees, that if a tree is "placed in a high and exposed situation, where it is much kept in motion by winds, the new matter which it generates will be deposited chiefly in the roots and lower parts of the trunk". Since then, many researchers have described the growth responses of trees to wind movement, with the majority of studies examining above ground responses. Schwendener (1874) and Mezger (1893) formulated a hypothesis that the cambium in the stem produces new wood in a way that equalises the stress on the outer surface – the *uniform stress hypothesis*. Although this still generates debate (Morgan and Cannell, 1994), many authors (reviewed by Telewski 1995) have described increased above ground cambial growth as a direct response to mechanical stress.

Plant growth responses to wind movement and other mechanical stress, termed 'thigmomorphogen-

* FAX No: 0131 445 514. E-mail: B.Nicoll@forestry.gov.uk

esis' by Jaffe (1973, 1980), Jaffe and Telewski (1984) and Telewski and Jaffe (1981, 1986a,b,c), include reduced height growth, greater stem taper, and development of smaller branches. There is, however, less information on the growth of roots in response to stress. Tree roots are known to move in the soil as the tree sways in the wind, for example, Hintikka (1972) described spruce roots in a clay soil lifting the surface by 13.6 mm for 3° stem displacement, and by 3.1 mm for 1° of stem movement. Conifers are known to allocate a larger proportion of total biomass below ground when they experience increased wind loading (Fritzsche, 1933; Neilson, 1990; Nicoll and Ray, 1996). The mass is also allocated differently in root systems of trees exposed to high wind speeds, compared to sheltered trees. For example, exposed forest trees are reported to allocate a larger proportion of root biomass to the lee side relative to the prevailing wind direction (Nicoll and Ray, 1996; Nicoll et al., 1995), and seedlings and newly planted trees show similar behaviour with increased thickening of roots on both windward and leeward sides, i.e. the roots along the axis of flexing (Stokes et al., 1995, 1997a).

Büsgen et al. (1929) described eccentric diameter growth of shallow spruce (*Picea* sp.) roots producing what they called a 'T-girder form' in cross section. They proposed that development of this shape resulted from root movement induced by stem swaying, as an extension of the uniform stress hypothesis to below-ground parts of the tree. Rigg and Harrar (1931) found similar root shapes on eight coniferous tree species growing on a sphagnum bog. Roots were circular in cross section when young, but older roots had increasingly more growth on the upper and lower surfaces, producing shapes of three types: 'the oval, the T-girder and the I-beam'.

Nicoll and Ray (1996) described a method to analyse the cross sectional shape of structural roots, quantifying the tendency toward development of T- and I-beam shapes. This method applied to shallow rooted Sitka spruce trees on an exposed upland site revealed that development of the T-beam shape was more pronounced in the buttressed part of the root adjacent to the stem on the leeward side of the tree relative to the prevailing wind. These trees also allocated more mass to the roots on the lee side. The I-beam shape was most developed on the windward side, particularly further out from the stem, past the buttressed region. The development of buttresses and eccentric root shapes was greatest on trees with the shallowest root plates, i.e. those trees that would have experienced the most wind movement (Eichkorn and Wolfart, 1991; Hintikka, 1972; Rizzo and Harrington, 1988). Ruth and Harris (1979) also described the largest supporting buttresses as being on the lee side of western hemlock (*Tsuga heterophylla* (Raf.) Sarg.) and Sitka spruce (*Picea sitchensis* (Bong.) Carr.) trees. This development on gymnosperm species contrasts with observations that buttresses develop predominantly on the windward side of angiosperm tree species (Senn, 1923 reanalysed by Baker, 1973; Grace, 1977; Henwood, 1973; Navez, 1930).

Reduction of root secondary thickening has been induced in a number of experiments by preventing stem sway (Fayle, 1968, 1976; Jacobs, 1939 1954; Wilson, 1975). Seedlings of Sitka spruce and European larch (*Larix decidua* (Mill.) grown in a wind tunnel responded to wind flexing by increasing the diameter on the windward and leeward sides, compared to non-moving controls (Stokes et al., 1995). Young (2-year-old) Sitka spruce trees that were flexed back and forward in a machine during the growing season increased the vertical (compared to horizontal) diameter of lateral roots on the axis of flexing (Stokes et al., 1997a). These studies indicate that there is a localised response to stress that results in a reallocation of assimilates between, along and around individual roots.

The localised root responses occur along-side a whole-tree response to movement, in which a greater proportion of assimilates are allocated to the lower stem (Jacobs, 1939) and root system (Nicoll and Ray, 1996), at the expense of the upper stem and branches (Telewski, 1995). What remains unclear is if tree responses, and in particular the root responses, to wind action are related to general patterns of movement throughout the tree's life, or if the tree responds proportionally to varying levels of stress as it grows in a variable wind climate. This paper tests the hypothesis that each year, roots thicken by an amount that is related to the wind movement experienced in that year. The method employed to test this hypothesis is a dendrochronological analysis of structural root growth rings in comparison with the wind records for the corresponding years.

Dendrochronology techniques have been used before to analyse root growth (Fayle, 1976; Jacoby et al., 1995; Krause and Eckstein, 1993; Schulman, 1945), but there have been no correlations of root growth ring widths with measured wind intensity from corresponding years. However, Urban et al. (1994) used root dendrochronology to demonstrate a release in structural root and stem growth following an increase in wind exposure of 120-year-old white spruce (*Picea gluaca* (Moench) Voss) trees. Measurement of root growth rings is sometimes reported to be difficult due to the unevenness of growth around the root, and occasional missing rings (for example, Schulman, 1945). However, in sections removed from the root systems of the 46-year-old Sitka spruce trees described by Nicoll and Ray (1996), growth was characteristically greatest on the upper side of structural roots, regardless of shape, with clear, wide, well differentiated growth rings suitable for measurement. As the upward (epinastic) thickening of spruce roots close to the stem is believed to be an adaptive response to wind movement (Nicoll and Ray, 1996), this part of the root is particularly suitable for a dendrochronological investigation of ring width in response to wind movement. Chronologies constructed from root ring measurements in four quadrants around the exposed trees used in the Nicoll and Ray (1996) study, at four distances from the tree centre, are here compared with records of wind speed and direction.

Methods

Annual growth ring measurements

Thirty-nine trees were randomly selected from an upland plantation (Crookburn, Kershope; 55°06′ N, 2°47′ W) of 46-year-old Sitka spruce that had grown on an exposed hillside with a strong prevailing wind from the south-west (248° from north). Soil at this site varied between peaty gley and surface water gley, with a winter water-table close below the surface (0.25–0.6 m). This shallow water-table had restricted root depth, and root plate thickness varied between 0.3 and 0.8 m. Complete details of the tree excavation method and initial measurements are given in Nicoll and Ray (1996) and Ray and Nicoll (1998). In the current study we used sections cut from the four largest roots on each of ten trees selected randomly from the original sample. Sitka spruce trees usually have between 3 and 11 structural roots, with most having at least four (Coutts, 1983; Nicoll et al., 1995); and so the four largest roots were expected to have the most structural importance, and hence be most adapted to wind action. Sections were cut at 0.5, 0.75, 1.0, and 1.25 m from the tree centre. The direction from the tree centre relative to north, and the distance out from the stem centre, was recorded for each root section, and a mark indicated its original vertical orientation. Annual growth ring widths were measured from the outer surface of each section to the biological centre. Growth rings were often indistinguishable around the base and sides of root sections. The rings were measured by placing the section on a motorised movable table, under a microscope. The position was recorded electronically each time a growth ring was advanced under the microscope cross-hairs, producing a set of ring widths for each section.

Constructions of root growth ring chronologies

Variation in growth ring widths can be separated into low-frequency variation that results from changes related to tree age, and high frequency variation that represents large year to year changes in ring width due to environmental factors (Cook and Kairiukstis, 1990; Fritts, 1970). To construct chronologies, growth curves were fitted through the data, and low frequency variation was removed to leave only the high frequency variation. We constructed 'standard' chronologies from our data using the 'ARSTAN' program (Cook, 1986), a subroutine of ITRDBLIB (Grissino-Mayer et al., 1992). Chronologies were based on mean growth from ten trees with separate chronologies for root sections at each of the four distances from the tree centre in four quadrants around the trees: north-east, south-east, south-west and north-west. There were, therefore, 16 separate chronologies. The south-west sector represented the windward side of the tree relative to the prevailing wind, and the north-west sector represented the leeward side.

Climate data

The closest meteorological station to the experiment site at Crookburn that had been in place for the life of the study trees was the UK Met Office station, 30 km to the north-east, at Eskdalemuir. This station has similar elevation and aspect to the experiment site. Climate data: monthly mean temperature and rainfall, monthly mean of hourly mean wind speed and direction, and monthly mean of hourly maximum wind speed, were available from the Eskdalemuir station. To test the validity of using this data to represent wind climate on the Crookburn experiment site, an anemometer on a 10 m mast was installed at Crookburn in February 1996, and wind speed and direction were monitored for 12 months. Mean daily wind speed and direction were compared between Crookburn and Eskdalemuir using linear regression. There was a highly

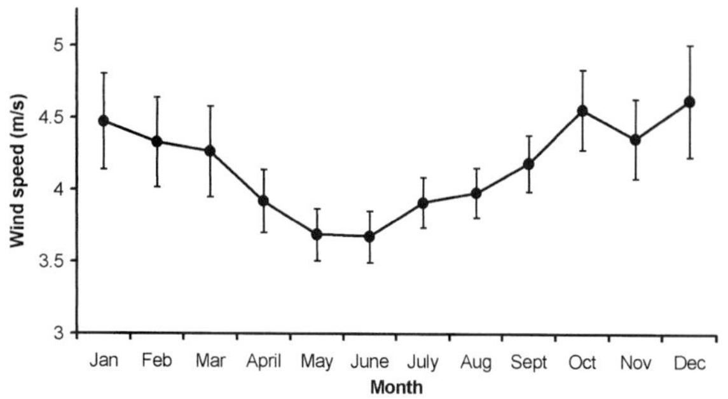

Figure 1. Monthly average wind speeds based on hourly means from Eskdalemuir Met station for the 47 years from 1947 to 1993. Error bars are 95% confidence intervals.

significant linear regression between wind speeds at the two sites ($P<0.001$), and between wind directions at the two sites ($P<0.001$), giving us confidence that data from Eskdalemuir represented the experiment site.

During the life of the study trees, over 70% of the maximum hourly wind speeds were from the south-west, with least wind (less than 3%) from the south-east. Monthly maximum and average hourly wind speeds over the same period were highest in December, falling to lowest in June (Figure 1). Monthly mean soil temperature over the life of the study trees varied from 3 °C in February to a maximum of 14 °C in August. Monthly mean air temperature varied between 1 °C in January and February, and 13 °C in July. Monthly average precipitation over the life of the trees, was lowest in May, with 75 mm, and highest in January, with 145 mm.

Statistical analysis

In this preliminary analysis of the data, each chronology was correlated with mean air temperature and rainfall, and with mean and maximum wind speed for each month in the corresponding years over the 47 years up to 1993. Positive and negative correlations were classed as being significant when $P<0.05$.

Results

The correlation matrix shown in Table 1 relates each of the growth ring chronologies with average hourly wind speeds, measured for each month in the corresponding years. There were significant correlations with average hourly wind speed for 15 of the 16 chronologies. Most of these were positive, except with data from the leeward sector at 1.25 m, where there were significant negative correlations. Most significant correlations in Table 1 are with mean wind in January to February, and July to October. The correlation matrix of the same chronologies against maximum wind speed (Table 3) shows most significant correlations in the windward and leeward sectors, but very few significant correlations in the perpendicular quadrants. In the leeward quadrant, the only positive correlations were for 0.75 m, while all significant correlations at 1.0 m and 1.25 m were negative. On the windward side of the tree, most correlations were positive. Most significant correlations in Table 2 were with maximum wind in February to March, and June to September.

Discussion

Dendrochronological techniques are commonly used to relate annual growth rings in tree stems to climate (for example, Fritts, 1970; Larsen and MacDonald, 1995; Pan et al., 1997), but few studies have related wind action to ring widths. Fayle (1976) found that restricting wind movement of young, c. 3.4 m tall, Scots pine (*Pinus sylvestris* L.) trees reduced the widths of new growth rings in the lower stem and adjoining parts of horizontal roots. In stems, reaction wood and eccentric rings appear to be developed concurrently, and may be promoted by the same stimulus (Wilson and Gartner 1996). Roots of conifer species do not normally develop reaction (compression) wood unless also exposed to light (Fayle, 1976), but root wood on the lee side of *Pinus pinaster* (Ait.) trees does have some properties similar to compression wood (Stokes et al., 1997b).

Stokes and Mattheck 1996 found that compressive strength of lateral roots of Norway spruce (*Picea abies*) and poplar species (*Populus nigra* and *Populus italiensis*) increased along the root to a maximum at between 0.5 and 1.0 m from the stem, before decreasing along the remaining root. This variation in development of root wood properties may be a response to wind induced mechanical stress, and appears to correspond with the pattern of assimilate allocation in the trees described here.

Although, in our study, there were few significant correlations between wind speed and stem chronologies, there were many positive correlations with root chronologies. The analysis indicated an overall increase in allocation of assimilates to the structural roots in response to wind. Similarly, Jacobs (1939)

Table 1. Spearman Rank Correlations of root chronologies with mean wind speed in each month of the corresponding year. + indicates a significant ($p<0.05$) positive correlation, − indicates a significant ($p<0.05$) negative correlation

Quadrant	1 (Leeward quadrant)				2				3 (Windward quadrant)				4			
Distance (m)	0.5	0.75	1.0	1.25	0.5	0.75	1.0	1.25	0.5	0.75	1.0	1.25	0.5	0.75	1.0	1.25
January		+	+			+	+			+						
February	+	+		−	+	+	+		+	+			+	+		
March				−												
April				−			+									
May																
June							+									+
July		+		−	+	+							+			
August		+				+	+						+		+	
September		+		−		+									+	+
October		+		−			+							+		
November																
December																

found an increase in secondary thickening of structural roots of free swaying radiata pine compared to restrained trees. This has also been observed on younger trees; for example, Stokes et al. (1997a) described an increase in root mass:shoot mass ratio in 2-year-old Sitka spruce that had been subjected to mechanical flexing.

Allocation to structural roots on the lee side was correlated more with maximum wind speeds than mean wind speed, and appeared to be at the expense of growth in roots further from the tree that would have had less of a structural role. Increased thickening of structural roots on the lee side of trees relative to the prevailing wind direction has been observed in previous studies (Nicoll et al., 1995; Nicoll and Ray, 1996). Development of windward roots correlated well with maximum gusts in the corresponding year. The few significant correlations between wind maximum gusts and thickening of roots in the two sectors perpendicular to the prevailing wind direction indicate that adaptive changes in allocation within root systems resulted predominantly from flexing of structural roots in the vertical plane, rather than twisting. In wind tunnel and mechanical flexing studies, Stokes et al. (1995, 1997a) also found more growth of windward and leeward (flexed) roots, compared to perpendicular (i.e. twisted) roots.

The amount of secondary thickening in any part of the root system in any one year appears from this study to be proportional to the amount of bending stress experienced at that point. This observation is compatible with, but not the same as the uniform stress hypothesis, in which stress is kept constant over the surface as the tree thickens. More research is required to find the shape of the dose response curve for conifer root response to mechanical stress.

The time of year that the roots experience stress may also be important: significant correlations between mean wind speed and root growth ring chronologies were mostly from wind movement during January and February, and July to October. As Sitka spruce roots are not expected to undergo growth in upland Britain during January and February (Coutts and Nicoll, 1990), the correlations at this time of year may result from a stored signal or resource. The growth promoter most commonly associated with plant responses to stress is ethylene, but as this is unlikely to persist long enough to be the signal involved here, a precursor or other growth promoter could be responsible. Alternatively, as Sitka spruce shows positive net photosynthesis throughout the winter while the temperature remains above −5 °C (Ludlow and Jarvis, 1971; Neilson et al., 1972), one possible mechanism is the storage of assimilates during this period in ray parenchyma adjacent to the dormant cambium particularly in regions of the structural roots that experience the most stress. The stored assimilates would enhance radial growth in these regions when the cambium becomes active in the spring (Wargo, 1979; Zimmerman, 1964). Although this hypothesis remains to be

Table 2. Spearman Rank Correlations of root chronologies with maximum wind speed in each month of the corresponding year. + indicates a significant ($p<0.05$) positive correlation, − indicates a significant ($p<0.05$) negative correlation

Quadrant	1 (Leeward quadrant)				2				3 (Windward quadrant)				4			
Distance (m)	0.5	0.75	1.0	1.25	0.5	0.75	1.0	1.25	0.5	0.75	1.0	1.25	0.5	0.75	1.0	1.25
January		+								+						
February		+		−					+	+						
March		+		−						+			+			
April				−												
May																
June		+		−					+	+						+
July				−				−	+							
August														−		−
September		+		−												
October																
November																
December				−												

tested, it is backed up by the observation that during spring and early summer while most assimilates are being used in shoot growth and stem thickening, there appeared to be little influence of wind on root development. Lundqvist and Valinger (1996) and Valinger et al. (1994, 1995) demonstrated similar responses above-ground in Scots pine; tree stems increased diameter growth after being subjected to mechanical loading in either the dormant or growing season. There would be benefits to trees responding to wind movement in the 'dormant' season – the highest wind speeds are usually experienced in the autumn and winter, and the loads on the tree can be increased by snow and ice accumulation in the crown. The tree would therefore adapt its morphology to resist to the largest loads that it experiences.

References

Baker H G 1973 Appendix. *In* A structural model of forces in buttressed tropical rainforest trees. Ed. K. Henwood. Biotropica 5, 83–93.

Büsgen M, Münch E and Thomson T 1929 The structure and life of forest trees. Chapman and Hall, London, UK.

Cook E R 1986 Guide for program ARSTAN. University of Arizona. pp 11.

Cook E R and Kairiukstis L A 1990. Methods of Dendrochronology. Kluwer Academic Publishers, Dordrecht, the Netherlands.

Coutts M P 1983 Root architecture and tree stability. Plant Soil 71, 171–188.

Coutts M P and Nicoll B C 1990 Growth and survival of shoots, roots, and mycorrhizal mycelium in clonal Sitka spruce during the first growing season after planting. Can. J. For. Res. 20: 861–868.

Eichkorn T and Wolfart A 1991 Untersuchungen über windinduzierte Wurzelverletzungen an Nadelbaümen in Südwestdeutschland. Allg. Forst-u. J. -Ztg 162, 83–92.

Fayle D C F 1968 Radial growth in tree roots. Tech. Rep. No. 9. Faculty of Forestry, University of Toronto, Toronto, Ontario

Fayle D C F 1976 Stem sway affects ring width and compression wood formation in exposed root bases. For. Sci. 22: 193–194.

Fritts H C 1970 Tree-ring analysis and reconstruction of past environments. *In* Tree-ring analysis with special reference to Northwest America. Eds J H G Smith and J Worrall. The University of British Columbia, Faculty of Forestry. Bulletin 7, 92–98.

Fritzsche K 1933 Sturmgefahr und Anpassung. Tharanther Forstliches Jahrbuch. pp 1–94.

Grace J 1977 Plant Response to Wind. Academic Press, New York.

Grissino-Mayer H, Holmes R and Fritts H C 1992 International tree-ring data bank program library manual. Laboratory of tree-ring research, University of Arizona, Tucson.

Henwood K 1973 A structural model of forces in buttressed tropical rain forest trees. Biotropica 5, 83–93.

Hintikka V 1972 Wind-induced root movements in forest trees. Commun. Inst. For. Fenn. 76, 1–56.

Jacobs M R 1939 A study of the effect of sway on trees. Commonw. For. Bur. Bull. No. 26. Commonwealth Forestry Bureau, Canberra, Australia.

Jacobs M R 1954 The effect of wind sway on the form and development of *Pinus radiata* D. Don. Aust. J. Bot. 2: 35–51.

Jacoby G, Carver G and Wagner W 1995 Trees and herbs killed by an earthquake ~300 yr ago at Humboldt Bay, California. Geology 23: 77–80.

Jaffe M J 1973 Thigmomorphogenesis: the response of plant growth and development to mechanical stimulation. Planta 114, 143–157.

Jaffe M J 1980 Morphogenetic responses of plants to mechanical stimuli or stress. Bioscience 30, 239–243.

Jaffe M J and Telewski F W 1984 Thigmomorphogenesis: callose and ethylene in the hardening of mechanically stressed plants. *In*

Phytochemical Adaptations to Stress. Eds B N Timmerman, C Steelink and F A Loewus. pp 79–95. Plenum Press, New York.

Knight T A 1803 Experiments on the descent of sap in trees. Phil. Trans. Roy. Soc. London 93: 277–289.

Krause C and Eckstein D 1993 Dendrochronology of roots. Dendrochronologia 11, 9–23.

Larsen C P S and MacDonald G M 1995 Relations between tree-ring widths, climate, and annual area burned in the boreal forest of Alberta. Can. J. For. Res. 25, 1746–1755.

Ludlow M M and Jarvis P G 1971 Photosynthesis in Sitka spruce (*Picea sitchensis* (Bong.) Carr.). I. General characteristics. J. Appl. Ecol. 8, 925–953.

Lundqvist L and Valinger E 1996 Stem diameter growth of Scots pine trees after increased mechanical load in the crown during dormancy and (or) growth. Ann. Bot. 77, 59–62

Mezger K 1893 Der Wind als massgebender Faktor für das Wachtsum der Bäume. Mundener Forstliche Hefte 3, 35–86.

Morgan J and Cannell M G R 1994 Shape of tree stems – a re-examination of the uniform stress hypothesis. Tree Physiol. 14, 49–62.

Navez A 1930 On the distribution of tabular roots in *Ceiba*. Proc. Natn. Acad. Sci. 16, 339–344.

Nielsen C C N 1990 Einflüsse von Pflanzenabstand und Stammzahlhaltung auf Wurzelform, Wurzelbiomasse, Verankerung sowie auf die Biomassenverteilung im Hinblick auf die Sturmfestigkeit der Fichte. University of Göttingen Faculty of Forestry and Niedersächsischen Forstlichen Versuchsanstalt, Thesis 100, J.D. Sauerländer's Verlag, Frankfurt am Main. 279 p.

Neilson R E, Ludlow M M and Jarvis P G 1972 Photosynthesis in Sitka spruce (*Picea sitchensis* (Bong.) Carr.) II. Response to temperature. J. Appl. Ecol. 9, 721–745.

Nicoll B C and Ray D 1996 Adaptive growth of tree root systems in response to wind action and site conditions. Tree Physiol. 16, 899–904.

Nicoll B C, Easton E P, Milner A D, Walker C and Coutts M P 1995 Wind stability factors in tree selection: distribution of biomass within root systems of Sitka spruce clones. *In* Wind and Trees. Eds M P Coutts and J Grace. pp 276-292. Cambridge University Press, Cambridge, UK.

Pan C, Tajchman S J and Kochenderfer J N 1997 Dendroclimatological analysis of major forest species of the central Appalachians. For. Ecol. Manage. 98, 77–87.

Ray D and Nicoll B C 1998 The effect of soil water-table depth on root-plate development and stability of Sitka spruce. Forestry (*In press*).

Rigg G B and Harrar E S 1931 The root systems of trees growing in sphagnum. Am. J. Bot. 18, 391–397.

Rizzo D M and Harrington T C 1988 Root movement and root damage of red spruce and balsam fir on subalpine sites in the White Mountains, New Hampshire. Can. J. For. Res. 18, 991–1001.

Ruth R H and Harris A S 1979 Management of western hemlock – Sitka spruce forests for timber production. U.S.D.A. Forest service, General technical report PNW-88.

Schulman E 1945 Root growth-rings and chronology. Tree-ring Bulletin 12, 2–5.

Schwendener S 1874 Das mechanische Princip im anatomischen Bau der Monocotyleden. Wilhelm Engelmann Verlag, Lepzig. 179 p.

Senn G 1923 Über die Ursachen der Brettwurzelbildung bei der Pyramiden-Pappel, Verh. Nat. Forsch. Ges. Basel 35: 405.

Stokes A and Mattheck C 1996 Variation of wood strength in tree roots. J. Exp. Bot. 47: 693–699.

Stokes A, Fitter A H and Coutts M P 1995 Responses of young trees to wind and shading: effects on root architecture. J. Exp. Bot. 46, 1139–1146.

Stokes A, Nicoll B C, Coutts M P and Fitter A H 1997a Responses of young Sitka spruce clones to mechanical perturbation and nutrition: effects on biomass allocation, root development, and resistance to bending. Can. J. For. Res. 27, 1049–1057.

Stokes A, Martin F, Sacriste S and Foucard T 1997b Adaptation of tree roots to wind loading – the relationship between mechanical behaviour and wood formation. *In* Plant biomechanics 1997: Conference proceedings. Eds G Jeronimidis and J F V Vincent. pp 339–346.

Telewski F W 1995 Wind-induced physiological and developmental responses in trees. *In* Wind and Trees. Eds M P Coutts and J Grace. pp 237–263. Cambridge University Press, Cambridge, UK.

Telewski F W and Jaffe M J 1981 Thigmomorphogenesis: changes in the morphology, and chemical composition induced by mechanical perturbation of 6-month-old *Pinus taeda* seedlings. Can. J. For. Res. 11, 380–387.

Telewski F W and Jaffe M J 1986a Thigmomorphogenesis: Field and laboratory studies of *Abies fraseri* in response to wind or mechanical perturbation. Physiol. Plant. 66, 211–218.

Telewski F W and Jaffe M J 1986b Thigmomorphogenesis: Anatomical, morphological and mechanical analysis of genetically different sibs of *Pinus taeda* in response to mechanical perturbation. Physiol. Plant. 66, 219–226.

Telewski F W and Jaffe M J 1986c Thigmomorphogenesis: The role of ethylene in the response of *Pinus taeda* and *Abies fraseri* to mechanical perturbation. Physiol. Plant. 66, 227–233.

Urban S T, Lieffers V J and MacDonald S E 1994 Release in radial growth in the trunk and structural roots of white spruce as measured by dendrochronology. Canadian Journal of Forest Research 24, 1550–1556.

Valinger E, Lundqvist L and Sundberg B 1994 Mechanical stress during dormancy stimulates stem growth of Scots pine seedlings. For. Ecol. Manage. 67, 299–303.

Valinger E, Lundqvist L and Sundberg B 1995 Mechanical bending stress applied during dormancy and (or) growth stimulates stem diameter growth of Scots pine seedlings. Can. J. For. Res. 25, 886–890.

Wargo P M 1979 Starch storage and radial growth in woody roots of sugar maple. Can. J. For. Res. 9, 49–56.

Wilson B F 1975 Distribution of secondary thickening in tree root systems. *In* The Development and Function of Roots. Eds J G Torrey and D T Clarkson. pp 197–219. Academic Press, New York.

Wilson B F and Gartner B L 1996 Lean in red alder (*Alnus rubra*): growth stress, tension wood, and righting response. Can. J. For. Res. 26, 1951–1956.

Zimmerman M H 1964 The relation of transport to growth in dicotyledonous trees. *In* The Formation of Wood in Forest Trees. Ed M H Zimmerman. pp 289–301. Academic Press, New York.

A. Stokes (ed.), The Supporting Roots of Trees and Woody Plants: Form, Function and Physiology, 227–238.

Responses of sap flow rate along tree stem and coarse root radii to changes of water supply

N. Nadezhdina and J. Cermak
Institute of Forest Ecology, Mendel University of Agriculture and Forestry, Zemedelska 3, 61300 Brno, Czech Republic

Key words: changes of soil moisture, radial pattern of sap flow, root severing, stem breakage, root-soil resistance, rapid changes

Abstract

Measuring the sap flow pattern along stem radii may help to understand relationships between layers of conducting sapwood and paths of water supplied by roots. Experiments were performed with specimens of four tree genera (*Ficus, Tilia, Pinus* and *Picea*) applying sap flow meters installed in tree stems and coarse roots. Roots were treated by naturally occurring changes in soil moisture (watering by rain and then drought stress) and by blocking the source of water (by the cutting of absorbing roots) or sink of water (tree crown) by breaking the stem. In the watering experiments sap flow increased significantly in the outer sapwood (60–80% of stem xylem radius, r_x) of the fig-tree when soil water supply increased after rain. Hot weather then caused drying-out of upper soil layers and a decrease in sap flow in the outer stem xylem. A similar situation occurred in the stem of the pine tree when a period of good water supply characterized by a higher flow in outer sapwood (80–100% of r_x) was followed by a period of hot and dry weather, which caused decreasing water content in surface soil layers. In response, sap flow decreased in outer sapwood, whilst no change in flow was observed in the deeper sapwood. The situation returned to its previous state after heavy rains. In the source-severing experiments (*Tilia*), sap flow in stem and roots increased immediately after treatment (which characterized changes in driving forces and the elimination of root–soil resistance) and than gradually decreased following the depletion of stored water. Severing experiments also indicated, that surface roots were presumably connected with the outer layers of sapwood, while deep roots supplied water uniformly over most of sapwood. In the sink-severing experiments (*Picea*), where the tree was winched until failure, sap flow decreased continuously in all roots, mostly in parts of sapwood where the highest flow was recorded before the treatment. Largest deceleration occurred in both the 'compressed' and 'tension' roots (with a time shift of around 4 min) followed by that in the root perpendicular to the direction of winching.

Introduction

To study complex root systems of large trees is a rather difficult and time consuming task, which was based usually on manual excavation methods (Jenik, 1978; Carlson and Harrington, 1987; Strong and La Roi, 1983). Only recently such methods were dramatically improved when using pressurized water or supersonic air (Rizzo and Gross, 1999) and when other methods such as ground penetrating radar were used in coarse root studies (Cermak et al., 1997, 1999; Hruska et al., 1998). All such methods describe the root structure in different details. In contrast, the present study considered description of functions in terms of water relations as related to rough parameters of architecture of complex root systems. The study was based on our previous experience in this field (Cermak and Kucera, 1990a; Cermak et al., 1980), when we estimated tree rooting depth and extent of rooted area from measurements of sap flow and soil water content.

We focus here on water transport in different parts of conductive tissues, i.e., the radial pattern of sap flow. The methodically oriented study considered diffuse porous broad-leaved and coniferous tree species growing under contrasting soil and climatic conditions. Untouched roots were treated by watering, and other roots by severing separately from the source of water (soil) and from its sink (crown). The aim of this study is to obtain information about the range of responses of sap flow in different xylem depths of stem and roots to rapid changes of natural soil water supply and to abrupt major perturbations of conductive systems: sudden severing of coarse roots from their anchored parts in the soil and of the upper stem with crown from the stem base with roots. The suitability of such approaches in field studies of root properties were thus evaluated. This study presents the first rough observations aimed at characterizing possible range of results for more detailed studies to be carried out later.

Material and methods

Experimental sites and sample trees

The fig tree (*Ficus carica* L., dbh=15.9 cm) was studied near the town of Radicondoli, Italy (altitude 550 m). The tree was situated very close to the wall of an old abandoned building at the place subjected to runoff of water from the large roof during rains. The south facing slope at the site was typical with loamy soil containing high to a very high percentage of stones, with surface layers easily subjected to desiccation. Mean annual and seasonal temperatures were 11.3 and 15.6°C, precipitation was 621 and 540 mm, respectively.

Scots pine (*Pinus sylvestris* L., dbh=28.6 cm) was sampled in Brasschaat, Belgium (see Cermak et al., 1998). The experimental plot was a pine plantation, on a slight north-facing slope, at an altitude of 16 m. Soil was moderately wet and sandy with a distinct humus and/or iron B-horizon, umbric regosol or haplic podzol in the F.A.O. classification (Baeyens et al., 1993). The groundwater depth normally ranges between 1.2 and 1.5 m and might be lower due to non-edaphic circumstances. The mean (over 28 years) annual, and growing season temperatures for the region were 9.76 and 13.72°C, and precipitation was 767 and 433 mm, respectively.

Lime (*Tilia cordata* L., diameter at breast height, dbh=13.8 cm) and spruce (*Picea abies* (L.) Karst., dbh=22.6 cm) were sampled in Sobesice forest district, Czechia at an altitude of 360 m, on a slight southward-facing slope. Soils were illimerized on weathered granodiorite with a cover of loess to a depth of 30–40 cm, and subject to desiccation. The mean annual temperature was 7.8°C, and the mean annual precipitation 580 mm, of which 360 mm was over growing seasons (Vasicek, 1984). The lime tree was growing in a closed stand, whilst the spruce tree grew near-by at the newly exposed edge of forest, opened up by the previous year's clear-cutting on the southern side.

Sap flow measurement

Sap flow rate was measured by the heat field deformation method (Nadezhdina and Cermak, 1998), applying dataloggers made by Environmental Measuring Systems & UNILOG, Brno, Czech Republic. Sensor applied for sap flow measurement consisted of the linear heater (insulated resistance wire inserted in the hypodermic needle) and two pairs of such needles with six thermocouples in each of them arranged in equal distances (with steps from 5 to 15 mm in different sensors). Stainless steel hypodermic needles 1.2 mm in outer diameter were used in both cases. One pair of needles with thermocouples was installed symmetrically placing one end above and the other (reference end) below the heater. The other pair of similar needles was installed so that its upper end was at a certain tangential distance from the heater and its lower reference end was below at the same level as the reference end of the symmetrical pair of needles.

The sensors were installed along trunk or root radii reaching from cambium to the pith. Thermocouples, marked as 'shallow' measured sap flow in the outer sapwood (closer to cambium), while 'deep' in the inner sapwood, closer to hardwood or pith. More points of sap flow along radii were obtained, when the needles were radially shifted during measurements over short periods of stable weather. Depth of the conducting xylem within sapwood and corresponding conducting area was estimated from the radial patterns of sap flow, taking into account the point, where the sap flow approached zero (this point was found using extrapolated part of curve in cases when sapwood was deeper then the length of sensors). The sap flow rate was measured per unit segment (i.e., 1 cm wide segment of xylem from the heater in the tangential direction), and data were routinely converted to flow density per sapwood area $Q_{w\ ar}$ in (g cm^{-2} h^{-1}).

Root treatment

During watering/drying experiments, sap flow in the stem of the fig tree was measured over a series of days in September with rather hot weather, when soil moisture was depleted rapidly from upper soil layers (data were measured each minute and the mean calculated and recorded for every 15 min). These days were followed by three rainy days, when the upper layers of soil were watered with a total precipitation of 20 mm (this amount was higher due to additional water coming as runoff from the roof of a neighboring building). Sap flow in the stem of the pine tree was measured during 2 months. Fine weather with periodic precipitation in July was followed by 3 weeks of hot weather without rains in August. A week-long period of rainy days followed with a total precipitation of 70 mm, when upper soil layers again became wetter.

The severing study consisted of two experiments, when sap flow was measured and recorded every minute. Fine roots, as the source of water, with corresponding coarse root branches, were severed from the rest of coarse roots where installed sensors were at 20 cm from the stem in the lime tree during the first experiment in May. Here the sap flow was measured (Exp. 1a) in the big first order coarse root (R1) (using the sensor S1), whilst two second order roots were severed by a sharp hand-saw. One large deep second order root remained functioning (Fig. 4a). The surface second order root was severed first and the intermediate second order root was severed later (after 10 min), (Exp. 1b) The same sensor S1 was used for measuring sap flow in the other two big first order coarse roots (R2 and R4) when they were completely severed, one after the other (Exp. 1c). Sap flow in the two opposite sides of the trunk in the same lime tree was measured (sensors S2 from West and S3 from North-East), whilst two big coarse roots (R2 and R3) were severed (Fig. 5a, b).

The sink of water was severed from coarse roots in the second experiment in spruce (Exp. 2). Here the sap flow was measured in three large coarse roots at 30 cm from the stem during fine weather at early afternoon in September. Sampled roots were oriented almost exactly to north, east and south. A steel cable was fastened to the stem at a height of 6 m above ground and the stem was pulled to the south by a winch for 5 min until it failed.

Results and discussion

Sap flow in stems during differing root water supply

Whole xylem cross section area was found to be conducting in the fig tree, with the peak flow along the radius occurring at a depth of 22 mm below the cambium (i.e., 27% of xylem radius) around noon (Fig. 1). Peak sap flow density in the stem reached a maximum

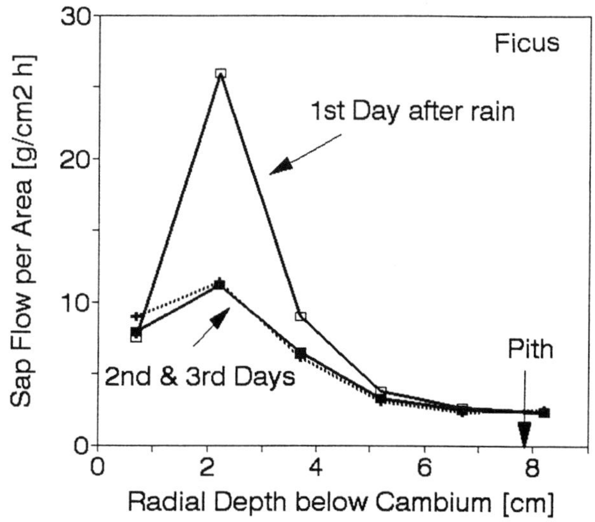

Figure 1. Changes in radial pattern of sap flow rate in the stem of a fig tree (*Ficus carica* L., dbh=15.9 cm) subjected to short-term natural watering during several days of rainy weather followed by hot and dry days. Radial pattern of sap flow is shown from S–E side of the stem, where the majority of branches were well-developed (there was a house wall on the opposite side and no branches).

around 26 g cm^{-2} h^{-1} during a fine day after several rainy days. However, the flow decreased during following days with continuous hot weather (Fig. 2). The flow decreased mostly at the depth of the original peak flow, while no changes in flow occurred in the outermost layers of sapwood near the cambium, nor in the deeper layers of sapwood towards the pith (see Figs 1 and 2).

Sap flow density in pine under good soil water supply during fine weather in July was characterized by a radial pattern, where the peak flow was at a depth of 15 mm below the cambium (i.e., about 10% of xylem radius). During 3 weeks of hot weather with no precipitation over the study period, the bare sandy soil started to dry from its upper layers. The beginning of drought stress in pine was reflected by decreasing sap flow mostly in outermost sapwood, whilst there was no decrease in flow in deeper sapwood layers (Fig. 3). The situation changed after a week-long period of rainy days, when the flow in the outermost layers of sapwood increased almost to its original level on the first sunny day, and the flow in deeper sapwood did not change significantly at the same time.

Sap flow in stems and roots after source-severing

Radial pattern of sap flow in the main first order root in lime tree was strongest in the outer sapwood (down

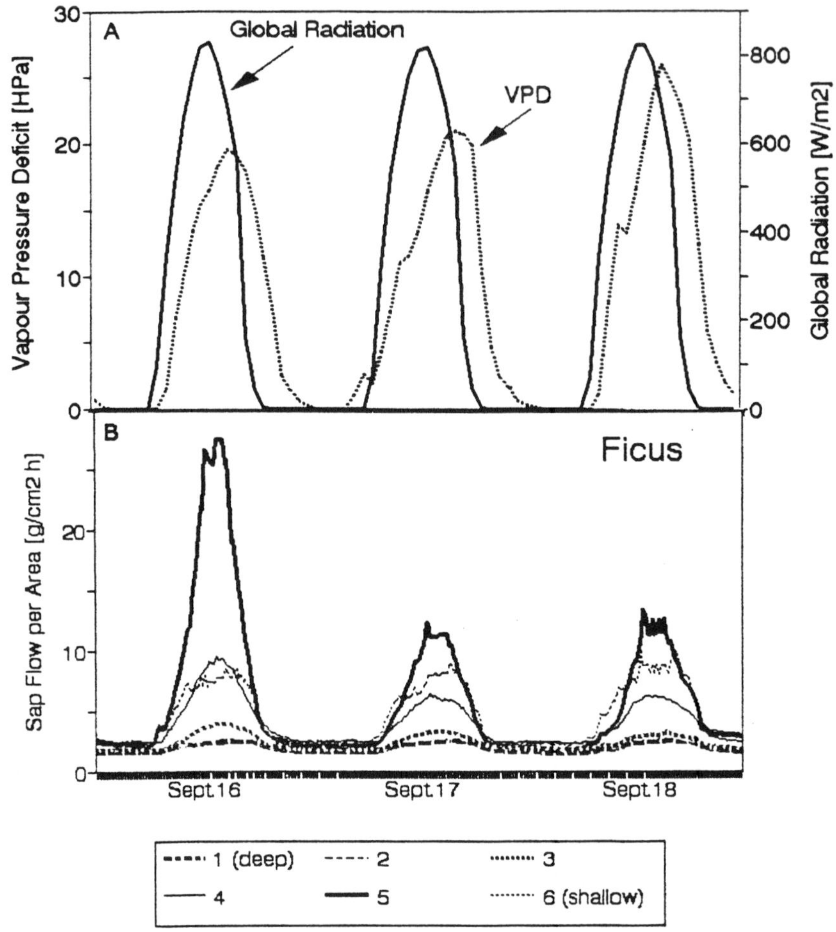

Figure 2. Diurnal courses of meteorological parameters (**A**) and sap flow in different depths below the cambium in the stem of a fig tree (see Fig. 1) after several days of rainy weather followed by a period of hot and dry days (**B**).

to the depth of about 2 cm below the cambium) and gradually decreased towards the pith. Severing one root branch situated in the surface soil layers (down to about 10 cm) caused a decrease in flow in the outer sapwood, whilst no change in flow occurred in the deeper sapwood. Severing the second root branch situated in the intermediate soil layers (between about 10 and 30 cm) caused an almost equal decrease in flow over the whole sapwood. The remaining flow, which was slightly higher in the outer sapwood, took place over the whole sapwood and characterized soil water supply from the remaining root branch in the deeper soil layers (Fig. 4).

In studies relating sap flow in stems to particular roots, it is important to situate the sensors exactly to appropriate sectors of conducting xylem, because continuous flow exists between one soil sector and the root bundle present in that soil sector as well as the stem sector corresponding to the root bundle (Cabibel, 1994). When the sensor was installed in the correct place in lime, a more pronounced response of sap flow to root severing was observed (Fig. 5c). When the sensor was placed slightly outside the conductive pathway coming from the experimental root, the sap flow decreased less in the outer sapwood after root severing, because some flow from neighboring intact roots was detected by the sensor simultaneously (Fig. 5d). This feature also confirms the sensitivity of flow rate to sectorial positioning at the stem base.

Sensors should be placed close to particular roots or in root buttresses if their specific behavior is to be studied, because there may be a large variation of sap flow around trunks (Cermak and Kucera, 1990b). Our results indicate that sensors located too high in

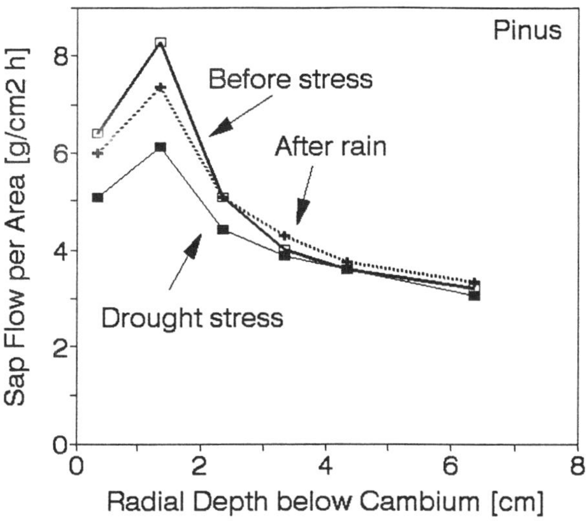

Figure 3. Radial pattern of sap flow rate in the stem of a pine tree (*Pinus sylvestris* L., dbh=28.6 cm). Curves characterize a period of good soil water supply (before stress, white squares) and after, when tree was subjected to partial drought stress (over 3 weeks of hot and dry weather, black squares), which was followed by a several day long rainy period. The situation returned again to the previous state in the first fine day after heavy rains.

stems do not necessarily measure the flow from the particular treated roots, but may reflect mostly wider sectors around trees. Variation around stems decreases with height, at least in some conifer species (Cermak and Kucera, 1990b), which means that the width of tree sectors covered by a single sensor increases when sensors are moved upwards, when water conducting paths from individual roots can overlap each other.

Sap flow in roots after sink-severing

Mean sap flow density was 2.0 g cm^{-2} h^{-1} in all spruce roots before winching. In roots from the north and east sectors sap flow was 3.0 g cm^{-2} h^{-1} and only 1.0 g cm^{-2} h^{-1} in roots growing towards the south. These densities correspond to the size of the crown projected surface area oriented in the different directions. The flow was slightly lower in the outer xylem layers and slightly higher in the inner layers, indicating deeper functional sapwood in the southern root. There were small differences in flow along radial patterns in the eastern and northern roots, and functionally sapwood seemed slightly shallower there.

The level of flow did not change in any of the roots during the first several minutes when the stem was winched to the south. No signs of tension in the stem nor in well-anchored roots were observed with the exception of a few quiet cracking sounds in the stem, after the bending angle of the upper stem reached 30°. However, an immediate response of sap flow occurred in all roots, when the stem broke at a height of 2.2 m, after it had reached an angle of 45°. The response time of sap flow in all roots was in the range of 4–8 min after stem breakage.

Sap flow changed dramatically after stem failure in the southern root (held in compression during winching) particularly in the deeper xylem layers, less in the medium layers and with very little response in the outer shallow xylem layers (Fig. 6). In the eastern root (perpendicular to the direction of winching) sap flow decreased less rapidly in the outer xylem, compared to the southern root, while it remained practically the same in the deeper xylem layers (Fig. 6). Sap flow started to decrease very little at first and then more rapidly up to 4 min after stem failure in the northern root (held in tension). Sap flow was nearly the same in all xylem layer, with less pronounced changes in flow in the outermost and innermost layers (Fig. 6).

Irrespective of different responses of flow in outer and inner xylem layers in all roots during the sink-severing experiment, the drop was always largest in layers where the flow was highest before stem treatment. These results indicate, that such layers were connected with the corresponding most active parts of the tree crown. Therefore sap flow decreased due to an abrupt loss of water potential gradient which was interrupted simultaneously over the whole cross-section of the stem. Thus, the remaining sap flow which persists for a few minutes after stem breakage, can reflect water saturating deficits in remaining xylem tissues between the sensors in roots and the broken end of the stem, i.e., in this case, thus about 2.7 m of the conducting pathway only (this is the maximum length, taking into account that some part of the xylem must be blocked by air entry from the open ends of tracheids at the broken part of stem).

Deceleration of sap flow at a certain distance from the cut end was slower when compared to, when the stem was cut just above the sap flow sensor (Cermak and Kucera, 1991). The largest deceleration of flow occurred in the root from the compressed southern side of the stem followed by that in the 'intermediate' eastern root. These two roots both showed peaks in deceleration of sap flow density, the first 3 min after stem failure, (Fig. 7). The magnitude of these peaks was always lower (about 25%) in the eastern root compared to the southern root. The peak of flow deceleration in the root from the tension northern side

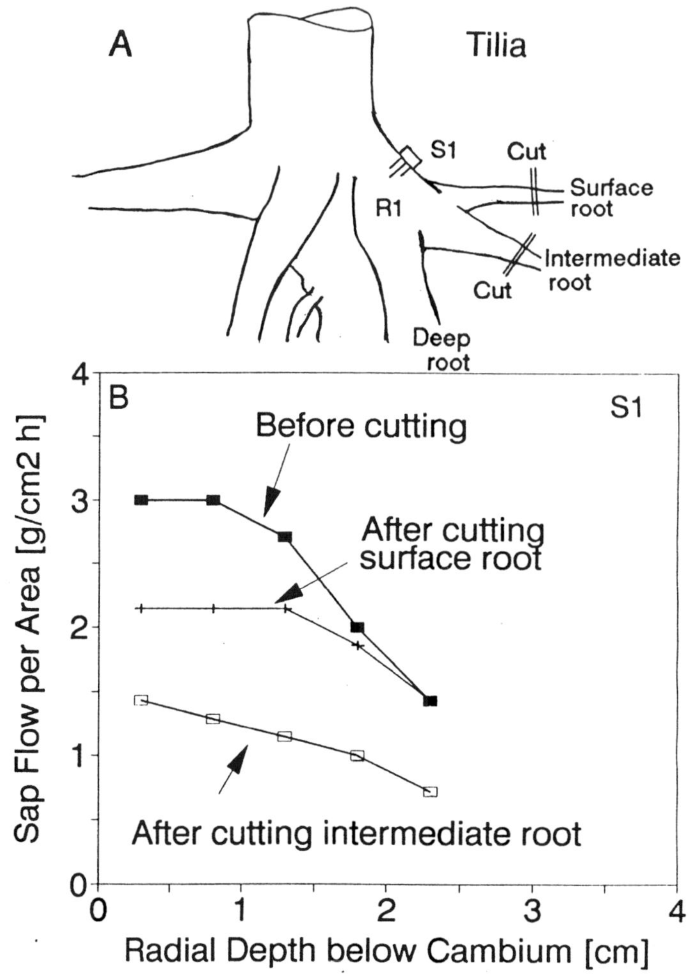

Figure 4. Scheme of the lime tree (*Tilia cordata*, dbh=13.8 cm) (**A**) and changes in radial pattern of sap flow rate after root severing (**B**). Sap flow was measured in a big coarse root (R1) just above ramification (at the point S1). First the second order surface root was removed and than the intermediate root, whilst the deep root was left intact.

of stem was of the same magnitude as that one from the compressed side, but took place 4 min after stem failure (Fig. 7). In all cases, highest deceleration occurred in xylem layers, where the maximum sap flows were recorded before stem breaking. There are almost no published data available to compare the above described responses of sap flow in different xylem depths to similar root and stem treatments, although the behaviour of trees during bending has been subjected to detailed mechanical studies (Stokes, 1999). It is difficult to explain the observed behavior clearly, especially when only one sample tree was used in this experiment. Nevertheless the results show the types of responses which could be expected in further experiments.

Estimating the rooting depth from sap flow measurements

A fast response of sap flow in the outer xylem layers (i.e., when taking into account 60–100% of xylem radius, r_{xyl}) to rather short-term changes of soil moisture was typical in experiments with fig and pine sample trees and applied drying and watering treatments. The results indicate, that outer xylem layers in tree-stems were mostly supplied by surface roots (Figs. 1 and 3). During severing experiments, the remaining flow in higher order roots characterized deep roots and was more uniformly distributed over the whole sapwood when the surface second order roots were removed (see Fig. 4).

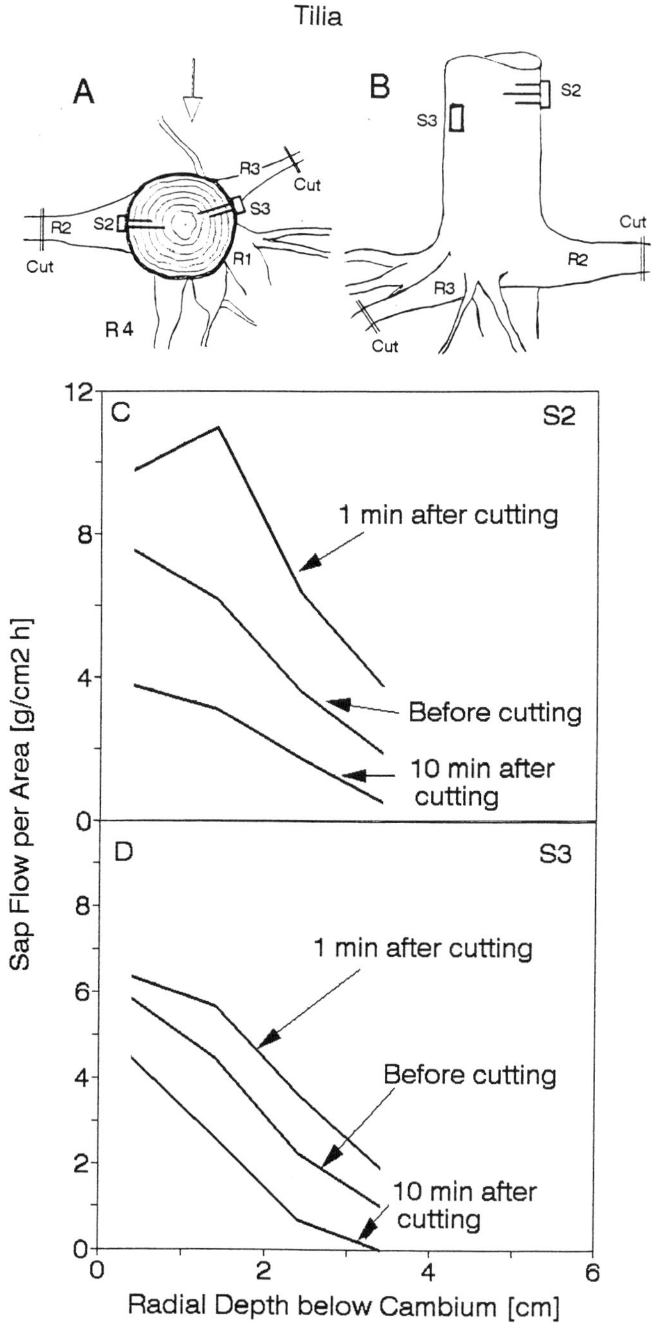

Figure 5. Scheme of the lime sample tree (the same tree as at the Fig. 4 with continuing numbering of roots and sensors). Position of roots and sap flow sensors is marked in the parts (**A**,**B**). Two coarse roots R2 and R3 were severed whilst sap flow was measured at points S2 and S3. Changes in radial pattern of sap flow in the stem (sensor S2) after severing a big coarse root R2 are shown at the part (**C**) and changes in radial pattern of sap flow in the stem (sensor S3) after severing a big coarse root R3 are shown at the part (**D**).

Trees of different species can be supplied by water equally from surface or deep roots (or both), or differently by different parts of their root systems, which depends on soil water and root distribution and overall site characteristics (Cermak and Prax, 1999; Cermak et al., 1980; Howard et al., 1996; Lott et al.,

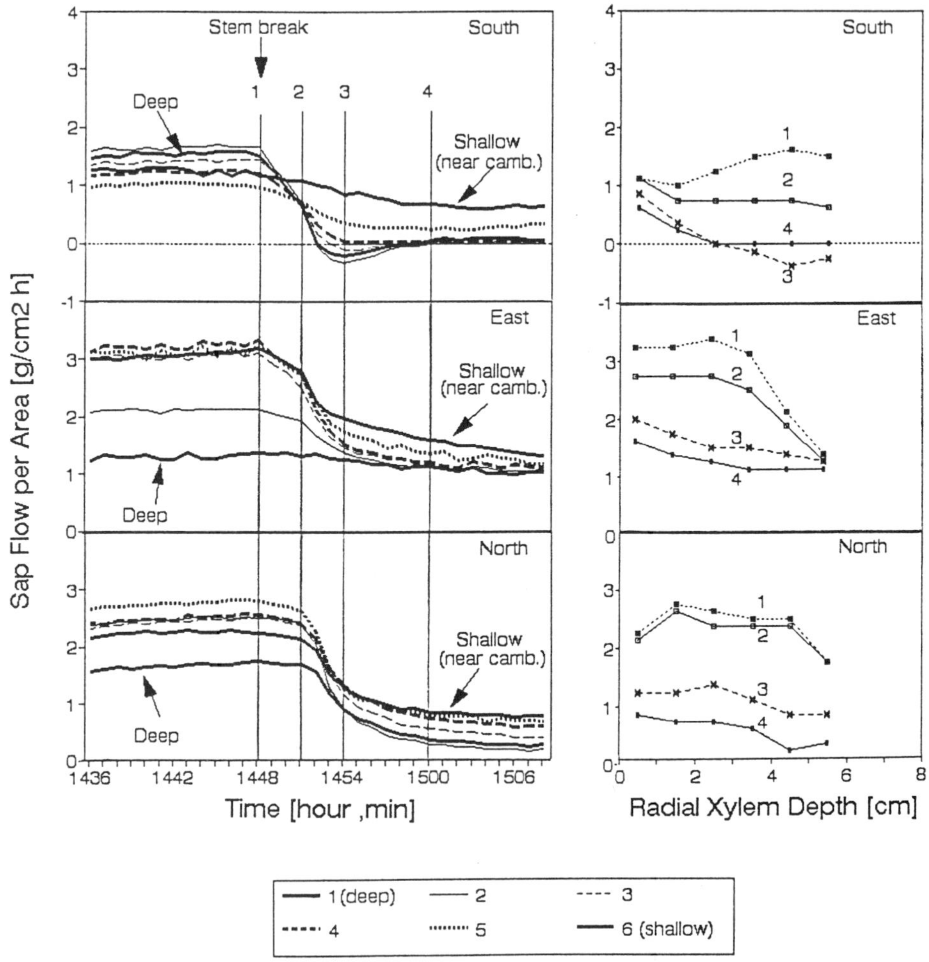

Figure 6. Response of sap flow rate in the coarse roots during the sink-severing experiment (by stem failure) in spruce. Individual roots of different orientations (south, compressed side; east, perpendicular side; and north, tension side) are shown one above the other. Sap flow at different radial depths over time are presented on the left, and radial patterns of flow at different time periods are shown (1–4) on the right.

1996). Our results presented here and those of previous studies (Cermak et al., 1984, 1992; Cermak and Nadezhdina, 1998) confirm that sapwood is not functionally homogenous over its total cross-section area. This may be associated with long-term development of xylem conduits (Mark and Crews, 1973), but may also be attributed to non-homogenous root and leaf distribution (Cermak and Kucera, 1990). Outer sapwood seems most important for water transport especially under conditions favorable for tree growth, when the majority of water comes from surface soil layers and where there are also more fine roots assuring nutrient supply (Clemensson-Lindell and Persson, 1995). However, whole sapwood area (even when working only partially) assures tree survival under less favorable conditions, when only deep roots can supply water.

The hydraulic resistance

Two types of sap flow responses to source-severing were observed. Sap flow usually increased first for a very short time (around 1 min, shorter time intervals were not possible to record with the given equipment) immediately after source-severing and then gradually decreased in an exponential pattern (Fig. 8). During the source-severing experiment (Exp. 1a), sap flow slightly increased in the large root R1 over the first minute when the small surface and intermediate roots were severed, whilst one bigger deep root remained functioning. Sap flow at different rates remained in

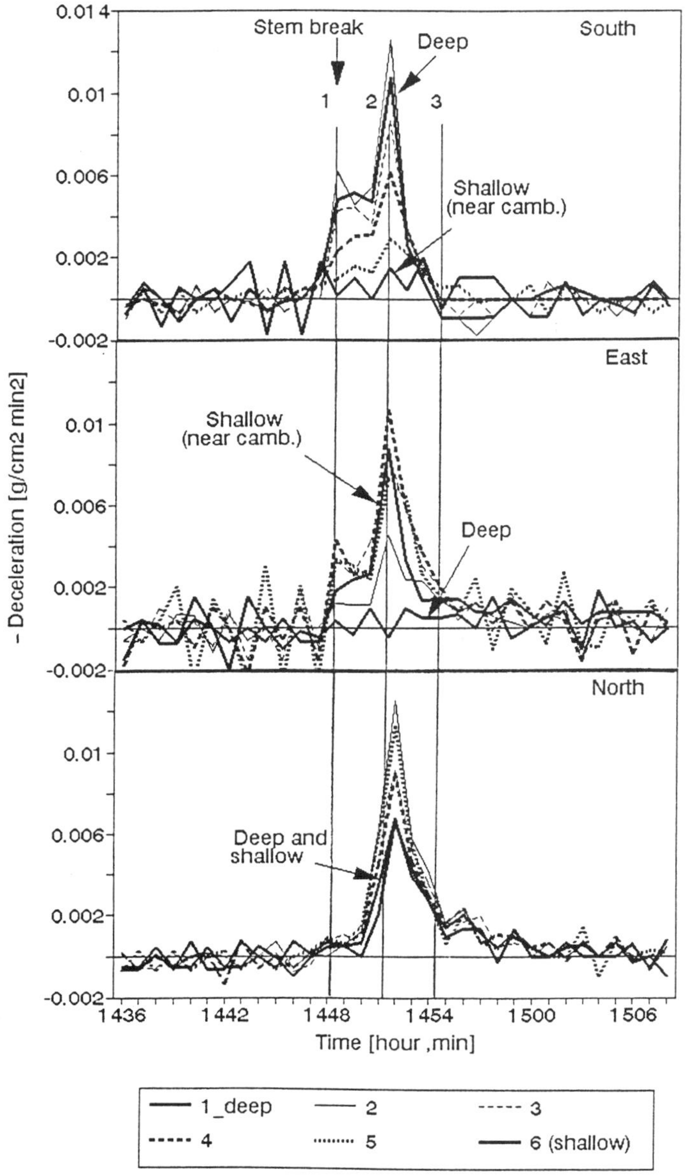

Figure 7. Deceleration of sap flow in the coarse roots of spruce in different radial depths during the sink-severing experiment. Individual roots of different orientations (south, compressed side; east, intermediate side; and north, tension side) are presented.

certain sapwood depths after source-severing. During the source-severing experiment (Exp. 1b), when roots R2 and R4 were removed completely, sap flow responded in a similar way: first, from abruptly increased and then gradually decreased over the whole sapwood. A similar reaction was also observed in the stem after source-severing experiment (Exp. 1c), but sap flow did not decrease equally over the whole sapwood due to additional flow coming from other roots along the same pathway in the stem (see Fig. 5).

It is known that sap flow (Q_w) in SPAC can be described as

$$Q_w = \mathrm{d}\psi_w/(R_{r+s} + R_{xyl} + R_{\text{leaves}}), \tag{1}$$

where $\mathrm{d}\psi_w$ is a gradient of water potential between leaves and soil; R_{r+s}, R_{xyl} and R_{leaves} are resistances of root–soil, xylem and leaves, respectively. $\mathrm{d}\psi_w$ increased and R_{spac} decreased abruptly at the first moment after source-severing due to changing water potential of the lower end of conductive pathway (re-

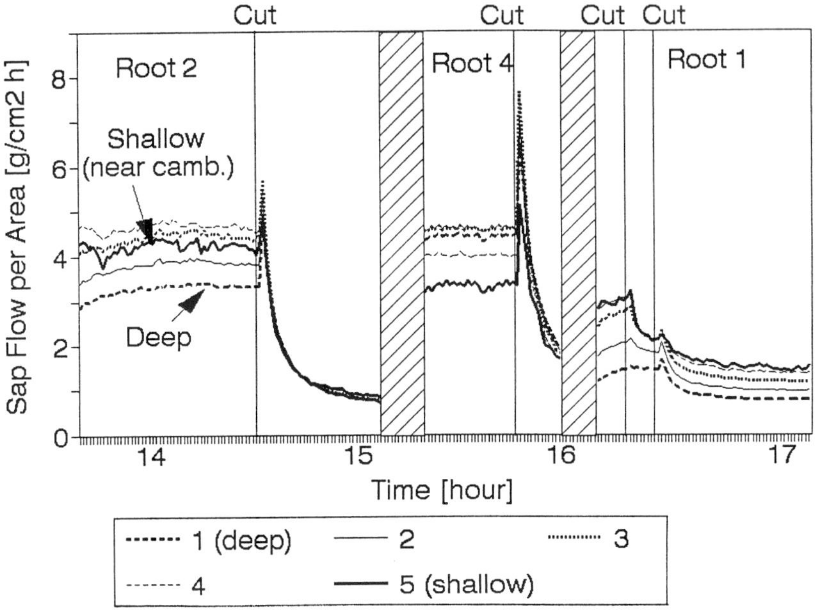

Figure 8. Immediate response of sap flow in large first order roots (R1, R2 and R4) in lime measured in different layers of sapwood (see scheme of roots at the Fig. 5A). Roots R2 and R4 were removed completely, while two small surface second order roots were severed from R1 subsequently, and one deep root remained functioning.

placing a usually higher negative value of ψ_w at the root surface by a lower positive pressure of the open atmosphere) and eliminating the root–soil resistance. Such resistance itself appears to be a major flux resistance, as it was characterized, e.g., for *Pinus contorta* (Running, 1980). Running showed that R_{r+s} under both high and low flux conditions in pine trees contributed more than one-half of the total resistance in the soil–plant–atmosphere continuum (R_{spac}). Roberts (1977) estimated that R_{r+s} represent 47–59% of R_{spac} in *Pinus sylvestris* tree. Both these reasons (increase of $d\psi_w$ and decrease of R_{spac}) should result in an abrupt increase of sap flow, which corresponds to our observations.

The second reaction of sap flow during the source-severing Exp. 1 (gradual decrease in flow) should correspond to a decreasing amount of water in conductive elements within the remaining tissues between the cut end of stem and the sensor below it, which was replaced by air. Integrated sap flow over the time period of decreasing flow correspond to differences in xylem water content before and after severing and mainly characterize total lumen volume of presumably large vessels (Cermak et al., 1984).

According to the above relationship (Eq. (1)) no peak flow can be expected in the sink-severing experiment 2, after cutting the stem from the rest of the tree equipped with sap flow sensors. The situation differed significantly from the source-severing experiment, because R_{r+s} remained the same, but R_{xyl} of stem and R_{leaves} disappeared (with the exception of that in the remaining part of stem) and the gradient of water potentials $d\psi_w$ changed from high values in the foliage direction to lower values in the direction of roots (when a high negative leaf water potential was replaced by a low positive atmospheric pressure above the sensors). This may cause blockage of flow or even reversed (downwards) movement, as observed in experiments when introducing dye into the xylem through a hole opened to the atmosphere during studies of conducting pathways and velocity (Cermak and Kucera, 1990b; Cermak et al., 1992). The flow should stop completely within seconds, when the cut is made just above the sensor (Cermak and Kucera, 1991). The delay in the blockage of flow in the roots of the winched spruce tree should correspond to the water capacity of the 2.7 m long conducting pathway between the cut end of stem and the sap flow sensors, which contains about 120 cm^3 of available water per cm^2 of sapwood.

Considering the above equation (Eq. (1)), we can assume with a reasonable simplification, that when taking into account directly measurable water absorption of roots (through the corresponding sap flow) we can estimate from Exp. 1 some parameters of root functional size. This situation is similar to that which exists in the aboveground parts of trees. It was shown in forest stands, that sap flow in branches, or leaf layers at a certain depth of the canopy corresponds to appropriate actual leaf areas weighted by the corresponding mean daily relative irradiances in the same layer, under non-limiting soil moisture conditions (Cermak, 1989). Analogous we can assume, that magnitude of sap flow rates in stems, buttresses or upper ends of root branches could be approximately interpreted in quantitative terms of actual functional size of roots.

Conclusions

First observations in several broad-leaf and coniferous species indicate, that measurement of the sap flow rate at the stem base by appropriately installed sensors allows an estimation of root distribution around trees (according to cardinal points) as well as according to soil depths. When considering the depth, this implies that surface roots supply water mostly to the surface layers of sapwood, whilst deep roots supply water more uniformly over the whole sapwood.

The possibility of studying the behavior of intact roots in situ by such a function-based, non-destructive technique may significantly help in quantitative field studies of tree root systems. A further application may be the identification of potentially damaging tree roots growing near buildings or sewerage pipes in an urban environment.

In experimental studies allowing a destructive approach, immediate responses of sap flow to source-severing may be applied to characterize some root properties (as the magnitude of the root/soil resistance to water flow, their water capacity, etc.). Response of sap flow to sink-severing seems to be useful for studies of the impact of mechanical stress on trees.

Acknowledgement

The study was technically assured by the Czech project VS 96077. The authors wish to express their best thanks to Dr Antonio Raschi from IATA, Firenze, Italy, to Prof. Reinhart Ceulemans from UIA, Antwerpen, Belgium, for their valuable support of this study when working on their related projects and to Prof. Radamír Ulrich for his help with tree winching.

References

Baeyens L, Van Slycken J and Stevens D 1993 Description of the soil profile in Brasschaat. Internal research paper. Institute for Forestry and Game Management, Geraardsbergen, Belgium, 17 pp.

Cabibel B 1994 Continuity of water transfer in the soil-root-trunk system. The case of fruit trees. Agronomie 14, 503–514.

Carlson W C and Harrington C A 1987 Cross-sectional area relationships in root systems of loblolly and shortleaf pine. Can. J. For. Res. 17, 556–558.

Cermak J, Huzulak J and Penka M 1980 Water potential and sap flow rate in adult trees with moist and dry soil as used for the assessment of the root system depth. Biol. Plant. (Praha) 22, 34–41.

Cermak J, Jenik J, Kucera J and Zidek V 1984 Xylem water flow in a crack willow tree (*Salix fragilis* L.) in relation to diurnal changes of environment. Oecologia (Berlin) 64, 145–151.

Cermak J 1989 Solar equivalent leaf area as the efficient biometrical parameter of individual leaves, trees and stands. Tree Physiology 5, 269–289.

Cermak J and Kucera J 1990a Water uptake in healthy and ill trees under drought and hypoxia and non-invasive assessment of the effective size of root systems. *In* Proc. COST 612 Workshop 'Above and belowground interactions in forest trees in acidified soils'. Ed. H Persson. pp 185–195. Simlangsdalen, Sweden.

Cermak J and Kucera J 1990b Scaling up transpiration data between trees, stands and watersheds. Silva Carelica 15, 101–20.

Cermak J and Kucera J 1991 Extremely fast changes of xylem water flow rate in mature trees, caused by atmospheric, soil and mechanical factors. *In* Proc. CEC International Workshop 'Methodologies to assess the impacts of climatic changes on vegetation: analysis of water transport in plants and cavitation of xylem transport in plants and cavitation of xylem conduits' Eds A Raschi and M Borghetti. 10 pp. 29–31 May 1991, Firenze, Italy.

Cermak J and Nadezhdina N 1998 Sapwood as a scaling parameter – defining according to xylem water content or radial pattern of sap flow? Ann. Sci. For. 55, 509–521.

Cermak J and Prax A 1999 Water balance of the floodplain forests in southern Moravia considering rooted and root-free compartments under contrasting water supply and its ecological consequences. Tree Physiol. (*In press*).

Cermak J Cienciala E Kucera J Lindroth A and Hallgren J-E 1992 Radial velocity profiles of water flow in stems of spruce and oak and response of spruce tree to severing. Tree Physiol. 10, 367–380.

Cermak J, Hruska J, Martinkova M, Krejzar T, Sustek S, Nadezhdina N and Tatarinov F 1997 Report on the situation of fully grown trees and their impact on constructions in the area of municipal bath of Zabrdovice (in Czech). Institute of Forest Ecology, Mendel Univ., Brno. 49 pp.

Cermak J, Riguzzi F and Ceulemns R 1998 Scaling up from the individual trees to the stand level in Scots pine: 1. Needle distribution, overall crown and root geometry. Ann. Sci. For. 55, 63–88.

Cermak J, Hruska J, Martinkova M and Prax A 1999 City tree roots and survival near houses analyzed using sap flow and ground penetrating radar technique. Plant Soil 219, 103–116.

Clemensson-Lindell A and Persson H 1995 The effect of nitrogen addition and removal on Norway spruce fine-root vitality and distribution in three catchment areas at Gardsjon. For. Ecol. Manage. 71, 123–131.

Howard S B, Ong C K, Black C R and Khan A A H 1996 Using sap flow gauges to quantify water uptake by tree roots from beneath the crop rooting zone in agroforestry systems. Agrofor. Syst. 35, 15–29.

Hruska J, Cermak J and Sustek S 1999 Mapping tree root systems with ground penetrating radar. Tree Physiol. 19, 125–130.

Jenik J 1978 Roots and root systems in tropical trees: morphogenic and ecologic aspects. *In* Tropical Trees as Living Systems. Eds P B Tomlinson and M H Zimmermann. pp 323–349. Cambridge University Press, Cambridge.

Lott J E, Khan A A H, Ong C K, Black C R, Topa M A, Rygiewicz P T and Cumming J R 1996 Sap flow measurements of lateral tree roots in agroforestry systems. Tree Physiol. 16, 995–1001.

Mark W R and Crews D L 1973 Heat-pulse velocity and bordered pit condition in living Engelmann spruce and lodgepole pine trees. For. Sci. 19, 291–296.

Nadezhdina N E and Cermak J 1998 The technique and instrumentation for estimation the sap flow rate in plants. Patent pending (PV-1587-98), Czech Republic (U.S. Patent and Trademark Rec. No. 69055, 1997).

Rizzo D M and Gross R 1998 Distribution of *Armillaria* on pear root systems and a comparison of root excavation techniques, DPSS 87, 293–299.

Roberts J 1977 Use of tree-cutting techniques in study of water relations of mature *Pinus sylvestris* L. 1. Technique and survey of results. J. Exp. Bot. 28(104), 751–767.

Running S W 1980 Field estimates of root and xylem resistances in *Pinus contorta* using root excision. J. Exp. Bot. 31(121), 555–569.

Stokes A 1999 Strain distribution during anchorage failure in *Pinus pinaster* Ait. at different ages and tree growth response to wind-induced root movement. Plant Soil 217, 17–27.

Strong W L and La Roi G H 1983 Root system morphology of common boreal forest trees in Alberta, Canada. Can. J. For. Res. 13(6), 1164–1173.

Vasicek F 1984 The characteristics of biogenocenoses. *In* Ecophysiological and Ecomorphological Studies of Individual Trees in the Spruce Ecosystem of the Drahanska Vrchovina Uplands (Czechoslovakia). Ed. F Vasicek. Folia Universitatis Agriculturae, Fac. Silviculturae, Agr. Univ. Brno.

A. Stokes (ed.), The Supporting Roots of Trees and Woody Plants: Form, Function and Physiology, 239–244.

Water conductance and osmotic potential of papaya (*Carica papaya* L.) roots as influenced by drought

Thomas E. Marler
College of Agriculture & Life Sciences, University of Guam, UOG Station, Mangilao, Guam 96923, USA

Key words: *Carica papaya*, osmometry, root hydraulic conductivity, water deficit

Abstract

Three studies were conducted with papaya seedlings to determine the influence of drought stress on root hydraulic conductance and osmotic potential (ψ_π). In Experiment 1, 'Sunrise' papaya seedlings were exposed to three successive drying cycles of 6 days each, with 2 days of well-watered conditions between them. After 24 h of re-hydration at the end of each drying cycle, water conductance was determined using positive pressure of 0.5 MPa. 'Sunrise' and 'Known You 1' papaya seedlings were used to determine osmotic relations in Experiments 2 and 3. In Experiment 2, the roots of the papaya seedlings were split and planted in containers that were comprised of four compartments. When plants were well-established, water was withheld from three compartments for one-half of the plants (1/4), and the remainder of the plants continued to receive water in all four compartments. Following 6 weeks, plants were harvested and ψ_π of root tips in all compartments was measured. In Experiment 3, treatments consisted of well-watered plants, of plants exposed to a rapid onset of water stress for 7 days, or of plants exposed to a slow onset of water stress for 1 month. The last treatment was imposed by adding back one-half of the water used during each preceding day. All plants were re-watered and allowed to re-hydrate for 2 days before determining ψ_π of root tips, young leaves, and stem tips. In Experiment 1, water conductivity ranged from 14 to 17 ml per g per MPa per hour, and there were no differences among the treatments. In Experiment 2, ψ_π of root tips in split-root compartments receiving water ranged from −0.6 to −0.7 MPa, and ψ_π of root tips in compartments from which water was withheld ranged from −1.0 to −1.1 MPa. In Experiment 3, leaf and stem tip ψ_π ranged from −1.6 to −2.4 MPa, and root ψ_π ranged from −0.6 to −0.9 MPa. The results of this study indicate that adjustment of root and ψ_π maintenance of root system water conductivity, characteristics of some plant which tolerate drought by postponing tissue dehydration, are expressed by papaya plants under some conditions of water deficits.

Introduction

Mechanisms by which plants compensate for water limitation are numerous. Papaya plants experiencing water deficits exhibit responses categorized as dehydration postponement (Marler and Mickelbart, 1998; Marler et al., 1994). Plants which tolerate drought by postponing dehydration of tissues express reduction of water loss, adjustment of osmotic potential, and maintenance of water uptake (Turner, 1986b). A profound reduction of water loss by leaf gas exchange was one of the initial responses of container-grown papaya plants during drying cycles (Marler et al., 1994).

The influence of drought stress on osmotic relations and maintenance of water uptake have not been sufficiently studied for papaya. Numerous cycles of drying and re-wetting did not lead to osmotic adjustment of leaf or root tissue (Marler et al., 1994). However, osmotic adjustment is more likely to occur when drought stress develops slowly (Turner and Jones, 1980), and rapid dehydration resulted from the methods employed during the previous studies with papaya (Marler et al., 1994). The role of papaya root hydraulic conductivity for maintaining water uptake during water deficits has not been demonstrated.

The purpose of this research was to further study the influence of drought stress on osmotic relations and water conductivity of papaya roots. For studying osmotic relations, we imposed a slow water depletion on plants by adding a portion of the daily water use back and by using partial root volume irrigation methods.

Materials and methods

Experiment 1

'Sunrise' papaya seeds were planted in 360-ml containers filled with a 50% peat, 50% sand substrate on 25 January 1993. The plants were grown in full sun with daily watering and weekly fertilizing with a complete nutrient solution (Excel 15N-2.2P-12.5K, Grace-Sierra, Milpitas, CA). The solution was based on a 7.5 mM nitrogen content. The plants were arranged in a randomized complete block design with four blocks containing 10 plants each. Within each block, plants were randomized into two treatments. One half of the plants in each block were designated to receive drought treatments, and the other half of the plants were well-watered throughout.

One block was selected to measure root conductance at the initiation of the study. Root conductance was determined using a PMS 650 pressure chamber (PMS Industries, Corvallis, OR, USA). The containers in which the plants were growing were small enough to fit into the pressure chamber. Rooting substrate was completely saturated with a 1 mM solution of $CaCl_2$ by raising the level of the solution from the bottom to avoid trapping air.

The pressure chamber was filled with the $CaCl_2$ solution. One plant was inserted into the chamber until the intact root system within the container was in the solution. The stem was cut at 70 mm above the collar, and was carefully inserted through the lid of the pressure chamber. Pressure within the chamber was increased slowly using compressed N_2 until the desired pressure of 0.5 MPa was reached within the chamber. The cut stump was attached to a graduated pipette by tubing. When exudate evolution reached a homogeneous rate, the volume of exudate which was expressed during a 10-min period was recorded. The root system was then removed from its container, carefully washed free of its rooting substrate, and dried at 70°C for 48 h before weighing.

The initial drying cycle was begun on 3 April 1993 with the remaining three blocks of plants. Following 6 days with no added water, the plants were severely wilted. All plants were re-watered and fertilized on 9 April 1993. One of the blocks was randomly selected for measurements on 10 April 1993. Measurements were begun at 07:00 h, and were as described above.

The plants in the two remaining blocks which received the initial drying cycle were subjected to a second drying cycle beginning 11 April 1993. All plants were watered and fertilized on 17 April 1993. One of the two remaining blocks was selected for measurements following two drying cycles. Methods were conducted as described above.

The remaining plants which had received two drying cycles were subjected to a third drying cycle beginning 18 April 1993. The plants were re-watered on 24 April 1993, and measurements as described above were conducted on 25 April 1993.

Experiment 2

'Sunrise' and 'Known You 1' papaya seeds were planted in nursery trays with 51-mm diameter cells on 13 December 1995. Rooting substrate was 60% peat and 40% perlite, and trays were located in full sun. The seedlings were watered as needed, and fertilized weekly with the complete nutrient solution described above.

Plants were removed one at a time from the nursery trays and prepared for planting in split-root containers on 20 February 1996. The bottom 10-mm of media was carefully washed from the roots, and a razor blade was used to split the exposed taproot into four homogeneous quadrants. The top portion of the root system which was undisturbed was placed into a sleeve with an inside diameter of 38 mm and a volume of 226 ml. The sleeve was placed on top of a split-root container with four sections. The top dimensions of each section were 100×100 mm, and the volume of each section was 730 ml. The sleeve was situated so that each quadrant of the taproot was positioned to extend into one of the compartments of the split-root container. The containers were filled with a 60% peat, 40% perlite medium, and watered thoroughly. The entire procedure from the time the medium was rinsed from the bottom half of a root system until the procedure was completed was about 3 min.

The plants were placed under shade screen of 20% sunlight transmission for 2 days, followed by 50% sunlight transmission for 2 days, and 75% sunlight transmission for 3 days. Thus, shade was provided for 1 week following the procedure. The experiment was conducted in full sun thereafter. Watering was as needed, and fertilizing continued to be weekly.

Two watering treatments consisted of withholding water from three of the four compartments (1/4) on 22 March 1996, or continuing to water all four compartments (4/4). The plants were arranged in two separate experiments, one for each variety. Within each variety, the treatments were arranged in a completely randomized design with six replications.

The drying cycle of the three compartments for the 1/4 plants was relieved on the morning of 1 May 1996, when all compartments were re-watered thoroughly. The plants were placed under shade screen of 20% sunlight transmission to provide conditions of low water demand. During pre-dawn on 3 May 1996, the plants were carefully removed from the split-root containers, and the media was washed from the roots with gentle water pressure. The terminal 10 mm of root tip from at least five roots per compartment were removed, rinsed in de-ionized water, sealed in plastic bags, and frozen. Roots with 1–2 mm diameter were selected.

The frozen tissue was thawed to determine ψ_π of expressed solution as previously described (Marler and Zozor, 1996). A Model 5500 vapor pressure osmometer was used to measure ψ_π (Wescor, Logan, UT, USA).

Experiment 3

'Sunrise' and 'Known You 1' papaya seeds were planted as described for Experiment 2 on 15 March 1996. Seedlings were transplanted into 2.6-liter containers on 26 April 1996, and were maintained until treatments were begun on 5 June 1996.

Three treatments consisted of plants which were well-watered throughout the study, plants exposed to a slow dehydration for 4.5 weeks, and plants exposed to rapid dehydration of 1 week. The plants were arranged as a two variety × three treatment factorial experiment in a randomized complete block design, with six replications.

Water was withheld from all plants receiving the slow dehydration treatment on 5 June 1996. The plants were weighed on each successive day, and half of the water lost during each day was added to the rooting substrate. The plants in the remaining two watering treatments were well-watered until 1 July 1996, when water was withheld from the plants receiving the rapid dehydration treatment. The plants receiving the slow dehydration treatment were wilted during the second half of each photoperiod for the last week of the dehydration period, and the plants receiving the rapid dehydration treatment were wilted during the last day of the dehydration period.

All plants were fully re-watered on the morning of 8 July 1996, and shade screen with 20% sunlight transmission was placed over the experimental area. The plants were maintained for 2 days under these conditions to allow re-hydration of plant tissues.

Samples of plant tissue were obtained during pre-dawn on 10 July 1996. The three youngest leaves were removed from each plant, rinsed in de-ionized water, sealed in plastic bags, and frozen. The terminal 5 mm of stem tip was cut from each plant, rinsed, sealed and frozen. Roots samples were prepared as described for Experiment 2. Osmotic potential was measured after thawing plant tissue as described in Experiment 2.

Data analysis

Water conductance was considered for an entire root system in Experiment 1, and then standardised to a unit root weight basis as water conductivity. Within each of the sampling dates, data were subjected to analysis of variance. The response to successive drying cycles was determined with regression analysis, using the number of drying cycles as the independent variable.

Data from Experiment 2 were analyzed separately for each variety. Two comparisons were subjected to a standard *t*-test. First, ψ_π of the watered compartment of 1/4 plants was compared with that of the 4/4 plants. Second, ψ_π of the dry compartments of 1/4 plants was compared with that of the watered compartment of 1/4 plants.

Data in Experiment 3 were subjected to analysis of variance as a 2 variety × 3 treatments × 3 plant tissues factorial experiment in a randomized complete block design.

Results

Experiment 1

Water conductance of the root system of 'Sunrise' plants receiving the drying and re-wetting cycles was relatively unchanged throughout three of these cycles (Fig. 1). There was a linear increase in conductance of the root system of well-watered plants for each successive cycle. As a result, water conductance of well-watered plants was greater than that of drought-stressed plants following two and three drying cycles ($P \leq 0.01$). However, when conductance was standardized to a unit root weight basis, there was no difference among any of the sampling dates and treatments for water conductivity (Fig. 1).

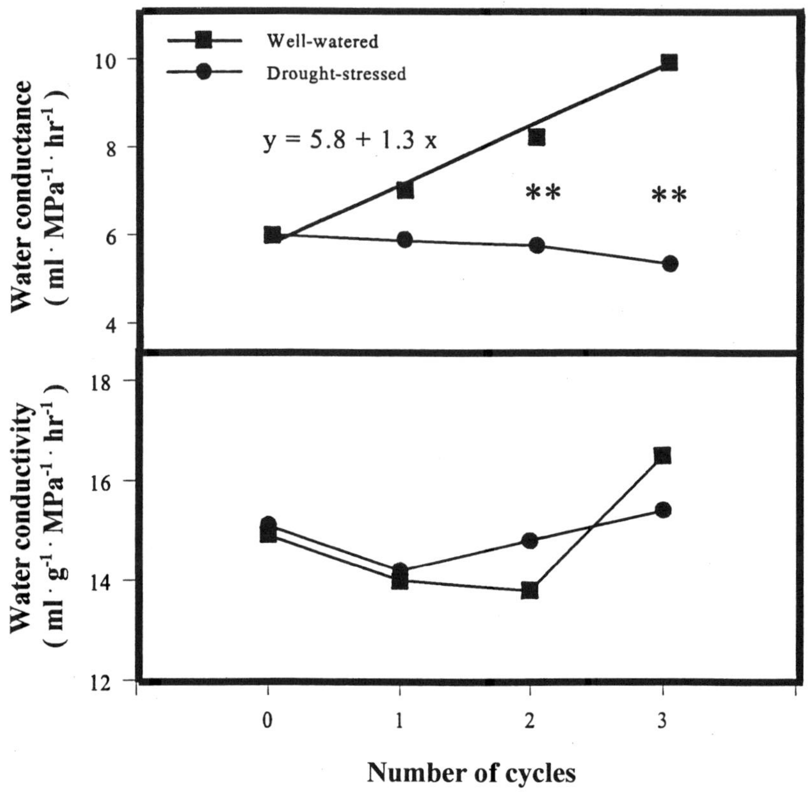

Figure 1. Water conductance (top) and conductivity (bottom) of 'Sunrise' papaya. **indicates difference ($P \leq 0.01$) between treatments for water conductance. Conductance increased linearly for well-watered plants. There were no differences between treatments for any of the sampling periods for water conductivity (n=5).

Experiment 2

After 6 weeks of withholding water from three of four compartments in split-root containers, there was no difference in ψ_π of 'Sunrise' papaya roots from the watered compartment of 1/4 plants and ψ_π of roots from the 4/4 plants (Fig. 2). However, ψ_π of roots from the dry compartment of 1/4 plants was ~0.4 MPa more negative than ψ_π of roots from the watered compartments of 1/4 plants ($P \leq 0.05$). Results were similar for the 'Known You 1' plants (Fig. 2).

Experiment 3

Analysis of variance for ψ_π results indicated that the block effect was not significant and no first- or second-order interactions occurred among the plant tissues, varieties, and treatments. Furthermore, ψ_π did not vary among the treatments or varieties main effects. In contrast, ψ_π differed among the plant tissues ($P \leq 0.01$). Leaf and stem tip ψ_π were about three times more negative than root tip ψ_π (Table 1).

Table 1. The influence of 4.5 weeks of slowly developing dehydration (slow) and one week of rapid dehydration (rapid) on osmotic potential (MPa) of the youngest three leaves, 5-mm of stem tip, and root tips of papaya seedlings. Data are means of 'Known You 1' and 'Sunrise' plants, since there were no differences between varieties (n=12).

Tissue	Treatment			Mean[a]
	Well-watered	Slow dehydration	Rapid dehydration	
Leaves	−1.94	−1.97	−1.77	−1.89a
Stems	−1.79	−1.69	−1.98	−1.82a
Roots	−0.61	−0.62	−0.63	−0.62b

[a]Mean separation within column by Duncan's multiple range test, $P \leq 0.01$.

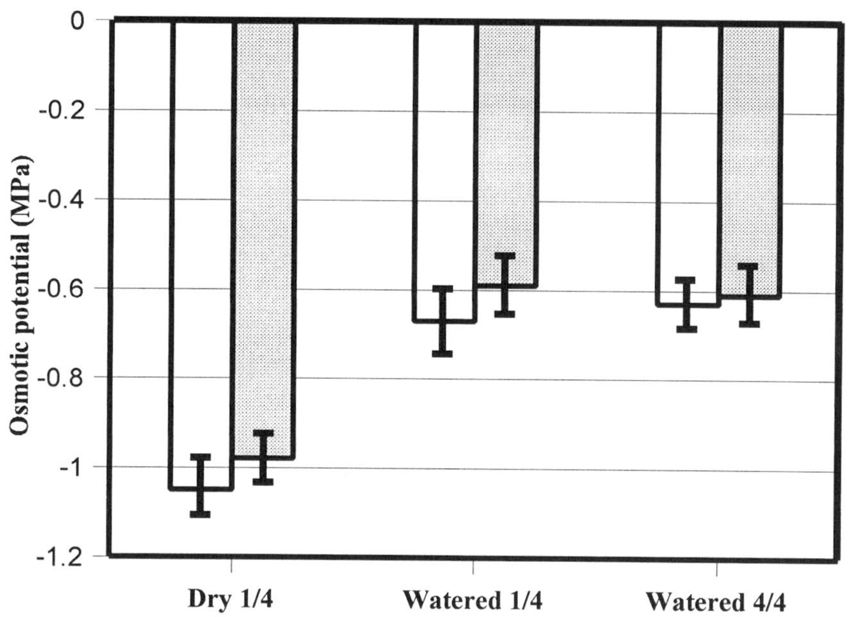

Figure 2. Osmotic potential of 'Sunrise' (open bars) and 'Known You 1' (shaded bars) papaya roots grown in split-root containers with water withheld from three compartments (1/4) or with all four compartments watered (4/4). There was no difference between watered 1/4 and 4/4 roots, according to *t*-test. Osmotic potential of dry 1/4 roots was more negative ($P \leq 0.01$) than that of watered 1/4 roots, according to *t*-test (n=6).

Discussion

Root systems subjected to water deficits often exhibit a decrease in water conductance (Kramer, 1983; Turner and Kramer, 1980; Wiersum and Harmanny, 1983). In this study, water conductance of intact roots from papaya plants experiencing cycles of withholding water was below that of continuously watered plants after two cycles (Fig. 1). However, this was exclusively due to continued root growth of the well-watered plants, while root growth ceased on the plants experiencing cycles withholding water. Thus, these data indicate that temporary water deficits did not elicit a reduction in water conductivity of papaya roots.

Stomatal conductance of papaya leaves is highly sensitive to water deficits. Although stomatal conductance declines during water deficits, upon re-watering there is an immediate return to pre-stress levels (Marler et al., 1994). Insensitivity of water conductivity of the papaya root system to water deficits may be one reason for this rapid recovery of leaf gas exchange upon re-watering.

The capacity to adjust ψ_π of plant tissues during water deficits is an important mechanism of drought tolerance (Turner, 1986a). In this study, slowly developing water stress resulted in severe leaf wilting throughout the photoperiod less than 4 weeks after withholding adequate water from papaya plants. However, more than 4 weeks of water deficits did not lead to any changes in leaf, stem tip, or root tip ψ_π when compared with continuously watered plants. In this study and in previous reports (Marler et al., 1994), ψ_π of plant tissues did not differ between well-watered plants and plants experiencing rapidly developing water stress. Roots typically have higher ψ_π than leaves (e.g., Parker and Pallardy, 1988), and in this study papaya root ψ_π was about 1.2 MPa higher than leaf ψ_π.

Adjustment of root ψ_π in response to localized water deficits did occur in this study when papaya roots were trained in split-root containers. Watering a quarter of the rooting volume in this system led to a 0.4 MPa decrease of ψ_π for root tips within the dry compartments, when compared with root tips within the watered compartment (Fig. 2). Papaya roots which receive no direct irrigation but are connected to a portion of a root system which receives water are able to maintain relative water content which does not differ from the portion which is watered (Marler and Discekici, 1996). Reduction of tissue ψ_π may play a role in achieving this homogeneity of relative water content in a heterogeneous system. The capacity for root tissues to adjust ψ_π may affect plant response to water deficits by influencing root growth, partitioning patterns among root and canopy tissues, and root-produced plant growth regulators (Turner, 1986a).

Dehydration postponement of plant tissues is achieved during periods of water deficits through reduction of water loss, adjustment of ψ_π, and maintenance of water uptake (Turner, 1986b). Previous reports indicated that drought-stressed papaya plants exhibit a strict control of leaf water loss (Marler and Mickelbart, 1998; Marler et al., 1994). The results of this study indicate that adjustment of root tissue ψ_π and maintenance of root system hydraulic conductivity are also expressed by papaya plants under some conditions of water deficits.

Acknowledgment

This research was supported by the US Dept of Agriculture under CSRS special grant agreement no 93-34135-8828, managed by the Pacific Basin Administrative Group.

References

Kramer P J 1983 Water Relations of Plants. Academic Press, Orlando, FL, USA, 489 pp.

Marler T E and Discekici H M 1996 Partial root volume irrigation of papaya plants in split root containers. HortScience 31, 685.

Marler T E and Mickelbart M V 1998 Drought, leaf gas exchange, and chlorophyll fluorescence of field-grown papaya. J. Am. Soc. Hortic. Sci. 123, 714–718.

Marler T E and Zozor Y 1996 Salinity influences photosynthetic characteristics, water relations, and foliar mineral composition of *Annona squamosa* L. J. Am. Soc. Hortic. Sci. 121, 243–248.

Marler T E, George A P, Nissen R J and Andersen P C 1994 Miscellaneous tropical fruits. *In* Handbook of Environmental Physiology of Fruit Crops. Vol II. Sub-tropical and Tropical Crops. Eds B Schaffer and PC Andersen. pp 199–224. CRC Press, Boca Raton, FL, USA.

Parker W C and Pallardy S G 1988 Leaf and root osmotic adjustment in drought-stressed *Quercus alba, Q. macrocarpa*, and *Q. stellata* seedlings. Can. J. For. Res. 18, 1–5.

Turner N C 1986a Adaptation to water deficits: a changing perspective. Austral. J. Plant Physiol. 13, 175–190.

Turner N C 1986b Crop water deficits: a decade of progress. Adv. Agron. 39, 1–51.

Turner N C and Jones M M 1980 Turgor maintenance by osmotic adjustment: a review and evaluation. *In* Adaptation of Plants to Water and High Temperature Stress. Eds N C Turner and P J Kramer. pp 87–103. Wiley, New York, USA.

Turner N C and Kramer P H 1980 Adaptation of Plants to Water and High Temperature Stress. J Wiley, New York, USA, 482 pp.

Wiersum L K and Harmanny K 1983 Changes in the water-permeability of roots of some trees during drought stress and recovery, as related to problems of growth in urban environment. Plant Soil 75, 443–448.

A. Stokes (ed.), *The Supporting Roots of Trees and Woody Plants: Form, Function and Physiology*, 245–258.

Effect of water stress on root meristems in woody and herbaceous plants during the first stage of development

D. Chiatante*, A. Di Iorio, L. Maiuro and S.G. Scippa
Università degli Studi del Molise, Dipartimento di Scienze Animali, Vegetali e dell'Ambiente, Campobasso, Italy

Key words: elongation, meristems, *Pinus nigra*, *Pisum sativum*, root, water stress

Abstract

We investigated the effect of water stress on the root system architecture of pine saplings and pea seedlings during the first stage of development. Attention was focused on meristematic tissue situated at the root tip because of the leading role played by the tissue in the planning of root system architecture. The data showed that both species are extremely sensitive and that plants arrest their growth immediately during water stress treatment. When stress treatment was not intense, both species recovered growth but presented modifications in the root system architecture. In pine saplings, the modification in root system architecture was the consequence of fine root meristems not recovering from water stress. The saplings survived by producing new lateral meristems from the cortical tannin zone above the fine root tip. In the case of pea seedlings, the meristematic tissues in the primary root arrested proliferation during water stress although they recovered when the event occurred during the first hours of germination. The response was different when water stress was enforced on older seedlings. In this case, root meristems never completely recovered their proliferation despite the increase in proline content observed in the cells. The modification of root system architecture in pea seedlings depended on the arrest of primary root elongation and the formation of new root laterals. As regards the primary roots, water stress treatment induced along the axis the formation of irregular 'swellings' in the cortical zone above the meristematic zone. Anatomical investigations suggested that such swellings may have derived from the changes in elongation direction of derivatives. The formation of new laterals was observed in hydroponic cultures when water stress treatment was enforced slowly and prolonged for a long time. The production of new lateral meristems may have been a similar response of woody and herbaceous plants to water stress conditions. It is not known whether these new meristems present characteristics of resistance to water stress.

Introduction

Plants experience many kinds of environmental stresses. The cumulative effects of the stresses determine species distribution, reproductive success in a given environment, and, in the case of crop plants, productivity (Cushman, 1990). Environmental stresses come in many forms, yet the most prevalent stresses (salt, cold, drought) have in common their effect on plant water status (Bohnert et al., 1995). Therefore, it is not surprising that the reactions of plants to many stresses partly overlap (Bartels and Nelson, 1994).

We have been studying the influence of water stress on root system architecture during the first period of plant development. The studies are conducted indifferently with a woody (*Pinus nigra* L.) and a herbaceous (*Pisum sativum* L.) plant because small, juvenile woody plants have physiological responses linked to environmental variations similar to those observed in herbaceous plants (Crocker et al., 1998). Moreover, the relationship between water stress effect and plant response is better studied when comparing parameters among a range of species varying widely in tolerance (Ludlow, 1989).

It has been suggested that the growing regions in a plant are exposed to water stress more than differentiated regions (Plant, 1995). Such differences could be primarily due to anatomical features that reg-

* FAX No: +39 874404678. E-mail: chiatant@hpsrv.unimol.it
Plant and Soil is the original source of publication of this article. It is recommended that this article is cited as: *Plant and Soil* **217:** 159–172, 1999.

ulate water movement in the two regions (Matsuda and Ryan, 1990). Our attention focused on the effect of water stress on the activity of root meristems because of the planning role played by the tissue in the construction of the root system. Another reason for investigating the tissue was due to the absence of any protection against water loss, which makes it particularly exposed even when the water stress lasts for a few minutes. In this regard, cytological investigations conducted in roots of woody plants indicated that water stress may damage the cellular membranes of root tips as revealed by electrolyte leakage measurements (McKay, 1992). A little more is known about the effect of water stress on root meristems in herbaceous plants. At different developmental stages, the occurrence of cellular damage such as lack of protoplasm, formation of membrane vesicles, and coagulation of cytoplasmic content has been demonstrated (Blackman et al., 1995). However, to the best of our knowledge, no investigation in woody and herbaceous plants has focused on the effect of water stress on the proliferative activity of root meristems. Similarly, no investigation has addressed the first stage of plant development when the effect of water stress on meristematic tissue could have a considerable effect on future root architecture.

In the present study, all the experiments were conducted with very young plants. The stress condition was enforced by withholding watering or by leaving the plants in a glass chamber under a condition of low relative humidity. Water stress in hydroponic cultures was used with pea seedlings to keep a close control on the value of water potential and to make repeated measurements of root elongation without damaging the tissues. Cytological and anatomical effects of water stress on meristematic root tissue were investigated.

Materials and methods

Water stress of pine saplings

Two-year-old *Pinus nigra* L. saplings were collected with bare roots from a nursery at the beginning of their vegetative season. Saplings were lifted from the nursery bed with their roots still included in a clump of soil to prevent damage during transportation to the laboratory. The soil was then gently washed away from the bare roots by dipping the clump of soil in water. The saplings were separated, taking care not to damage the roots, and re-planted in an iron box (10 cm wide, 30 cm long and 40 cm deep) after recording the image of each root system on a computerised scanner system. At the bottom of each box, there was an iron grille to enable excess water to percolate and to be sucked up by means of a small plastic tube left in the soil during re-planting operations. The boxes were maintained in a water bath at 22 °C and 16-hr daily light of 300 μmol m^{-2} s^{-1} photosynthetically active radiation. The saplings were grown under such conditions for 6 weeks before applying water stress treatments by means of variable watering regimes (7- or 10-day intervals). The same watering regime was applied to four boxes for a total of 20 saplings. Control saplings were maintained on a 3-day-interval watering regime; 20 saplings were never watered during the treatments. At the end of the stress treatment, which lasted for 3 weeks, all the saplings were returned to the initial 3-day-interval watering regime and left to grow for another 3 weeks. After 3 weeks the boxes were emptied, saplings were removed without damaging the roots, and image analysis of the root systems was carried out with a computerised scanner system. Relative moisture content (RMC) was calculated on saplings before and after water stress by weighing all the root system before and after drying at 105 °C for 24 hr (RMC= fresh weight-dry weight/fresh weight đ 100).

Water stress treatments of pea seedlings

Dry seeds of *Pisum sativum* L. cv Lincoln were surface sterilised with 15% (v/v) sodium hypochlorite for 30 min, rinsed in running tap water for 1 hr, left at room temperature for 4 hr, and placed on moist vermiculite until use at 24 °C in the dark. After 3 days an aliquot of pea seedlings was transferred to dry vermiculite. Batches of 20 pea seedlings from all the treatments were collected at regular intervals and used for primary root elongation measurements. Alternatively, the seedlings grown in moist vermiculite for 3 days were placed for two days on a filter paper in a glass chamber in the presence of water or a 20% glycerol solution. After 2 days of treatment in the glass chamber containing glycerol, an aliquot of seedlings were returned in a glass chamber in the presence of water and were left to recover from water stress. As regards hydroponic cultures, 3-day-old pea seedlings germinated in moist vermiculite were rinsed in tap water and their roots immersed in Hoagland's medium or alternatively in Hoagland's medium containing polyethylene glycol 6000 (PEG). PEG concentration was calculated according to Michel and Kaufman (1973) to produce a final external water potential of –1.7 m Pa. Care was taken to ensure that all the primary root axes were immersed in the solution, and seedlings having the same length of the root axis were chosen for the same treatment. In water stress experiments where a slower rate of drying was to be tested, seedlings grown in moist vermiculite for 3 days were transferred every 24 hr to solutions containing increasing concentrations of PEG 6000 to give according to Michel and Kaufman (1973) values of water potential respectively of –0.5, –0.8, –1.0, –1.5, and –1.7 M Pa. In order to reverse the inhibition of root elongation exerted by PEG 6000, seedlings were rinsed in tap water several times before transferring them to Hoagland's medium. In all the experiments, the primary root elongation value was obtained by measuring the distance between the root tip and cotyledons with a ruler. In some seedlings, the primary root axis was detached by a hand scalpel and used to establish the RMC and to make cytological and anatomical preparations.

Mitotic index

Meristematic root tips (0–1 cm) were fixed in absolute ethanol-glacial acetic acid (3:1) and then prepared according to the standard Feulgen procedure, with a hydrolysis time of 10 min in 1 N HCl at 60 °C. The stained roots were squashed between the slide and cover-slip with a drop of acetic acid. The slides were frozen on dried ice, the cover-slip was removed, and the slides were washed (15 min each washing) in 100%, 95%, 70%, 50%, 35%, and 0% cold absolute ethanol. The slides were then dried and mounted with immersion oil and xylene (1:1 v/v) and viewed with a Leitz Orthoplan microscope to determine the mitotic index. The mitotic index was obtained by counting the percentage of cells in the mitotic phase. At least 1000 cells per root tip were counted.

Proline content in root meristems of pea seedlings

The proline content in root meristems of pea seedlings was determined by HPLC. The root meristem extract was obtained by homogenising the tissue in TRIS buffer, pH 7.8. The homogenate was filtered through cheese-cloth and centrifuged at 30,000 × g for 30 min. The pellet was discarded and the supernatant used as crude extract. The sample was prepared by adding 100 μl of crude extract to 200 μl of borate buffer (0.2 m boric acid solution adjusted to pH 8.5 with 30% KOH solution). After addition of 40 μl of OPA reagent (5 mg/ml), the solution was mixed for 1 min. Then, 20 μl of FMOC reagent (3 mM in acetone) was added and mixed for 3 min. Excess FMOC reagent was removed by 20 μl of heptyl-amine reagent. We used a Kontron MT2 HPLC system (Kontron Instruments, Milan, Italy), a Kontron 422 intelligent pump, a Kontron 482 column thermostat (the column temperature was 40 °C), a Kontron 465 intelligent autosampler, and a Kontron 465 spectrophotometer with excitation wavelength at 265 nm and with emission wavelength at 315 nm. The OPA reagent was first added to the column to remove the primary amino acid influence at the retention time. The proline was eluted (40 °C) by a reverse phase C8 column (20 × 2.1 cm) under isocratic conditions at a flow rate of 1 ml/min. The eluent phase consisted of 20 mM acetate buffer, pH 7.2 (60%), acetonitrile (20%), and methanol (20%).

Histological preparation of roots

Roots of pea seedlings were excised by hand with a scalpel, fixed by 2% glutaraldehyde for 24 hr and dehydrated in ethanol. Initial paraffin infiltration was obtained by leaving the specimens for 24 hr in a solution containing absolute ethanol and paraffin at the ratio 1:1 (v/v). The specimens were then transferred to a fresh solution containing absolute ethanol and paraffin (1:3, v/v) for 24 hr. Final infiltration was obtained by leaving the specimens in liquid paraffin for 24 hr before hardening in the oven. Specimens were cut by a Leitz 1512 microtome, and 1–2 μm sections were collected, mounted on a slide and stained with hematoxylin and eosin. Slides were observed by a Leitz Aristoplan microscope and photographed.

Results

Effect of water stress on roots of pine saplings

At the end of water stress treatments in a growth chamber, we classified as 'recovering saplings' those showing the presence of new green needles. According to this parameter, we calculated the percentage of survival. We found that 57% saplings maintained at the 7-day interval of watering survived treatments. The survival percentage fell to 10% and 2% when saplings were maintained respectively under a 10-day and a 0-day interval of watering (Figure 1). At the beginning and at the end of water stress experiments, images of the root system for each sapling had been

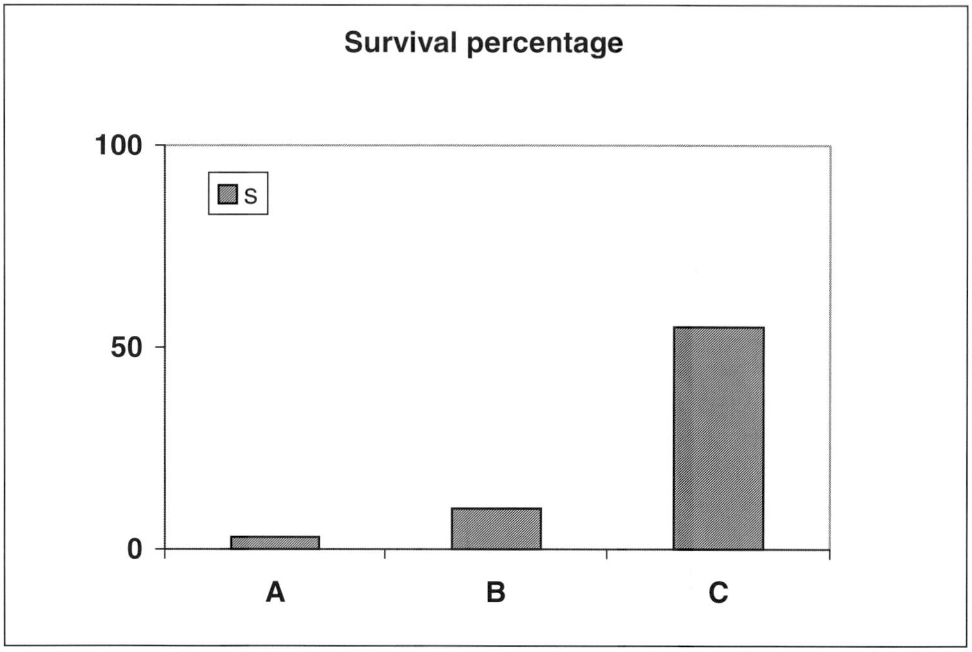

Figure 1. Survival percentage of pine saplings under different watering regimes. Pine saplings lifted from the nursery bed are re-planted in the growth chamber. After a period of re-growth, the plants are divided in aliquots and stressed by different watering regimes: A=0-day interval; B=10-day interval; C=7-day interval; control=3-day interval.

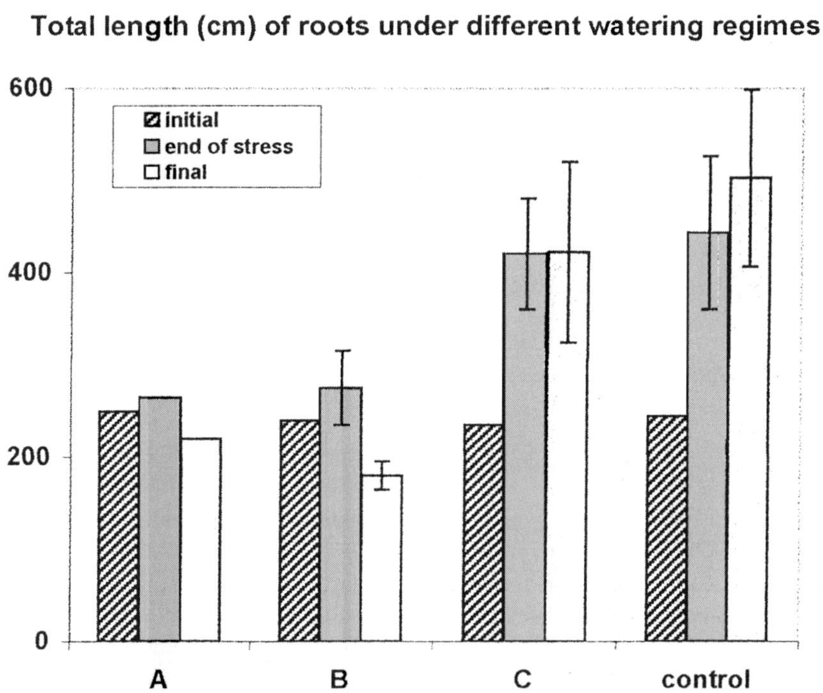

Figure 2. Total root length in pine saplings under different watering regimes. Watering regimes are described in Figure 1. The different columns refer to measurements made at re-planting time, at the end of stress treatment, and at the end of the experiment.

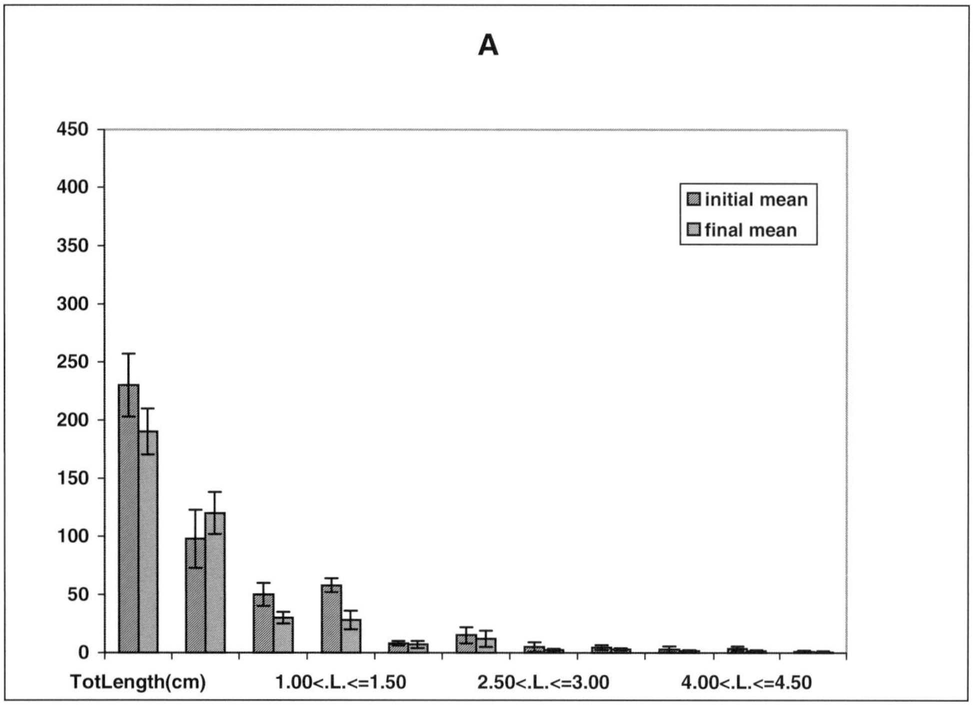

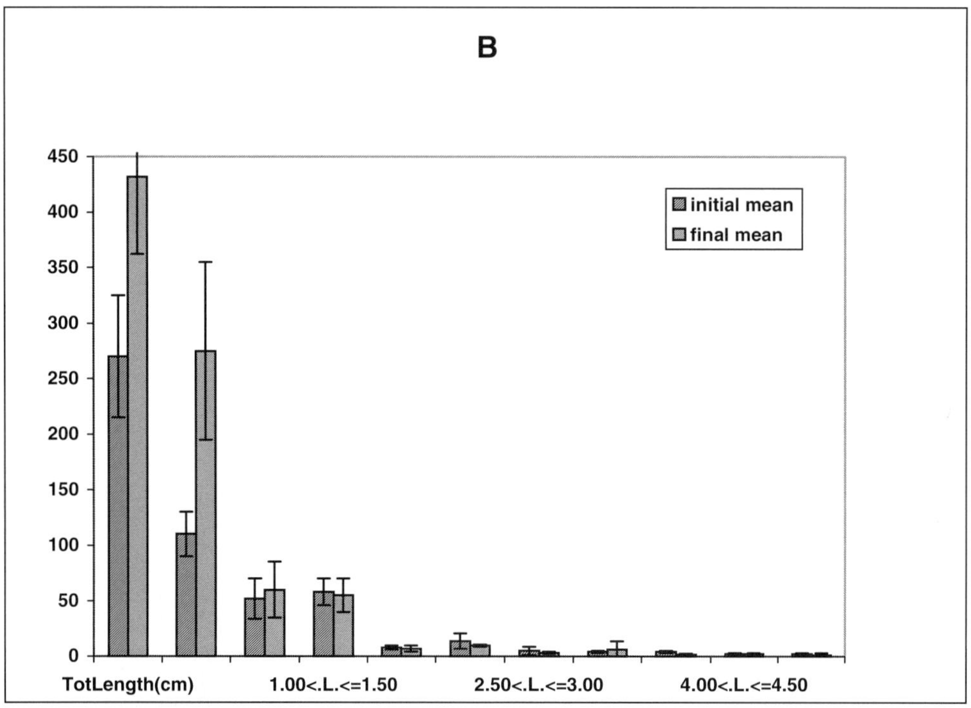

Figure 3. Computerized analysis of root system of pine saplings under different watering regimes. Root systems of saplings under a 10-day (A) or a 7-day (B) interval of watering regime are analyzed. The analysis is done before (initial mean = ▨) and after (final mean = ▨) the water stress treatment. The parameters obtained from the analysis and reported in A and B refer to the total length of root system (Totlength) and to the length of each category formed by roots with the same diameter (0–0.5 mm; 0.50–1.00 mm; 1.00–1.50 mm; 1.50–2.00 mm; 2.00–2.50 mm; 2.50–3.00 mm; 3.00–3.50 mm; 3.50–4.00 mm; 4.00–4.50 mm; 4.50–...mm). The values reported are the mean values of 20 root systems for each treatment. The bars indicate the standard deviations.

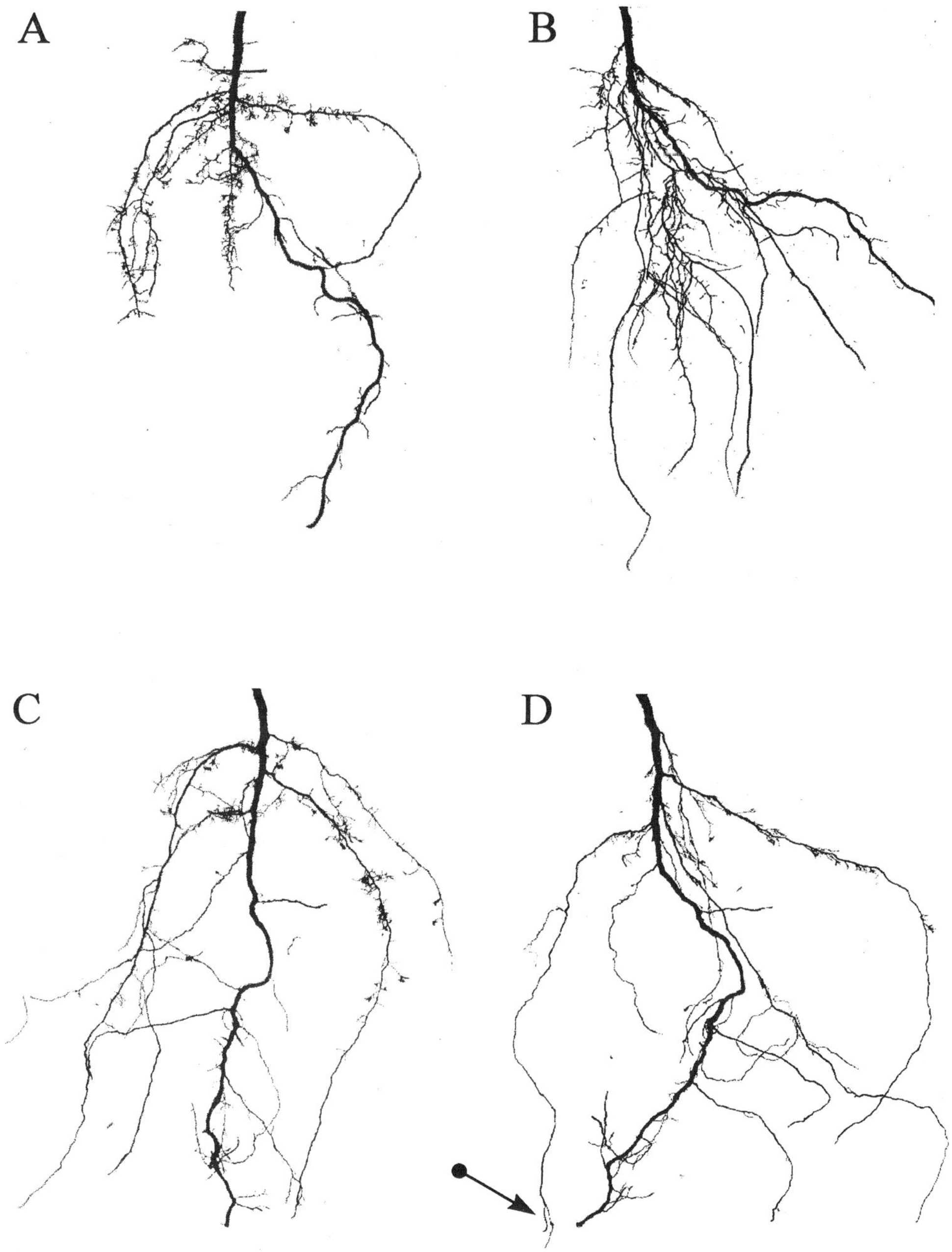

Figure 4. Images of root systems obtained by computerized scanning system before and after water stress treatment. A and C are the images of two different root systems of pine saplings recorded before re-planting them in growth chamber. The image in B represents the root system of the same pine sapling shown in A recorded at the end of the 7-day interval treatment. The image in D represents the root system of the same pine sapling shown in C recorded at the end of the 10-day interval treatment.

taken by a computerized scanner system and stored. Stored images were analyzed by a specific software program in order to obtain many parameters: total length, length of roots divided in categories according to their diameter, number of tips, number of forks and crossings. The comparison of such parameters indicated that saplings at the 7-day watering interval increased the total root length (Figure 2C) mainly for the increase in length of fine roots having a diameter <0.5 mm (Figure 3B). Saplings at the 10-day watering interval showed no increase in root length (Figure 2B), with the exception of a new lateral which was observed in a single sample (arrow in Figure 4D). Furthermore, comparative analysis of all the images indicated that the fine roots present at the beginning of the experiment with a diameter <0.5 mm did not elongate while in the growth chamber (compare images in Figure 4). Saplings under severe stress (10-day watering interval) did not show emission of new laterals (Figure 4C, D). Furthermore, they presented a significant decrease in the total length of the root system (Figure 2B), which was considerable in the categories of roots with a diameter of 0.5–1.5 mm (Figure 3A).

In order to explain the decrease in total length in the root systems as shown by computer analysis of pine saplings treated at the 10-day watering interval, we comparatively examined all the root system images recorded before and after re-planting the saplings in the boxes. The analysis revealed that all pine saplings collected at the end of the experiment had lost their fine ecto-mycorrhizal roots (compare all images in Figure 4). This fact indicated that these roots did not survive the re-planting procedure. Such an event was also confirmed in control saplings, thereby indicating that loss of fine ecto-mycorrhizal roots was independent of the watering regimes and was due to an unknown re-planting effect. The initial presence of these fine ecto-mycorrhizal roots with their numerous and typical short ramifications must have certainly influenced the computerized analysis by giving a higher total length value in the root system. In particular, it was clear that the initial length value in the 0.5 mm category of roots during our computerized analysis should have been considerably influenced. Such findings justified the decrease in total length value of the root system as shown by computerized analysis in some samples at the end of the experiment. The event was disguised in saplings whose root system had started to grow again during the experiment. In fact, in such saplings the decrease in the total length value of the root system or the decrease in length value in the 0.5 mm category of fine roots could have been balanced only by the growth of new laterals.

Microscopic analysis of the root systems showed that saplings showing re-growth in the growth chamber during the treatments produced new laterals, which always emerged from the dark zone of existing roots where cortical cells were blackened by tannins in their cell wall (McKenzie and Peterson, 1995). The aspect of these new emerging laterals was white or yellow. The presence of the white meristematic zone at the tip of an old fine root producing new laterals was never observed in our experiments. Ecto-mycorrhizal infection was never observed in new growing laterals, at least after a superficial microscopic analysis.

Studies were conducted on root meristems of saplings treated by different watering regimes to examine their proliferative activity. In this regard, we tried to calculate the mitotic index of meristematic root tips. The experiments failed because the tissue excised from the old root tips was difficult to squash under the glass slides. The mitotic index could be obtained only with the meristematic tissue excised from root tips of new young laterals or from old roots of control plants.

Effect of water stress on roots of pea seedlings

We chose pea seedlings to investigate the effect of water stress on the root system of a herbaceous plant during the first stage of development. Preliminary experiments had shown that pea seedlings germinated in moist vermiculite for 3 days and immediately arrested primary root elongation when transferred to dry vermiculite. The same type of arrest in primary root elongation was observed when pea seedlings were transferred on filter paper to a glass chamber in the presence of a 20% glycerol solution, thereby generating inside the chamber a low value of relative humidity. In both cases, the water stress reduced the elongation of the primary root and arrested the emission of laterals (not shown). In addition to the arrest of proliferation in meristematic tissue, we also observed mophological alterations in the primary root axis. In particular, we observed a considerable reduction in the primary root axis diameter. When water-stressed seedlings were returned to normal growth conditions, primary root elongation never fully recovered to the same value of controls, although the seedlings immediately showed clear signs of recovery in shoots. In fact, the shoots immediately began to grow again. Furthermore, when we examined the root system during recovery, we often observed the occurrence of a

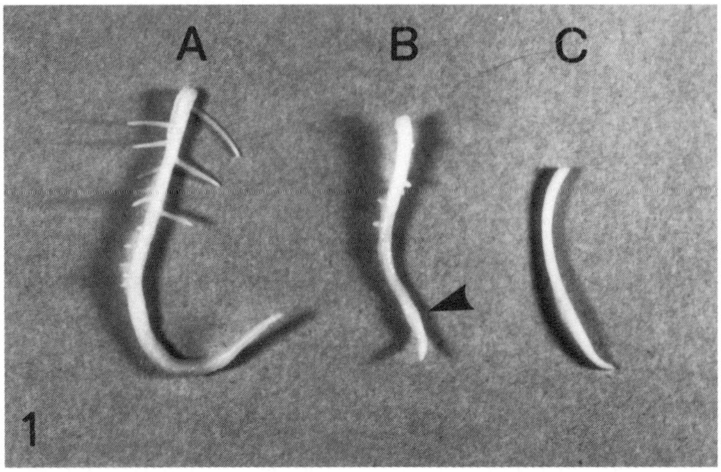

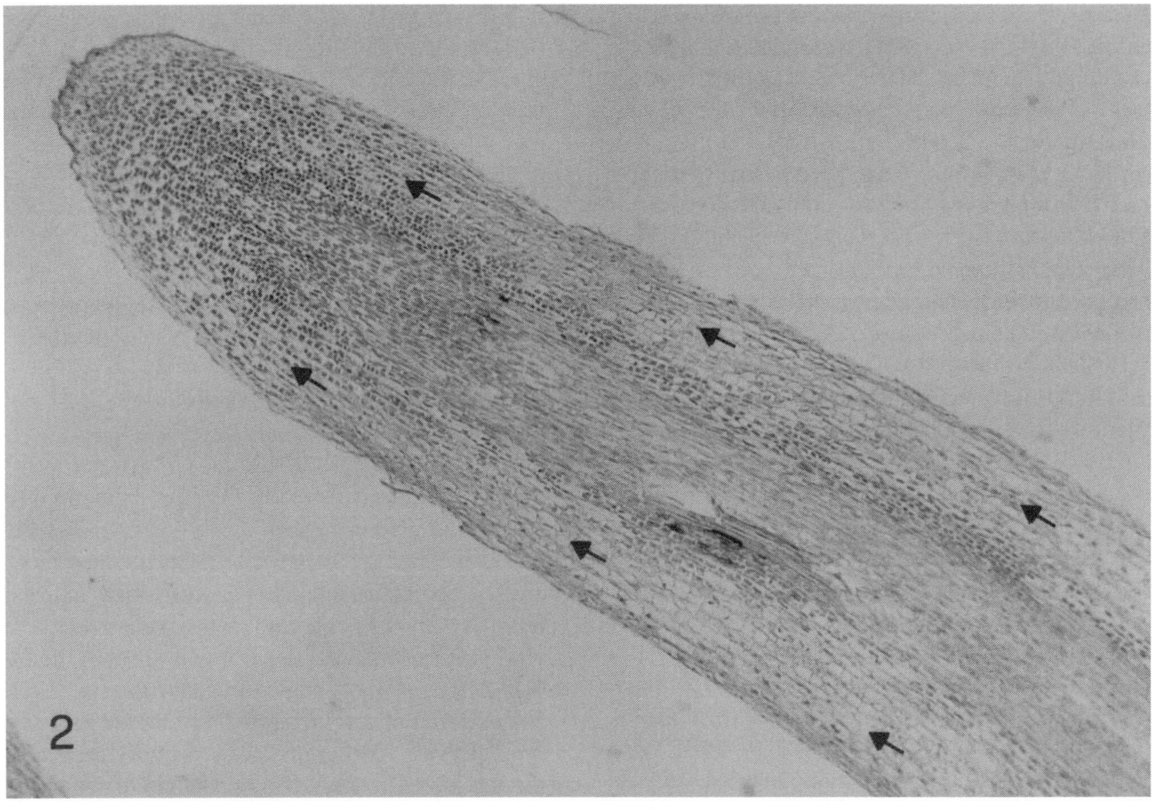

Figure 5. Effect of water stress on the primary root axis of pea seedlings. 1A=control seedling are grown for 3 days in moist vermiculite and then transferred on dry filter paper for 2 days in a glass chamber in presence of water. 1B=pea seedling are grown for 3 days in moist vermiculite, transferred on dry filter paper for 2 days in a glass chamber in presence of a solution containing 20% glycerol solution, and eventually returned for 2 days to the glass chamber in presence of water; 1C=pea seedling are grown for 3 days in moist vermiculite, transferred on filter paper for 4 days in a glass chamber in presence of 20% glycerol solution. 2=Low magnification (63×) of a longitudinal section of the primary root in a control seedling grown as described above (1A).

'swelling' zone above the meristematic root tip (arrow in Figure 5.1B). The swellings zones remained visible along the primary root axis.

These preliminary observations induced us to closely study the dynamics of such events. We decided to use a hydroponic system because it allows a close control of the value of water stress intensity to be applied to the seedlings. More importantly, with hydroponic cultures it is possible to visually follow the events taking place in the root system during the treatments. Repeated measurements of elongations on the same root were possible without damaging the tissues. A PEG concentration giving –0.5 M Pa water potential was effective in reducing root growth (Figure 6B). A PEG concentration giving –1.7 M Pa water potential considerably affected root growth and showed a good survival percentage. The best survival percentage was observed in experiments where the highest PEG concentration (–1.7 M Pa) was achieved slowly (see Materials and methods). At the end of 24 hr treatments with –1.7 M Pa water potential, statistical analysis of primary root length measurements showed a 25% inhibition in comparison with the control (Figure 6A, C). In stressed seedlings, we also found a 40% drop in moisture content (RMC) in the primary roots with respect to the control, which could explain the inhibition of elongation and the arrest of cell divisions. The arrest of cell proliferation in the meristematic tissue was demonstrated by the absence of mitosis after 24 hr of PEG treatment. The fact that PEG affected roots was also confirmed by the increase in proline concentration in meristematic tissue with respect to the control (data not shown). The water stress treatment induced in pea seedlings the formation of swelling zones along the root axis similar to those shown by treatments in the glass chamber (arrows in Figure 6C). When we prolonged the stress for several days, in a few cases we observed the production of new laterals, although the primary root continued to show morphological alterations along its axis.

Roots washed at the end of stress treatments and returned to a PEG-free medium showed a rapid increase in the RMC value. The increase was 80% of the control value after 12 hr. Contemporaneously, root-axis diameter returned to its normal value and proline concentration in the meristematic tissues returned to a normal level (not shown). Elongation in the primary root axis never fully recovered to the control value. Furthermore, the primary root axis of recovering pea seedlings still presented the swelling regions (Figure 6D).

In order to understand the reason for the appearance of swollen regions along the primary root axis during the water stress treatments, anatomical investigations were carried out. Microscopic analysis of 1-μm-thick fixed sections of primary root axes presenting these characteristic swellings (Figure 7A) revealed that at a short distance from the meristematic tip the derivative cells had changed their usual direction of elongation (compare Figure 5.2 with Figure 7A). In fact, the direction of elongation of many derivative cells present in the swollen region (Figure 7B) no longer appeared parallel with the longitudinal axis of the root as in the derivative cells of the control seedlings (Figure 5.2). In particular, some cells even showed an elongation direction clearly perpendicular to the main root axis (arrows in Figure 7B).

The considerable implications of the morphological modifications of the primary root on the overall root system architecture required the exclusion of possible artefacts. In order to exclude artefacts, we treated pea seedlings with sodium butyrate, which is a compound known to be a strong inhibitor of meristematic cell activity (Chiatante et al., 1986). In this case, the treatments consisted of leaving the pea seedlings in Petri dishes on paper soaked with a medium solution containing sodium butyrate. We observed an arrest of primary root elongation similar to that induced by water stress. Furthermore, during the recovery from sodium butyrate treatment, we observed the appearance of swellings along the primary root axis above the meristematic region similar to those observed during the recovery from water stress (Figure 8A, B). Analysis of the direction of elongation in derivative cells after butyrate treatment showed that in swelling zones these cells presented an elongation direction perpendicular to that of the main root axis.

Discussion

It has been shown that damage to tissues may occur in the root tip of woody plants during water stress (McKay, 1992, 1997). Data presented herein indicate that water stress affects pine saplings by arresting the elongation of fine roots with a diameter of 0.5 mm or less. We report also the occurrence of a reduction in length of the white zone of such roots. As the white root zone includes the elongation zone and the meristematic zone (McKenzie and Peterson, 1995), such a reduction may be explained only by the loss of activity in the meristems, followed by differentiation of all the

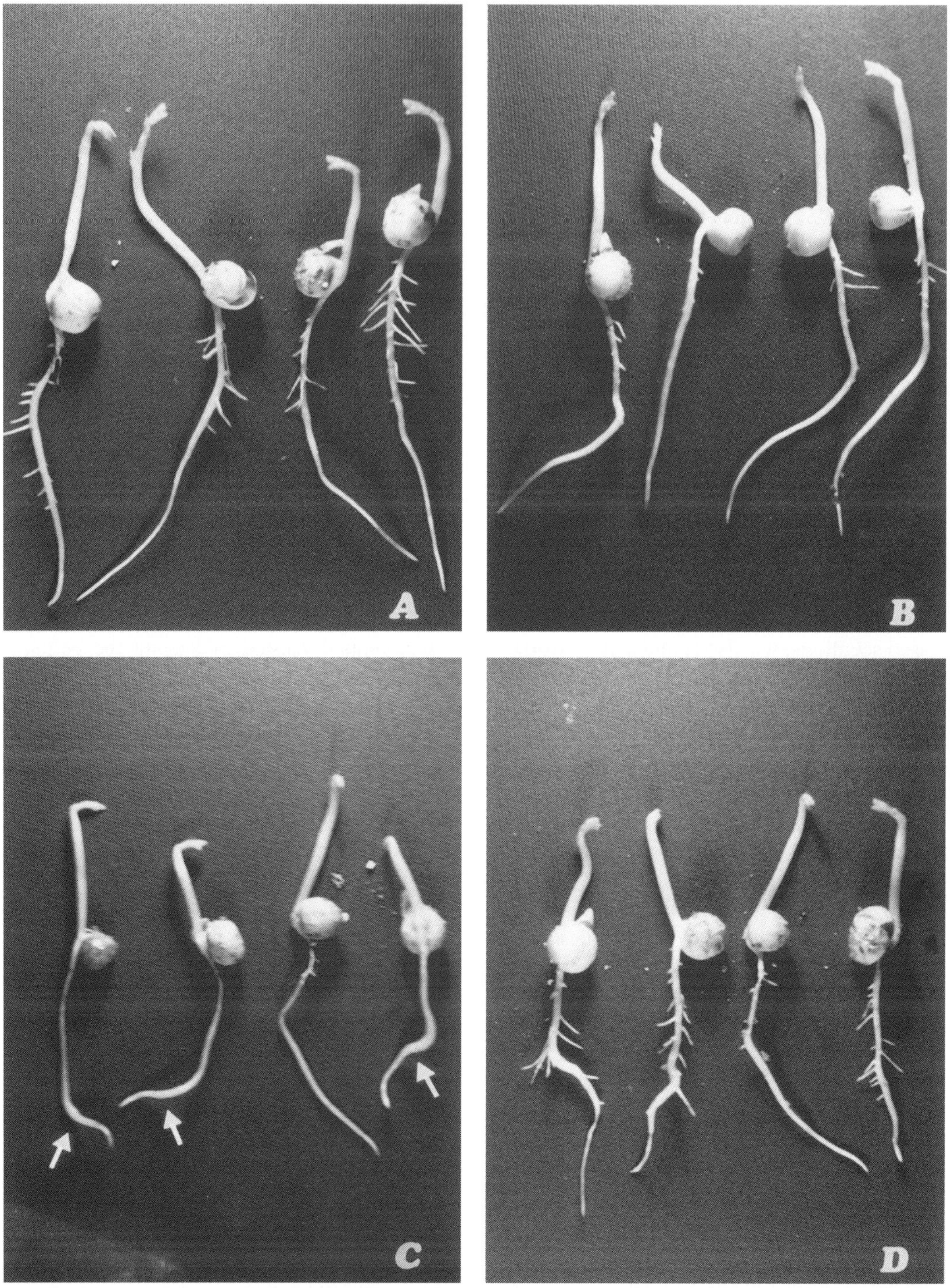

Figure 6. Pea seedlings stressed in hydroponic culture in presence of high PEG concentration. A=control seedlings germinated in vermiculite for 3 days and grown in hydroponic culture for 5 days; B=seedlings germinated in vermiculite for 3 days and grown in hydroponic culture for 5 days in presence of a solution containing PEG at –0.5 M Pa water potential; C=seedlings germinated in vermiculite for 3 days and transferred every 24 h in 5 different solutions containing increasing concentration of PEG (maximum concentration produced –1.7 M Pa water potential); D=seedlings treated as described in C then washed and recovered for 3 days in hydroponic culture containing water.

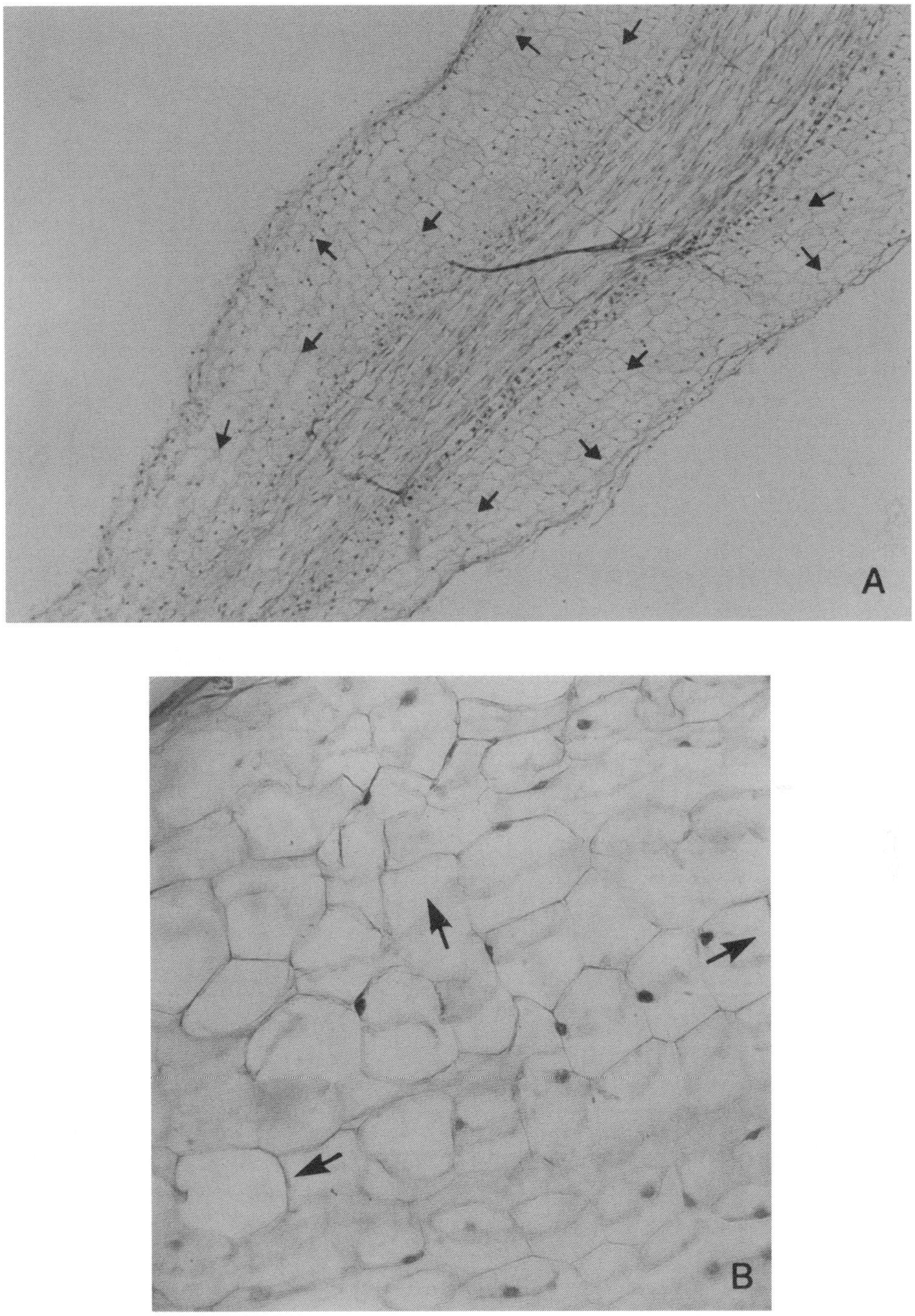

Figure 7. Histological investigation of the swollen zone in the root axis of a pea seedling. Water stress treatment are described in Figure 6D. A=low magnification (250 ×) of the swollen zone including cortical cells with an irregular direction of elongation (compare the arrows). B=high magnification (1000 ×) of the same swollen zone shown in A.

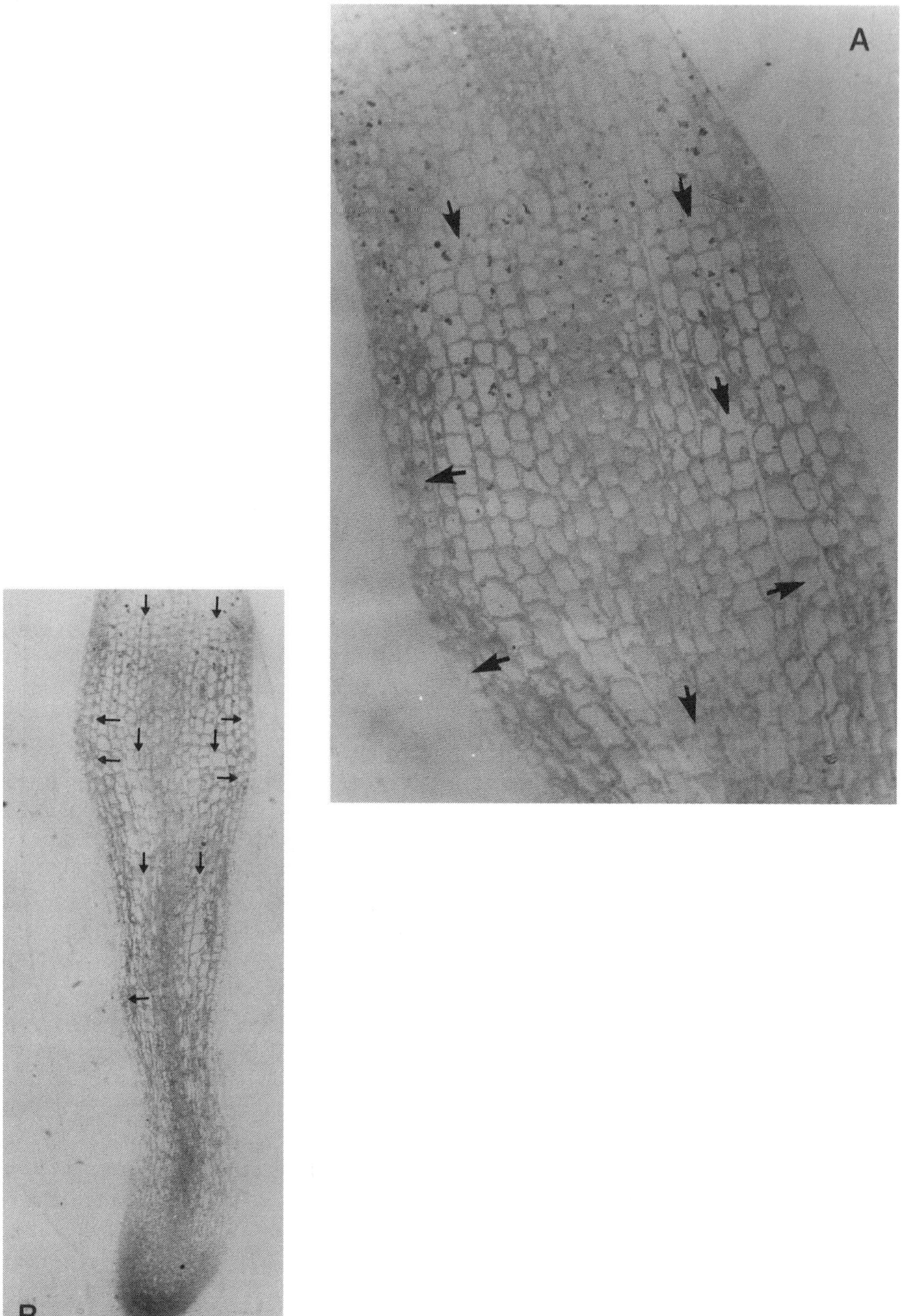

Figure 8. Aspect of the primary root axis after treatment with sodium butyrate. A=high magnification (400×) of the swollen zone where the cortical cells show an irregular direction of elongation (compare the arrows). B=low magnification (63×) showing the overall aspect of the swollen zone present in the cortical region of the primary root axis.

derivative cells. Therefore, our observations suggest that meristems are extremely sensitive to water stress, which would explain why all the fine roots present at the beginning of the water stress treatment did no resume elongation during recovery. At the same time, such observations lead us to conclude that re-growth of pine saplings at the end of water stress is dependent on the emission of new laterals. This was demonstrated by the increase in a number of parameters such as forks, crossings and tips after computerized analysis of root system images taken from saplings before and after water stress.

Another interesting indication emerging from our experiments with woody plants was that fine roots of pine saplings suffer from water stress during handling and transportation from the nursery bed to the open fields or to growth chambers. Also in this case, the indication emerges from a comparative analysis of root systems images taken before and after re-growth. In fact, re-growth depends on the emission of new laterals.

The root system of young herbaceous plants is also affected by water stress. Elongation of the primary root is arrested as a consequence of cell proliferation inhibition in root meristems. The proline content increase in meristematic cells probably is not sufficient to avoid (Jones and Mullet, 1995) the loss of water and the arrest of cell proliferation. For this reason, we confirm that, at least for the effect on root meristems, juvenile woody plants have physiological responses linked to environmental variations similar to those observed in herbaceous plants (Croker et al., 1998).

More difficult to explain is the appearance of a swollen zone above the meristematic zone along the primary root axis of pea seedlings. In this regard, our anatomical and cytological investigations suggested that the swelling may have been due to the combination of two factors: a) the loss of longitudinal orientation of cortical derivatives during elongation; b) the possible death of some meristematic initials and a consequent displacement of cortical derivatives from their usual distribution in parallel lines of cells. The hypothesis appears to be supported by the results of Saab et al. (1993), which indicated that the capacity of primary root cells to maintain elongation at low water potential apparently decreases as cells are displaced away from the apex. In their experiments, maintenance of root cell elongation was associated with the expression of specific polypeptides (Saab et al., 1993). Sharp et al. (1993) reported that an external water potential of about −1.6 M Pa (close to that used in our experiments) inhibits root elongation rate and induces an increase in the radial diameter of cortical cells in the elongation zone during recovery. The authors suggested that inhibition of elongation during water stress changes the quantitative and qualitative accumulation pattern of polypeptides in the elongation region of the root. All these hypotheses point to a different sensitivity to water stress conditions of meristematic cells. In this regard, it is known that quiescent cells in meristematic tissue are generally more resistant to environmental extremes, and such an effect is apparent at the level of root anatomy (Feldman, 1998).

We can reasonably exclude that swellings in our roots may have been induced by a toxic effect of PEG or by a reduction in oxygen content in the PEG solution during hydroponic culturing. In fact, we used a vigorous aeration in hydroponic cultures, and the control seedlings showed no arrest in root elongation or the occurrence of swellings. Swellings were observed during recovery even when pea seedlings were stressed in a glass chamber by growing them on paper and in the presence of low humidity. Furthermore, we observed a swollen region in roots whose proliferation was affected by sodium butyrate, which is known to be an inhibitor of meristematic cells (Chiatante et al., 1986).

The use of hydroponics with pea seedlings enabled us to observe the production of new laterals during prolonged water stress treatments. The production and elongation of new laterals under stress conditions suggest that herbaceous plants may produce new tissues adapted to live under water stress conditions. These new tolerant meristems could represent the response of pea seedlings to adapt to water stress conditions. Unfortunately, we completely ignore the molecular events which make new meristems tolerant to water stress. There is a large body of evidence regarding leaves of herbaceous species being resistant to desiccation, whereas there is scarce information regarding roots. In woody species, the literature indicates that roots are not damaged even after 120 hr of desiccation (McKay, 1997 and references therein), but there are no indications regarding the specific tolerance of root meristematic tissue during such an event. We have not yet used hydroponic cultures with woody species, but we cannot exclude that woody species may produce new meristems similarly to pea seedlings. In this regard, we do not know whether the new root meristems of laterals produced in the root system of pine saplings and pea seedlings during recovery have an increased tolerance to water stress with respect to old meristems. To answer to this question, biochemical studies have been programmed which will study the protein composition of these different meristems.

Acknowledgements

The authors acknowledge MIRAAF (Ministero Italiano per le Risorse Agricole, Alimentari e Forestali) project: 'Biotecnologie Vegetali', MURST (60%), and EC (Fair project no. CT95-0497 'Plant vitality and dormancy') for supporting this research. We thank B. Johnston for editing the manuscript.

References

Bartels D and Nelson E D 1994 Approaches to improve stress tolerance using molecular genetics. Plant Cell Environ. 17, 659–667.

Blackman S A, Obendorf R L and Leopold A C 1995 Desiccation tolerance in developing soybean seed: the role of stress protein. Physiol. Plant. 93, 630–638.

Bohnert H J, Nelson D E and Jensen R G 1995 Adaptation to environmental stresses. Plant Cell 7, 1099–1111.

Chiatante D, Levi M and Sparvoli E 1986 Arrest of the cell cycle induced by dibutyryl-cAMP and sodium butyrate in root meristems of lettuce. Caryologia 39, 143–150.

Crocker J L, Witte T W and Angé R M 1998 Stomatal sensitivity of six temperate, deciduous tree species to non-hydraulic root-to-shoot signalling of partial soil drying. J. Exper. Biol. 49, 761–774.

Cushman J C 1990 Gene expression during adaptation to salt stress. *In* Anatomy; A Key Factor Regulating Plant Tissue Response to Water Stress. Ed. F Katterman. pp 173–204. Academic Press, The Hague, Netherlands.

Feldman L J 1998 Not so quiet quiescent centers. Trends Plant Sci. 3, 80–81.

Jones J T and Mullet J E 1995 A salt- and dehydration-inducible pea gene, Cyp 15a, encodes a cell-wall protein with sequence similarity to cysteine proteases. Plant Mol. Biol. 28, 1055–1065.

Ludlow M M 1989 Strategies in response to water stress. *In* Structural and Functional Responses to Environmental Stresses: Water Shortage. Eds H K Kreeg, H Richter and T M Hinckley. pp 269–281. Academic Press, The Hague, The Netherlands.

Matsuda K and Rayan A 1990 Anatomy: a key factor regulating plant response to water stress. *In* Anatomy; A Key Factor Regulating Plant Tissue Response To Water Stress. Ed. F Katterman. pp 63–72. Academic Press, The Hague, The Netherlands.

McKay H M 1992 Electrolyte leakage from fine roots of conifer seedlings: a rapid index of plant vitality following cold-storage. Can. J. For. Res. 22, 1371–1377.

McKay H M 1997 A review of the effect of stresses between lifting and planting stock quality and performance. New Forests 13, 379–399.

McKenzie B E and Peterson C A 1995 Root browning in *Pine banksiana* Lamb and *Eucaliptus pilularis* Sm. 1. Anatomy and permeability of the white and tannin zones. Bot. Acta 108, 127–137.

Michel B E and Kaufman M R 1973 The osmotic potential of polyethylene glycol 6000. Plant Physiol. 51, 914–916.

Plant Z 1995 Sensitivity of crop plants to water stress at specific developmental stages: reevaluation of experimental findings. Israel J. Plant Sci. 43, 99–111.

Saab I N, Ho D and Sharp R 1993 Role of endogenous ABA in root and shoot growth at low-water potentials: spatial distribution of gene expression and elongation rate. *In* Plant Responses to Cellular Dehydration During Environmental Stress. Eds T J Close and E A Bray. pp 285–286. American Society of Plant Physiologists.

Sharp R E, Voetbory G S, Saab I M and Bernstein N 1993 Role of abscisic acid in the regulation of cell expression in roots at low water potential. *In* Plant Responses to Cellular Dehydration During Environmental Stress. Eds T Close and E A Bray. pp 57–66. American Society of Plant Physiologists.

A. Stokes (ed.), The Supporting Roots of Trees and Woody Plants: Form, Function and Physiology, 259–267.

Influence of soil temperature on root freezing tolerance of Scots pine (*Pinus sylvestris* L.) seedlings

Eva Stattin[1,*] and Anders Lindström[2]
[1]*Swedish University of Agricultural Sciences, Department of Forest Yield Research, S-776 98 Garpenberg, Sweden and* [2]*Dalarna University, S-776 98 Garpenberg, Sweden*

Abstract

The influence of soil temperature on the root freezing tolerance of one-year-old containerized Scots pine (*Pinus sylvestris* L.) seedlings was investigated. In addition, the TTC and electrolyte leakage methods were evaluated in terms of their suitability for use in detecting damage to roots caused by freezing. In mid-August, seedlings were placed in three thermostat-controlled soil beds in a greenhouse with an initial soil temperature of 14.3 °C. Soil temperature was lowered in two of the soil beds, resulting in temperatures of 10.7 and 5.3 °C respectively. Each soil temperature, i.e. 14.3, 10.7 and 5.3 °C was maintained for eight weeks. Starting in early September, damage to roots induced by artificial freezing was estimated biweekly by measuring electrolyte leakage, triphenyl tetrazolium chloride (TTC) reduction and potential root growth in a three-week cultivation test. In addition, the root freezing tolerance of seedlings placed outdoors was tested. Measurements showed that these seedlings were exposed to soil temperatures ranging from 13.0 °C in mid-August to 0.5 °C in November. Generally, the development of root freezing tolerance was more pronounced for seedlings exposed to lower (0.5 and 5.3 °C) soil temperatures compared with those exposed to higher (10.7 and 14.3 °C) ones. Root freezing tolerance was highest among the seedlings placed outdoors which were also exposed to the lowest soil temperatures registered in the study. To examine the effect of a temporary warm period, the soil temperature in one treatment was increased from 5.4 °C to 13.9 °C, maintained at the latter temperature for two weeks in October and then lowered to 5.7 °C. Root freezing tolerance was reduced by exposure to the warmer soil temperature. However, after four weeks at the colder soil temperature, the tolerance of the seedlings had returned to the level measured prior to exposure to the warm soil temperature. Methods based on the measurement of root electrolyte leakage and TTC reduction were both found to have limitations when used to detect root freezing damages in containerized seedlings.

Introduction

In autumn, before seedlings are ready for frozen storage, containerized seedlings placed outdoors may be exposed to root-zone temperatures cold enough to injure their roots (Desjardins and Chong, 1980; Lindström, 1986; Tinus, 1982). Despite this risk, seedlings must remain outdoors until they are hardy enough to be placed in frozen storage, as indicated by shoot (Colombo, 1990) and root (Lindström and Nyström, 1987; Lindström and Stattin, 1994) freezing tolerance tests. The development of root freezing tolerance is likely influenced by both the thermoperiod and photoperiod (Johnson and Havis, 1977). In Scots pine (*Pinus sylvestris* L.) (Lindström, 1995; Smit-Spinks et al., 1985) and several other species (Bigras and D'Aoust, 1992, 1993; Bigras et al., 1989) root freezing tolerance appears to be influenced mainly by the thermoperiod. It would be of practical value to minimize the period during which seedlings are exposed to unpredictable outdoor temperature conditions. One way to achieve this would be to artificially hasten the development of root freezing tolerance in autumn. Such a treatment would reduce the risk of root freezing injury outdoors as well as indoors during frozen storage.

Root freezing tolerance can be assessed by subjecting seedlings to freezing temperatures and then measuring their vitality. Plant vitality can be measured as

* FAX No: +46 22 526 100. E-mail: evas@sluger.slu.se
Plant and Soil is the original source of publication of this article. It is recommended that this article is cited as: *Plant and Soil* **217**: 173–181, 1999.

root growth capacity (RGC) (Stone, 1955), triphenyl tetrazolium chloride (TTC) reduction (Steponkus and Lanphear, 1967), the proportion of decaying roots (Havis, 1976), leakage of ninhydrin-reactive compounds (Siminovitch et al., 1962), and electrolyte leakage (Dexter et al., 1930). If freezing tests are to be used to measure storability (Colombo, 1990) the methods used to evaluate seedling vitality after freezing must be rapid and reliable.

The aims of this study were (i) to determine how various soil temperature regimes during autumn influence the development of root freezing tolerance in one-year-old containerized Scots pine seedlings and (ii) to evaluate two methods for detecting lethal root damage caused by low temperature exposure i.e. the measurement of root electrolyte leakage and TTC reduction.

Materials and methods

Plant material

In 1992, Scots pine seedlings (provenance 61°59′ N; 13°49′ E; 540 m) were grown in 50-mL multipot containers (Hiko V50, Sweden, 881 seedlings m^{-2}) filled with peat substrate (Närkes miljöproduktion AB, Sweden) mixed with perlite (11–15%) and 1 kg m^{-3} of dolomitic lime at a commercial forest tree nursery in central Sweden (60°49′ N; 16°38′ E). The seedlings were grown in a greenhouse for the first seven weeks, whereupon they were moved outdoors and placed on pallets elevated above the ground. On August 17, the seedlings were transported to the Swedish University of Agricultural Sciences at Garpenberg (60°15′ N; 16°15′ E) and placed outdoors. From September 17 onwards, the seedlings were kept in a thermostatically controlled bed that did not allow the soil temperature to fall below −4 °C. Fertilizer was supplied from early April, when seedlings were three weeks old, until early October. The total nitrogen supply varied between 40 and 55 g N m^{-2} depending on when the seedlings were tested.

Soil temperature treatments

On August 19, seedlings were placed in a greenhouse room in three soil beds in which the container-soil temperature was thermostatically controlled. Soil in the beds was kept cool by passing coolant (65% ethanol) through 1.25 cm diam. polyvinyl chloride (PVC) tubing that had been laid out along the bottom of the beds at 10-cm intervals. A 2-cm-thick layer of sand covered the tubing. Soil temperature was regulated manually by adjusting the thermostats (Danfoss, RT 13, Denmark).

Four different soil temperature treatments and a control were included (Table 1). In three of them the soil temperature was kept constant for eight weeks (treatment B, C, and D), and the fourth simulated a temporary warm period in late autumn (treatment E). The control consisted of seedlings exposed to ambient temperatures outdoors (treatment A).

In all treatments, calibrated copper-constantan thermocouples attached to an automatic datalogger (Campbell CR10WP, UK) were used to measure container-soil temperatures at half depth and the air temperature at plant level. The accuracy of the measurements was ±0.2 °C. Temperatures were recorded each minute and stored as 60-min mean values from August 19 until November 24 (Table 1). The initial soil temperature was 14.3 °C in treatments B, C and D and was kept at this temperature for eight weeks in treatment B. On September 2, the soil temperature was reduced by approximately 5 °C in treatment C and D and was kept at this temperature (10.7 °C) for eight weeks in treatments C. On September 16, the soil temperature was reduced again in treatment D and maintained at this temperature (5.3 °C) for eight weeks. The soil temperature in treatment E was increased from 5.4 °C (E1) to 13.9 °C in October (E2) and kept at this temperature two weeks, whereupon it was lowered to 5.7 °C (E3). The day/night air temperature was 20/15 °C from August 19 until September 2, when it was lowered to 15/15 °C, and it was lowered again on October 28 to 10/10 °C. All seedlings were exposed to the natural photoperiod.

Artificial root freezing and evaluation of root damage

Starting on September 2, seedlings were evaluated biweekly for root freezing tolerance. Randomly selected seedlings were transferred into bottom-sealed test tubes, and the root parts of the seedlings were immersed in an ethanol-water freezing bath. Environmental conditions and techniques used during the freezing tests were similar to those described by Lindström and Stattin (1994). Test temperatures ranged from −5 to −25 °C at 5 °C intervals and 18 seedlings from each treatment were exposed to each test temperature. Minimum temperature was selected at each sampling date with the purpose of killing all roots (Table 1). In addition, 18 control seedlings per treat-

Table 1. Mean soil temperatures between sampling dates for treatment A-E. Roots were exposed to freezing temperatures (5 °C intervals from −5 °C down to stated minimum temperature) and freezing damage was estimated for treatments on indicated (•) sampling dates

Treatment	Mean soil temperature (°C) for indicated periods Aug. 19		Sep. 2		Sep. 16		Sep. 30		Oct. 14		Oct. 28		Nov. 11		Nov. 25
A		13.0	•	9.6	•	10.1	•	5.4	•	2.8	•	0.5	•	0.6	•
B		14.3	•	14.3	•	14.3	•	14.3	•						
C		14.3		10.7	•	10.7	•	10.7	•	10.7	•				
D		14.3		11.6		5.3	•	5.3	•	5.3	•	5.3	•		
E		14.3		11.6		5.4		5.4		13.9	•	5.7	•	5.7	•
						E1		E1		E2		E3		E3	
Minimum temp. at artificial freezing, °C			−10		−15		−20		−20		−25		−25		−25

ment were kept in darkness at 2 °C. Damage to roots after freezing was evaluated by assessing survival in cultivation tests, and by measuring the reduction of TTC to formazan and root electrolyte leakage.

For each treatment – test temperature combination the cultivation tests consisted of three replicates, with three seedlings per replicate. After thawing, the seedlings in a replicate were planted in a mixture of peat and sand (50% v/v) according to a completely randomized design and cultured for three weeks in a growth chamber. The cultivation tests were performed according to the method described by Mattsson (1986). At the end of the cultivation test, seedlings were classified as either *dead* (seedlings with no new roots) or *living* (seedlings producing new roots), and LT_{50}-values (the lethal root freezing temperature that killed 50% of the seedlings) were estimated by linear interpolation.

Measurements of TTC reduction to formazan and root electrolyte leakage were made on another three replicates, with three seedlings per replicate. In the TTC test, in which the level of TTC reduction to formazan indicates tissue vitality (Steponkus and Lanphear, 1967), root samples representing the whole root system (both fine and woody roots, (approx. 70 mg/seedling) were analyzed. The roots were put in test tubes, and 6 ml of 0.6% (w/v) TTC in 0.06 *M* Na_2HPO_4-KH_2PO_4 and 0.05% (v/v) detergent (Tween 20, Kebo AB, Sweden) were added to the tubes. No vacuum was needed for infiltration. Samples were incubated at 30 °C for 20 h. Formazan was extracted from the samples in 96% (v/v) ethanol in an 80 °C water bath. The absorbances were recorded at 520 nm and expressed as percentage of the value obtained with the unfrozen control. TTC data from November 11 were excluded owing to improper handling.

In connection with the root electrolyte leakage measurements it became obvious on the first sampling date (September 2) that electrolytes were being lost from freeze-injured roots while they were being rinsed under running water; thus electrolyte leakage results from this date were excluded. From the second sampling date onwards the preparation of roots was standardized as follows: The peat was carefully removed from the roots by rinsing for one minute under running, cold, 5 °C tap water. Subsequently any remaining weed roots and peat residues, etc., were removed by hand. Roots were pooled, as in the TTC measurements, with three replicates and three seedlings per replicate. Roots from one replicate were dip-rinsed five times in 1 L of fresh tap water. Finally, the roots were dip-rinsed in three beakers containing 0.25 L de-ionized water. Tap water and de-ionized water were replaced after processing three replicates. A 100-mg root sample was cut off from the distal quarter below the center of each root system. Neither the main root nor roots greater than 2 mm in diameter were included in the sample. Root samples from one replicate were put in a glass bottle using forceps, whereupon the bottle was filled with 18 mL de-ionized water and sealed. The samples were thoroughly shaken and then incubated for 24 h in darkness at a temperature of 16 °C. The samples were then taken into room temperature and thoroughly shaken, whereupon their electrical conductivity (EC) was measured (= EC_{frozen}). EC measurements were made using a conductivity meter (model CDM 83 Conductivity Meter, Radiometer,

Table 2. Root freezing temperature, °C, resulting in the death of 50% (LT_{50}-value) of outdoor-stored (A) Scots pine seedlings and seedlings exposed indoors to different soil temperatures (B–E) (Table 1). Survival was assessed in a 3-week cultivation test, and LT_{50}-values were obtained by linear interpolation. $N = 9$

Treatment	Soil temp. (°C)[1]	Soil temp. kept as stated between		Lethal (LT_{50}) root freezing temp., °C, at sampling date						
		start	end	Sep. 2	Sep. 16	Sep. 30	Oct. 14	Oct. 28	Nov. 11	Nov. 25
A	Outdoors			−7.2	−7.5	−10.5	−17.1	−17.5	< −25	< −25
B	14.3	Aug. 19	Oct. 14	−7.5	−8.7	−7.9	−12.5			
C	10.7	Sep. 2	Oct. 28		−11.8	−9.4	−13.2	−12.5		
D	5.3	Sep. 16	Nov. 11			−9.5	−16.3	−16.9	−19.2	
E1	5.4	Sep. 16	Oct. 14							
E2	13.9	Oct. 14	Oct. 28					−12.1		
E3	5.7	Oct. 28	Nov. 25						−13.7	−17.5

[1]Mean soil temperature for the indicated period. For treatment A see Table 1.

Denmark) with a CDC 304T measuring probe with built-in temperature compensation. The samples were then autoclaved for 10 min at 1.2 bars pressure and allowed to cool to room temperature. After shaking, the conductivity was remeasured (= EC_{boiled}). Relative root electrolyte leakage was calculated as $EC_{frozen}/EC_{boiled} * 100$.

Statistical analysis

Differences in seedling survival after root freezing between treatment A and treatments B-E were analyzed statistically using χ^2 test (Zar, 1974). Survival data were pooled for all freezing temperatures (−5 to −25 °C at 5 °C intervals) by sampling date to increase the number of observations per cell. All χ^2 tests were made as pairwise comparisons with 18–45 seedlings, depending on the sampling date, in each pooled treatment (df = 1).

Seedling survival in the cultivation tests after the artificial freezing was examined for intervals of EC_{boiled}-values, TTC-values in% of control seedlings, and TTC absorbance at 520 nm. Differences in survival between the interval closest to 50% survival and other intervals were tested using the χ^2 test. These tests were based on pairwise comparisons (df = 1).

Results

Influence of soil temperature in autumn on root freezing tolerance

Root freezing tolerance increased successively during autumn. For treatment A (outdoors) LT_{50}-values, based on survival in a three-week cultivation test after artificial root freezing, fell from −7.2 °C in early September to below −25 °C in mid-November (Table 2). A linear regression displayed high correlation between the LT_{50}-values of seedlings stored outdoors and length of exposure to soil temperature treatment ($r^2 = 0.93$, df = 6, $p = 0.0003$). A high correlation was also obtained when the LT_{50} was regressed on mean soil temperature during the fortnight before artificial freezing ($r^2 = 0.91$, df = 6, $p = 0.0005$).

For treatment D (5.3 °C), LT_{50}-values fell from −7.5 °C in early September to −19.2 °C in mid-November (Table 2). It was only on the last sampling date (November 11) that these seedlings showed significantly lower survival after root freezing compared with seedlings stored outdoors (Table 3). By that date, seedlings kept outdoors had been exposed to a colder mean soil temperature over the fortnight before artificial freezing (0.5 °C, Table 1).

In late October, the temporary increase in soil temperature from 5.4 °C to 13.9 °C (treatment E2) resulted in significantly lower survival after root freezing compared with seedlings exposed to outdoor temperature conditions (Table 3). Later, after the soil temperature had been lowered and kept at 5.7 °C for two weeks (November 11) survival was still lower compared with that of seedlings placed outdoors and seedlings exposed to the 5.3 °C soil temperature treatment ($p = 0.090$). After four weeks at 5.7 °C, the roots had reached the level of freezing tolerance measured prior to the onset of the 14-day warm period.

For treatments B (14.3 °C) and C (10.7 °C) there were only a slight increase in root freezing tolerance

Table 3. Survival,%, of Scots pine seedlings in a three-week cultivation test following artificial root freezing. Data for all freezing temperatures were pooled (Table 1) for each sampling date. Survival of seedlings from the different indoor soil temperature treatments (B-E) was compared with the survival of seedlings stored outdoors (A) using χ^2 test, p-values within parentheses. N = 18–45, df = 1

Treatment	Soil temp. (°C)[1]	Soil temp. kept as stated between		Survival, %, at sampling date						
		start	end	Sep. 2	Sep. 16	Sep. 30	Oct. 14	Oct. 28	Nov. 11	Nov. 25
A	Out-doors			44.4	33.3	38.9	72.2	60.0	97.8	88.9
B	14.3	Aug. 19	Oct. 14	50.0 (0.739)	44.4 (0.402)	30.6 (0.458)	50.0 (0.053)			
C	10.7	Sep. 2	Oct. 28		59.3 (0.056)	33.3 (0.624)	55.6 (0.141)	40.0 (0.058)		
D	5.3	Sep. 16	Nov. 11			36.1 (0.808)	66.6 (0.609)	57.8 (0.830)	64.4 (<0.001)	
E1	5.4	Sep. 16	Oct. 14							
E2	13.9	Oct. 14	Oct. 28					37.8 (0.035)		
E3	5.7	Oct. 28	Nov. 25						46.7 (<0.001)	60.0 (0.002)

[1]Mean soil temperature for the indicated period. For treatment A see Table 1.

during the 8-week period of exposure (Tables 2 and 3). In a linear regression, including treatments B-E, length of exposure to soil temperature treatment and mean soil temperature the fortnight before artificial freezing explained about 75% of the LT_{50} variation (df = 14, $p = 0.0002$). Of these two factors, length of exposure to soil temperature treatment was best correlated with the LT_{50}-values ($r^2 = 0.65$).

Evaluation of root freezing damage by using methods based on electrolyte leakage and TTC

The relative root electrolyte leakage increased as roots were exposed to lower freezing temperatures (Table 4). This result was consistent for all sampling dates and treatments (data not shown). The increase in relative root electrolyte leakage can be attributed to the fact that as freezing temperature decreased EC_{boiled}-values decreased as well, whereas EC_{frozen}-values were unaffected (Figure 1). EC_{boiled}-values over 400 μS cm^{-1} g^{-1} root fresh weight corresponded with high survival in the cultivation test. Only 3.9% of the seedlings with EC_{boiled}-values over 400 μS cm^{-1} g^{-1} died in the cultivation tests, whereas mortality was 99.1% for seedlings with EC_{boiled}-values $\leq$200 μS cm^{-1} g^{-1} (Table 5). For intermediate EC_{boiled}-values (201 to 400 μS cm^{-1} g^{-1}) mortality was 50% (Table 5).

TTC-values over 60% of unfrozen control values were associated with high survival in the cultivation test (Table 5); i.e. 93.7% of the seedlings with TTC-values above this level survived in the cultivation tests. Of the seedlings with TTC-values less than 60% of the unfrozen control, mortality was 76.7%.

Absorbance at 520 nm over 1.5 corresponded with high survival in the cultivation test (Table 5). Of 333 seedlings with absorbance values exceeding 1.5, only 3.3% were classified as dead, whereas mortality was 61.8% for seedlings with absorbance values less than 1.5.

Discussion

Influence of soil temperature on root freezing tolerance

Root freezing tolerance has been shown to vary with species (Lindström and Nyström, 1987; McKay, 1994), provenance (Lindström and Nyström, 1987; McKay, 1994), seedling age (Lindström and Nyström, 1987; Coursolle et al., 1998), photo- period (Colombo et al., 1995) and thermoperiod (Smit-Spinks et al., 1985). Our study was restricted to a single provenance, and the soil temperature treatments were not replicated in different growth chambers. Thus, these

Table 4. Relative root electrolyte leakage (%) for unfrozen pine seedlings and pine seedlings frozen at −5, −10, −15 and −20 °C on September 30 1992. $N = 3$

Treatment	Soil temp. (°C)[1]	Soil temp. kept as stated between		Relative root electrolyte leakage, %, after exposure to				
		start	end	+2 °C	−5 °C	−10 °C	−15 °C	−20 °C
A	Out-doors			16.6	19.8	29.2	39.5	43.7
B	14.3	Aug. 19	Oct. 14	17.4	23.2	35.4	44.9	48.1
C	10.7	Sep. 2	Oct. 28	21.3	24.7	31.0	44.9	46.0
D	5.3	Sep. 16	Nov. 11	23.7	24.0	23.5	30.1	61.7

[1]Mean soil temperature for the indicated period. For treatment A see Table 1.

restrictions need to be considered when interpreting the results since they might have been different if other provenances, seedling ages or thermoperiods had been used.

In this study, the root freezing tolerance of the Scots pine seedlings placed outdoors gradually increased during autumn. This pattern is in accordance with earlier findings regarding *Pinus sylvestris* (Lindström and Nyström, 1987; Smit-Spinks et al., 1985) as well as for other woody species (cf. Johnson and Havis, 1977; Mityga and Lanphear, 1971; Pellett and White, 1969).

Differences in root freezing tolerance were small between treatment A and D at occasions when outdoor soil temperature were similar to the soil temperature in the 5.3 °C soil bed (treatment D). However, seedlings placed outdoors (treatment A) rapidly increased their root freezing tolerance in early November, after exposure to 0.5 °C soil temperature during two weeks. During this period LT_{50}-values for seedlings stored outdoors decreased from −17.5 °C to below −25 °C, whereas the decrease was less pronounced for seedlings in the 5.3 °C soil bed (from −16.9 °C to −19.2 °C). Root freezing tolerance was greater at low soil temperatures (treatment A and D) than at higher ones (treatment B and C). These findings are supported by the results of Lindström and Stattin (1994) who found that exposure of spruce seedlings to subzero temperatures (−4.9 to −4.1 °C) resulted in higher root freezing tolerance during late autumn compared with seedlings exposed to temperatures around and above zero (−0.6 to 4.7 °C). Thus, it can be concluded that low temperature exposure is essential for the development of maximum root freezing tolerance. The strong influence of temperature on root freezing tolerance was also demonstrated in treatment E2 where root freezing tolerance was reduced as the mean soil temperature increased from 5.3 °C to 13.9 °C. Our results provide additional evidence of the important role that soil temperature plays in the development of root freezing tolerance and are in accordance with earlier studies regarding *Pinus sylvestris* (Smit-Spinks et al., 1985), *Picea glauca* (Bigras and D'Aoust, 1993), *Juniperus chinensis* (Bigras et al., 1989), and *Taxus cuspidata* (Mityga and Lanphear, 1971).

Seedlings that had been exposed to 13.9 °C for two weeks (treatment E) did not regain their initial level of freezing tolerance until they had been exposed to a low (5.7 °C) soil temperature for four weeks. To our knowledge this is the first study showing that a long period of cold exposure is needed in order for seedlings to redevelop root freezing tolerance once it has been reduced by warm temperatures. Several studies have shown that the freezing tolerance of conifer roots (Smit-Spinks et al., 1985; Zhao et al., 1995) and shoots (Aronsson, 1975; Bigras and D'Aoust, 1993) gradually increases over a period of several weeks in autumn. Furthermore, Lindström and Stattin (1994) found that root freezing tolerance continued to slowly develop in Scots pine and Norway spruce seedlings kept in cold (2 °C) or frozen (−4 °C) storage. The importance of the duration of cold-temperature exposure on the development of root freezing tolerance is further demonstrated by the results obtained with seedlings exposed to a soil temperature of 5.3 °C (D); i.e. root freezing tolerance successively increased during the course of the eight-week treatment. These results indicate that it is a slow process to develop adequate root freezing tolerance. Moreover, at higher soil temperatures (10.7 and 14.3 °C) the increase in root freezing tolerance during the eight-week treatments was negligible, implying that there is an interaction between duration of exposure and temperature.

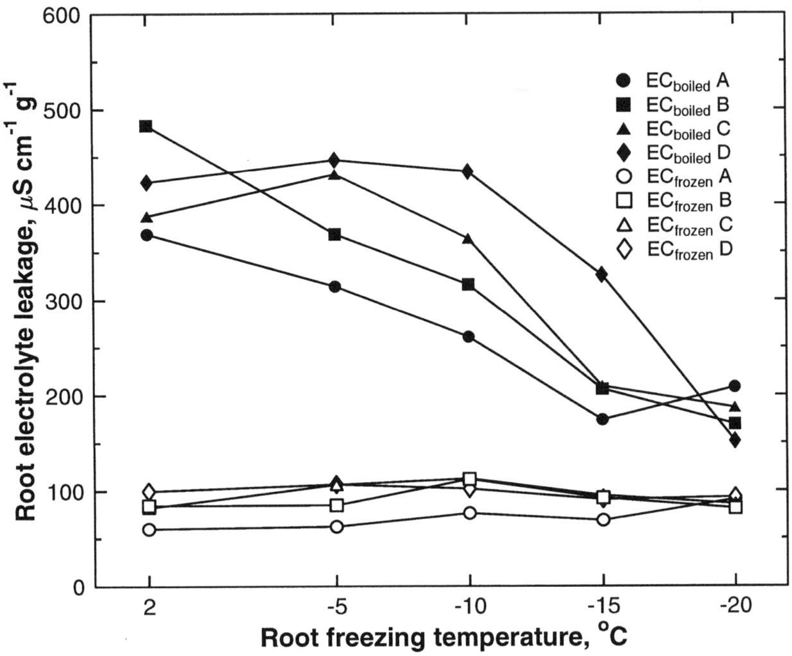

Figure 1. Electrolyte leakage from roots of Scots pine seedlings after freezing (EC_{frozen}) and boiling (EC_{boiled}). Samples are from treatments A–D (Table 1). Roots were frozen at −5, −10, −15 and −20 °C on September 30. $N = 3$.

Root electrolyte leakage as an indicator of freezing damage

A decrease in subzero temperature resulted in an increase in relative root electrolyte leakage, as reported earlier for *Picea mariana* (Bigras and Calmé, 1994). It is expected when using relative root electrolyte leakage, as an estimate of freezing damage, that the electrolyte leakage increases with increasing damage as shown by Dexter et al. (1932). However, in our study the leakage values after autoclaving (EC_{boiled}) decreased as roots had been exposed to lower temperatures whereas leakage values after freezing (EC_{frozen}) were not influenced by freezing temperature. This means that electrolytes were lost before or during the preparation of freeze injured roots. Our results support Wilner's (1959) hypothesis that electrolytes may diffuse into the soil from injured root tissue and then be washed away while preparing the roots for electrolytic tests. Under practical conditions the EC_{boiled}-value could be used as an indicator of seedling root damage. However, the correlation between survival after root freezing and electrolyte leakage after autoclaving must be further investigated before a standard method can be developed because (i) electrolyte levels may vary over the year (Dexter et al., 1932; Lapins, 1961) and (ii) we were not able to correctly distinguish between living and dead seedlings in cases where EC_{boiled}-values were intermediate.

TTC as an indicator of freezing damage

Steponkus and Lanphear (1967) defined the killing point for *Hedera helix* stem and leaves as the temperature at which the TTC reduction to formazan were 50% of the amount of TTC reduction of undamaged control seedlings. This level was later confirmed for roots regarding *Taxus cuspidata* (Mityga and Lanphear, 1971), *Pinus sylvestris* (Lindström and Nyström, 1987), and *Picea abies* (Lindström and Mattsson, 1989). If the killing point is defined as 50% survival, our results agree with these earlier findings. In practice, for instance, in nursery management, a stricter limit might be more appropriate. In this study TTC-values over 60% corresponded with high survival (>89%) in the cultivation test. If the method is to be used on a routine basis in nurseries, a group of undamaged seedlings would be needed to obtain a reference value.

Earlier studies have shown that the amount of formazan produced is correlated with the severity of freezing injury in shoots of *Prunus persica* (Yadava et al., 1978) and roots of *Picea abies* (Lindström and

Table 5. Survival of unfrozen and artificially frozen Scots pine seedlings in a 3-week cultivation test. Survival was assessed for intervals of EC_{boiled}, TTC% of unfrozen control, and TTC absorbance. Artificial freezing of roots was performed at two-week intervals between September 2 and November 25. Freezing temperatures ranged between −5 and −25 °C at 5 °C intervals (Table 1). Differences in survival were assessed using χ^2 tests. In these pairwise comparisons (df = 1), the interval within which survival was closest to 50% was compared with each of the other intervals

Evaluation method	Number of seedlings			χ^2 test, p-value
	Living	Dead	Total	
EC_{boiled}, $\mu S\ cm^{-1}\ g^{-1}$ root fresh weight				
0–200	1	107	108	<0.0001
201–400	231	228	459	–
401–600	362	15	377	<0.0001
601–800	9	0	9	0.0032
TTC,% of unfrozen control				
20.0–39.9	29	196	225	<0.0001
40.0–59.9	65	114	179	–
60.0–79.9	96	12	108	<0.0001
80.0–99.9	78	3	81	<0.0001
100.0–119.9	52	2	54	<0.0001
120.0–139.9	9	0	9	0.0001
140.0–159.9	18	0	18	<0.0001
TTC, absorbance at 520 mm				
0–0.749	27	180	207	<0.0001
0.750–1.499	169	137	306	–
1.500–2.249	225	9	234	<0.0001
2.250–2.999	88	2	90	<0.0001
3.000–3.749	9	0	9	0.0076

Mattsson, 1989) and *Pinus sylvestris* (Lassheikki et al., 1991). Although our results support these findings, formazan production levels in our study differed substantially from those cited in the previous reports. This discrepancy may reflect differences in absorption between (i) species, (ii) injury levels (iii) type of tissues analyzed or (iv) sampling dates. Further research aimed at determining the effects of various factors on formazan formation will have to be carried out before it will be possible to use the amount of formazan produced as an indicator of freezing damage in practice.

Acknowledgements

This study was supported by grants from the Swedish Council for Forestry and Agricultural Research (SJFR) and the Swedish Forestry Research Foundation (SSFf). We thank Dr W. Livingston (University of Maine), Dr I. Ekberg (Swedish University of Agricultural Sciences), Dr A. Mattsson, Claes Hellqvist and Prof A. Persson (Dalarna University) for their helpful comments during the preparation of this manuscript, and the staff at the Department of Forest Yield Research, Garpenberg, for their cooperation.

References

Aronsson A 1975 Influence of photo- and thermoperiod on the initial stages of frost hardening and dehardening of phytotron-grown seedlings of Scots pine (*Pinus sylvestris* L.) and Norway spruce (*Picea abies* (L.) Karst.). Stud. For. Suec. 128. 20 p.

Bigras F J and Calmé S 1994 Viability tests for estimating root cold tolerance of black spruce seedlings. Can. J. For. Res. 24, 1039–1048.

Bigras F J and D'Aoust A L 1992 Hardening and dehardening of shoots and roots of containerized black spruce and white spruce seedlings under short and long days. Can. J. For. Res. 22, 388–396.

Bigras F J and D'Aoust A L 1993 Influence of photoperiod on shoot and root frost tolerance and bud phenology of white spruce seedlings (*Picea glauca*). Can. J. For. Res. 23, 219–228.

Bigras F J, Paquin R, Rioux J A and Therrien H P 1989 Influence de la photopériode et de la température sur l'évolution de la tolérance au gel, de la croissance et de la teneur en eau, sucres, amidon et proline des rameaux et des racines de genévrier (*Juniperus chinensis* L. 'Pfitzerana'). Can. J. Plant. Sci. 69, 305–316.

Colombo S J 1990 Bud dormancy status, frost hardiness, shoot moisture content, and readiness of Black spruce container seedlings for frozen storage. J. Amer. Soc. Hort. Sci. 115, 302–307.

Colombo S J, Zhao S and Blumwald E 1995 Frost hardiness gradients in shoots and roots of *Picea mariana* seedlings. Scand. J. For. Res. 10, 32–36.

Coursolle C, Bigras F J and Margolis H A 1998 Frost tolerance and hardening capacity during the germination and early developmental stages of four white spruce (*Picea glauca*) provenances. Can. J. Bot. 76, 122–129.

Desjardins R L and Chong C 1980 Unheated environments for overwintering nursery plants in containers. Can. J. Plant. Sci. 60, 895–902.

Dexter S T, Tottingham W E and Graber L F 1930 Preliminary results in measuring the hardiness of plants. Plant Physiology 5, 215–223.

Dexter S T, Tottingham W E and Graber L F 1932 Investigations of the hardiness of plants by measurement of electrical conductivity. Plant Physiology 7, 63–78.

Havis J R 1976 Root hardiness of woody ornamentals. HortScience 11, 385–386.

Johnson J R and Havis J R 1977 Photoperiod and temperature effects on root cold acclimation. J. Amer. Soc. Hort. Sci. 102, 306–308.

Lapins K 1961 Artificial freezing of 1-year-old shoots of apple varieties. Can. J. Plant Sci. 41, 381–393.

Lassheikki M, Puttonen P and Räsänen P K 1991 Planting performance potential of *Pinus sylvestris* seedlings as evaluated by root growth capacity and triphenyl tetrazolium chloride reduction methods. Scand. J. For. Res. 6, 91–104.

Lindström A 1986 Outdoor winter storage of container stock on raised pallets – effects on root zone temperatures and seedling growth. Scand. J. For. Res. 1, 37–47.

Lindström A 1995 Overwintering of containerized conifer seedlings – A risky venture. *In* L'hivernage de plantes ligneuses et des vivaces ornamentales. Proc. IQDHO conf. Trois-Rivières, Canada 1995. pp 167–177.

Lindström A and Mattsson A 1989 Equipment for freezing roots and its use to test cold resistance of young and mature roots of Norway spruce seedlings. Scand. J. For. Res. 4, 59–66.

Lindström A and Nyström C 1987 Seasonal variation in root hardiness of container-grown Scots pine, Norway spruce, and lodgepole pine seedlings. Can. J. For. Res. 17, 787–793.

Lindström A and Stattin E 1994 Root freezing tolerance and vitality of Norway spruce and Scots pine seedlings; influence of storage duration, storage temperature, and prestorage root freezing. Can. J. For. Res. 24, 2477–2484.

Mattsson A 1986 Seasonal variation in root growth capacity during cultivation of container grown *Pinus sylvestris* seedlings. Scand. J. For. Res. 1, 473–482.

McKay H M 1994 Frost hardiness and cold-storage tolerance of the root system of *Picea sitchensis*, *Pseudotsuga menziesii*, *Larix kaempferi* and *Pinus sylvestris* bare-root seedlings. Scand. J. For. Res. 9, 203–213.

Mityga H G and Lanphear F O 1971 Factors influencing the cold hardiness of *Taxus cuspidata* roots. J. Amer. Soc. Hort. Sci. 96, 83–86.

Pellett N E and White D B 1969 Soil-air temperature relationships and cold acclimation of container grown *Juniperus chinensis* 'Hetzi'. J. Amer. Soc. Hort. Sci. 94, 453–456.

Siminovitch D, Therrien H, Wilner J and Gfeller F 1962 The release of aminoacids and other ninhydrin-reacting substances from plant cells after injury by freezing; a sensitive criterion for the estimation of frost injury in plant tissues. Can. J. Bot. 40, 1267–1269.

Smit-Spinks B, Swanson B T and Markhart III A H 1985 The effect of photoperiod and thermoperiod on cold acclimation and growth of *Pinus sylvestris*. Can. J. For. Res. 15, 453–460.

Steponkus P L and Lanphear F O 1967 Refinement of the triphenyl tetrazolium chloride method of determining cold injury. Plant Physiol. 42, 1423–1426.

Stone E C 1955 Poor survival and the physiological condition of planting stock. For. Sci. 1, 90–94.

Tinus R W 1982 Successful overwintering of container-grown seedlings. *In* Proceedings of the 1981 Intermountain Nurserymen's Association Meeting. Can. For. Serv. Edmonton, Alberta Inf. Rep. NOR-X-241, 26–41.

Wilner J 1959 Note on an electrolytic procedure for differentiating between frost injury of roots and shoots in woody plants. Can. J. Plant. Sci. 39, 512–513.

Yadava U L, Doud S L and Weaver D J 1978 Evaluation of different methods to assess cold hardiness of peach trees. J. Amer. Soc. Hort. Sci. 103, 318–321.

Zar J 1974 Biostatistical analyses. Prentice-Hall, Inc., Englewood Cliffs, N.J.

Zhao S, Colombo S J and Blumwald E 1995 The induction of freezing tolerance in jack pine seedlings: The role of root plasma membrane H+-ATPase and redox activities. Physiol. Plant. 93, 55–60.

A. Stokes (ed.), The Supporting Roots of Trees and Woody Plants: Form, Function and Physiology, 269–276.

Sucrose metabolism, growth and transplanting stress in sweetgum seedling taproots and stems

Shi-Jean S. Sung* and Paul P. Kormanik

Tree Root Biology, Southern Research Station, Forest Service, United States Department of Agriculture, 320 Green Street, Athens, GA 30602-2044, USA

Key words: *Liquidambar styraciflua*, pyrophosphate-dependent phosphofructokinase, stress, sucrose synthase, transplanting

Abstract

One-year-old nursery-grown bare-root sweetgum (*Liquidambar styraciflua* L.) seedlings were lifted and transplanted into a nearby nursery bed or a cleared forest field in late January 1994. Seedlings remained in the same bed for the second year were the nontransplanted controls. Seedlings growing in beds were watered regularly and those in field received only rain. Specific activities of sucrose synthase (SS), pyrophosphate-dependent phosphofructokinase (PPi-PFK), ATP-dependent PFK (ATP-PFK), fructokinase (FK), and glucokinase (GK) were determined in the xylem-side cambial tissues of control and transplanted seedling taproots and stems from April through November 1994. Stem and taproot SS activity of control seedlings exhibited similar seasonal patterns with stem SS higher than root most of the time. Low SS activity was observed during early leaf expansion in the spring with resumption of taproot SS activity occurring 2 weeks later than stem SS. Seedling SS activity peaked in June and September and decreased sharply near leaf abscission in fall. Seasonal patterns of PPi-PFK in control seedlings were similar to those of SS except that PPi-PFK peaked only in June. Taproots of nontransplanted seedlings had similar level of PPi-PFK to that of stems. From late April to mid-June, there were 35–75% and 20–45% decreases in SS and PPi-PFK activities in transplanted seedlings, respectively, as compared with the controls. Moreover, seedlings transplanted into the forest field had much less growth and lower activities of SS and PPi-PFK than seedlings transplanted into the nursery beds. Activities of ATP-PFK and FK were generally lower in transplanted seedlings than controls.

During spring and early summer, some transplanted seedling stems had tip dieback or black mottling and scars or both. These seedlings usually had small reddish leaves, little new root growth, and very low SS and PPi-PFK activities as compared with healthy transplanted seedlings of the same site. Some of the black mottling/scare areas on discolored seedlings were infected with *Botryosphaeria dothidea* (Moug.:Fr.) Ces. De Not. At the end of the first year, about 10% mortality was observed with seedlings transplanted into the nursery bed or the forest site. No further mortality occurred on the forest site for the following 3 years. It is concluded that transplanting stress, based on sucrose metabolism, lasted at least through late June for sweetgum seedlings planted in January.

Abbreviations: ATP-PFK – ATP-dependent phosphofructokinase; FK – fructokinase; GK – glucokinase; PPi-PFK – pyrophosphate-dependent phosphofructokinase; SG – sweetgum; SS – sucrose synthase

Introduction

Sucrose is the major translocated form of carbohydrates in trees (Zimmermann and Brown, 1971). The central roles of sucrose in plant cell growth, storage, and gene regulation have been investigated intensively in recent years (for review, see Koch, 1996). For example, in addition to being the starting and end points of sucrolysis and sucroneogenesis, sucrose can modulate genes for starch synthesis, storage protein synthesis, sucrose cleavage, and nitrate

* FAX No: +1-706-559-4291.
E-mail: ssung/srs_athens@fs.fed.us

reductase (Koch, 1996, and references cited therein). In the sucrose metabolic pathway, sucrose synthase (SS) and pyrophosphate-dependent phosphofructokinase (PPi-PFK) are active in actively growing and storing tissues and inactive in dormant tissues (Sung et al., 1989; Xu et al., 1989). Activities of the alternative enzymes to SS and PPi-PFK, such as invertases and ATP-dependent PFK, usually remain at low levels regardless of the physiological status of plants (Sung et al., 1989; Xu et al., 1989). Sucrose synthase has been identified as a stress indicator for *Pinus taeda* seedlings and trees (Sung et al., 1993b, 1996) and for *P. jeffreyi* trees (Otrosina et al., 1996). It took transplanted loblolly pine seedlings 8 months to resume their cambial growth and SS activities similar to those of the nontransplanted ones (Sung et al., 1993b).

In contrast to conifers, most hardwoods do not appear to suffer as much from transplanting stress. It was reported that transplanted sweetgum (SG, *Liquidambar styraciflua* L.) seedlings have delayed root growth compared to leaf growth (Kormanik, 1986b). Moreover, 41–89% of transplanted SG seedlings had mid-June stem dieback during the first year of transplanting (Kormanik, 1986a). It is critical for transplanted seedlings to establish, as quickly as possible, root–soil contacts for water and nutrient uptake. These processes require energy from stored carbohydrate reserves, current photosynthates, or both (Johnsen et al., 1988; Philipson, 1988). Moreover, before leaves become photoautotrophic, they live off sucrose. Elongation and diameter growth of stems also require sucrose. Seasonal sucrose metabolism was reported in 2- and 3-year-old nursery grown, nontransplanted SG seedlings (Sung et al., 1989). No reports, however, have been made on the biochemical characteristics of transplanting stress experienced by SG seedlings in controlled environments or on forest sites. The objectives of this study were: (1) to follow seasonal patterns of SS, PPi-PFK, ATP-PFK, FK, and GK in nontransplanted and transplanted SG seedling taproots and stems; and (2) to relate SS activity and growth with transplanting stress or other stresses in SG seedlings.

Materials and methods

In April 1993 stratified sweetgum seeds of mixed seed lots were sown in a raised concrete block nursery bed ($18.3 \times 1.2 \times 1.2$ m) at the Whitehall Experiment Forest (Athens, GA, USA). Seedlings were grown at a density of 66 m^{-2} following the nursery protocol of Kormanik (1986b) except that soil P level was 100 ppm and no vesicular-arbuscular-mycorrhizae were added to the beds. In late January 1994, seedlings from half of the bed were lifted, tagged, and measured prior to transplanting to a nearby nursery bed at the Whitehall Experiment Forest or to a cleared forest field on the Savannah River Site, National Environmental Research Park (New Ellenton, SC). A total of 225 seedlings were transplanted into the nursery bed at a density of 33 m^{-2}. Another 300 seedlings were planted in the forest field at 3×3 m spacing. Seedlings remained in the same nursery bed for the second year were thinned to 33 m^{-2} and served as the nontransplanted controls. Thinning was accomplished by cutting off every other seedlings at the root collar level in later January. The controls and transplanted seedlings in the nursery beds were water regularly. Transplanted seedlings in the forest field received only rain. Seedling condition was evaluated for the presence of stem tip dieback or disease in August, 1994. First year seedling survival was assessed in December 1994. In December 1996 and July 1998, growth measurements and mortality were recorded for plants in the forest field.

The nursery seedlings were lifted periodically from April through November of 1994 for enzyme activities. From June to August, discolored transplanted seedlings also were lifted. Because of the variable cambial SS and PPi-PFK activities among first-order lateral roots of young sweetgum trees during active growing seasons (Sung et al., 1993a), only taproot cambial tissues were analyzed for enzyme activities. At each sampling time, four nontransplanted, four healthy transplanted, or eight discolored transplanted were lifted from nursery beds or the field. Diameter and weight growth of each lifted seedling was recorded. Protocols for sampling and extracting taproot and stem cambial tissues and for enzyme assays and protein determination were based on the procedures of Sung et al. (1993b). Briefly, after bark peeling, xylem-side cambial tissues were scraped into a weight boat. Tissues from two seedlings were composited into a stem or a root sample. Duplicate samples were collected and analyzed each time. Samples were immediately frozen with liquid nitrogen and extracted with extraction buffer, at a 10:1 ratio (weight:volume), containing 200 mM Hepes/NaOH (pH 7.8), 3 mM Mg acetate, 5 mM dithiothreitol (DTT), 1% (w/v) soluble polyvinylpyrrolidone-40, and 10% (v/v) glycerol. After centrifugation at 34 $000 \times g$ for 20 min at 4°C, supernatant was analyzed for enzyme assays and protein determination. Changes of OD at 340 nm in all enzyme assays were monitored with a Beckman DU-70 spectrophotometer at 25°C. Sucrose synthase was assayed with 100 mM sucrose, 0.5 mM UDP, and 1 mM PPi as substrates. Pyrophosphate-dependent PFK was assayed with 10 mM fructose-6-P and 1 mM PPi. Enzyme specific activities between duplicate samples varied less than 15%. Average values are reported here.

Results

Nontransplanted SG seedlings started bud break and new root growth near the end of March. Bud break and new root growth of transplanted seedlings began near the end of April. There existed a clear seasonal pattern for SS specific activity in nontransplanted SG stem cambial tissues (Fig. 1A). Stem SS activity was at the lowest level in early April when mean leaf size was about 8 cm^2 (Fig. 1A). Near the end of April, stem SS activity increased more than 4-fold and peaked in June followed by decreases in July and August. The second peak SS activity was observed in September. By November SS activity was low (Fig. 1A). Sweetgum leaf abscission occurred in early December at this nursery. Generally, taproot cambial SS activity tracked that of the stem throughout the year with delayed resumption of activity in the spring (Fig. 1A). Except for May, taproot SS was lower than stem SS. The SS activity in transplanted seedling taproots and stems was at much lower level than the nontransplanted controls from April to June except for SS in transplanted stems sampled in May (Fig. 1B). Transplanting related losses of SS activity were more obvious and lasted longer with taproots than with stems. Maximal loss of SS was 75% for taproots in June and 60% for stems in April. In August, stem and taproot SS in transplanted seedlings were greater than controls (Fig. 1). Nevertheless, transplanting did not alter the spatial pattern of SS activity. Transplanted taproot had lower SS than stem except in November (Fig. 1).

Peak PPi-PFK activity in nontransplanted and transplanted seedling stems and taproots was greater than SS in corresponding organs (Figs. 1 and 2). In contrast to SS, taproot had similar PPi-PFK activity to that of stem in nontransplanted seedlings (Fig. 2A). Nevertheless, the seasonal patterns of seedling taproot and stem PPi-PFK activity (Fig. 2A) were similar to those of SS (Fig. 1A) with only one peak PPi-PFK activity in June. Losses of PPi-PFK activity, about 20–45%, in transplanted seedlings occurred from April through June (Fig. 2B). Sweetgum seedling taproots and stems did not exhibit any seasonal or spatial patterns for ATP-PFK, FK, and GK activities (data not shown). Furthermore, transplanting related losses of ATP-PFK, FK, and GK activities in taproot and stem cambial tissues were minimal compared to those of SS or PPi-PFK (Fig. 3 vs. Figs. 1 and 2).

Table 1. Growth (mean±S.E.) and survival of sweetgum seedlings transplanted into a nursery bed or a forest field. Seedlings (1–0) were lifted and transplanted in late January 1994

	Nursery bed	Forest field
January 1994		
Seedling number	225	300
First-order later root number	9.3±0.3	9.0±0.3
Height (cm)	94.5±1.1	91.3±1.2
Root collar diameter (mm)	11.9±0.2	11.7±0.1
December 1994		
Survival (%)	94.0	88.7
December 1996		
Survival (%)[a]	–	89.1
Height (m)	–	2.4±0.04
Diameter at breast height (mm)	–	21.3±0.7
July 1998		
Survival	–	88.4
Height (m)	–	4.9±0.04
Diameter at breast height (mm)	–	55.3±1.1

[a] All nursery bed seedlings were discarded in December 1994. Some dead seedlings in the field sprouted by December 1996.

In this study, about 25% of seedlings transplanted to the nursery bed or the forest field had both stem dieback and black mottling or scars on stems or branches. Another 25% of transplanted seedlings had black mottling or scars on stems or branches. Some of the discolored areas were verified to be infected with the fungus *Botryosphaeria dothidea* (Moug.:Fr.) Ces. De Not (WJ Otrosina, personal communication). Most of the discolored seedlings had small, reddish purple colored leaves. Only a few nontransplanted controls had the above-mentioned symptoms in 1994. At the end of the first year after transplanting, nearly 90% of seedlings survived (Table 1). The dead seedlings generally had fewer first-order lateral roots (FOLR) at transplanting than those survived: 6.9±0.8 (mean±S.E.) vs. 9.3±0.3 for nursery beds

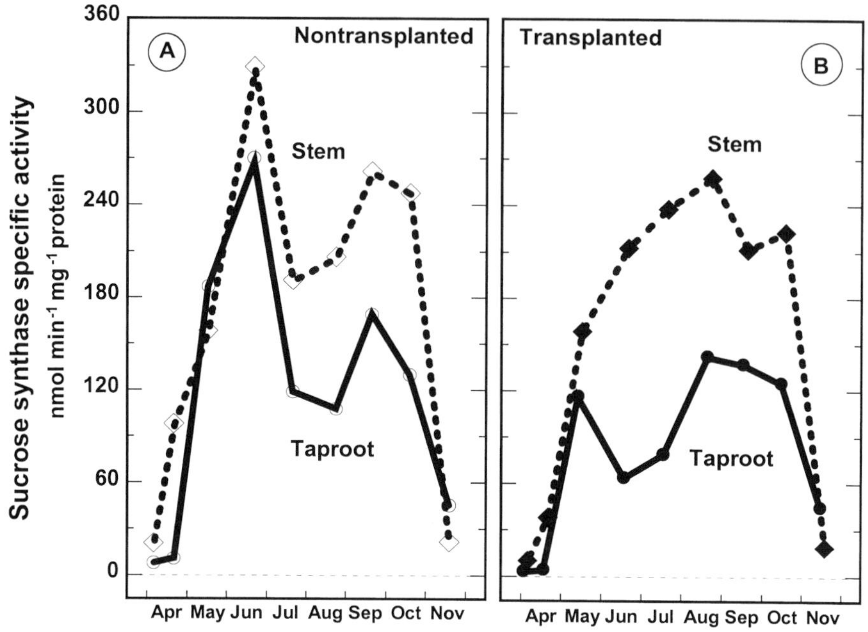

Figure 1. Seasonal patterns of sucrose synthase specific activity in taproot and stem cambial tissues of (A) nontransplanted and (B) transplanted sweetgum seedlings during 1994. Seedlings were lifted and transplanted into a nearby nursery bed in late January 1994. Seedlings remained in the same nursery bed were the nontransplanted controls.

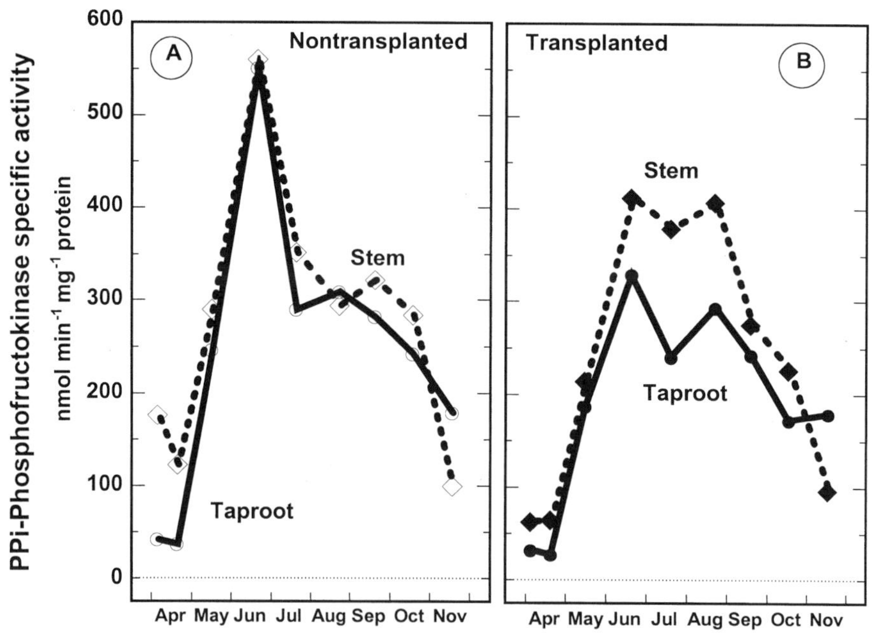

Figure 2. Seasonal patterns of pyrophosphate-dependent phosphofructokinase specific activity in taproot and stem cambial tissues of (A) nontransplanted and (B) transplanted sweetgum seedlings during 1994. Seedlings were lifted and transplanted into a nearby nursery bed in late January 1994. Seedlings remained in the same nursery bed were the nontransplanted controls.

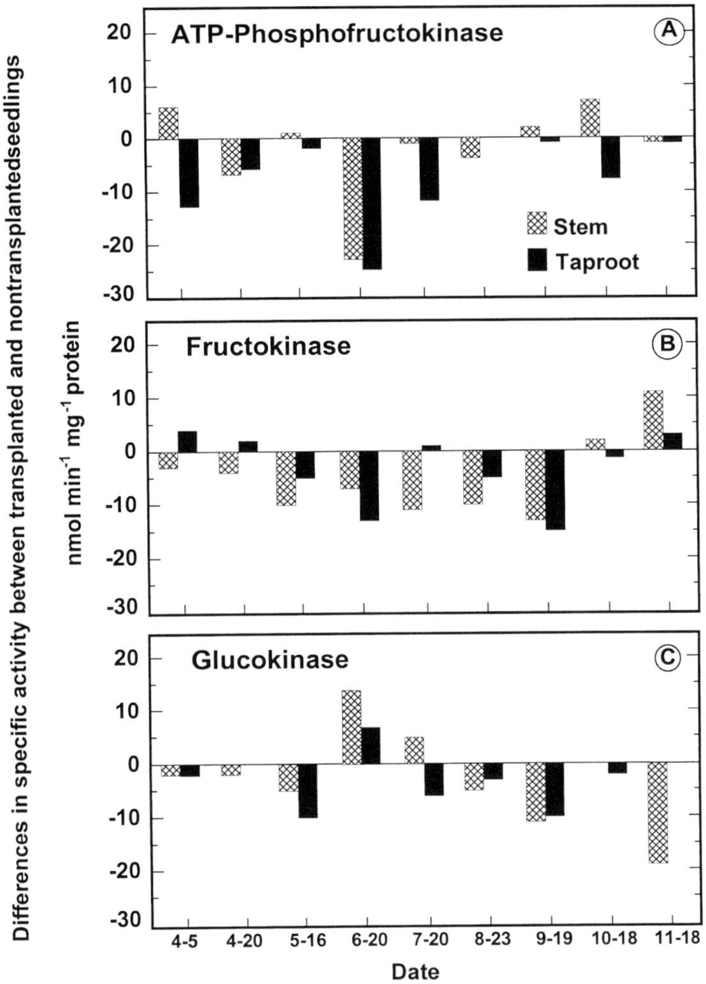

Figure 3. Differences in (A) ATP-dependent phosphofructokinase, (B) fructokinase and (C) glucokinase specific activities between transplanted and nontransplanted sweetgum seedlings during 1994. Seedlings were lifted and transplanted into a nearby nursery bed in late January 1994. Seedlings remained in the same nursery bed were the nontransplanted controls.

and 8.2±0.8 vs. 9.0±0.3 for the forest site. No further mortality occurred for the forest field plants for the following 3 years (Table 1).

Diameter and weight growth of healthy or discolored transplanted SG lifted in the months of June through August were presented in Table 2. Enzyme activities of these seedlings were presented in Table 3. In general, the discolored seedlings had fewer FOLR and smaller diameter at transplanting than the healthy ones (Table 2). The discolored seedlings had little root, stem or leaf growth but greater amounts of stem dieback, compared to the healthy seedlings of the same site (Table 2). Seedlings in the forest field grew much less than those transplanted into the nursery bed. Moreover, SS activity of healthy seedlings in the forest field remained low even in August when transplanted seedlings in the nursery beds had greater SS activity than controls (Table 3, Fig. 1A). Generally, SS and PPi-PFK activities in discolored seedling taproots and stems were much lower than those of the healthy seedlings (Table 3). Again, activities of ATP-PFK, FK, and GK were constant regardless of seedling condition or transplanting site.

Discussion

Results in Figure 1 and Tables 2 and 3 indicated that there exists a close relationship between SS activity

Table 2. Growth of healthy and discolored sweetgum seedlings in a nursery bed or a forest field during the first year after transplanting. Seedlings in the nursery bed were watered regularly but those in the forest field received rain only

Date	Site	At transplanting		Ater transplanting (fresh weight, g)							
		FOLR[a] (#)	RCD (mm)	Seedling Condition[b]	RCD (mm)	New root[c]	Total root	Stem/ branch	Dieback stem	Leaf	Seedling
6-3	Forest	10	13.1	Healthy	14.1	6.5	68	60	2.6	34	162
		6	10.9	Discolored	11.0	0.1	34	26	7.1	4	64
6-20	Nursery	7	11.1	Healthy	13.4	8.8	60	77	0.8	62	199
		5	10.5	Discolored	10.9	1.1	30	27	9.2	17	74
7-20	Nursery	11	11.3	Healthy	18.3	31.1	173	192	0	234	599
		5	10.0	Discolored	13.5	2.4	61	57	10.0	52	170
8-10	Forest	9	11.6	Healthy	15.5	–	69	58	0.4	40	167
		7	10.2	Discolored	13.4	–	57	42	0.6	17	116
8-23	Nursery	9	12.9	Healthy	20.7	–	216	320	0.5	290	826
9-19	Nursery	17	12.4	Healthy	25.9	–	347	534	0.1	396	1277
10-18	Nursery	12	13.4	Healthy	26.1	–	276	441	0	244	961
11-18	Nursery	12	12.6	Healthy	28.5	–	304	593	7.4	203	1100

[a]FOLR and RCD: first-order lateral root number and root collar diameter of seedlings lifted from the nursery bed in January 1994. Values presented were means from four healthy or eight discolored seedlings.
[b]Same seedlings lifted at each sampling date were analyzed for enzyme activities. Discolored seedlings had black mottling or scars on stems or branches. Some discolored areas were infected with *Botryosphaeria dothidea*. Leaves of discolored seedlings were generally small and reddish purple in color.
[c]New lateral roots were non-suberized and white in color. They were not distinguishable from the original roots by August.

Table 3. Sucrose metabolic pathway enzyme activities in the cambial tissues of taproots and stems of healthy and discolored sweetgum seedlings during the first year after transplanting. Seedlings in the nursery bed were watered regularly but those in the forest field received rain only

Date	Site	Seedling condition[a]	Taproot enzyme specific activity (nmol min^{-1} mg^{-1} protein)					Stem enzyme specific activity (nmol min^{-1} mg^{-1} protein)				
			SS[b]	PPi-PFK	ATP-PFK	FK	GK	SS	PPi-PFK	ATP-PFK	FK	GK
6-3	Forest	Healthy	4	183	30	38	28	62	264	21	34	19
		Discolored	6	76	46	24	12	1	68	7	14	16
6-20	Nursery	Healthy	65	329	33	51	50	213	413	35	58	48
		Discolored	48	225	30	50	34	94	346	29	49	21
7-20	Nursery	Healthy	80	241	24	52	27	238	380	46	50	33
		Discolored	16	170	18	40	21	35	163	16	47	26
8-10	Forest	Healthy	27	151	30	51	23	65	188	25	40	18
		Discolored	8	40	23	23	22	1	20	35	19	11

[a]Same seedlings were measured for growth as presented in Table 2. Two healthy or four discolored seedlings were composited into a sample. Duplicate samples were analyzed. Discolored seedlings had black mottling or scars on stems or branches. Some discolored areas were infected with *Botryosphaeria dothidea*. Leaves of discolored seedlings were generally small and reddish purple in color.
[b]SS, sucrose synthase; PPi-PFK, pyrophosphate-dependent phosphofructokinase; ATP-PFK, ATP-dependent PKF; FK, fructokinase; GK, glucokinase. Values presented were means of two samples.

and tissue growth: low SS activity in early spring prior to full leaf expansion, increasing SS activity as more leaves exporting sucrose in the summer and early fall, and low SS activity toward leaf abscission. Furthermore, healthy transplanted seedlings had higher SS level than the discolored seedlings of the same site. And, transplanted seedlings receiving regular watering had higher SS activity and grew more than those planted in the forest field and received only rain. These lines of evidence further support results of studies on tree and crop species: that SS activity is closely related to growth (Otrosina et al., 1996; Ross and Davies, 1992; Sun et al., 1992; Sung et al., 1989, 1993b, 1994, 1996; Xu et al., 1989). In other words, SS activity can be used as a stress indicator for plant tissues. In this study, transplanting stress only lasted through June with sweetgum seedlings transplanted into nursery beds and watered regularly. However, most plantation establishment for trees does not have irrigation facility. Thus transplanting stress can be longer, as in the case of forest field seedlings, depending on weather conditions. The enhancing effects of irrigation on photosynthesis rate and fertigation on leaf area, as reported by Samuelson (1998), were assessed beginning in June. It is not known the effect of fertilization on sucrose metabolism of these transplanted SG.

Activity of PPi-PFK more or less tracked that of SS in sweetgum seedlings of different environments. However, this enzyme is not as responsive to growth/stress as SS (Figs. 1 and 2, Table 3). Activities of the other sucrose metabolic pathway enzymes remained constant and were only slightly affected by stress (Fig. 3, Table 3). These results were similar to studies with other crops and trees of different environments where ATP-PFK and GK were identified as the maintenance enzymes for being constant and low in activity (Sung et al., 1989, 1993b; Xu et al., 1989).

Stem dieback and delayed resumption of root growth (as compared to that of leaf) was reported for transplanted sweetgum seedlings (Kormanik, 1986a,b). In fact, initial stem dieback was considered 'common' for sweetgum plantation by foresters. It is not clear whether stem dieback is directly associated with infection by *B. dothidea*. Since no discolored symptom was observed with SG seedlings at transplanting, there exist two possibilities. The first is that *B. dothidea* infected and stayed latent during the first year in the nursery and became aggressive later when transplanted sweetgum are under other stress such as drought. In other words, it was possible that transplanting exacerbated the discolored/disease problem. The second possibility is that the inferior seedlings, those with fewer FOLR and smaller diameter, were infected more often by the disease in the field. Discolored seedlings sampled in this study for growth and enzyme activity have fewer FOLR and smaller diameter at transplanting as compared to healthy transplanted seedlings. Kormanik (1986a) reported that SG seedlings with greater FOLR number also had greater diameter at lifting from the nursery. These seedlings outperformed those with fewer FOLR in the field. Similarly, in this study dead transplanted seedlings had fewer FOLR, at transplanting, than those survived. More study is needed to correlate extents of stem discoloration or stem dieback to seedling FOLR. No *B. dothidea* related symptoms have been observed in trees of the forest field at least for the fourth year after transplanting.

Acknowledgements

This research was funded by the Department of Energy grant DE-AI09-76SR00870 via Savannah River Natural Resource Management and Research Institute.

References

Johnsen K H, McConnell S P, Regelbrugge J C and Seiler J R 1988 Hydraulic conductance of roots present at the time of lifting and newly regenerated roots of 2+0 eastern white pine seedlings. Tree Planters' Notes 39, 5–8.

Koch K E 1996 Carbohydrate-modulated gene expression in plants. Annu. Rev. Plant Physiol. Plant Mol. Biol. 47, 509–540.

Kormanik P P 1986a Lateral root morphology as an expression of sweetgum seedling quality. For. Sci. 32, 595–604.

Kormanik P P 1986b Development of vesicular-arbuscular mycorrhizae in a young sweetgum plantation. Can. J. For. Res. 15, 1061–1064.

Otrosina W J, Sung S S and White L M 1996 Effects of subsoiling on lateral roots, sucrose metabolizing enzymes, and soil ergosterol in two Jeffrey pine stands. Tree Physiol. 16, 1009–1013.

Philipson J J 1988 Root growth in Sitka spruce and Douglas-fir transplants: dependence on the shoot and stored carbohydrates. Tree Physiol. 4, 101–108.

Ross H A and Davies H V 1992 Sucrose metabolism in tubers of potato (*Solanum tuberosum* L.). Effects of sink removal and sucrose flux on sucrose-degrading enzymes. Plant Physiol. 98, 287–293.

Samuelson L J 1998 Influence of intensive culture on leaf net photosynthesis and growth of sweetgum and loblolly pine seedlings. For. Sci. 44, 308–316.

Sun J, Loboda T, Sung S S and Black C C 1992 Sucrose synthase in wild tomato *Lycopersicon chmieleswskii* and tomato fruit sink strength. Plant Physiol. 98, 1163–1169.

Sung S S, Kormanik P P, Xu D P and Black C C 1989 Sucrose metabolic pathways in sweetgum and pecan seedlings. Tree Physiol. 5, 39–52.

Sung S S, Kormanik P P and Black C C 1993a Understanding sucrose metabolism and growth in a developing sweetgum plantation. *In* Proceedings 22nd Southern Forest Tree Improvement Conf. pp 114–123. Atlanta, GA. June 1993.

Sung S S, Kormanik P P and Black C C 1993b Vascular cambial sucrose metabolism and growth in loblolly pine (*Pinus taeda*) in relation to transplanting stress. Tree Physiol. 12, 243–258.

Sung S S, Shieh W J, Geiger D R and Black C C 1994 Roles of sucrolytic enzymes in developing *Phaseolus vulgaris* L. fruits. Plant Cell Environ. 17, 419–426.

Sung, S S, Kormanik P P and Black C C 1996 Temporal and spatial aspects of root and stem sucrose metabolism in loblolly pine trees. Tree Physiol. 16, 1003–1008.

Xu D P, Sung S S and Black C C 1989 Sucrose metabolism in lima bean seeds. Plant Physiol. 89, 1106–1116.

Zimmermann M N and Brown C L 1971 Trees Structure and Function. Springer-Verlag, Berlin 316 pp.

NEW METHODS IN ROOT RESEARCH

A. Stokes (ed.), The Supporting Roots of Trees and Woody Plants: Form, Function and Physiology, 279–285.

A method for predicting the possible site of failure in trees during mechanical loading

Method for predicting tree failure

Alexia Stokes[1,*], Michael Drexhage[2] and Daniel Guitard[3]
[1]*Laboratoire de Rhéologie du Bois de Bordeaux, CNRS/INRA. Université Bordeaux I, B.P.10, Domaine de l'Hermitage, 33610 Cestas Gacinet, France;* [2]*University of Göttingen, Institute of Silviculture, Büsgenweg 1, 37077 Göttingen, Germany;* [3]*Laboratoire Energétique et Phénomène de Transfére, LEPT-ENSAM, Esplanade des Arts et Métiers, Université Bordeaux I, 33400 Talence, France (*Corresponding author)*

Abstract

In order to determine if it is possible to predict the site and mode of failure under mechanical loading, different tree species were subjected to bending tests. Four species were examined: European beech, English oak, Douglas fir and Norway spruce. All trees were of a similar age and size. Strain gauges were placed on the surface of the trunk and along the lateral roots. Strain gauges measure deformations on the surface of a material during loading. Trees were winched horizontally using increments of force of 0.25 kN, and strain was measured in each of the strain gauges for each increment of force applied. Resistance of species to mechanical failure was in the order: beech, oak, Douglas fir and spruce. Beech required over three times as much force to cause stem breakage, compared to spruce which generally broke in the root system. Oak also broke in the trunk whilst Douglas fir usually broke at the root/stem joint. In those species which broke in the trunk, strain was always found to be higher in the stem than in the roots during winching. However, in Douglas fir and Norway spruce, strain was always highest at the root/stem joint and in the root system, respectively. Root system architecture may influence the mode of failure as beech, oak and Douglas fir all have deeper growing roots than spruce, and they all broke above ground level. However, a more detailed study of tree architecture and wood properties would be needed to verify this hypothesis. It appears that through the use of winching tests and strain measurement, it may be possible to predict how and where a tree might fail under mechanical loading. More species and trees of different ages and in different states of health would need to be examined using this method.

Introduction

An increased awareness of the importance of tree safety in urban areas has led to the development of new techniques and tools for assessing tree health and potential mechanical failure in the last ten years (Mattheck, 1995, 1998; Mattheck and Breloer, 1995; Wessolly, 1998). Trees in urban areas are often subjected to much greater environmental stresses than forest trees, e.g. drought, pollution, soil compaction under roads or paving and mechanical damage, especially to root systems through trenching (Nicoll and Armstrong, 1998). Therefore, the risk of wounding and decay entering through a wound and attacking an unhealthy tree can be high. Hence, non-invasive and visual methods for determining tree safety are highly desirable, so as to not encourage the spread of pathogens throughout the tree.

Mattheck (1995, 1998) has pioneered the use of portable devices to measure wood strength in samples bored from the tree. Using a database, in which many data have been collected from a large range of tree species at different sites and ages, Mattheck (1995, 1998) can predict if the wood is sound, or if it has been attacked by fungi. The thickness and volume of wood remaining around a decayed centre can be a further indication if a tree is likely to fail or not. Despite Mattheck's success over the last few years, his methods have received criticism from a German counterpart, Wessolly (1998). Wessolly claims that by drilling holes to take wood samples, any heart-rot present in the tree may spread to the surface via the holes, thus endangering a previously safe tree.

Wessolly (1998) proposes that his own method, which involves winching the tree sideways, provides a good indicator of tree safety. Inclinometers are attached at the tree base and elastometers at different heights up the trunk. Measurements of strain and force needed to pull the tree over a few degrees during winching are made. From these measurements, the bending stiffness can be calculated (Brüchert and Spatz, 1999). Such a technique may provide the user with information about the global mechanical stability of the tree, but is expensive and cannot always give indications about localised zones of weakness. These zones, which often exist due to the presence of decay within the tree, may be a potential failure site if the tree is subjected to wind loading. Although results using Wessolly's method (1998) for assessing tree safety should be analysed with caution, a similar cheaper method could be envisaged to predict the mode of tree failure.

The mode of failure of a tree depends on its external shape and internal wood properties as well as local climatic conditions (Coutts, 1983, 1986; Fraser and Gardiner, 1967; Petty and Swain, 1985). In trees which fail in the trunk, the amount of stem taper is thought to be important in determining the site of stem breakage (Peltola and Kellomäki, 1993). Poor wood quality and the presence of compression wood in conifer stems has also been thought to increase the likelihood of stem failure (Dunham and Cameron, 1999). Root system asymmetry can also contribute to tree instability (Coutts et al., 1999, this volume). However, although root system architecture is known to be important with regards to tree stability, few studies consider the rooting type present when examining tree stability and safety in different species (Crook and Ennos, 1996, 1997; Stokes and Mattheck, 1996). Three basic and inherent types of root system exist in adult trees: heart, tap and sinker. Heart root systems consist of both vertical and lateral roots, tap root systems are comprised of one large vertical root with smaller horizontal lateral roots, whilst sinker root systems possess few vertical roots and have large horizontal lateral roots (Köstler et al., 1968). Some species also have root systems which are a combination of two of these types. The shape of a root system can also be very much influenced by its local environment as well as by silvicultural practices, hence root anchorage may also be weakened (Fraser and Gardiner, 1967; Lindstrom and Rune, 1999, this volume). As information about the contribution of tree architecture to stability is relatively scanty, it is vital to investigate the role of root system shape with regards to anchorage failure and stem breakage. It could be considered that deeper rooting species, e.g. oak and Douglas fir are better anchored than shallowly rooted species, e.g. spruce, but no quantitative studies have yet been carried out.

In a preliminary study, winching tests, similar to those proposed by Wessolly (1998) and Brüchert and Spatz (1999), were carried out on four tree species, in order to determine if it is possible to predict the mode and site at which a tree will fail when mechanically loaded. Two broadleaf sp. with tap and heart root systems, and two conifer sp. with heart and sinker root systems were chosen in order to investigate if rooting type influenced the type of failure mode (anchorage failure or stem breakage). Trees were pulled over until uprooting or root or stem breakage occurred, the site of failure was noted and the relationship between strain and failure mode investigated.

Materials and methods

Study Site

Four tree species (Table 1) were chosen for the study from a small area of forest in the Göttinger Wald, Lower Saxony, northern Germany (longitude 10° 06'24" E, latitude 51° 36'30" N, 250–300 m a.s.l.). The soil in the Lower Saxony regions was a typical gleyic luvisol (FAO) with a silty loam/clay texture and is developed from newly weathered sandstone. Water is removed slowly so that the soil remains wet at shallow depths for long periods (Drexhage and Lupke, 1995). Three trees per species were chosen and were approximately 25 years old. Trees chosen were comparable in both height and size at diameter breast height (DBH) (Table 2).

Soil samples were taken from the soil around each tree immediately after the winching tests had been carried out, in order to determine if soil water content influenced soil shear strength and hence tree anchorage. The samples were weighed when fresh and then oven-dried for three days at 80 °C and dry weight measured.

Winching Tests

Winching tests were carried out on trees in order to determine if a relationship existed between tree mechanical behaviour and the site of failure during maximum loading. A hand winch was attached to the base of an anchoring tree at a distance of several metres from the experimental tree (Table 2). The

Table 1. The inherent rooting type of the four different species studied at 25 years of age (Köstler et al., 1968)

Species	Type of root system
European beech (*Fagus sylvatica* L.)	Heart
Douglas fir (*Pseudotsuga mexziesii* (Mirb.) Franco)	Heart
English oak (*Quercus robur* L.)	Tap
Norway spruce (*Picea abies* (L.) Karst.)	Sinker

winch cable was attached to the experimental tree in the crown at a height of a few metres (Table 2), rather than at the base of the tree so that strain in the tree during winching would imitate that under wind loading as best possible. The tree stem was protected with a rubber sling at the point of attachment to avoid damage by the metal cable during winching. A force transducer was attached between the winch and the winch cable and could measure force applied to the tree up to 10 kN.

Two lateral roots per tree were excavated, one in the direction away from the winch 'counter-winchward' (CW) and the second towards the winch 'winchward' (WD). In order to measure movement in these roots during mechanical loading, plastic backed strain gauges (Kyowa 10 mm gauge length, 120 ohm resistance) were used to estimate longitudinal strains. Strain gauges convert longitudinal deformations of a metal element into an electrical signal and must be connected to a strain indicator (Kyowa SD-10) in quarter bridge mode, via a switching box (Vishay), in order for the electrical signal to be converted into micro-deformations (μstrain). Bark was removed with a chisel at DBH and at three or four intervals along the length of a root, starting from the stem-root joint (Table 2). The distance of strain gauges along the lateral roots was approximate, as the exact position depended on root straightness and proximity to daughter branches. Care was taken not to damage the surface fibres of the wood during removal of the bark. Strain gauges were glued to the wood where the bark had been removed, using Loctite 401 multi-usage glue, which took approximately 15 minutes to dry.

The initial values of each strain gauge were recorded before winching commenced. The trees were then pulled sideways using increments of force of 0.25 kN. Strain was measured in each of the gauges after each increment of force had been applied. The tree was winched sideways until failure occurred. The site of failure was noted.

Results

Several different failure mechanisms were observed and much variability was found between species with regards to the bending moment (M_b = force x length of lever arm) required to cause mechanical failure. A mean M_b of 22.5 kNm was needed to cause stem breakage in beech (Figures 1a, b), compared to only 7.5 kNm which resulted in failure of the root system in spruce trees (Figure 2, Table 3). Both oak and beech broke in the stem whereas Douglas fir failed at the root/stem joint (Figure 3). Soil humidity was slightly variable and did not appear to influence soil shear strength and hence anchorage (Table 3). Soil water content (21%) was highest in oak and lowest in Douglas fir (8.0%), both of which failed in the trunk or stem/root joint compared to spruce (10.6%) which failed in the root system. Analysis of variance of mean M_b within and between species, using soil humidity as a variable, was not significant, and would need more data for a fuller analysis.

The amount of strain experienced by the tree during winching varied between species and appeared to be related to the mode of failure. Strain in the trunks of beech and oak was over twice that in the CW roots and both these species failed in the trunk (Figure 1a, b). In Douglas fir, strain was found to be greatest at the root-stem joint (Figure 2). However, in the roots of spruce, strain was found to be over two times greater than that experienced in the trunk (Figure 3). Remarkably, the area where the tree experienced the highest strains during initial winching, on the CW side, was always the region where the tree failed under maximal loading. These results were not repeated on the WD side of the tree, which was held in compression under loading. Spruce roots broke during anchorage failure in two trees, and the highest strain was noted in a CW root at 64 cm from the stem base, where the root broke (Figure 3). Surprisingly, the strain gauge at the root-stem joint showed a negative value, which could

Table 2. Tree diameter and height data and height of winch cable (lever arm) in the stem, along with distance between the experimental tree and the anchoring tree. The distance of strain gauges along the lateral roots are representative of the real position, which depended on root straightness and proximity to daughter branches

Species	DBH (cm)	Tree height (m)	Distance of experimental tree from anchoring tree (m)	Height at which cable was attached in the experimental tree (m)	Distance of gauges along lateral roots from stem centre (m)
Beech	15.7 ± 1.5	13.3 ± 0.4	13.7 ± 2.9	5.2	0.10, 0.30, 0.50, 0.65
Douglas fir	15.5 ± 2.1	12.3 ± 0.7	18.4 ± 1.2	5.2	0.12, 0.35, 0.55, 0.75
Oak	15.3 ± 1.8	11.8 ± 0.6	13.6 ± 2.3	4.3	0.15, 0.40, 0.60
Spruce	15.0 ± 0.1	13.4 ± 0.1	19.6 ± 4.9	6.0	0.15, 0.45, 0.90

Table 3. Soil humidity, maximum bending moment (M_B) and site of failure in the tree

Species	Soil humidity (%)	Maximum M_B around the tree base (kNm)	Site of failure
Beech	14.3 ± 3.2	22.5 ± 3.1	At a height of 1.2 ± 0.5 m up the trunk from stem base
Douglas fir	8.0 ± 0.5	13.9 ± 3.6	Two trees failed at the stem/root joint and one tree at a height of 0.9 m up the stem
Oak	21.6 ± 1.5	16.1 ± 6.0	At a height of 1.8 ± 0.5 m up the trunk from stem base
Spruce	10.6 ± 4.3	7.5 ± 1.0	Two trees failed in the root system and one tree at a height of 1.5 m up the stem

be attributed to the CW root being pulled upwards, thus placing the gauge under compression (Figure 3). Douglas fir broke at the stem-root base in two trees, and strain in this area was significantly greater than in either the roots or the trunk on the CW side of the tree (Figure 2). Beech and oak both broke in the trunk and strain was always highest on the CW side of the trunk compared to that in the roots.

Discussion

This pilot study of non-invasive methods for determining possible zones of weakness in trees has provided interesting results, despite the low numbers of trees examined, and the fact that all the trees were of a similar size and growing in a forest rather than an urban environment. Furthermore, the different failure mechanisms appear to be linked to root system shape, although a study of root architecture would be needed to verify this hypothesis.

The order of resistance to stem breakage or anchorage failure, of the four species were: beech, oak, Douglas fir and spruce. Beech was over three times more resistant to failure than spruce, whereas oak and Douglas fir were only twice as resistant. Although further stability characteristics should be considered e.g. crown area and architecture, it is rather unfortunate that the species which appears to be most frequently damaged during a storm is that which is most widely planted in Northern Europe, where windthrow causes huge economic losses every year (Nielsen, 1995). However, new management regimes, involving better planting, thinning and felling techniques, are currently being implemented in an effort to reduce windthrow of spruce sp. (Gardiner and Quine, 1999). However, as Douglas fir appears to be more windfirm than spruce, and provides high quality timber, it could be argued that from the stability point of view, this species be used to replace spruce in the windiest parts of Northern Europe.

Beech and oak trees broke in the stem during mechanical failure, compared to Douglas fir and spruce, which generally failed at the tree base or in the root system. Beech and oak both have much stronger wood than either Douglas fir or spruce (Anon, 1987), therefore the stem should be more resistant to breakage. However, in static tests, if trees are well anchored by the root system, and the moment required to resist overturning is greater than that needed to cause

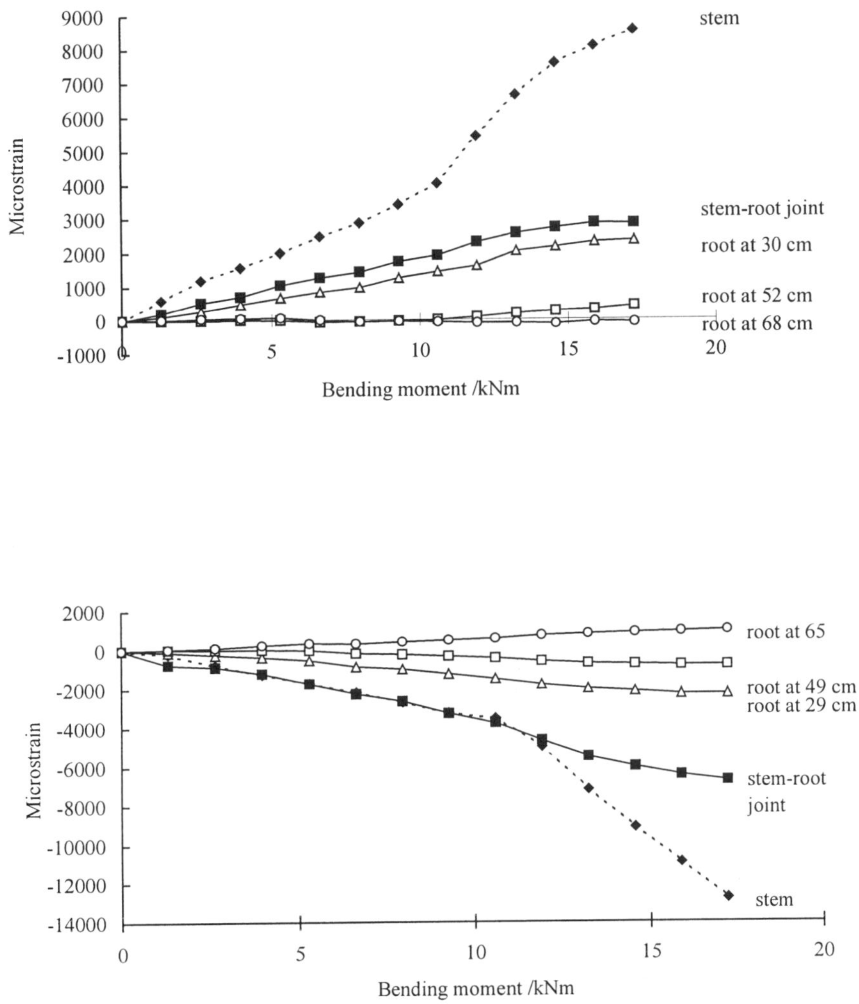

Figure 1. (a) Strain on the counter-winchward (tension) side and (b) on the winchward (compression) side of the tree, increases with bending moment in beech, as the tree was pulled sideways. Strain was greatest in the stem (where failure occurred) compared to the stem-root base and the lateral roots.

stem failure, the trunk will break (Putz et al., 1983). Therefore, beech and Douglas fir (heart type root systems) and oak (tap root system) must all be better anchored than spruce, which failed in the root system. Although neither shoot nor root biomass was considered in this study, root biomass is thought to be one of the most important factors determining anchorage strength, along with root system asymmetry and root strength (Coutts et al., 1999, this volume; Nicoll et al., 1995; Putz et al., 1983; Stokes and Mattheck 1996; Stokes et al., 1997). Therefore, a more complete study of overturning resistance, root/shoot biomass and architecture would provide much information about how anchorage is affected by tree form.

Many studies have been carried out on the anchorage strength and failure of spruce trees, which generally overturn during winching due to the poorly anchored shallow root plate (Blackburn et al., 1988; Coutts, 1983, 1986; Fraser and Gardiner, 1968) identified different components of root anchorage in Sitka spruce and determined that the CW roots, held in tension, were the most important in resisting overturning. Coutts, (1983, 1986) measured the displacement of soil in the soil-root plate during winching, and found that most displacement occurred at a distance of 75 cm from the stem base, on the CW side of the tree, as found with roots from Norway spruce in this study. In a study on root-wood strength, Stokes and Mattheck (1996) found that wood strength increased at a dis-

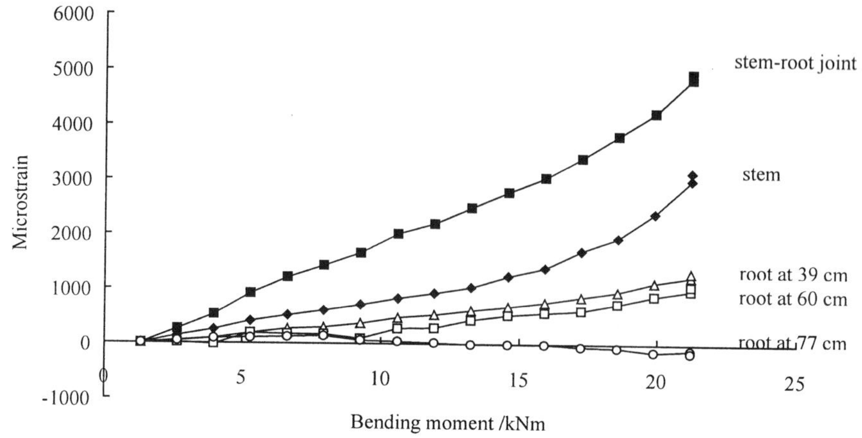

Figure 2. Strain on the counter-winchward side of the tree, increases with bending moment in Douglas fir, as the tree was pulled sideways. Strain was greatest at the stem- root base (where failure occurred) compared to the stem and lateral roots.

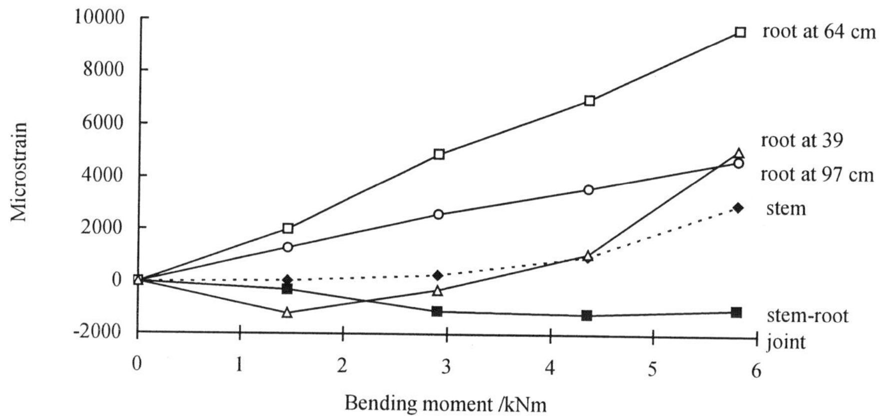

Figure 3. Strain on the counter-winchward side of the tree, increases with bending moment in spruce, as the tree was pulled sideways. Strain was greatest in the lateral roots at a distance of 64cm from the stem base (where failure occurred) compared to the stem and stem-root base.

tance of 50–100 cm from the stem-base in Norway spruce. Nicoll and Ray (1996) also examined this zone in Sitka spruce and determined that on the windward side of the tree, secondary thickening was increased on the upper and lower sides of the root compared to further along the lateral root. As roots in this zone will be susceptible to rupture, it appears that adaptive growth can occur to prevent rupture in localised regions of the tree.

The results showing that the site of highest strain on the CW side of the tree during winching is the probable site of failure are rather surprising, as trees are such heterogeneous structures. If the same results were obtained with a homogenous structure made of an isotropic material, such a response could be expected. However, wood is not homogenous, therefore strain should differ throughout the tree, depending on the material properties present. These results may be explained by the way in which the tree behaves globally under loading; if well-anchored, the roots are not likely to move very much, compared to the stem (as in the case of beech and oak). However, if the roots are poorly anchored, the root system will pivot and sway during loading. It would be interesting and instructive to carry out a similar study in which strain gauges are placed along the length of the stem in beech and oak, in order to determine if in regions of highest strain, failure occurs under maximal loading. Similar studies have been carried out on rubber trees (*Hevea brasiliensis*) and it was found that when strain gauges were placed at four different heights up the tree, the tree broke nearest the gauge which indicated the highest strain values (T. Fourcaud, personal communication).

Although a small number of trees were examined in this study, and all were healthy and of a similar age and size, the results indicate that the method used might be suitable for predicting the possible site of failure in trees during mechanical loading. This possibility was shown in a similar study (Stokes, 1999, this volume) in which Maritime pine, containing grafting tissue in the stem (considered as mechanically weak wood), broke at this point during loading, as indicated by the strain gauges. Further research should be carried out on trees of different ages and sizes, as well as trees with decay in the stem and roots, in order to test the method more thoroughly. If viable, such a method, where the trunk only needs to be winched sideways very slightly, would be useful in determining possible zones of weakness in trees; such regions could then be investigated more carefully, using one of the portable devices proposed by Mattheck (1995, 1998).

Acknowledgements

This study was carried out with aid from the European Commission (project no. AIR3-CT93 1269). Thanks are due to P. Heinze who helped with the fieldwork. The investigation site was put to our disposal by the Forest Office Radolfshausen.

References

Anonymous 1987 Wood Handbook: Wood as an Engineering Material. U.S. Department of Agriculture, Washington D.C. Agriculture Handbook 72.

Blackburn P, Petty J A and Miller K F 1988 An assessment of the static and dynamic factors involved in windthrow. Forestry 61: 235–244.

Brüchert F and Spatz H-C 1999. The mechanics of Norway spruce (*Picea abies* L. Kurst) mechanical properties of standing trees from different thinning regimes. For. Ecol. Managemt. (*In press*).

Coutts M P 1983 Root architecture and tree stability. Plant Soil 71, 171–88.

Coutts M P 1986 Components of tree stability in Sitka spruce on peaty, gley soil. Forestry 59, 173—97.

Coutts M P, Nielsen C C N, Nicoll B C 1999 The development of symmetry, rigidity and anchorage in the structural root systems of conifers. Plant Soil 217, 1–15.

Crook M J and Ennos A R 1996 The anchorage mechanics of deep-rooted larch, *Larix europea* × *L. japonica*. J. Exp. Bot 47, 1509–1517.

Crook M J and Ennos A R 1997 The increase in anchorage with tree size of the tropical tap rooted tree *Mallotus wrayi*, King (Euphorbiaceae). *In* Plant Biomechanics. Eds G Jeronimidis, J F V Vincent. pp 31–36. Centre for Biomimetics, Reading, UK.

Drexhage M and v. Lüpke, B 1995 Aspects of sustainability by afforestation of agricultural set-aside areas: development of roots and root/shoot-ratios. EU-project TREEARCH, 3. Progress Report. Edited by I Denker and C N Nielsen, 63–85.

Dunham R and Cameron A 1999 Crown, Wood and stem properties of wind damaged and undamaged Sitka spruce For. Ecol. Mangemnt. (*In press*)

Fraser A I and Gardiner J B H 1967 Rooting and stability in Sitka spruce. Forestry Commission Bulletin No. 40. HMSO, London.

Gardiner B A., Quine C 1999 Management of forests to reduce the risk of abiotic damage – a review with particular reference to the effects of strong winds. For. Ecol. Managemt (*In press*).

Köstler J N, Brückner E and Bibelriether H 1968, Die Wurzeln der Waldbäume. Verlag Paul Parey, Hamburg & Berlin.

Lindström A and Rune G, 1999 Root deformation in plantations of container-grown Scots pine trees: effects on root growth, tree stability and stem straightness. Plant Soil 217, 29–37.

Mattheck C 1995 The internal optimization of wood. Springer-Verlag, Berlin.

Mattheck C 1998 Design in Nature – Learning from Trees. Springer-Verlag, Berlin.

Mattheck C and Breloer H 1995 The body language of trees. A handbook for failure analysis. HMSO, London.

Nicoll B C, Easton E P, Milner A D, Walker C, Coutts M P 1995 Wind stability factors in tree selection: distribution of biomass within root systems of Sitka spruce clones. *In* Wind and Trees. Eds M P Coutts, J. Grace. pp 276–292. Cambridge University Press, UK.

Nicoll B C and Ray D 1996 Adaptive growth of tree root systems in response to wind action and site conditions. Tree Physiology 16, 899–904.

Nicoll B C and Armstrong A 1998 Development of *Prunus* root systems in a city street: pavement damage and root architecture. Arboric. J. 22, 259–270.

Nielsen C C N 1995 Recommendations for stabilisation of Norway spruce stands based on ecological surveys. *In* Wind and Trees. Eds M P Coutts and J Grace. pp. 424–435. Cambridge University Press, UK.

Peltola H and Kellomäki S 1993 A mechanistic model for calculating windthrow and stem breakage of Scots pines at stand edge. Silva Fennica 27, 99–111.

Petty J A and Swain C 1985 Factors influencing stem breakage of conifers in high winds. Forestry 58: 75–85.

Putz F E, Coley P D, Lu K, Montalvo A and Aiello A 1983 Uprooting and snapping of trees: structural determinants and ecological consequences. Can. J. For. Res. 13: 1011–1120.

Stokes A 1999 Strain distribution during anchorage failure of *Pinus pinaster* Ait. at different ages and tree growth response to wind-induced root movement. Plant Soil, 217, 17–27.

Stokes A and Mattheck C 1996 Variation of wood strength in roots of forest trees. J. of Exp. Bot. 47, 693–699.

Stokes A, Nicoll B C Coutts M P and Fitter A H 1997 Responses of young Sitka spruce clones to mechanical perturbation and nutrition; effects on biomass allocation, root development and resistance to bending. Can. J. For. Res. 27, 1049–1057.

Wessolly L and Erb M 1998 Handbuch der Baum Statik und Baum Kontrolle. Patzer Verlag, Berlin-Hannover.

A. Stokes (ed.), The Supporting Roots of Trees and Woody Plants: Form, Function and Physiology, 287–293.

Review of in situ shear tests on root reinforced soil

J. E. Norris and J. R. Greenwood
Department of Civil and Structural Engineering, The Nottingham Trent University, Burton Street, Nottingham NG1 4BU, UK

Key words: in situ shear strength, root reinforced soil, shear box

Abstract

Bioengineering and the quantification of the effects of vegetation in civil engineering are still relatively new concepts in the United Kingdom. This paper reviews the development and use of in situ shear box techniques to measure the shear strength of root reinforced soil. Details of variation in design and the test procedures adapted by researchers are described and problems outlined. The review has led to the development of a 150 × 150 × 100 mm in situ shear box at Nottingham Trent University for testing root reinforced Gault Clay on a bioengineering demonstration site on the M20 motorway at Maidstone, UK.

Introduction

Bioengineering in the United Kingdom is a relatively new concept. Vegetation has been used in the past as a means of reducing the visual impact of civil engineering works and enhancing the quality of the landscape. It has been recognised that vegetation has contributed to slope stability but to an unquantifiable degree.

Recent work in Europe (Coppin and Richards, 1990; Mulder, 1991; Tobias, 1995), New Zealand (Phillips and Watson, 1994), Japan (Endo, 1980) and North America (Ziemer, 1981) has provided information on the engineering benefits of vegetation which must now be further developed and applied.

In this paper, previous work on in situ shear testing of root reinforced soil is reviewed to assist in the development of suitable apparatus for use on a Gault Clay bioengineering demonstration slope in the United Kingdom (Greenwood et al., 1996).

Review of in situ shear testing methods

The principle of in situ shear testing is described in BS5930. The sample of ground is subjected to direct shear using a stress system similar to the laboratory shear box (BS1377). However, this system has mainly been developed for testing in situ rock but not root-reinforced soil. A number of researchers have therefore devised and developed in situ shear testing apparatus for this purpose.

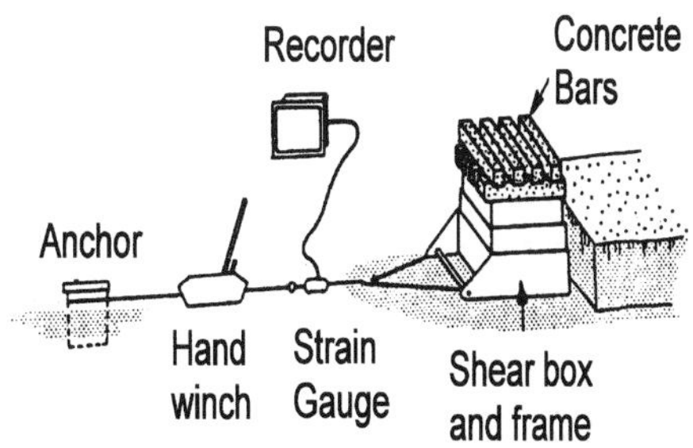

Figure 1. Schematic illustration of the in situ shear apparatus used by Endo and Tsuruta (1969) and Endo (1980) (modified from Endo 1980).

The first reference to testing of root reinforced soil was Endo and Tsuruta in 1969 who designed apparatus to measure the contribution of small *Alnus glutinosa* roots to the strength of homogeneous nursery soil in Japan. O'Loughlin (1972) later used a slightly modified design for the study of old-growth forests of coastal British Columbia, Canada. Endo reused his shear apparatus in 1980 to measure the strength of *Betula japonica* Sieb. and *Alnus japonica* Steud. tree roots.

The design of Endo and Tsuruta (1969) shearing apparatus (Figure 1) comprises a 500 mm wide × 500 mm long × 300–600 mm adjustable height shearing case made from 3-mm thick iron plates, in which the soil block is fully enclosed on all four sides; an outer frame to hold the shearing case and to act as a carrier; a wooden load stand; concrete bars for additional load; a hand winch of 2.5 tons capacity and a set of strain gauges and recorder attached by 10-mm wire ropes. The apparatus measures the in situ shear strength by pulling the shear case and surcharge load by continually levering the hand winch. The pulling force generated is monitored continuously using the strain gauge and recorder.

In 1981, Ziemer developed a shear box to measure the strength of mature *Pinus contorta* on coastal sands in California. Ziemer's shear box apparatus differed markedly from Endo and Tsuruta's original design in that instead of enclosing the soil block on four sides, two of the sides of the box remained open to allow for the continuous nature of horizontal roots through the soil block and into the surrounding soil. A base plate was also used. In this instance, shearing is along two vertical parallel planes and not along the base of the shear box and the shear box is pushed as opposed to being pulled as in Endo and Tsuruta's shear apparatus. Ziemer's apparatus consists of an open sided shear box made of steel plates bolted together on the top, bottom, front and back of the soil block. A mechanical jack extends at a rate of 12.7 mm/min for approximately 7 min and a proving ring mounted between the mechanical jack and the shear box is used to measure the shearing force.

Wu et al. (1988) successfully used Ziemers design for their study of soil–root systems in tree species Western hemlock, Alaska cedar and Silver maple at four test sites in Oregon, Alaska (× 2) and Ohio. The particular tree species used in this study were found to have many horizontally aligned roots so Ziemers design and test method proved fruitful although Wu et al. modified the dimensions of the shear box to suit their particular needs.

O'Loughlin (1981) adapted Ziemers design to test Beech forest in New Zealand. The shear box remained open on two sides but the base plate was removed so that the base of the soil block was kept in continuity with the surrounding soil (Figure 2). O'Loughlin reduced the dimensions of the shear box to 300 mm wide × 300 mm long × 150 mm high. The shear box is powered by a horizontally mounted CBR (California Bearing Ratio) test jack, which is turned at a constant rate of 1 rps. The jacking mechanism forces the soil block forward at 13 mm/min. Shear stresses are recorded using a dial gauge and proving ring.

In 1986, Abe and Iwamoto designed a large scale apparatus to test *Cryptomeria japonica* D. DON (sugi) a widely planted tree in Japan. This large scale shear apparatus shows a combination of design features from Endo and Tsuruta and Ziemer's apparatuses. Abe and Iwamoto's shear apparatus was of a much larger scale to any of the other previous in situ shear boxes, being 1 m wide × 1 m long × 0.3–1 m adjustable height. The apparatus comprises the following parts (Figure 3): (1) shear box made of steel plates 3 mm thick capable of holding the soil block together during the test; (2) oil jack and pump used to produce the shear load with a maximum capacity of 20 tons and 500-mm long stroke; (3) a 5-ton capacity load cell to measure the shearing load; (4) displacement meters to 0.02 mm accuracy; (5) digital strain recorder; (6) a steel bar for additional load (50 kg in weight, 1 m in length).

The apparatus works by pushing the shear box forward. The shear method is incrementally stress controlled by increasing the shear load by 100 kgf every 20 min.

Reported advantages of this system are: (i) whole live tree root systems could be sheared (shearing area 1 m^2); (ii) the displacement can be observed from the side of the block; (iii) both the acting forms of the roots in the soil and soil block movement can be observed by exposure after the test and the results can be expressed by the Coulomb equation (Abe and Iwamoto 1986).

Clark (1992) during research into the principles of bioengineering in East Nepal developed an in situ shear box to measure the root strength of Nepalese grasses. The design criteria required the shear box to be portable, for use on slopes steeper than 30° to the horizontal and to simulate shallow translational failures to a maximum depth of 300 mm. Initial design features included a large metal shear box, 500 mm wide × 750 mm long × 70 mm high, which was pulled by a hand winch to apply the strain. Sand bags were used as additional normal load. This design bears some resemblance to Endo and Tsuruta's original shear apparatus. This first design encountered problems during strain application. The back edge of the shear box created a passive failure with the shear plane, which penetrated the surface forcing the box to ride up and tilt. To overcome this Clark attempted to divide the shear box into a series of three smaller units (250 × 500 mm) using vertical grouser plates. These in turn had their problems and were subsequently abandoned due to the plates reducing the effectiveness

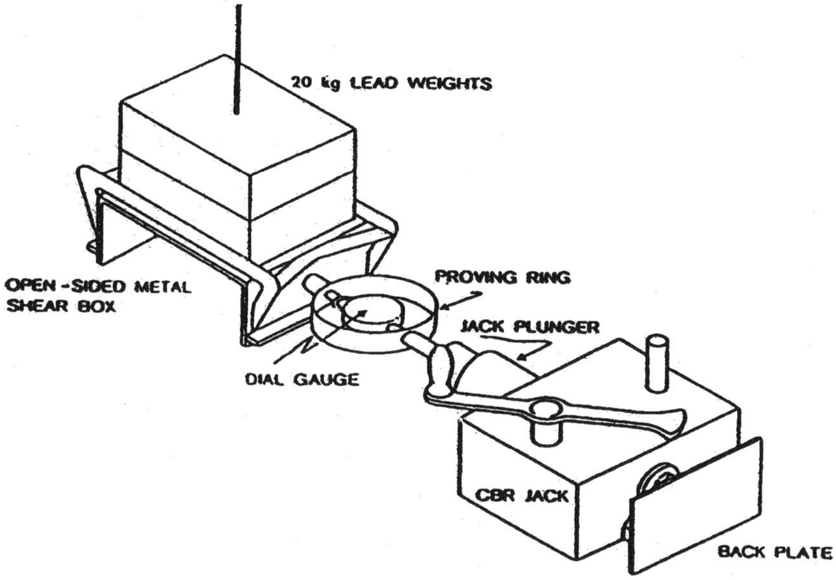

Figure 2. Field shear box testing equipment of O'Loughlin (1981).

of root reinforcement in the soil block as they cut the root mass/ball.

A revised design of a four-sided steel frame 250 mm wide × 250 mm long × 100 mm high pushed by a mechanical jack was finally used. The shear box was displaced at a rate of 1 mm/s. This design works in much the same way as Abe and Iwamoto's large scale direct shear apparatus although on a much smaller scale. This shear box was not without its limitations as the maximum displacement of the apparatus was 35 mm and an estimated 100 mm displacement was required to pull or break the roots and the front edge of the shear box lifted off the ground during jacking. Clark also felt that having digital recorders to measure and record displacement, shear force and time would have greatly improved the operators ability to observe the movements within the shear box.

Tobias (1995) also used the same principle method as Endo and Tsuruta to measure the shear strength of grassland (Table 1) in Switzerland. The shear box had dimensions of 500 mm wide × 500 mm long × 150 mm high. These dimensions reduced marginal effects such as roots being cut by the pushing of the metallic shear frame and also kept the whole apparatus reasonably manageable.

Yatabe et al. (1996) briefly discusses in situ direct shear tests on weathered granite soil with roots of Chinese cedar, fir shrub, bamboo and grass. Their apparatus reflects the BS 5930 in situ shear apparatus as it uses jacking systems for both normal and shear stress.

The main features of each of the referenced in situ shear box tests on root reinforced soil are summarised in Table 1.

Discussion

As described above only a few researchers have reported attempts to use an in situ shear box to determine the contribution of roots to soil strength. This may be due to the lack of standard commercially available in situ shear testing apparatus and the difficulties in preparing the soil block and setting up of the apparatus prior to testing.

The problems in the design of an in situ shear box may have been partly alleviated in more recent work through researchers highlighting the problems of their apparatus for example Clark (1992) encountered problems with tilting of the shear box during shearing, limited displacement of the shear box and difficulties in observing the test. Tobias (1995) stated that "it is indispensable to watch carefully what is happening during the test . . . ".

The two distinct shear box designs of Endo and Tsuruta (1969) and Ziemer (1981) measure the shear resistance of root reinforced soil in quite different ways. Ziemer in his open sided shear box provided for horizontally aligned roots normal to the shear plane,

Table 1. Review of in situ shear box testing relating to root reinforced soil

Author	Location	Vegetation type tested	Shear box dimensions ($w \times l \times h$	Comments
Endo and Tsuruta (1969)	Japan	*Alnus glutinosa*	500 × 500 × 300–600 mm (adjustable height) (Figure 1)	Roots 1–2 mm diameter pulled from the soil block during shearing.
O'Loughlin (1972) in Ziemer (1981)	British Columbia, Canada	*Pseudotsuga menziesii* *Thuja plicata* *Tsuga heterophylla*	Modified design of Endo and Tsuruta (1969)	Roots 5 mm in diameter pulled out of the soil block instead of shearing along the shear plane.
Endo (1980)	Japan	*Betula japonica* Sieb. *Alnus japonica* Steud.	500 × 500 × 300–600 mm (Figure 1)	A simple design but made possibly impractical by the addition of two kinds of concrete sticks to provide additional load. Vertically aligned roots may pull out or break.
Ziemer (1981)	California	*Pinus contorta*	Open sided shear box 600 × 300 × 300 mm. Open side is 300 mm^2.	Designed so that the shear box shears along two parallel vertical planes. Shear rate 1.27 cm/min. Roots lying horizontally and aligned normal to the shear planes are undisturbed by the shear box and continue into the soil either side of the box.
O'Loughlin (1981), O'Loughlin et al. (1982)	Maimai experimental catchment, New Zealand	Beach forest	Open sided and base shear box 300 × 300 × 150 mm (Figure 2)	Similar design to Ziemer (1981) although on a reduced scale. Shear rate 1.3 cm/min. During shearing, shear surfaces develop at the sides and base of the soil block; roots either stretch and break or pull out at these surfaces. Reviewed by Phillips and Watson (1994).
Abe and Iwamoto (1986)	Chiyda Experimental Branch Station nursery, Forestry and Forest Products Research Institute, Ibaraki Prefecture.	*Cryptomeria japonica* D. DON (sugi)	Large-scale direct shear box 1 × 1 × 0.3–1 m (adjustable height) (Figure 3)	Shear loading rate 100 kgf per 20 min. A large tree root present in the soil can easily be sheared by the oil jack with a uniform pressure.

Table 1. Continued

Author	Location	Vegetation type tested	Shear box dimensions ($w \times l \times h$	Comments
Wu et al. (1988)	Siuslaw National Forest Mapleton, Oregon	Western hemlock	Undrained or quick direct shear tests	Apparatus designed by Ziemer (1981). The dimensions of this box represent a compromise
	Tongass National Forest Alvin Bay, Alaska	Western hemlock Alaska cedar	$300 \times 600 \times 300$ mm	since increasing the length of the box would increase the number of roots that would be cut off
	Tongass National Forest Peterson Cr., Alaska	Western hemlock		and reducing the length would reduce the resistance developed by the root inside the box.
Clark (1992)	Ohio State University Columbus, Ohio	Silver maple		
	Ghopa Camp, Nepal	*Pennisetum Purpeneum* (Napier grass)	$250 \times 250 \times 100$ mm	It was noted that the four sided shear box had tilting problems during shearing.
	Karkichap, Nepal	*Seteria anceps* (local Nepalese fodder grass)		
	Ambote, Nepal	*Themeda* sp., *Neyraudia* sp., *Cymbopogan Microtheca*		
	Dhankuta, Nepal	*Cymbopogan Microtheca*, Imperata.		
Tobias (1995)	Zurich, Switzerland	Grasses (*Alopecurus geniculatus, Poa pratensis Agrostis stolonifera, Festuca rubra, Lolium multiflorum, Festuca pratensis*	$500 \times 500 \times 150$ mm	Marginal effects of roots cut by pushing in the metallic frame are minimised with these dimensions. Tests interpreted as unconsolidated and undrained.
Yatabe et al. (1996)	Japan	Chinese cedar, fir, shrub, bamboo, grass	Direct shear apparatus $300 \times 300 \times 120$ mm	Normal stress applied by air cylinder, shear stress applied with a hydraulic jack.

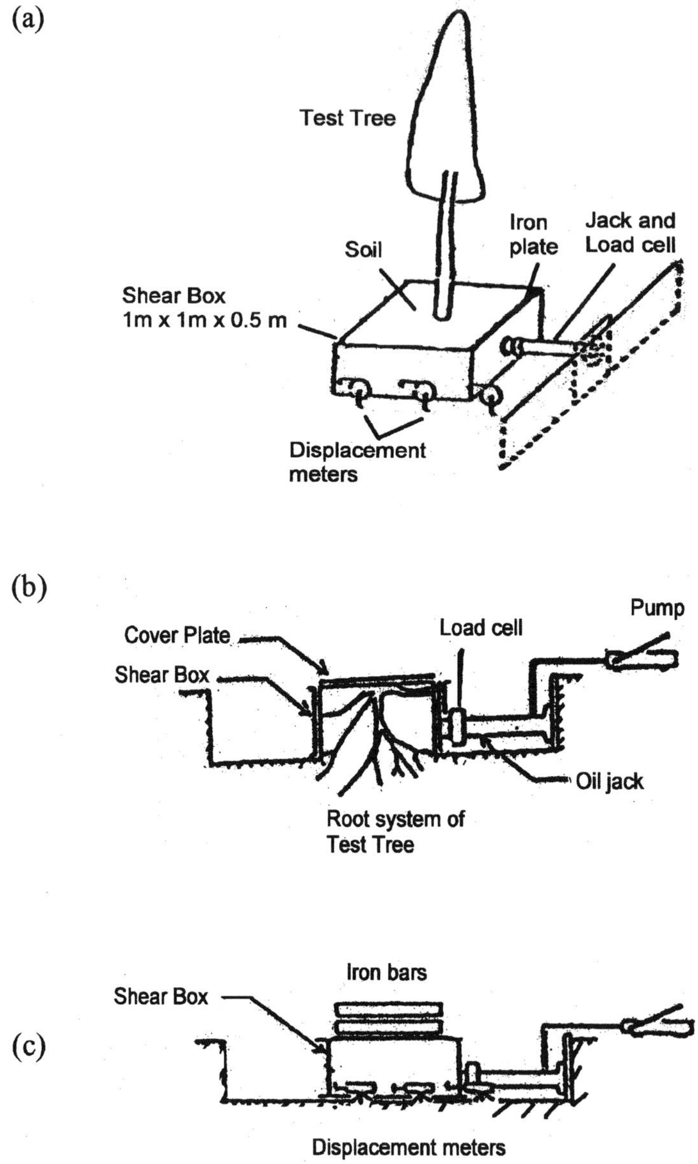

Figure 3. Large-scale direct-shear apparatus (modified from Abe and Iwamoto 1986).

which protrude undisturbed through the soil block and into the soil on either side of the shear box. Hence, these roots can only break when shearing takes place. However, Endo and Tsuruta's closed shear box roots can only extend undisturbed in the vertical direction, therefore these roots are able to break or pull out of the soil. When O'Loughlin (1981) removed the base plate of Ziemers shear apparatus roots could then be sheared in all directions (both horizontally and vertically) – although this probably creates more unknowns in the analysis.

As each author has used slightly different dimensions for the shear box and also used either a push or pull method of shearing the soil block this raises questions about repeatability between methods.

With more experimental work being carried out the results of these tests will gain credibility for application in geotechnical engineering and in the future, a database of root strength, should evolve for major plant species planted in particular soil types.

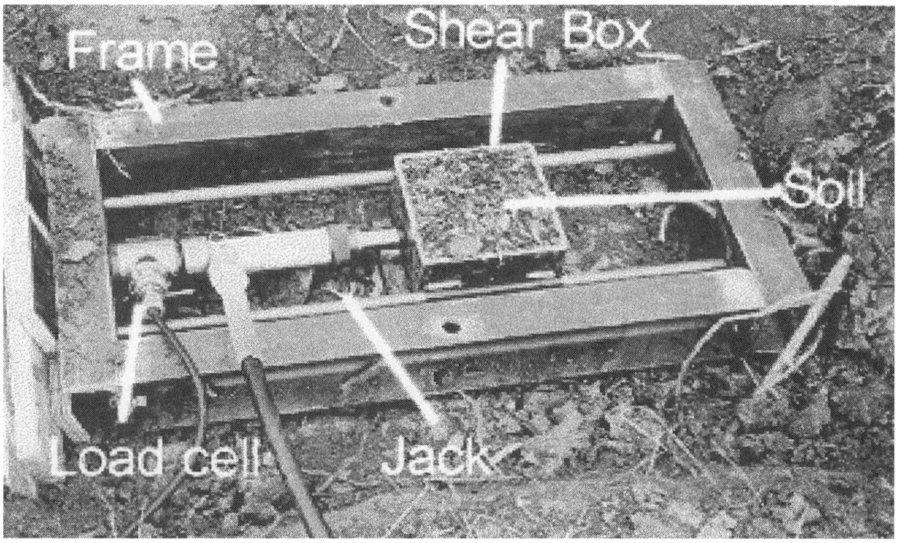

Figure 4. In situ shear box apparatus, developed by Nottingham Trent University, in operation on the bioengineering demonstration site, M20 motorway, Maidstone, UK.

Nottingham Trent University's in situ shear apparatus

This review of in situ shear boxes has assisted in the development of an in situ shear box at Nottingham Trent University. Whilst large shear boxes are desirable to overcome the local discontinuities effects with a soil/root system. It was recognised that the large shear boxes introduced more problems in terms of control and handling of equipment and the actual testing procedure. It was therefore concluded that a relatively small (150 × 150 mm) portable shear box apparatus is most appropriate, as it permits a greater number of tests. It is also hoped to include a means of anchoring larger roots within the shear box to obtain an indication of the pull out forces generated during shear.

A push jacking mechanism is favoured as opposed to the pulling mechanism of Endo and Tsuruta, for reasons outlined by Clark (1992). Clark also had tilting problems with the push method so to prevent tilting the shear box is designed inside a steel frame that sits on two runners. The steel frame also enables vertical roots protruding out of the ground to be clamped in position therefore simulating the trees resistance to shearing and also to measure the tensile force required to pull the root out of the ground.

The shear box is 150 mm wide × 150 mm long × 100 mm high made of steel plates. A hydraulic jack and pump provide the shearing force, which in turn is measured by a 10 kN load cell. Figure 4 shows the in situ shear box apparatus in operation on the bioengineering demonstration site on the M20 motorway Maidstone UK.

More information and results of the in situ shear testing and bioengineering trial are provided in Greenwood et al. (1999).

Acknowledgements

This work is funded by a Research Enhancement Fund grant from The Nottingham Trent University. CIRIA (the Construction Industry Research and Information Association) is thanked for providing a suitable field site and for supporting bioengineering research in the UK.

References

Abe K and Iwamoto M 1986 Preliminary experiment on shear in soil layers with a large direct shear apparatus. J. Jpn For. Soc. 68, 61–65.

BS1377 1990 Methods of Test for Soils for Civil Engineering Purposes. HMSO, London.

BS5930 1995 (Draft edition) Code of Practice for Site Investigations. HMSO, London.

Clark J E 1992 Principles of bioengineering with reference to East Nepal. Ph.D dissertation, Silsoe College, Cranfield University, UK.

Coppin N J and Richards I G 1990 Use of Vegetation in Civil Engineering. CIRIA Butterworths, London.

Endo T 1980 Effect of tree roots upon the shear strength of soil. Jpn Agric. Res. Quart. 14, 112–115.

Endo T and Tsuruta T 1969 The effect of tree roots upon the shearing strength of soil. Annual Report of the Hokkaido Branch, Forest Place Experimental Station, Sapporo, Japan, 167–182 (In Japanese with English summary).

Greenwood J R, Morgan R P C and Short J 1996 Bio-engineering: a field trial at Longham Wood Cutting, M20 Motorway. CIRIA, Special Publication 128, London.

Greenwood J R, Morgan R P C and Coppin N 1999 Bioengineering: the Longham Wood Cutting Field Trial. CIRIA, London.

Mulder H F H M 1991 Assessment of landslide hazard. Faculty of Geographical Sciences University of Utrecht, The Netherlands.

O'Loughlin C L 1972 An investigation of the stability of the steepland forest soils in the Coast Mountains, southwest British Columbia, Ph.D dissertation, University of British Columbia, Vancouver, BC, Canada.

O'Loughlin C L 1981 Tree roots and slope stability. What's new in Forest Research. Forest Research Institute Publication No. 104, Rotorua, New Zealand.

O'Loughlin C L, Rowe L K and Pearce A J 1982 Exceptional storm influences on slope erosion and sediment yield in small forest catchments, North Westland, New Zealand. 84–91. *In* E M O'Loughlin and L J Brens First National Symposium on Forest Hydrology, Melbourne, 1982 Eds Institution of Engineers, Australia, National Conference Publication 82/6.

Phillips C J and Watson A J 1994 Structural tree root research in New Zealand: a review. Landcare Research Science Series No. 7. Manaaki Whenua Press, Lincoln, Canterbury, New Zealand.

Tobias S 1995 Shear strength of the soil root bond system. Vegetation and Slopes, Thomas Telford.

Wu T H, Beal P E and Lan C 1988 In situ shear test of soil root systems. J. Geotech. Eng. 114, 1376–1394.

Yatabe R, Yagi N, Yokota K, Enoki M, Kobori S and Mukaitani M 1996 Stability analysis of slope reinforced by roots network. 835-840. *In* Earth Reinforcement. (Eds H Ochiai, N Yasufuku and K Omine). Balkema, Rotterdam.

Ziemer R 1981 Roots and shallow stability of forested slopes. Int. Assoc. Hydrol. Sci. 132, 343–361.

A. Stokes (ed.), The Supporting Roots of Trees and Woody Plants: Form, Function and Physiology, 295–304.

An experimental study on the mechanical behaviour of the maize root system: analysis of the soil/root bond and determination of rigidity modulus

Sandrine Denis[1], Fabrice Bremand[1] and Yannick Hebert[2]
[1]*LMS, Université de Poitiers, CNRS UMR 6610, SP2MI - Téléport 2 - BP 179, 86960 Futurscope Cedex, France*,*
[2]*INRA - Station d'Amélioration des Plantes Fourragères, 86600 Lusignan, France*

Key words: maize, rigidity modulus, root/soil bond, root system

Abstract

The main goal of this work is concerned with the experimental measurement of mechanical parameters of maize roots, such as rigidity modulus and the root/soil bond in order to numerically simulate the mechanical behaviour of the root system in soil. For the determination of rigidity modulus, non-contact and non-disturbing optical methods of strain measurement have major advantages over classical extensometers due to the complexity of material. Amongst all these techniques, the tracking of two markers constitutes the best choice. Since a maize root is not rectilinear, a tensile test introduces two important perturbations given by the transverse and the out-of plane displacements. To avoid these non-suitable strains, we have used two CCD cameras allowing the determination of the spatial co-ordinates of the two markers. Thus longitudinal strain can be easily determined. With stress evaluated from a special testing machine, the rigidity modulus for three genotypes of maize is shown. In order to evaluate root/soil bond, 430 tests were carried out to measure the tensile force required to pull out each root from soil on three different genotypes. Correlations were made between plant anchorage, age, length and branching of roots, and also soil moisture.

Introduction

Lodging of maize (Zea mays L.) by wind and storms may cause much damage to the plants. This phenomenon is principally due to anchorage failure by the root system. As such damage results in decreasing yield and quality of the harvest, selection for increasing root lodging resistance is an important objective.

As the maize plant is pulled sideways, up to an angle of 10°, leeward roots move down and windward roots upwards (Ennos et al., 1993, Hébert, unpublished data). Furthermore, little movement of the soil is observed around the base of the stem: soil on the windward side moves upwards and soil on the leeward side downwards. At angles of about 10°, leeward roots buckle (Ennos et al., 1993, Hébert, unpublished data). The same behaviour has been observed for wheat (Ennos, 1991). These soil movements show that the root/soil bond is an important factor of the anchorage of root system. Anchorage depends, on the geometrical data of roots, i.e. length and branching (Easson et al., 1995; Stokes et al., 1996); and also, on soil moisture and the substrate composition (Easson et al., 1995).

The goal of this study is to evaluate the influence of the main parameters governing the embedding of the root system in the soil. The genetic and environmental variations in root morphology are now well known (Duparque and Pellerin, 1990; Guingo and Hébert, 1997; Guingo et al., 1998; Hébert et al., 1992; Zuber, 1968), and the importance of root number, root diameter and spreading direction have been clearly demonstrated. Nevertheless, a better performance of selection procedures still remains dependent upon an integrated view of the variations of each trait involved. At the same time, we carry out experimental analysis on the mechanical behaviour of the root system for various maize genotypes. The correlation between

* FAX No: (33)(0)549496504.
E-mail: bremand@ims.univ.poitiers.fr

Table 1. The genetic material used in the experimental designs

Genotype	Source
W117 × EP1	Experimental hybrid
F2 × F1772	Experimental hybrid
DK300	Sockalb

both approaches will allow us to develop a model using the Finite Element Method (F.E.M.) of a lateral pushing test on a maize plant in soil (for an angle inferior to 5°), in order to investigate the mechanical behaviour of each genotype. Therefore, it is necessary to determine some mechanical properties, e.g. rigidity modulus of roots and stem and the root/soil bond, i.e. to determine the behaviour of leeward roots to being pushed into the soil and the behaviour of the windward roots to being pulled outwards from the soil, as the plant is lightly pulled sideways.

In this paper, we present the experimental techniques used for the evaluation of rigidity modulus of each maize root and the soil-bond interface as the root is pulled out from the soil. For small angles, the role of the windward roots is as important as that of the leeward roots. Therefore, we have decided to study initially the mechanical behaviour of the windward roots in soil.

Hence, the geometrical data will allow us to carry out a mesh of maize plant structure using F.E.M. and the boundaries conditions will be determined using the mechanical parameters calculated during the pull-out tests.

Materials and methods

Plant material and experimental design

One maize variety and two experimental hybrids (Table 1) were chosen due to their different lodging behaviour and their different root morphology i.e. number of roots, angular spread and root diameter. The level of resistance to lodging is high for DK300, whilst the two experimental hybrids are sensitive to this phenomenon but they have very different root morphological traits.

Plant material was sown, at a seed density of 95 000 seeds/ha, in a field divided into two replicates, where the soil composition was silt (25%), clay (15%), sandy and organic material. Only one of the middle two rows was considered for the measurements. Standard irrigation was provided.

Figure 1. The root system of mature maize plant. The adventitious roots of inter-node 8 emerge above the soil.

120 plants (20 for each genotype and for each replication) were selected at random. For each variety, only inter-nodes 6, 7, 8 (when existing), noted E6, E7 and E8, have been studied (Figure 1) because of the important role of these inter-nodes in the mechanical anchorage of root system (Guingo and Hébert, 1997; Hébert et al., 1992). These roots are called prop roots. The function of the other inter-nodes is to absorb water and nutrients from soil.

For each inter-node which emerges above ground, a drawing was made in order to record the angular position of each root and the day of appearance.

Rigidity modulus determination

Hooke's law states that the studied material must be homogeneous and isotropic, which is not the case in maize roots. Therefore, it is preferable to use the term rigidity modulus rather than Young's modulus.

Method

Tensile tests were performed on maize root specimens which were selected depending on certain geometrical conditions. First of all, the most rectilinear part of the root, close to the stem must be chosen because this portion of root is the most sensitive to mechanical constraints. Furthermore the root length should be at least

50 mm and its radius R must be almost constant in order to determine the stress $\sigma = F/\pi R^2$, where F is the tensile force. If the roots were too curved, only a combined tension bending test could be performed, which would not give a correct rigidity modulus. Since this complex material is non-homogeneous with a surface which can present roughness or other irregularities (Figure 1), classical extensometers as strain gauges are not suitable. The choice of optical methods of strain measurement appears very important because such methods can performed non-contact and non-disturbing evaluation. Furthermore, these techniques have the advantage to allow us to measure the local strains, i.e. on a short area, and not to measure the global strain, i.e. on the whole surface of the specimen. Several methods are available i.e. optical diffraction of a laser beam by a grating marked on the surface (Brémand and Lagarde, 1986), numerical spectral analysis by Fast Fourier Transform (FFT) (Dupré et al., 1993) or white light speckle by numerical correlation (Sutton et al., 1986), but they need to either stick or to engrave a grid on the surface which is not possible with maize roots (Denis, 1996). To obtain the longitudinal elongation only two points should be identified which is the tracking of the two markers method (Brémand et al., 1995). These markers are drawn with black ink onto the root. With the help of an imaging system, this method consists of recording images of the specimen for each loading step. Thus, the x and y co-ordinates are obtained by the calculation of the centre of gravity of each marker (Brémand et al., 1992).

During the tests, different displacements were noted. The longitudinal displacement is due to the tensile force, whereas the transverse and the out-of-plane displacements come from the irregular geometry of maize roots. The latter induces variations in the enlargement ratio leading to false strains. To avoid these variations, a stereoscopic experimental set-up is used.

In order to quantify the transverse and the out-of-plane displacements, two identical CCD cameras are used (Denis et al., 1998). They are located perpendicularly (Figure 2) in such a manner that the transverse displacements of one camera corresponds to the out-of-plane displacements of the other. The two CCD cameras have identical lenses with a focal length f.

The physical referential is determined as follows:

- X-axis is parallel to the line passing by the focal points $I_1(L, -L, 0)$ and $I_2(-L, -L, 0)$ of each camera,
- Y-axis goes from cameras to specimen,
- Z-axis is vertical,
- the origin O is defined as the intersection of the two optical axis which are located at 45° related to Y-axis.

The physical co-ordinates (x, y, z) of a point M (Figure 3) are obtained by triangulation with the two corresponding points M_1 and M_2 (images of M, respectively, by CCD1 and CCD2). Let us call a_1 and b_1 the co-ordinates of M_1 and a_2 and b_2 the co-ordinates of M_2, these quantities are expressed in pixels and are obtained from the calculation of the gravity centre by a classical interpolation (Brémand et al., 1992).

The x and y co-ordinates of M can be determined by the calculation of the intersection of the two straight lines (A_1, I_1) and (A_2, I_2) and these values are obtained according to the lengths f, L and the x-components a_1 of M_1 and a_2 of M_2 where A_1 and A_2 are the intersection between each optical axis and the two straight lines (M_1, I_1) and (M_2, I_2). The last co-ordinate z is obtained from x, y previously calculated and the lengths f and L.

Strain measurement

In order to measure longitudinal strain, the tracking of two markers method has been chosen. The stereoscopic method gives knowledge of the three-dimensional co-ordinates of markers P_1 and P_2. Let (x_1, y_1, z_1) be the co-ordinates of P_1 and (x_2, y_2, z_2) the co-ordinates of P_2, at time t. At time t=0, these co-ordinates were (x_{10}, y_{10}, z_{10}) and (x_{20}, y_{20}, z_{20}). Hence the displacement $\vec{U}$ of the point P_1 has for components:

$$\begin{cases} u_{x_1} = x_1 - x_{10} \\ u_{y_1} = y_1 - y_{10} \\ u_{z_1} = z_1 - z_{10} \end{cases} \text{(This is the same for } P_2\text{).} \quad (1)$$

The strain tensor ϵ is obtained by deriving the displacement $\vec{U}$ (Germain, 1983). Only the longitudinal strain ϵ_{zz} is of interest. Thus

$$\epsilon_{zz} = \frac{1}{2}\left(2^* \frac{\partial u_z}{\partial z} + \left(\frac{\partial u_x}{\partial z}\right)^2 + \left(\frac{\partial u_y}{\partial z}\right)^2 + \left(\frac{\partial u_z}{\partial z}\right)^2\right) \quad (2)$$

where $\frac{\partial u_z}{\partial z} = \lim_{\Delta z \to 0}\left(\frac{u_{z_1} - u_{z_2}}{\Delta z}\right)$, with $\Delta z = z_{10} - z_{20}$ which leads to

$$\epsilon_{zz} = \frac{1}{2}\left[2^*\left(\frac{u_{z_1} - u_{z_2}}{\Delta z}\right) + \left(\frac{u_{x_1} - u_{x_2}}{\Delta z}\right) + \left(\frac{u_{y_1} - u_{y_2}}{\Delta z}\right) + \left(\frac{u_{z_1} - u_{z_2}}{\Delta z}\right)^2\right] \quad (3)$$

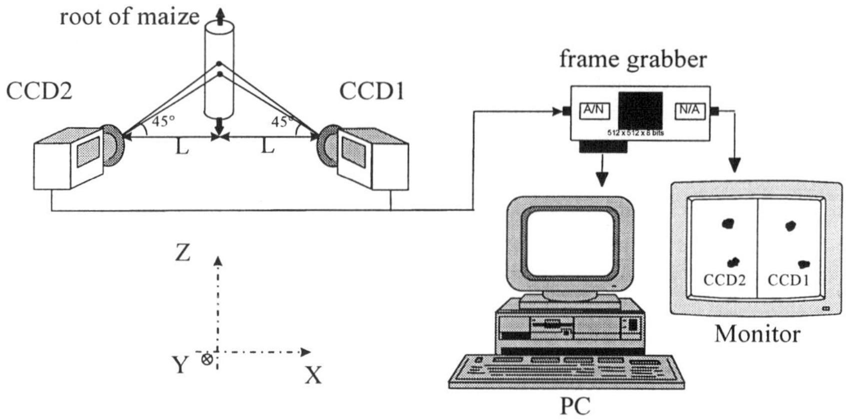

Figure 2. Experimental apparatus allowing us to determine rigidity modulus: two CCD cameras located perpendicularly and a PC used for images acquisition.

when $\Delta z \to 0$.

Usually it can be thought that the error made on the location of one marker is one tenth of a pixel. For these experiments (with $L = 370$ mm and $f = 50$ mm), an accuracy of around one hundredth of millimetre has been obtained for each co-ordinate. As the distance Δz around 5 mm, the strain accuracy is estimated to 10^{-3} which is more than enough for the purposes of these experiments.

Experimental apparatus

In order to perform these tests, a specially designed low charge testing machine was used. The stepper motor was controlled by a PC. The loading was applied at an imposed displacement speed of 1 mm/min and was recorded by an acquisition card at the regular interval: $\Delta_t = 2$ s, chosen by the user (Figure 4). A second PC was used for images acquisition. This PC contained a frame grabber Matrox PIP 1024 (512*512 pixels, 8 bits). This card received pictures of the two CCD cameras with the same time interval $\Delta_t = 2$ s.

Root–soil interactions

A portable device has been specially made for the study of the cohesion between root and soil. This machine allows us to directly measure the tensile force required to pull out each root from soil, *in situ*. This force is based on the linear response of a spring when submitted to a force F; its length variation Δ is given by $F = k^*\Delta$, where k represents stiffness (Figure 5). The displacement $\Delta_{\max}$, corresponding to the maximum tensile force, was measured by a displacement sensor. The value was read using a voltmeter. This device is capable of measuring tensile forces in the range 0–300 N with an accuracy around 0.1 N. Anchorage was quantified by measuring the maximum pull out force.

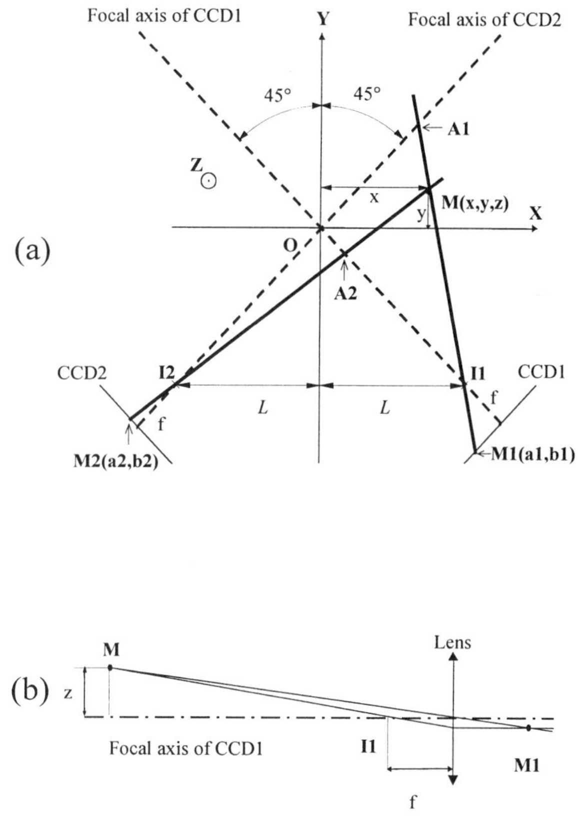

Figure 3. Stereoscopic setup from the top (a) and from the side (b) for one of the two CCD cameras.

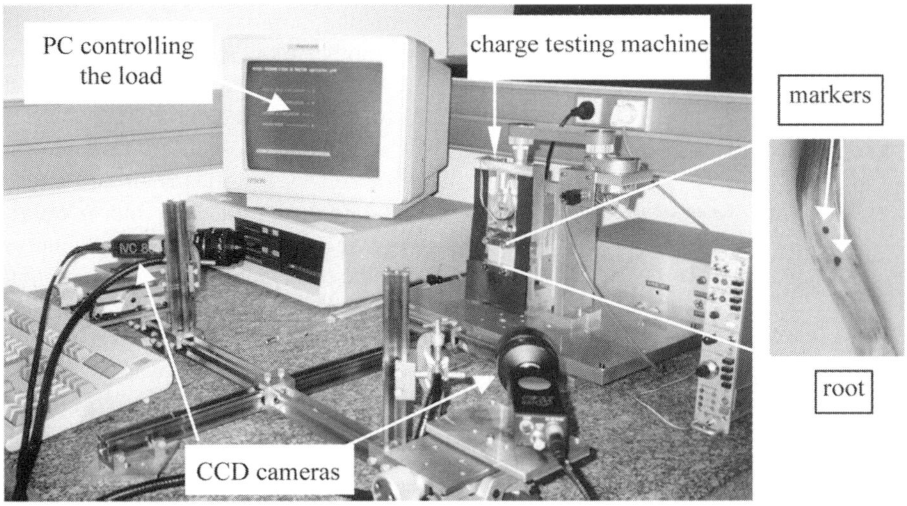

Figure 4. View of experimental device including CCD cameras, for the determination of the three-dimensional co-ordinates of the two markers and the load obtained at an imposed displacement.

Table 2. Tensile behaviour of root

Behaviour coefficient	Behaviour
1	Failure of root due to the slipknot
2	Failure of root
3	Breaking of the cortex / inner stele bond
4	Intact root

The principle was to isolate a root from the maize root system, without disturbing the soil, in order to not damage the soil-root interface. Then a tensile test was performed on the axis of the root, until the breaking of the root or the deterioration of the link between soil and root was obtained. This value corresponds to the breaking load of the vegetal material or uprooting force. Furthermore, for each root, observations were made on soil moisture, on branching, on the tensile behaviour of root and on the approximate length of the part of root useful for plant anchorage. Maize roots are conical and highly tapered along the upper part. The function of the tapered part of the root is principally to absorb water from soil and not prevent the plant from being uprooted. Only the first part of the root is necessary for plant anchorage. Five different levels of branching were identified (Figure 6) and coefficients were assigned depending on root tensile behaviour (Table 2).

Results and Discussion

Results

Rigidity modulus determination

One hundred tests were performed. For each, the stress σ versus the longitudinal strain ϵ was plotted. Two representative experimental curves are presented in Figure 7. Curves of $\sigma = f(\epsilon_{ccd1})$ and $\sigma = f(\epsilon_{ccd2})$ were also plotted, where ϵ_{ccd1} and ϵ_{ccd2} are respectively the longitudinal strains obtained by CCD1 and CCD2 (Figure 7). Due to the non homogeneous nature of maize roots, experimental curves were not completely linear. In some cases the longitudinal strains obtained by one of the two cameras were irregular and negative (Figure 7). This is the result of out of plane displacements in which the specimen moves to the opposite direction from the CCD cameras. Since the curves obtained by CCD1 and CCD2 are very different, a stereoscopic method for the study of maize roots is therefore required. This method results in a curve obtained from a complex function of CCD1 and CCD2 signals (Figure 7b), but does not lead to the average between CCD1 and CCD2 as shown on Figure 7a. It can also be remarked that the new curve is smoother than those obtained from the CCD1 and CCD2.

In order to obtain the rigidity modulus for each curve $\sigma = f(\epsilon)$, a linear interpolation passing through the origin is calculated by the least squares method consisting of minimising the normal distance between the experimental results and their fitting curve.

Figure 5. The experimental device used to pull out a root from soil.

For each inter-node of each variety, we notice a scatter in the rigidity modulus (Figure 8) due to the nature of the material, i.e. the morphological variations between roots of the same inter-node. For a few roots, the rigidity modulus is greater than 300 MPa whilst the majority is about 150 MPa.

Furthermore, there are no significant differences between varieties. The mean obtained for the inter-node 7 of the W117 × EP1 is lightly higher than averages of the others genotypes, but this difference is due to one test, where the rigidity modulus reaches 600 MPa.

Root–soil interactions

430 tensile tests were carried out on the three genotypes in two replicates with different soil moistures. First, a part of the field was watered until the soil was saturated, simulating soil at the time of root lodging. In another plot, soil was watered in order to have a moist soil. In the latter, irrigation was stopped 24 hours before tests (called dry soil).

Results of the pull-out tests are shown in Table 3, which gives the means of the pull-out limit F and standard errors for each inter-node of each genotype. Significant differences among genotypes were found. For the three genotypes, the maximum pull-out force is obtained for inter-node 6 and is especially noticeable in W117 × EP1. However, on inter-node 7 of this genotype, pull-out load was the weakest. The root system of the F2 × F1772 was most resistant to uprooting as the pull-out limit obtained for E7 is the highest. Furthermore, this genotype has an additional inter-node E8, with a mean equal to that of inter-nodes E6 and E7.

In order to investigate the relationships between the different parameters such as soil moisture, age, length, branching and the tensile behaviour of roots, regressions were calculated for each inter-node of each variety versus the pull out force (Figure 9).

The regressions were not significant, but showed that peak load increases with age, length and branching ($y = 3.283^* \, x_{age} - 29.953$, $R^2_{age} = 0.309$; $y = 0.221^* \, x_{length} + 13.899$, $R^2_{length} = 0.168$; $y = 11.038^* \, x_{branching} + 18.643$, $R^2_{branching} = 0.094$). This indicates the older the root is, the longer and more branched it is. Consequently, we also noticed that tensile force was correlated to root age, probably due to an increase in branching.

No significant correlations were observed between the level of branching and pull out limit, and between tensile behaviour and pull out limit, for the different types of soil. However, for an equal level of branching, the tensile force is higher for dry soil than wet soil. For example, concerning the inter-node 7 of F2 × F1772 and for the level 4 of branching, a pull-out limit equal to (70.67 ± 14.36 N) was obtained for a moist soil and was equal to (98.81 ± 35.99 N) for a dry soil. When the soil is wet, the soil–root bond is weaker which leads to a lower pull out force. Furthermore in dry soil, the root generally breaks before the deterioration of soil. In the case where the behaviour coefficient is equal to 3, the cortex–inner stele bond appears weaker than the soil–root interface.

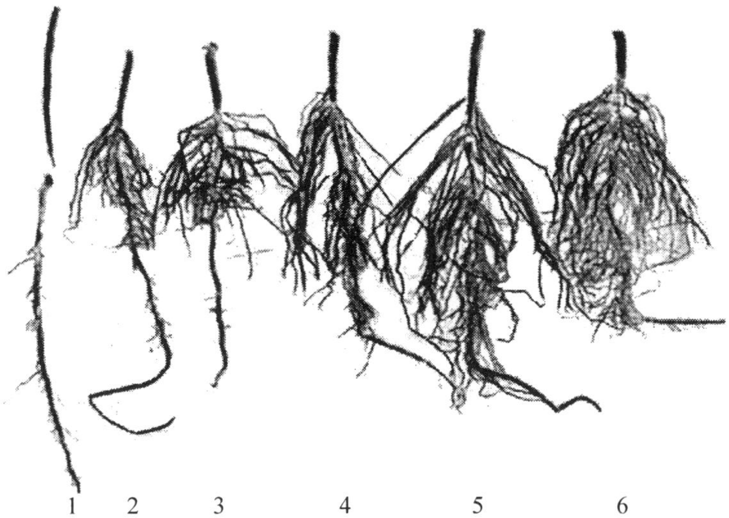

Figure 6. Roots were classified according to their branching pattern.

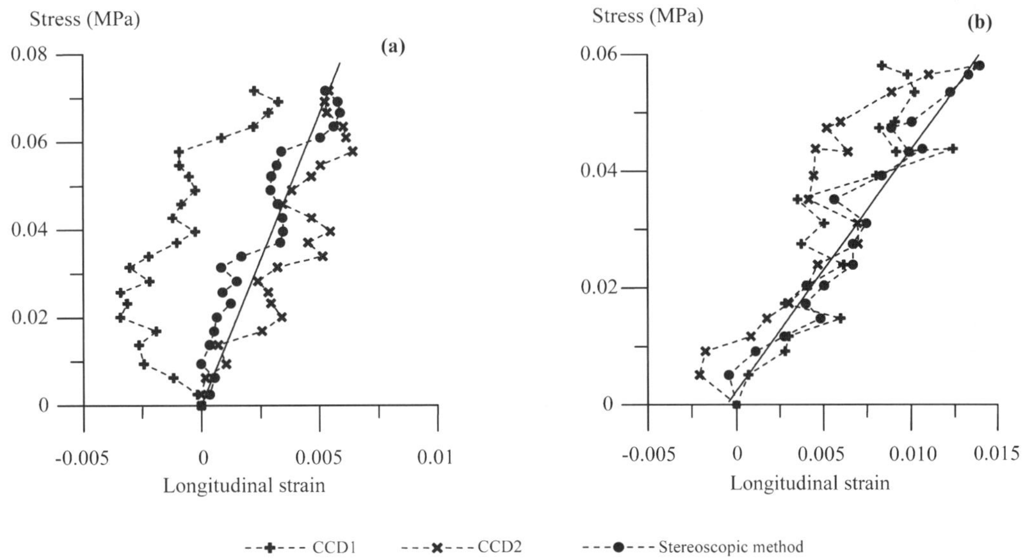

Figure 7. The variations of the stress with the longitudinal strain obtained by the CCD1 and CCD2 and by stereoscopic method. The slope of the 'stereoscopic' curve is plotted (continuous line) in order to determine the rigidity modulus.

Discussion

Rigidity modulus determination

The use of a non-contact and a non-disturbing optical method allows us to test plant material. The study on the mechanical behaviour of a maize root system required the tracking of two markers associated with a stereoscopic method to determine the three spatial co-ordinates of the points for each loading step. Co-ordinates were obtained with an accuracy of around 10^{-2} mm and longitudinal strain accuracy is estimated to 10^{-3}.

An important number of tests was carried out in order to determine a rigidity modulus of each inter-node of each variety. In fact, the uniqueness of specimens, morphological and anatomical differences of roots, on a same inter-node of a maize plant implies that we have obtained wide variation in the results. The rate of

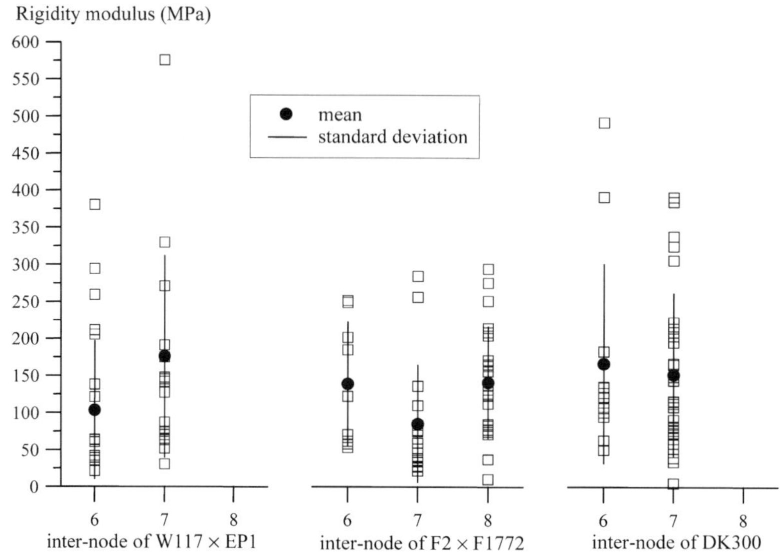

Figure 8. The relationship between the rigidity modulus and inter-nodes for the three genotypes.

Table 3. Tensile force required to pull maize root from soil for each inter-node of each genotype

	W117 × EP1	F2 × F1772	DK300
Inter-node 6	114.044 ± 31.188 N	74.04 ± 26.31 N	81.949 ± 34.534 N
Inter-node 7	24.698 ± 19.784 N	72.772 ± 21.462 N	50.612 ± 23.799 N
Inter-node 8		59.066 ± 18.169 N	

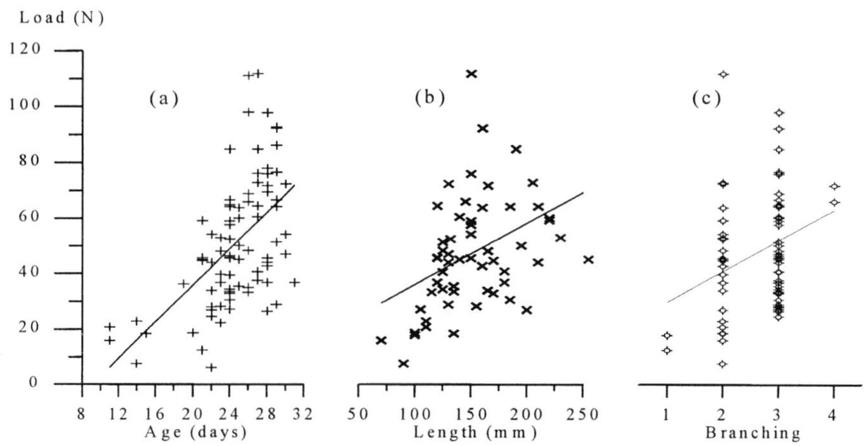

Figure 9. For one variety, the DK300, the relationship between the tensile force required to pull maize root from soil and three parameters: (a) age, $y = 3.283^* x - 29.953$, $R^2 = 0.309$; (b) length, $y = 0.221^* x + 13.899$, $R^2 = 0.168$; (c) level of root branching, $y = 11.038^* x + 18.643$, $R^2 = 0.094$.

lignified cell walls measured by Hébert (unpublished data) also shows scatters in results due to root age. For each genotype, the rigidity modulus of each inter-node are similar.

The rigidity moduli of the roots for each inter-node, measured in this paper is higher than that determined by Ennos et al. (1993) which was 72 ± 37 MPa. It may be due to the different manner of

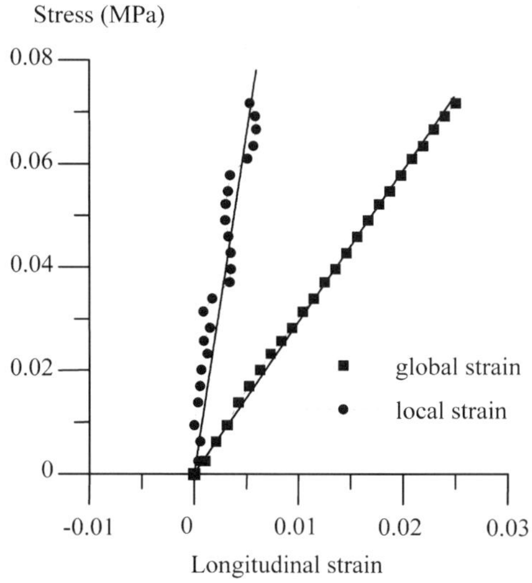

Figure 10. Variations in stress with longitudinal strain in local and global strains.

measuring strain. The tracking of two markers allows measurement of local strains, whereas Ennos et al (1993) determined the global strains, i.e. they has measured the displacement of roots held between the jaws of an Instron machine, during tensile tests, and then they deduced the global strains using equation (3) , where Δz, which is the initial distance between the two clamps, and does not tend towards zero.

In order to compare results obtained by these two methods, the local strains and the global strains were simultaneously measured during a tensile test (Figure 10). The rigidity modulus obtained by local strains is higher than this obtained by global strains, owing to the fact that global strains are greater than local strains. This result is also due to sliding of the clamps. Therefore, another advantage of our method is that it takes into account any sliding of the specimen in the Instron jaws.

Furthermore, the use of the stereoscopic method, allows the determination of the correct rigidity modulus, by taking into account the irregular geometry of maize roots. Thus the mean of rigidity modulus of each inter-node will be used in order to numerically simulate the mechanical behaviour of a maize root system in soil.

Root–soil interactions

The conception of a portable device has allowed us to directly examine the soil–root bond in the field, without disturbing the soil by measurement of the tensile force required to pull maize roots out of soil.

For the three maize varieties, the pull-out limit was correlated with the length of roots; long roots produce large resisting load to uprooting which is in agreement with observations by Stokes et al. (1996). The pull out tests also indicated that the greater the branching, the higher the tensile force (Easson et al., 1995; Stokes et al., 1996). The two parameters, length and branching of root, are directly correlated with the age of the root, hence the pulling resistance of the root increased with the age of the root.

The increase of soil moisture content induces a decrease in the tensile force. For the roots extracted from the wetter soil, the pull-out limit was lower than for the drier soil. Furthermore, in the case of dry soil, the soil–root bond is stronger than the cortex–inner stele bond; hence, the cortex–inner stele bond fails before the root soil–bond. We obtained the inverse phenomenon, when the soil is wet. These two observations support the findings of Easson et al. (1995) during the study of the tensile properties of wheat roots.

The main interest of this study was the determination of the maximum tensile force required to pull out a maize root for a particular type of soil and for a certain rate of branching.

The knowledge of these two mechanical parameters will allow us to commence a three-dimensional model using the Finite Element Method, which consists of two parts:

1. The design of a mesh, to define the characteristics of the material and to choose the behaviour of the model. A mesh of maize shoot and roots, was determined according to geometrical and morphological data of each variety, i.e. length, radius and number of roots, radial angle and insertion angle of roots. We have decided to define the behaviour of the model as a linear isotropic elastic behaviour, hence emphasizing the necessity of measuring the rigidity modulus of roots.

2. We must choose the manner to apply load and boundaries conditions to the mesh. The evaluation of the tensile force, required to pull-out roots from soil, allows us to apply to the mesh boundaries conditions. However, during the pull out tests, we must nevertheless measure a further parameter, the displacement of the root versus the tensile force. Thus the conception of a new portable device is in progress in order to carry out further relevant and reliable pull out tests.

Acknowledgement

We would like to thank Region Poitou-Charentes for their financial support.

References

Brémand F and Lagarde A 1986 A new method ofoptical strain measurement with applications. pp 686–694. Proceeding of the 1986 SEM Spring Conference on Experimental Mechanics. New Orleans, U.S.A.

Brémand F, Dupré J C and Lagarde A 1992 Non-contact and non-disturbing local strain measurement methods. Part. I: Principle. European J. Mech. 11A (3), p 349–366.

Brémand F, Dupré J C and Lagarde A 1995 Mesure des déformations sans contact par analyse d'images. pp 171–178. Photomécanique 95. Cachan, Eyrolles, France.

Denis S 1996 Etude des propriétés mécaniques du système racinaire et de ses organes chez le maïs. DEA de Mécanique, Université de Poitiers, France.

Denis S, Belkadi S and Brémand F 1998 Mesure des déformations par stéréoscopie sur surface quelconque – Application aux racines de maïs. pp 73-80. Photomécanique 98, Marne la Vallée, France.

Duparque A and Pellerin S 1990 Some morphological characters of maize root and shoot systems and their relationships with resistance to root lodging. p 43. Scaife A. Proceedings of the First Congress of the European Society of Agronomy, 5–7 Dec. ESA, Colmar, France.

Dupré J C, Brémand F and Lagarde A 1993 Numerical spectral analysis of a grid. Application to strain measurements. Opt Lasers Engrg. 18(3), 159–172.

Easson D L , Pickles S J and White E M 1995 A study of the tensile force required to pull wheat roots from soil. Ann. Appl. Biol. 127, 363–373.

Ennos A R 1991 The Mechanics of Anchorage in Wheat Triticum aestivum L. II: Anchorage of mature wheat against lodging. J. Exp. Bot. 42(245), 1607–1613.

Ennos A R, Crook M J and Grimshaw C 1993 The anchorage mechanics of maize, *Zea mays*. J. Exp. Bot. 44(258), 147–153.

Germain P 1986 Mécanique Tome 2. Ecole Polytechnique, Edition Marketing.

Guingo E and Hébert Y 1997 Relationships between mechanical resistance of the maize root system and root morphology, and their genotypic and environmental variation. Maydica 42, 265–274.

Guingo E, Hébert Y and Charcosset A 1998 Genetic analysis of root traits in maize. Agronomie 18, 232–243.

Hébert Y, Duparque A and Pellerin S 1990 Verse en végétation: la variabilité génétique du système racinaire et de l'appareil aérien du maïs et ses conséquences sur la séléction. *In* Colloque la vie du maïs, Physiologie et production du maïs. pp 215–225.

Hébert Y, Barrière Y and Bertholeau J C 1992 Root lodging resistance in forage maize: Genetic variability of root system and aerial part. Maydica 37, 173–183.

Stokes A, Ball J, Fitter A H, Brain P and Coutts M P 1996 An experimental investigation of the resistance of model root systems to uprooting. Ann. Bot. 78, 415–421.

Sutton M A, Chenz M, Peters W H, Chao Y J and McNeill S R 1986 Application of optimized digital correlation method to planar deformations analysis. Image Vis. Comput. 4(3), 143–150.

Zuber M S 1968 Evaluation of corn root systems under various environments. p 1-9. Proc 23rd Ann. Corn and Sorghum Industry Res. Conf. 10–12 Dec. Chicago, U.S.A.

A. Stokes (ed.), The Supporting Roots of Trees and Woody Plants: Form, Function and Physiology, 305–311.

Distribution of *Armillaria* on pear root systems and a comparison of root excavation techniques

D. M. Rizzo[1] and R. Gross[2]
[1]*Department of Plant Pathology, University of California, Davis, CA 95161, USA and* [2]*Dendro Tech, PO Box 766, Calistoga, CA 94515, USA*

Key words: Armillaria mellea, irrigation, pears, *Pyrus*, root disease, root excavation:-pressurized water-supersonic air, root grafts

Abstract

Armillaria root disease (causal agent *Armillaria mellea sensu stricto*) is a major disease of fruit and nut crops in California, USA. An aspect of Armillaria root disease in orchard crops that has not been extensively studied is the spatial distribution of the fungus on the roots of mature trees and how this translates to symptom expression and yield loss. Root systems of Bartlett pear trees at two commercial orchards were excavated utilizing one of two methods: pressurized water or supersonic air. At both sites, trees were selected to represent a range of crown conditions (from declining to apparently healthy) in and around Armillaria infection centers. Root systems were excavated to a minimum diameter of 2 m and a minimum depth of 1 m. At both sites, numerous independent infections by Armillaria were observed on the roots of diseased trees. Discrete lesions were scattered on root surfaces, ranging in size from several square centimeters to large lesions covering hundreds of square centimeters. The root system excavations revealed a lack of root contact or root grafts between adjacent pear trees. Rhizomorphs were prevalent on root surfaces and in the soil at both sites. Both methods resulted in loss of very fine roots (<0.5 mm diam). Fine (0.5–1 mm diam) and small roots (1–5 mm diameter), however, were left intact. Epiphytic rhizomorphs were generally not affected by the excavation techniques. In comparing the advantages and disadvantages of the two systems, we concluded that the hydraulic system gives the better final product for root disease work. Larger volumes of soil are easily removed and the root surface is clean and easy to interpret. However, the supersonic air system will probably prove more useful to researchers. This technique is easier to master and waste disposal is much simpler.

Introduction

Armillaria root disease is an important disease of fruit and nut crops in many places throughout the world. In California, the most important hosts include almonds, pears, walnuts, grapes, and citrus (Ogawa and English 1991; Smith 1941). In temperate growing areas, the primary causal agent of the disease is *Armillaria mellea* (Vahl) Kummer. The fungus attacks the roots of plants, killing the cambium and decaying the woody tissues. Symptoms of the disease include poor shoot growth, premature yellowing and dropping of leaves, dieback of branches, and eventually death of the tree. Spread of the pathogen is primarily by growth of fungal mycelium through susceptible host tissue and root to root contact, or through the formation of rhizomorphs (root-like aggregations of fungal tissue). The disease occurs in distinct foci, known as infection centers, that expand over time. The fungus can potentially survive for long periods in woody debris in the soil leading to long-term establishment on a site. Replants are often killed leading to areas of the orchard/vineyard that are unprofitable for production.

Research on *Armillaria* in orchard crops has generally concentrated on chemical controls (particularly in California), to limited extent on interactions of chemicals with potential biological controls (e.g., Munnecke et al., 1973, 1976, 1981; Ohr et al., 1973) and finally on resistant rootstocks (Guillaumin et al., 1989). Basic biological studies on *Armillaria* (such as pathogen distribution, spread, virulence, and inoculum survival) have received relatively limited attention in orchard crops as compared to forest ecosystems (but see Kable, 1974).

One aspect of *Armillaria* root disease in orchard crops that has not been extensively studied is the spatial distribution of the fungus on the roots of mature trees and how this translates to symptom expression and yield loss. It is generally assumed that *Armillaria* initially infects the peripheral parts of the root system and slowly colonizes the roots until the root collar is reached. A better understanding of colonization patterns of root systems can add to our knowledge of pathogen spread. In addition, development of post-plant chemical treatments will benefit from information on the distribution of the pathogen on individual root systems.

Examination of the root systems of mature trees is the most difficult task in *Armillaria* research. Traditionally, root excavations have been done by hand or using a backhoe or similar heavy machinery. Hand excavation is time consuming and very labor intensive; it useful in shallow forest soils, but of limited value in examining deep-rooted orchard species. Use of a backhoe allows for examination of a greater part of the root system; however, it is very destructive to the root system. Our objectives in this pilot study were to (1) examine the roots of *Armillaria*-infected pear trees to determine the spatial distribution of the pathogen on individual root systems, and (2) compare two methods of non-destructive root excavation: pressurized water and supersonic air. These two objectives are part of a larger study to determine the effects of *Armillaria* root disease on the productivity of pear orchards.

Materials and methods

Study sites

Two pear orchards in Lake County, California, were chosen for the root excavations. Orchard 1 was a Bartlett pear orchard currently planted on a 4×4 m spacing. The orchard was initially established in the late 1800s on French domestic rootstock. Some of these original trees are still found in the orchard, but numerous interplantings have been made with various rootstocks (*Pyrus betulaefolia*, Winter Nelis, and OH×F). Orchard 1 was located adjacent to a stream; the soil was generally a clay loam with gravel streaks running through it. The orchard was flood irrigated until 1971 when solid set sprinklers were installed. Mortality due to Armillaria root disease was first noted by the grower in the late 1970s. Two infection centers approximately 100 m apart were chosen for excavation.

Orchard 2 was also a Bartlett pear orchard and was located approximately 15 km from Orchard 1. The orchard was originally planted in 1878 on a 10×8 m spacing on French domestic rootstock. Many of the original trees are still found in the orchard. The orchard was interplanted in 1962 on various rootstocks (*P. betulaefolia*, Winter Nelis, Old Home) resulting in the current 4×8 m spacing. The soil is a mixture of sandy loam and clay loam, and unlike site 1 is not located near an existing streambed. The orchard was flood irrigated until 1975, when solid set sprinklers were installed. Crown symptoms and mortality were first noted in the mid-1980s by the grower, although it was not recognized as Armillaria root disease at the time.

All isolates from both orchards were identified as *Armillaria mellea sensu stricto* (Rizzo et al., 1998). In addition, within each of the excavated portions of the orchards all Armillaria isolates were found to be members of the same somatic incompatibility group and, therefore, putatively the same genetic individual (Rizzo et al., 1998).

Root system excavations

At Orchards 1 and 2, 30 and 10 trees, respectively, were chosen for extensive root excavation to determine the spatial distribution of *Armillaria* on the root systems of individual pear trees. At both sites, trees were selected to represent a range of crown conditions (from declining to apparently healthy) in and around *Armillaria* infection centers. Prior to excavation, tree crowns were visually rated on a scale of 1–5 (1, healthy, full crown with good shoot growth; 2, reduced top shoot growth; 3, most shoots with reduced growth; 4, dieback evident, leaves small and lighter in color; 5, dead).

Root systems were excavated utilizing one of two methods. At orchard 1, pressurized water (Gross, 1995; Lindsey et al., 1995) was used to remove soil from the roots of nine trees in 1997. Prior to excavation, the soil surrounding the trees was irrigated to saturation. During excavation, water was delivered to the roots with a hand-held automatic pressure controlling fire-fighting nozzle (TFT brand, Mid-Matic model), supplied by a 4.45-cm diameter fire hose at 5.3 kg/cm^2 (75 psi) (Fig. 1). The soil–water slurry was removed by a Vactor brand model 1645 mobile industrial-vacuum conveyor through a 1.2-m long, ri-

Figure 1. Hydraulic excavation method using a hand-held automatic pressure controlling fire-fighting nozzle supplied by a 4.45-cm diameter fire hose at 5.3 kg/cm^2 (75 psi). The soil–water slurry was removed by a Vactor brand model 1645 mobile industrial-vacuum conveyor through a 1.2-m long, rigid plastic tip on flexible tubing (20 cm diam. and 30 m long).

Figure 2. Excavation using supersonic air method. The air was delivered through an engineered air jet nozzle and a 3.2-cm diameter hose at 93.5 hl/min at 7 kg/cm^2 (100 psi).

gid plastic tip on flexible tubing (20 cm diam and 30 m long) (Fig. 1).

Supersonic air was used to excavate root systems at Orchard 2 in 1997 and an additional 21 trees at Orchard 1 in 1998. The air was delivered through an engineered air jet nozzle and a 3.2-cm diameter hose at 93.5 hl/min at 7 kg/cm^2 (100 psi). The procedure utilizes the soil pores for air penetration which causes localized soil structural failure. The air stream penetrates the soil, filling soil pores until the soil fails in shear and exploded away in the direction of least resistance. Soil moisture was maintained at field capacity for excavation; this allowed for soil pores to be maintained while lubricating soil particles for enhanced rate of excavation. In the 1997 experiment, the soil was removed during excavation by vacuum as described above (Fig. 2). In 1998, a vacuum was not used; the soil was simply displaced out of the hole with the airstream.

Figure 3. Excavated root systems at Orchard 1.

Root systems were excavated to a diameter of approximately 2 m from the stem and a depth of approximately one meter (Figs. 1–3). The area between adjacent trees was also excavated to examine root overlap (Fig. 3). Root systems were examined periodically during the excavation process to ascertain that critical information was not lost. Soil pits were examined for the presence of inoculum. Following soil removal, root systems were individually dissected using hand tools and photographed. The exposed roots were examined for signs of *Armillaria*. The presence of various host responses (e.g., callus formation, cracking and discoloration of bark), as well as extensively decayed roots, allowed for the location of *Armillaria* infections. Fungal samples were removed from lesions, decayed portions of the root system, and rhizomorphs; isolations were made in the laboratory as described previously (Rizzo et al., 1998).

Results and discussion

Root excavations

Following root system excavations, *Armillaria* was detected on 17 of 30 trees at Orchard 1, and five of 10 trees at Orchard 2. All trees in which *Armillaria* was not detected on the root system were located beyond the border of the infection center as delimited by trees with declining crowns. Of the trees colonized by *Ar-*

millaria, most (16 of 22) had discrete lesions scattered on the surfaces of large (10–20 mm diam.) and very large roots (>20 mm diam.) (root system terminology follows Sutton and Tinus (1983)). These lesions ranged in size from several square centimeters to large lesions covering hundreds of square centimeters. On six of the 16 trees, numerous lesions appeared to coalesce to almost completely colonize the root system; however, we could still discern discrete lesions with uninfected tissue between the lesions. Discrete lesions were not commonly found on roots less than 10 mm diam. These roots tended to be completing colonized by the fungus. *Armillaria* was not observed on roots less than 1 mm diam. On five of the infected trees, lesions (up to 20×30 cm) were found primarily in the root crown area (i.e., at or above the soil line) rather than on the root system. On only one tree, at the border of an infection center, was *Armillaria* found solely at the periphery of the root system. A single decayed root (40 mm diam) was found on this tree approximately 3.25 m from the root crown area. In general, infection did not appear to proceed from distal portions of the root system towards the root collar. Instead, infections were initiated at scattered points throughout the root system.

The root system excavations revealed a lack of root contact or root grafts between adjacent pear trees. Roots from adjacent trees were generally separated by 30–60 cm. Only one instance of transmission of *Armillaria* via root to root contact was observed in the 40 trees excavated. This was the only root to root contact observed regardless of colonization by *Armillaria*. Root excavations of 10 trees in a third pear orchard (as part of an unrelated study) also did not reveal any instances of root to root contact.

Rhizomorphs were prevalent on root surfaces and in the soil at both sites. Rhizomorphs could be traced from lesions into the soil and could also be found on the walls of the excavation pit throughout the soil profile (Fig. 4). Extension of rhizomorphs from diseased roots through the soil appeared to be limited to 30–60 cm. This would be sufficient to bridge the gap between root systems of adjacent trees and between different roots on the same tree. Rhizomorphs have not been considered to be an important part of the epidemiology of *Armillaria* in California orchards (e.g., Anonymous, 1991; Raabe, 1966, 1972), most likely due to the dry conditions encountered in most growing areas in the state. However, during the 1970s, sprinkler irrigation systems replaced flood irrigation in many pear orchards, thereby facilitating more frequent irrigations. Increased irrigation may have led to conditions that are much more conducive to rhizomorph formation and growth than had generally occurred in California orchards.

Figure 4. Rhizomorph (arrow) and an old, decayed pear root (arrow) within the soil profile.

Large pieces of dead woody roots were found in several of the root pits at Orchard 1 (Fig. 4). Most likely, these roots were from pear trees previously killed by *Armillaria*, although some roots may have been from oaks which occupied the site prior to the establishment of the orchard. Most of the roots had *Armillaria*-type decay (i.e., white, spongy rot) and appear to have served as primary inoculum sources during development of the current infection centers. Much of the spread of *Armillaria* infection centers in the two orchards could be accounted for by movement of the fungus, via rhizomorphs, from primary inoculum rather than secondary spread between trees. Growth of rhizomorphs through the soil from primary inoculum could explain the occurrence of lesions restricted to the root collar of some trees.

We attempted a quantitative measure of the impact of *Armillaria* on the root system by measuring the percentage of the diameter of the root collar that was impacted by *Armillaria* and measuring the sizes of most lesions. These measurements were used to estimate the amount of root system colonized by mycelial fans of *Armillaria*. Root systems were then rated on the percentage of the root system impacted by the fungus (0, 1–20, 21–40, 41–60, 61–80, 81–100%). As expected there was a highly significant correlation ($r = 0.92$; $P < 0.001$) between the condition of the crown and the estimated root impact (Fig. 5). It was difficult, however, to determine the relative importance of various types of infection on individual trees. For

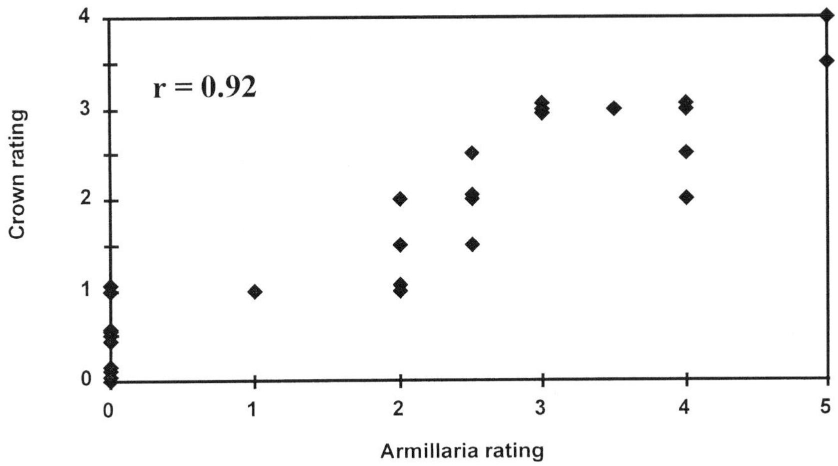

Figure 5. Correlation between crown symptoms and colonization of root systems by *Armillaria*. Root systems were rated on the percentage of the root system impacted by the fungus (0, 1–20, 21–40, 41–60, 61–80, 81–100%). Tree crowns were visually rated on a scale of 1–5 (1, healthy, full crown with good shoot growth; 2, reduced top shoot growth; 3, most shoots with reduced growth; 4, dieback evident, leaves small and lighter in color; 5, dead).

example, several trees had over 60% of their root system impacted by *Armillaria* with only some reduced growth (crown rating 2). It should be pointed out that *Armillaria* initially colonizes the cambium of trees; following cambial colonization then fungus then begins to decay the xylem. Therefore trees, that had only recently been colonized would have the fungus confined to the cambium which may not restrict water movement to the crown. It is possible that at this stage *Armillaria* infection does not impact the host to any great extent. This is a testable hypothesis that we are currently examining.

Pears have traditionally been considered to be among the most resistant orchard crops to Armillaria root disease (Raabe, 1966, 1979; Thomas, 1934). Based on our observations and previously published work (Thomas, 1934), it appears that pear trees may tolerate infection on peripheral parts of the root system for many years. Despite apparent resistance, reports of Armillaria-infected pear trees have increased throughout the pear growing regions of northern California since about 1990. Much of the increased disease incidence may be explained by changes in environment as discussed above; changes in irrigation regimes have led to conditions that are much more conducive to rhizomorph growth. In addition, horticultural factors may be contributing to the observed distribution of Armillaria on pear root systems. Growers tend to plant pear trees deep in the soil; in this study, we found the main lateral supporting roots on all trees being initiated 30 cm or greater below the soil line. According to growers, this was done to prevent suckering on older types of rootstocks and to provide windfirmness. In recent years, however, there does not seem to be any real biological reason, other than tradition, for this practice. Because the root collar is found well below the soil line on these trees, we would suggest that this makes the trees much more susceptible to death from Armillaria infection. This is very similar to what has been found with other orchard crops where trees seem to tolerate root infections to some extent, but are quickly killed when infection occurs at the root collar (Guillaumin, 1977; Morrison et al., 1991; Munnecke et al., 1976). Because of the prevalence of root collar infections that occurred separately from root infections in our study, one potential way to lessen the risk of mortality of pear trees by Armillaria may be to plant trees with the root collar predominantly at or above the soil line. Our findings could also have an impact on the efficacy of post-plant chemical applications. If rhizomorphs are the primary means of spread then soils drenched with fungicide treatments may prevent initial infection on young trees or older trees that have not been infected.

Comparison of root excavation methods

Both excavation methods allowed for observation of large portions of pear tree root systems with little or no damage to healthy structural roots. Severely decayed roots were susceptible to breakage during excavation; this was particularly true with the hydraulic excavation system. During removal of waste soil and/or water, the plastic tip of the vacuum hose occasionally caused abrasion of the bark on the larger roots. Both methods resulted in loss of very fine roots (<0.5 mm diam.); fine (0.5–1 mm diam.) and small roots (1–5 mm diam.), however, were left intact. Epiphytic rhizomorphs were generally not affected by the excavation techniques. The loss of the very fine roots was not a problem in this study, because Armillaria only colonizes larger woody roots.

Although damage to roots was minimal with both methods, there were advantages and disadvantages to using each excavation method. The hydraulic system allowed for quick removal of large volumes of soil; we estimated excavation rates of up to 8 m^3 of soil per hour during our studies. The hydraulic delivery system had an effective range of excavation of up to 3 m. This allowed us to easily cut trenches between trees and excavate to depths of greater than 1 m. Excavation utilizing pressurized water left a clean root surface that allowed for easy visualization of lesions and other defects that made for easier diagnosis of Armillaria.

There were some disadvantages to using pressurized water to excavate roots. While the roots themselves are clean, the surrounding soil becomes muddy and slippery and requires caution when moving around. Availability of enough water can also be a problem. The method generates a considerable amount of waste; our estimate is approximately a 10-fold increase in waste volume over the amount of soil removed. Depending on the soil type and location of the field site, waste disposal can be a problem. Movement of the soil–water slurry off-site may require special permits. In our situation, the soil–water slurry (approximately 80 m^3) was dumped down one of the orchard rows; the water soaked into the ground leaving the soil spread over an approximately 50 m long area. Because of this, soil from another source was needed to replace the soil removed during excavation. Finally, considerable expertise is required to operate and maintain the equipment in an efficient and safe manner.

The supersonic air system had some distinct advantages over pressurized water. First, minimal expertise is required to operate the equipment. It was relatively easy for a newcomer to master the technique. The waste generated in this method is limited to the amount of soil actually excavated. In excavating 10 trees at Orchard 2, we generated approximately 7 m^3 of soil. This soil was easily replaced around the sample trees. Unlike water, the air supply is basically unlimited.

The use of air to excavate roots is slower than using water. We estimated that approximately 2 m^3/h of soil could be removed by the technology currently available. The technique is dirty and leaves a fine film of clay particles on the surface of the roots. For our, study, this required using a small hand-held sprayer to wash the roots and allow us to see lesions and other root discoloration. Some physical limitations of the system include heating of the tools to high temperatures which slows down the excavation work. In addition, the tools generate very high noise levels that require operators and observers to wear ear protection.

Overall, we found both of the non-destructive sampling methods to be very useful in examining the spatial distribution of Armillaria on the root systems of mature trees. In addition to the data that can be obtained, these excavation techniques have the advantage that growers are much more willing to allow healthy trees to be excavated. In comparing the advantages and disadvantages of the two systems, we conclude that the hydraulic system gives the better final product for root disease work. Larger volumes of soil are easily removed, and the root surface is clean and easy to interpret. In the long run, however, the supersonic air system will probably prove more useful to researchers. This technique is easier to master, more adaptable to varied research situations, and waste disposal is much simpler.

Acknowledgement

The cooperation of Broc Zoller and Nick Ivicevich in setting up the field plots is greatly appreciated. Cameron Whiting and Rachel Elkins assisted during the root excavations. We thank Task Force Tips, Inc. (Valparaiso, Indiana) for comprehensive and detailed technical assistance and nozzles. EZ-Pup (Ontario, Canada) is acknowledged for supplying the fatigue reducing nozzle to shoulder device. Concept Engineering Group, Inc., (Pittsburgh, PA) developed the supersonic soil excavation device. This study was supported by funding from the California Pear Advisory Board, University of California Division of Agriculture and Natural Resources, and by a cooperative agreement with United States Department of Agriculture.

References

Anonymous 1991 Integrated pest management for apples and pears. University of California, Statewide Integrated Pest Management Project, Division of Agriculture and Natural Resources, Oakland, California. Publication 3340.

Gross R 1995 Construction application of hydraulic soil excavation. *In* Trees and Building Sites. Eds G Watson and D Neeley. pp. 177–184. International Society of Arboriculture, Savoy, IL, USA.

Guillaumin J J 1977 Apricot root rot, *Armillariella mellea* (Vahl) Karst. EPPO Bull. 7, 125–135.

Guillaumin J J, Pierson J and Grassely C 1989 The susceptibility of different Prunus species used as stone fruit rootstocks to *Armillaria mellea* (sensu stricto). *In* Proceedings of the Seventh International Conference on Root and Butt Rots. Ed., D J Morrison. pp. 197–207. Forestry Canada, Pacific Forestry Centre, Victoria, British Columbia, Canada.

Kable P F 1974 Spread of *Armillariella* sp. in a peach orchard. Trans. Br. Mycol. Soc. 62, 89–98.

Lindsey P A, Gross R and Milano R 1995 An investigation to assess the impact of street infrastructure improvement on the roots of adjacent cork oak trees. *In* Trees and Building Sites. Eds G Watson and D Neeley. pp. 22–32. International Society of Arboriculture, Savoy, IL, USA.

Morrison D J, Williams R E and Whitney R D 1991 Infection, disease development, diagnosis, and detection. *In* Armillaria Root Disease. (Eds C G Shaw III and G A Kile. pp. 62–75. USDA For. Serv. Agric. Handb. No. 691.

Munnecke D E, Kolbezen M J and Wilbur W D 1973 Effects of methyl bromide or carbon bisulfide on *Armillaria* and *Trichoderma* growing on agar medium and relation to survival of *A. mellea* in soil following fumigation. Phytopathology 63, 1352–1357.

Munnecke D E, Kolbezen M J, Wilbur W D and Ohr H D 1981 Interactions involved in controlling *Armillaria mellea*. Plant Dis. 65, 384–389.

Munnecke D E, Wilbur W D and Darley E F 1976 Effect of heating or drying on *Armillaria mellea* and *Trichoderma viride* and the relation to survival of *A. mellea* in soil. Phytopathology 66, 1363–1368.

Ogawa J M and English H 1991 Diseases of temperate zone tree fruit and nut crops. University of California, Division of Agriculture and Natural Resources, Oakland, California. Publication 3345.

Ohr H D, Munnecke D E and Bricker J L 1973 The interaction of *Armillaria mellea* and *Trichoderma* spp. as modified by methyl bromide. Phytopathology 63, 965–973.

Raabe R D 1966 Testing plants for resistance to oak root fungus. Calif. Agric. 20, 12.

Raabe R D 1972 Variation in pathogenicity and virulence in single-spore isolates of *Armillaria mellea*. Mycologia 64, 1154–1159.

Raabe R D 1979 Resistance or susceptibility of certain plants to Armillaria root rot. University of California Publication No. 2591. 11 pp.

Rizzo D M, Whiting E C and Elkins R B 1998 Spatial distribution of *Armillaria mellea* in pear orchards. Plant Dis. 82, 1226–1231.

Smith R E 1941 Diseases of Fruits and Nuts. California Agricultural Extension Service Circular No. 120.

Sutton R F and Tinus R W 1983 Root and root system terminology. Forest Science Monograph 24. Society of American Foresters, Washington, DC, USA.

Thomas H E 1934 Studies on *Armillaria mellea* (Vahl) Quel., infection parasitism and host resistance. J. Agric. Res. 48, 187–218.

A. Stokes (ed.), The Supporting Roots of Trees and Woody Plants: Form, Function and Physiology, 313–324.

An approach to minirhizotron root image analysis

T. Vamerali*, A. Ganis, S. Bona and G. Mosca
Dipartimento di Agronomia Ambientale e Produzioni Vegetali, Agripolis, Università di Padova, Via Romea 16, 35020 Legnaro, Padova, Italy

Key words: Image analysis, minirhizotron, roots, soil type, sugarbeet

Abstract

Minirhizotrons speed up research on root demography, but image quality often hampers standardization of the image processing method. A simple procedure working on the blue band of colour images was tested on fibrous roots of sugarbeet (*Beta vulgaris* var. *saccharifera*). With respect to green and red, the blue band allows better detection of roots when their luminance is very similar to that of the background. The method makes use of an exponential algorithm of contrast stretching, which takes luminance frequency distribution into account. Based on a single threshold level, the procedure includes skeletonization. A minimum segment length was adopted to discriminate roots from extraneous objects. Although a specific minimum root length (MRL) value was calculated for each soil type, results show that a single value can be applied, indicating that this method can be profitably used for processing large samples of images.

Abbreviations: ANS – artificial neural systems; MRL – minimum root length; RGB – red-green-blue; RLD – root length density

Introduction

Minirhizotrons inspected with a miniature video camera allow continuous and non-destructive observations of roots, leading to the acquisition of a large number of root images (Upchurch and Ritchie, 1983). Although images may be analysed directly in the field, more often they are recorded on a video tape, for easier analysis in the laboratory later.

Cheng et al. (1991) developed techniques for evaluating root demography and dynamics based on manual tracing of roots on transparent sheets. Hendrick and Pregitzer (1993) measured root length by applying an interactive procedure, drawing individual roots manually with a PC mouse. This method increases precision, but it often requires more time than manual counting.

Since the ordinary method of counting roots manually (Upchurch, 1987) is tedious, very time-consuming and not always correlated with root length, direct digitalization of images in the field and application of automatic image-analysis techniques can be very helpful in speeding up root research. When dealing with transparent-wall root images, the main problem is to recognize roots with respect to the background, which often includes extraneous objects (lighter soil particles, droplets of water, insects). The only way to discriminate roots against extraneous objects is to consider all the features. Traditional methods of feature recognition are global techniques like thresholding, which produces a binary image, light objects (roots, extraneous particles) becoming white and the background being converted into black. Outlining algorithms are useful local techniques but, when used alone, they are unable to identify roots exclusively and accurately. So expert systems or techniques based on certain parameters, such as the length/width ratio of root segments, seem necessary to identify roots over objects having similar light intensity (Stockman et al., 1990). Artificial neural systems (ANS) were also introduced by Nater et al. (1992), allowing ac-

* E-mail: vamerali@agripolis.unipd.it
Plant and Soil is the original source of publication of this article. It is recommended that this article is cited as: *Plant and Soil* **217:** 183–193, 1999.

curate identification of roots through a training set. Nevertheless, the ANS model requires computer implementations and may cause a substantial decrease in accuracy when applied to images on which it has not been trained.

Therefore, there is a need to work out a fast, sufficiently accurate procedure to analyse large numbers of minirhizotron root images. This paper presents some results of the image analysis of sugarbeet fibrous roots, based on a simple, rapid, automatic procedure applied to a large set of images and tested with different soil types, so that possible effects of varying image background were taken into account. A preliminary study was also carried out to ascertain whether the automatic procedure is able to take advantage of the additional information offered by coloured images.

Materials and methods

Image acquisition

Sugarbeet (*Beta vulgaris* var. *saccharifera*, cv. Monodoro, Novartis) was sown on March 5 1997 at the 'Lucio Toniolo' experimental farm of Legnaro (Padova) in four lysimeters filled with the following soil types: sandy, clay, silty loam, organic. Each lysimeter was 2 m long, 2 m wide and 1.5 m deep. Plant density was 10 plants m^{-2}, with 0.45-m row spacing (4 rows per lysimeter). Soil was fertilized with 120-150-100 kg ha^{-1} of N, P_2O_5 and K_2O respectively; water was supplied by a drip system to minimize drought stress, and the crop was kept free of weeds manually.

Each lysimeter was equipped with four 45°-angled plexiglass minirhizotrons, placed with their open tops at one end of each plant row, 8 cm from the row centre. The minirhizotrons had a 1.5-m long underground portion and a 50-mm inside diameter, allowing 1 m of soil depth to be explored. 24-bit red, green and blue (RGB) colour images were taken on May 28 by a camera (XL-6 Pulnix, CA, USA) moving along each minirhizotron. The images, with a viewed soil-tube interface area of 32 mm (horizontal)×24 mm (vertical), were captured every 2.5 cm of tube length (i.e., 1.77 cm of soil depth) and recorded by a personal computer equipped with a frame-grabber (Screen Machine 2, FAST, München, Germany) (Figure 1). The resulting digitized TIFF-format images had a width of 736 pixels, a height of 560 pixels, and a size of 1.209 Mb. Each pixel in a digitalized colour image was represented by a luminance value for each of the three bands, ranging from 0 to 255. In a grey-scale image, 0 corresponds to black and 255 to white.

In order to assess the root-background contrast over time, two images were taken from the central portion of each tube on 6 observation dates: May 28, June 11, July 2, July 16, August 8 and August 29, thus providing a sample of 192 images.

Image analysis

Luminance profiles for the three RGB bands were extracted from root segments and near-background segments in the images of the six observation dates. Output data were evaluated as averages of the profile line, in order to verify whether a single band for all images could be more suitably processed. Subsequently, the following processing steps were performed on the set of images acquired on May 28, using image-analysis software (KS 300, Carl Zeiss Vision GmbH, Eching bei München, Germany):

(1) Extraction of the blue-band grey-scale (Figure 2a).

(2) Radiometric enhancement. Although linear contrast stretching can be used (Heeraman et al., 1993) in order to highlight roots over the background, preliminary studies indicated that this method is not suitable for the following steps. An exponential function weighted on the luminance frequency distribution of the image pixels was therefore used (Figure 2b). The equation was in the following form:

$$y = a\exp(bx) + c \tag{1}$$

where y is the output luminance of a pixel P of input luminance L_P; x=arcsin(H^d); H is the integral of the relative frequency distribution function of luminance, calculated over the range $L_P \div L_{MAX}$; and L_{MAX} is the maximum possible value for luminance. a, b, c and d are empirical parameters. Thus, input luminance L_P may be converted to different output values depending on the image luminance distribution.

(3) Bias due to the luminance gradient produced by non-uniform lighting was reduced with a local smoothing operation based on the average background luminance value of image boxes sized 50×50 pixels.

(4) Irregular noise due to light soil particles, water droplets, etc. was reduced by applying a median filter on a box sized 7×7 pixels. This filter preserves the edges among objects (Pratt, 1991) and therefore should not affect foreground–background discrimination.

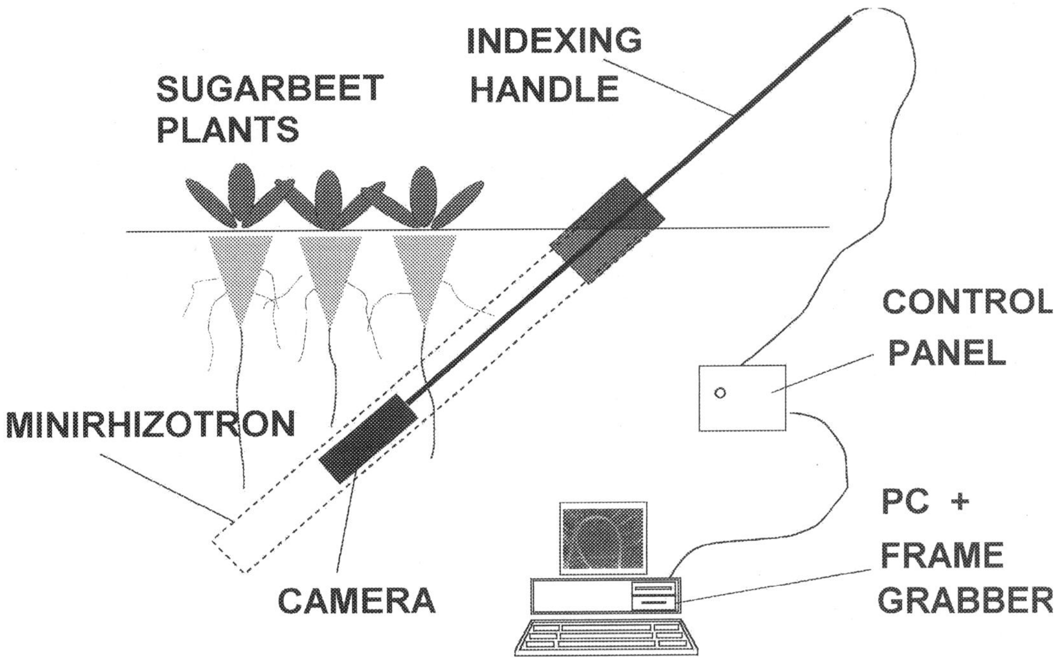

Figure 1. Sketch of experimental device.

(5) A grey value of 127 was assumed as threshold level, thus providing binary images. Subsequently, root skeletons were extracted (Figure 2c) by an operation of repeated boundary erosion of foreground objects, yielding 1-pixel wide line segments with the same topology as the original objects. The kind of erosion adopted here only works across the major axis of an object, leaving its end pixels unchanged. All individual segments obtained by skeletonization were automatically counted and measured.

A sample of 40 images for each soil type was used to estimate the minimum root length (MRL) to be considered in order to classify a segment as root. For this purpose, linear regression was calculated for each soil between data obtained by manually tracing the roots with the PC mouse, and automatic measurements of root length. After this calibration, the whole set of available images was automatically analysed (Figure 2a–c). The actual root length was estimated for each soil from the sum of root skeletons longer than the MRL (Figure 2d) and the respective regression equations.

Results and discussion

Foreground–background contrast

The red band was the most intense in overall images (Figure 3), as well as in soil and roots separately. Since the intensity of the three bands was nearly proportional in soil and roots, the red band turned out to give the greatest contrast throughout the growing season ($P \leq 0.01$). Significant interaction was found between observation date and band (Figure 4). Therefore, images to be analysed interactively (e.g., by tracing roots with the PC mouse or some other device) or even automatically without any stretching enhancement, may be profitably processed in the red band.

Soil luminance varied only weakly with time, depending on soil moisture and on the light intensity of the camera, neither of which was constant, whereas root luminance progressively decreased with time, due to senescence. Therefore, the contrast decreased significantly for the three bands throughout the season from May to August (Figure 4), increasing difficulties in root recognition. Similar results were reported by Smit and Zuin (1996) for Brussels sprouts and leek roots observed under white light with grey-scale images.

The hierarchy among the three bands in terms of root-background contrast observed throughout the sea-

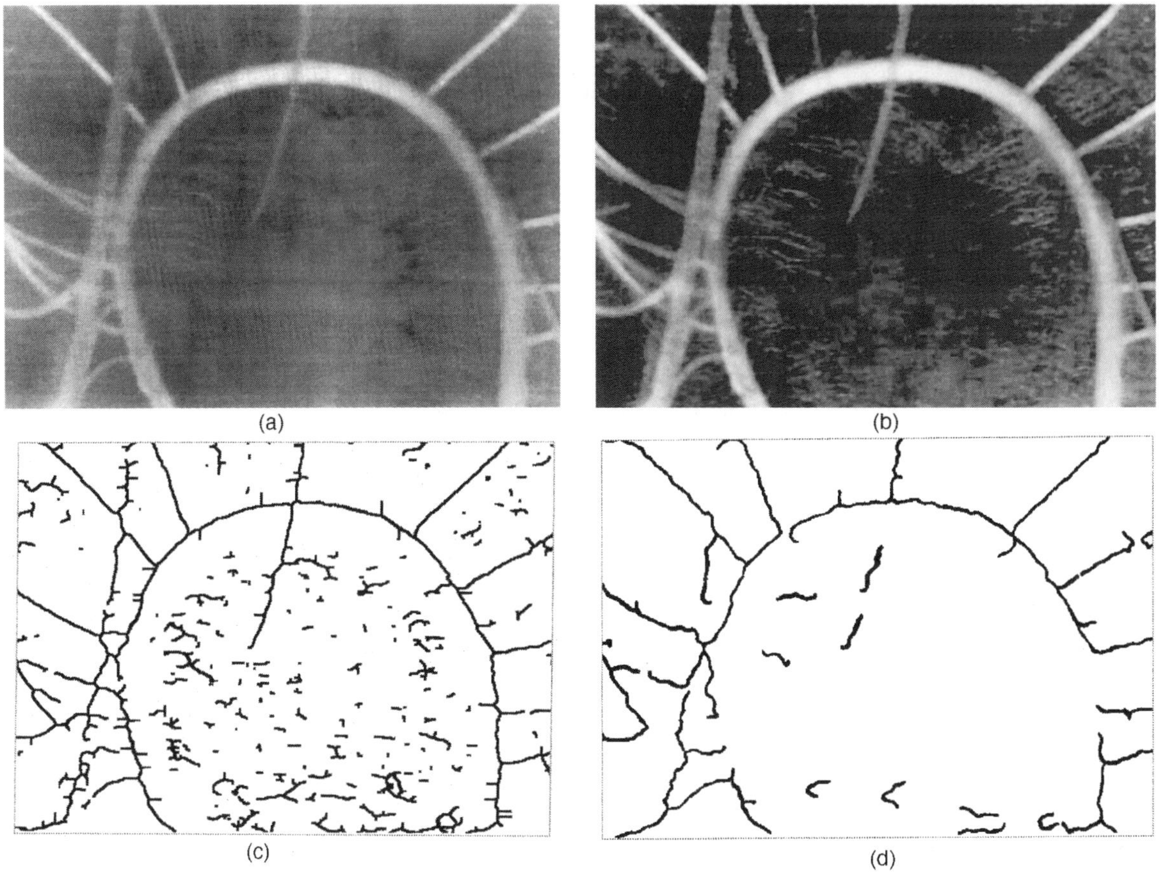

Figure 2. Main steps of image processing. a) original image (grey-scale of blue band); b) image after contrast enhancement (exponential algorithm); c) image after skeletonization; d) root skeletons filtered with MRL.

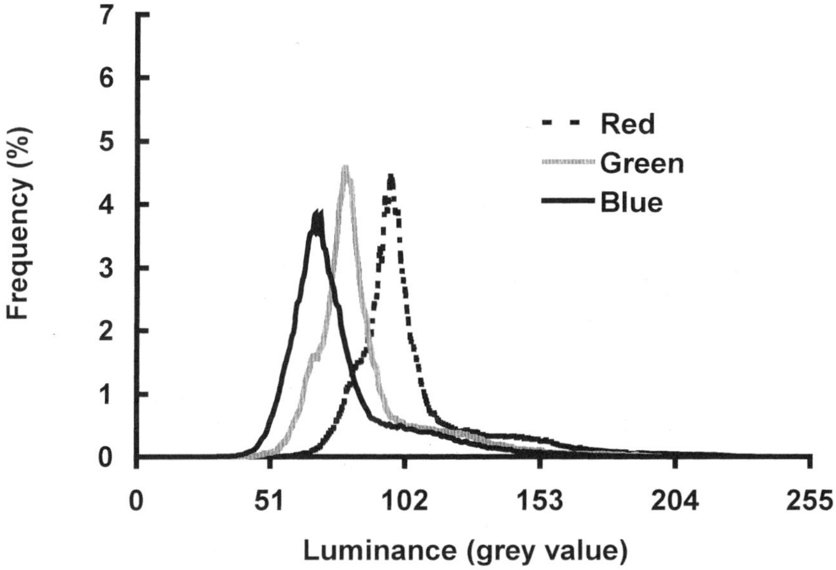

Figure 3. Luminance distribution for three channels (red, green and blue) in a true colour minirhizotron root image (May 28, 1997).

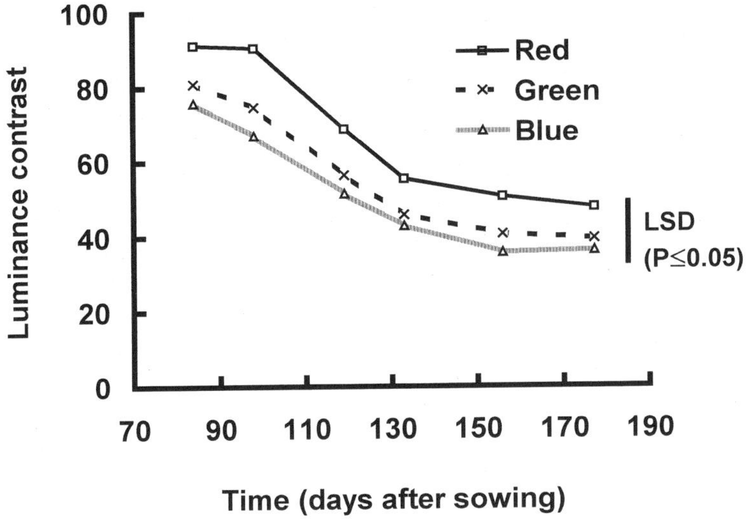

Figure 4. Absolute luminance contrast (difference in grey value) between roots and background for three channels (red, green and blue) in true colour minirhizotron images throughout the growing season: values are means of four soil types. 'Observation date×Colour band' interaction significant at $P \leq 0.05$, LSD=15.9.

son also appears from a comparison of different soil types (Figure 5). Organic soil showed the highest contrast ($P \leq 0.05$); the other soils had similar values. As a general rule, this suggests that it is better to use dark substrates when dealing with root observations by means of transparent-wall methods. Thus, when comparisons among treatments are performed and a large number of images are analysed, even the use of artificial dark substrates can make automatic root recognition easier and faster, although absolute values of root density may not be considered realistic.

Choosing which band to process

As explained above, large differences either in mean RGB luminance or in a single band (in our case, red) between foreground and background are helpful in both interactive and threshold-based image analysis methods. However, some root regions may have mean RGB luminance values very close to that of the background, so that root luminance may slightly exceed that of the background in one band but be lower in another. This is due to the different colours of roots and soil, and is useful when the exponential stretching method (see Experimental) is applied, as will be discussed below. In order to examine possible differences in colour between roots and background, the mean Blue/Red and Blue/Green ratios (i.e., the ratio of blue-band luminance to that of the other two bands) were calculated from root and background arrays of pixels. These arrays were the same as those used for the extraction of luminance profiles.

As overall mean, background and foreground were different in hue (Table 1), the Blue/Red ratio of roots being greater ($P \leq 0.01$) than that of soil (0.731 vs. 0.714). The difference in hue was ascribed mainly to the first observation date and associated with all soil types, except organic.

No more information could be retrieved from the green band, as the Green/Red ratio values were not significantly different ($P > 0.05$).

When exponential stretching was applied, the hierarchy of luminance between a foreground and a background region was much more important than their absolute difference. Because of the different Blue/Red ratios, when the mean RGB luminance was similar, the right hierarchy of luminance (foreground>background) was sometimes set in the blue band but inverted in the red band (foreground<background). This caused loss of continuity of root skeleton segments when the red band was processed, producing fragments shorter than the MRL. This drawback, which may increase with root age as a consequence of the decreasing root-background contrast, was reduced by processing the

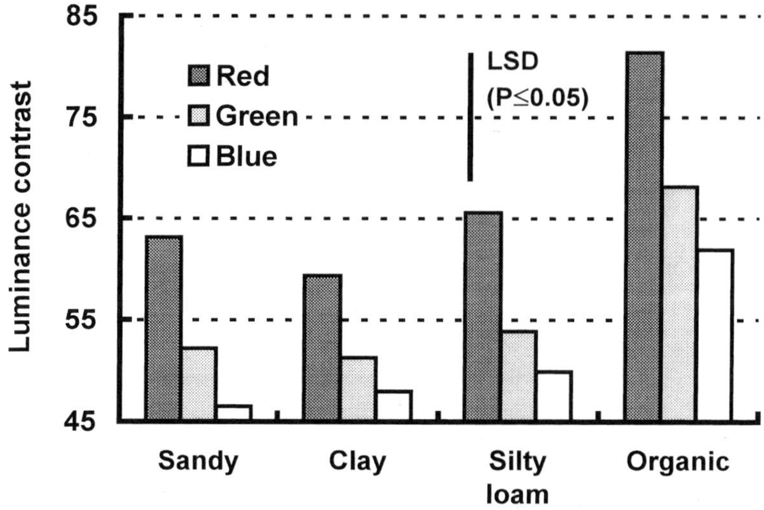

Figure 5. Absolute root luminance contrast over background (difference in grey value) as a function of colour band and soil: means of entire growing season. 'Soil type×Colour band' interaction significant at $P \leq 0.05$, LSD=13.0.

Table 1. Blue/Red luminance ratio calculated from pixel arrays. Means with same letter within a root-background couple are not different at $P \leq 0.05$ (LSD test). 'Observation date×Soil type' interaction significant at $P \leq 0.05$, LSD=0.04

Date	Soil type								All soils	
	Sandy		Clay		Silty loam		Organic			
	Root	Background	Root	Background	Root	Background	Root	Background	Root	Background
May 28	0.698	0.655	0.779	0.688	0.739	0.663	0.788	0.761	0.751a	0.692b
June 11	0.664	0.656	0.739	0.703	0.666	0.633	0.784	0.806	0.713a	0.700a
July 2	0.721	0.715	0.743	0.723	0.717	0.702	0.775	0.797	0.739a	0.734a
July 16	0.713	0.690	0.733	0.703	0.739	0.715	0.766	0.783	0.738a	0.723a
August 8	0.671	0.667	0.699	0.692	0.697	0.685	0.758	0.795	0.706a	0.710a
August 29	0.728	0.683	0.727	0.730	0.728	0.724	0.764	0.772	0.737a	0.728a
Mean	0.699a	0.678b	0.737a	0.707b	0.714a	0.687b	0.773a	0.786a	0.731a	0.714b

blue band. These observations were supported by some preliminary trials of image analysis, performed by processing the three bands separately in order to compare their effectiveness in root length estimation. As roots with small dark regions were generally more easily recognized as a whole in the blue than in the red and green bands, the blue band was chosen for processing in the following part of this experiment.

Image processing

When applied to transparent-wall images, typically formed of a relatively small area occupied by bright roots and a much larger area by darker background, exponential radiometric enhancement (see Equation 1) produced a drastic reduction in the radiometric values of the background and those of minor root parts (Figure 6); a large fraction of the resulting values fell in a range which was narrower and closer to zero. Instead, most of the root pixels were converted into grey values close to the maximum (i.e., 255), and only a few root pixels were left, with other bright objects, in the intermediate luminance interval (near 127). Most of these objects were formed of bright soil particles, water droplets and spots caused by uneven diffusion of light through the minirhizotron wall. They tended to be circular so that, after skeletonization, most of them appeared as short irregular segments and were later removed with the MRL method, discussed below.

A minimum root length (Table 2) was determined for each soil type by maximizing the coefficient of

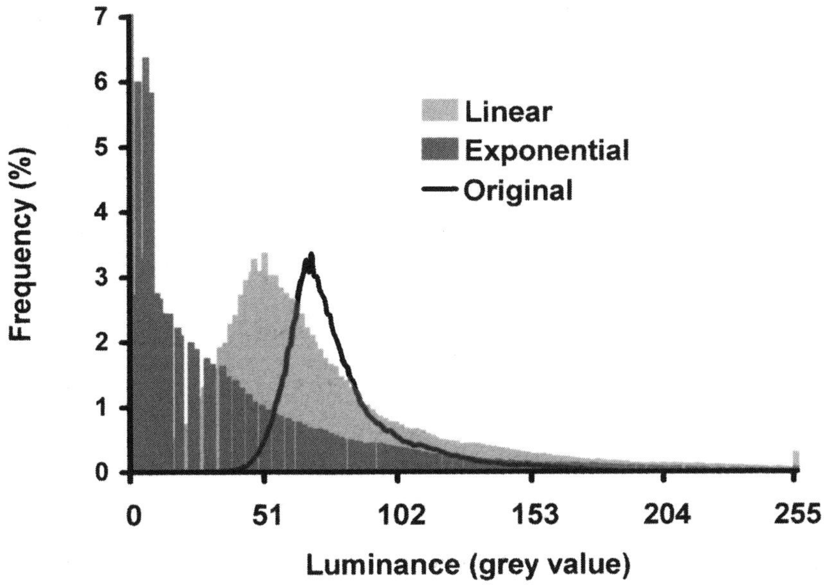

Figure 6. Luminance distribution of blue band in a minirhizotron root image before ('original') and after linear and exponential stretching enhancement.

Table 2. Minimum root length (MRL) calculated for relationships between manual and automatic measurements

Soil type	MRL (mm)
Sandy	2.80
Clay	2.97
Silty loam	2.08
Organic	2.11
All	2.80

determination (R^2) for regression (Figure 7) between manual and automatic measurements (see Experimental); however, a significant R^2 (0.738, Figure 8) could also be obtained by applying a single MRL (2.80 mm) to all soil types.

Comparisons among soils did not reveal any differences in the maximum depth reached by the root system, because it exceeded the length of the minirhizotrons.

Differences in root length density (RLD) values in the depth interval 50÷100 cm were found among soils (Table 3; Figure 9), the largest being associated with organic soil and the lowest with sandy soil. From these results and from other experiments (data not shown), both growth and root depth appear to be positively related to N content in soil. Klepper (1987) found reduced growth of sugarbeet fibrous roots in sandy soil; for the same crop, growth of both tap and lateral roots is reported as directly related to thermal time (Jaggard et al., 1996). It is likely that the lower brightness of organic soil (on average, a luminance value of 77 against 92 for other soils; $P \leq 0.05$) favoured soil heating, at least in early growth stages, thus accumulating this effect with that of N content. In our sandy soil, the lower availability of N probably played a more important role: as reported by Jaggard et al. (1996), soil fertility strongly increases the density of lateral roots.

The main problem of the minirhizotron method is image quality (Smucker et al., 1987). Background homogeneity and colour are major determinants for quality, as well as the behaviour of a minirhizotron towards light and the uniformity of lighting. In particular, we encountered some difficulties with bright extraneous objects, especially with sandy soil. The MRL method cannot completely overcome this inconvenience, but partial elimination of short root segments cannot be avoided either.

Organic soil showed a good fit of automatic measurements with manual ones, owing to good contrast with roots, and also helped in foreground recognition. However, the best fit was found in clay soil, due to the greater homogeneity of its background.

With all soils tested, the method described here causes a loss in root length, the amount depending on background: in our case, it was minimal with silty loam and maximal with sandy soil (see regression

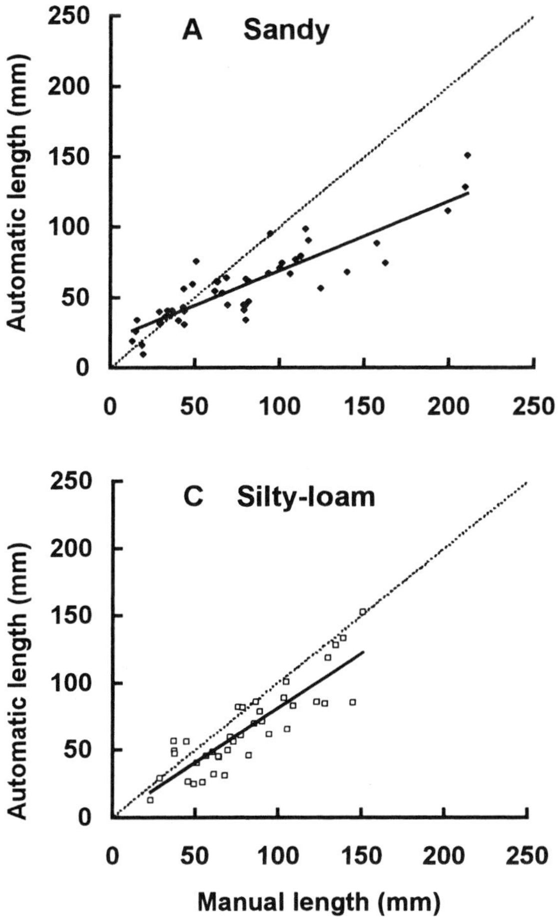

Figure 7. Relationships between manual and automatic measurements of root length. The equations are as follows: Sandy: y=0.4934 (±0.040) x+19.389 (±3.722); R^2=0.772. Clay: y=0.7071 (±0.045) x+8.420 (±4.433); R^2=0.866. Silty-loam: y=0.8119 (±0.076) x+0.049 (±6.643); R^2=0.743. Organic: y=0.7651 (±0.060) x+13.070 (±5.327); R^2=0.817. Values in parentheses are standard errors.

curves, Figure 7). The variability of loss in length measurements means that this method cannot be profitably used with a single image, in which case local techniques using various radiometric threshold levels and feature-based rules are more reliable.

As pointed out by Smit and Zuin (1996), when fine roots die they do not always completely disappear and no exact criterion is available for classifying them as dead or alive. When transparent-wall methods, like minirhizotrons, are used, only morphological features can be considered as age indicators. The most important one is colour (Cheng et al., 1991; Hendrick and Pregitzer, 1993): the young living roots of many plants are white (unpigmented), whereas brown and dark roots are classed as dead. When roots appear white and only slight senescence is noticeable, the images can be analysed by simple thresholding, and a single luminance threshold can be adopted in the foreground–background partition (Casarin et al., 1991; Nater et al., 1992).

Smit and Vamerali (1998) have often observed permanent shrinking of root diameter in potato, which implies loss of contact with soil particles and consequently loss of root functionality (Veen et al., 1992). Therefore, in root turnover studies, when a root starts to darken, judgement about its viability may seem quite arbitrary and prone to human error when manual or interactive procedures are applied. Identification of roots through the automatic method tested here requires a minimum root-background contrast. Although this method was tested on white roots, when applied to root turnover studies, the exact relationship between root luminance (under constant camera light intensity) and functionality should be known. If

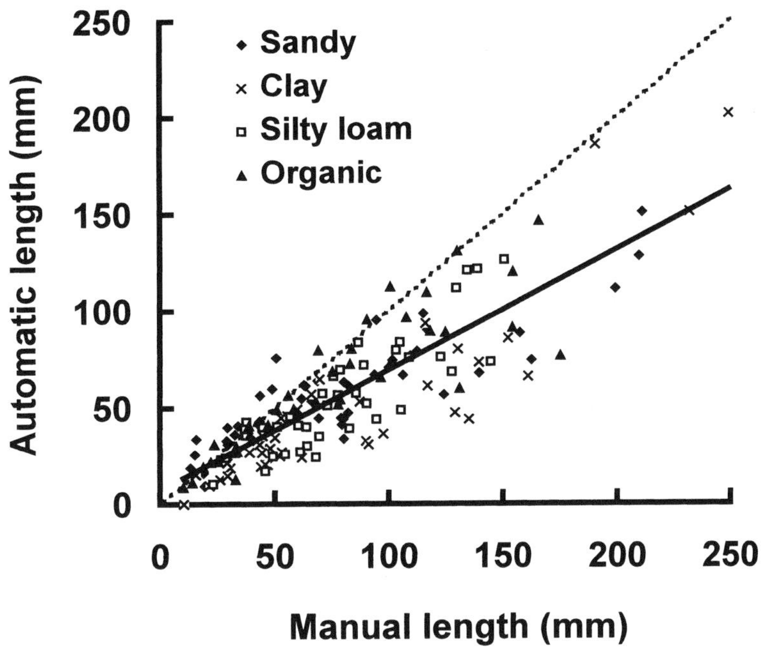

Figure 8. Relationship between manual and automatic measurements of root length for four soils pooled: y=0.622 (±0.029) x+7.216 (±2.676); R^2=0.738. Values in parentheses are standard errors.

Table 3. Root length density (RLD), soil nitrogen and organic matter contents (before sowing) in different soil types. Means with same letter are not different at $P \leq 0.05$ (LSD test)

Soil type	RLD 0–50 cm ($cm\ cm^{-2}$)	RLD 50–100 cm ($cm\ cm^{-2}$)	Total soil N (‰)	Organic matter (%)
Sandy	0.879a	0.939b	0.3	0.61
Clay	0.747a	1.217ab	1.4	2.46
Silty loam	0.989a	1.254ab	0.9	1.56
Organic	0.883a	1.464a	2.0	3.19

contrasted (visible) roots are considered to be alive (Huck and Taylor, 1982), this method can measure the length of living roots at any time. It is therefore a non-destructive dynamic procedure for root estimation and provides data which can be used in root dynamic studies for both growth and uptake models.

When large image samples are to be processed in root demography studies, the above results indicate that the approach described here is suitable for automated processing after proper adjustment to the specific characteristics of the experimental soil. The whole analysis is very fast: sequential loading, processing and data storing take about 30 s per image, for a PC with a 100-Mhz processor and 32 Mb of RAM. The method requires great stocks of images, but then any minirhizotron-based technique requires them; the need for greater numbers of samples with respect to the soil-core method is due to interference between minirhizotron and root arrangement. Using angled minirhizotrons, in sugarbeet a lower root density was detected in the first 40–50 cm of soil depth with respect to the next 50 cm. Since most of the literature reports the opposite trend for this crop (Brown and Biscoe, 1985; Jaggard et al., 1996), the use of minirhizotrons presumably leads to under-estimation of root density in the upper soil layers (Upchurch and Ritchie, 1983; Majdi et al., 1992). Thus, as an instrument for quantitative root observations (root length and density, branching), minirhizotrons have some drawbacks (McMichael and Taylor, 1987), but the

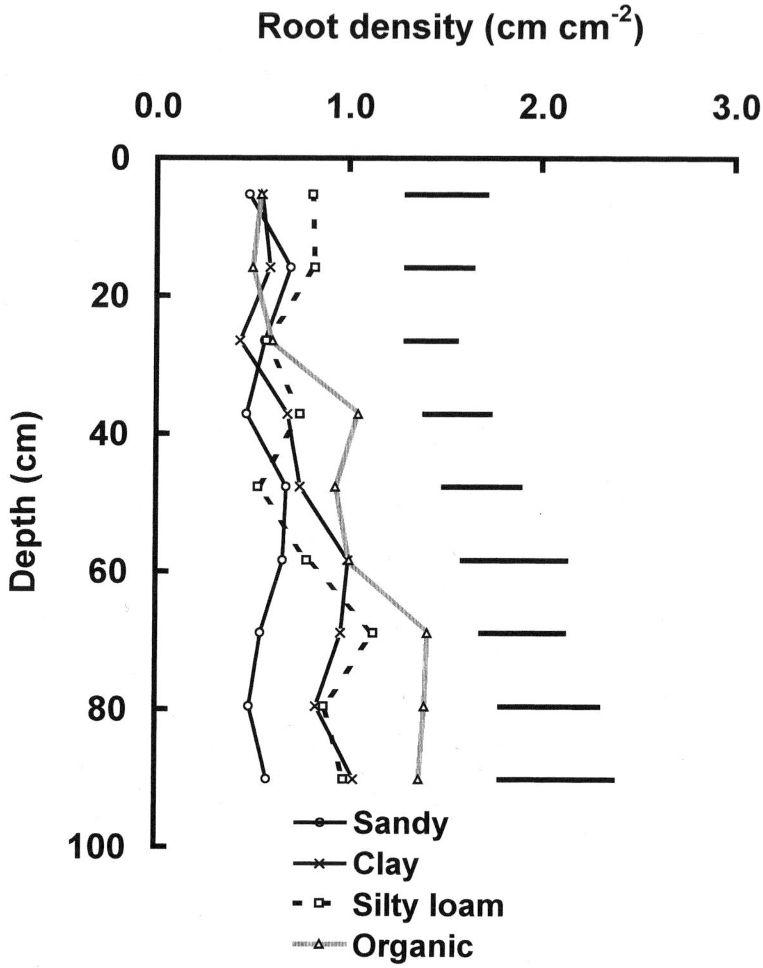

Figure 9. Patterns of root length density (automatic measurements): sugarbeet in different soil types. Horizontal bars: LSD values ($P \leq 0.05$) at different depths.

method described here proved quite practical, and only good PC memory capacity is required.

Conclusions

Colour images from minirhizotrons are more suitable than grey-scale images for automatic analysis. The advantage of analysing colour images may be increased by using cameras equipped with blue or, for some species, UV light, the latter case profiting by root fluorescence in a short-wavelength interval of visible radiation. Thus, extracting the blue channel from a digitized frame allows better recognition of roots, particularly young ones. Automatic analysis of a large set of images can be performed, based on an exponential algorithm of contrast stretching. Reduction of non-root objects can be made when a background-specific MRL is applied, but good accuracy can be achieved even by fitting a single MRL to varying conditions.

Acknowledgements

Financial support for this research ‘Architecture of root systems’ (ASiR) from the Italian National Research Council (CNR) is gratefully acknowledged. We are also grateful to Adriano Massignan for help with installation and sampling of minirhizotron tubes, and to Gabriel Walton for revision of the English text.

References

Brown K F and Biscoe P V 1985 Fibrous root growth and water use of sugar beet. J. Agric. Sci., Cambridge 105, 679–691.

Casarin M, Jacquey S, Fouere A and Girardin P 1991 Digital picture processing applied to the evaluation of plant root dynamics. *In* Plant roots and their environment: proceedings of an ISSR-symposium. Eds B L McMichael and H Persson. Proc. of ISSR Symposium, 21–26 Aug., 1988, Uppsala, Sweden. Elsevier Science Publishers B.V., Amsterdam, The Netherlands, pp 570–575.

Cheng W, Coleman D C and Box Jr J E 1991 Measuring root turnover using the minirhizotron technique. Agric. Ecosyst. Environ. 34, 261–267.

Heeraman D A, Crown P H and Juma N G 1993 A color composite technique for detecting root dynamics of Barley (*Hordeum vulgare* L.) from minirhizotron images. Plant and Soil 157, 275–287.

Hendrick R L and Pregitzer K S 1993 Patterns of fine root mortality in two sugar maple forests. Nature 361, 59–61.

Huck M G and Taylor H M 1982 The rhizotron as a tool for root research. Adv. Agron. 35, 1–35.

Klepper B 1987 Origin, branching and distribution of root systems. *In* Root Development and Function. Eds P J Gregory, J V Lake and D A Rose, pp 103–124. Cambridge University Press.

Jaggard K W, Dunham R J, Brown K F and Werker A R 1996 Season and soil type effects on sugar beet root development. *In* Proceedings of 59th IIRB Congress, 13–15 Feb., 1996. pp 213–228.

Majdi H, Smucker A J M and Persson H 1992 A comparison between minirhizotron and monolith sampling methods for measuring root growth of maize (*Zea mays* L.). Plant and Soil 147, 127–134.

McMichael B L and Taylor H M 1987 Applications and limitations of rhizontrons and minirhizotrons. *In* Minirhizotrons observation tubes: methods and applications for measuring rhizospere dynamics. Ed. H M Taylor. ASA Special Publication no. 50, Madison, WI, pp 1–14.

Nater E A, Nater K D and Baker J M 1992 Application of artificial neural system algorithms to image analysis of roots in soil. I. Initial results. Geoderma 53, 237–253.

Pratt W K 1991 Digital Image Processing. Ed. J Wiley & Sons, Inc., New York.

Smit A L and Zuin A 1996 Root growth dynamics of Brussel sprouts (*Brassica oleracea* var. *gemmifera*) and leeks (*Allium porrum* L.) as reflected by root length, root colour and UV fluorescence. Plant and Soil 185: 271–280.

Smit A L and Vamerali T 1998 The influence of potato cyst nematodes (*Globodera pallida*) and drought on rooting dynamics of potato (*Solanum tuberosum* L.). Europ. J. Agronomy 9, 137–146.

Smucker A J M, Ferguson J C, Debruyn W P, Belford L R and Richie J T 1987 Image analysis of video recorded plant root systems. *In* Minirhizotron observation tubes: methods and applications for measuring rhizosphere dynamics. Ed. H M Taylor. ASA Special Publication no. 50, Madison, WI, pp 67–80.

Stockman G, Smucker A J M and Halgren R 1990 Processing images of complex root systems. *In* Abstr., Conf. Analytical Methods for Quantifying Root and Soil Dynamics. St. Louis, MO, August 13–14, p 18.

Upchurch D R 1987 Conversion of minirhizotron-root intersections to root length density. *In* Minirhizotron observation tubes: methods and applications for measuring rhizosphere dynamics. Ed. H M Taylor. ASA Special Publication no. 50, Madison, WI, pp 51–65.

Upchurch D R and Ritchie J T 1983 Root observations using a video recording system in mini-rhizotrons. Agron. J. 75, 1009–1015.

Veen B W, Van Noordwijck M, De Willigen P, Boone F R and Kooistra M J, 1992 Root-soil contact of maize, as measured by a thin-section technique. III. Effects on shoot growth, nitrate and water uptake efficiency. Plant Soil 139, 131–138.

A. Stokes (ed.), The Supporting Roots of Trees and Woody Plants: Form, Function and Physiology, 325–334.
© 2000 Kluwer Academic Publishers. Printed in the Netherlands.

A field method of determining NH_4^+ and NO_3^- uptake kinetics in intact roots: Effects of CO_2 enrichment on trees and crop species

Hormoz BassiriRad[1,*], Stephen A. Prior[2], Rich J. Norby[3] and Hugo H. Rogers[2]
[1]*Department of Biological Sciences, University of Illinois at Chicago, Chicago, IL 60607, USA;* [2]*USDA-ARS National Soil Dynamics Laboratory, P.O. Box 3439, Auburn, AL 36831-3439, USA and* [3]*Oak Ridge National Laboratory, Bldg. 1059, P.O. Box 2008, Oak Ridge, TN 37831-6422, USA*

Key words: CO_2 enrichment, field method, intact roots, NH_4^+, NO_3^-, trees, uptake kinetics

Abstract

Models describing plant and ecosystem N cycles require an accurate assessment of root physiological uptake capacity for NH_4^+ and NO_3^- under field conditions. Traditionally, rates of ion uptake in field-grown plants are determined by using excised root segments incubated for a short period in an assay solution containing N either as a radioactive or stable isotope tracer (e.g., $^{36}ClO_3$ as a NH_4^+ analogue, $^{14}CH_3NH_3$ as an NO_3^- analogue or $^{15}NH_4^+$ and $^{15}NO_3^-$). Although reliable, this method has several drawbacks. For example, in addition to radioactive safety issues, purchase and analysis of radioactive and stable isotopes is relatively expensive and can be a major limitation. More importantly, because excision effectively interrupts exchange of compounds between root and shoot (e.g., carbohydrate supply to root and N transport to shoot), the assay must be conducted quickly to avoid such complications. Here we present a novel field method for simultaneous measurements of NH_4^+ and NO_3^- uptake kinetics in intact root systems. The application of this method is demonstrated using two tree species; red maple (*Acer rubrum*) and sugar maple (*Acer saccharum*) and two crop species soybean (*Glycine max*) and sorghum (*Sorghum bicolor*). Plants were grown in open-top chambers at either ambient or elevated levels of atmospheric CO_2 at two separate US national sites involved in CO_2 research. Absolute values of net uptake rates and the kinetic parameters determined by our method were found to be in agreement with the literature reports. Roots of the crop species exhibited a greater uptake capacity for both N forms relative to tree species. Elevated CO_2 did not significantly affect kinetics of N uptake in species tested except in red maple where it increased root uptake capacity, V_{max}, for NH_4^+. The application, reliability, advantages and disadvantages of the method are discussed in detail.

Introduction

Soil physical and chemical properties such as bulk solution concentration, diffusion and mass flow determine the availability of mineral nutrients at the root surface (Nye and Tinker, 1977). However, the rate of ion entry into the plant is regulated by root physiological uptake kinetics (Barber, 1984; Chapin, 1980; Clarkson, 1985; Schenk, 1996) which plays an important role in determining overall plant nutrient status. Root nutrient uptake kinetics vary widely among species and genotypes and this variation may explain differences in competitive ability in nutrient acquisition among co-occurring species (Chapin et al., 1986; Lambers and Poorter, 1992) influencing the dynamics of ecosystem nutrient cycling. Despite its significant role in regulating plant and ecosystem nutrient dynamics, measurement of root uptake kinetics in field-grown plants remains a difficult task.

Currently, the excised root technique (Epstein, 1972) is the most common field method for measuring root uptake kinetics. Briefly, in this method fine active roots are obtained from soil cores or other soil excavation techniques, washed and cut into 5 to 10 mm

* FAX No: +1 (312) 413 2435. E-mail: hormoz@uic.edu
Plant and Soil is the original source of publication of this article. It is recommended that this article is cited as: *Plant and Soil* **217:** 195–204, 1999.

long segments and are further divided into subsamples. Each subsample is then placed in a buffered uptake solution for a period of up to two hours. The uptake solution contains a radioactive or stable isotope tracer of the nutrient of interest (e.g., ^{32}P and ^{15}N are often used to measure P and N uptake). Rates of uptake in the saturable ranges of the substrate (labeled ion) concentration are determined by analyzing tracer accumulation in the tissue. Often a minimum of five or six data points are required to construct the uptake rate vs. concentration curve for calculation of kinetic parameters V_{max} and K_m.

Although excised root techniques can produce relatively reliable estimates of root physiological uptake kinetics, their application in field research is subject to a number of limitations. For example, since excised roots are cut from the endogenous carbohydrate supplies, uptake measurements have to be conducted rapidly following the excision to avoid carbohydrate exhaustion in the root tissues. In most cases, this problem is minimized by exogenous addition of carbohydrates (e.g., glucose or sucrose) into the uptake solution. However, as excised roots can readily absorb and metabolize soluble sugars (Bryce and Ap Reese, 1985), results may be compromized particularly in studies where the treatment is expected to affect nutrient uptake via root carbohydrate status.

Excised root experiments also require use of isotopic tracers as a labeling tool. There are several drawbacks associated with such a requirement. First, all radioactive tracers such as ^{13}N and ^{32}P pose a potential health hazard. Therefore, their use and waste disposal requires special training and care. Second, for N uptake studies, only a few radio tracers are available, but each offer a unique disadvantage. For example, ^{13}N has a half life of roughly 10 min. which limits its utility for field experiments. On the other hand, the use of analogues such as ClO_3 (synthesized with ^{36}Cl) for NO_3^- and CH_3NH_3 (labeled with ^{14}C) for NH_4^+ uptake studies have recently been criticized for not reliably reflecting the uptake behavior of NO_3^- and NH_4^+ (Jackson and Bloom, 1994; Kosola and Bloom, 1994, 1996; Siddiqi et al., 1992). More recently, ^{15}N is widely used as a tracer in N uptake studies. ^{15}N has very little or no hazard potential and as such is an attractive tracer of N. Nevertheless, because tissue ^{15}N analysis is fairly expensive its utility particularly in large-scale experiments may be cost prohibitive.

Here we describe a novel method of simultaneously determining kinetics of NH_4^+ and NO_3^- uptake of intact root systems in the field. The application of the method is demonstrated using two tree and two crop species. Since the method uses intact root systems, the possibility of carbohydrate exhaustion is avoided. Furthermore, because NH_4^+ and NO_3^- uptake rates are calculated from the depletion of these ions, (1) changes in the solution concentration can be analyzed by a range of conventional analytical methods for NH_4^+ and NO and (2) there is no requirement for use of tracers. We are particularly interested in examining how projected doubling of atmospheric CO_2 concentration will affect root uptake capacities of these inorganic N forms in intact roots of field grown plants. The knowledge of species differences in root N uptake kinetics and their relative responses to increases in atmospheric CO_2 is critical in predicting ecosystem productivity and potential changes in species composition. To date, there is little information about the potential impact of high CO_2 on kinetics of root N uptake in tree species (BassiriRad et al., 1997a, b) and to our knowledge this is the first report of such measurements in intact field-grown plants.

Materials and methods

Growth conditions

This work was conducted in the summer of 1996. Two national sites involved in CO_2 research were used: (1) Oak Ridge National Laboratory's Global Change Field Research Facility on the National Environmental Research Park in Oak Ridge, TN and (2) USDA-ARS National Soil Dynamics Laboratory in Auburn, AL. The work involving the tree species was conducted in twelve open-top chambers (OTC) at Oak Ridge. These OTCs were constructed (Rogers et al., 1983) at the facility on soils classified as Captina silt loam (fine-silty, siliceous, mesic Typic Fragiudult) with moderate-to-medium granular structure and medium internal drainage. The chambers were 3.0 m in diameter and 2.4 m high. An additional 1.2 m panel was installed at the beginning of the third growing season (1996) to accommodate the height growth of the seedlings. The chambers were modified to operate at either ambient temperature or 4 degree C above ambient, in combination with ambient or elevated (+30 Pa) atmospheric CO_2 partial pressure (Norby et al., 1997). A randomized complete block design was used with four treatments (the factorial combinations of two CO_2 levels with two temperature regimes) in each of three blocks. In the current investigation only the ambient temperature chambers were used.

One-year-old red maple (*Acer rubrum* L.) and sugar maple (*Acer saccharum* Marsh.) seedlings from a commercial nursery were planted directly into the soil within the chambers in spring, 1994. Additional seedlings (not used in the current study) were planted in spring, 1995, for a total of 10 plants of each species per chamber. The sides and tops of the OTCs were covered with 73% shade cloth to avoid unnatural levels of side light to these shade-tolerant maple seedlings. When roots were sampled in this experiment in June, 1996, the average height and basal diameter across the treatments were 2.1 m and 20 mm for red maple and 2.4 m and 18 mm for sugar maple. Trees in elevated CO_2 had an estimated aboveground dry mass, based on basal area times height, that was 76% greater in red maple and 23% higher in sugar maple compared to trees in ambient CO_2.

The experiments on crop species was also conducted in OTCs (Rogers et al., 1983) that were almost identical to those used at Oak Ridge and were part of an ongoing CO_2 experiment initiated in June 1992. Carbon dioxide levels inside the chambers were set at partial pressures of ambient or ambient plus 36 Pa. The experimental site was a soil bin (2 m deep, 6 m wide, and 76 m long) which was uniformly filled with surface soil of a Blanton loamy sand (loamy, siliceous, thermic Grossarenic Paleudult) that had been continuously fallow for over 25 years (Batchelor, 1984). A more detailed description of the soil status prior to initiation of the study, fertilizer and lime amendments during the study, and subsequent soil analysis results have been reported previously (Reeves et al., 1994). All experimental plots were managed under no-till conditions since 1992.

Crop species tested here were soybean (*Glycine max* (L.) Merr.; 'Stonewall') and grain sorghum (*Sorghum bicolor* (L.) Moench.; 'Savanna 5'). Plants were grown from seed to maturity in OTCs at two CO_2 levels. Seeds were sown on 9 May in 1996. Soybean seeds were inoculated with commercial Rhizobium (Nitragin Co.) prior to planting. Plants were thinned for uniformity to a final density of 30 plants m^{-2} for soybean and 26 plants m^{-2} for sorghum. To ensure adequate plant establishment, fertilizer N was broadcast at a rate of 34 kg N ha^{-1} to both the grain sorghum and the soybean shortly after planting. In the grain sorghum, an additional 67 kg N ha^{-1} was applied 30 days after planting. Weed control during the growing season was done by hand. In the off season, weed control was done both by hand and by use of glyphosate (N-[phosphonomethyl] glycine). At the time of uptake studies, plants were near the end of their vegetative growth.

The Auburn experiment was a split-plot design with three replicate blocks (6 m by 25 m). Whole plot treatments (plant species) were randomly assigned to half of each block. Subplot treatments (CO_2 levels) were randomly assigned to two open top chambers (3 m dia.) within each whole plot. Nutrient uptake experiments on these species were also performed in June of 1996 within a week of the tree species.

Root NH_4^+ and NO_3^- uptake kinetics

Experimental procedures for uptake experiments were similar for both sites. Net influx of NH_4^+ and NO_3^- for the target roots were determined by the depletion technique. Higher order laterals, containing a large number of fine active roots, were carefully dug out to a shallow soil. In all cases, sufficient suitable roots were present in the top 5 cm. Excavation of fine roots proceeded rapidly and cautiously until enough root material was exposed. Fine roots less than 2 mm in diameter with either white or light brown color were considered active in N uptake. Soil particles were carefully and thoroughly washed off the fine intact roots with distilled water making sure to avoid any physical damage. Immediately after rinsing, fine roots, still attached to the plant, were placed into a 2 mL microfuge tube containing 1.5 mL solution of NH_4NO_3 at one of the following concentrations: 25, 50, 75, 100, 150 and 200 μM. Uptake solutions also contained a phosphate buffer composed of 1 mM, KH_2PO_4 and 0.05 mM, K_2HPO_4 which maintained the pH at 6.0. More recent work in our lab indicates that rates of NH_4^+ and NO_3^- uptake do not differ significantly in the presence or absence of this phosphate buffer. Therefore, the use of phosphate buffer is not critical. The solutions were well aerated prior to uptake assay.

Roughly 5 to 10 mg of root dry weight were used in crop species and 10 to 20 mg in tree species. The amount of roots placed in each uptake tube was determined empirically. The amount of root necessary for this method may vary significantly for different species and functional groups. The ratio of root to solution volume has to be adjusted so that there is a detectable depletion during the uptake period, but to avoid a complete exhaustion of NH_4^+ or NO_3^- concentration prior to the completion of the incubation period. After immersing the roots into the uptake solution, the microfuge tubes were covered tightly with a piece of parafilm to reduce evaporation. All other exposed root parts were covered with wet paper towels throughout the uptake period. The incubation period lasted thirty to sixty minutes at which time the uptake assay was terminated by removing the roots from individual assay solutions. At that point, the portion of roots immersed in the assay solution was cut from its lateral and oven dried to be used in the uptake rate calculations. The microfuge tubes containing the N depleted assay solutions were capped tightly and frozen in liquid nitrogen until they were analyzed for NH_4^+ and NO_3^-.

Concentrations of $NH_4^+$and NO_3^- were simultaneously determined on an automated HPLC system modified after Goyal and Huffaker (1986). Briefly, The HPLC system used consisted of the following: two LC pumps, an SIL-10A autosampler, an SPD-10A variable wavelength UV detector, an RF-10A^{XL} fluorescence detector, and an SCL-10A system controller all from Shimadzu Scientific Instruments (Columbia, MD), a home-made heat reaction coil, a Valco (Houston, TX) electrically actuated nine port eight position stream selection valve, and a manual three port two position selection valve from Upchurch Scientific (Oak Harbor, WA). The autosampler was equipped with a 50 μL sample loop. Data collection and analysis from the HPLC system was performed using a software, CLASS-VP, which was also provided by Shimadzu. From each sample only 10 μL volume was needed to analyze for both NH_4^+ and NO_3^- concentrations. NO_3^- was isocratically eluted from a strong anion exchange Zorbax column (4.6×12.5 mm, 5 μm, Mac-Mod analytical, Chads Ford, PA) using 50 m*M* phosphate buffer (pH 3.0) at a flow rate of 1 mL/min. NO_3^- was detected by UV at 210 nm. The same sample was then derivatized the on line addition (via a PEEK^{TM} tee-connector) of a 100 m*M* phosphate buffer (pH 7.0) containing 3.5 m*M* o-phthaldialdehyde (OPA) and 5 m*M* 2-mercapthoethanol flowing at a rate of 1.5 mL/min. This reaction was carried out in a 3 meter coil of a microbore Teflon tubing maintained at 63 °C. The OPA-derivitized NH_4^+ was detected by fluorescence using emission and excitation wavelengths set at 410 and 470 nm, respectively.

The detection limit of this system for either NH_4^+ or NO_3^- is in the 10^{-9} M ranges. Because water uptake by roots will contribute to changes in NH_4^+ and NO_3^- concentration, this quantity was measured for each tube and taken into account in net uptake calculations. Net uptake rates (μmol g^{-1}dw h^{-1}) vs. concentration data were fitted to a Michaelis-Menten function (based on a Hanes plot transformation described in Price and Stevens, 1989) to estimate V_{max} and K_m.

Statistics

The results of each experiment were analyzed separately. Two-way Analysis of Variance (ANOVA) was used to test for the main effect of CO_2 on derived variables (K_m and V_{max}) for NH_4^+ and NO_3^- uptake. Values represent means calculated from three replications. Within treatments, each replicate represents a different open-top chamber. Data were tested for normal distribution and were log or square transformed as necessary. Differences were considered significant at $P<0.05$.

Results

Figure 1 shows representative net uptake rates of both N forms in the concentration ranges of 0 to 200 μM. All species exhibited a detectable net uptake of both NH_4^+ and NO_3^- in this range. Although roots of all species were able to absorb both N forms, NH_4^+ was always taken up at a greater rate (Figure 1). In fact, for a given species and a given external concentration, net NH_4^+ uptake rate was always greater (often two fold) than net NO_3^- uptake (Figure 1). There was no significant effect of CO_2 on species preference for N form however, the relative preference for NO_3^- compared to NH_4^+ was substantially higher in crop species than in tree saplings (Figures 1 and 2). When rates of net NO_3^- uptake were calculated as a proportion of total N uptake and averaged over both CO_2 treatments we found that in soybean and sorghum an average of 40% of N was taken up as NO_3^- whereas this proportion averaged for the tree species accounted for only 28% of total N uptake. The preference for NH_4^+ compared to NO_3^- is also illustrated in Figure 2 where absolute V_{max} values for NH_4^+ are substantially greater than NO_3^- in all species.

A two-way analysis of variance, performed separately for each experiment, indicated that CO_2 enrichment did not significantly affect V_{max} values of NH_4^+ and NO_3^- in all species tested. While tree species exhibited a significant CO_2 by species interaction for maximum NH_4^+ uptake capacity, there was no significant interaction term for crop species for either N form (Table 1). When each species is analyzed separately

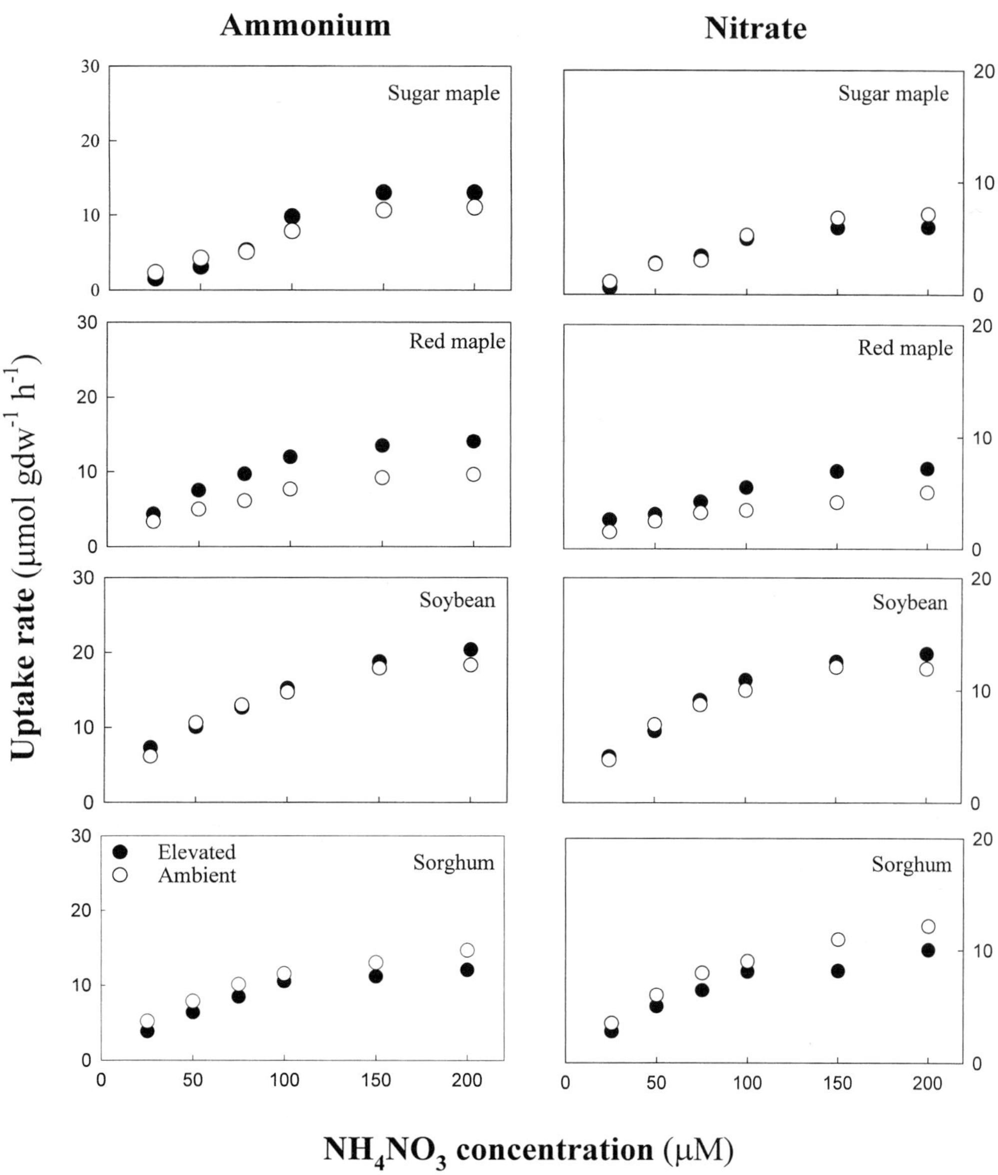

Figure 1. Root net uptake rate of NH_4^+ and NO_3^- from a wide range of external concentration of NH_4NO_3 for representative plants of two tree (red and sugar maple) and two crop species (soybean and sorghum) at ambient and elevated CO_2 levels. Rates are measured on intact field grown roots and are expressed on gram dry weight basis. Note the differences in scales of the Y axes between NH_4^+ and NO_3^-.

using a one-way ANOVA with CO_2 as the main factor, we found that CO_2 had no significant effects on either NH_4^+ or NO_3^- uptake capacity except in red maple (Figure 2). In this species, high CO_2 significantly increased V_{max} for NH_4^+, but not for NO_3^- indicating that the significant interaction term observed in Table 1 was largely due to a positive V_{max} response of red maple to CO_2 enrichment. There was no significant CO_2 effect K_m for either N form (data not shown).

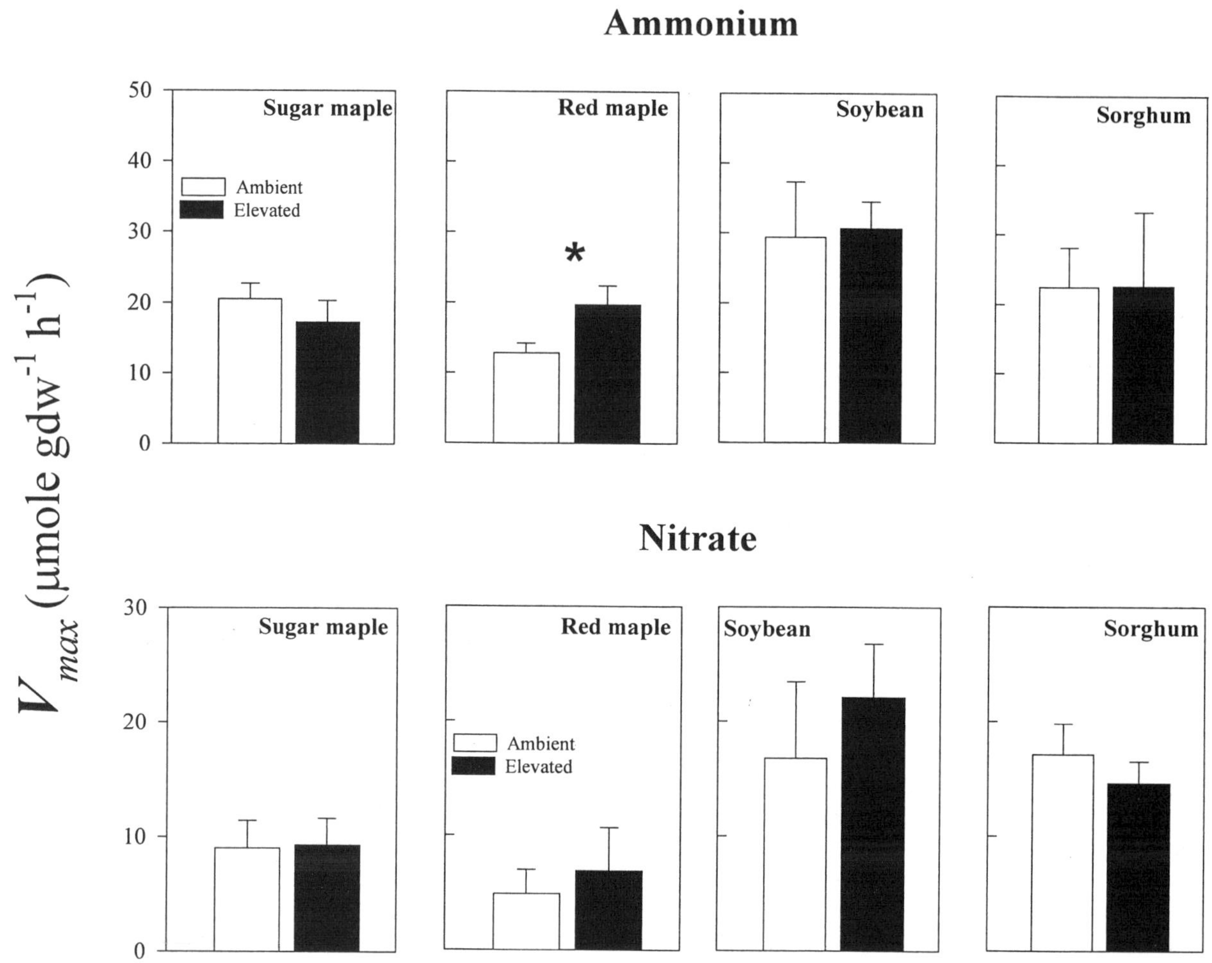

Figure 2. Maximum net uptake rate, V_{max}, of NH_4^+ and NO_3^- by intact roots of red maple, sugar maple, soybean and sorghum grown under field conditions at either ambient or elevated CO_2 levels. Each bar represents a mean of three replicates for each species. Note the differences in scales between NH_4^+ and NO_3^- . The CO_2 effect for each species is analyzed separately using a one-way ANOVA and significant differences are denoted by $* = P < 0.05$.

Discussion

Reliability, advantages and disadvantages of the method

Numerous laboratory studies have previously used the depletion technique to evaluate root NO_3^- and NH_4^+ uptake kinetics, but we are not aware of any studies that have applied this approach for an *in situ* determination of N uptake kinetics. Rennenberg and colleagues described an elegant field method for determining *in situ* net uptake rates of NH_4^+ and NO_3^- in spruce and beech trees (Gessler et al., 1998; Rennenberg et al., 1996). Although their method is similar to ours in terms of using intact roots, they determined net uptake rates of these ions at only a single external concentration of NH_4^+ and NO_3^-.

There are strong reasons why the method described here produces reliable information about an important root characteristic that is critical for a better understanding of the N dynamics in natural and managed ecosystems. The absolute values of net NH_4^+ and NO_3^- uptake rates as well as V_{max} values for both N forms reported here are qualitatively and quantitatively within the range of those reported by others. However, it must be noted that a direct comparison of the kinetic values reported here and those from the literature should be viewed with caution as experimental protocols and genotypic characteristics of plant materials differ substantially from one lab to an-

Table 1. Significant effects in two-way analysis of variance (ANOVA) for data presented in Figure 2. Data from each site are analyzed separately

Source	NH_4^+			NO_3^-		
	df	MS	Fcalc	df	MS	Fcalc
			Oak Ridge-Tree species			
Species	1	19.89	3.26	1	31.95	4.19
CO_2	1	9.28	1.52	1	3.79	0.51
Sp* CO_2	1	78.69	12.88**	1	2.36	0.31
Error	8	6.11		8	7.63	
			Auburn-Crop species			
Species	1	167.90	3.00	1	38.30	1.95
CO_2	1	1.30	0.02	1	5.90	0.30
Sp* CO_2	1	1.00	0.02	1	45.70	2.33
Error	8	56.01		8	19.64	

*Significant at $P<0.05$; ** Significant at $P<0.01$.

other. Additionally, a direct comparison may also be difficult due to inconsistency in expression of uptake rate units. Nevertheless, in our study the mean V_{max} value for sugar maple, 21.4 μmol g^{-1} h^{-1}, fell within the range of the values reported for the same species by Rothstein et al., 1996. In fact, V_{max} values reported for the tree species tested in the present study are in good agreement with the values generally reported for deciduous trees (Chapin et al., 1986; Gessler et al., 1998; Lajtha, 1994). Our kinetic results for crop species (Figure 2) are also within the range of values reported for soybean (Chaillou et al., 1994; Delhon et al., 1996; Saravitz et al., 1998), grasses (Goyal and Huffaker, 1986; Hole et al., 1990; Lee and Rudge, 1986; Lycklama, 1963; Macduff and Jackson, 1992; Siddiqi et al., 1989; Smart and Bloom, 1998) and other herbaceous species (Høgh-Jensen et al., 1997; Lainé et al., 1993; Macklon et al., 1990; Wieneke, 1992).

Our results are also in general agreement with other reports demonstrating a differential root uptake capacity for NH_4^+ as opposed to NO_3^- in most native and crop species (Blacquiere, 1987; Haynes and Goh, 1978; Marschner et al., 1991). We found that all species tested here had a distinct preference for NH_4^+ over NO_3^- but this preference was more pronounced for tree species compared to crop species (Figures 1 and 2). The greater preference for NH_4^+ vs. NO_3^- is almost a universal root characteristics in tree species and is often associated with an adaptation to forest soils that are relatively low in NO_3^- content (Cole, 1981; Haynes and Goh, 1978; Kronzucker et al., 1997; Marschner et al., 1991). Root preference for NH_4^+ uptake is not limited to tree species since most crop species also prefer NH_4^+ compared to NO_3^- (Glass and Siddiqi, 1995; Lee and Drew, 1989) when both N forms are equally present in the root medium roots.

There are a number of advantages in using the method described here as opposed to excised root techniques for determining root physiological uptake capacity for N. The most important advantage in this field method is that the target roots remain attached to the plant during uptake measurements. Therefore, unlike the excised root method, much of exchange of compounds between root and shoot remains uninterrupted. This is an important consideration particularly since the delivery of carbohydrates and nitrogenous compounds to the root and/or removal of absorbed N from the root play a critical role in net rates of NH_4^+ and NO_3^- uptake (Glass and Siddiqi, 1995; Pate and Layzell, 1990; Saravits et al., 1998). Uptake kinetics measurements of intact roots as described here also has the advantage of not requiring major equipment or radioactive/stable isotope tracers at the field site.

Although we have successfully applied this method to assess kinetics of N uptake in both tree and crop species, its application may be subject to some limitations. These limitations have also been discussed by Rennenberg et al. (1996). For example, because root excavation can damage fine roots, this procedure is most suitable for field situations where either digging is carried out with relative ease (e.g. coarse soils) or a substantial portion of the fine roots appear near the soil surface. In the present study both of these conditions were met but attempts are underway in our lab to improve the application of this method to more diverse soil conditions. It is also important to note that under field conditions most plant roots are mycorrhizal. Both endo- and ectomycorrhizae affect net N uptake rates (Boxman and Roelofs, 1988; Eltrop and Marschner, 1996; Rygiewicz et al., 1984) but at present this method does not separate the relative contribution of the mycorrhizae to overall N uptake. This problem is not unique to our method, but we suggest that until further quantification of root vs. mycorrhizal contribution, the term 'root system' rather than 'root' uptake kinetics be used.

CO_2 response

Because CO_2 enrichment generally increases carbohydrate levels in roots (BassiriRad et al., 1996a; Norby, 1994; Tschiplanski et al., 1993), metabolically regulated processes such as specific root N uptake rate are expected to be stimulated under high CO_2. In addition, because energy requirements associated with the uptake and assimilation of NO_3^- is considerably greater than NH_4^+ (Blacquiere, 1987; Glass and Siddiqi, 1995; Haynes and Goh, 1978) we expected a greater positive effect of CO_2 on NO_3^- than NH_4^+. However, with the exception of red maple, CO_2 did not significantly affect uptake kinetics of either N forms in species tested (Table 1 and Figure 2). Even though these results are inconsistent with our expectations they are not entirely surprising. Using potted seedlings we have shown elsewhere (BassiriRad et al., 1997a, b) that high CO_2 increased, decreased or had no significant effect on NO_3^- uptake kinetics depending upon species tested. In those studies the species dependent nature of the N uptake responses may have resulted from differences in patterns of C partitioning. If a greater proportion of the additional carbon resulting from growth under high CO_2 is partitioned into root growth then high CO_2 can still lead to an increase N uptake on a whole plant basis while uptake rates per unit root mass remains unchanged. In fact, in a different pot experiment, we found that CO_2 enrichment downregulated NH_4^+ uptake capacity in two conifer seedlings, yet the whole plant N uptake increased in response to CO_2 (BassiriRad et al., 1996b). In that case, increased biomass allocation to fine roots under high CO_2 counteracted the inhibitory effect of CO_2 on root N uptake capacity. Other mechanisms involved in regulation of root system uptake of N such as changes in root morphology, increased mycorrhizal hyphae and decreased plant demand by higher nitrogen use efficiency may be simultaneously operative in determining the magnitude and the direction of kinetic changes in response to CO_2. Finally, in most plants root kinetics of N uptake is regulated by demand which is likely to exhibit a seasonal pattern (Gessler et al., 1998; Glass and Siddiqi, 1995). Recently, in a hydroponic experiment using soybean and sunflower, we observed that root N uptake kinetics responses to CO_2 enrichment was highly dependent on the stages of development and root age (unpublished). Therefore, we suggest that a 'one point in time' determination is not adequate and more measurements of root N uptake kinetics are necessary to draw valid conclusions about possible effects of CO_2.

Acknowledgements

We wish to thank Barry and Tammy Dorman for their technical assistance in field data collection and two anonymous reviewers for their valuable comments. This work was funded by an interagency program TECO through the Office of Health and Environmental Research, United State Department of Energy contract # DE-FG02-97ER62332 and the National Science Foundation Division of Environmental Biology grant number NSF IBN-9728785.

References

Barber S A 1984 Soil Nutrient Bioavailability. A Mechanistic Approach. John Wiley and Sons, New York. 398 p.

BassiriRad H, Thomas R B, Reynolds J F and Strain B R 1996a Differential responses of root uptake kinetics of NH_4^+ and NO_3^- to enriched atmospheric CO_2 concentration in field-grown loblolly pine. Plant Cell Environ. 19, 367–371.

BassiriRad H, Griffin K L, Strain B R and Reynolds J F 1996b Effects of CO_2 enrichment on growth and root $^{15}NH_4^+$ uptake rate of loblolly pine and ponderosa pine seedlings. Tree Phys. 16, 957–962.

BassiriRad H, Griffin K L, Reynolds J F and Strain B R 1997a Changes in root NH_4^+ and NO_3^- absorption rates of loblolly and ponderosa pine in response to CO_2 enrichment. Plant Soil 190, 1–9.

BassiriRad H, Reynolds J F, Virginia R A and Brunelle M H 1997b Growth and Root NO_3^- and PO^3-_4 uptake capacity of three desert species in response to atmospheric CO_2 enrichment. Aust. J. Plant Physiol. 24, 353–358.

Batchelor J A Jr. 1984 Properties of bin soils at the national tillage machinery laboratory. In house report. National Soil Dynamics Laboratory, Auburn, AL.

Blacquiere T 1987 Ammonium and nitrate nutrition in *Plantago lanceolata* and *P. major* ssp. *major*. II. Efficiency of root respiration and growth. Comparison of measured and theoretical values of growth respiration. Plant Phys. Biochem. 25, 1775–1785.

Bloom A J, Sukrapanna S S and Warner R L 1992 Root respiration associated with ammonium and nitrate absorption and assimilation by barley. Plant Phys. 99, 1294–1301.

Boxman A W and Roelofs J G M 1988 Some effects of nitrate versus ammonium nutrition on the nutrient fluxes in *Pinus sylvestris* seedlings. Effects of mycorrhizal infection. Can. J. Bot. 66, 1091–1097.

Bryce J H and Ap Rees T 1985 Effects of sucrose on the rate of respiration of *Pisum sativum*. J Plant Phys. 120, 363–367.

Chaillou S, Rideout J W, Raper C D Jr. and Morot-Gaudry J-F 1994 Responses of soybean to ammonium and nitrate supplied in combination to the whole root system or separately in a split-root system. Phys. Plantarum 90, 259–268.

Chapin F S III 1980 The mineral nutrition of wild plants. Ann. Rev. Ecol. Syst. 11, 233–260.

Chapin F S III, Van Cleve K and Tryon P R 1986 Relationship of ion absorption to growth rate in taiga trees. Oecologia 69, 238–242.

Clarkson D T 1985 Factors affecting mineral nutrient acquisition by plants. Ann. Rev. Plant Physiol. 36, 77–115.

Cole D W 1981 Nitrogen uptake and translocation by forest ecosystems. *In* Terrestrial Nitrogen Cycles. Processes, Ecosystem Strategies and Management Impacts. Eds F E Clark and T Rosswall. Ecological Bulletin. pp 219–232. Swedish Natural Science Research Council, Stockholm.

Delhon P, Gojon A, Tillard P and Passama L 1996 Diurnal regulation of NO_3^- uptake in soybean plants IV. Dependence on current photosynthesis and sugar availability to the roots. J. Exp. Bot. 47(300), 893–900.

Eltrop L and Marschner H 1996 Growth and mineral nutrition of non-mycorrhizal and mycorrhizal Norway spruce (*Picea abies*) seedlings grown in semi-hydroponic sand culture. I. Growth and mineral nutrient uptake in plants supplied with different forms of nitrogen. New Phytol. 133, 469–478.

Epstein E 1972 Mineral Nutrition of Plants: Principles and Perspective. John Wiley and Sons, New York. 104 p.

Gessler A, Schneider S, Von Sengbusch D, Weber P, Hanemann U, Huber C, Rothe A, Kreutzer K and Rennenberg H 1998 Field and laboratory experiments on net uptake of nitrate and ammonium by the roots of spruce (*Picea abies*) and beech (*Fagus sylvatica*) trees. New Phytol. 138, 275–285.

Glass A D M and Siddiqi M Y 1995 Nitrogen absorption by plant roots. *In* Nitrogen Nutrition in Higher Plants. Eds H S Srivastava and R P Singh. pp 21–56. Associated Publishing Co., New Delhi, India.

Goyal S S and Huffaker R C 1986 A novel approach and a fully automated microcomputer-based system to study kinetics of NO_3, NO_2, and NH_4 transport simultaneously by intact wheat seedlings. Plant Cell Environ. 9, 209–215.

Haynes R J and Goh K M 1978 Ammonium and nitrate nutrition of plants. Biol. Rev. 53, 465–510.

Høgh-Jensen H, Wollenweber B and Schjoerring J K 1997 Kinetics of nitrate and ammonium absorption and accompanying H^+ fluxes in roots of *Lolium perenne* L. and N_2- fixing *Trifolium repens* L. 20, 1184–1192.

Hole D J, Emran A M, Fares Y and Drew M C 1990 Induction of nitrate transport in maize roots and kinetics of influx, measured with nitrogen-13. Plant Physiol. 93, 642–647.

Jackson L E and Bloom A J 1994 Assessment of methylammonium as an analog for ammonium in plant uptake from soil. Plant Soil 164, 195–202.

Kosola K R and Bloom A J 1994 Methylammonium as a transport analog for ammonium in tomato (*Lycopersicon esculentum* L.). Plant Physiol. 105, 435–442.

Kosola K R and Bloom A J 1996 Chlorate as a transport analog for nitrate absorption by roots of tomato. Plant Physiol. 110, 1293–1299.

Kronzucker H J, Siddiqi M Y and Glass A D M 1997 Conifer root discrimination against soil nitrate and the ecology of forest succession. Nature 385, 59–61.

Lainé P, Ourry A, Macduff J, Boucaud J and Salette J 1993 Kinetic parameters of nitrate uptake by different catch crop species: effects of low temperatures of previous nitrate starvation. Physiol. Plant. 88, 85–92.

Lajtha K 1994 Nutrient uptake in eastern deciduous tree seedlings. Plant Soil 160, 193–199.

Lambers H and Poorter H 1992 Inherent variation in growth rate between higher plants: A search for physiological causes and ecological consequences. Adv. Ecol. Res. 23, 187–261.

Lee R B and Drew M C 1989 Rapid reversible inhibition of nitrate influx in barley by ammonium. J. Exp. Bot. 40, 741–752.

Lee R B and Rudge K A 1986 Effects of nitrogen deficiency on the absorption of nitrate and mamonium by barley plants. Ann. Bot. 57, 471–486.

Lycklama J C 1963 The absorption of ammonium and nitrate by perennial ryegrass. Acta Bot. Neerl. 12, 361–423.

Macduff J H and Jackson S B 1992 Influx and efflux of nitrate and ammonium in Italian ryegrass and white clover roots: comparisons between effects of darkness and defoliation. J. Exp. Bot. 43(249), 525–535.

Macklon A E S, Ron M M and Sim A 1990 Cortical cell fluxes of ammonium and nitrate in excised root segments of *Allium cepa* L.; Studies using ^{15}N. J. Exp. Bot. 41(224) 359–370.

Marschner H, Häussling M and George E 1991 Ammonium and nitrate uptake rates and rhizosphere pH in non-mycorrhizal roots of Norway spruce [*Picea abies* (L.) Karst.] Trees 5, 14–21.

Norby R J 1994 Issues and perspectives for investigating root responses to elevated atmospheric carbon dioxide. Plant Soil 165, 9–20.

Norby R J, Edwards N T, Riggs J S, Abner C H, Wullschleger S D and Gunderson C A 1997 Temperature-controlled open-top chambers for global change research. Global Change Biol. 3, 259–267.

Nye P H and Tinker P B 1977 Solute Movement in the Soil-Root System. pp 243-289. University of California Press, Berkeley, California.

Pate J S and Layzell D B 1990 Energetic and biological cost of nitrogen assimilation. *In* Biochemistry of Plants, Vol. 16, Intermediary Nitrogen Metabolism. pp 1–42. Academic Publishing, San Diego.

Price N C and Stevens L 1989 Fundamentals of Enzymology, pp 143–144. Oxford University Press, Oxford.

Reeves D W, Rogers H H, Prior S A , Wood C W and Runion G B 1994. Elevated atmospheric carbon dioxide effects on sorghum and soybean nutrient status. J. Plant Nutr. 17(11), 1939–1954.

Rennenberg H, Schneider S and Weber P 1996 Analysis of uptake and allocation of nitrogen and sulphur compounds by trees in the field. J Exp. Bot. 47(303), 1491–1498.

Rogers H H, Heck W W and Heagle A S 1983 A field technique for the study of plant responses to elevated carbon dioxide concentrations. J. Air Poll. Control Assoc. 33, 42–44.

Rothstein D E, Zak D R and Pregitzer K S 1996 Nitrate deposition in northern hardwood forests and the nitrogen metabolism of *Acer saccharum* marsh. Oecologia 108, 338–344.

Rygiewicz P T, Bledsoe C S and Zasoski R J 1984 Effects of ectomycorrhizae and solution pH on [^{15}N]nitrate uptake by coniferous seedlings. Can. J. For. Res. 14, 893–899.

Saravits C H, Devienne-Barret F, Raper, C D Jr., Chaillou S and Lamaze T 1998 Nitrate uptake rate by soybean and wheat plants determined by external nitrate concentration and shoot-mediated demand. Int. J. Plant Sci. 159(2), 305–312.

Schenk M K 1996 Regulation of nitrogen uptake on the whole plant level. Plant Soil 181, 131–137.

Siddiqui M Y, Glass A D M, Ruth T J and Fernando M 1989 Studies of the regulation of nitrate influx by barley seedlings using $^{13}NO_3^-$ Plant Physiol. 90, 806–813.

Siddiqi M Y, King B J and Glass A D M 1992 Effects of nitrite, chlorate, and chlorite on nitrate uptake and nitrate reductase activity. Plant Physiol. 100, 644–650.

Smart D R and Bloom A J 1998 Investigations of ion absoption during NH_4^+ exposure I. Relationship between H^+ efflux and NO_3^- absorption. J. Exp. Bot. 49(318), 95–100.

Tschaplinski T J, Norby R J and Wullschleger D S 1993 Responses of loblolly pine seedlings to elevated CO_2 and fluctuating water supply. Tree Physiol. 13, 283–296.

Wieneke J 1992 Nitrate fluxes in squash seedlings measured with ^{13}N. J. Plant Nut. 15(1), 99–124.

FINE AND WOODY ROOT CHARACTERISTICS, ULTRASTRUCTURE AND RHIZOSPHERE

A. Stokes (ed.), The Supporting Roots of Trees and Woody Plants: Form, Function and Physiology, 337–346.

Adaptive tactics and characteristics of tree fine roots

Hans Persson
Department of Ecology and Environmental Research, Swedish University of Agricultural Sciences, Box 7072, S-750 07 Uppsala, Sweden

Key words: fine root biomass, fine-root chemistry, forest trees, live/dead ratios, nutrient availability, root distribution, root length, site quality, stem density, water availability

Abstract

Tree fine roots form an integrated network in the forest soil, decreasing in density from the soil surface downwards. Fine roots are opportunistic and exploitive in their growth habits, adapting rapidly to climatic variation and to changes in mineral nutrient and water supply. There is a substantial variation in fine-root biomass (live), necromass (dead) and the live/dead ratios in tree stands depending on tree age and site quality. As uptake organs they sustain the growth processes of forest trees and are sensitive indicators of the growth response of forest trees to soil resources. Mixed tree stands may have more fine roots than pure stands, stand density being an important factor determining the quantity of fine roots. Increasing fine-root biomass and decreasing live/dead ratios with age is to a great extent overshadowed by site factors favouring growth such as increased water and mineral nutrient availability. At the stand level, nitrogen fertilisation may cause a shift in carbon investments on a relative basis from below- to aboveground tree structures. A reverse relationship is often observed between the amount of fine roots and nutrient availability.

Background

Despite differences in morphology between the root systems of different tree species resulting from site quality, stem density and stand age, root systems have many characteristics in common. Tree roots extend into new areas by the means of fast growing lateral roots originally formed by the taproot (the primary root). One important feature of tree root systems is their woody framework, which normally undergoes extensive secondary thickening. The large-diameter woody roots of the forest trees provide structural support and prevent wind movements. A second important feature of tree root systems in mature forest stands is their dense network of fine roots, which function in the acquisition of available soil resources. Tree root systems are divided into coarse (structural roots) roots and fine roots on the basis of diameters (Flower-Ellis and Persson, 1980; Vogt and Persson, 1991). Studies of the growth of structural fractions of a forest tree – branches, stems, large-diameter roots, etc. – may be a prerequisite for a deeper understanding of how the tree allocates its resources with age, while studies of the growth of leaves (or needles) and fine roots are of a more immediate functional importance.

As a forest stand ages, a considerable weight fraction will accumulate in large-diameter roots (Ovington, 1957; Persson et al., 1995), but carbon budgets suggest that the annual investment in the maintenance and turnover of the fine roots is higher (Ågren et al., 1980; Linder and Axelsson, 1982). Many large-diameter roots in forest trees survive for a long time and may die with the tree itself, while the fine roots exhibit a high rate of death and renewal (Persson, 1983). The form, growth direction and distribution of coarse roots tend to be under genetic control (Köstler et al., 1968), this has to be altered and modified by the soil environment to a certain extent. The quantity of fine roots, the growth dynamics, the distribution patterns and fine-root numbers is strongly influenced by local environmental conditions (Persson, 1996; Persson & Ahlström, in press).

The distribution of fine-root biomass is closely connected with changes in root length, number of root tips, surface area and volume (Persson, 1980). Mycorrhizal root tips distributed on the fine roots are

Table 1. Amount of live fine roots (DW g m^{-2}), live/dead ratios (g/g) in forest stands of different tree species and age in natural forest stands (no numbers) or in stands subjected to different environmental conditions: control and thinned; increasing distance from the tree stem (0.5, 1.0, 1.5 m), control, drought and ammonium sulphate application; control, increased and decreased N and S deposition; dry, moderate and wet sites; control and carbonate lime; and control and dolomite; control, carbonate lime in two doses, dolomite, peat ash and wood ash; increasing site quality (I–IV); low and high productivity sites. N.D., no data, only live fine roots recorded

Treatments, etc./species	Age	Diameter (mm)	Biomass (g m^{-2})	Live/dead ratio (g/g)	References
Natural stands					
Acer rubrum	2–3	<2	5–58	9.3–28.9	Fredericksen & Zedaker, 1995
Betula parpyrifera	50–60	<1	273	N.D.	Braekke & Kozlowski, 1977
Fagus sylvatica	90	<1	118	0.6	van Praag et al., 1988
Mixed hardwood	80	<0.5	73–175	0.5–2.9	Burke & Raynal, 1994
Picea abies	26	<2	492	2.8	Majdi & Persson, 1993
Picea abies	35	<1	134	0.3	van Praag et al., 1988
Picea abies	40	<2	276	0.7	Roerig Hansen & Thomsen, 1991
–	45	<2	301	0.8	
–	66	<2	725	0.4	
Picea abies	115	<2	510	2.0	Puhe et al., 1986
–	106	<2	350	2.1	
–	80	<2	360	1.1	
Picea abies	85–95	<2	110–222	1.6–3.0	Ulrich et al., 1984
Pinus taeda	2–3	<2	44–86	4.4–25.9	Fredericksen & Zedaker, 1995
Pinus sylvestris	20	<2	26	0.4	Persson, 1983
–	120	<2	123	1.9	
Pinus sylvestris	100	<2	145–256	0.9–1.2	Persson, 1987
Pinus resinosa	38	<1	196	N.D.	Braekke & Kozlowski, 1977
Robinia pseudoacacia	2–3	<2	16–45	2.3–13.3	
Pinus, Acer	2–3	<2	35–54	>2.5	Fredericksen & Zedaker, 1995
Pinus, Robinia	2–3	<2	34–101	>5.3	
Thinned					
Pinus radiata					
control	12	<1	138	0.3	Santantonio & Santantonio, 1987
thinned	12	<1	55	0.1	
Increasing tree distance					
Tectonia grandis					
0.5 m	2–20	<1	10–73	N.D.	Singh & Srivasta, 1984
1.0 m	2–20	<1	8–17	N.D.	
1.5 m	2–20	<1	3–34	N.D.	
Water, N-fertilisation					
Picea abies – control	28	<1	273	0.4	Persson et al., 1995
drought	28	<1	142	0.2	
ammonium sulphate	28	<1	265	0.4	
Picea abies – control	80–85	<1	119–220	0.5–2.1	Persson & Ahlström, submitted
increased N, S dep.	80–85	<1	120–296	0.4–2.5	
decreased N, S dep.	80–85	<1	142–269	0.6–2.4	
Pseudotsuga menziesii					
dry	70	<1	253	0.2	Santantonio & Hermann, 1985
moderate	170	<1	350	0.4	
wet	120	<1	315	0.8	

Table 1. Continued

Treatments, etc./species	Age	Diameter (mm)	Biomass (g m^{-2})	Live/dead ratio (g/g)	References
Fertilised except N					
Picea abies – control	31	<1	168	N.D.	Clemensson-L. & Persson, 1993
carbonate lime	31	<1	288	N.D.	
Picea abies – control	69	<1	257	N.D.	Clemensson-L. & Persson, 1993
carbonate lime	69	<1	250	N.D.	
Picea abies – control	48–53	<1	182	N.D.	Clemensson-L. & Persson, 1993
carbonate lime	48–53	<1	175	N.D.	
Picea abies – control	55	<1	179	N.D.	Clemensson-L. & Persson, 1993
carbonate lime low	55	<1	149	N.D.	
carbonate lime high	55	<1	169	N.D.	
dolomite lime	55	<1	132	N.D.	
peat ash	55	<1	161	N.D.	
wood ash	55	<1	118	N.D.	
Pinus sylvestris – control	38	<1	122	N.D.	Clemensson-L. & Persson, 1993
carbonate lime	38	<1	160	N.D.	
Increasing site quality					
Cedrus deodara-I	33	<2	1414	N.D.	Ares & Peinemann, 1997
II	33	<2	260	N.D.	
IV	33	<2	5583	N.D.	
Cupressus sempervirens-I	30	<2	424	N.D.	Ares & Peinemann, 1997
IV	24	<2	1254	N.D.	
Pinus halepensis-I	24	<2	305	N.D.	Ares & Peinemann, 1997
II	25	<2	273	N.D.	
III	33	<2	668	N.D.	
IV	23	<2	900	N.D.	
Pinus radiata-I	26	<2	649	N.D.	Ares & Peinemann, 1997
II	26	<2	291	N.D.	
IV	19	<2	1042	N.D.	
Pseudotsuga menziesii					
low productivity	11–163	<2	2–90	2.2–8.6	Vogt et al., 1983
high productivity	13–160	<2	6–55	0.8–9.4	

morphologically and functionally different from the rest of the root system. The mycorrhizal sheet comprises 20–30% of the volume of the root tips, is a significant part of dry weight and insulates the root tissue from the soil (Harley and Smith, 1983). Due to the intimate contact between the fine roots (and mycorrhiza) in the root–soil interface, the chemical concentration in fine roots is a direct and reliable indicator of prevailing soil conditions. Although deficiency symptoms in forest trees are reflected in low nitrogen/mineral nutrient ratios in the fine-root chemistry, few attempts have been made to explain effects of nutrient imbalances from fine-root data (Persson et al., 1995).

In the past, researchers have arbitrarily chosen root diameter size classes to describe what have been called fine roots varying from <1 to 10 mm (Vogt and Persson, 1991). For most forest trees, roots <1 mm in diameter consist of ramifications with mycorrhizal root tips, morphologically very distinctive from the rest of the root system. In the present paper we deal with 'fine roots' in diameter fractions <0.5, <1 or <2 mm, avoiding tendentious names related to resource acquisition such as 'feeding' roots or 'conducting' roots. Fine root data with a diameter >2 mm were excluded from the data sets, and only data calculated directly on an area basis were included (Table 1).

Available information suggests substantial fine root production with a seasonal pattern different from needle or leaf production (Ågren et al., 1980; Persson, 1983). Growth dynamics of fine roots may vary considerably between different forest types and tree species at varying latitude, and under changing climatic conditions from one year to another (Persson, 1996). Therefore, attempts to correlate fine-root development in forest trees with environmental factors have often yielded inconclusive results (see Santantonio and Hermann, 1985). Fine root characteristics such as the amount of fine roots and the live/dead ratio seem to depend on a variety of abiotic and biotic factors specific for different forest stand (Persson, 1996). As a result of human impacts such as soil acidification, elevated aluminium/base cation ratios in the soil solution, high nitrogen/mineral nutrient ratios, etc., an increased sensitivity of the root systems to environmental stress may be found such as drought, windbreak, nutrient shortage, etc. (see Persson et al., 1995).

An increasing amount of information is so far available in literature on tree fine roots and their distribution. The primary objective of the present paper is to evaluate our present knowledge on fine-roots-development in forest ecosystems based on live fine root data (biomass) or standing crop estimates separated into live (biomass) and dead tissues (necromass). Data on fine root standing crop, with no separation into live, or live and dead fractions in most cases have been excluded. There seem to be no simple relationships between the fine-root biomass (live) and fine-root necromass (dead) at any given time and for any forest site. Minirhizotron data were not included since it is not possible to separate living and dead fine roots correctly through the transparant platic tubes and the data are not possible to convert to dry weight basis (Vogt and Persson, 1991). The lack of fine-root biomass data for different forest types and for important tree species is apparent (cf. Table1). Thus, much published data have been collected in conifer sites located in cold temperate or boreal zones. It was found impossible to include data reported in literature only in diagrams. The sequential sampling of soil cores or small soil blocks/monoliths, from which the fine roots are excavated, is the most common method of quantifying fine root standing crop in forests (Vogt and Persson, 1991). The data sets of fine root biomasss and live/dead ratios (Table 1) were collected/calculated from investigations using this technique.

Table 2. Average amount of fine roots in relation to the distance to the sampling trees on five sampling occasions (May, July, September and October 1996 and May 1997). Sampling was carried out on the Aubure *Fagus sylvatica* and *Picea abies* stands to a depth of 30 cm. n=25/n=100. The age of the *F. sylvatica* stand was 150 and the *P. abies* stand 100 years old. The different letters indicate significant differences between the two stands (Students t-test, $P > 0.05$)

Species/distance (m)	Biomass ($g\ m^{-2}$)	Necromass ($g\ m^{-2}$)	Live/dead ratio ($g\ g^{-1}$)
Fagus sylvatica			
1.5	75	107	1.3
2.5	82	117	0.8
3.4	93	103	1.3
4.5	81	114	1.0
1.5–4.5	*83*	*110a*	*1.1a*
Picea abies			
1.5	54	60	1.4
2.5	53	64	1.3
3.4	59	56	1.8
4.5	60	44	2.9
1.5–4.5	*57b*	*56b*	*1.8b*

Factors affecting fine-root distribution in forests

Spatial variation

In most forest soils the amount of fine roots diminish rapidly from the soil surface, or near-surface layers, downward (Persson et al., 1995; Stone and Kalisz, 1991). Most data reported in literature (Table 1) are based on total amount of fine roots from the soil surface to depths without any roots. The density of fine roots (dry weight, length, etc., per unit soil volume) is strongly affected by water and nutrient availability and competition among plants with roots in the same soil volume (Persson, 1996). The live/dead ratio, here used as an indicator of vitality of the fine roots, generally decreases with depth (see, e.g., Puhe et al., 1986). The most 'vital' fine roots are to be found in the uppermost organic soil horizon. The vitality of fine roots seems to be strongly influenced by the prevailing soil conditions (Clemensson-Lindell and Persson, 1995).

The lateral roots from different trees grow over and into one another in the different soil layers, forming an integrated network of the expanding fine roots. Fine roots are usually distributed in a fairly patchy pattern, independently of the distance to the nearest tree (Persson, 1980). The amount of fine roots excavated from core samples taken with increasing distances of

1.4, 2.5. 3.4 and 4.5 m from the tree stems in nearby *Fagus sylvatica* and *Picea abies* stands developed on similar soil types, confirmed that there were no significant decreases with distance from the investigated trees (Table 2). Nor did the live/dead ratio of the fine roots in the latter stand change significantly with distance from the trees.

Mixed forest stands may have substantially more fine roots than pure stands (Köstler et al., 1968). Thus, a mixture of *Pinus taeda* and *Robinia pseudoacacia* in a young plantation, had a higher fine-root biomass than pure stands (Table 1; Fredericksen and Zedaker, 1995). A more substantial fine-root biomass in mixed stands may be attributed to an intensive utilisation of the total soil volume by the fine roots. But, in the case of the present *Pinus–Robinia* stand, more N availability due to N fixation by *Robinia* may be the cause of the increase in the fine-root biomass, since there were no parallell increase in fine-root biomass in the *Pinus–Acer* treatment. The differences in fine-root distribution between pure and mixed stands may have important implications with regards to resource competition and other interference interactions.

Tree stem density seems to be an important factor determining the amount of fine roots present in different forest stands at any time. Santantonio and Santantonio (1987) showed (Table 1) that fine root quantity differed considerably between two similar *Pinus radiata* stands (55 g m^{-2} vs. 138 g m^{-2}) with different tree densities (166 vs. 567 stem ha^{-1}). But since the stand with lower stem density (60% reduction in basal area) had been thinned 2 years before the sequential root sampling was carried out (above-ground biomass was 59 vs. 134 , it is possible that the fine roots were still affected by the thinning operation. The live/dead ratio in the thinned versus the unthinned stand (0.3 vs. 0.1) suggest that the newly formed live fine roots in the thinned site were either still in a vigorous growth phase, or that dead fine roots were sloughed off after the thinning operation and not yet completely decomposed.

Species and age

The age of the forest stands (Table 1) varied from 2 to 170 years old for the coniferous tree species *Picea abies, Pseudotsuga menziesii, Pinus radiata, P. halepensis* and *P. sylvestris*. There is no evidence of age dependency in the data. The general trend in the data – increasing fine root biomass and decreasing live/dead ratios with age – seemed to a great extent be overshadowed by other factors. The fine-root biomass for *Picea abies* in different stands (28–106 years old) varied between 110 and 510 g m^{-2}, for *Pseudotsuga menziesii* (11–170 years old) between 2 and 350 g m^{-2} and for the different *Pinus* species (2–120 years old) between 26 and 1042 g m^{-2}. The live/dead ratio for *Picea abies* varied between 0.3 and 2.8, for *Pseudotsuga menziesii*, between 0.2 and 9.4 and for the different *Pinus* species between 0.4 and >5.3, respectively. Highest live/dead ratios were found in 2–3-year-old *Pinus taeda* stands which initially constituted only live fine roots (Fredericksen and Zedaker, 1984).

Lyr and Hoffmann (1967) suggested a high variability of root growth when compared to shoot growth in temperate forest trees, with site conditions substantially influencing the control of fine root growth. Clear differences (H. Persson, unpublished data) in fine-root distribution were found in two nearby *Picea abies* and *Fagus sylvatica* forest stands on similar site conditions (Table 2). It was evident that the *F. sylvatica* fine roots were more deeply rooted than *P. abies* fine roots. The amount of fine roots was higher in the *F. sylvatica* stand, while the live/dead ratio was higher in the *P. abies* stand.

High live/dead ratios (Table 1, Fredericksen and Zedaker 1995) were found in very young (2–3 years old) forest plantations, characterised by active growing fine roots and yet a very low necromass accumulation. Low live/dead ratio (0.4 vs. 1.9) was found in a young versus an old *Pinus sylvestris* stand (20 vs. 120 years old). Low live/dead ratios in the young stand were found also for some ground-layer vegetation species, *Calluna vulgaris* and *Vaccinium vitis-idaea*. The young stand was characterised by a high fine-root growth rate resulting in a high fine-root necromass accumulation (Persson, 1983). Most mature forest stands (Table 1) showed increasing live/dead ratio with age, probably also related to that fine roots in mature forest stands probably stays alive for a longer period of time than in young stands (Persson, 1993). The life/dead ratios also seem to be influenced by fluctuations in soil water and mineral nutrient availability (see below).

Since substantial seasonal variation is generally to be found in fine root biomass and necromass, there is an obvious limitation in data from investigations only carried out on one sampling occasion during the season. The data sets in Table 1 however include data from a variety of forest types with sequential sampling of fine roots carried out repeatedly during the season (e.g., Braekke and Kozlowski, 1977; Persson, 1983;

Persson et al., 1995; Santantonio and Santantonio, 1987; Vogt et al., 1983).

Water and temperature

Water availability influences mineral nutrient availability and the total carbon allocation patterns in trees. Water and mineral nutrient uptake is then closely related to the degree of penetration by the fine-root tips, i.e., the fine root growth and the rate of penetration by the mycorrhizal fungi. A high growth rate means that the root tips and mycorrhizal fungi penetrate a large soil volume, which increases the accessibility to water reserves and contributes to the maintenance of a stable water regime in the trees. Data from one field experiment including drought (two-thirds of the precipitation was excluded by a roof during the growth period) indicate (Table 1; Persson et al., 1995) reduced fine-root growth in the upper parts of the soil profile, whereas the deeper parts were more substantially penetrated by fine roots. In dry sites, the whole soil profile appears to be more occupied by fine roots than wet sites (Persson et al., 1995).

Besides water availability, temperature can influence the plant performance indirectly by affecting the water uptake and or/ the absorption of nutrients (Bowen, 1991). Many reviews have been written on the relationships of low and/or high temperature on plant functions such as utilisation of photosynthesis, carbon partitioning, water and ion uptake (McMichael and Burke, 1996). Soil moisture stress during the summer and low temperature in winter seem to be the factors, which limit fine root growth most (Lyr and Hoffman, 1967). Because of the fluctuations in soil temperature, roots seldom function under optimal temperature conditions in the field. Soil temperature also affects the degree of suberization, with implications for water and ion uptake (Persson, 1978). Henrick and Pregitzer (1993) found no significant temporal differences in fine root standing crop in two well-stocked *Acer saccharum* forests (one southern and one northern site at a distance of 80 km) located in the northern Lower Peninsula of Michigan. The southern site was exposed to higher temperature during the growth period.

The fact that forest trees are well adapted to water stress was shown by a field experiment in a 20-year-old *Pinus sylvestris* stand (Puhe et al., 1986). During a period of about 1 year of artificial drought, there was no significant influence on the fine-root development when compared with trees on control plots. A sufficient water uptake by some parts of the root system may provide the necessary water for the whole tree. Therefore, some roots of the same root systems can force their way through dry zones, when water uptake and supply is guaranteed by other roots.

Table 3. The vertical distribution down to 60 cm of the mineral soil of living fine roots (<2 mm in diameter) in two mature *Pinus sylvestris* stands (Ih 0 and Ih 5) in Central Sweden with different ground water tables. The soil water table was at its highest 1–3 m in Ih 0 and 10 m level in Ih 5. Data from Persson (1992)

Soil horizons/depth (cm)	The amount of living fine roots ($g\ m^{-2}$)	
	Ih 0	Ih V
LFH (6-7)	79	99
Mineral soil		
0–10	59	72
10–20	37	15
20–30	18	2
30–40	20	1
40–50	6	1
50–60	1	1
Total soil profile	221	189

Deans (1979) reported a seasonal influence of soil temperature and moisture on the rate of fine-root growth: as soil temperature increased, root growth increased. But this relation was overridden by low soil moisture during the later part of the growth period. In forest soils, fine roots grow towards more humid zones. In two nearby mature *Pinus sylvestris* stands on similar soil conditions, but with varying depth of the soil water table (Persson, 1982; Table 3) the fine roots were more deeply distributed in the stand with the highest soil water table. The fine roots in the latter case evidently penetrated more deeply in the soil profile due to increased water availability.

Live and dead fine roots (Santantonio and Hermann, 1985; Table 1) in dry, moderate and wet habitats of *Pseudotsuga menziesii* were sampled sequentially during a 3-year period. The amount of live fine roots on the dry plots was low compared with the other two sites. The most striking observation was, however, that the amount of dead fine roots (necromass) increased in dry in conditions. Thus, the live/dead ratio decreased from 0.78 in wet, to 0.43 in moderate and to 0.24 in dry conditions. Optimum amounts of fine roots should be expected in the forest soil at times of high water (and nutrient) availability, with loss of fine roots when conditions become adverse (Deans, 1979; Persson, 1980).

Nitrogen

Data on fine-root growth in response to plant nutrients in the soil profile are available from several nitrogen manipulated experimental areas (e.g., Ahlström et al., 1988; Persson, 1980; Persson and Ahlström, 1991, 1994, in press). Most of those data were obtained by using ingrowth cores; few data are available from investigations using the soil coring technique. The general impressions from those field experiments are that N fertilisation increases fine-root growth activity, but only locally in soil volumes directly affected by the nitrogen fertilisers. The nitrogen availability in itself does not determine the rate of fine-root exploitation as much as the total mineral nutrient available, e.g., the 'site quality' (see below).

Data from one field experiment with ammonium sulphate application (NS) show an increased fine root growth in the upper parts of the soil profile (Table 1; Persson et al., 1995). However, the data for the total soil profile did not indicate any changes in the amount of live fine roots. The live/dead ratio for the NS treated plots did not significantly differ from the control plots. Nevertheless, in the NS plots a reduced 'vitality' was indicated, in fine roots sorted into three vitality classes (Clemensson-Lindell and Persson, 1995). The reason for the morphological changes in the fine roots may be caused by alterations in soil pH and increased amounts of Al^{3+} in the soil solution imposed by the NS application. Experimentally decreased N and S deposition in catchment areas in SW Sweden (a roof experiment) showed an increased amount of live fine roots and live/dead ratios (Table 1; Persson and Ahlström, submitted).

At the stand level, nitrogen fertilisation may cause a shift in carbon investments on a relative basis from below- to aboveground tree structures (Ågren et al., 1980; Linder and Axelsson, 1982; Santantonio and Hermann, 1983). Carbon allocation to foliage/needles is increased when the allocation to fine roots is decreased (Gower and Vitousek, 1989). From a teleological point of view, it may be argued that fine roots on forest sites with good N supply, do not need so much fine roots. A decreased fine-root growth rate does not mean that the fine roots are 'less vital' or damaged. Root damage per se seems to be related to unbalanced mineral nutrient conditions characterised by low availability of most elements, except for nitrogen (Persson et al., 1995). Leaching of base cations, in combination with high nutrient requirements related to the increased growth rate caused by the high nitrogen load, enhance the risk of nutrient imbalances in the forest trees.

Nitrogen in excess of requirements within the forest ecosystem may result in both acidification and eutrophication of the soil, both processes increasing the risk of forest damage (needle loss or fine-root loss – see Marschner et al., 1991; Schulze, 1989). In damaged forest stands there is a clear relationship between the needle loss of the forest trees and the loss of live fine roots (Puhe et al., 1986; Ulrich et al., 1984). Obviously there is a declining trend in the amount of living fine roots with increased needle loss, which is also shown in a decreasing live/dead ratio for the fine roots. Whether forest damage is due to nutrient imbalances or a function of high nitrogen and sulphur concentration in the soil solution causing soil acidification (and Al^{3+} blockading the uptake of important nutrients from the soil solution) is still not yet finally settled.

Other mineral nutrients

Experimental treatments such as liming and compensatory fertilisation of mineral nutrients except nitrogen (Table 1; Clemensson-Lindell and Persson, 1993; Persson and Ahlström, 1991) may result in no effect, or both negative and positive effects, on the growth and development of fine roots, depending on the soil type, the fertilisation agents and the dose at which they are applied. The investigations from *Picea abies* stands in South Sweden (Persson and Ahlström, 1994) clearly demonstrate that element concentrations in fine roots (Ca, P, Mg and K) are strongly affected by changes in soil nutrient status after liming, ash application, etc. At the same time, there is a general increase in nitrogen concentration in the fine roots, which might be related to increased levels of nitrogen in the soil released by enhanced nitrogen mineralization. It is well known that liming on nitrogen-rich soils enhances nitrate formation (Staff et al., 1996).

It is reasonable to assume that the lime application in itself does not influence fine-root growth, but the increased nitrogen availability by higher nitrogen mineralisation after liming. In temperate forests nitrogen is quantitatively the most important nutrient strongly limiting forest growth (Tamm, 1991). Liming carried out at high doses in combination with N fertilisation was shown to have a strong negative effect on the development of fine roots, but also to extensive changes in plant, animal and microbial communities (Persson and Ahlström, 1991; Staaf et al., 1996). In forest stands with reduced amount of fine roots or less vital fine roots due to soil acidification, it is likely that this negative effect is much stronger (Persson and Ahlström, 1991).

Site quality

Fine root growth is correlated to environmental conditions favourable for photosynthesis, although local site conditions tends to cause variation in this relationship (Vogt et al., 1986). Traditionally, forest researchers have attributed differences in wood yield between sites to differences in 'site quality'. The mineral nutrient balances, soil texture and water holding capacity in the forest soil determine site quality. While effects of site conditions can give useful leads, they are difficult to interpret because of the confounding of variables (such as soil type, slope, past management, etc.) and a lack of information on processes controlling growth. Soil organic matter, its structure and rate of decomposition, has a strong control over the availability of nutrients, and in turn, the growth rate of forest stands. There is plenty of evidence that the proportion of carbohydrates produced by trees that is allocated to fine-root growth is higher in poor growth conditions than where water and nutrients are less limiting (Gower, 1987).

A reverse relationship between the amount of fine roots and nutrient availability has been observed in a variety of forest stands (Nadelhoffer et al., 1985; Gower and Vitouch, 1989). The implication is that the trees maintain a large heavily ramified mycorrhizal fine root system in poor sites versus growth in above-ground tree structures. A study by Ares and Peinemann (1997) was conducted to determine amounts and distribution of *Cedrus deodara*, *Cupressus sempervirens*, *Pinus halepensis* and *P. radiata* fine roots in stands of different 'site quality' in the temperate region of Argentina (Table 1). Differences in the amount of fine-roots between poor versus better sites of similar age indicate decreasing amount of fine roots with increasing site quality (Table 1). Soil organic matter content furthermore provides a good correlation with root density. Observation on single profiles indicated that at low organic matter content ($<1\%$) the root density notably decreased. There was a negative correlation between root density and clay content that suggested that fine roots decreased as clay content increased.

Concluding remarks

Within similar climatic conditions fine root growth is limited by important nutrients appearing in low concentration in relation to nitrogen. There is an overall lack of information concerning the fine root development in fertilised forests from which nitrogen or other mineral nutrients have been eliminated as growth-limiting factors. New pathways of carbon and mineral nutrient cycling require further assessment. Few attempts have been made to quantify the cost in terms of carbon used by the mycorrhizae. Few attempts have also been made to include mycorrhizal acquisition for water and nutrients in the perspective of carbon investments by the forest trees. Although it may be difficult from investigations in forest ecosystems to sort out which specific factor is most important for fine root growth, a greater proportion of the total net primary production may go to fine roots and mycorrhizas, when site conditions are less favourable. Although mycorrhizal fungi infect most fine roots, the respiration costs of mycorrhizal versus non-mycorrhizal fine roots are largely unknown. The number of mycorrhizal ramifications is increasing with increasing stand age (Persson, 1984). Mycorrhizas may perform a functional role as effective nutrient absorbing organs even at relative high nitrogen concentrations in the soil water (Högberg, 1989).

Site quality should be regarded as an integrated characteristic of total water and mineral nutrient availability. The growth of fine roots may differ fundamentally from that of shoots by being far more opportunistic and exploiting. In mixed forest stands, shallow-rooted and more deeply rooted fine roots from different forest trees may serve to exploit the mineral soil horizons more efficiently (Köstler et al., 1968). The location of large proportion of fine roots near the soil surface suggests that forests stands are vulnerable to disturbances by management practices.

With increasing age of the forest stand, the total soil profile is more completely occupied by fine roots. Nevertheless, the majority of tree fine roots, are located in the top 20 cm of the soil horizon in most forest stands (Persson et al., 1995; Vogt et al., 1981). Fine roots are adapting to the thinned conditions by decreased growth, and a decreased live/dead ratio (Santantonio and Santantonio, 1987; Table 1). There is a general increase in the total standing crop (biomass+necromass) and live/dead ratios of fine roots with increasing age (Persson, 1984). When competing with other plants, trees must invest sufficiently in root structures to provide both mechanical support and to be able to compete for soil resources. When managing forest stands, considerations should be given to practices that reduce any adverse effect on the root systems. Since the wood production is among the silviculturist's highest priorities, the ability of the root system to supply enough water and nutrients is of fundamental importance.

References

Ågren G, Axelsson B, Flower-Ellis J G K, Linder S, Persson H, Staaf H and Troeng E 1980 Annual carbon budget for a young Scots pine. Ecol. Bull. 32, 307–313.

Ahlström K, Persson H and Börjesson I 1988 Fertilization in a mature Scots pine (*Pinus sylvestris* L.) stand: effects on fine roots. Plant Soil 106, 179–190.

Ares A and Peinemann N 1992 Fine-root distribution of coniferous plantations in relation to site in southern Buenos Aires, Argentina. Can. J. For. Res. 22, 1575–1581.

Bowen G D 1991 Soil temperature, root growth, and plant function. *In* Plant Roots: the Hidden Half. Eds Y Waisel, A Eshel and U Kafkafi. pp 309–330. Marcel Dekker, New York.

Burke M and Raynal D J 1994 Fine root phenology. Production, and turnover in a northern hardwood forest ecosystem. Plant Soil 162, 135–146.

Braekke F H and Kozlowski T T 1977 Distribution and growth of roots in *Pinus resinosa* and *Betula papyrifera* stands. Medd. Norsk Inst. Skogforsk. 33, 437–451.

Clemensson-Lindell A and Persson H 1993 Long-term effects of liming on the fine-root standing crop of *Picea abies* and *Pinus sylvestris* in relation to chemical changes in the soil. Scand. J. For. Res. 8, 384–394.

Clemensson-Lindell A and Persson H 1995 The effects of nitrogen addition and removal on Norway spruce fine-root vitality and distribution in three catchment areas at Gårdsjön. For. Ecol. Manage. 71, 123–131.

Clemensson-Lindell A and Persson H 1993 Fine root vitality in a Norway spruce stand subjected to various nutrient supplies. Plant Soil 168–169, 167–172.

Deans J D 1979 Fluctuations of the soil environment and fine root growth in a young sitka spruce plantation. Plant Soil 52, 195–208.

Flower-Ellis J G K and Persson H 1980 Investigations of structural properties and dynamics of Scots pine stands. Ecol. Bull. 32, 125–138.

Fredericksen T S and Zedaker S M 1995 Fine root biomass, distribution, and production in young pine-hardwood stands. New Forests 10, 99–110.

Gower S T 1987 Relations between mineral nutrient availability and fine root biomass in two Costa Rican tropical wet forest. Biotropica 19, 252–257.

Gower S T and Vitousek P M 1989 Effect of nutrient amendments on fine root biomass of a primary successional tropical forest in Hawaii. Oecologia 81, 566–568.

Grier C C, Vogt K, Keyes M and Edmonds R 1981 Biomass distribution and above- and below-ground production in young and mature *Abies amabilis* zone ecosystems of the Washington Cascades. Can. J. For. Res. 11, 155–167.

Harley J L and Smith S E 1983. Mycorrhizal Symbiosis, Academic Press, London, 483 pp.

Hendrick R L and Pregitzer K S 1993 The dynamics of fine root length, biomass, and nitrogen content in two northern hardwood ecosystems, Can. J. For. Res. 23, 2507–2520.

Högberg P 1989 Growth and nitrogen inflow rates in mycorrhizal and non-mycorrhizal seedlings of *Pinus sylvestris*. For. Ecol. Manage. 28, 7–17.

Köstler J N, Bruckner E and Bibelrieter H 1968 Die Wurzeln der Waldbäume. Verlag Paul Parey, Hamburg, 284 pp. (in German).

Linder S and Axelsson B 1982 Changes in carbon uptake and allocation patterns as a result of irrigation and fertilization in a young *Pinus sylvestris* stand. *In* Carbon Uptake and Allocation in Subalpine Ecosystems as a Key to Management. Ed. R H Waring. pp 38–44. For. Res. Lab., Oregon State University, Corvallis, OR, USA.

Lyr H and Hoffmann G 1967 Growth rates and growth periodicity of tree roots. Int. Rev. For. Res. 2, 181–236.

Majdi H and Persson H 1993 Spatial distribution of fine roots, rhizosphere and bulk-soil chemistry in an acidified *Picea abies* stand. Scand. J. For. Res. 8, 147–155.

Marschner H, Häussling M and Eckhard G 1991 Ammonium and nitrate uptake rates and rhizosphere pH in non-mycorrhizal roots of Norway spruce (*Picea abies* (L.) Karst.). Trees 5, 14–21.

McMichael B L and Burke J L 1996 Temperature effect on root growth. *In* Plant Roots: the Hidden Half. 2nd edn. Eds Y Waisel, A Eshel and U Kafkafi. pp. 383–396. Marcel Dekker, New York.

Nadelhoffer K J, Aber J D and Melilo J M 1985 Fine roots, net primary production, and soil nitrogen availability: a new hypothesis. Ecology 73, 1139–1147.

Ovington J D 1957 Dry-matter production by *Pinus sylvestris* L. Ann. Bot. NS 21, 288–414.

Persson H 1978 Root dynamics in a young Scots pine stand in Central Sweden. Oikos 30, 508–519.

Persson H 1980 Fine-root dynamics in a Scots pine stand, with and without near optimum nutrient and water regimes. Acta Phytogeogr. Suec. 68, 101–110.

Persson H 1980 Spatial distribution of fine root growth, mortality and decomposition in a young Scots pine stand in Central Sweden. Oikos 34, 77–87.

Persson H 1982 Changes in the tree and dwarf shrub fine-roots after clear cutting in a mature Scots pine stand. Swed. Con. For. Proj., Techn. Rep. 31, 18 pp.

Persson H 1983 The distribution and productivity of fine roots in boreal forests. Plant Soil 71, 87–101.

Persson H 1984 The dynamic fine roots of forest trees. *In* State and Change in Forest Ecosystems – Indicators of Current Research. Ed. Ågren, G. Dept. Ecol. and Env. Res., Res. Rep. 13, pp 193–204.

Persson H 1988 Root growth and root processes – effects of air pollution in forest stands. *In* Relationships between above and below ground influences of air pollutants on forest trees. Eds J Bervaes, O Mathy and P Evers. Air Pollution Research Report 16, pp 180–188.

Persson H 1996 Fine-root dynamics in forest trees. Acta Phytogeogr. Suec. 81, 17–23.

Persson H and Ahlström K 1991 The effects of forest liming on fine-root growth. Water Air Soil Pollut. 54, 305–375.

Persson H and Ahlström K 1994 The effects of alkalizing compounds on fine-root growth in a Norway spruce stand in SW Sweden. J. Environ. Sci. Health 29, 803–820.

Persson H and Ahlström K 2000 Effect of nitrogen deposition on tree roots in boreal forests. *In* Going Underground – Ecological Studies in Forest Soils. Eds N Rastin and J Bauhus (*In press*).

Persson H and Ahlström K Submitted Fine-root response to nitrogen in three nitrogen manipulated Norway spruce catchment areas during a five year period.

Persson H, von Firks Y, Majdi H and Nilsson L O 1995 Root distribution in a Norway spruce (*Picea abies* (L.) Karst.) stand subjected to drought and ammonium-sulphate application. Plant Soil 168–169, 161–165.

Persson H, Majdi H and Clemensson-Lindell A 1995 Effects of acid deposition on tree roots. Ecol. Bull. 44, 158–167.

Puhe J, Persson H and Börjesson I 1986 Wurzelwachstum und Wurzelschäden in Skandinavischen Nadelwäldern. Allge. Forst Zeitschr. 20, 488–492 (in German).

Roehrig Hansen H B and Thomsen L 1992 Rodundersögelser og biomassemålinger på tre danske rödgranlokaliteter. Report Laboratoriet for Ökologi og Miljoelaere, Lyngby, Denmark, 111 pp. (in Danish).

Santantonio D and Santantonio E 1987 Effect of thinning on production and mortality of fine roots in a *Pinus radiata* plantation on a fertile site in New Zealand. Can. J. For. Res. 17, 919–927.

Santantonio D and Hermann R K 1985 Standing crop, production, and turnover of fine roots on dry, moderate and wet sites of mature Douglas-fir in Western Washington. Ann. Sci. 42, 113–142.

Schulze E D 1989 Air pollution and forest decline in a spruce (*Picea abies*) forest. Science 244, 776–783.

Singh K P and S K Srivasta 1984 Spatial distribution of fine root mass in young trees (*Tectonia grandis*) of varying girth sizes. Pedobiologia 27, 101–170.

Staaf H, Persson T and Bertills U (Eds) 1996 Skogsmarkskalkning. Naturvårdsverket, Rep. 4559, 290 pp. (in Swedish).

Stone E L and Kalisz P J 1991 On the maximum extent of tree roots. For. Ecol. Manage. 46, 59–102.

Tamm C O 1991 Nitrogen in terrestrial ecosystems. Ecol. Stud. 81, 1–116.

Ulrich B, Pirouzpanah D and Murach D 1984 Beziehubgen zwischen Bodenversaurung und Wurzelentwicklung von Fichten mit unterschiedlich starken Schadesymptomen. Forstarchiv 55, 124–134 (in German).

van Praag H J, Sougnez-Remy S, Weissen F and Carlett G 1988 Root turnover in a beech and a spruce stand of the Belgian Ardennes. Plant Soil 105, 87–103.

Vogt K A and Persson H 1991 Measuring growth and development of roots. *In* Techniques and Approaches in Forest Tree Ecophysiology. Eds J P Lassoie and T M Hinckley. pp 477–501. CRC Press, Boca Raton, FL.

Vogt K A, Edmonds R I and Grier C C 1981 Seasonal changes in biomass and vertical distribution of mycorrhizal and fibrous-textured conifer fine roots in 23- and 180-year-old subalpine *Abies amalis* stands. Can. J. For. Res. 11, 223–229.

Vogt K A, Moore E E, Vogt D J, Redlin M J and Edmonds R L 1983 Conifer fine root mycorrhizal root biomass within the forest floors of Douglas-fir stands of diffrent ages and site productivities. Can. J. For. Res. 13, 429–437.

Vogt K A, Grier C C, Meir C E and Edmonds R L 1986 Overestimation of net root production: a real or imaginary problem. Ecology 67, 577–579.

A. Stokes (ed.), *The Supporting Roots of Trees and Woody Plants: Form, Function and Physiology*, 347–355.

Pine root structure and its potential significance for root function

Carol A. Peterson*, Daryl E. Enstone and Jeff H. Taylor
Department of Biology, University of Waterloo, Waterloo, ON, Canada N2L 3G1

Key words: condensed tannins, cork, metacutis, passage cells, *Pinus*

Abstract

Actively growing roots of pouch-grown *Pinus banksiana* Lamb. are known to have three anatomically distinct zones, i.e., white, condensed tannin, and cork (in order of increasing distance from the root tip). Roots of pouch and pot-grown *Pinus taeda* L., and field-grown *P. banksiana* also develop these three zones. The terminal region of a dormant root resembles the condensed tannin zone, with the addition of a suberized metacutis partially surrounding the apical meristem. White roots are anatomically suited for efficient ion uptake due to the presence of a living cortex. The condensed tannin zones of both growing and dormant roots have a dead cortex but retain passage cells in their endodermal layers, through which some ion uptake could occur. The effect of the maturation from white to condensed tannin zone on water uptake is difficult to predict, but some uptake would occur through the endodermal passage cells. In the young cork zone, no ion and little water absorption should occur. The discrepancies between results of separate anatomical and physiological investigations of tree roots need to be resolved by correlative studies incorporating both approaches in individual experiments.

Introduction

The morphology of pine roots is complex and heterorhizic, and individual roots change anatomically along their lengths (see Wilcox, 1964 and McCrady and Comerford, 1998). A detailed analysis of wall modifications and viability of cells in each of the discrete zones of the root is necessary to begin to understand the anatomy of the system as a whole, and should serve as one of the useful predictors of function.

The variation in structure of any individual root is a consequence of its growth and maturation pattern (Esau, 1977). The root is formed by an apical meristem and consequently grows at its tip. Thus, there is a gradient in age along the length of the root, with the youngest area being at the apex and the oldest near the shoot. Because changes in structure occur with age and environment, the anatomy of the root is non-uniform along its length. During primary growth in pine, striking changes occur in the xylem, endodermis and remainder of the cortex. Secondary growth, brought about by divisions in the vascular and cork cambia, result in further, massive anatomical changes. When the roots become dormant, the cells of the apical meristem cease dividing, and further modifications occur at the tip.

In addition to their own innate variability, pine roots usually associate with ectomycorrhizal fungi. A general description of mycorrhizae on several members of the Pinaceae can be found in Brundrett et al. (1990). The significant delivery of ions and water to pine roots by this symbiont is outside the scope of this article. The reader is referred to George and Marschner (1996) for a review of ectomycorrhizae on tree roots. Similarly, the consequences of this symbiosis for nutrient uptake directly from the soil by the roots are considered elsewhere (Taylor and Peterson, 1999).

Tree roots, like plant roots in general, function to absorb nutrient ions and water from the soil. The movement of these materials can be divided into two main processes, i.e., radial transport, which involves passage through a series of cells between the soil and the xylem, and axial transport, which occurs through the conducting elements of the xylem (see Steudle

* E-mail: cpeterso@uwaterloo.ca
Plant and Soil is the original source of publication of this article. It is recommended that this article is cited as: *Plant and Soil* **217**: 205–213, 1999.

and Peterson, 1998). There is also a slower axial movement through the nonvascular cells. The present review concerns the anatomy of pine roots as it affects ion and water movement through the radial path.

Development of pine roots

According to McKenzie and Peterson (1995a,b), roots of pouch-grown *P. banksiana* Lamb. can be divided into three zones. The youngest (most apical) is the 'white zone'; adjacent to this is the 'tannin zone' (herein referred to as the condensed tannin zone), which appears tan or brown macroscopically; adjacent (basipetal) to the tannin zone is the 'cork zone' (Figure 1). The root was divided into these three zones because each had a distinctive internal structure which should have consequences for ion and water uptake (McKenzie and Peterson, 1995a,b). It supercedes the earlier classification of white roots as 'nonsuberized' and brown roots as 'suberized' (e.g., Kramer and Bullock, 1966). The current analysis of pine root structure can be extended to *P. taeda*, grown either in pouches or in pots with an artificial soil mix (Enstone and Peterson, unpublished data) and also to field-grown *P. banksiana* (Taylor and Peterson, 1999). Maturation of the three zones is compressed toward the root tip in the more slowly growing soil roots.

White zone
As the term implies, this zone of the root appears white to the naked eye. In cross sections proximal to the root cap, the following regions can be identified – cortex (the innermost layer of which is the endodermis), pericycle, xylem, and phloem (Figure 2A). A typical epidermis is usually absent, since it tends to be sloughed early in root development (Mirov, 1967).

Condensed tannin zone
With the exception of the white tips, pine roots are brown. The youngest (most apical) part of the root which becomes macroscopically brown is the condensed tannin zone. Here, the cortical cells have died and deposited brown-colored, condensed tannins (proanthocyanidins) in their walls (Figure 1 and 2B). These cells fail to accumulate fluorescein in their cytoplasm which, by itself, is not a sure indicator of cell death. However, when treated with berberine hemisulfate and then potassium thiocyanate, crystals of berberine thiocyanate form in the walls and also protrude into the lumena of the cells. The latter crystal growth is not seen in living cells (McKenzie and Peterson, 1995a). The diameter of the root usually shrinks slightly, due to collapse of the outer cortical cells (Figure 2B).

Cork zone
Secondary growth is initiated by divisions of the vascular and cork cambia, which in roots typically arise in the stele (Esau, 1977). This is the case in *P. banksiana* and *P. taeda*, so that the cork cells are produced interior to the endodermis (Figure 1 and 2C). The ultrastructure of root cork cells has not been studied, but if they are like their counterparts in shoot bark, they will be dead and possess walls with many layers of suberin (suberin lamellae) on all their wall surfaces (Sitte, 1962). Dye permeability studies suggest that root cork cells have a Casparian band-like structure in their primary walls (McKenzie and Peterson, 1995b).

Endodermis
The innermost layer of the cortex, the endodermis, is highly modified. Its component cells go through four stages of development as described by Van Fleet (1961). The cells are considered mature when they have formed a Casparian band in their radial and transverse walls (Figure 2A), and this is known to coincide with lignification of the first tracheids in *Pinus resinosa* Ait. (Wilcox, 1964). It is generally assumed that the presence of a Casparian band prevents the movement of water and ions through the walls at this point. The former has not been tested experimentally in any system due to technical difficulties, but there is evidence from several types of experiments that the Casparian bands of the endodermis drastically reduce ion movement. A compartmental elution study (Baker, 1971), a root pressure probe study (Peterson et al., 1993), and ultrastructural studies of lanthanum and uranyl ion distribution (Nagahashi et al., 1974; Robards and Robb, 1972) all in herbaceous species, indicated that the Casparian bands of the endodermis acted as an apoplastic barrier to ions. More recently, direct evidence of an endodermal apoplastic barrier to calcium and magnesium ions in *Picea abies* roots has been provided by Kuhn et al. (1999) using LAMMA (laser microbeam mass analysis) and SIMS (secondary ion mass spectrometry) detection of stable mass isotopes in freeze-substituted samples personal communication (personal communication).

In young pine roots, the endodermis enters a second stage of development in which suberin lamellae are deposited on all wall inner surfaces (Figure 2A,B). The development of these lamellae tends to be

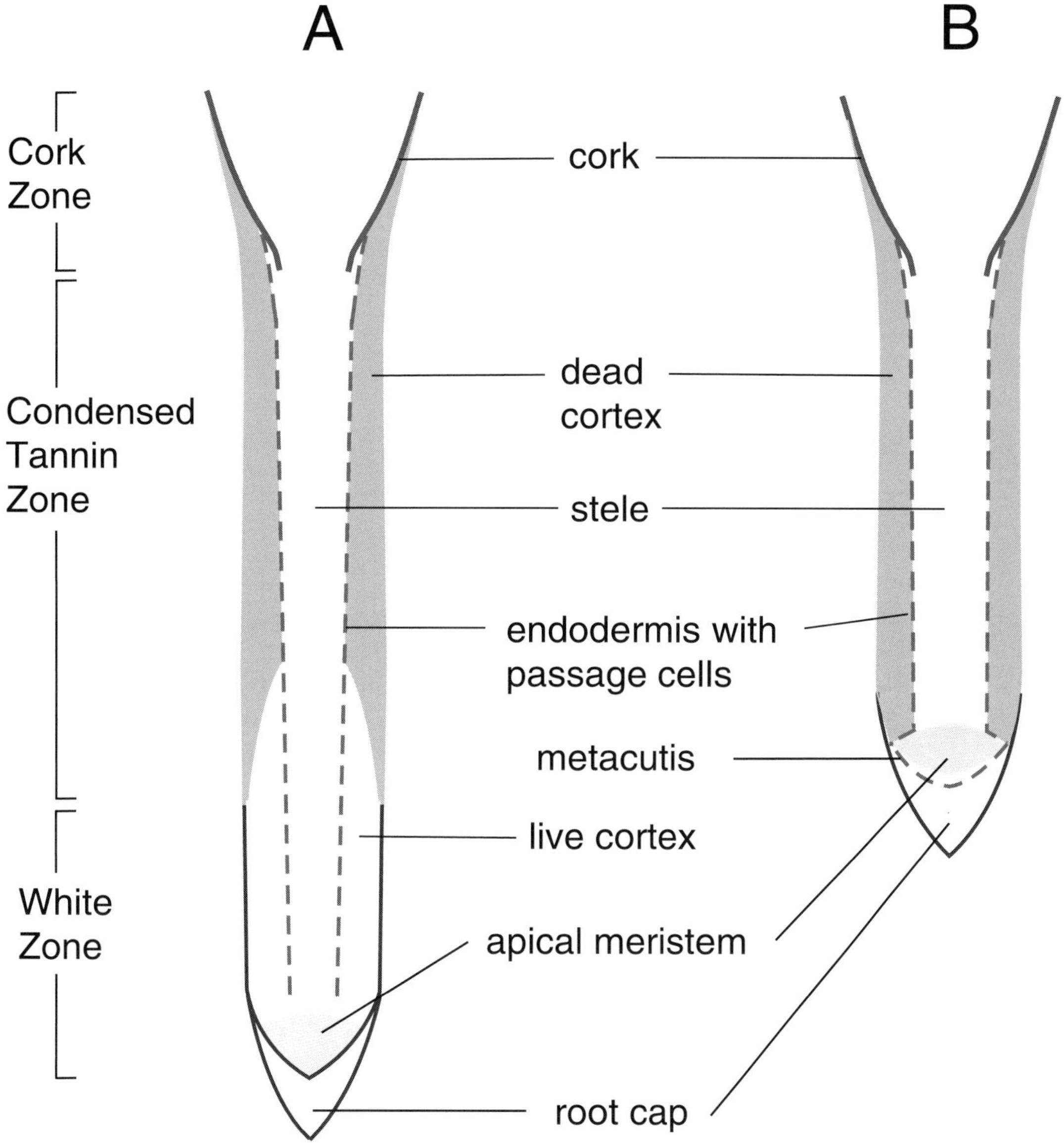

Figure 1. Growing (A) and dormant (B) roots of pine, illustrating various anatomical features as seen in median, longitudinal section.

delayed in those endodermal cells nearest the xylem poles, and the lamella-less cells are known as passage cells. The general functions of passage cells have been reviewed by Peterson and Enstone (1996).

In the white zone of a pouch-grown *P. taeda* root, the endodermis has only Casparian bands in the youngest zone examined (10 mm from the root tip). Suberin lamellae begin to develop 10–20 mm from the tip so that at the point where the condensed tannin zone commences, about 80% of the endodermal surface has suberin lamellae. This percentage remains relatively constant until the onset of cork cell maturation, at which time almost all endodermal cells develop lamellae (Enstone and Peterson, unpublished data).

Dormant roots

Roots that have ceased to extend as a result of inactivity of the apical meristem are referred to as dormant roots. Their basic anatomy was first elucidated by Müller in 1906, and further discussed by Plaut (1918). Their most striking feature is the presence of a metacutized (or metacutinized) layer that resembles an acropetal extension of the fully differentiated endo-

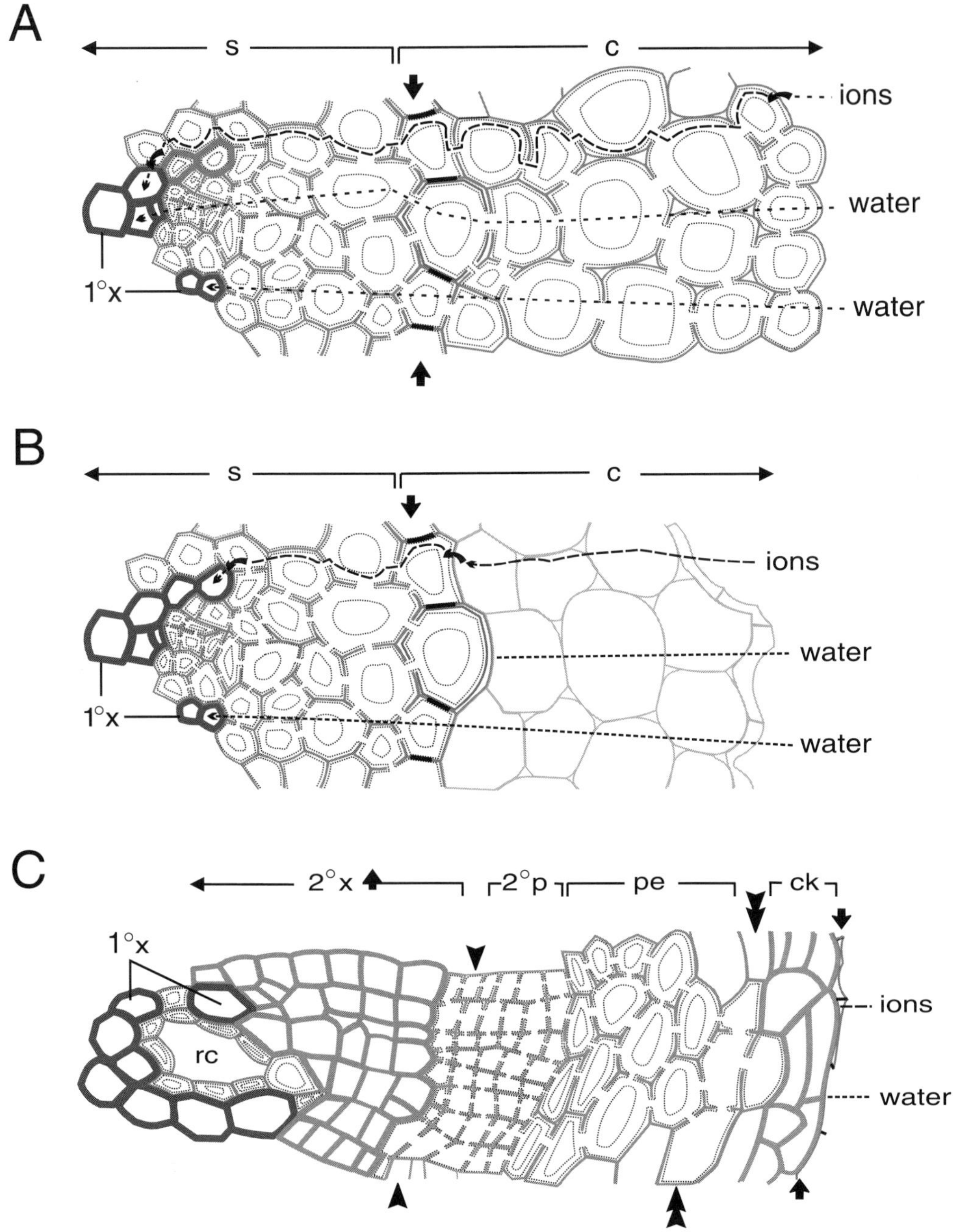

Figure 2. Pine root cross sections showing the major pathways of radial movement of ions and water from the soil solution across the cortex (c) into the stele (s) and finally into the tracheids of the primary xylem (1° x). Both plasmalemma and tonoplast membranes are shown by dotted lines. Steps in which ions cross a membrane actively are indicated by curved arrows. In the endodermis (arrows), suberin lamellae are drawn as black lines inside the walls, and Casparian bands by blackened primary walls. The central cell of the endodermis has a suberin lamella; the upper and lower endodermal cells do not and thus are passage cells. After secondary growth, a resin canal (rc) forms in the xylem, secondary xylem (2° x) and secondary phloem (2° p) are produced by the vascular cambium (single arrowhead). The pericycle (pe) enlarges, and cork cells (ck) form from the cork cambium (double arrowheads). Remnants of the endodermis (arrows) remain external to the cork. (A) White zone. Water is indicated as moving through the transcellular path, except at the endodermal cell with the suberin lamellae, where its movement is symplastic. (B) Condensed tannin zone. Dead cortical cells are assumed to be waterlogged, allowing diffusion of water and ions across the cell lumena. (C) Cork zone.

dermis to encompass the apical meristem (Figure 1). This metacutized layer is derived from cells of the cortex and young root cap, and involves lignification and suberization of the walls. The layer of cells outside the metacutized layer dies, becoming similar to the cortex of the condensed tannin zone (Wilcox, 1954).

Dormancy is generally a temporary state through which roots pass, and the stimuli that initiate it are complex. In the field, an increase in the number of dormant roots correlates with a decrease in the available soil moisture (Taylor and Peterson, 1999). In fact, root dormancy has been induced by lowering the soil moisture to a critical level (Leshem, 1970). However, dormancy occurs even in roots that are maintained under constant, ideal conditions (Wilcox, 1964; Enstone, personal communication), suggesting that root dormancy may be a programmed event. In field-grown roots, metacutized roots are often totally brown (Wilcox, 1964). This is brought about by development of the condensed tannin zone to the tip (Taylor and Peterson, 1999). Wilcox (1964) occasionally found winter-dormant, metacutized apices that were white, which he attributed to a peripheral location of the metacutis in these cases; that is, there were no significant external tissues to become brown. When the apical meristem resumes activity, breaking the dormancy, the root extends through the metacutized layer and develops a new white tip that is typical of growing roots (Wilcox, 1968).

The radial paths of ion and water movement

The potential for the developmental changes described above to influence ion and water uptake by the root can be understood by a brief consideration of these physiological processes, the details of which have been studied mainly in herbaceous species. The specific cells through which materials move in the radial path are the cortical parenchyma, endodermis, pericycle and stelar parenchyma (Figure 2A). There are three pathways through such a series of cells, i.e., apoplastic (through the cell walls), symplastic (through the cytoplasm of individual cells and plasmodesmata linking the cells), and transcellular (through the walls, cytoplasm and vacuoles, i.e., repeatedly through membranes). Because ions and water move differently through the cells, they will be considered separately below. More extensive reviews of ion and water movement in roots are available (Clarkson, 1996; Steudle and Peterson, 1998, respectively).

Ions

It is an oversimplification to group all ions together, since most are transported across membranes by specific carriers (see Nissen, 1996). However, the schemes in Figure 2 can be applied to most essential ions. In the white zone (Figure 2A), ions initially diffuse into the root through the outermost walls of the cortex and then may be actively transported across the plasmalemma (the outermost membrane of the cell) if suitable membrane proteins are present. Once in the cytoplasm, ions are assumed to diffuse from the cytoplasm of one cell to the next through plasmodesmata until reaching the pericycle or stelar parenchyma. Here, they move from the cytoplasm into the wall system. The mechanism of ion release into the wall has been controversial. An older hypothesis that ions are passively released from the cells due to low oxygen tension in the stele has been supplanted by the idea of a controlled release. In barley, the ultrastructure of the cells involved (Läuchli et al., 1974) and the concentration gradients of potassium across the root (Läuchli, 1971) were early indications that release from the symplasm in the stele might be an active process. Later, Clarkson et al. (1984) and Clarkson and Hanson (1986) demonstrated regulation of xylem sap pH in onion and that this was associated with potassium release into the transpiration stream. Finally, two outwardly rectifying potassium channels were identified in the stelar parenchyma cells of maize roots (de Boer and Wegner, 1997). The movement of potassium through these channels is powered by a proton ATPase pump in the membrane. This raises the possibility that the release of other essential ions is also controlled, but further experiments are necessary to determine the mechanism(s) of how each passes through the membrane. Once in the walls of the pericycle and stelar parenchyma, ions would diffuse along their concentration gradient to the lumena of the tracheary elements. The impermeability of the Casparian band to ions and the tight connection of the band with the plasmalemma means that ions take a predominantly symplastic route for at least part of their journey into the xylem, and thus the composition of the xylem sap can be controlled.

Water

The majority of water movement in the root is driven by gradients in water potential. Under normal conditions, a tension in the xylem is transmitted through the continuum of water within the root, and water is drawn inward (Figure 2). Its movement is passive and some water will be moving through all three pathways (apoplastic, symplastic, and transcellular). The extent to which an individual pathway is used depends on its resistance relative to those of the other pathways. The resistance depends on the specific resistance of the pathway and on the area of the pathway. In the case of materials moving from the soil to the xylem, the relevant areas would be those occurring in sequential, tangential longitudinal sections. It is not possible to distinguish between transcellular and symplastic movement experimentally, and they are grouped together as the cell-to-cell pathway.

To complicate matters still further, the resistances of the pathways can vary. The resistance of the transcellular pathway is lowered by aquaporins, membrane-spanning proteins through which water can diffuse (see Maurel, 1997 and Schäffner, 1998). The number of proteins in the membranes could be altered by changing their rate of insertion, assuming a constant turnover. Similarly, opening and closing of plasmodesmata would influence water flow in the symplastic path. Suberin lamellae increase the resistance to water flow in the transcellular path and Casparian bands have the same effect for the apoplastic path. With all these variables, it is perhaps not surprising that it has been difficult to discover, either by experimentation or modelling, the fraction of water moving in the three pathways.

Pine root structure and absorptive function

With a basic understanding of root structure and the processes of ion and water absorption, we can now consider the potential for root development to influence these processes.

Ions

As described earlier, movement through the symplast is a requirement for ion uptake, specifically in the endodermis. Thus, it is necessary for most ions to pass from the wall system through membranes into living cells. In Figure 2A (the white zone), the active uptake step is shown as occurring in an outermost cortical cell. However, ions may diffuse through the apoplast as far as the Casparian bands of the endodermis, so that the ions could be taken into the symplast in any of the cortical cells. The white zone, with its living cortex, has a large plasmalemma surface with space for many protein carriers and thus has a great potential for ion uptake. In Figure 2A, the development of a suberin lamella in an endodermal cell is assumed not to sever its plasmodesmata, so that the inward, symplastic diffusion of ions into the pericycle is not impeded. In herbaceous plants, endodermal plasmodesmata often remain intact although they are constricted by suberin lamellae (Clarkson et al., 1971; Hass and Carothers, 1975; Warmbrodt, 1985), but it is not known whether or not this is the case for pine.

Death of the cortical cells in the condensed tannin zone drastically reduces the membrane area available for ion uptake (Figure 2B). The dead cortical cells with modified walls were found to be permeable to berberine, a fluorescent dye (molecular weight 336.4 Da) indicating that ions (which are smaller) should diffuse through this area (McKenzie and Peterson, 1995a). However, to cross the endodermis, the ions need to be in the symplast, and the only site for their entry into this compartment is the outer membrane of the passage cells. Therefore, the presence of endodermal passage cells in the condensed tannin zone is critical for ion uptake in this region of the root. As yet, there have been no direct measurements of ion uptake in this discrete root zone.

During early secondary growth, cork cells form a continuous cylinder adjacent to the endodermis. The sites for ion uptake into the symplast are reduced to nil and it would seem to be a safe prediction that this zone would be incapable of absorbing any ions. However, experimental results with older roots show otherwise (Hay et al., 1986; Sougnez-Remy et al., 1993; Van Praag et al., 1993; Van Rees and Comerford, 1990; see review by Comerford et al., 1994). These conflicting results may be resolved by a correlated anatomical and physiological study.

In metacutized roots, the condensed tannin zone extends to the root tip, but the endodermis still retains passage cells. Thus, even dormant roots have some potential for ion absorption.

Water

The effects of root development on water uptake is even more difficult to predict than for ions. The white zone is analogous to a herbaceous plant root. In maize, it has been shown that when the endodermis has only Casparian bands, the major resistance to water uptake occurs in the living parenchyma of the cortex (Peterson and Steudle, 1993). This would be the situation in the young white zone of the root which has Casparian bands but no suberin lamellae. Development of suberin lamellae in progressively more endodermal cells would increase the resistance to water flow in the transcellular pathway in older regions of the white zone. However, these static changes in root anatomy may be compensated for by aquaporins.

It is also difficult to predict the effect of condensed tannin zone development on water uptake. Although the number of passage cells remains rather constant in this zone, death of the cortical cells, and abrasion of the outer layers would drastically decrease resistance to water flow. On the other hand, deposition of condensed tannins in the walls has an as yet untested effect.

Assuming that root cork has the same physiological properties as stem cork, maturation of this tissue should greatly increase resistance in both the apoplastic pathway and cell-to-cell pathway, and thus water movement in this root zone should be drastically curtailed. Experimental results, however, have not borne out this prediction. Measurable amounts of water are taken up by roots in secondary growth (Kramer and Bullock, 1966; McFall et al., 1991). Some of these discrepancies may be ascribed to differences in root age. In young, pouch-grown roots, the first layer of mature cork forms a complete barrier to the movement of dyes, but this is not the case in older roots. Addoms (1946) describes dye movement through dead lateral roots, cracks and lenticels in the bark. The question of permeability of older cork zone roots to water should be addressed by a correlated anatomical and physiological study. It is important (and difficult) not to injure the roots by handling during permeability tests so that rapid, artifactual movement through injuries is prevented (Peterson and Moon, 1993).

In dormant roots, the passage cells of the condensed tannin zone presumably are a relatively low-resistance pathway for water entry into the stele. Dormant roots would be capable of water uptake. The metacutis, which partially surrounds the meristem, resembles cork in structure (Leshem, 1970), and Mager (1913) proposed that metacutization protects the root from water loss. However, Leshem (1970) found dormant roots to be more susceptible to water loss than growing, white tips. The function of the metacutis remains unknown.

Conclusions and future directions

The work of McKenzie and Peterson (1995a,b) on pine root anatomy clearly needs to be extended to older roots and to the whole heterorhizic system. We are as yet at an early stage of understanding pine root system anatomy, having studied only young seedling roots. However, it is encouraging that the three zones first identified in pouch-grown roots also apply to pot and field-grown trees.

Formation of the condensed tannin zone in pine roots (which may be typical of tree root structure in general) is curious. What is the adaptive advantage of this developmental pattern? Is the energetic cost of maintaining a viable cortex some distance from the tip too high? Is it a mechanism for limiting the growth of mycorrhizae, which are nourished only by living cortex? Is it part of a defence strategy against soil pathogens? Is the root already preparing for a dormant phase in which some absorptive capacity is maintained?

From a consideration of pine root anatomy and ion uptake, it is clear that the presence of passage cells in the endodermis is important for the absorptive capacity of the condensed tannin zone. This zone is of interest throughout the year, but is particularly important when the root is dormant since, in that condition, the white zone is replaced by the condensed tannin zone. Ultrastructural aspects of the metacutis are unknown.

It would be of interest to study the effects of environment on the formation of suberin lamellae in pine, as exposure to air accelerates their formation in both the endodermis and exodermis of herbaceous plant roots (Van Fleet, 1961; Enstone and Peterson, 1998), as does salinity for the exodermis (Reinhart and Rost, 1995).

While a knowledge of anatomy is essential to understanding root function, it is only part of the picture. Less visible changes such as the activities of aquaporins and ion carriers, porters, and channels may override the predictions presented above. Experimental work is essential to move from the realm of speculation to the realm of reality. As a step in this direction, Dr. Steve Hallgren (Oklahoma State University) is undertaking to test the hydraulic conductivities of the white, condensed tannin, and cork zones in loblolly pine roots. It would be instructive to do the same for a variety of essential ions.

Acknowledgements

This research was supported by a University of Waterloo Interdisciplinary Grant awarded to Steve Hallgren (Oklahoma State University) and CAP, by a United States Department of Agriculture/National Research Initiative Competitive Grants Program

(USDA/NRIGCP) grant number 96-35100-3162 awarded to Steve Hallgren, and by an NSERC-Industrial Scholarship awarded to JHT (Industrial partner, Mikro-Tek, Timmins, Canada).

References

Addoms R M 1946 Entrance of water into suberized roots of trees. Plant Physiol. 21, 109–111.

Baker DA 1971 Barriers to the radial diffusion of ions in maize roots. Planta 98, 285–293.

de Boer AH and Wegner LH 1997 Regulatory mechanisms of ion channels in xylem parenchyma cells. J. Exp. Bot. 48, 441–449.

Brundrett M, Murase G and Kendrick B 1990 Comparative anatomy of roots and mycorrhizae of common Ontario trees. Can. J. Bot. 68, 551–578.

Clarkson D T 1996 Root structure and the sites of ion uptake. *In* Plant Roots – The Hidden Half. Eds Y Waisel, A Eshel and U Kafkafi. pp 483–510. Marcel Dekker, Inc. New York.

Clarkson D T and Hanson J B 1986 Proton fluxes and the activity of a stelar proton pump in onion roots. J. Exp. Bot. 37, 1136–1150.

Clarkson D T, Robards A W and Sanderson J 1971 The tertiary endodermis in barley roots: fine structure in relation to radial transport of ions and water. Planta 96, 292–305.

Clarkson D T, Williams L and Hanson J B 1984 Perfusion of onion root xylem vessels: a method and some evidence of control of the pH of the xylem sap. Planta 162, 361–369.

Comerford N B, Smethurst P J and Escamilla J A 1994 Nutrient uptake by woody root systems. New Zealand Journal of Forestry Science 24, 195–212.

Enstone D E and Peterson C A 1998 Effects of exposure to humid air on epidermal vitality and suberin deposition in maize (*Zea mays* L.) roots. Plant Cell Environ. 21, 837–844.

Esau K 1977 Plant Anatomy, Second Edition. John Wiley & Sons, Inc., New York. pp 215–255.

George, E and Marschner, H 1996 Nutrient and water uptake by roots of forest trees. Z. Pflanzenernähr. Bodenk. 159, 11–21.

Haas D L and Carothers Z B 1975 Some ultrastructural observations on endodermal cell development in *Zea mays* roots. Amer. J. Bot. 62, 336–348.

Hay M J M, Dunlop J and Hopcroft D H 1986 Phosphate uptake and anatomy of unthickened and secondarily thickened adventitious roots of field-grown white clover (*Trifolium repens* L.). New Phytol. 103, 659–668.

Kramer P J and Bullock H C 1966 Seasonal variations in the proportions of suberized and unsuberized roots of trees in relation to the absorption of water. Amer. J. Bot. 53, 200–204.

Kuhn A J, Schröder W H and Bauch J 1999 Kinetics of Ca and Mg entry into mycorrhizal spruce roots. Planta. (*In press*).

Läuchli A, Kramer D, Pitman M G and Lüttge U 1974 Ultrastructure of xylem parenchyma cells of barley roots in relation to ion transport to the xylem. Planta 119, 85–99.

Läuchli A, Spurr A R and Epstein, E 1971 Lateral transport of ions into the xylem of corn roots. II. Evaluation of a stelar pump. Plant Physiol. 48, 118–124.

Leshem B 1970 Resting roots of *Pinus halepensis*: structure, function and reaction to water stress. Bot. Gaz. 131, 99–104.

MacFall J S, Johnson G A and Kramer P J 1991 Comparative water uptake by roots of different ages in seedlings of loblolly pine (*Pinus taeda* L.). New Phytol. 119, 551–560.

Mager H 1913 Versuche über die Metakutisierung. Flora 106, 42.

Maurel C 1997 Aquaporins and water permeability of plant membranes. Ann. Rev. Plant Phys. Plant Mol. Bio. 48, 399–429.

McCrady R L and Comerford N B 1998 Morphological and anatomical relationships of loblolly pine fine roots. Trees 12, 431–437.

McKenzie B E and Peterson C A 1995a Root browning in *Pinus banksiana* Lamb. and *Eucalyptus pilularis* Sm. 1. Anatomy and permeability of the white and tannin zones. Bot. Acta 108, 127–137.

McKenzie B E and Peterson C A 1995b Root browning in *Pinus banksiana* Lamb. and *Eucalyptus pilularis* Sm. 2. Anatomy and permeability of the cork zone. Bot. Acta 108, 138–143.

Mirov N T 1967 The Genus Pinus. The Ronald Press Company, New York. pp 363–369.

Müller H 1906 Über die Metakutisierung der Wurzelspitze und über die verkorkten Scheiden in den Achsen der Monokotyledone. Bot. Zeit. 64, 53–84.

Nagahashi G, Thomson W W and Leonard R T 1974 The Casparian strip as a barrier to the movement of lanthanum in corn roots. Science 183: 670–671.

Nissen P 1996 Uptake mechanisms. *In* Plant Roots – The Hidden Half. Eds Y Waisel, A Eshel and U Kafkafi. pp 4511–4527. Marcel Dekker, Inc., New York.

Peterson C A and Moon G J 1993 The effect of lateral root outgrowth on the structure and permeability of the onion root exodermis. Bot. Acta 106, 411–418.

Peterson C A and Steudle E 1993 Lateral hydraulic conductivity of early metaxylem vessels in *Zea mays* L. roots. Planta 189, 288–297.

Peterson C A and Enstone D E 1996 Functions of passage cells in the endodermis and exodermis of roots. Physiol. Plant. 97, 592–598.

Peterson C A, Murrmann M and Steudle E 1993 Location of the major barriers to water and ion movement in young roots of *Zea mays* L. Planta 190, 127–136.

Plaut M 1918 Über die Morphologischen und Mikroskopischen Merkmale der Periodizität der Wurzel sowie über die Verbreitung der Metakutisierung der Wurzelhäube im Pflanzenreich. Festscher. 100-jähr. Best kgl. Wurttemb. Landw. Hochschule. 129–151. Hohenheim.

Reinhardt D H and Rost T L 1995 Salinity accelerates endodermal development and induces an exodermis in cotton seedling roots. Enviro. Exp. Bot. 69, 1170–1178.

Robards A W and Robb M E 1972 Uptake and binding of uranyl ions by barley roots. Science 178, 980–982.

Schäffner A R 1998 Aquaporin function, structure, and expression: are there more surprises to surface in water relations? Planta 204, 131–139.

Sitte P 1962 Zum Feinbau der Suberschichten im Flaschenkork. Protoplasma 54, 555–559.

Sougnez-Remy S, Waterkeyn L and Van Praag H J 1993 L'absorption-translocation spécifique d'éléments nutritifs en relation avec la structure anatomique des radicelles de hêtre (*Fagus sylvatica*) et d'épicéa (*Picea abies*). II. Observations relatives à la structure anatomique des radicelles. Belg. J. Bot. 126, 184–190.

Steudle E and Peterson C A 1998 How does water get through roots? J. Exp. Bot. 49, 775–788.

Taylor J H and Peterson C A 1999 Morphometric analysis of root anatomy in *Pinus banksiana* Lamb. during a 3-month field study. Trees, submitted. Van Fleet D S 1961 Histochemistry and function of the endodermis. Bot. Rev. 27, 165–220.

Van Praag H J, Sougnez-Remy S and Weissen F 1993 L'absorption-translocation spécifique d'éléments nutritifs en relation avec la structure anatomique des radicelles de hêtre (*Fagus sylvatica*) et d'épicéa (*Picea abies*). I. Mesures de l'absorption spécifique. Belg. J. Bot. 126, 175–183.

Van Rees K C J and Comerford N B 1990 The role of woody roots of slash pine seedlings in water and potassium absorption. Can. J. For. Res. 20, 1183–1191.

Warmbrodt R D 1985 Studies on the root of *Zea mays* L. – structure of the adventitious roots with respect to phloem unloading. Bot. Gaz. 146, 169–180.

Wilcox H 1954 Primary organization of active and dormant roots of noble fir, *Abies procera*. Amer. J. Bot. 41, 812–821.

Wilcox H 1964 Xylem in roots of *Pinus resinosa* Ait. in relation to heterorhizy and growth activity. *In* Formation of Wood in Forest Trees. Ed. M Zimmerman. pp 459–478. Academic Press, Inc., New York.

Wilcox H 1968 Morphological studies of the root of red pine, *Pinus resinosa*. I. Growth characteristics and patterns of branching. Amer. J. Bot. 55, 247–254.

A. Stokes (ed.), *The Supporting Roots of Trees and Woody Plants: Form, Function and Physiology*, 357–370.

Tracheids in white spruce seedling's long lateral roots in response to nitrogen availability

M.J. Krasowski[1,*] and J.N. Owens[2]
[1]*British Columbia Ministry of Forests, Research Branch, Red Rock Research Station, 18435 Forest Nursery Rd., Prince George, BC V8N 5Y7, Canada and* [2]*School of Forest Biology, Department of Biology, University of Victoria, P.O. Box 1700 Victoria, BC V8W 2Y2, Canada*

Abstract

This study examined how the availability of inorganic nitrogen (N) modified the anatomical characteristics of white spruce (*Picea glauca* (Moench) Voss) roots related to their hydraulic properties. Seedlings were grown for one growing season in 4 L capacity pots filled with sand under one of three N levels: low (10 ppm), medium (50 ppm) and high (125 ppm). First order lateral roots with intact tips were sampled from dormant seedlings in October. Root segments were collected from 4, 10, and 14 cm distances above the root tip for fixation and sectioning and for maceration. Additional specimens were collected from the 4 and 14 cm distances for maceration and scanning electron microscopy of xylem pits. Root diameter and surface area occupied by the xylem in root cross sections increased basipetally in all treatments but exceptions were found. Higher N-levels significantly increased root diameter and surface area occupied by the xylem. In the two higher N treatments secondary root development was more advanced near the root tip than in the low N treatment. There was a strong positive correlation between root diameter and cross-sectional root area occupied by the xylem (30–50% of the root cross section) but not in portions with little secondary development. Non-conducting space within the xylem occupied 10–13% of its cross-sectional surface. Tracheids of the primary xylem were larger, had larger lumens but thinner cell walls than those of the secondary xylem. Low N treatment seedling tracheids had smaller total cross-sectional area, less lumen, and less cell wall surface area than the two other N treatments. Tracheid diameter means were between 19–20 μm in the high and medium N treatments, and 15.2 μm in the low N treatment. The range was 4.5–51.3 μm. Tracheid length was not significantly affected by N. The average tracheid was about 1000 μm long, and the range was 110–3530 μm. Pit-border diameters ranged between 4.1–20.6 μm (average 10–11 μm) and were not affected by the N treatment. Pit aperture diameters were within 0.62–10.2 μm range (average between 3–4 μm) and were also not significantly affected by the N treatment, although tracheids from the medium N-treatment roots tended to have larger apertures. The pit border diameter equals that of the margo while the aperture size should be similar to that of the torus of the pit membrane. If the capacity for axial water transport in spruce roots is affected by N, it would be by its impact on conduit diameter and, possibly on the pit-membrane pore sizes but not by changes to conduit length and to the size of the pit membrane surface area.

Introduction

Water uptake by terrestrial plants is determined by root tissue characteristics (Moreshet and Huck, 1991). Xylem anatomy is, therefore, an important factor affecting water flow through roots (Tyree and Karamanos, 1981). While xylem development in relation to conductive capacity of roots has been extensively studied in maize (McCully and Canny, 1988; Wang et al., 1991; Wenzel et al., 1989; St. Aubin et al., 1986), the development of xylem water conducting capacity in conifers has not been broadly researched. Anatomical studies of Wilcox (1954, 1962, 1964) emphasized the development of primary tissues, including the xylem.

* FAX No: +(250) 963 3436.
E-mail: mkrasowski@gems1.gov.bc.ca

Plant and Soil is the original source of publication of this article. It is recommended that this article is cited as: *Plant and Soil* **217:** 215–228, 1999.

The most detailed coverage of conifer root primary xylem characteristics is probably that represented by Wilcox's (1962) study on *Pinus resinosa* where not only the developmental pattern was described but also characteristics of the primary xylem tracheids, including their pits. Xylem differentiation in roots of *A. alba* was described by Rypacek et al. (1976) but it did not extend beyond primary development.

There have been few reported studies on wood formation in roots. Woods (1991) noted that the common understanding of root growth is almost invariably synonymous with root elongation. In his review on cambial activity in roots, this author stated that in spite of the coverage of the subject in works by Fayle (1968), Philipson et al. (1971), Yeoman (1976) and Zimmerman (1964), only the general principles of applicable processes had been described while research on cambial growth in roots of individual species still remained to be carried out. Among the published reports on cambial growth and wood formation in roots there were several studies of conifer roots. These included the description of seasonal patterns of cambial activity in several species (Liphshitz et al. 1984; Lobzhanidze and Kandelaki, 1972), and the impact of soil moisture on that process (Coutts, 1982b; Pompart-Dasse and David, 1974). The properties of conifer root tracheids in general and as influenced by environmental or cultural factors have been described in only a few cases. Pompart-Dasse and David (1974) found pine root tracheids shorter on wet and on dry sites, compared to those produced on sites with intermediate soil moisture. Manwiller (1972) compared the dimensions of root and stem tracheids of southern pines finding the former longer, larger in cell and lumen diameter, but having thinner cell walls (Woods, 1991).

Unlike the properties of wood in roots, stem wood characteristics have been studied frequently both in relation to the mechanical quality of wood and to its water conducting capacity. The size and shape of xylem conduits are influenced by genetic factors (Burdon and Low, 1992; Ford, 1993) tree age (Connor and Lanner, 1990; Dinwoodie, 1963; Pande et al. 1995), the season of initiation, and the growth rate (Antal and Micko, 1991; Burdon and Low, 1992; Dodd and Fox, 1990; Tavita, 1992), position on the bole (Tavita 1992; Saranpää, 1994; Dinwoodie 1961), environmental factors (Antonova and Stasova, 1993; Larson, 1960; Wodzicki, 1960, 1964, 1971), spacing (Persson et al. 1995; Yang and Hazenberg, 1994), nutrition (Harvey and van den Driessche, 1997), variation in the size of the tree crown (Funada et al., 1990), the level of toxic trace elements in the plant's environment (Hagemeyer et al., 1994), and different site, climate and fertilization conditions (Sandercock et al., 1995).

Anatomical properties of wood conduits determine their hydraulic efficiency (Tyree et al., 1994). Cross-walls in the xylem increase the resistance to axial flow (Clarkson, 1991; St. Aubin et al., 1986; Sanderson et al., 1988) and large, open xylem elements would offer little resistance to water flow (Passioura, 1988). McCully and Canny (1988) considered the diameter of the largest open xylem element(s) in a given root segment to be the most important anatomical characteristic related to its hydraulic conductivity. This property would change with the distance from the root tip as a result of the progressing xylem maturation. Conifer tracheids, although empty a short distance above the root tip (Wilcox, 1954, 1962), are closed by cell walls at both ends and tracheid to tracheid communication is through pits in their side walls (Bailey and Tupper, 1918; Esau, 1977). Because tracheid length is between 1–3 mm, water has to pass through up to 1000 tracheids per meter of wood length (Tyree et al., 1994). Short tracheids decrease wood hydraulic conductivity (Pothier et al., 1989). Wide conduits are more efficient in water transport than narrow conduits (Zimmerman, 1983). However, they are also more prone to dysfunction by the breakage of the continuity of water column under increased tension and to embolism-caused cavitation (see the discussion in Tyree et al., 1994 and Zimmerman 1983). Recently, the loss of hydraulic conductivity in response to raised xylem tension or cavitation in conifers has been attributed more to the anatomy of the tracheid pit membrane than to the conduit size. The variation in tracheid diameter is apparently small (Sperry and Sullivan, 1992; Cochard, 1992). The decline in hydraulic efficiency of xylem under high tension can result not only from cavitation but also from plugging pit apertures by pressure-displaced toruses (Sperry and Tyree, 1990). This subject is discussed in depth by Sperry and Tyree (1990), Tyree and Sperry (1989), and Tyree et al. (1994). Pit membrane elasticity determines its functioning under increased and relaxed tension. The size of margo pores reflects on its water permeability affecting hydraulic conductivity of tracheids and their vulnerability to cavitation. Although cavitation has been reported to occur mainly in the stem xylem (Crombie et al., 1985; Jackson et al., 1995; Tyree and Dixon, 1983), it may also occur in the roots (Byrne et al., 1977).

The purpose of this work was to examine the properties of root xylem in white spruce seedlings grown under different levels of nitrogen availability.

Materials and methods

Plant material and fertilizer regimes

A single local white spruce seed source (British Columbia Ministry of Forests registered seed lot number 6866, Central Plateau, elevation 940 m) was used. Seeds were sown in March 1996 into over two hundred small containers[1] filled with peat/vermiculite mix and germinated in a greenhouse. Forty five uniform germinants were individually transplanted into 4 L capacity pots (150×150×180 mm) containing unsterilized masonry sand and grown in a common greenhouse culture until May 5, 1996. N-treatments began then and overhead watering was terminated. Fifteen seedlings were randomly assigned to each of three N-treatments: (1) low N (5–10 ppm)[2], (2) medium N (50 ppm), and (3) high N (125 ppm) applied as NH_4NO_3 at pH 5.2 twice a week until September. All other macronutrients and all micronutrients were supplied at the same concentrations required to prevent their potential deficiency (R. van den Driessche, 1996, person. comm.). Nutrient stock solutions were mixed and diluted to desired concentrations at the time of application to prevent precipitation. Seedlings received 500 mL of the nutrient solution in each application. Additional watering was carried out as necessary and every 2–3 weeks the pots were flushed with water to minimize the build up of salt residues.

Sampling and measurements

In October 1996, three seedlings from each treatment were randomly sampled. Five first-order laterals longer than 15 cm with intact, unpruned root tips were sampled from each seedling from three positions along the root length: 140 mm, 100 mm, and 40 mm above the root tip. Two approximately 6 mm long root segments were collected from each position. The distal (toward the root tip) segment from each pair was fixed in formalin-acetic-acid fixative (FAA), dehydrated in an alcohol series, and embedded in paraffin (Johansen, 1940), then sectioned transversely at 6 μm and stained with safranin and hematoxylin. Proximal segments were put into marked vials, macerated as described by Johansen (1940), then stained with 1% aqueous safranin. After rinsing, external tissues were removed with fine needles. Macerated tracheids were spread in Farrant mounting medium on glass slides and secured with cover slips.

[1] Spencer-Lamaire Industries Ltd., Edmonton, Alberta, Canada.

[2] N-deficiency symptoms were noticed in July and N was increased to 10 ppm.

Light microscopy

Root cross sections were viewed using microscope-mounted video camera. On-screen measurements were taken using the Optimas 3 software[3]. Depending on the size of sections, ×1, ×4 and ×10 objectives (with appropriate calibration of the display) were used. The smallest and largest section diameters, section surface area and area occupied by the xylem were measured. When it was visually estimated that the primary xylem occupied more than a quarter of the total xylem area, separate measurements were made of the area occupied by the primary and by the secondary xylem. Otherwise, the total area occupied by the xylem was measured. Resin ducts and wide rays were excluded.

To evaluate the amount of non-conducting space in the xylem, detailed measurements were taken under a ×40 objective. Areas between 8,000–10,000 μm^2 (5,000–6,000 μm^2 in sections with little secondary xylem) were randomly subsampled from the xylem cross section by free-hand drawing the cursor and enclosing the subsampled area. Two to four such areas were subsampled, depending on the size of the section and whether the primary and secondary xylem were measured separately or not (see previous paragraph). All individual tracheids within each sampling area were measured by tracing the cross-cursor around the outer cell perimeter, then around the inner perimeter. The total area of each tracheid section and lumen area were determined and the difference between the two equaled the area occupied by the cell wall. Partially-enclosed tracheids were also measured but such measurements were marked so they could be distinguished from whole-cell sections. Cells other than tracheids were not measured. Tracheid radii (TR) were calculated from measured tracheid surface areas (TA) measurements as follows:

$$TR = \frac{\sqrt{\frac{TA}{3.14}} + \frac{\sqrt{TA}}{2}}{2} \tag{1}$$

[3] BioScan, Edmonds, WA, USA.

and lumen radii (LR) were calculated identically but lumen areas were used instead of TA. It was assumed that the formula would approximately cancel out calculation errors caused by shape variation (errors of calculations based on the same area would be −12.8% and +11.4% if applied to a rectangle and a circle, respectively)

Average cell wall thickness was calculated for each tracheid as:

$$TR - LR \quad (2)$$

Tracheid widths (walls included) and lengths were measured on macerated wood under a compound microscope equipped with a calibrated eyepiece. Initially, 100 intact tracheids were sequentially sampled starting from a random point on the slide but this was too time consuming so widths were measured on 50 and lengths on 30 tracheids in most root-segment samples. The width of each tracheid was measured in three places, excluding tapered ends and the mean was taken as an average width.

Scanning electron microscopy

Additional three roots were sampled from three seedlings per treatment but only from two positions on the root (4 and 14 cm from the root tip). These samples were macerated as described above, gradually dehydrated and transferred from 100% alcohol into hexamethydyldisilazane (a substitute for critical-point drying). The woody root portions were spread over the surface of an aluminum stub covered with adhesive paste and sputter-coated with gold. Pit boarders and outer diameters of apertures were measured with a ruler directly on the screen of a JEOL JSM-35 scanning electron microscope. Scanning electron microscope images of pits and pit apertures suffered from distortion caused by the viewing angle dependent on the tilt of the tracheid relative to the position of the electron detector. Assuming that the pits were perfectly round, the largest observed diameter of the ellipse representing a distorted view of a round object would equal the true diameter of that object. Thus, only these dimensions were measured and analyzed.

Statistical analyses

Data were analyzed using the SAS[4] 6.12 statistical package. All but one analysis of variance (ANOVA) were performed according to a split-plot design with the nitrogen treatment (N) being the main plot factor. Individual trees (T) were replicates of N and were treated as experimental units for the main plot. Roots were subsampling units and were nested in the higher order factors. Xylem development class (see the Result section) was treated as a split-plot factor. When differences between the primary and secondary xylem were considered, the xylem type was a split-split plot factor and it was subsampled. Multiple comparisons were examined using the Least Significant Difference method; a Bonferroni correction was also used to control the overall experimentwise type I error rate (Milliken and Johnson, 1992). Diagnostic checking was conducted with the Univariate Procedure in SAS using studentized residuals (each residual divided by its standard error). The underlying assumption of normality was confirmed when: (i) the distribution of the studentized residuals was bell-shaped and centered near zero, (ii) both estimated skewness and kurtosis had values between 2 and −2, and (iii) more than 95% of the studentized residuals fell between −2 and 2. When these conditions were not fulfilled, the data were transformed using the Rank Procedure in SAS and the diagnostic tests were repeated (Peter K. Ott, 1996–98; personal communications[5]).

[4] SAS Institute, Cary, North Carolina, USA.

Results

Root sections

Root sections (130) varied in the advancement of the secondary xylem development. To account for this variation, four classes of xylem development were arbitrarily established: (i) **class 1** with the primary to secondary xylem area ratio $\leq$1; (ii) **class 2** with the ratio of 1$\geq$5, (iii) **class 3** with the ratio 5$\geq$20, and (iv) **class 4** with the ratio >20. Classes 1 and 2 were distinguished on the basis of a precise measurement while classes 3 and 4 were distinguished subjectively. Examples of specimens representing each class of the secondary xylem development are shown in Figures 1–4.

Generally, the distribution of secondary xylem development classes indicated less cambial activity in the low N treatment than in the other two treatments. In most roots, the advancement of secondary xylem development progressed in the basipetal direction but

[5] Peter, K. Ott, biometrician, British Columbia Ministry of Forests, Research Branch, Victoria, B.C.

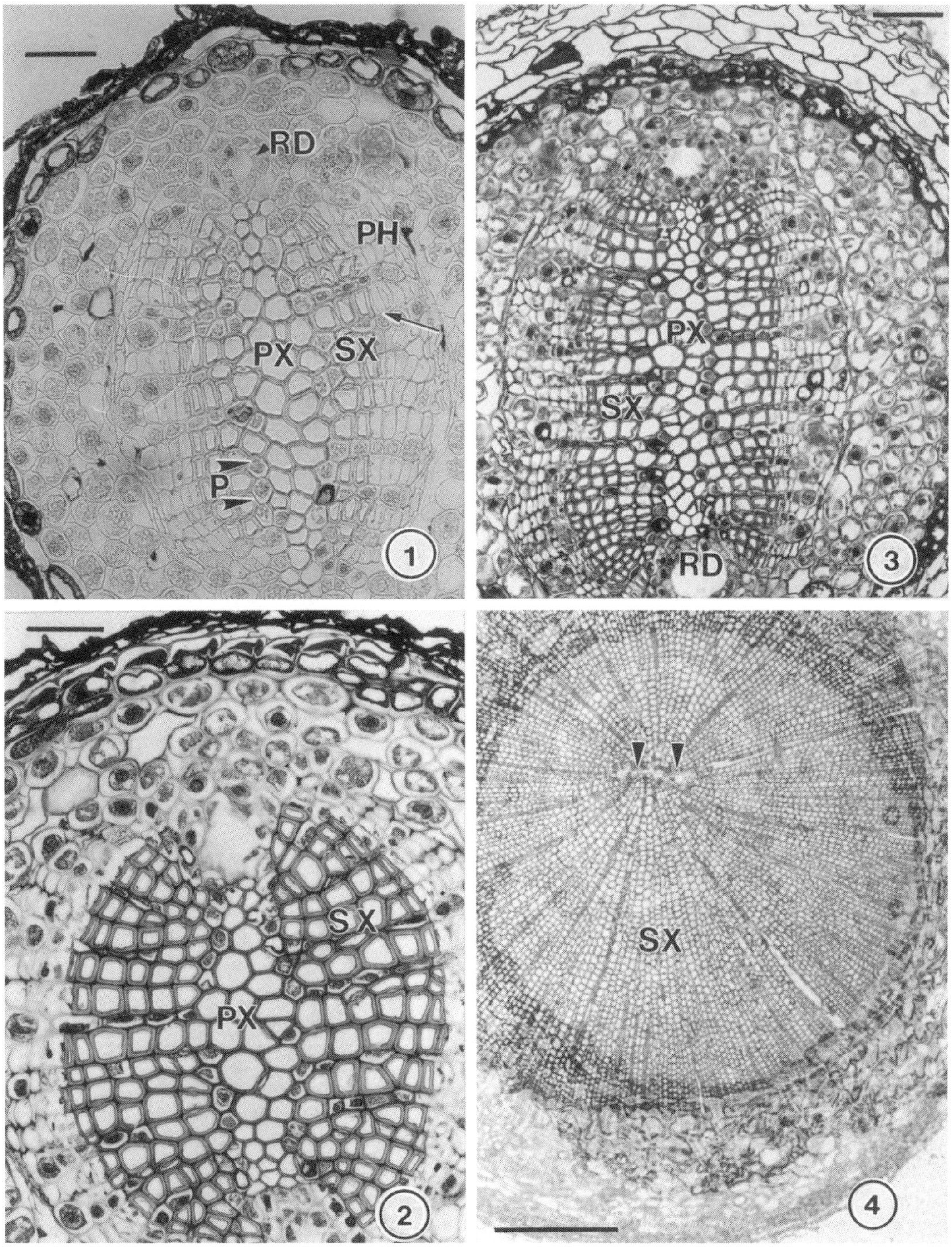

Figures 1–4. Transverse sections of long lateral roots of white spruce seedlings. The sections represent the four arbitrarily established classes of the advancement in secondary xylem development. **Figure 1.** A section showing xylem development referred to as class 1 in the text. The figure shows large cells of the primary xylem (PX) and smaller cells of the secondary xylem (SX) arranged in files on both sides of the primary xylem. Secondary phloem (PH) is separated from the secondary xylem by vascular cambium (arrow). Also shown are: a resin duct (RD) and parenchyma (P). Bar=50 μm. **Figure 2.** A section showing xylem development referred to as class 2 in the text. Bar=50 μm. **Figure 3.** A section showing xylem development referred to as class 3 in the text. No separate measurements of the primary and secondary xylem were made for specimens in this class. Bar=50 μm. **Figure 4.** A section showing xylem development referred to as class 4 in the text. The primary xylem located in the center of the section (arrowheads) constitutes only a small portion of the entire xylem area and was not separately measured. Note the variation in secondary-xylem cell sizes. Bar=10 μm.

Figure 5

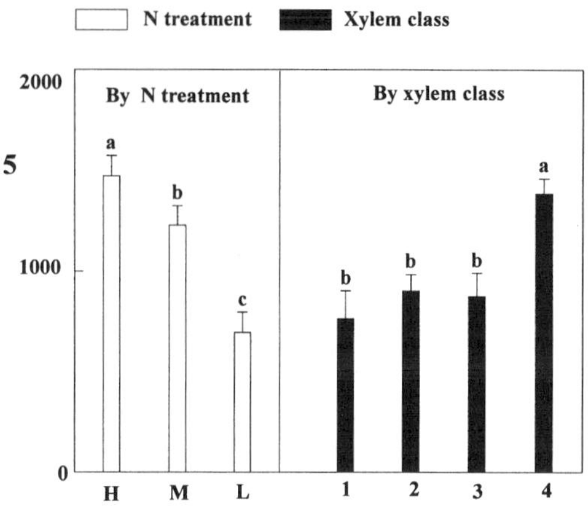

Figure 6

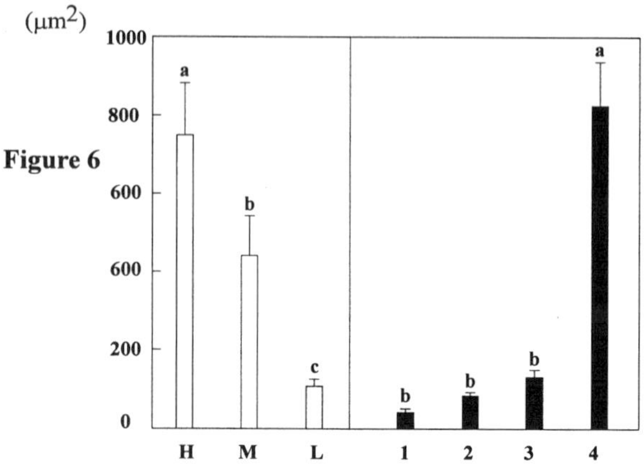

Figure 7

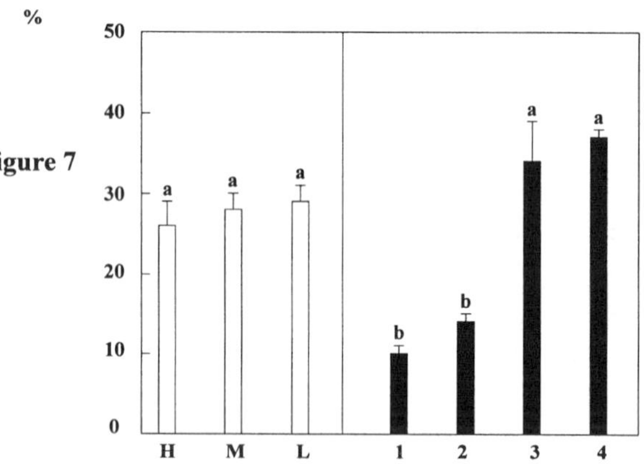

Figures 5–7. Bar graphs showing mean root diameters, mean xylem surface areas on root cross sections and the mean proportions between the xylem surface area and the total cross sectional root area in roots of one-year-old white spruce. The means are shown by N treatment (H=high N, M=medium N, L=low N treatments and by the xylem development class (1–4 as shown on Figures 1–4 and explained in the text). Capped bars represent standard errors of the means. **Figure 5.** Mean root diameters measured on transversely-sectioned specimens. Individual root diameters represented the average between the greatest and the smallest diameter of each transverse root section. **Figure 6.** Mean xylem surface areas measured on cross sectioned root specimens. **Figure 7.** Mean percent of the total cross sectional root area occupied by the xylem.

in some roots secondary growth was more advanced near the root tip than in more proximal locations.

The variation in root diameter among the segments sampled from different trees, roots, and sampling positions on the root was large and ranged from about 0.5 mm to over 3 mm. Seedlings grown under the high N treatment had the thickest roots (Figure 5). All three treatments differed significantly (p 0.04) from each other (α=0.05) with roots in the low N treatment being about half the size of those from the two other treatments. With some exceptions, root diameter was generally greater near the base than near the root tip. The ANOVA showed a significant effect of the xylem development class on root diameter. Class 4 root segments had the largest diameter (Figure 5) while root diameters in the other classes did not significantly differ from each other. The effect of the xylem class on root diameter was consistent across all three N treatments.

Not surprisingly, there was a strong positive correlation between root diameter and the area occupied by the xylem on root cross sections. The strongest correlation was in the xylem classes 4 (Pearson correlation coefficient 0.98) and 3 (0.93) but it was weaker in the classes 2 (0.47) and 1 (0.67). The greatest xylem areas were in roots of the high N treatment seedlings and the smallest in the low N treatment (Figure 6). All three N treatments differed significantly (p 0.05) from each other. Xylem area in the xylem class 4 (Figure 6) was significantly (p 0.0001) greater from those in the other three classes. Generally, the area occupied by the xylem constituted less than half of the cross-sectional root area. This proportion was unaffected by N (Figure 7).

N-treatment significantly affected the size of transverse sections (p 0.0001) of individual tracheids and areas occupied by the cell walls (p 0.0016) and by lumens (p 0.0004) on these transverse sections. Tracheids in the low N-treatment were significantly smaller in terms of their cross sectional area, lumen area, and cell wall area than those in the other two treatments (Table 1). However, N did not significantly change the proportion between areas occupied by cell walls and lumens. On average, cell walls and lumens occupied about the same proportion of the tracheid transverse surface area. The shape of these sections and that of the lumens could be best described as intermediate between elliptical and square or rectangular. Cell walls were thicker in tracheid corners.

Tracheids from plants grown under low N had significantly thinner cells walls (p 0.05) and smaller lumen radii (p 0.004) than plants grown under medium and high N (Table 1). Cell walls were generally thicker in specimens from basal positions and in those from xylem classes 3 and 4 than xylem classes 1 and 2 (Table 1). This indicated that cell wall thickness increased with progressing development of secondary xylem. Non-conducting spaces (parenchyma, intercellular spaces) occupied between 5–30% of the xylem area with a large variation throughout the same root cross section. Because of this variation, a mean of all replicates per section was calculated to obtain the proportion of the area of conducting elements to the total area occupied by the xylem for each individual root section. The non-conducting areas occupied averages of 10%, 11% and 13% of the xylem area in the low, medium and high N treatments, respectively, with no significant treatment effect.

On average, tracheids of the primary xylem had significantly (p 0.02) larger transverse areas than those of the secondary xylem (Figure 8). The data where unbalanced because, due to developmental differences, the primary xylem was measured on more sections in the low N treatment that in the other two N treatments. Cell wall areas were only somewhat greater in tracheids of the primary xylem but lumens were much larger in the primary than in the secondary xylem (Figure 8). This explains the significant (p 0.0001) difference in the proportions of the cell wall surface to the whole tracheid cross sectional surface between the two xylem types (Figure 9). Thus, average primary xylem cells were larger, had larger lumens, but thinner cell walls than the secondary xylem cells as the area occupied by the cell walls was stretched around a greater circumference.

Root macerations

One hundred and thirty four macerated root segments were examined. The shortest measured tracheid was 110 μm long while the longest was 3530 μm. Tracheids in the medium N treatment were longer than those in the other two treatments (Table 1) but this effect of N was not significant. The xylem development class (Table 1) had a significant effect on the length of tracheids and so did its interaction with the N treatment (p 0.03). In the medium N treatment, the longest tracheids were in the xylem class '1' containing the greatest percentage of primary-xylem tracheids. In the other two treatments tracheids in the class '1' were the shortest. There is no obvious explanation of this interaction.

Individual tracheids varied in width somewhat as their shapes were not always regular. The narrowest tracheid (cell wall included, a mean of three width measurements) was 4.57 μm while the widest tracheid was 51.3 μm wide. They both occurred in the low N treatment specimens. While extreme width differences occurred in all three N treatments, most of the tracheids from the same sample tended to be of similar width. Mean widths of tracheids in the high and medium N treatments were greater than those in the low N treatment (Table 1) and a significant treatment effect (p 0.001) on tracheid width was detected. The xylem development class effect was also significant (p 0.02) with the class 4 being the only one significantly different from the other three and having wider tracheids (Table 1).

The pits

When viewed under the light microscope, the pit-halves were reasonably round in shape of borders and apertures (Figure 10). The maceration technique used in this study separated tracheids by dissolving the middle lamellae and causing the cells to separate. In this process, most of the primary walls, including areas where they formed pit membranes, were lost exposing pit borders and pit apertures (Figures 11–12; Figure 14). The images shown on these figures represent halves of bordered-pit pairs or the tracheid halves of the half-bordered pits. Occasionally, more or less intact pit membranes were observed (Figures 13, 15) and these showed great variation in pit-membrane pore sizes.

Table 1. Morphological characteristics of individual tracheids from first order lateral roots of white spruce seedlings and their standard errors (in brackets). Length and width were measured on macerated specimens, other characteristics on root cross sections

	Nitrogen treatment			Xylem development class			
Variable	Low	Medium	High	1	2	3	4
Length (μm)	1019a	1155a	991a	1039de	1110c	1069cd	1023e
	(11)	(12)	(11)	(31)	(13)	(15)	(9)
Width (μm)	15.2b	19.8a	19.6a	17.6d	17.4d	16.8d	18.2c
	(0.10)	(0.11)	(0.12)	(0.27)	(0.13)	(0.16)	(0.09)
Transverse area (μm^2)	199b	284a	290a	239cd	258c	219d	257c
	(9)	(9)	(8)	(12)	(10)	(10)	(8)
Cell wall area	100b	136a	138a	111d	120cd	111d	129c
	(3)	(3)	(1)	(6)	(4)	(3)	(4)
Cell wall thickness (μm)	2.26b	2.58a	2.56a	2.37d	2.38d	2.47c	2.58c
	(0.04)	(0.04)	(0.04)	(0.09)	(0.05)	(0.07)	(0.05)
Lumen area (μm^2)	96b	148a	153a	128c	138c	107d	128c
	(7)	(7)	(6)	(12)	(10)	(8)	(6)
Lumen radius (μm)	5.09b	6.10a	6.22a	5.69c	5.99c	5.90c	5.50c
	(0.15)	(0.15)	(0.14)	(0.33)	(0.18)	(0.26)	(0.17)

Means with the same letters are not different from each other at α=0.05 (within the group by which they were tested).
Tracheid length and widths were measured on macerated tracheids. All other characteristics shown in this table were measured on transverse root sections or calculated (cell wall thickness and lumen radius) from these measurements as explained in the text.

The smallest measured width of a pit border was 4.1 μm while the greatest was 20.6 μm. The smallest pit aperture was 0.62 μm while the greatest was 10.2 μm. Although many pits on adjacent tracheids were of similar sizes (Figure 11), there was a considerable variation in pit sizes both within the same tracheid and among tracheids, with the pits located near the tips being often much smaller than those in the mid-tracheid portion, as shown on Figure 12. Although the N treatment did not significantly affect the pit border (p 0.14) and pit aperture (p 0.12) dimensions, the p values were not far from the arbitrarily chosen α=0.05 level. There was little difference in pit border means among the treatments (11.7, 11.8, and 10.2 μm, in the high, medium, and low N treatments, respectively). The pit aperture mean was greater in the medium N treatment than in the other two (4.2, 3.7, and 3.3 μm, in the medium, high, and low treatments, respectively). The correlation between pit border and pit aperture dimensions was not strong (Pearson's correlation coefficient=0.48).

Discussion

This study has described the characteristics of water conducting tissues from roots of young white spruce seedlings. Effects of nitrogen availability on these characteristics were examined as well as the advancement of the secondary root development under different N supply. The results reported here may contribute to the improvement of the understanding of the growth and functioning of conifer roots and the mechanism underlying the efficiency of water transport through the root xylem. Anatomical details of the secondary xylem were compared for roots and stems by Patel (1965) but only older trees of angiosperm species were included in that report. Vessel diameters reported in that study were determined from sectioned material. The wide array of studies on conifer xylem from stems, discussed in the introduction, represented a variety of technical approaches to these measurements. Tracheid lengths and widths were measured on either sectioned or macerated specimens. In estimations of tracheid cross-sectional characteristics, a diameter measurement was usually reported and cell wall thickness determined by linear rather than area measurements. The study of Antonova and Stasova (1993) represented a different approach and included measurements of cross sectional surface area.

There are some important differences and potential problems related to the technical approaches used in the estimation of tracheid dimensions. Maceration appears to be a technique suitable for measurements of tracheid widths and lengths as the tracheids are separated along their entire length and the individual

Figure 8

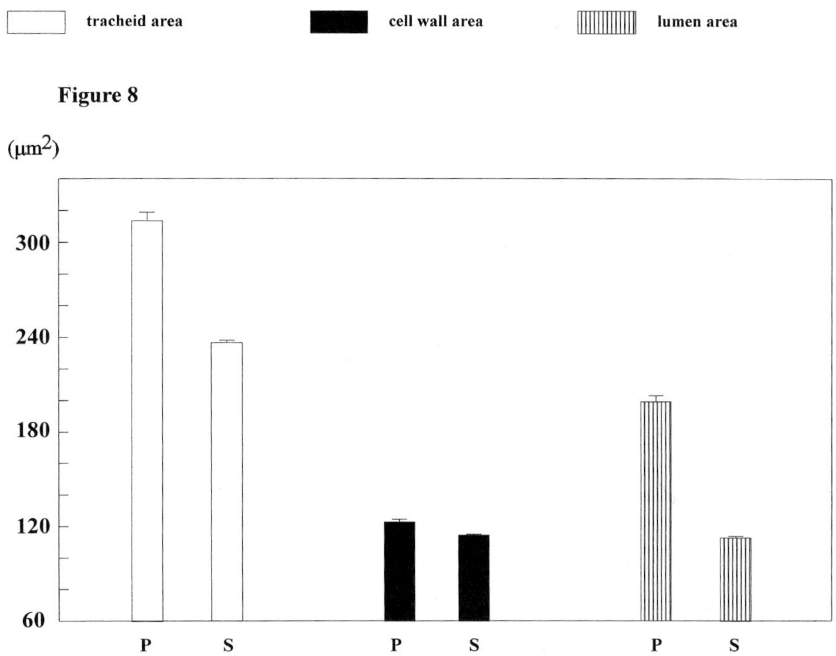

Figure 9

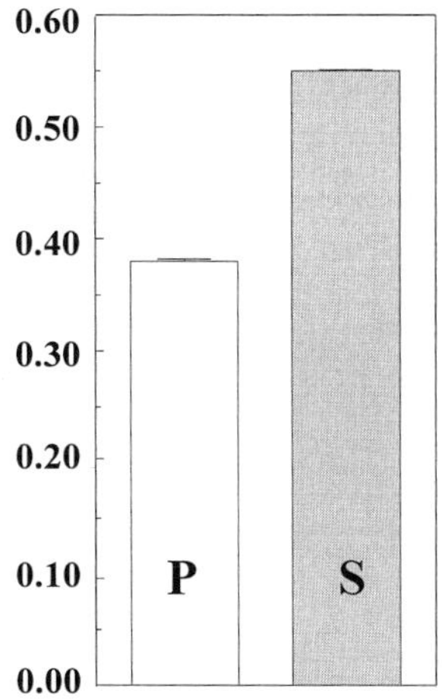

Figures 8–9. Mean cross sectional tracheid areas, their mean cell wall and lumen areas, and the proportion of the cell walls in primary and secondary xylem of cross sectioned tracheids from the roots of one-year-old white spruce seedlings. Capped bars represent standard errors of the means. **Figure 8.** Mean cross sectional tracheid areas (blank bars), their mean cell wall (black-filled bars) and lumen (line-patterned bars) areas of primary and secondary tracheids. P=primary xylem, S=secondary xylem. **Figure 9.** Mean proportions of cell walls in tracheid cross sections of the primary (blank bar marked 'P') and the secondary (shaded bar marked 'S') xylem.

variation in their thickness (width) can be estimated. In the current study, a mean of three measurements was taken to estimate the average tracheid diameter. It was noted that some tracheids were more regularly shaped than others, some had both ends tapered but more commonly one end was pointed while the other end was blunt. Tracheids were neither perfectly oval or square in their cross sections. Their shapes were more or less rectangular and, depending on which side the tracheid was placed on the slide, its wider or narrower side could be exposed to the view.

Measuring cross-sectioned specimens introduces a different problem caused by the taper of the tracheid ends. It is not known whether one has measured the diameter of a narrow tracheid or a narrow end of a wide one unless serial sections are also examined. The

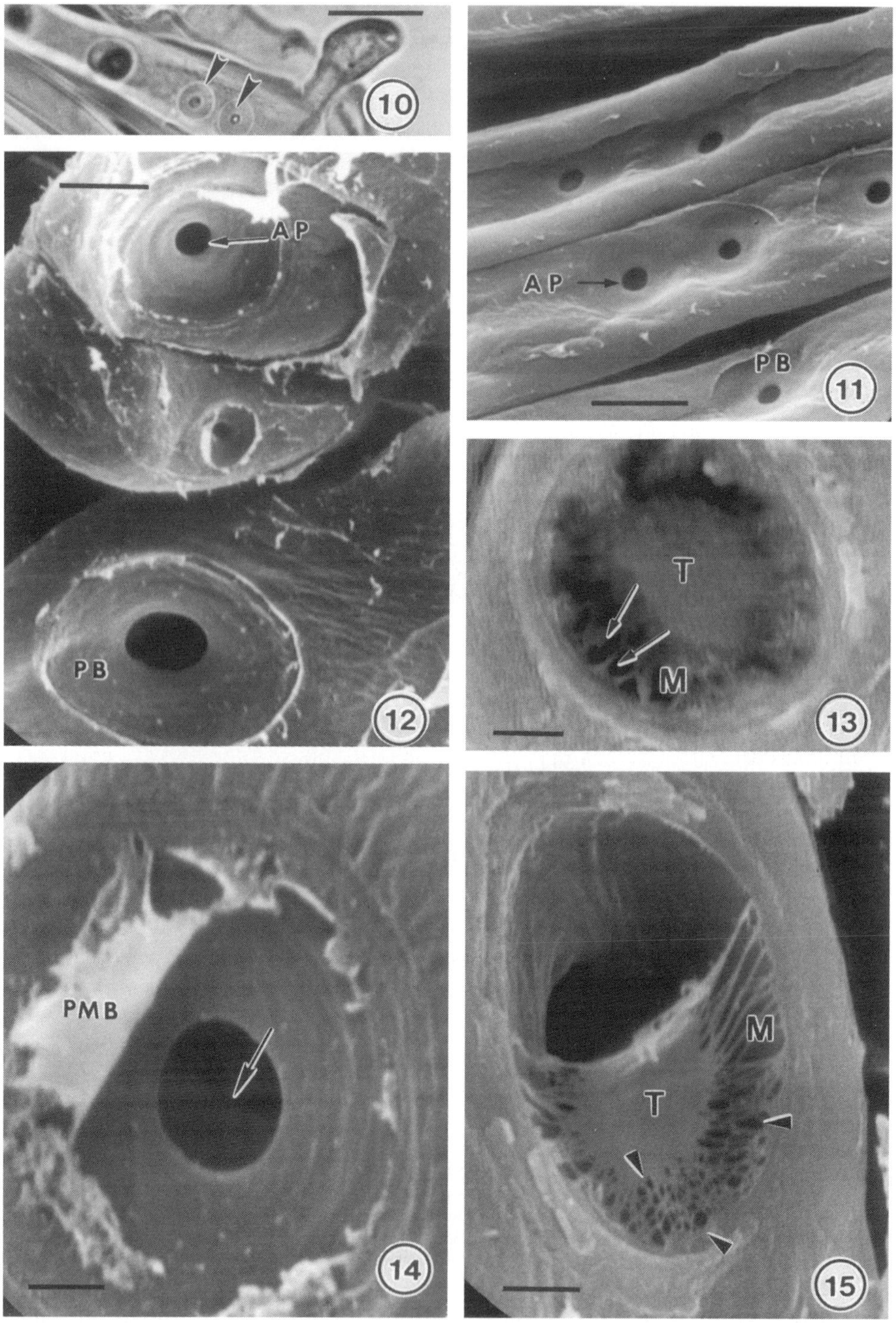

Figures 10–15. Light and scanning electron micrographs of bordered pits and pit membranes of tracheids from macerated segments of the roots of one-year-old white spruce seedlings. **Figure 10.** Phase-contrast view of the pits in tracheid cell walls showing pit borders (arrowheads) and pit apertures (small circles in the centre of the bordered regions). Bar=50 μm. **Figure 11.** Scanning electron micrograph of tracheids showing uniform size pit borders (PB) and pit apertures (AP). Note the elliptical shape of the borders and apertures, a distortion caused by the viewing angle. Bar=5 μm. **Figure 12.** Scanning electron micrograph of tracheids illustrating the range of pit border and pit aperture sizes in tracheids from the same region of the same root. Bar=10 μm. **Figure 13.** Scanning electron micrograph of a pit membrane showing the torus (T), and the margo (M) with pores (arrows) of various sizes formed by the cellulose microfibrils of the primary wall. Bar=150 μm, Bar=2 μm. **Figure 14.** Scanning electron micrograph showing remnants of the pit membrane (PMB) almost completely removed by specimen maceration and an unobscured view of the aperture (arrow). Note that the specimen is tilted relative to the viewing angle and the thickness of the cell wall on the side of the pit aperture can be discerned. Bar=2 μm. **Figure 15.** Scanning electron micrograph showing a partially torn pit membrane with torus (T) and margo (M) with pit membrane pores (arrowheads) of various shapes and sizes. Bar=2 μm.

same is true for measurements of cross sectional areas. When lumens are measured, the same problem is encountered as the lumens taper when the tracheid tapers while the cell wall thickness does not change much. Consequently, the proportion of the cell wall to lumen area and the cell wall to tracheid cross sectional area is greater at the tapered ends than in the middle. Perhaps such measurements should be made on longitudinal sections to avoid some of these problems. However, longitudinal sections through thin roots are difficult to interpret. The proportion of tracheids from the primary and secondary xylem seen on different longitudinal sections could be difficult to estimate. The primary xylem tracheids measured in this study were often wider than those of the secondary xylem so their proportion on different longitudinal cross sections could impact on the results of measurements. The problems discussed above pertained to this study as much as to the other reported studies. Comparing the mean tracheid widths measured on macerated specimens with tracheid cross sectional areas measured on sectioned specimens showed similar effects of the N treatment on tracheid size. Lumen radius of a macerated tracheid could be calculated by deducting the cell wall thickness from half of tracheid's width. When these radii were compared to those calculated from transverse sections, the former means were greater than the latter by 0.25, 1.22, and 1.02 μm in the low, medium, and high N treatments, respectively. This underestimation of lumen radii in sectioned specimens could be related to some tracheids being sectioned through their narrow, tapering ends while in measurements on macerated specimens these narrow ends were avoided.

Tracheid lengths measured in this study generally corresponded to the range listed by Tyree et al. (1994) which was 1–3 mm. The mean lengths of white spruce root tracheids were about 1 mm. Although tracheids about 3.5 mm long were observed, there were many more tracheids shorter than 1 mm than those longer than 3 mm. There was a large background variation in the size of tracheids, regardless of whether it was the width, the length, or cross sectional area that was measured. These characteristics varied among roots and among trees, indicating that they could be related to growth behavior of individual roots and to genetic factors peculiar to individual trees. Over seven thousand tracheids were individually measured (not counting those only partially enclosed in the subsampled area), on root cross sections. It could be expected that the effects of the imposed treatments on tracheid size, if strong, would override the error caused by sectioning through some tapering parts of the tracheids. Mean diameters of tracheids in roots of one-year-old spruce seedlings were greater than the effective lower limit on conduit diameter in trees which, according to Tyree et al. (1994) would be about 10 μm. They were, on average, almost twice that size but that included the cell wall. Some much larger tracheids were observed and it was obvious that the size of these conduits increased mainly due to the increasing lumen diameter while the cell wall thickness increased less. The proportion between the areas occupied by the cell wall and the lumen was not significantly affected by the N treatments. Thus, the larger the tracheid's external width the larger its potential conductive capacity would be and the less of its diameter would be taken by the cell wall. Large in transverse dimensions tracheids of the primary xylem had much thinner cell walls than the usually smaller cells of the secondary xylem. Tracheids with thicker cell walls tended to occur in the more advanced xylem development classes 3 and 4. It must be noted that much fewer primary xylem than secondary tracheids were measured.

The area occupied by the xylem constituted usually less than half of the root cross-sectional area. In roots with more advanced secondary development, the correlation between these variables was very strong so estimates of the approximate partition of the root cross sectional area to the xylem could be accurately made from external root dimensions. Such estimates would be much less accurate for roots with less advanced secondary development. Results of this study indicated that the cambial activity of roots in white spruce seedlings was stimulated by the availability of inorganic nitrogen. This was expected and confirmed the results reported by Coutts and Philipson (1976) and Philipson and Coutts (1977) for *Picea sitchensis*. The observed increase in root diameters and cross sectional surface areas, both total and occupied by the xylem, was achieved by both an enlargement of the tracheid size, that is its diameter and cross sectional area, and by an increased number of cells produced by the vascular cambium. This was demonstrated by an increased number of root sections with advanced secondary xylem development in the two higher N treatments, compared to the low N. However, the difference in cambial activity and tracheid enlargement was significant only when a near-deficiency treatment was compared to the approximately optimal N supply (considered to be the 'medium', 50 ppm, N-treatment, R. van den Driessche, personal communication, 1997) and much less or not at all when the medium and high N treatments were compared. Therefore, the limit for the effective stimulation of radial root growth in white spruce roots would not be much above the 50 ppm N supplied twice a week. First-initiated secondary xylem tracheids were narrow, thus early initiation and differentiation of the secondary xylem did not cause an immediate significant increase in root diameter.

The distribution of sections representing different stages of the secondary xylem development in roots of young white spruce seedlings suggested that there was a general trend for increasing number of secondary xylem cell layers in the basal direction which resulted in a basipetally increasing root diameter. This was not surprising. However, the fact that less advanced segments were sometimes found at proximal locations than near the root tip showed that the regulation of cambial activity could vary along the root length and could be independent of the position of the cambium relative to the root tip. This supports the statements of Woods (1991) regarding the behavior of the root cambium in older trees where annual growth rings did not correspond to the real age of the root and differed in numbers in different root segments, not necessarily in the manner corresponding to that observed in the growth of stems. The fact that less secondary xylem could occur further away from the root tip than near the root tip could be interpreted that in young roots in which there are numerous open primary and secondary xylem elements the size of the conducting area may not impose a serious limitation to the axial transport of sap through the roots. The large dimensions of the primary xylem elements might represent a developmental adjustment enhancing the conductive capacity of root portions close to the root tip where the secondary xylem has not yet developed. It should be kept in mind that the examined roots were collected from dormant seedlings without any white root tips.

Addoms (1946) observed absorption of weak aqueous dye solutions by older, suberized roots of yellow poplar, sweet gum, and shortleaf pine. She observed that these roots did not appear to absorb unless younger branch roots were excised and the wounds sealed. This suggested that the absorption by older roots would occur only if the necessity arose, perhaps due to increased xylem tension caused by the inability to feed the transpirational stream by the normally absorbing branch roots that had been removed. Kramer (1946) found that roots that varied from 3.2 to 17.0 mm in diameter absorbed water even through many layers of cork. He remarked that water absorption through non-elongating, older roots must be of vital importance to overwintering trees. Chung and Kramer (1975), supported this conclusion by quantifying the absorption of water and radioactive phosphorus by loblolly pine roots. Krasowski et al. (1996) reported efficient water absorption and transport through brown, non-elongating roots of dormant white spruce seedlings at soil temperatures barely above freezing but at low foliage water content indicative of high tension in the xylem. Thus, the anatomical observations reported in the current study support the notion of high capacity for axial transport of water in brown non-elongating roots beginning short distances above their root tips.

Peterson and Steudle (1993) calculated the hydraulic conductivity of early metaxylem vessels in *Zea mays* L. They postulated that the radial movement of water through the non-pitted portions of the cell wall could be substantial and that it should not be ignored in detailed analysis of water flow through roots. Their calculations included a number of assumptions that could introduce an error that might account for 8–31% of the contribution to the transport through the wall to the total lateral water transport through vessels. Unfortunately, their methods of calculation cannot be readily applied to evaluate the lateral hydraulic conductivity of conifer tracheids. This is because a single tracheid can be laterally connected to a number of other tracheids and to parenchyma cells which makes it difficult to estimate how many pits conduct water into the tracheid and how many conduct it out of that tracheid. A large variation among different tracheids can be expected in this respect.

The distance between pit borders is that occupied by the margo. Since the torus must fit the pit aperture, it can be presumed that its size is closely related to the size of the pit aperture. N treatments did not significantly alter these properties, although the pit aperture sizes tended to be greater in the optimal medium N treatment than in the other two treatments. If the capacity for axial water transport in spruce roots is affected by N, it would be by its effects on conduit diameter and, possibly, the pit-membrane pore sizes but not by changes to the size of the pit membrane surface area or to changes to conduit length.

Results of the study reported here could be of use in work concerned with modeling the axial flow of xylem sap in the roots. The differences in root anatomy reported in this study may be the function of growth rate differences in different N treatments. It is possible that other factors that affect growth rates could produce similar responses.

Acknowledgement

Technical assistance of Adam (Oli) Caputa (Red Rock Research Station) is greatly appreciated. We also thank Peter Ott (biometrician, Research Branch, British Columbia Ministry of Forests) for statistical help and Karen Dale (Red Rock Research Station) for secretarial help. Special thanks to Dr R van den Driessche for advice on nutrition. This study was supported by the NSERC Industry/University Partnership Grant to J N Owens and by the British Columbia Ministry of Forests. It is a portion of dissertations submitted to the University of Victoria by the first author in partial fulfillment of the requirements for the degree of a doctor of philosophy.

References

Addoms R M 1946 Entrance of water into suberized roots of trees. Plant Physiol. 21, 109–111.

Antal M and M M Micko 1991 Variation of wood density and tracheid length of white spruce. Drevársky Výskum 128, 35–41.

Antonova G F and V V Stasova. 1993 Effects of environmental factors on wood formation in Scots pine stems. *In* Trees: structure and function, vol. 7, pp 214–219. Springer Verlag.

Bailey I W and W W Tupper 1918 Size variation in tracheary cells: I. A comparison between the secondary xylems of vascular cryptogams, gymnosperms, and angiosperms. Amer. Acad. Arts Sci. Proc. 54, 149–204.

Burden R D and C B Low 1992 Genetic survey of *Pinus radiata*. Wood properties: variation, heritabilities, and interrelationships with other traits. N.Z. J. For. Sci. 22, 228–245.

Byrne G F Begg, J E and Hansen G K 1977 Cavitation and resistance to water flow in plant roots. Agric. Meteorol. 18, 21–25.

Charlton W A 1991 Lateral root initiation. *In* Plant roots. The hidden half. Eds Y Waisel, A Eshel and U Kafkafi. pp 103–128. Marcel Dekker, Inc., New York, Basel, Hong Kong.

Chung H H and Kramer P J 1975 Absorption of water and P through suberized and unsuberized roots of loblolly pine. Can. J. For. Res. 5, 229–235.

Clarkson D T 1991 Root structure and sites of ion uptake. *In* Plant roots. The hidden half. Eds Y Waisel, A Eshel and U Kafkafi. pp 417–453. Marcel Dekker, Inc., New York, Basel, Hong Kong.

Cochard H 1992 Vulnerability of several conifers to air embolism. Tree Physiol. 11, 73–83.

Connor K F and R M Lanner 1990 Effects of tree age on secondary xylem and phloem anatomy in stems of Great Basin bristlecone pine (*Pinus longaeva*). Amer. J. Bot. 77, 1070–1077.

Coutts M P 1982 The tolerance of tree roots to waterlogging: 5. Growth of woody roots of Sitka spruce and lodgepole pine in water logged soil. New Phytol. 90, 467–476.

Coutts M P and Philipson J J 1976 The influence of mineral nutrition on the root development of trees. I. The growth of Sitka spruce with divided root systems. J. Exp. Bot. 27, 1102–1111.

Crombie D S, Hipkins M F and Milburn J A 1985 Gas penetration of pit membranes in the xylem of rhododendron as the cause of acoustically detectable sap cavitation. Plant Physiol. 12, 445–453.

Dinwoodie J M 1961 Tracheid and fibre length in timber – a review of literature. Forestry 34, 125–144.

Dinwoodie J M 1963 Variation in tracheid length in *Picea sitchensis* Car. Department of Scientific and Industrial Research. Forest Products Research Special Report No. 16. Her Majesty's Stationery Office, London.

Esau K 1977 Anatomy of seed plants. J Wiley and Sons, New York.

Fayle D C F 1968 Radial growth of tree roots. Fac. For., Univ. Toronto Tech. Rep. 9, 1–183.

Ford E D 1993 Potential measures for use in developing selection procedures for high timber density in juvenile wood of *Picea sitchensis*: II. Cell dimensions and wood histology. J. Sustainable Forestry 1, 43–67.

Funada R, Kubo T and Fushitani M 1990 Early and late wood formation in *Pinus densiflora* trees with different amounts of crown. IAWA Bulletin n.s., 11, 281–288.

Hagemeyer J, Hepper T and Breckle S W 1994 Effects of Cd and Zn on the development of annual rings of young Norway spruce (*Picea abies*) plants. *In* Trees: structure and function, vol. 8, pp 223–227. Springer Verlag.

Harvey H P and van den Driessche R 1997 Nutrition, xylem cavitation, and drought resistance in hybrid poplar. Tree Physiology 17, 647–654.

Jackson G E, Irvine J and Grace J 1995 Xylem cavitation in Scots pine and Sitka spruce saplings during water stress. Tree Physiol. 15, 783–790.

Johansen D A 1940 Plant microtechnique. McGraw-Hill, New York.

Kramer P J 1946 Absorption of water through suberized roots of trees. Plant Physiol. 21, 37–41.

Krasowski M J, Letchford T, Caputa A, Bergerud W A and Ott PK. 1996. The susceptibility of white spruce seedlings to overwinter injury and their post-injury field responses. New For. 12, 261–278.

Larson P R 1960 A physiological consideration of the springwood-summerwood transition in red pine. For. Sci. 6, 110–122.

Liphschitz N, Lev-Yadun S, Rossen R and Waisel Y 1984 The annual rhythm of activity of the lateral meristems (cambium and phellogen) in Pinus halapensis mill. and Pinus pinea L. IAWA Bull. n.s. 5, 263–274.

Lobzhanidze E D and Kandelaki A A 1972 The periods of cambial activity in the stems and roots of Picea orientalis of different generations in uneven-aged stands (in Georgia). Tr. Tbilis. Inst. Lesa 19/20, 202–214. (In Russian).

Manwiller F G 1972 Tracheid dimensions in rootwood of southern pine. Wood Sci. 5, 122–124.

McCully M E and Canny M J 1988 Pathways and processes of water and nutrient movement in roots. Plant Soil 111, 159–170.

Milliken G A and Johnson D E 1992 Analysis of messy data. Volume 1: Designed Experiments. Van Nostrand, Reinhold, New York.

Moreshet S and Huck M G 1991 Dynamics of water permeability. *In*: Plant roots. The hidden half. Eds Y Waisel, A Eshel and U. Kafkafi. pp. 605–626. Marcel Dekker, Inc., New York, Basel, Hong Kong.

Pande P K, Rao R V, Agrawal S P and Singh M 1995 Variation in the dimensions of tracheid elements of *Pinus Caribaea* Var. *Bahamensis*. J. Trop. For. Prod. 1, 117–123.

Passioura J B 1988 Water transport in and to roots. Annu. Rev. Plant. Physiol. Plant Mol. Biol. 39, 245–265.

Patel R N 1965 A comparison of the anatomy of the secondary xylem in roots and stems. Holzforschung (Berlin) 19, 72–79.

Persson B, Persson A, Ståhl E G and Karmlats U L 1995 Wood quality of *Pinus sylvestris* progenies at various spacings. For. Ecol. Manag. 76, 127–138.

Peterson C A and Steudle E 1993 Lateral hydraulic conductivity of early metaxylem vessels in *Zea mays* L. roots. Planta 189, 288–297.

Philipson J J and Coutts M P 1977 The influence of mineral nutrition on the root development of trees: II. The effect of specific nutrient elements on the growth of individual roots of Sitka spruce. J. Exp. Bot. 28, 864–871.

Philipson W R, Ward J M and Butterfield B G 1971 The vascular cambium. Chapman & Hall, London.

Pomparat-Dasse M and David R 1974 Influence des facteurs edaphiques de trois sites caractéristiques de la forêts de Gascogne sur les variations de la longueur des tracheides de la racine et al la tige de *Pinus maritimes* poussant dans chacun des sites: relation avec l'activité cambiale. C.R. Hebd. Seances Acad. Sci. 279, 169–174.

Pothier D, Margolis H A, Poliquin J and Waring R H 1989 Relation between the permeability and the anatomy of jack pine sapwood with stand development. Can. J. For. Res. 19, 1564–1570.

Rypacek V, Hauck O and Vyslouzil J 1976 Conducting tissue differentiation in seedlings of some forest tree species. Scr. Fac. Sci. Nat. Univ. Purkynianane 6, 103–112.

Sandercock C F, Sands R, Ridoutt B G, Wilson L F and Hudson I 1995 Factors determining wood microstructure in eucalyptus. *In* Eucalypt plantations: Improving fibre yield and quality. Eds M B Potts, N M G Borralho, J B Reid, R N Cromer, Tibbits W N and Raymond C A. Proc. CRC-IUFRO Conf. Hobart, 19–24 Feb. pp 128–135. Cooperative Research Centre for Temperate Hardwood Forestry, Hobart, Australia.

Sanderson J, Whitbread F C and Clarkson D T 1988 Persistent xylem cross-walls reduce the axial hydraulic conductivity in the apical 20 cm of barley seminal root axes: implications for the driving force of water movement. Plant Cell Environ. 11, 247–256.

Saranpää P 1994 Basic density, longitudinal shrinkage and tracheid length of juvenile wood of *Picea abies* (L.) Karst. Scand. J. For. Res. 9, 68–47.

Sperry J S and Sullivan J E M 1992 Xylem embolism in response to freeze-thaw cycles and water stress in ring-porous, diffuse-porous, and conifer species. Plant Physiol. 100, 605–613.

Sperry J S and Tyree M T 1990 Water-stress induced xylem embolism in three species of conifers. Plant Cell Environ. 13, 427–436.

St. Aubin G, Canny M G and McCully M E 1986 Living vessel elements in the late metaxylem of sheated maize roots. Ann. Bot. 58, 577–588.

Tavita Y L 1992 Influence of sampling position and growth rate on the wood properties of Benguet Pine (*Pinus kesiya* Royle ex. Gordon). FPRDI J. 21, 31–41.

Tyree M T and Sperry J S 1989 Water stress induced cavitation and embolism in some woody plants. Physiol. Plant. 66, 397–405.

Tyree M T and Dixon M A 1983 Cavitation events in *Thuja occidentalis* L. Plant Physiol. 72, 1094–1099.

Tyree M T and Karamanos A J 1981 Water stress as an ecological factor. *In* Plants and their atmospheric environment Eds J Grace, E D Ford and P G Jarvis. pp 237–261. Blackwell Scientific, Oxford.

Tyree M T, Davis S D and Cochard H 1994 Biophysical perspectives of xylem evolution: is there a tradeoff of hydraulic efficiency for vulnerability to dysfunction? IAWA J. 15, 335–360.

Wang X L, Canny M J and McCully M E 1991 The water status of the roots of soil-grown maize in relation to the maturity of their xylem. Physiol. Plant. 82, 157–162.

Wenzel C L, McCully M E and Canny M J 1989 Development of water conducting capacity in the root systems of young plants of corn and some other C-4 grasses. Plant Physiol. 89, 1094–1101.

Wilcox H 1954 Primary organization of active and dormant roots of noble fir, *Abies procera*. Amer. J. Bot. 41, 812–821.

Wilcox H 1964 Xylem in roots of *Pinus resinosa* Ait. in relation to heterorhizy and growth activity. *In* Ed. M Zimmerman. pp 459–478. Formation of wood in forest trees. Academic Press, Inc., New York.

Wilcox H 1962 Growth studies of the root incense cedar, *Libocedrus decurrens*. I. The origin and development of primary tissues. Amer. J. Bot. 49, 221–236.

Wodzicki T J 1960 Investigation on the kind of *Larix polonica* Rac. wood formation under various photoperiodic condition. Acta Soc. Bot. Pol. 29, 713–730.

Wodzicki T J 1964 Photoperiodic control of natural growth substances and wood formation in larch (*Larix decidua* D.C.) J. Exp. Bot. 15, 584–599.

Wodzicki T J 1971 Mechanism of xylem differentiation in *Pinus sylvestris* L.J. Exp. Bot. 22, 671–687.

Woods F W 1991 Cambial activity of roots. *In* Plant roots. The hidden half. Eds Y Waisel, A Eshel and U Kafkafi. pp 149–160. Marcel Dekker, Inc. New York, Basel, Hong Kong.

Yang K C and Hazenberg G 1994 Impact of spacing on tracheid length, relative density, and growth rate of juvenile wood and mature wood in *Picea mariana*. Can. J. For. Res. 24, 996–1007.

Yeoman M M Ed. 1976 Cell division in higher plants. Academic Press, New York.

Zimmerman M H Ed. 1964 The formation of wood in forest trees. Academic Press, New York.

Zimmerman M H 1983 Xylem structure and the ascent of sap. Springer-Verlag, Berlin, 192 p.

A. Stokes (ed.), The Supporting Roots of Trees and Woody Plants: Form, Function and Physiology, 371–375.

The efficiency of the water conducting system of tomato roots

Hydraulic conductivity of tomato roots

A. Eshel*, Ilona Shick and Y. Waisel

*Department of Plant sciences, The George S. Wise Faculty of Life Scienced, Tel Aviv University, Tel Aviv 69978, Israel (*Corresponding author)*

Abstract

Hydraulic conductivity of the root system of aeroponically grown tomato plants, was investigated. Two perpendicular diameters of every xylem element in a single basal root and all laterals connected to it were measured in microscopical cross sections. Only 1.6% of the 2100 elements analyzed were found to be circular. Therefore hydraulic diameter of the xylem elements was calculated according to an equation that takes into account their deviation from the circular shape. The sum of the hydraulic diameters of all the elements identified within each cross section was used for estimation of the conducting capacity. An increase of the conducting capacity in the basipetal direction was observed, resulting from an increase in the number of xylem vessels and in their diameter. A certain level of redundancy was found in the structure of the vascular tissues of this tomato root. The changes in the conducting capacity were gradual, but constrictions were found at some points along the main axis. Comparing the conducting capacity above and below certain branching points along the axes indicated excess of capacity in proximal zones. This apparent excessive investment in conducting elements is explained by the need to preserve hydraulic conductivity even when many of the supplying vessels are incapacitated by embolism.

Introduction

Conduction of water is one of the most important functions of plant roots. Unlike the uptake of ions or the production of hormones which take place mostly in the younger portions of roots, water transport is performed by roots throughout their life. The water conducting tissues are comprised of dead xylem elements whose hydraulic conductivity is determined predominately by their diameter and length.

The hydraulic architecture of root systems has not been analyzed in such details as that of the shoots. But we assume that the same principles of efficiency in terms of the structure and function of the branching system should apply (Fitter, 1996; Nobel, 1996). There are several ways to determine the overall hydraulic conductivity of plant root systems (Tyree et al., 1995). However, one must remember that root systems are comprised of various root types, of diverse lengths, of several diameter classes, and of various physiological traits (Eshel and Waisel, 1996; Waisel and Eshel, 1991). Thus, understanding the hydraulic architecture of a root system requires the quantitative analysis of all its individual constituents, categorization of all its tributaries, and estimation of the contribution of each constituent to the total water supply of the whole plant. Another important parameter that should be considered in this context is the efficiency of the water conducting structures. Here we used the ratio between the sum of the hydraulic conductivity of a root segment and all its laterals to the hydraulic conductivity of the next proximal segment, as a measure of the efficiency. In the following we report such an analysis of a whole portion of a root system of a tomato plant, extending from the stem base to the tips of all the terminal rootlets.

Materials and methods

Tomato plants (*Lycopersicon esculentum* Mill. cv. Daniella) were grown with their roots in the aeroponic chamber of the Sarah Racine Root Laboratory (Waisel, 1996).The 6 m deep root chamber was intermittently sprayed with a fine mist of 50% Hoagland's nutrient solution, supplemented by 5 *mM* $MgSO_4$ and 5 *mM* $CaCl_2$. The solution was sprayed for 20 seconds, every minute. The temperature of the root chamber was 20

± 2 °C and the temperature of the greenhouse, on top of the root chamber, was maintained at 24 ± 2 °C.

Roots were taken from 2 months old tomato plants at the flowering stage. One of the main basal roots, 0.9 m long bearing 10 second order laterals was analyzed. The laterals had a combined length of 4.9 m and bore scores of third and fourth order laterals. The examined root was excised from the plant, spread in water on a large tray and mapped. It was then divided into 120 segments representing all levels of branching. Samples of each of the segments were fixed in FAA, dehydrated in ethanol, imbedded in paraffin, sectioned with a rotary microtome, stained with Safranine-Fast green and analyzed by light microscopy. Each cross section was photographed, using a Leitz microscope with a digital camera.

Analysis of the microscopic images included measurements of the root diameter and of the diameter of the vascular cylinder, counting of all xylem vessels, and measuring two perpendicular diameters of every vessel. The hydraulic diameter of every vessel (*Dh*) was calculated according to the equation of Lewis and Boose (1995) (1), for capillaries with elliptic cross section.

$$Dh = \frac{2a3b3}{a2 + b2} \tag{1}$$

In this equation *a* and *b* are the short and long diameters of the elliptic capillary. The equivalent hydraulic diameter of any root cross section (*Ds*) was calculated by summation of the *Dh* values of all the xylem vessels identified in that cross section:

$$Ds = \sum_{i=1}^{N} Dhi \tag{2}$$

where *N* is the number of xylem vessels identified in the root cross section, and *Dhi* is their respective hydraulic diameter. Assuming uniform viscosity of the sap, the quantitative contribution of each root to the water conductivity of the whole root system, was considered to be proportional to *Ds*.

Results and discussion

The Hagen-Poiseuille equation for flow that is commonly used for calculating xylem conductivity (Nobel, 1991; Zimmermann, 1983), is based on the assumption that the individual vessels function like ideal circular capillary tubes. Evaluation of the water conductivity that is based on such an assumption

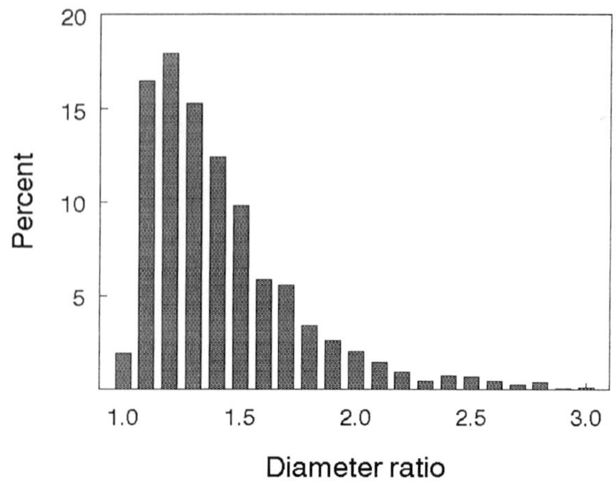

Figure 1. Distribution of xylem vessels in tomato roots according to the ratio of two perpendicular diameters.

introduces errors when used for estimation of water flux through non-circular conduits. Therefore, we have used the equation for capillaries with elliptical cross sections, i.e., Equation (1) (Lewis and Boose, 1995). We are not the first ones to consider the deviation of xylem vessels from the ideal circular shape when estimating their conductivity. Melchior and Stueudle (1993) used a different formula for elliptical cross sections when estimating the hydraulic conductivity from microscopical sections of onion roots. However, there are no reports that specify to what extent the xylem vessels of each species deviate from the circular shape.

In order to evaluate the elliptic shape of the conducting elements we measured two perpendicular diameters of 2110 xylem vessels in roots of all branching levels. Only 1.6% of these vessels were circular, having a diameter ratio of unity. All other vessels were elliptical, as indicated by their higher diameter ratio. Most vessels had diameter ratio between 1.1 and 2.0 (Figure 1). In that range of diameter ratio the hydraulic diameter calculated according to Equation (1) differs by up to 13% from the commonly used value of r^4. A total of some 5% of the vessels had a diameter ratio between 2 and 3, and few vessels that were severely compressed had a higher diameter ratio reaching 4.2. However, those were vessels with a small average diameter, and we estimate that their contribution to the overall conductivity of the roots should be small. There were no significant differences in the distribution of xylem vessels by diameter ratio among the branching levels I to IV, indicating the deviation from the circular shape is inherent characteristic of this plant.

Table 1. Distribution of the number of vessels of in six size classes according to the hydraulic diameter [*Dh*], and % of large and of small vessels, along the main axis of a tomato root system. The most abundant size class in each segment is highlighted

Root segment	Hydraulic diameter [*Dh*] (μm^4)						Vessel number	% vessels	% vessels
	10^4	10^5	10^6	10^7	10^8	10^9	$10^4<$	$<10^6$	$10^8<$
1-base	0	0	4	9	**14**	12	39	0	66.7
2	0	1	13	8	**18**	12	52	1.9	57.7
3	0	0	8	6	**15**	10	39	0	64.1
4	0	2	**13**	8	10	4	37	5.4	37.8
5	0	**9**	**8**	7	7	1	32	28.1	25.0
6	0	4	**30**	14	10	2	60	6.7	20.0
7	0	7	**24**	12	6	2	51	13.7	15.7
8	0	**23**	13	5	4	0	45	51.1	8.9
9	0	**10**	10	3	4	0	27	37.0	14.8
10	0	**17**	7	1	4	0	29	58.6	13.8
11	0	**27**	4	7	1	0	39	69.2	2.6
12	0	**14**	5	4	0	0	23	60.9	0
13-apex	3	**10**	0	1	0	0	14	92.9	0

The variation of the root diameter, and the diameter of the stele, along the main root axis is presented in Figure 2. The difference between the root diameter and the diameter of the stele represents the thickness of the cortex. As the cortex varied along the root, the external root diameter could not serve as a good measure of its water conducting capacity. The diameter of the stele increased gradually in the basipetal direction. Widening of the vascular cylinder was not uniform along the root, and abrupt increases in certain points were found.

The number of vessels and their diameter increased considerably from the root tip towards its base (Table 1). Such a change in xylem vessel diameter with increasing distance from the root apex of the tomato differs from what was described before by Fahn (1964) for roots of *Retama*, by Pate et al. (1995) for *Banxia prionotes,* and by Zimmermann (1983), and by Aloni and Griffith (1991) for shoots and roots of grasses. The present report is the first one to describe the variation of vessel diameter along an undisturbed aeroponically grown root. Therefore, it is too early to determine whether the discrepancy between our results and those reported for soil grown plants results from genetic differences between the studied plant species, or from the particular growth conditions.

The hydraulic diameter of each of the root segments, was estimated according to Equation (2). This parameter reflects the potential for water conductivity of the root at the examined point. If the gradient of the water potential along the root system is uniform, the hydraulic conductivity of the main root, at each point of branching, should equal the combined conductivity of the segments connected distal to that point.

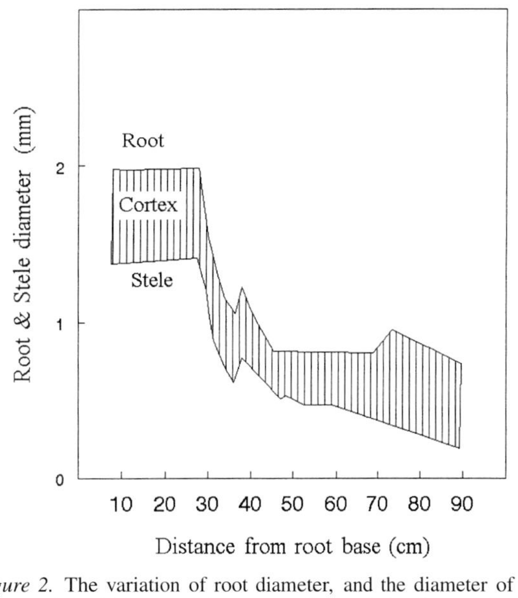

Figure 2. The variation of root diameter, and the diameter of the stele, and the thickness of the cortex, along a 0.9 m long main axis of a basal root of tomato.

Table 2. Changes in the utilization of conducting capacity along the main axis of a tomato root

Hydraulic diameter of the main axis segment (*Ds*) (μm^4)	THe sum of hydraulic diameters of proximal branches (ΣDs) (μm^4)	% utilisation
$3.60 \cdot 10^9$	$3.40 \cdot 10^9$	94%
$3.40 \cdot 10^9$	$2.50 \cdot 10^9$	74%
$2.20 \cdot 10^9$	$1.07 \cdot 10^9$	49%
$1.00 \cdot 10^9$	$4.93 \cdot 10^8$	49%
$4.90 \cdot 10^8$	$8.45 \cdot 10^8$	172%
$7.80 \cdot 10^8$	$7.40 \cdot 10^8$	96%
$6.40 \cdot 10^8$	$2.97 \cdot 10^8$	46%
$2.30 \cdot 10^8$	$2.40 \cdot 10^8$	104%
$2.00 \cdot 10^8$	$1.20 \cdot 10^8$	60%
$9.00 \cdot 10^7$	$7.35 \cdot 10^7$	82%
$7.00 \cdot 10^7$	$2.15 \cdot 10^7$	31%
$2.00 \cdot 10^7$	$5.60 \cdot 10^6$	28%

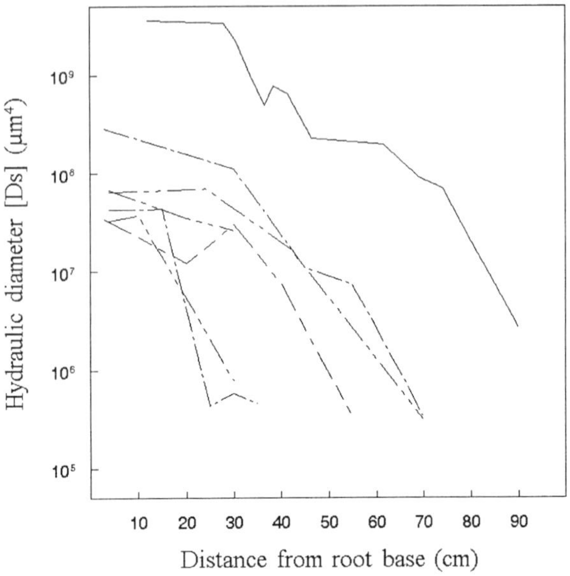

Figure 3. The variation along the root of root hydraulic diameter (Ds) of the main axis (solid line) and six second order laterals (dashed lines)

The *Ds* value of every root segment was compared with the sum of *Ds* values of the distal segments attached immediately next to it. This served as a measure of the utilisation efficiency of the conducting tissue. The data presented in Table 2 show such comparisons for 12 branching points along the main axis. In half of them utilisation was 60% or lower and only one point showed the inverse situation. The results of such analysis for 20 branching points along the laterals attached to this axis were similar. Among these points the utilisation in five was less than 20%, in additional five utilisation was between 30% and 60%, eight more exhibited utilisation between 70% and 130%. Two points indicated sharp constrictions where the sum of the distal segments was twice and three times larger than that of the proximal axis. Thus, the presented data indicate that there is a considerable degree of redundancy in the structure of the water conducting tissue. In half the points less than 60% of the capacity were being utilized. Still in many segments, the *Ds* of the main root was nearly equal to the sum of the distal segments. Several points of constriction, where *Ds* of the main axis was significantly smaller than that of the segments immediately distal to it, were identified.

The calculated hydraulic diameter *Ds* changed along the main root by five orders of magnitude (Figure 3), from a value of $2.6 \cdot 10^6$ μm^4 near the main root apex, to $3.6 \cdot 10^9$ μm^4 at the root's base. Along second order laterals *Ds* changed by two orders of magnitude. Variations in hydraulic conductivity along roots were reported in very few cases before. Melchior and Steudle (1995) found that the axial hydraulic conductivity of onion roots changed by two or three orders of magnitude along the apical 200 mm segments. Pate et al. (1995) reported that the specific hydraulic conductivity of a *Banksia prionotes* root increased 20 fold along a 130 cm section of a single sinker root. Thus, the rate of change in hydraulic conductivity along the tomato root is not uncommon.

The *Dh* value of every xylem vessel is proportional to the 4th power of its radius. Thus, when calculated on the basis of the vessel diameter, most of the hydraulic conductivity of roots is attributed to a few large vessels, and the contribution of the many narrow ones is negligible. However, the validity of upscaling this rule from individual capillaries to the entire root should be reevaluated in view of the hazard of cavitation of xylem elements. Roots of hardwoods were reported to be more vulnerable to xylem cavitation than the stems (Sperry and Ikeda, 1997). Under such conditions some, or all of the large xylem elements, may not be able to conduct water at all. This is a very severe danger because embolism is considered to be more persistent in the larger capillaries than in small ones (Yang and Tyree, 1992). In such cases when the few large vessels, that are responsible for most of the conducting capacity, are incapacitated by embolism or by a damage caused by pathogens, the plant relies on conductivity of the narrow elements for its water supply. Therefore, judging the potential water conducting capacity by vessel diameter alone may seriously underestimate the contribution of the narrow xylem elements to the hydraulic conductivity of the roots of field grown plants.

The observed changes in the hydraulic diameter along the axis, indicate that there are sites of constriction along tomato roots. We suggest that such constrictions should be of value for tomato plants under water stress conditions by prevention of the spreading of embolism. Constrictions may thus play an important role in the water conducting system of tomatoes, by preserving the continuity of the xylem water column, during periods of high water stress.

The high degree of redundancy in the structure of the water conducting system of the tomato plant reflects the overruling influence of the long-term safety considerations over those of the current efficiency. The same principle was shown by Eshel (1992) to govern overall bean root growth, when subjected to various water and oxygen levels. The excess of *Ds* in many segments of tomato roots and the constrictions that are found in other segments along the root both point to that same explanation.

Acknowledgement

This investigation was supported by the BMBF (The German Ministry for Sciece Research) and the University of Bielefeld. Special thanks are due to the indefatigable Prof. S.W. Breckle for the coordination of this German- Egyptian-Israeli and Jordanian research project.

References

Aloni R and Griffith M 1991 Functional xylem anatomy in root-shoot junctions of six cereal species. Planta 184, 123–129.

Eshel A 1992 Response of plant roots to water and oxygen stresses. Biel. Okol. Beitr. 6, 23–32.

Eshel A and Waisel Y 1996 Multiform and multifunction of various constituents of one root system. *In* Plant Roots: The Hidden Half. Second edition revised and expanded. Eds Y Waisel, A Eshel and U Kafkafi. pp 175–191. Marcel Dekker Inc., New York, USA.

Fahn A 1964 Some anatomical adaptations of desert plants. Phytomorphology 14, 93–102.

Fitter A 1996 Characteristics and functions of root systems. *In* Plant Roots: The Hidden Half. Second edition revised and expanded. Eds Y Waisel, A Eshel and U Kafkafi. pp 1–20. Marcel Dekker Inc., New York, USA.

Lewis A M and Boose E R 1995 Estimating volume flow rates through xylem conduits. Amer. J. Bot. 82, 1112–1116.

Melchior W and Steudle E 1993 Water transport in onion (*Allium cepa* L.) roots. Changes of axial and radial hydraulic conductivity during development.

Nobel P S 1991 Physicochemical and Environmental Plant Physiology. Academic Press Inc., San Diego, CA. USA. 635 p.

Nobel P S 1996 Ecophysiology of roots of desert plants, with special emphasis on agaves and cacti. *In* Plant Roots: The Hidden Half. Second edition revised and expanded. Eds Y Waisel, A Eshel and U Kafkafi. pp 823–844. Marcel Dekker Inc., New York, USA.

Pate J S, Jeschke W D and Aylward M J 1995 Hydraulic architecture of the dimorphic root systems of South-West Australian species of Proteaceae. J. Exp. Bot. 46, 907–915.

Sperry J S and Ikeda T 1997 Xylem cavitation in roots and stems of Douglas-fir and white fir. Tree Physiol. 17, 275–280.

Tyree M T, Patino S, Bennik J and Alexander J 1995 Dynamic measurements of root hydraulic conductance using a high pressure flowmeter in the laboratory and in the field. J. Exptl. Bot. 46, 83–94.

Waisel Y 1996 Aeroponics: A tool for root research. *In* Plant Roots: The Hidden Half Second edition revised and expanded. Eds Y Waisel, A Eshel and U Kafkafi. pp 239–245. Marcel Dekker Inc., New York, USA.

Waisel Y and Eshel A 1991 Multiform behavior of various constituents of one root system. *In* Plant Roots: The Hidden Half. Eds Y Waisel, A Eshel and U Kafkafi. pp 39–52. Marcel Dekker Inc., New York, USA.

Yang S and Tyree MT 1992 A theoretical model of hydraulic conductivity recovery from embolism with comparison to experimental data on *Acer saccharum*. Plant, Cell Env. 15, 633–643.

Zimmermann M H 1983 Xylem Structure and the Ascent of Sap. Springer Verlag, Berlin, Germany. 143 p.

A. Stokes (ed.), The Supporting Roots of Trees and Woody Plants: Form, Function and Physiology, 377–389.

Root structure and arbuscular mycorrhizal colonization of the palm *Serenoa repens* under field conditions

Jack B. Fisher[1,*] and K. Jayachandran[2]
[1]*Fairchild Tropical Garden, 11935 Old Cutler Rd., Miami, FL 33156, USA and Department of Biological Sciences, Florida International University, Miami, FL 33199, USA and* [2]*Dept. of Environmental Studies and Southeast Environmental Research Center, Florida International University, Miami, FL 33199, USA*

Key words: arbuscular mycorrhizae, palms, root anatomy, root architecture, *Serenoa*, soil profiles

Abstract

Serenoa repens (Bartr.) Small is a palm native to the southeastern USA. It is a common understory plant in pine communities on both acid sands and alkaline limestone. Roots have only primary growth and range in thickness from 8.0 mm (first order roots from the stem) to 0.8–2.9 mm (ultimate roots of third to fifth order). The thickest roots occur at soil depths >20 cm; fine roots (<1.2 mm) occur at all depths (1–60 cm). Some second and third order roots are negatively geotropic and grow up to the mineral soil surface. The epidermis of all roots has a thick, eventually lignified outer wall. Except for the thinnest, all roots have a single-layered, thick-walled exodermis, which is first suberized and later lignified. Root hairs are never present. A hypodermis composed of several layers of lignified cells (up to 8-cells-thick) is next to the exodermis and forms the outer cortex. Radial series of thin walled and slightly lignified cells sporadically occur in the outer cortex of the thinnest roots, but there are no passage cells in the exodermis, which is continuous. The remaining inner cortex is composed of unlignified parenchyma with air canals and a completely lignosuberized endodermis in old roots. Passage cells were seen the the endodermis of the some of the thinnest roots. Arbuscular mycorrhizal (AM) fungi occur in the outer one-third of the cortical parenchyma adjacent to the hypodermis. Fungal coils, arbuscules and vesicles are found most frequently in the thinnest roots, but also occur sporadically in all root orders. Cells a few mm from the apical meristem are sometimes colonized. At sites of appressoria, coils of AM hyphae occur within an epidermal cell and exodermal and hypodermal cells beneath. Intercellular hyphae with intracellular branch arbuscules (*Arum*-type) are common in the inner cortex. There is evidence of a dieback of the highest order roots during the winter dry season. Profiles of soil and roots have the highest density of AM spores in the surface 10 cm layer. Total AM spore density ranged from 130 to 1100 spores per 50 g soil in different samples. *Glomus* spp. dominated followed by *Gigaspora* spp. The findings are related to a more general understanding of growth and AM colonization in long-lived roots of tropical woody monocotyledons. Palm roots, in particular, are slow growing and are protected by massive hypodermal layers.

Introduction

Palms are woody perennial monocotyledons, whose roots are known to be mycorrhizal (St. John, 1988; Zona, 1996). Mycorrhizal colonization promotes palm growth on nutrient poor native soil (Janos, 1977) and is potentially significant in the ecology of wild palms and in the cultivation of ornamental palms. Although palm roots lack secondary growth, they are long-lived and often woody and massive in size. There are few descriptions of palm root architecture and little information on their arbuscular mycorrhizae (AM) association, possibly because palm roots are difficult to study because of lignified tissues and silica bodies, as well as their predominantly tropical distribution (Tomlinson, 1990).

* FAX No: +1 (305) 665-2840. E-mail: fisherj@fiu.edu
Plant and Soil is the original source of publication of this article. It is recommended that this article is cited as: *Plant and Soil* **217:** 229–241, 1999.

We report on root structure and mycorrhizae of *Serenoa repens* (Bart.) Small, the saw palmetto, a prostrate understory palm ranging from warm temperate to subtropical climates in the southeastern United States. The palm is common both in pine rockland vegetation on shallow alkaline soils over limestone and in pinelands on deep acidic sandy soil. It is often the dominant understory plant beneath various species of *Pinus*. *Serenoa* is also significant in producing fruit that has been shown to be of medicinal value (Bennett and Hicklin, 1998) and is becoming significant in local economies.

This study will provide a foundation for future experimental research on palm mycorrhizae. Therefore, the pathway of fungal penetration and relationship of root anatomy to the presence of AM are emphasized.

Materials and methods

Habitat

The research site is a remnant of undisturbed natural pine rockland vegetation in Miami-Dade County, Florida. The original overstory of *Pinus elliottii* Engelm. var. *densa* Little & Dorman was eliminated by a bark beetle outbreak in 1993 following Hurricane Andrew, five years before this study. Although dead pine trunks remain, there is little seedling regeneration. The following plants grow within 10 m of the collection site: *Bursera simaruba* (L.) Sarg., *Myrica cerifera* L., *Pithecellobium keyense* Britton ex Britton & Rose, *Quercus virginiana* Mill., *Rhus copallinum* L. The site was chosen because *Serenoa* plants grow in an area of deep sand and do not have other native palms (*Sabal palmetto* (Walter) Lodd. ex Schult. & Schult. f. and *Coccothrinax argentata* (Jacq.) L.H. Bailey) nearby which might complicate root sampling.

Soil

The site has shallow sandy soils over a bed of oölitic limestone. The surface 10 cm is a white sand; pH = 6.2–6.8; available P = 5.9–8.8 mg/kg; total P = 28–35 mg/kg (Bray and Kurtz, 1945). The soil at 20–50 cm depth is a yellow-orange sand; pH = 6.3–8.2; available P = 11.5–17.3 mg/kg; total P = 12–35 mg/kg.

Collection of roots and soil

Field study: A trench was dug perpendicular to a single prostrate palm trunk, starting about 20 cm behind its distal portion in contact with the soil, in late

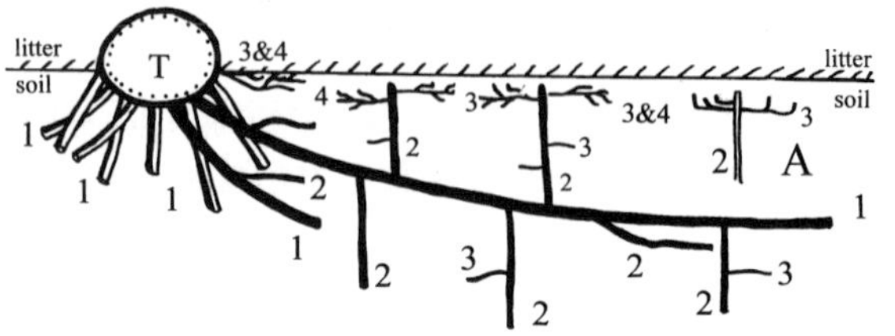

Figure 1. A. Diagram of root architecture. Soil surface and layer of litter (hatched) are indicated. Root orders (1–4) and horizontal trunk (T) are labeled. Order 2 roots may be horizontal or vertical (growing either up or down). Inset A shows a vertical Order 2 with its Order 3 & 4 roots becoming erect at their tips).

December 1997. This region of trunk had fully mature adventitious roots. Samples of soil (10×10×10 cm) and roots were cut from the trench wall and used for analyses of spores and root densities. The soil profile was sampled 0-50 cm deep and up to 120 cm from the trunk. The newly exposed trench wall was washed gently with a water spray to expose roots in natural positions.

Soil samples (1 litre volume) from the side of the trench were wet sieved down to 32 μm particle diameter, and fungal spores were separated and counted by wet sieving (Brundrett et al. 1996). Root density was determined by washing roots free of soil in 1 litre soil samples, sorting into four diameter classes, and measuring total root length per soil volume for four root diameter classes using a morphometric video analysis system ('AgVision', Decagon Devices Inc., Pullman, WA).

Greenhouse study: Seedlings were first grown in sterilized soil mix for 12 months. They were transplanted into 2-liter pots with fresh, unsterilized native sandy soil taken from the field trench and contained wild *Serenoa* root fragments. After 5 months, roots were collected and used for histological study. Anatomical descriptions are based on both roots from wild, field-grown mature plants and roots of potted, greenhouse-grown seedlings colonized by AM fungi.

Microtechnique

Root anatomy was examined in both fresh and FAA-fixed roots using hand sections stained with aqueous toluidine blue. Histochemical stains were used following the procedures of (Perumalla et al., 1990): HCl-phloroglucinol for lignin; sudan IV for suberin; fluorescence of berberine sulfate and aniline blue combination (Brundrett et al., 1988), and natural auto-

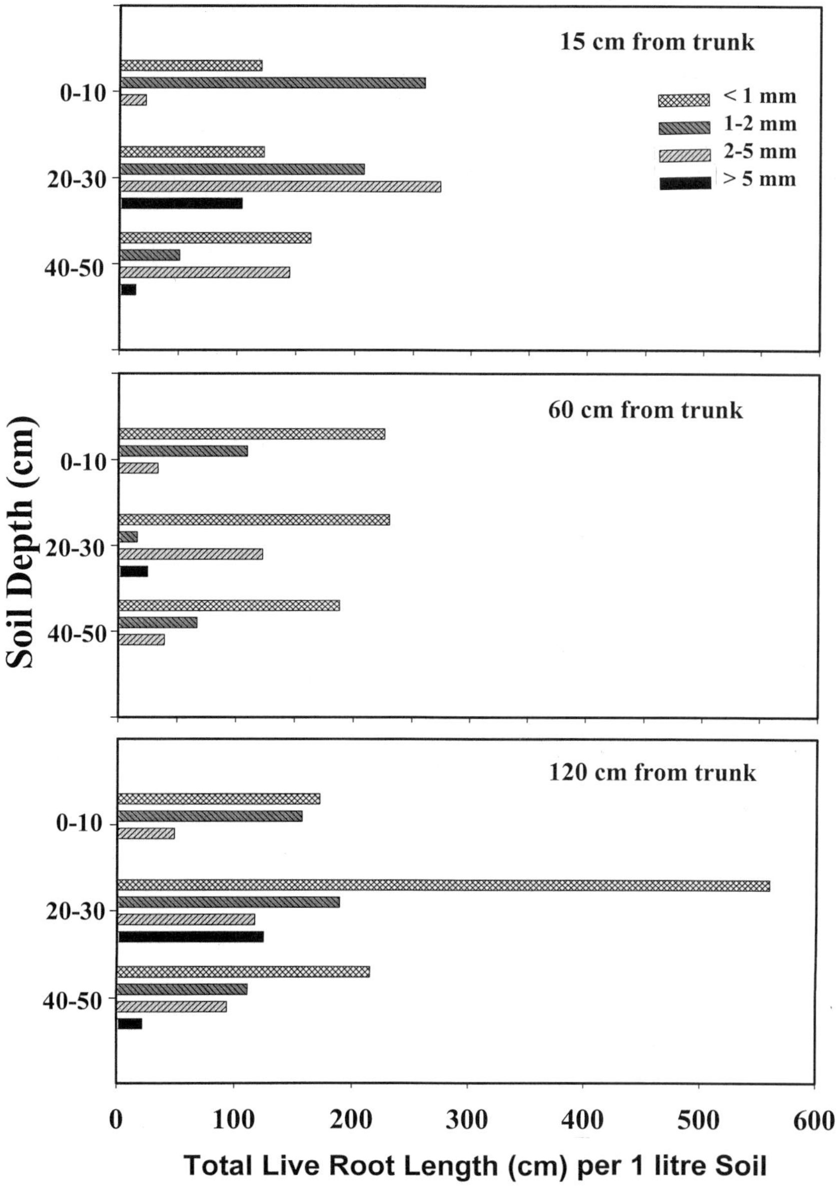

Figure 2. Total length of live roots found in 1 litre of soil in the profile of a trench from one plant. Bar graphs show four different root diameter classes at each of three soil depths at three distances from the trunk (above each other) along the trench. Bars are root diameter classes from thinnest (top) to widest (bottom): <= 1 mm; >1=2 mm; >2<5 mm; >=5 mm; respectively.

fluorescence for cell walls using a Leitz Ortholux II fluorescence system with wide band UV and narrow band blue filters.

For observing AM fungi, roots were cleared in KOH, bleached with NH_4OH-H_2O_2, and stained with trypan blue in acidic glycerol (Brundrett et al., 1996). The cortex of most roots was difficult to clear and observe through the thick lignified hypodermis. Whenever possible, roots were cut into longitudinal or transverse slices before processing.

Results

General mature architecture of root system and profile of root distribution

The stem of *Serenoa* is typically a horizontal axis ca. 8–15 cm in diameter. The plant produces vegetative lateral buds and eventually forms a cluster of prostrate trunks radiating from the original seedling site (Fisher and Tomlinson, 1973).

A generalized diagram of the root system of a mature, field-grown palm is shown in Figure 1. All roots are adventitious in origin on the lower one-third of the circumference of the horizontal trunk which is below the level of leaf litter. These roots are the thickest and are designated Order 1. They initially grow downwards or obliquely, and then become horizontal at soil depths >15 cm when not constrained by surface rocks (Figure 2). The living roots have a reddish surface color and are common in the 20–30 cm deep profile. The entire architecture is distorted by either very shallow soils or by localized deep pockets of sand which effectively act like deep flower pots.

Order 2 roots are either positively gravitropic, horizontal, or negatively gravitropic. Some erect Order 2 roots were more than 30 cm long. Order 3 roots are usually horizontal and eventually bear Order 4 roots as ultimate short feeder roots. This is especially noticable as the soil is removed where vertical Order 2 roots end within 1–2 cm of the soil surface, at the boundary of the compact humus layer and the loose leaf litter. The distal end of this non-growing, erect root may have a root cap or shows an aborted or damaged apex. Order 3 roots are clustered in the few cm behind the erect end. There is an obvious mat of horizontal fine roots (Orders 3 and 4) starting 6–7 cm below the soil surface and extending down to the 20–30 cm depth, as seen in Figure 2 especially in the fine roots at 120 cm distance. In some locations, Order 3 and 4 roots tend to bend up and end as an erect short root at the soil surface (inset A in Figure 1). Generally, the four diameter classes presented in Figure 2 represent Orders 1–4, respectively. However, root fragments are difficult to accurately classify. Roots <1 mm include both Order 4 and 5.

Root 'branch angles' were not measured. However, we observed lateral roots arise at a wide range of angles, both with respect to gravity and to the orientation of the parent root. These different angles are show in Figure 1.

In vigorous root systems, Order 5 roots may occur, but they are short and often swollen or tubercle-like. Order 5 roots are more common in potted seedlings and in vigorous roots within deep natural sand pockets. When roots are sampled at the start of the dry season (December 1997), there are many detached dead or short and shriveled Order 5 roots which were lost and not counted. In the early rainy season (June 1998) there are many white and growing Order 4 and 5 roots, which were not seen when the soil samples were collected earlier (December). This observation suggests a seasonal die-back that may start in the drier surface soil layer and is reflected in the greater number of finest roots below 10 cm (Figure 2).

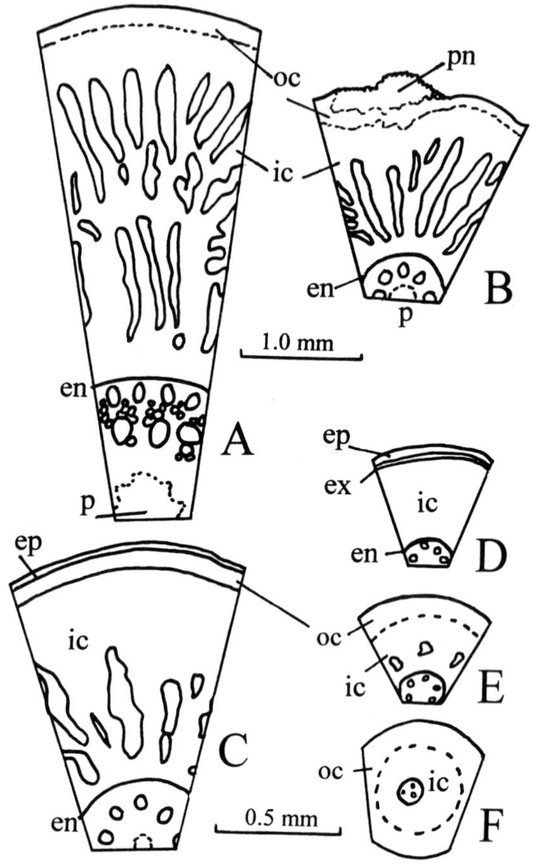

Figure 3. Serenoa roots, different orders in transverse section to show organization of tissues. **A**. Order 1 old root after epidermis is lost; **B**. Order 2 including a pneumathode region; **C**. Order 3; **D**. Order 4 with little lignification of the outer cortex; **E**. Order 4 with poorly defined exodermis and some lacunae in the cortex; **F**. Order 5 without a clearly defined exodermis. Abbreviations: en – endodermis, ep – epidermis, ex – exodermis, ic – inner cortex, oc – outer cortex, p – pith, pn – pneumathode. Scale line for A & B = 1.0 mm; for C–F = 0.5 mm.

Between the old dead leaf bases of the trunk, roots sometimes occur which are the size of Orders 2 and 3, apparently arising directly from the trunk. The ultimate roots are mainly Order 4 and sometime Order 5. However, it is unclear whether these are the only feeder roots or whether Order 3 are also feeder roots. Ultimate roots are very brittle and fragile; they easily detach during extraction. Thin roots, equivalent in diameter to Order 3 or 4, may also arise occasionally on Order 1 and 2 roots, respectively (Figure 1).

Table 1. Characteristics of different orders of roots of *Serenoa*

Features	Order 1	Order 2 Erect	Order 2 & 3	Order 4 & 5
Diameter (mm)	6.0 – 8.0	3.2 – 5?	1.4 – 4.0	0.87 – 1.18
Surface color	Reddish brown	Reddish brown	Brown	White to brown
Pneumathodes	Present	Present	Present	Rare
Exodermis	Present	Present	Present	Present but not always thickened
Outer cortex (= hypodermis)	Lignified	Lignified	Lignified	Sometimes unlignified
Inner cortex width (mm)	2.68	1.00	0.17 – 1.0	0.21 – 0.45
Inner cortex lacunae	Present	Present	Present	Present or absent
Endodermis	Present	Present	Present	Present ± passage cells
Stele diameter (mm)	2.0 – 2.13	0.93	0.51 – 0.60	0.10 – 0.18
Number phloem poles	28	18	12–13	4
Pith parenchyma	Present	Present	Present or absent	Absent

Anatomy and development roots of different orders

Although roots contain only primary tissues, significant changes occur during aging. The diversity of root size and histology for different Orders are summerized in Table 1 and diagramed in Figure 3.

Root caps remain attached after roots stop growth and their apical meristems have matured. The root surface shows regular, transverse rings of darkened, dead cells which represent root cap remains, perhaps due to rhythmic elongation. These dark collars of dead, pigmented root cap cells give palm roots a characteristic appearance. The root cap is formed by a distal layer and border cells of the apical meristem according to von Guttenberg (1968), although we did not confirm this. The root cap cells extend back to varying distances, sometimes forming a visible tubular sheath covering short longitudinal segments of the root. In inactive roots, the root cap is small and suberized, forming a restrictive cap. It seems that when a root resumes growth, the new tissues push through the suberized cap which is left behind, and a slight constriction in root diameter persists.

The epidermis (also called the 'rhizodermis') is one-celled thick, sometimes with remnant root cap fragments to the outside (Figure 4). All roots lack root hairs. The epidermal cells are papillose with a thickened but initially unsuberized and unlignified outer cell wall. In larger roots, the epidermis becomes lignified (Figure 8).

All root orders also exhibit an irregular longitudinal pattern of white rings of surface eruptions. These are a result of proliferation of subepidermal cells and a rupturing of the epidermis (Figure 3B). The proliferated cells are spherical, thin-walled, highly suberized, and with many intercellular spaces (Figure 9). Thick-walled exodermal and outer cortical cells appear to enlarge and may undergo one or two internal paradermal cell divisions. These lenticel-like structures are called pneumathodes (Seubert, 1997) because of their presumed function in gas exchange. They are most common in roots near the surface and on erect Order 2 roots.

The cell layer adjacent to the epidermis is first distinguished from the epidermis as an exodermis by the thickening of cell walls and development of suberin in the anticlinal walls. With age, all exodermal cell walls become equally thickened and both suberized and lignified (Figures 4, 7). We found no evidence of thin-walled or unsuberized passage cells.

The multilayered nature of the secondary wall and quality of staining with toluidine blue clearly distinguish these exodermal cells from adjacent outer cortical cells which are thick-walled with fewer or no wall layers (Figures 4, 5). Except for the highest order roots, these outer cortical cells mature into a hard, lignified hypodermis (Figures 4–8, 18). The thinnest Order 4 and 5 roots tend to lack a clear hypodermis and have a weakly developed exodermis (Figure 13), but some old Order 5 have a lignified hypodermis. In old roots, the epidermis disintegrates leaving the exodermis as the outer boundary with the soil.

The thickest roots have radially aligned lacunae in the inner cortex and variable number of cell layers between the lacunae and the endodermis. Thicker roots are aerenchymous but hard due to the thick shell of the hypodermis (Figure 3A–C).

As root tissues mature, endodermal cells opposite the protophloem poles are first to differentiate with the development of first suberin in the inner one-third of the radial cell walls followed by lignification of radial (Figures 12, 14) and then inner tangential walls (Figures 10, 11, 15). Differentiation of the endodermal cells progresses toward the xylem poles (Figure 11) and passage cells occur in young roots. Eventially, all endodermal cells become lignified and have thickened inner and thin outer tangential walls (Figure 10), although passage cells are found irregularly in the smallest roots (Figure 15).

The stele or vascular core has a one-layered pericycle and multiple protoxylem and protophloem poles, depending upon the diameter of the roots (Table 1; Figure 10). Wide metaxylem vessels remain immature with cytoplasm at a distance of several cm behind the apical meristem in wide roots. At these positions, the hypodermis, endodermis and central region of the stele are fully lignified. The stele center of thin roots has sclereids. Widest roots (> 2–3 mm) have a well-differentiated pith composed of large, thin-walled parenchyma surrounded by thick-walled, smaller sclereids (Figure 3).

Mycorrhizal fungal colonization

The root surface is covered with a variety of non-septate fungal hyphae. Narrow, golden-brown hyphae anastomose in the valleys of the epidermal cells thus following the pattern of anticlinal epidermal walls. Wider unpigmented hyphae, which stain darkly with trypan blue, occur on the epidermal surface and on root cap remnants. Various forms of appressoria occur, ranging from a hyphal plexus with swelling on the epidermal surface (Figure 22) to an absence where a hypha enters directly into the epidermal cell with no external modification (Figures 18–20). The hyphae tend to form a coil within one (rarely two or three) adjacent epidermal cells and coils or loops within the adjacent exodermal and hypodermal cells (Figures 24–29). In surface view, there is no consistent pattern of appressoria on the root, which is consistant with a uniform epidermis and an exodermis that has no passage cells.

Beneath each appressorium, a hypha invades an epidermal cell, an adjacent exodermal cell and then a radial series of outer cortex cells, which may have thick walls or may be part of the sporadic series of thin-walled cells. It is unclear if these cells were penetrated before or after adjacent exodermal and hypodermal cell walls were suberized and lignified. Intercellular hyphae were not observed in the compact cells of the outer cortex (Figures 16, 17). Hyphae travel mainly in the longitudinal intercellular spaces of the inner cortical parenchyma. Adjacent cortical cells, positioned both longitundinally and transversely to the root axis, are invaded by coils (Figures 21, 23), hyphal branches with arbuscules, and vesicles. Vesicles appear to be both inter- and intracellular.

In all root orders, hyphae and arbuscules are concentrated in the peripheral parenchyma cells of inner cortex (Figures 16, 17). AM hyphae are variously distributed, either completely around the circumference, in a sector when present in thin roots (orders 4–5), or in the peripheral half to three-quarters of the inner cortex. Fungi rarely occur within 2–3 parenchyma cells of both the endodermis and the thick-walled outer cortical cells. In thicker, higher order roots (Orders 1–3), limited AM fungi occur in the outer one-third or one-fourth of the inner cortex, to the exterior of the radial air lacunae.

AM fungi are never observed in the stele or endodermal cells. In old roots, hyphae (some septate and possibly saprophytic) grow on the surface of the lacunae and enter cells of the inner cortex adjacent to the endodermis but do not penetrate endodermal cells.

Because of the thick, lignified layer of exodermal and outer cortical cells, whole root segments can not be reliably cleared and stained to document presence of internal AM. Thick roots clear and stain well after they are cut into longitudinal slices to expose the cortex. Thin roots are problematic, especially when old and sclerified. Thick cross sections that are processed like whole roots are stained most reliably, but colon-

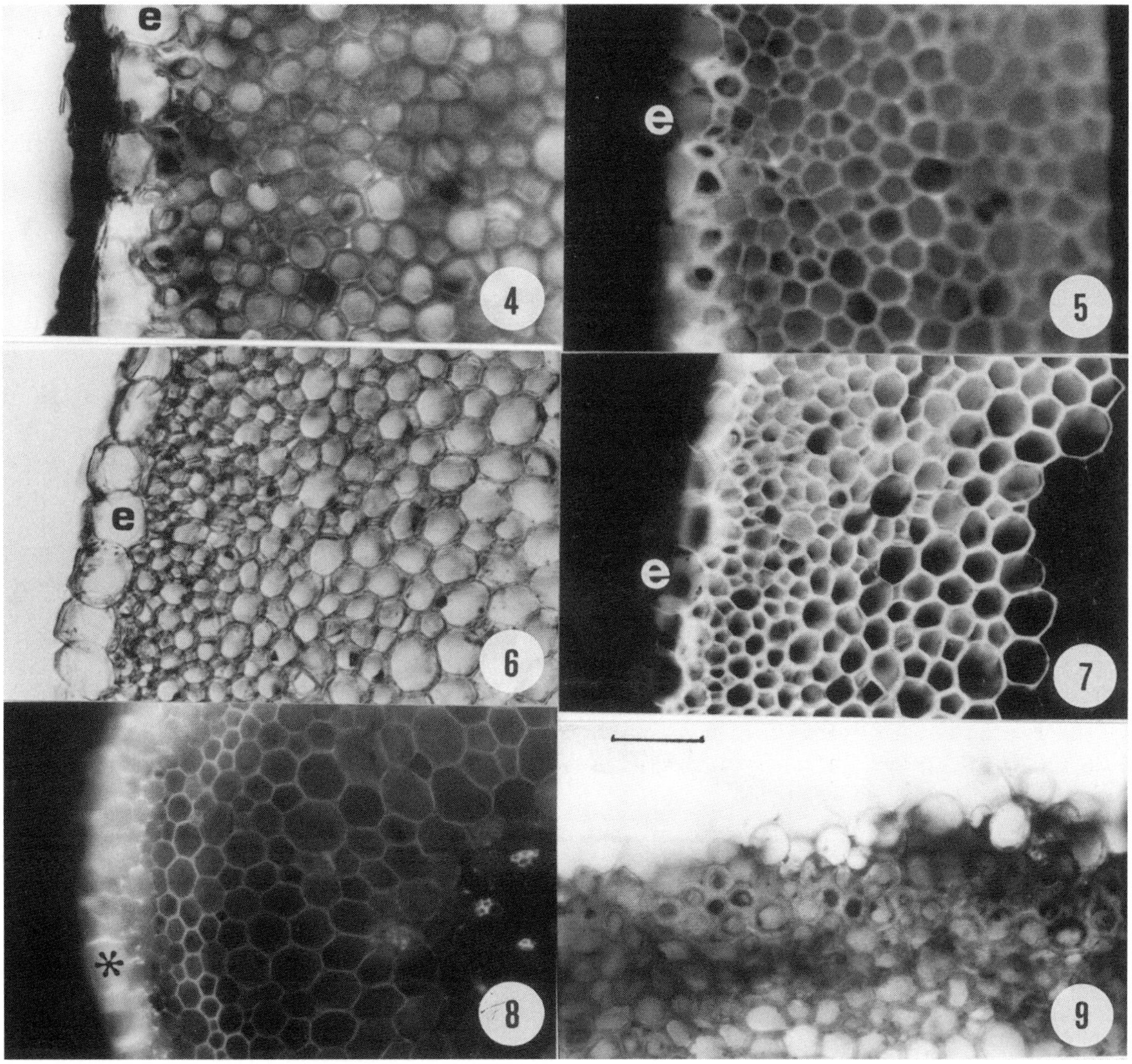

Figures 4–9. Roots in transverse section; all stained with berberine-analine blue. 4. Young Order 1 with epidermis on left and outer cortex on right; dark layer on the epidermis is the crushed remains of root cap; viewed with white light. 5. Same section viewed UV light; thick walls of exodermis are obvious, with no fluorescence of root cap and little for the epidermis. 6. Old Order 2 with exodermis and outer cortex similar in wall thickness when viewed with white light. 7. Same section viewed with UV light; walls of epidermis (inner walls only), exodermis and outer cortex all lignified. 8. Young Order 3 viewed with UV light; showing lignified epidermis and exodermis, inner cortex with some autofluorescence; endodermis with only casparian strips. 9. Surface of pneumathode region of Order 2 with surface dead cells and underlying thick-walled suberized cells; viewed with white light. Figures 4–9 same magnification. Scale line in Figure 9 = 50 μm. Asterisk (*) – lignified epidermis, e – epidermis.

ization along root length cannot be easily followed. Our inability to clear the exodermal/hypodermal layer prevents the use of the grid point method to determine percent colonization, which is standard practice for other plant species.

Soil characterization and fungal spore density in profiles

The soil is typical of a sand pocket in the pine rockland community. The substrate is oölitic limestone that forms visible outcrops and generally has a shallow overlayer of sandy soil with a thin surface layer of organic litter, depending upon the length of time since the last burn of this fire climax community. Leaf litter

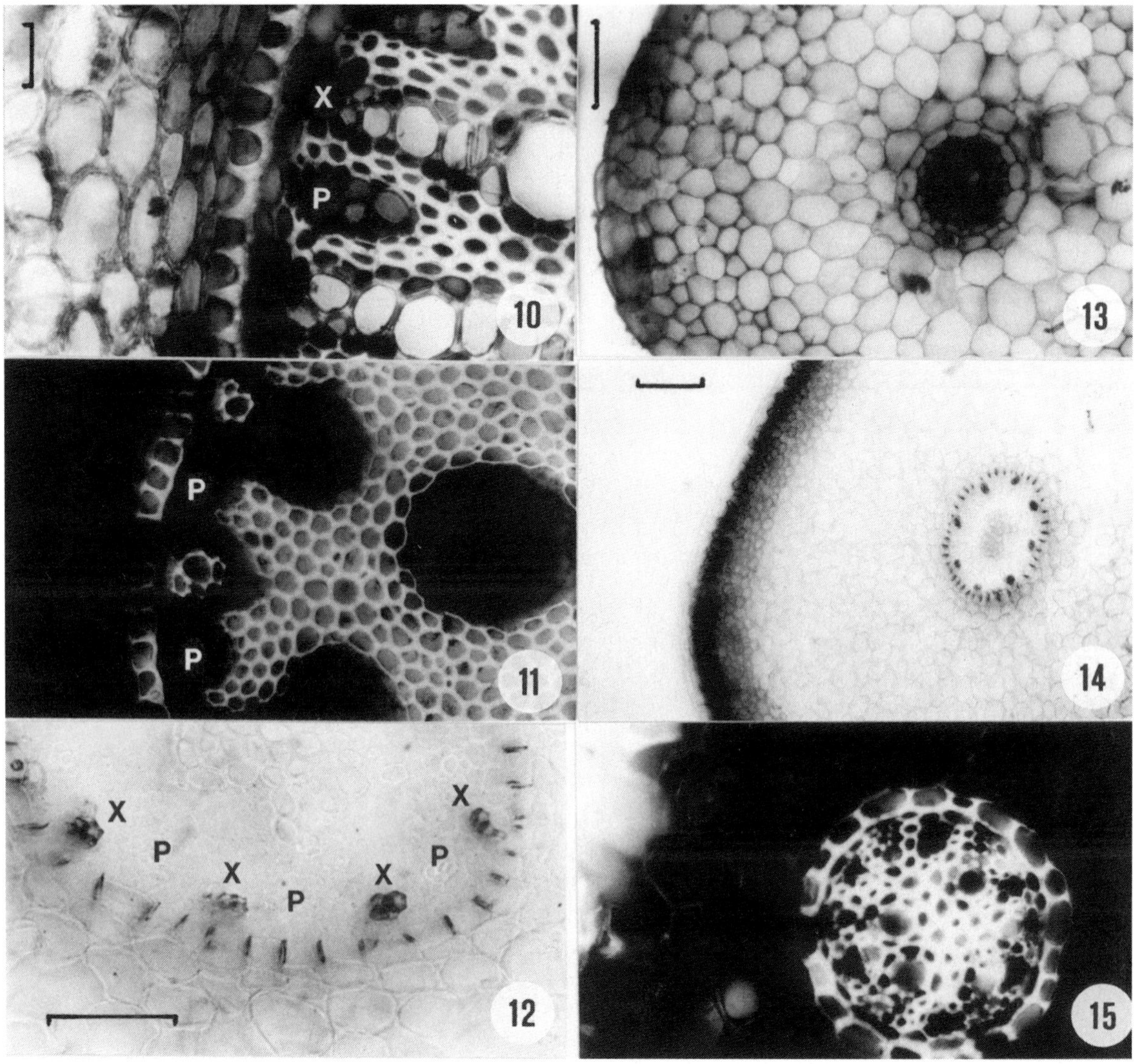

Figures 10–15. Root steles in transverse section; all stained with berbeine-analine blue, except for Figure 13 which is stained with toluidine blue. 10. Mature Order 1 with thick-walled lignified endodermis; viewed with white light. 11. Young Order 1 with developing endodermis that is thick-walled only opposite phloem poles; viewed with UV light. 12. Young Order 3 or 4; endodermis with only tangential walls lignified; lignin stained darkly at xylem poles (x); viewed with white light. 13. Mature Order 5 lacking a defined exodermis but with a clear endodermis; viewed with white light. 14. Young Order 3 or 4 with lignified outer cortex and radial walls of endodermis; lignin stained darkly; viewed with white light. 15. Mature Order 4 or 5, with lignified endodermis; viewed with UV light. Abbreviations: p – phloem pole, x – xylem pole. Scale lines = 200 μm in Figure 10; 50 μm in Figure 13; 100 μm in Figure 14; and 50 μm in Figures. 11, 12, 15.

and twigs reach a depth of 2 cm. The top soil layer is a white sand which changes to yellow-orange sand at about 20–30 cm depth. The pH is neutral to slightly acidic and is low in available phosphorus.

Frequency of AM fungal spores is presented in Figure 30. The highest spore counts were near the soil surface and nearest to the trunk, decreasing with depth and distance from the trunk. AM spore number (Figure 30) does not show a close correlation with root length distribution (= root density), especially with the thinnest size class at the 20–30 cm depth (Figure 2).

Preliminary observations with the dissecting microscope indicated that spores collected on the 250 μm sieve were mostly the genus *Gigaspora* or *Scutelospora.* These spores were typically large with a bulbous base. Spores smaller than 250 μm were mostly *Glomus* spp. We did not observe obvious *Acaulospora, Entrophospora*, or *Sclerocystis*.

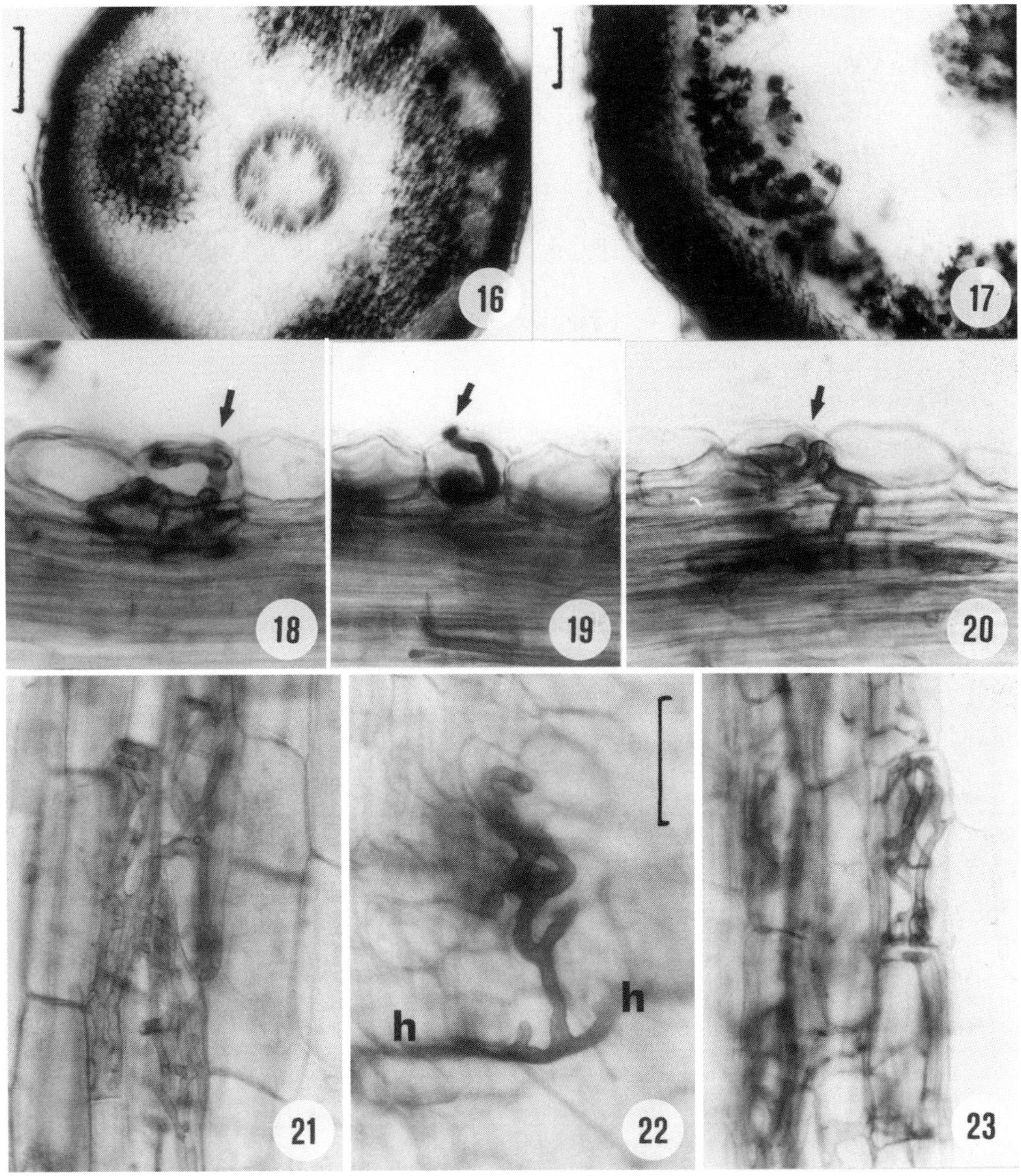

Figures 16–23. Roots in thick transverse and longitudinal sections; all stained with trypan blue and viewed with white light. 16. Order 3 with patchy distribution of AM fungi in cortex. 17. Order 3 or 4 with AM fungi in peripheral region of outer cortex. 18–20. Hyphal penetration of three different epidermal cells; in longitudinal view with epidermis at top. 21. Cortical cells with intracellular hyphae. 22. External hypha at site of appressorium (microscope focused above the epidermal surface). 23. Cortical cells with inter- and intracellular hyphae. Scale line = 200 μm in Figure 16; 100 μm in Figure 17; and 50 μm in Figures 18–23. Arrow – site of hyphal penetration, h – hyphae external to root and continuing out of focus.

Discussion

Root structure

The architecture and distribution of roots within the soil profile are similar to those described for African oil palm (*Elaeis guineensis*) by Purvis (1956) and Jourdan and Rey (1997), although the finest roots (presumably feeder roots) of *Serenoa* occur more evenly in all soil depths (0–50 cm) rather than mainly in the surface 20 cm reported for *Elaeis* (Tailliez, 1971) and

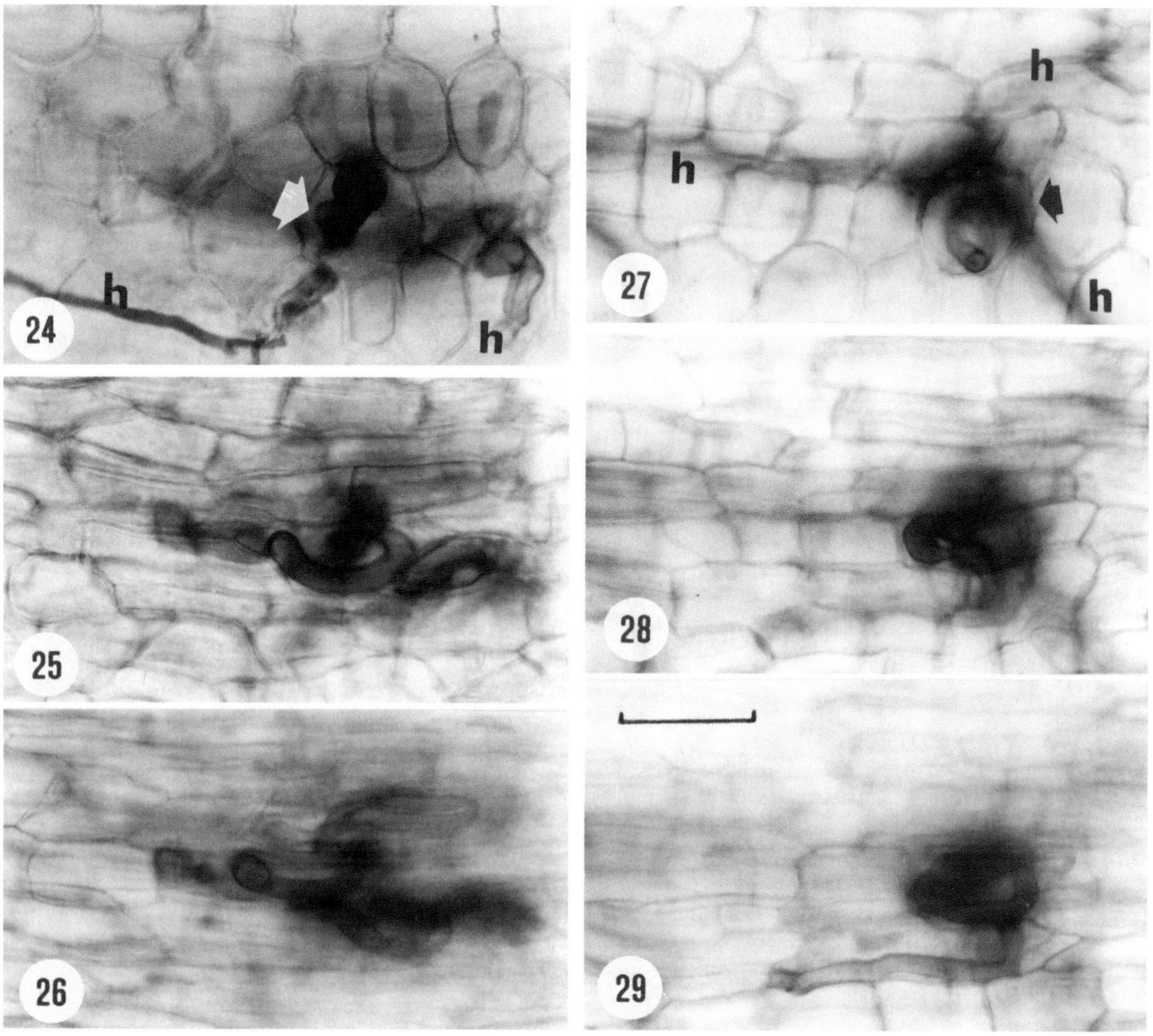

Figures 24–29. Roots in surface view; all stained with trypan blue and viewed with white light. 24–26. Three images of the same appressorium at focal planes of the epidermis, exodermis and next cell below, respectively. 27–29. Three images of the same appressorium at focal planes of the epidermis, exodermis, and next cell below, respectively. All at same magnification. Scale line in Figure 29 = 50 μm. Arrow – site of hyphal penetration, h – hypha external to the root and continuing out of focus.

Bactris (Vandermeer, 1977; Ferreira et al., 1990; Ferreira et al., 1995). The more uniform distribution of fine roots in *Serenoa* may be due to the well-drained sandy substrate as opposed heavy clay. However, the distribution of feeder roots may vary during the growing season. We found many old and senescing Order 4 and 5 roots in December (end of wet season) and many new, growing higher order roots in June (start of wet season), suggesting an annual turnover of feeder roots. This observation must be substantiated by following feeder root growth and death throughout the year. Thus, the distribution of living (non-shriveled and solid) roots in Figure 2 may only represent the winter state of live roots, in which few or no roots < 1 mm were found. In *Elaeis*, the finest roots elongated more slowly in adult plants than in seedlings, i.e. short lateral roots in Table 3 versus 1 in Jourdan and Rey (1997). In addition, these workers found that Order 4 roots (= 'ultimate unbranched roots') were self-pruning, living for one month or less (Jourdan and Rey, 1997: Table 4). *Serenoa* may also have short-lived Order 4 and 5 roots which would account for the observation of white fine roots growing near the surface during the rainy season.

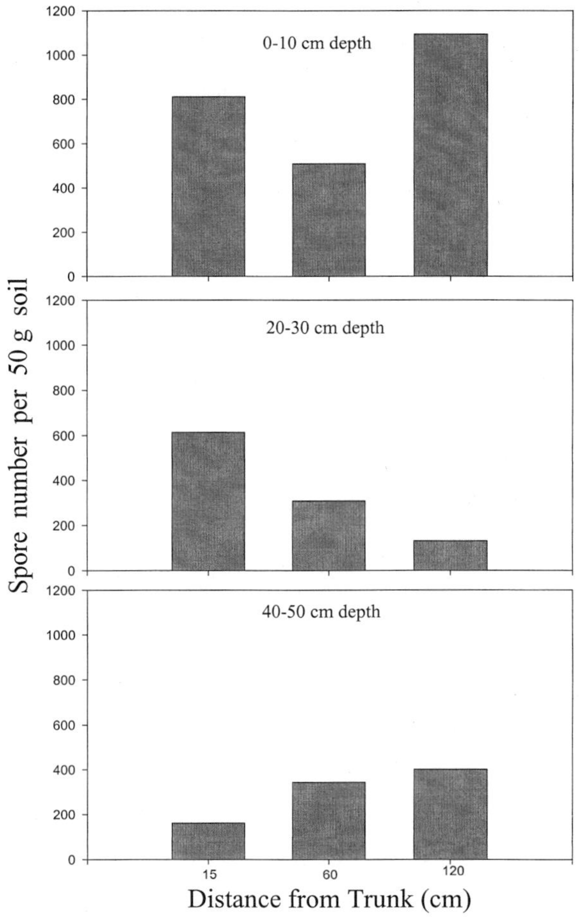

Figure 30. Spore number as bar graph at three distances from the trunk, repeated for the three soil depths. Only AM spores >32 μm diameter were counted per 50 g fresh soil taken from the same trench as in Figure 2.

Root anatomy conformed to the descriptions by Seubert (1997). Suberization of the exodermis occurred later than that of the endodermis and before maturation of the hypodermis, which is also lignosuberized (as defined by Peterson and Enstone, 1996). Throughout development and in old roots, the exodermis remains distinct from hypodermal cells in terms of cell size and cell wall staining. Therefore, we consider the exodermis as single layered, uniform, and without passage cells.

The regular production of dark rings of root cap remnants and periodic slight constrictions in root diameter indicate rhythmic extension growth or individual roots. We have no information on whether such periodicity is synchronized within the entire or a part of the root system.

The zone of epidermal absorption appears to be limited because ultimate roots (Orders 4 and 5) are usually short (<= 1 cm) and higher order roots (Orders 2 and 3) have thick exo- and hypodermal layers with all cells walls lignosuberized within a few cm of the root tip. This surface area limitation is especially probable because root hairs are absent. Water continuity between the soil and living cell membranes is probably maintained by mycorrhizal hyphae that pass through suberized cell walls of the outer cortex (exo- and hypodermis). However, we did not determine the extent of living external hyphae along the length of a root. In old thick roots, the epidermis is lost, and only the exodermis remains as the outer root-soil boundary.

Mycorrhizae

Although the species of AM fungi have not been identified, a description of colonization by these endosymbionts are useful in understanding the biology of palm roots and will serve as a foundation for future descriptive and experimental studies of AM in *Serenoa* and other palms. The approach of Merryweather and Fitter (1998) in identifying the species of AM fungi within a root system is needed for *Serenoa*. The presence and identity of spores (*Glomus*, *Gigaspora*, and *Scutelospora*) in adjacent soil does not confirm which AM fungi are present in colonized *Serenoa* roots.

The hyphae of AM fungi penetrate the root of *Serenoa* through a single epidermal cell and then form intracellular coils within epidermal, exodermal and hypodermal cells, similar to both *Arum*- and *Paris*-type mycorrhizae (Smith and Smith, 1997). The infection unit expands laterally after reaching the thinner-walled cells of the inner cortex. Intercellular hyphae predominate in this region where hyphae grow in the longitudinal intercellular spaces and lacunae. Thus, the bulk of colonizing root hyphae are of the *Arum*-type with arbuscules produced by hyphal branches that enter cortical cells. This form of root colonization is similar to that reported for wild temperate herbaceous monocots by Brundrett and Kendrick (1990b) and in cultivated *Allium porrum* (Brundrett et al., 1985). However, we did not observe hyphal projections (bobbits) described by Widden (1996) in some Liliaceae. In their literature review, Smith and Smith (1997: Table 1 Arecaceae) noted that both *Arum*- and *Paris*-types of AM have been reported in palms. Our observation of thin brown hyphae on *Serenoa* root surfaces that form arbuscules in adjacent epidermal cells needs to be expanded and documented.

We could not determine the state of the exodermis at the time of earliest hyphal penetration because the roots could not be cleared adequately. Continued hyphal penetration occurs at some distance (5 mm or more) from the root tip, but we do not know if the exodermis was fully suberized in this region. Brundrett and Kendrick (1990a, b) found that most AM hyphal penetration of the epidermis in wild plants occurred before the exodermis was fully suberized, i.e. before paradermal walls were suberized. They and we assume that fungal penetration cannot occur after the exodermis and hypodermis are suberized and lignified, but this must be verified in palms. We are currently trying to improve techniques so that these critical stages of hyphal penetration can be described better.

It now appears that AM are widespread among palms (St. John, 1988; Zona, 1996), and a few studies have clearly shown enhanced nutrient uptake and growth promotion by AM colonization in several palms (Janos, 1977; Blal et al., 1990; Clement and Habte, 1995). The native soil is sandy, well-drained, and low in phosphorus. Thus, AM could be ecologically significant in phosphorus and water uptake (Smith and Read, 1997). However, the role of AM in the ecology of native palms and the significance of AM for the horticulture of palms will only be clarified by future descriptive and experimental studies.

Acknowledgements

We thank: Mohamed Bakkar, David Janos, and Ted St. John for advice on staining AM; Meera Nair and Marianne Vanevic for technical assistance; and David Janos for helpful comments on the manuscript. The Montgomery Botanical Center permitted use of their pine rockland site. Southeast Environmental Research Center contribution number 90.

References

Bennett B C and Hicklin J 1998 Uses of *Seneroa repens* (Bartr.) Small (saw palmetto) in Florida. Econ. Bot. 52, 365–375.

Blal B, Morel C, Gianinazzi-Pearson V, Fardeau J C and Gianinazzi S 1990 Influence of vesicular-arbuscular mycorrhizae on phosphate fertilizer efficiency in two tropical acid soils planted with micropropagated oil palm (*Elaeis guineensis* Jacq.). Biol. Fertil. Soils 9, 43–48.

Bray R H and L T Kurtz 1945 Determination of total, organic and available form of phosphorus in soil. Soil Sci. 59, 39–45.

Brundrett M and Kendrick B 1990a The roots and mycorrhizas of herbaceous woodland plants. I. Quantitative aspects of morphology. New Phytol. 114, 457–468.

Brundrett M and Kendrick B 1990b The roots and mycorrhizas of herbaceous woodland plants. II. Structural aspects of morphology. New Phytol. 114, 569–479.

Brundrett M C, Enstone D E and Peterson, C A 1988 A berberine-analine blue fluorescent staining procedure for suberin, lignin, and callose in plant tissue. Protoplasma 146, 133–142.

Brundrett M C, Piche Y and Peterson, R L 1985 A developmental study of the early stages in vesicular-arbuscular mycorrhiza formation. Can. J. Bot. 63, 184–194.

Brundrett M, Bougher N, Dell B, Grove T and Malajczuk N 1996 Working with mycorrhizas in forestry and agriculture. Austral. Cent. Int. Agr. Res. Monograph 32, 1–374.

Clement C R and Habte M 1995 Genotype variation in vesicular-arbuscular mycorrhizal dependence of the pejibaye palm. J. Plant Nutrition 18, 1907–1916.

Ferreira S A do N, Clement C R, Ranzani G 1990 Contribuição para conhecimento do sistema radicular da pupunheira (*Bactris gasipaes* H.B.K.–*Guilielma gasipes* (H.B.K.) Bailey). I. Solo latossolo amarelo, textura média. Acta Amazonica 10, 245–249.

Ferreira S A do N, Clement C R, Ranzani G and Costa S de S 1995 Contribuição ao conhecimento do sistema radicular da pupunheira (*Bactris gasipaes* Kunth, Palmae). II. Solo latossolo amarelo, textura argilosa. Acta Amazonica 25, 161–170.

Fisher J B and Tomlinson P B 1973 Branch and inflorescence development in saw palmetto (*Serenoa repens*). Principes 17, 10–19.

Janos D P 1977 Vesicular-arbuscular mycorrhizae affect the growth of *Bactris gasipaes*. Principes 21, 12–18.

Jourdan C and Rey H 1997 Architecture and development of the oil-palm (*Elaeis guineensis* Jacq.) root system. Plant Soil 189, 33–48.

Meryweather J and Fitter A 1998 The arbuscular mycorrhizal fungi of *Hyacinthoides non-scripta*. I. Diversity of fungal taxa. New Phytol. 138, 117–129.

Perumalla C J, Peterson C A and Enstone D E 1990 A survey of angiosperm species to detect hypodermal Casparian bands. I. Root with a uniseriate hypodermis and epidermis. Bot. J. Linn. Soc. 103, 93–112.

Peterson C A and Enstone D E 1996 Functions of passage cells in the endodermis and exodermis of roots. Physiol. Plant. 97, 592–598.

Purvis C 1956 The root system of the oil palm: Its distribution, morphology, and anatomy. J. W. Afr. Inst. Oil Palm Res. 1, 60–82.

St. John TV 1988 Prospects for application of vesicular-arbuscular mycorrhizae in the culture of tropical palms. Adv. Econ. Bot. 6, 50–55.

Seubert E 1997 Root anatomy of palms. I. Coryphoideae. Flora 192, 81–103.

Smith S E and Read D J 1997 Mycorrhizal symbiosis. Second edition. Academic Press, San Diego.

Smith F A and Smith S E 1997 Tansley Review No. 96. Structural diversity in (vesicular)-arbuscular mycorrhizal symbioses. New Phytol. 137, 373–388.

Tailliez B 1971 The root system of the oil palm on the San Alberto plantation in Colombia. Oléagineux 26, 435–447.

Tomlinson P B 1990 The structural biology of palms. Oxford University Press, Oxford.

Vadermeer J 1977 Observations on the root system of the pejibaye palm (*Bactris gasipaes* H.B.K.) in Costa Rica. Turrialba 27, 239–242.

von Guttenberg H 1968 Der primäre Bau der Angiospermenwurzel. In Handbuch der Pflanzenanatomie. Bd. 8, Teil 3. Borntraeger, Berlin.

Widden P 1996 The morphology of vesicular-arbuscular mycorrhizae in *Clintonia borealis* and *Medeola virginiana*. Can. J. Bot. 74, 679–685.

Zona S 1996 *Roystonea* (Arecaceae: Acrecoideae). Flora Neotropica 71, 1–36.

A. Stokes (ed.), The Supporting Roots of Trees and Woody Plants: Form, Function and Physiology, 391–396.

Is maintenance respiration in roots a constant?

Tjeerd J. Bouma[1,2], David Bryla[2], Yadong Li[2] and David M. Eissenstat[2]
[1]Netherlands Institute of Ecology, Center for Estuarine and Coastal Ecology, P.O. Box 140, 4400 AC Yerseke, The Netherlands and [2]Department of Horticulture, Pennsylvania State University, 103 Tyson Building, University Park, PA 16802-4200, USA

Key words: acclimation, age, dark-decay, drought, droughted surface soil, maintenance, respiration, root, root age, woody roots

Abstract

Maintenance respiration is often considered to be a constant under constant temperature conditions. Here we report on two independent observations that cause us to question this assumption. Respiratory requirements of detached root segments from apple trees decreased with age. Respiration after a dark-decay period was lower for citrus roots in droughted surface soil than for similar roots in irrigated soil. Although we are well aware of the limitations with respect to interpreting these observations in terms of maintenance respiration, it is our opinion that present observations indicates that we need to re-think the methods by which root respiration and its maintenance component are commonly being studied. Based on present observations, we propose that (i) specific respiration for maintenance is *not* constant, (ii) the population of fine roots should *not* be regarded as one entity and (iii) root respiration *needs* to be studied on individual roots to improve understanding of physiological responses to changes in the soil environment.

Introduction

Present knowledge on respiratory responses of roots to the soil environment is still quite limited, especially regarding the maintenance component. Maintenance respiration was originally defined as any respiration not associated with growth (Pirt, 1975; Szaniawski, 1981; Thornley, 1970). Because of the large energy requirements of ion uptake in roots, Veen (1980) redefined maintenance for roots as any respiration not associated with growth and ion uptake. Depending on the stage of development, maintenance respiration may account for over 50% of the rate of total ATP production in roots (Van der Werf et al., 1988), with overall root respiration up to 50% of plant photosynthates produced daily (Lambers et al., 1996). Important energy-requiring maintenance processes in plants include protein turnover, maintaining ion gradients, and transport processes (Bouma and De Visser, 1993; Bouma et al., 1994, 1995; De Visser et al., 1992; Johnson, 1990; Penning de Vries, 1975). However, quantitative information on the specific costs of these processes and their rates in relation to environmental conditions are still too scanty to enable process-based simulation of the dynamics of maintenance respiration.

Simulation models of plant growth reflect this limited understanding of maintenance respiration, and values for it are often simply calculated as the product of plant biomass and a constant maintenance factor (Goudriaan and van Laar, 1994; Marcelis et al., 1998). Such model calculations may be based on experimental methods that yield a single value for specific maintenance costs per unit root biomass (mol ATP/O_2/CO_2 g^{-1} dry weight s^{-1}). For example, *multiple regression approaches* yield a single value for maintenance respiration even though measurements are typically made on plants grown under different environmental conditions or at different stages of development (e.g., Bouma et al., 1996; Van der Werf et al., 1988; Veen, 1980). Another drawback of this method is limitations involved in relating specific maintenance processes to regression components due to the correlative nature of the method (Bouma et al., 1996).

* FAX No: +31-113-573616;
E-mail: tbouma@cemo.nioo.knaw.nl

Alternative methods to assess specific maintenance costs (reviewed by Amthor, 1984) also suffer limitations, primarily due to uncertainties in underlaying assumptions. For example, the *'dark-decay' method* is based on the assumption that only the most critical tissue maintenance processes occur when carbohydrate supply is limited. Hence, maintenance costs are estimated from the reduced respiration of plants kept in darkness for a prolonged period of time. Although the basic assumption of the 'dark-decay' method has been criticized (Gary, 1989), it is still used (e.g., Thomas et al., 1993) because of lack of alternatives. Similar to the regression approach, the 'dark-decay' method does not relate respiratory costs to specific processes. However, in contrast to the regression approach, the 'dark-decay' method offers a relatively simple way to compare physiological responses. So despite its uncertainties, it may reveal some indication whether maintenance costs are constant or changing over time. A *budget approach* (i.e., $R_{\text{maintenance}} = R_{\text{total}} - R_{\text{growth}} - R_{\text{uptake}}$) may also be used to derive whether maintenance costs are constant (e.g., Peng et al., 1993). However, this approach is rather indirect, especially because respiratory costs associated with ion uptake are difficult to estimate. If specific costs for root maintenance are indeed variable, we need to reconsider not only current simulation models, but also standard research approaches for studying root respiration.

In the present study we discuss two observations. Although we realize that both observations have limitations with respect to the interpretation, it is our opinion that they do raise serious doubt about the assumption that specific costs for root maintenance are constant for (i) roots occurring under different physiological conditions and (ii) roots differing in age. Consequences for methods to study root respiration are discussed.

Methods

Relationship between maintenance respiration and root age

Twenty root boxes (rhizotrons) with a clear acetate window (0.55×0.30 m) were placed 0.5 m from the bole of 20-year-old apple trees (Red Chief Delicious on M26 rootstock). Trees were grown in an orchard at the Russell B. Larson Experimental Research Station of the Pennsylvania State University in Centre County, PA, USA (40°85′N, 77°83′W). Every 1 or 2 weeks, newly emerging roots were traced on the acetate windows. Different colored permanent markers were used for root tracings to indicate the date of appearance. After obtaining a population of roots of variable ages, individual roots were excavated by cutting through the acetate window. Roots were transported as quickly as possible to the laboratory in 1 mM $CaSO_4$ buffered with 5 mM MES (2-[*N*-morpholino]ethanesulfonic acid) adjusted to pH 5.5. Respiration rates were measured in a 3-ml cuvette (10-mm ID) filled with $CaSO_4$–MES buffer maintained at 25°C using a Clark-type oxygen electrode connected to a CB1D control box (Hansatech Instruments Ltd, Norfolk, UK). To eliminate respiratory costs associated with nutrient uptake, no nutrients were added to the solution. Re-uptake of efflux may have occurred, but could not be quantified.

Subsamples were used to determine respiratory costs for root construction (mmol CO_2 g root^{-1}) based on elemental composition (i.e., C, H, N and O content; McDermitt and Loomis, 1981). Specific root length (m g^{-1}) was obtained by dividing length by the weight of root segments.

Root maintenance respiration in wet and dry soil

Seeds of Volkamer lemon (*Citrus volkameriana* Tan. & Pasq.) were germinated in flats filled with vermiculite in a glasshouse at University Park, PA, USA (40°85′N, 77°83′W). After 4 weeks, seedlings were transplanted into 0.75-l pots filled with sterilized Chandler fine sandy soil (Typic quartzipsamment) collected from the Citrus Research and Education Center in Lake Alfred, FL, USA. Plants were watered daily and fertilized weekly using a modified Hoagland's solution (5 mM KNO_3, 5 mM $Ca(NO_3)_2$, 5 mM KH_2PO_4, 2 mM $MgSO_4$, 1 mM Fe as FeEDTA and micronutrients; Hoagland and Arnon, 1939). Greenhouse temperatures fluctuated between 20 and 35°C, and maximum midday photosynthetically active photon flux densities reached 1400 μmol m^{-2} s^{-1} on clear days and 500 μmol m^{-2} s^{-1} on days with dense cloud cover.

After 4 months, 12 uniform seedlings were transplanted into soil columns (60 cm long, 76 mm ID) constructed from plastic (PVC) pipe similar to those described by Bryla et al. (1997) (Fig. 1). The columns were filled with sterile Chandler soil. Soil columns were hydraulically and atmospherically separated into upper and lower compartments by a wax layer (1:9

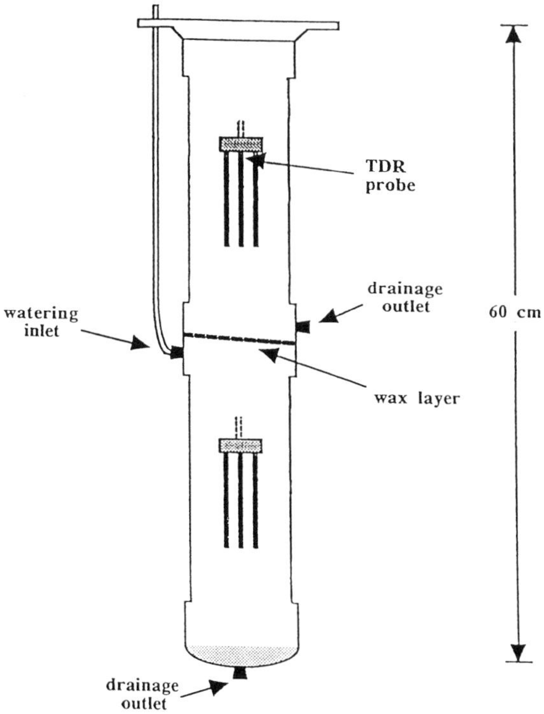

Figure 1. Split-pot design used to measure respiration of roots growing in droughted versus well-watered surface soil (adapted from Bryla et al., 1997).

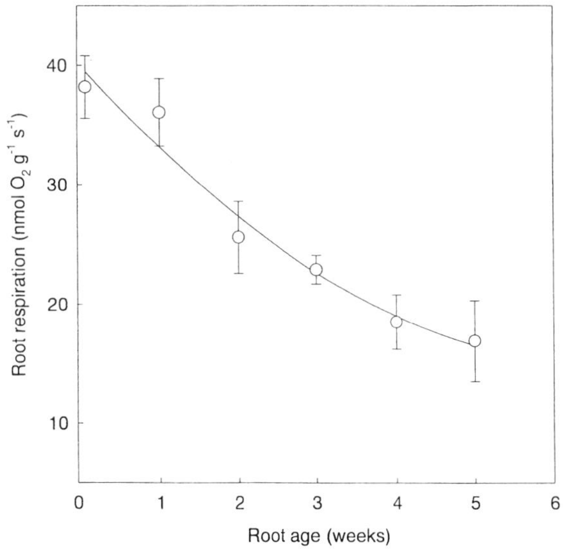

Figure 2. Root respiration rates (nmol O_2 g^{-1} s^{-1}) as a function of root age. Respiration was measured on detached root segments collected from 20-year-old apple trees (Red chief Delicious on M26 rootstock).

ratio (v/v) of hard paraffin wax and petroleum jelly). This enabled respiration measurements to be made on the upper and lower compartments separately. After 4 months, a sufficient number of roots penetrated the wax layers into the lower soil compartment so that roots could be irrigated in the lower compartment while roots in the upper compartment could be exposed to dry soil for a prolonged period of time. Half the plants were droughted (upper compartment only) 1 month prior to making respiration measurements; non-droughted compartments were watered daily. Measurements with time domain reflectometry probes (Fig. 1) indicated that the volumetric soil water content in droughted upper compartments were less than 0.3% within 2–3 days. One plant from the droughted treatment died and was not included in the study.

After 1 month of drought, respiration rates of the upper soil compartments were measured daily on both droughted and non-droughted plants. An earlier study showed that this was sufficient time for root respiration to acclimate to drought (Bryla et al., 1997). To measure respiration, a 0.75-l cylindrical PVC chamber (adapted from the Li-Cor LI-6000-09 soil respiration chamber) was attached to the top of the soil columns, and connected to a closed-system portable gas-exchange system (LI-6200, Li-Cor Inc, Lincoln, NE, USA). After 3 days of measurements, plants were placed in continuous darkness and measurements were continued for another 8 days, a technique known as the 'dark-decay' method.

At the end of the experiment, roots were extracted from the soil columns and rinsed with deionized water. Fine roots were separated from the woody taproot, dried at 70°C for at least 1 week, and weighed. Respiration rates were expressed per g fine root dry weight (i.e., on average 9.1±0.4 g in the irrigated upper compartments and 10.1±0.5 g in the droughted upper compartments).

Results and discussion

Relationship between maintenance respiration and root age

The root boxes enabled us to collect roots of known ages from mature field-grown apple trees, and measure their rates of respiration. Root respiration rates were found to decrease strongly with root age (Fig. 2). This cannot be ascribed to differences in growth because root extension only occurs at the root tip. Secondary growth seemed unlikely to occur as root carbon content (46±0.2% C), respiratory costs in root construction as calculated from elemental composition (9.8±0.1 mmol CO_2 g root^{-1}) and specific root length (5–40 m g^{-1}) were independent of root age. Also, respiration associated with ion uptake should have been negligible, because of the absence of nitrogen in the buffer solution. Hence, by definition, the decrease in apple root respiration with age must have been due to reduced maintenance requirements as the roots became older.

Age dependence of apple maintenance respiration is consistent with work in sugar maple. Excised root respiration was highly correlated to root N concentration (Pregitzer et al., 1998), and root N concentration is higher in daughter roots than their parent roots (Pregitzer et al., 1997). Other researchers have also found respiration to be correlated with N concentration under conditions where ion-uptake respiration was minimized (Burton et al., 1996; Reich et al., 1998; Ryan et al., 1996), which would be consistent with protein turnover representing a large fraction of maintenance respiration. Although we did not find any difference in N concentration among our roots, this may in part be related to different partitioning of N compounds in roots of different ages. If older apple roots partition N relatively more for structural defenses compared to younger roots, then this may not entail as high a maintenance cost as N used for assimilatory enzymes. In these same roots we find a higher V_{max} for P uptake in younger roots. Thus, young, highly absorptive roots may have higher protein requirements, causing faster maintenance respiration then that of older, less absorptive roots.

Summarizing, respiratory costs for maintenance of roots appear to decrease with age, but the nature of the energy-requiring maintenance processes is not fully understood. Our results and those of Pregitzer and colleagues show that the entire population of fibrous roots on a tree should not be regarded as one entity in respiration studies. Measuring (maintenance) respiration on a sample containing many roots will give an average value; the value will depend on the age structure of the sample. Average values may be sufficient for obtaining carbon budgets of individual plants. However, such average values are insufficient to provide insight into mechanisms underlaying soil exploitation by roots and optimal root behavior.

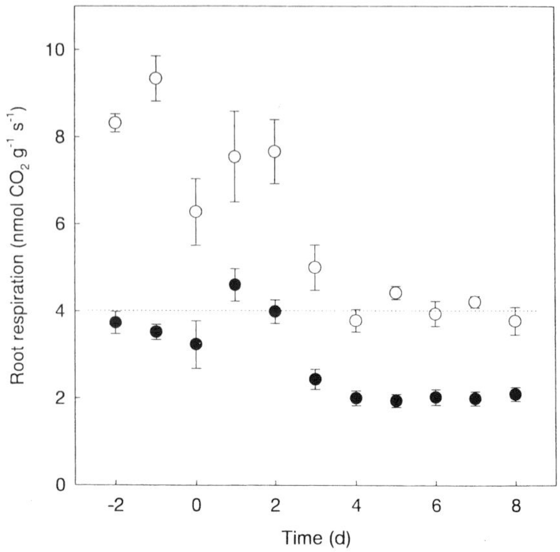

Figure 3. Root respiration rates (nmol CO_2 g^{-1} s^{-1}) of citrus seedlings in wet (open symbols) or dry (closed symbols) surface soil. For 3 days, measurements were made on plants exposed to normal cycles of day and night (i.e., up to day 0). Plants were then placed in darkness for 24 h day^{-1} ('dark-decay' treatment), and measurements were continued for another 8 days (i.e., after day 0).

Root maintenance respiration in wet and dry soil

Citrus seedlings were selected to quantify the effect of drought on maintenance respiration of roots, as citrus root mortality is not increased by a drought period of more than a 100 days (Bryla et al., 1997). This is in contrast to observations made on various other species, where dry soil can cause significant root mortality (e.g., Hayes and Seastedt, 1987; Klepper et al., 1973; Smucker et al., 1991). In citrus, root growth and ion uptake slow down or stop during prolonged periods of drought (Bryla et al., 1997; Espeleta and Eissenstat, 1998; Espeleta et al., 1999). Thus, respiration rates of roots in droughted soil are reduced because carbon demand for growth and nutrient uptake is limited. The question is, are the maintenance costs of such 'inactive' roots in droughted soil different from that of active roots in well irrigated soil? To answer this question, 1-year-old citrus seedlings with roots in droughted versus irrigated soil patches were placed in the dark for a period of 8 days.

We observed that respiration rates of the roots present in dry and in wet soil patches both decreased approx. 50% after 4 days of dark (Fig. 3). This decrease may be due to different maintenance requirements of roots from dry and wet soil patches or may be due to different maintenance requirements of taproots and fine roots, as the weight ratio between fine roots and taproot differed marginally between the droughted (0.59±0.03) and irrigated (0.74±0.09) upper compartments. Because of the 'black-box' nature of the 'dark-decay' method (e.g., Gary, 1989), no firm conclusions can be drawn on the processes underlying the decrease in respiration after 4 days of darkness. Similarly, we can also not account for the increase in respiration during the first 2 days of the dark period. However, irrespective of the nature of the underlying causes, present results do indicate that maintenance respiration is lower for roots that have been exposed to dry soil.

Conclusions and recommendations

Present observations indicate that the population of fine roots should not be regarded as one entity; maintenance respiration appears to be influenced by root age and abiotic conditions in the soil (e.g., drought). This brings into question models where specific costs for maintenance are regarded as constant. Maintenance respiration (R_m) needs to be divided into a component for basal requirements of 'passive' tissue (R_{mb}) and a component related to tissue activity (R_{ma}): $R_m = R_{ma} + R_{mb}$. In models, the need for such separation has been resolved by linking part of the maintenance respiration to tissue protein content (Goudriaan and van Laar, 1994; Marcelis et al., 1998). However, a well-formulated theoretical basis for calculating maintenance respiration, as presented for growth respiration (Penning de Vries et al., 1974), still does not exist. To gain an experimental basis for such quantitative understanding of maintenance respiration, we need to be able to relate respiration rates of single roots of specific age in well defined-soil patches to biochemical and physiological traits of such roots (e.g., membrane composition, protein content and enzyme activity). This requires techniques that allow quantification of respiration rates of single roots under a broad range of soil conditions. So which methods are presently available, and which alternatives may be studied?

Firstly, inserting a single root into a cuvette connected to an open circuit gas-exchange system allows accurate quantification of the effects of factors such as temperature, water status and root age on root respiration rates (e.g., Holthausen and Caldwell, 1980; Nobel and Palta, 1989; Nobel et al., 1991; Palta and Nobel, 1989a,b,c; Sisson, 1983). Apart from the relatively high costs involved in setting up an open gas-exchange system, this approach is elegant and reliable. *Secondly*, respiratory responses of individual roots may also be assessed on isolated root segments using a Clark-type oxygen electrode. The biggest disadvantage of this method is that roots have to be detached from the plant in order to make the measurements, which can affect their physiology (Rakonczay et al., 1997). Developing a system that allows electrode measurements on attached roots would be a great improvement. *Thirdly*, the concept of CO_2 trapping on single roots has potential for physiological and ecological field studies. Although less elegant than open gas-exchange systems, this method is promising as the low costs makes it widely applicable. The sensitivity of the respiration measurements may be customized by adjusting the concentration of the trapping and titration solution and the length of the measurement period. Preliminary tests in our laboratory indicated that the concept has potential, but the method needs more work on the details. The concept remains, however, limited to (i) experiments that do not require a high resolution over time and (ii) species from which the root respiration is not sensitive to soil CO_2 concentrations (Bouma et al., 1997a,b).

We wish that present observations will help to reevaluate present view of maintenance respiration of roots, and will lead towards other approaches. We do believe that future studies on carefully characterized individual roots in a well defined environment, will be an important step towards gaining a quantitative understanding of maintenance respiration of roots.

Acknowledgements

This research was supported by grants from the National Science Foundation (IBN-9596050) and United State Department of Agriculture (NRI 94-37107-1024).

References

Amthor J S 1984 The role of maintenance respiration in plant growth. Plant Cell Environ. 7, 561–569.

Bouma T J and De Visser R 1993 Energy requirements for maintenance of ion concentrations in roots. Physiol. Plant. 89, 133–142.

Bouma T J, De Visser R, Janssen J H J A, De Kock M J, Van Leeuwen P H and Lambers H 1994 Respiratory energy requirements and rate of protein turnover in vivo determined by the use of an inhibitor of protein synthesis and a probe to assess its effect. Physiol. Plant. 92, 585–594.

Bouma T J, De Visser R, Van Leeuwen P H, De Kock M J and Lambers H 1995 The respiratory energy requirements involved in nocturnal carbohydrate export from starch-storing mature source leaves and their contribution to leaf dark respiration. J. Exp. Bot. 46, 1185–1194.

Bouma T J, Broekhuysen A G M and Veen B W 1996 Analysis of root respiration of *Solanum tuberosum* as related to growth, ion uptake and maintenance of biomass. Plant Physiol. Biochem. 34, 759–806.

Bouma T J, Nielsen K L, Eissenstat D M and Lynch J P 1997a Estimating respiration of roots in soil: Interactions with soil CO_2, soil temperature and soil water content. Plant Soil 195, 221–232.

Bouma T J, Nielsen K L, Eissenstat D M and Lynch J P 1997b Soil CO_2 concentration does not affect growth or root respiration in bean or citrus. Plant Cell Environ. 20, 1495–1505.

Burton A J, Pregitzer K S, Zogg G P and Zak D R 1996 Latitudinal variation in sugar maple fine root respiration. Can. J. For Res. 26, 1761–1768.

Bryla D R, Bouma T J and Eissenstat D M 1997 Root respiration in citrus acclimates to temperatures and slows during drought. Plant Cell Environ. 20, 1411–1420.

De Visser R, Spitters C J T and Bouma T J 1992 Energy costs of protein turnover: theoretical calculation and experimental estimation from regression of respiration on protein concentration of full-grown leaves. *In* Molecular, Biochemical and Physiological Aspects of Plant Respiration. Eds H Lambers and L H W van der Plas. pp 493–508. SPB Academic Publishing, The Hague.

Espeleta J F and Eissenstat D M 1998 Responses of citrus fine roots to localized soil drying: a comparison of seedlings with adult fruiting trees. Tree Physiol. 18, 113–119.

Espeleta J F, Eissenstat D M and Graham J H 1999 Citrus root responses to localized drying soil: a new approach to study mycorrhizal effects on the roots of mature trees. Plant Soil 206, 1–10.

Gary C 1989 Temperature and time-course of the carbon balance of vegetative tomato plants during prolonged darkness: examination of a method of estimating maintenance respiration. Ann. Bot. 63, 449–458.

Goudriaan J and Van Laar H H 1994 Modeling Potential Crop Growth Processes. Textbook with Exercises. Current Issues in Production Ecology. Volume 2. Kluwer Academic Publishers, Dordrecht/Boston/London.

Hayes D C and Seastedt T R 1987 Root dynamics of tallgrass prairie in wet and dry years. Can. J. Bot. 65, 787–791.

Hoagland D R and Arnon D I 1939 The water-culture method for growing plants without soil. University of California, Agricultural Experimental Station Circular 347, Berkeley, CA, USA.

Holthausen R S and Caldwell M M 1980 Seasonal dynamics of root system respiration in *Atriplex confertifolia*. Plant Soil 55, 309–317.

Johnson I R 1990 Plant respiration in relation to growth, maintenance ion uptake and nitrogen assimilation. Plant Cell Environ. 13, 319–328.

Klepper B, Taylor H M, Huck M G and Fiscus E L 1973 Water relations and growth of cotton in drying soil. Agric. J. 65, 307–310.

Lambers H, Atkin O K and Scheurwater I 1996 Respiratory patterns in roots in relation to their functioning. *In* Plant Roots. The Hidden Half. Eds Y Waisel, A Eshel and K Kafkaki. pp 323–362. Marcel Dekker, New York.

Marcelis L F M, Heuvelink E and Goudriaan J 1998 Modeling biomass production and yield of horticultural crops: a review. Sci. Hort. 74, 83–111.

McDermitt D K and Loomis R S 1981 Elemental composition of biomass and its relation to energy content, growth efficiency and growth yield. Ann. Bot. 48, 275–290.

Nobel P S and Palta J A 1989. Soil O_2 and CO_2 effects on root respiration of cacti. Plant Soil 120, 263–271.

Nobel P S, Lopez F B and Alm D M 1991 Water uptake and respiration of root systems of two cacti: observations and predictions based on individual roots. J. Exp. Bot. 42, 1215–1223.

Palta J A and Nobel P S 1989a Root respiration for *Agave deserti*: influence of temperature, water status and root age on daily patterns. J. Exp. Bot. 40, 181-186.

Palta J A and Nobel P S 1989b Influence of water status, temperature, and root age on daily patterns of root respiration for two cactus species. Ann. Bot. 63, 651–662.

Palta J A and Nobel P S 1989c Influence of soil O_2 and CO_2 on root respiration for *Agave deserti*. Physiol. Plant. 76, 187–192.

Peng S, Eissenstat D M, Graham J H, Williams K and Hodge N C 1993 Growth depression in Mycorhizal citrus at high-phosphorus supply. Plant Physiol. 101, 1063–1071.

Penning de Vries F W T 1975 The costs of maintenance processes in plants. Ann. Bot. 39, 77–92.

Penning de Vries F W T, Burnsting A H M and Van Laar H H 1974 Products, requirements and efficiency of biosynthesis: a quantitative approach. J. Theor. Biol. 45, 339–377.

Pirt S J 1975 Principles of Microbe and Cell Cultivation. Blackwell Scientific Publications, Oxford, pp 63–74.

Pregitzer K S, Kubiske M E, Yu C K, and Hendrick R L 1997 Root architecture, carbon and nitrogen in four temperate forest species. Oecologia 111, 302–308.

Pregitzer K S, Laskowski M J, Burton A J, Lessard V C and Zak D R 1998 Variation in sugar maple root respiration with root diameter and soil depth. Tree Physiol. 18, 665–670.

Rakonczay Z, Seiler J R and Kelting D L 1997 Carbon efflux rates of fine roots of three species decline shortly after excision. Environ. Exp. Bot. 38, 243–249.

Reich P B, Walters M B, Tjoelker M G, Vanderklein D and Buschena C 1998 Photosynthesis and respiration rates depend on leaf and root morphology and nitrogen concentration in nine boreal tree species differing in relative growth rate. Funct. Ecol. 12, 395–405.

Ryan M G, Hubbard R M, Pongracic S, Raison R J and McMurtie R E 1996 Foliage, fine-root, woody tissue and stand respiration in *Pinus radiata* in relation to nitrogen status. Tree Physiol. 16, 333–343.

Smucker A J M, Nunez-Barrios A and Ritchie J R 1991 Root dynamics in drying soil environments. Belowground Ecol. 2, 4–5.

Sisson W B 1983 Carbon balance of *Yucca elata* during a hot and cool period in situ. Oecologia 57, 352–360.

Szaniawski R K 1981 Growth and maintenance respiration of shoot and roots in scot pine seedlings. Z. Pflanzenphysiol. 101, 391–398.

Thomas R B, Reid C D, Ybema R and Strain B R 1993 Growth and maintenance components of leaf respiration of cotton grown in elevated carbon dioxide partial pressure. Plant Cell Environ. 16, 539–546.

Thornley J H M 1970 Respiration, growth and maintenance in plants. Nature 227, 304–305.

Van der Werf A, Kooijman A, Welschen R and Lambers H 1988 Respiratory costs for the maintenance of biomass, for growth and for ion uptake in roots of *Carex diandra* and *Carex acutiformis*. Physiol. Plant. 72, 483–491.

Veen B W 1980 Energy cost of ion transport. *In* Genetic Engineering of osmoregulation. Impact on Plant Productivity for Food, Chemicals and Energy. Eds D W Rains, R C Valentine and C Holaender. pp 187–195. Plentum Press, New York, NY.

A. Stokes (ed.), The Supporting Roots of Trees and Woody Plants: Form, Function and Physiology, 397–404.

Is the root system the key to higher nitrogen uptake in rapeseed?

The case of a composite hybrid line.

T. Vamerali*, S. Bona, G. Mosca and P. Sambo
Dipartimento di Agronomia Ambientale e Produzioni Vegetali, Università di Padova, Agripolis, Via Romea 16, 35020, Legnaro, Padova, Italy

Key words: rapeseed, hybrid, nitrate uptake, nitrogen leaching, root

Abstract

The recently commercialised rapeseed CHL (composite hybrid lines) are reported to give an average increase of about 15% in grain yield compared with 'old' varieties in normal growing conditions in France. In Italy, this observation has not been confirmed, since CHL Synergy and a good-yielding variety Ceres generally have similar yields (average of 12 field trials). However, several differences in morphological and physiological characters have been identified in the two genotypes, especially at root level. Both shoot biomass and the amount of shoot nitrogen in Synergy were slightly higher compared with Ceres for most of the growing season, until the end of winter. As regards the root system, Synergy shows evident hybrid vigour, having faster growth and, especially in early stages, greater weight and total length than Ceres, together with higher specific root length. These traits are a prerequisite for the uptake of greater amounts of nitrogen, which may eventually be leached. Although an increase in plant density led to higher root length density in the first soil layer (0–30 cm) for both varieties, nitrate leaching of Synergy exceeded that of Ceres at the high plant density, thus indicating that this hybrid is more sensitive to competition for nitrogen uptake. The increase in nitrate uptake rate per unit root length (V_{max}) found for the hybrid compared with the 'old' variety was accompanied by an unfavourable increment of affinity ($1/K_m$), which may be responsible of a loss of efficiency at high plant density. The morphological and physiological modifications of the hybrid seem to be more useful for reducing nitrogen leaching during winter rather than for increasing grain yield.

Abbreviations: CHL – Composite hybrid lines; Km – Michaelis–Menten constant; PAR – Photosynthetic active radiation; RLD – Root length density; SRL – Specific root length; V_{max} – Maximum nutrient uptake rate.

Introduction

Rapeseed cultivation is still based on the use of varieties, and new varieties (hybrids) have recently been made available in order to improve yields. The first types of rapeseed hybrid are called CHL (composite hybrid lines). Higher grain yields in CHL are reported in certain growing conditions in France (CETIOM, 1997) and may vary from 10 to 20% with respect to that of older varieties. However, grain yield increases are not quoted for other environments, including that of Italy, and mechanisms through which better yields may be achieved have not been clarified.

In other plants high-yielding hybrids have been obtained by improving the agronomic efficiency of fertiliser (yield per unit of nutrient supplied) which encompasses both physiological efficiency (yield per unit of nutrient taken up) and efficiency of recovery from the soil (nutrient taken up per unit of nutrient supplied). In maize, for example, breeding has focused on physiological efficiency, so that an excess of nutrient supply with respect to absorbed amounts is required, thus raising problems of environmental quality (Craswell and Godwin, 1984).

In rapeseed, a positive correlation between nitrogen uptake and quantitative root development has not been mentioned in the literature but, in many plants, relatively few roots are needed to take up large amounts of nitrates, at least when water is not

* E-mail: vamerali@agripolis.unipd.it

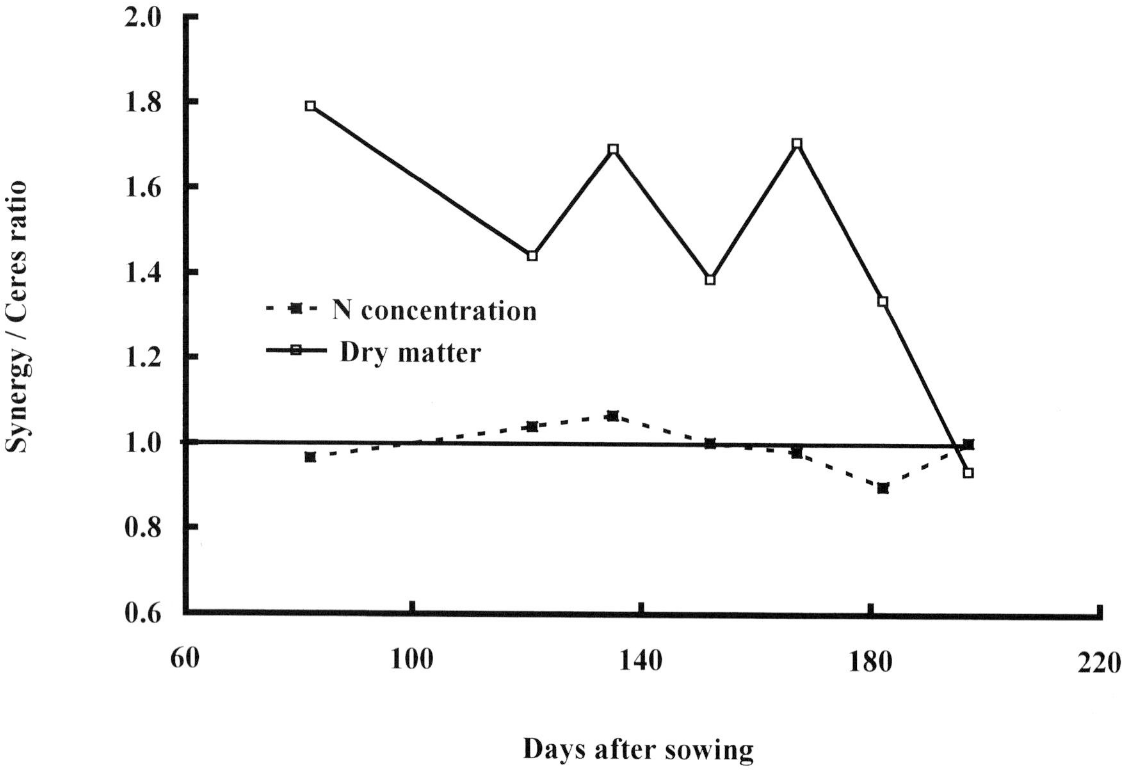

Figure 1. Open field experiment: shoot dry matter accumulation and its nitrogen concentration (ratio Synergy/Ceres) during growing season. Data from the two varieties were significantly different ($P \leq 0.05$).

a limiting factor (Barraclough 1986, 1989; Burns, 1980).

Soil nutrient availability for crop growth depends on several factors, including amount and chemical form of ions and their mobility (Marschner, 1989). Fertilisation, particularly with nitrogen, aims to supply nutrients to the rhizosphere at a constant rate, ensuring the highest potential of root uptake at any time. However, the flux of ions to the soil solution may not be kept permanently high as not all the soil is populated by roots, depleted areas may surround roots as well. Therefore, root uptake changes and soluble ions may be leached deeper into the soil. Nitrate leaching is one important factor contributing to degradation of ground-water (Duyinsveld et al., 1988) and agricultural policy (EU directives 676/91 and 2078/92) regulates water pollution from agricultural system, fixing a maximum nitrate content in drinking-water.

In order to reduce costs and to carry out ecological farming, it is therefore essential to study root system growth and N-uptake efficiency in relation to varying management strategies. Correct synchronisation and synlocalisation of root growth and nitrogen availability may improve the nitrogen use efficiency, thus reducing leaching, at least in the short term (Martinez and Giraud, 1990). Growing a specific crop during autumn and winter is a good technique to reduce nitrogen losses. Reau (1995), comparing various crops in set-aside management, found that rapeseed allowed greater reduction of N leaching than Italian rye-grass.

Rapeseed is known to take up large amounts of nitrogen, up to 300 Kg ha^{-1} (Gabrielle and Gosse, 1995), although in northern Italy nitrate leaching may significantly reduce nitrogen availability because of high rainfall during winter.

This paper presents some results on the nitrogen uptake of rapeseed and relationships with root development and efficiency, comparing a new hybrid (Synergy) and an old variety (Ceres).

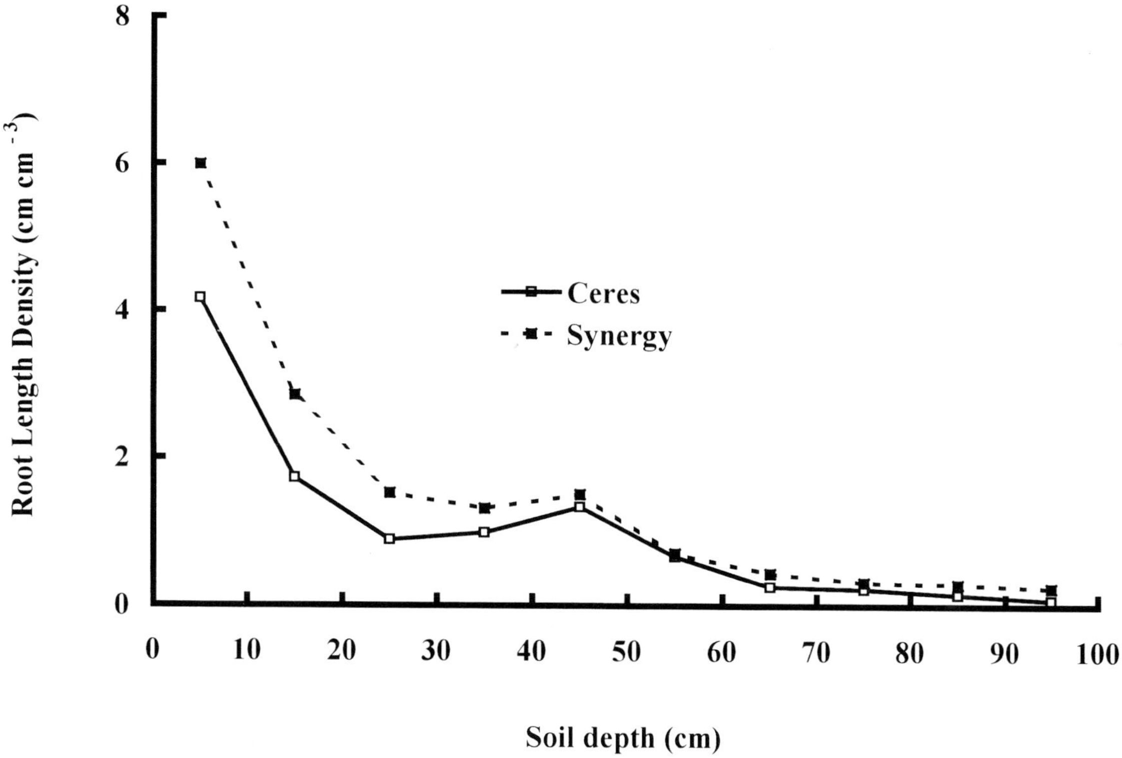

Figure 2. Open field experiment: root length density along soil profiles in two varieties, Ceres and Synergy. Measurements carried out at stage D2 (March 23, 1994). Data are significantly different for first three points only ($P \leq 0.05$).

Materials and methods

1993–94 open field experiment

The study was carried out on the experimental farm 'Lucio Toniolo' at Legnaro (Padova) during 1993–94. Rapeseed was sown on September 30, 1993 and harvested at physiological maturity on June 10, 1994. Two varieties (Synergy and Ceres) and two sowing densities (100 and 300 seeds per m^{-2}) were factorially combined in a randomised block design with 4 replicates. The total amount of nitrogen was 130 Kg ha^{-1}.

Shoot biomass and nitrogen content were measured every two weeks by harvesting one metre of the crop row, starting on December 20. Eight soil cores for each treatment and depth were sampled for $N–NO_3$ determination. During the cycle, cores were taken six times close to rows at three interval depths (0–20, 20–40 and 40–60 cm) and their nitrate content determined by the salicylic acid method (Haroon et al., 1975).

The soil core method (Ellis and Barnes, 1971) was used to measure root growth between depths of 0 and 100 cm. Samples were washed in water and, after manual removal of extraneous objects, root length was measured using a Root Length Scanner (Comair). Root investigations were carried out twice during the season, at flowering (stage D2 – March 23) and during grain filling (stage G4 – May 13).

Field-pot experiment

Forty-nine varieties of rapeseed were investigated to determine nitrogen accumulation in the biomass during winter. Four pots for each variety were buried in sandy-silty-loam soil, leaving the upper borders just protruding from soil; each pot was filled with the same soil as outside. As the pots had 4 holes in the bottom, a certain continuity between the soil in the pots and that outside the pot was guaranteed.

Seeds were sown on October 1, 1996 and seedlings were sampled on February 18, 1997. The aboveground parts of the plants were dried and weighed,

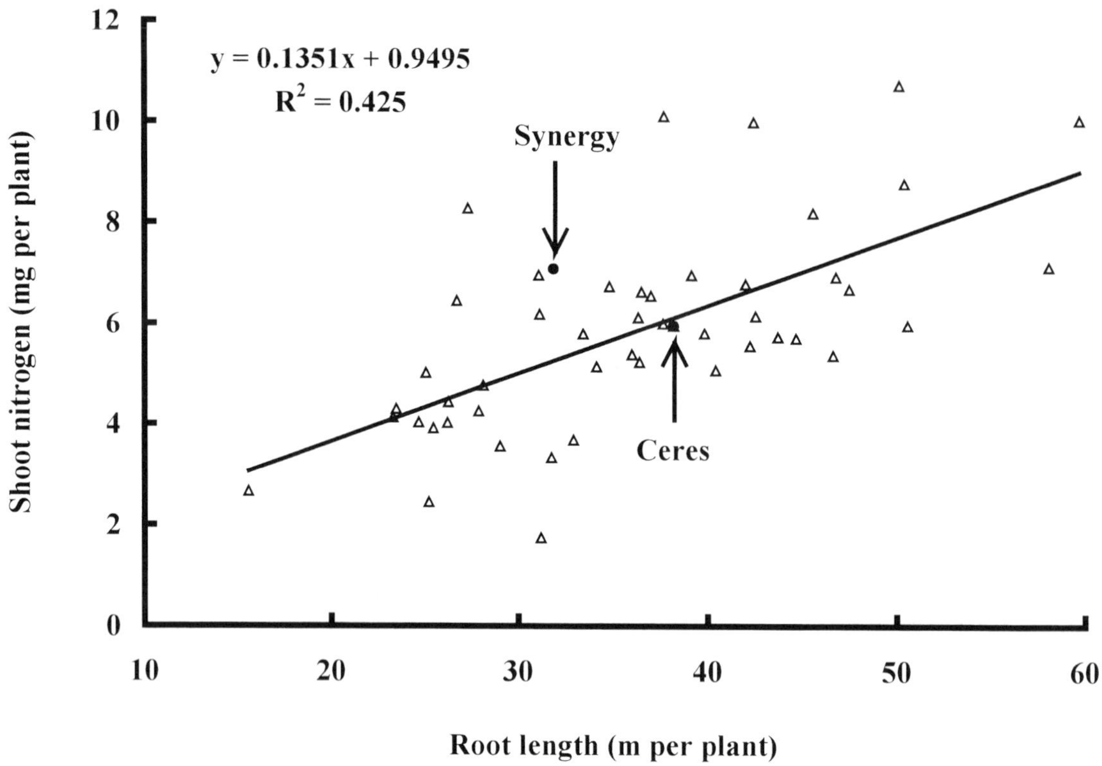

Figure 3. Field-pot experiment: relationship between root length and shoot nitrogen content. Measurements carried out on 49 varieties (markers are average of four replicates). Linear regression was significant at $P \leq 0.05$.

and nitrogen contents determined using the kjeldhal method. The pots were then extracted from the soil, most of the root system being collected. The roots were washed, measured (Root Length Scanner – Comair), dried and weighed and nitrogen contents were determined, as for the shoots.

Growth chamber experiment

In a growth chamber (PAR 400 μMol m^{-2} sec^{-1}), three-week-old rapeseed seedlings (stage of two leaves), previously germinated in a tray containing sandy soil, were used to determine the nitrate uptake parameters using the 'depletion method' (Claassen and Barber, 1974). Thirty seedlings with 3 replicates for each experiment were used after removal from sand. The seedlings were washed and placed into a tank containing 0.8 mMol N–NO_3 nutritive solution (modified Hogland) continuously flushed with air. Induction of enzymes responsible for nitrate uptake was achieved by keeping plants overnight in the solution. The following morning the plants were placed into small beakers containing 140 mL of the same nutritive solution. The beakers were completely covered with an aluminium sheet to prevent the plant roots from being subjected to light. Every 20 min from the beginning of the experiment, 1 mL of the solution was sampled. Nitrate concentration was calculated by measuring adsorbance at 210 nm (Norman et al., 1985). The experiment was repeated 8 times, yielding 24 replicates for each variety. $V_{\max}$ and K_m for nitrates using Michaelis–Menten equations (Wong, 1975) were calculated by plotting depletion curves.

Statistical analysis

Analysis of variance was applied to the set of available data and the Duncan test at $P \leq 0.05$ was used to separate means. Statistical analyses were performed with Statgraphics 6.0 software (Manugistics, Inc. 1993, Maryland 20852, USA).

Nitrate depletion curves had the following form:

$$I = \frac{V_{\max} * C}{K_m + C} \tag{1}$$

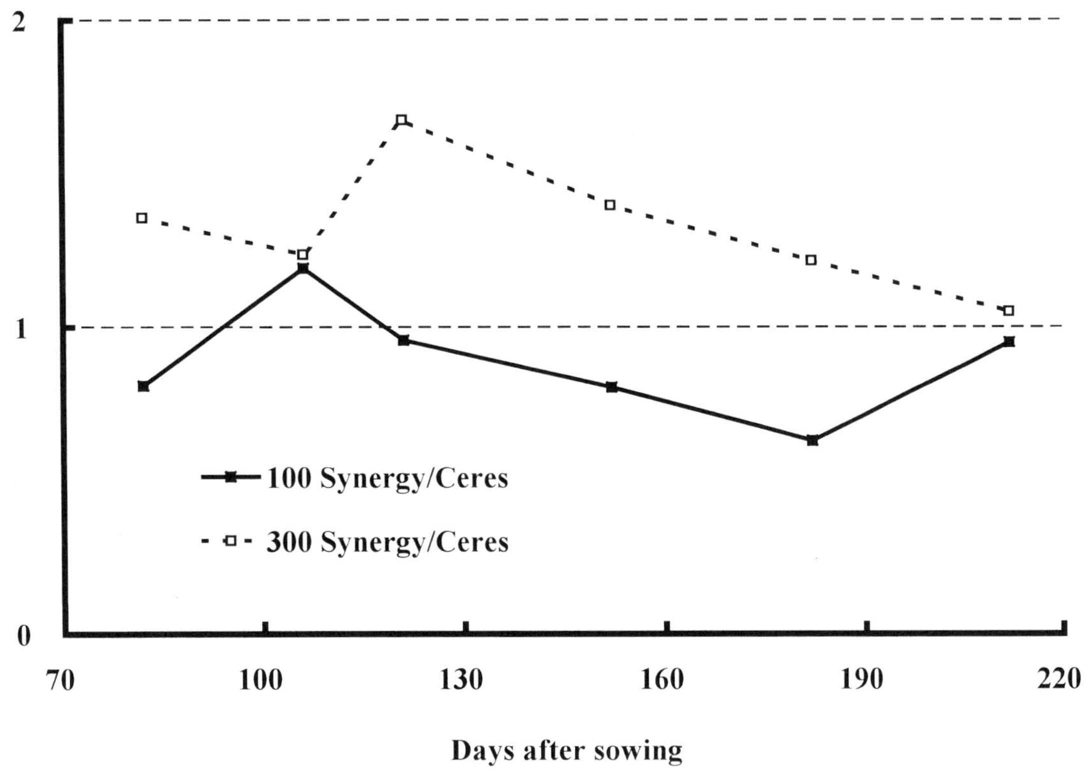

Figure 4. Open field experiment: soil nitrate contents (average of 0–60 cm layer) during growing season for Synergy divided by that of Ceres at two sowing densities, 100 (continuous line) and 300 (dotted line) seeds per square metre. Each point is average of 48 replicates.

where: I = nitrate uptake rate; C = nitrate concentration; $V_{\max}$ = maximum nitrate uptake rate; K_m = Michaelis–Menten constant.

Results and Discussion

Results

In the open field, the shoot biomass of CHL Synergy was greater than that of Ceres in the first part of the growing season (Figure 1) by an average of over 50% up to the end of stem elongation (190 days after sowing). During the second phase, the biomass was not significantly different and remained so until the end of the growing season. Since shoot nitrogen concentration did not differ between the two varieties (Figure 1), nitrogen contents reflected differences in biomass.

With regard to root growth, the CHL showed greater development, mainly in the top layers of the soil; this difference was evident up to stage D2, on March 3, 1994 (Figure 2). Thereafter, an almost similar root pattern was observed between depths of 0 and 100 cm for both varieties. Therefore, the root growth potential expressed by Ceres was very similar to that of Synergy, but only late in season.

The pot experiment yielded a positive significant correlation between root length and nitrogen uptake, data coming from 49 rapeseed varieties (Figure 3). Ceres and Synergy data points were in line with this statement. Linear regression analysis explained around 42% of data variation.

Figure 4 shows the effects of increased plant density on soil nitrate contents. Two sowing densities were compared, 100 (a common value in Italy) and 300 seeds per square metre. The continuous line represents the quantity of nitrates measured in the 0–60 cm soil layer for Synergy at a density of 100 seeds, divided by the amount of nitrates detected at the same density for Ceres. As for all the sampling dates, the curve remains below the value of 1, indicating that, at 100 seeds per square metre, Synergy takes up more nitrates from the soil than Ceres and thus confirming that, even in open field nitrogen uptake is related to quantitative development of the root system. Nevertheless, with increased seed density (Figure 4, dotted line) and consequently

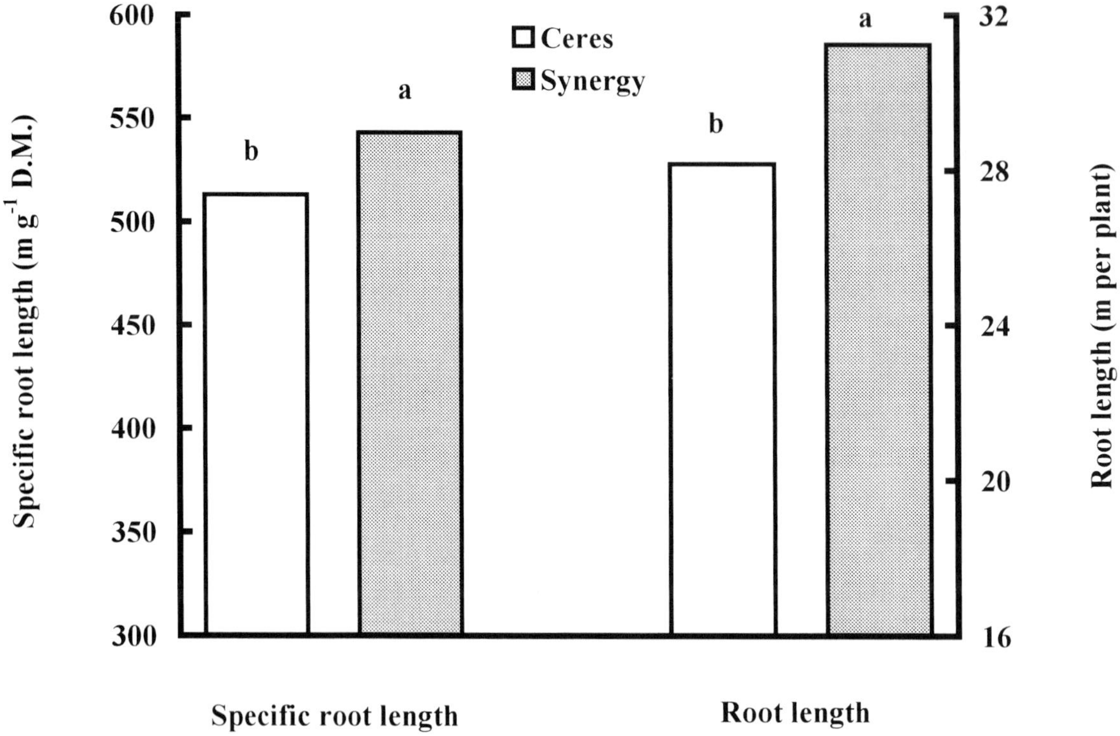

Figure 5. Growth chamber experiment: specific root length (SRL) and root length for Ceres and Synergy. Means (24 replicates) with the same letter are not different at $P\leq 0.05$ (Duncan test).

number of roots, Ceres was shown to tolerate higher intraspecific competition.

Even in the growth chamber experiment, the higher growth rate of Synergy was clear; in the early phases (seedlings) the root length of Synergy was higher than that of Ceres (Figure 5). This ability was partly due to the fact that this variety produced roots with lower biomass. The ratio between root length and root dry matter (SRL) was higher in Synergy (542 vs. 512 m g^{-1} of root d.w.).

The superiority of Synergy was also evident from a physiological standpoint. The uptake kinetic parameters, K_m and V_{max}, were more suitable for nitrate uptake, V_{max} being higher and K_m lower than the values for Ceres (Figure 6); therefore, the hybrid showed an enhanced ability to take up nitrates very quickly. Nevertheless, an increase of 10% in V_{max} was accompanied by an unfavourable increase (3%) of affinity (reciprocal of K_m) for nitrates.

Discussion

The new rapeseed hybrid Synergy seems to be more suitable for taking up nitrogen than the older Ceres, mainly because of a greater potential of biomass growth. Synergy can grow faster in the first part of its cycle, expanding both above- and below-ground parts thus exploiting all conditions for greater nitrogen accumulation.

Although at maturity the two varieties expressed almost similar root patterns, Synergy had much faster root growth than Ceres. This situation allowed the former to colonise the soil soon after sowing with a deeper and more extended root system, before winter rains began. Especially in north Italy, much rain falls in winter and may lead to considerable nitrogen leaching.

As the life-cycle of rapeseed develops between September and June, during a period in which the risk of nitrate leaching is very high (Carter et al., 1987), careful agronomic management must be applied, particularly when nitrogen fertiliser is given.

Since a generally good correlation between root length and nitrogen accumulation was found in 49 different varieties of rapeseed, one strategy for reducing the risk of nitrogen leaching is to increase the length and numbers of roots in the soil before winter. Higher root length density may simply be achieved by increas-

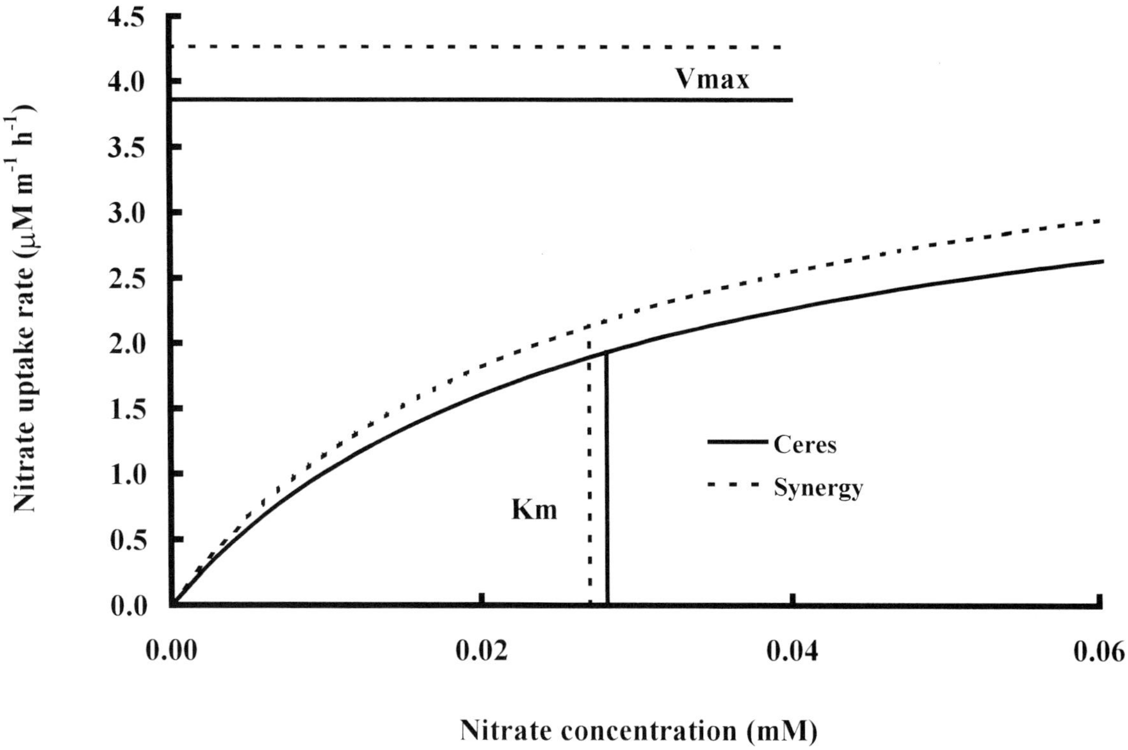

Figure 6. Growth chamber experiment: Michaelis–Menten kinetics for NO^3 uptake for Ceres and Synergy.

ing plant density, although no changes in grain yield may be expected because of morphological modifications of the shoot. At high plant densities, most of the seeds are inserted on the main stalk and just a little on side branches; the opposite situation is observed at low plant densities.

Although rapeseed hybrids are characterised by higher yields with respect to normal varieties in growing conditions in France (CETIOM, 1997), in north Italy we could not confirm this observation, since the yield of Synergy was not statistically different to that of Ceres (4.02 vs. 3.95 t ha^{-1} d.w.). Furthermore, new hybrids have the great drawback that their seeds are very expensive, even though they have been available for a few years on the market. As Synergy seeds cost more than Ceres seeds, and if nitrate uptake is the most important goal, then an increase in seed density in Ceres can partly compensate the lower root growth.

The advantage of Synergy is not only morphological, since kinetic parameters (V_{max} and K_m) are also more suitable than those of Ceres for nitrate uptake. The average increase in maximum uptake rate (V_{max}) and affinity ($1/K_m$) for nitrate were 10% and 3% higher, respectively. Therefore, this new hybrid can improve nitrogen uptake efficiency although, significant increases in grain yield cannot be expected.

The relative efficiency of Synergy is due to greater root expansion associated with better nitrate uptake parameters but, of these two factors, it is not possible to state which is the main one responsible for the higher growth rate. This probably depends on the better uptake parameters which initially induce a greater increase in biomass, leading to greater root expansion which, in turn, induces a greater uptake of nitrates, and so on. This auto-catalytic process is probably driven by improved nitrate uptake characteristics which are able to induce better performance before and during winter.

However, one question does remain: why does Synergy not maintain its growing performance after the end of winter? The answer to this question is not yet clear for the north Italian environment but, as reported for the our site for winter wheat, the main factor limiting high yields is the availability of nitrogen (Sinclair et al., 1993). At low plant densities, soil nitrate contents for Synergy during the winter were lower than for Ceres. This means that Synergy takes up more nitrogen than Ceres, but then it has less nitrogen for growth after the end of winter, while Ceres can partly use it later in the season. Therefore, the substantial equivalence in yield of these two varieties is probably due to nitrogen availability. The period of nitrogen uptake is different, but the quantity is similar (around 200 Kg ha^{-1}, roots excluded).

Agronomically, Synergy is not the best variety to grow because there is no real yield increase compared with Ceres. However, as regards environmental protection, a plant that is able to take up more nitrogen during winter is preferable, in order to limit nitrate leaching. Economically, the fact must also be borne in mind that seed costs are very different, and it is possible to take up the same amount of nitrates by increasing the seed dose of Ceres, thus saving money.

As has already occurred in the case of maize, the selection pressure applied by breeders has led to new hybrids characterised by an increase of nutrient uptake rates accompanied by an unfavourable increment of affinity for the substrate (Saccomani et al., 1984). As a consequence, the amounts of fertiliser exceed the need of the crop and thus raise environmental problems. In rapeseed, a further drawback may also be increased sensitivity to root competition for nutrients. When Synergy is sown at high densities, nitrate leaching is greater than that of Ceres. In these conditions, the amount of nitrate uptake per unit of root length is reduced, so that the hybrid cannot take advantage of its higher physiological efficiency for nitrogen uptake.

Conclusions

Although a new CHL Synergy did not give yield increases in north Italy compared with an 'old' variety, a greater accumulation of biomass in the winter period was found to be associated with a higher nitrogen uptake capability. These traits led to good efficiency in nitrogen recovery from the soil in an environment where most precipitation takes place in winter.

Synergy showed 'hybrid vigour' with faster root and shoot growth, especially in the early stages. Even the kinetic parameters for nitrate uptake were improved with respect to Ceres. Both root length and physiological efficiency are prerequisites for the uptake of greater amounts of nitrogen.

When rapeseed hybrids and hybrids in general are selected, careful breeding programs must be applied, aiming to accompany increased nutrient uptake rate with proportionally increased affinity.

Acknowledgements

Research supported by the Italian Ministry for Agricultural Policy (MiPA), Special Project PRiSCA, and National Research Council–Project ASiR.

References

Barraclough P B 1986 The growth and activity of winter wheat roots in the field: nutrient inflows of high yielding crops. J. Agric. Sci. 106, 53–59.

Barraclough P B 1989 Root growth, macronutrient uptake dynamics and soil fertility requirements of a high yielding winter oilseed rape crop. Plant Soil 119, 59–70.

Burns I G 1980 Influence of the spatial distribution of nitrate on the uptake of N by plants. A review and a model for rooting depth. J. Soil Sci. 31, 155–173.

Carter A D, Palmer R C and Monkhouse R A 1987 *In* Vulnerability of soil and ground water to pollutants. Eds W van Duijvenbooden and H G van Waegeningh. Proc. Infor., 38: 679–685.

CETIOM 1997 Colza d'hiver, variétés. Mai 12–13.

Claassen P and Barber J 1974 Method for characterizing the relation between nutrient concentration and flux into roots of intact plants. Plant Physiol. 54, 564–568.

Craswell E T and D C Godwin 1984 The efficiency of nitrogen fertilizers applied to cereals in different climates. Adv. Plant Nutr. 1, 1–56.

Duyinsveld W H M, Strebel O and Böttcher J 1988 Are nitrate leaching from arable land and nitrate pollution of groundwater avoidable? Ecol. Bull. 39, 116–125.

Ellis B B and Barnes B T 1971 Mechanical method for obtaining soil cores. Plant Soil 35, 209–212.

Gabrielle B and Gosse G 1995 Bilan environnemental du colza: premiers resultats experimentaux. OCL 6, 443–444.

Haroon M, Schrader L E and Young U L 1975 Rapid colorimetric determination of NO_3 in plant tissue by nitration of salicylic acid. Soil Sci. Plant An. 6, 71–80.

Marschner H 1989 Nutrients dynamics at the root-soil interface (rhizosphere). Z. Planzenernahr. Bodenk 152, 151–157.

Martinez J and Giraud G 1990 A lysimeter study of the effects of a ryegrass catch crop during winter wheat/maize rotation on leaching and on the following crop. J. Soil Sci. 41, 5–16.

Norman R J, Edberg J C and Stuck J W 1985 Determination of nitrate in soil extracts by dual-wavelength ultraviolet spectrophotometry. Soil Sci. Am. J. 49, 1182–1185.

Reau R 1995 Les risques de pollution par l'azote dans les cultures de colza. OCL 6, 445–449.

Saccomani M, Cacco G and Ferrari G 1984 Changes in the kinetic parameters of sulfate uptake in maize hybrids during the selection period 1930 through 1975. Maydica 24, 133–140.

Sinclair T R, Mosca G and Bona S 1993 Simulation analysis of variation among seasons in winter wheat yields in northern Italy. J. Agron. Crop Sci. 170, 202–207.

Wong J T 1975 Kinetics of enzyme mechanism. Academic Press, London, UK. 294 p.

A. Stokes (ed.), *The Supporting Roots of Trees and Woody Plants: Form, Function and Physiology*, 405–417.

Effects of liming on rhizosphere chemistry and growth of fine roots and of shoots of sessile oak (*Quercus petraea*)

M.R. Bakker, R. Kerisit, K. Verbist and C. Nys*
Equipe Cycles Biogéochimiques, INRA Nancy, 54280 Champenoux, France

Key words: fine roots, liming, *Quercus petraea* Liebl. M., rhizosphere, rhizotron, shoot:root

Abstract

Soil acidification can be detrimental to root growth and nutrient uptake, and liming may alleviate such acidification. In the following study, seedlings of sessile oak (*Quercus petraea* Liebl. M.) were grown in rhizotrons and subjected to liming (L) or gypsum (G) treatments and compared with the control (C). In order to study and interpret the impact of these calcium rich treatments on fine root development and tree growth, the following parameters were assessed: fine root biomass, fine root length, seedling development (height, diameter, leaves), seedling biomass, nutrient content of roots and seedlings, bulk soil and soil solution chemistry and rhizosphere soil chemistry. The results show that liming increased bulk soil pH, exchangeable Mg, Ca and the Ca/Al molar ratio, and decreased exchangeable Al, mainly in the A-horizon. Gypsum had a similar but smaller impact on exchangeable Al, Ca, H^+ and the Ca/Al molar ratio in the A-horizon, but reacted with depth, so that exchangeable Mn, Mg and Ca were increased in the B-horizon. In the rhizosphere, the general pattern was determined by the treatment effects of the bulk soil. Most elements were more concentrated in the rhizosphere than in bulk soil, except for Ca which was less concentrated after liming or gypsum application. In the B-horizon rhizosphere pH was increased by the treatments (L > G,C) close to the root tips. Furthermore, the length of the zone with a positive root-induced pH increase was greater for the limed roots as compared with both the other treatments. Fine root growth was stimulated by liming (L > G,C) both in terms of biomass and length, whereas specific root length was not obviously affected apart from the indication of some stimulation after liming at the beginning. The live:dead ratio of fine roots was significantly higher in the limed rhizotrons as compared to the control (G not assessed), indicating lower mortality (higher longevity). Shoot growth showed greater lime-induced stimulation (L > G,C) as compared to root growth. As a result the shoot:root ratio was higher in the limed rhizotrons than in the control (L > G,C). Liming induced a higher allocation of P, S, Mg, Ca and K to the leaves, stem and twigs. Gypsum showed similar effects, but was only significant for S. Liming increased the foliar Ca/Al ratio by both increasing foliar Ca and decreasing foliar Al, whereas gypsum did not clearly improve foliar nutrition. This study suggests that a moderate application of lime can be successful in stimulating seedling growth, but that gypsum had no effect on seedling growth. It can be concluded that this lime-induced growth stimulation is directly related to the improved soil fertility status, and the alleviation of Al toxicity and acid stress, resulting in better foliar nutrition. The impact of liming on fine roots, as a consequence, was not limited to a stimulation of the total amount of fine roots, but also improved the root uptake performance.

Introduction

Soil acidification has a negative effect on fine root development (Rost-Siebert, 1983; Marschner, 1991). High concentrations of Al, N or protons, and low cationic:Al and cationic:N ratios in the soil solution inhibit root growth, damage root tissues or lead to impaired uptake (Rost-Siebert, 1983; Göransson and Eldhuset, 1987; Ljungström and Nihlgård, 1995). A decrease in root growth results in a low specific root length (Olsthoorn et al., 1991) or even mortal-

* FAX No: +33 (0)3 83 394069. E-mail: nys@nancy.inra.fr
Plant and Soil is the original source of publication of this article. It is recommended that this article is cited as: *Plant and Soil* **217**: 243–255, 1999.

ity of parts of the root system. Where acidification is stratified according to soil depth, roots are often concentrated in the upper organo-mineral layers (A horizons), reflecting avoidance of the acid subsoil layers (Marschner, 1991) and the buffering of Al toxicity by organic matter. Liming has been advocated to counterbalance many of the undesirable effects of soil acidification (low pH, Ca and Mg concentrations, high Al concentrations) and their impact on fine root development (Hüttl and Zöttl, 1993). Effects of liming on fine root development are reported to be positive (Murach and Schünemann, 1985; Schüler and Zwick, 1992; Gobran et al., 1993), although some state the contrary, depending on site, soil type, form and doses of liming and the time lapse since liming (Persson and Ahlström, 1994; Persson et al., 1995; Staaf et al., 1996).

Earlier results from field sites on acidic soil, suggested that liming stimulated fine root development and tree growth of sessile oak (*Quercus petraea* Liebl. M.) (Bakker, 1999). The present paper aims to link the *in situ* observations of the field sites with the lime-induced changes in the rhizosphere. For this 2 series of rhizotron experiments have been carried out. In the first (indirect) series of experiments, the stock of nutrients at the beginning (soil, acorns, litter, lime treatments) and the amount of biomass formed during the experiment (leaves, branches, roots) were determined. In the second (direct) series, a chemical characterization of the rhizosphere soil was carried out at the end of the experiment, while measurements of rhizosphere pH were carried out during the experiment. It was expected that the lime or gypsum-induced changes in the bulk soil and bulk soil solution would react differently in the rhizosphere, thereby improving the uptake conditions, which would explain the increased tree growth.

Materials and methods

Experimental treatments

Seedlings of sessile oak were grown in rhizotrons in four successive experiments. Rhizotrons, measuring 55 (length)×40 (height)×3.5 cm (width), were filled with an acid brown forest soil (FAO Dystric Cambisol; USDA Dystrochrept) collected from the control plot of a liming experiment at La Croix-Scaille (French Ardennes). The oak stand of that trial showed low foliar K and very low foliar Mg and Ca nutrition. The A-horizon (average pH-H_2O 3.8) of 15 cm was placed on top of the B-horizon soil (average pH-H_2O 4.4) of 20 cm and both were re-constituted with a density equivalent to field conditions. A control (C) was compared soils treated with doses of $CaCO_3$ (Lime, L) or $CaSO_4$ (Gypsum, G) equivalent to 1.4 t ha^{-1} CaO. In order to accelerate the effect of these applications on root growth, half of the dose of lime or gypsum was mixed with the top A-horizon soil before filling the rhizotrons. The other half was spread on top of a shallow litter-layer of oak leaves. This situation (half of the applied calcium in form of lime incorporated into the soil) corresponds to a period of 20 months after liming if applied to the surface as shown in previous studies on the same soil (Belkacem and Nys, 1997).

Three of the experiments, each replicated 4 times, were carried out for 3–5 months, whereas the longest lasted 13 months (replicated twice only). The first three experiments (consisting of two short experiments and the long experiment) were carried out with small seedlings (age 4 months). For the fourth (short) experiment acorns (brought to simultaneous germination by pre-treatment) were used, as seedlings were not available any more. Both the acorns and seedlings have the same genetic origin (northeastern France). The seedlings from the laboratory nursery were also obtained from acorn germination and grown on a growth substrate that could be removed easily before replanting in the rhizotrons. Before replanting nearly all roots were pruned. In all rhizotrons either three to four seedlings were planted or three to four acorns (average dry weight 1.5 g) sown.

The rhizotrons were placed at an angle of about 30°, so as to induce fine root growth preferentially along one side. All experiments were carried out in the greenhouse. A summary of the initial situation is presented in Table 1 (soil, litter and acorns).

Sampling procedure

Seedling growth, root growth and soil analysis (bulk and rhizosphere) were determined at the end of each experiment. Seedlings were cut at the root collar and the following growth parameters were assessed: number of leaves, dry weight of leaves, stem diameter, stem weight and height of seedling. Sub-samples of leaves and stems were analysed for their element content. Dry weight was determined after drying at 65 °C to constant weight. Roots were sampled destructively on opening the rhizotrons and they were collected separately for both soil layers. The roots were separated from the soil by wet sieving with tapwater over

Table 1. Composition of soil layers A and B, litter and acorns used for the experiments with sessile oak (*Quercus petraea*) seedlings, presented as averages over n repetitions, respectively (* = missing value)

Stratum	*n*	pH-H_2O	pH-KCl	K	Ca	Mg	Na	Mn	H^+	Al^{3+}	BS (%)
							$cmol_c\ kg^{-1}$				
A-layer	3	3.79	2.88	0.22	0.91	0.27	0.06	0.14	1.58	4.23	21.6
B-layer	3	4.54	3.97	0.12	0.01	0.02	0.04	0.06	0.00	3.18	7.3
		S ($g\ kg^{-1}$)	P	K	Ca	Mg	Mn	Al			
Litter	2	1.60	0.89	1.32	6.96	0.62	1.24	0.53			
Acorns	10	3.40	*	87.7	7.41	4.40	0.98	0.00			

two consecutive sieves (4 and 2 mm mesh size), followed by floating the roots collected from the sieves in a bin filled with water. Dead roots were separated from living roots using general visible criteria: resilience, brittleness, colour of bark and xylem (Vogt and Persson, 1991), the reliability of which was confirmed by observations under a binocular microcope. Roots were separated into fine (<2 mm dia.) and coarse (>2 mm) roots. Dry weight of roots was assessed by drying at 105 °C to constant weight and length was determined by the line intersection method (Tennant, 1975). The element contents of fine roots were analysed on subsamples of the live roots, dried at 65 °C to constant weight.

Soil adhering to the roots was collected and formed the 'rhizosphere' soil fraction. For this, only roots of a comparable size (<2 mm) were used. Rhizosphere soil was obtained by gently shaking the fine roots until only small amounts of soil remained adhered to the roots (Hendriks and Jungk, 1981; Olsthoorn et al. 1991). Roots were air dried for some hours before the rhizosphere soil was brushed off. Bulk soil was sampled at random throughout every rhizotron, but not in the vicinity of roots. Soil solution was sampled during the experiment using rhizon soil moisture samplers (Eijkelkamp BV, Giesbeek, the Netherlands, reference 19.21.05). The rhizons (length 10 cm, outer dia. 2.3 mm) were inserted horizontally in both A and B-layers after boring small holes in the plexiglass. After installation and a contact time of several days, small vacuum-tubes were connected to the outer edge of the rhizons using injection needles and were left overnight in order to collect a sufficient sample volume (generally between 2 and 7 mL). Rhizosphere pH was determined using micro-electrodes (outer diameter of the tip of the electrode and reference electrode 0.5–1.0 mm) with agar of a pH comparable to the soil pH (pH around 4) as a contact medium between the electrode and the soil. During the pH measurements, the selected rhizotron was placed horizontally and opened. On suitable roots (measurable over a 4–5 cm length) small pieces of agar (size approx. 4×4 mm, thickness about 2 mm) were put at regular distances from the root tip before measuring the pH at these localities, with the reference electrode placed in the vicinity of the measurement localities (Gijsman, 1990).

Sample processing

Soil samples were air-dried and then sieved at 2 mm. 0.5 *M* NH_4Cl-exchangeable Al, Ca, Mg, K, Mn and Na (Trüby, 1989) were determined by ICP (emission spectrometry), exchangeable acidity by automatic titration (Trüby 1989), and pH–KCl and pH–H_2O on a 1:2.5 dilution basis with pH-electrodes. Concentrations were expressed as oven-dried (105 °C) weights. Cation Exchange Capacity (CEC) was defined as the sum of exchangeable Mn, Mg, Ca, Na, K, plus titrable Al^{3+} and H^+ (Espiau and Pedro, 1980). Base saturation was defined as the proportion of exchangeable Mn, Mg, Ca, Na and K in relation to total CEC. Soil solution analysis was performed by ICP (S, Mn, Mg, Al, Ca, Na, K) and colorimetry (Technicon Autoanalyzer) for nitrate and ammonium. For chemical analyses of foliage, stems and live fine-roots, subsamples were pretreated with peroxide (H_2O_2), then mineralized with $HClO_4$ and analysed by ICP. The acorns were mineralized by HCl after consecutive ignition up to 500 °C and then analysed by ICP.

Data analysis

All statistics were calculated with Unistat 4.0 software (Unistat Ltd, 1995). Distribution was tested for normality and wherever necessary values were log transformed to improve the distribution. The three short experiments were regrouped prior to analysis while the long experiment was analysed separately if this was feasible according to number of repetitions. The one experiment for which acorns were used in stead of pruned seedlings, was not analysed separately, as the genetic origin was comparable and the amount of roots formed was not different from that in the other experiments. The above-ground parameters, seedling growth and foliar composition were analysed per seedling by a Two-Way ANOVA for treatment and experiment effects for both the short experiments and the long experiment, followed by an ANOVA for treatment effects per individual experiment. The below-ground parameters, soil, solution and roots were analysed per rhizotron by a Three-Way ANOVA for effects of treatment, soil layer and experiment for the set of data of the three short experiments. This was followed by a Two-Way ANOVA for treatment and soil layer for each single experiment. For the long experiment, treatment and soil layer effects were only tested for the soil and fine root mineral concentration data, as the number of replications was too low elsewhere. As there was some variation in absolute mean values of the three short experiments with regards to shoot:root ratios and also the amount of mineral elements immobilized in the biomass of the plants (summed values per rhizotron), some further statistical analyses were carried out on weighted values. For this, the original values shoot:root values were divided by the mean value of the individual experiment. Similarly, for the total amount of elements immobilized per rhizotron, the values were weighted, dividing them by the number of trees, in order to avoid an effect due to the different number of seedlings. An ANOVA was then carried out on these standardized values (shoot:root and quantity of immobilized mineral elements). Wherever treatment effects were significant, the Student–Newman–Keuls test was used to determine the significantly different treatments.

Results

Plant growth

Table 2 presents the overall treatment effects on above-ground growth. Generally, the limed seedlings grew significantly more in the short experiments than the seedlings in the control and gypsum treatments. This was also true for each individual short experiment, and in the long experiment the tendency – although not significant – is the same. The lower seedling number in the $CaSO_4$ treatment is mainly the result of the experiment with acorns, and could suggest a lower seedling survival due to less favourable conditions for root development in the early stages of seedling establishment. Also, the seedling perfomance in the long experiment seems to decline for the $CaSO_4$ treatment as compared to the short experiments, but this is not significant. So, the lime treatment increased above-ground growth, whereas the gypsum treatment had no effect, or perhaps even a slightly negative effect.

The root system was analysed per rhizotron, thereby decreasing the number of replicate samples, since it was not feasible to attribute the roots to each single seedling. Table 3 presents the overall means for fine root biomass, fine root length and specific root length for the short and long experiments. Liming seems to have stimulated fine root length and both liming and gypsum applications appear to have increased specific root length in the A-layer, but this was not significant on an overall basis. In only one of the short experiments was this lime-induced stimulation of fine root length and specific root length significant at $p<0.05$. Similar to the observations on seedling growth, the $CaSO_4$ treatment tends to decline in the longest experiment relative to fine root biomass and length. About half of all root biomass and root length was formed in the A-layer and half in the B-layer.

There were no treatment effects on coarse roots (data not presented). The live/dead ratio of the fine roots (Figure 1), based on a few replicate samples in the long experiment (not assessed for gypsum), was significantly higher in the limed rhizotrons (7.2±1.2) than in the control (3.6±0.4) (values are means ± standard error).

In Table 4 the effects of the treatments on the foliar nutrition and fine root concentrations are presented. In the short experiments, liming significantly increased foliar Ca concentration and foliar Ca/Al ratio, while lowering Al concentration. Gypsum significantly increased N nutrition while significantly lowering Mg nutrition. In the long experiment the only significant effect was an increase of the Mg concentration in the foliage by liming. The element concentrations in the fine roots did not show as many significant treatment effects. Liming significantly increased Ca

Table 2. Effects of liming and gypsum treatments on seedling growth; presented are n = number of seedlings and mean values of seedling height (H in cm), seedling diameter (D in mm), number of leaves (Nl), dry weight of leaves (Wl in g seedling^{-1}), dry weight of stem and branches (Wst in g seedling^{-1}), total shoot dry weight (Wtot in g seedling^{-1}). The different letters indicate significant treatment effects at p<0.05 level (st. errors in italic)

Treatment	n	H		D		Nl		Wl		Wst		Wtot	
Short experiments (3 to 5 months)													
Control	35	13.2b	*0.9*	3.7b	*0.2*	14b	*1.0*	0.70b	*0.1*	0.40b	*0.1*	1.1b	*0.2*
$CaCO_3$	34	17.3a	*1.4*	5.4a	*0.5*	21a	*1.6*	1.4a	*0.2*	0.74a	*0.1*	2.1a	*0.3*
$CaSO_4$	32	13.6b	*1.0*	3.7b	*0.2*	14b	*1.2*	0.76b	*0.1*	0.45b	*0.1*	1.2b	*0.1*
Long experiment (13 months)													
Control	6	17.9	*4.0*	5.5	*0.3*	19	*3.1*	1.5	*0.3*	1.1	*0.3*	2.6	*0.6*
$CaCO_3$	6	19.2	*3.3*	6.2	*1.1*	28	*9.4*	1.7	*0.6*	1.2	*0.5*	2.9	*1.1*
$CaSO_4$	6	13.2	*0.8*	6.3	*0.3*	19	*5.5*	1.1	*0.3*	0.8	*0.2*	1.9	*0.5*

Table 3. Effects of lime and gypsum applications on fine root biomass (FRB sum of A and B layer in g rhizotron^{-1}), fine root length (FRL sum of A and B layer in m rhizotron^{-1}) and specific root length (per layer in m g^{-1}); with standard errors in italic (no significant overall effects for the short experiments; the long experiment was not tested due to low sample number)

Treatment	n	FRB		FRL		SRL (A)		SRL (B)	
Short experiments (3 to 5 months)									
Control	12	1.91	*0.33*	24.1	*3.5*	12.5	*1.8*	20.6	*3.7*
$CaCO_3$	12	2.24	*0.32*	34.5	*4.8*	15.0	*1.4*	20.5	*3.6*
$CaSO_4$	12	1.98	*0.39*	25.4	*3.7*	15.4	*2.9*	23.6	*4.8*
Long experiments (13 months)									
Control	2	2.48	*0.18*	37.0	*0.4*	14.9	*4.6*	16.8	*3.4*
$CaCO_3$	2	3.02	*0.41*	42.7	*7.9*	16.2	*1.7*	12.9	*0.2*
$CaSO_4$	2	1.90	*0.05*	32.8	*3.8*	16.0	*0.5*	19.2	*3.4*

concentration in the A-layer and the gypsum application resulted in significantly higher S concentrations in the B-layer and lower Mn concentrations in the fine roots of the A-layer. In the long experiments (data not shown) treatment effects were not tested (n=2), but mineral concentrations in the fine roots were in the same range as in the short experiments. The element concentrations of the coarse roots only revealed lower Al concentrations after liming and gypsum as compared to the control.

Table 5 presents the total mineral element content in the seedlings (shoot + root) in the 4th, short, experiment only (4 rhizotrons each). Since the number of seedlings was different for the treatments, the summed content values were divided by the number of seedlings. An ANOVA on these summed content values per seedling per rhizotron, showed that liming significantly increased total content of P, Mg, Ca and K as compared to the control. Both liming and gypsum increased S content as compared to the control. The percentage allocation to the shoot is significantly increased by liming for S, P and K as compared to the control (Table 6). Mg allocation was significantly higher after liming when compared to the gypsum treatment.

In Figure 2 the overall effects (for the three short experiments) on shoot:root (dry weight) ratio are presented (standardized values). For the non-standardized values, shoot:root ratios (means + standard errors) ranged from 1.24±0.17, 1.52±0.39 to 1.90±0.41 for the control, gypsum and liming treatments, respectively. The differences between the

Table 4. Effect of lime and gypsum applications on foliar mineral concentrations (in the short experiments based on 11, 10 and 8 seedlings for C, L and G, respectively, and in the long experiment on 6, 6 and 5 seedlings, respectively) and root mineral concentrations (based on 4 rhizotrons each for the short experiments; the long experiment is not presented) in g kg^{-1} dry weight (letters indicating significant differences at p<0.05; st. errors in italic)

	N		S		P		K		Ca		Mg		Mn		Al		Ca/Al	
	g kg^{-1}																(mol/mol)	
Foliage: short experiments (3 to 5 months)																		
Control	23.8b	*1.1*	1.9	*0.2*	1.2	*0.3*	10.1	*1.7*	5.2b	*0.3*	1.5a	*0.1*	4.2	*0.5*	0.2a	*0.0*	22.7b	*2.2*
$CaCO_3$	23.5b	*0.7*	1.7	*0.1*	1.3	*0.1*	7.9	*0.3*	7.0a	*0.7*	1.3a	*0.1*	4.0	*0.2*	0.1b	*0.0*	39.9a	*5.2*
$CaSO_4$	27.2a	*1.5*	2.0	*0.1*	0.8	*0.1*	6.0	*0.4*	5.3b	*0.4*	0.9b	*0.1*	4.4	*0.4*	0.2a	*0.0*	18.5b	*1.1*
Foliage: long experiment (13 months)																		
Control	28.4	*1.1*	1.9	*0.1*	1.0	*0.3*	7.0	*0.3*	3.21	*0.5*	1.4b	*0.1*	1.9	*0.1*	0.1	*0.0*	75.8	*11.7*
$CaCO_3$	28.1	*1.2*	2.0	*0.1*	1.6	*0.1*	7.9	*0.4*	4.72	*0.4*	1.7a	*0.1*	2.1	*0.0*	0.1	*0.0*	107	*8.3*
$CaSO_4$	30.8	*1.7*	1.7	*0.2*	1.5	*0.3*	6.9	*0.7*	3.92	*0.4*	1.3b	*0.2*	1.8	*0.3*	0.1	*0.0*	73.8	*3.9*
Fine-roots, A-layer (short experiments only)																		
Control	12.1	*1.3*	1.79	*0.3*	1.3	*0.3*	7.3	*0.5*	3.1b	*0.5*	1.4	*0.1*	0.6a	*0.0*	4.3	*0.7*	0.54	*0.15*
$CaCO_3$	16.9	*2.5*	1.88	*0.4*	1.5	*0.3*	7.4	*0.5*	4.4a	*0.2*	1.6	*0.1*	0.5ab	*0.1*	7.6	*1.6*	0.45	*0.08*
$CaSO_4$			2.78	*0.3*	1.5	*0.2*	7.1	*0.5*	3.3b	*0.1*	1.6	*0.2*	0.4b	*0.1*	6.5	*0.9*	0.36	*0.04*
Fine-roots, B-layer (short experiments only)																		
Control	14.4	*1.4*	1.7b	*0.2*	0.8	*0.1*	7.4	*1.8*	2.3	*0.3*	1.5	*0.1*	0.8	*0.0*	8.8	*1.0*	0.18	*0.02*
$CaCO_3$	15.2	*0.9*	1.5b	*0.1*	0.9	*0.1*	4.3	*2.2*	3.0	*0.3*	1.4	*0.1*	0.8	*0.1*	10.7	*1.7*	0.20	*0.02*
$CaSO_4$	16.2	*0.5*	2.6a	*0.2*	1.2	*0.3*	7.1	*0.2*	3.0	*0.3*	1.6	*0.1*	0.7	*0.1*	10.1	*0.6*	0.21	*0.02*

Table 5. Total amount of elements in the seedlings in mg seedling^{-1}, after five months of growth as a function of lime or gypsum treatments based on only one of the short experiments (small letters denoting significant treatment effects)

Treatment	S		P		K		Ca		Mg		Mn		Al	
Control	9.94b	*1.9*	9.29b	*2.5*	27.0b	*6.1*	16.0b	*4.1*	4.98b	*1.1*	8.69	*2.5*	8.91	*2.8*
$CaCO_3$	18.6a	*1.3*	24.5a	*2.0*	52.0a	*6.0*	42.3a	*5.4*	9.60a	*0.8*	15.2	*3.0*	14.2	*1.8*
$CaSO_4$	20.2a	*2.6*	13.3b	*2.6*	30.9b	*5.3*	23.4b	*4.4*	6.40ab	*1.2*	10.3	*2.2*	11.4	*2.8*

Table 6. Percentage allocation of nutrients to the shoot after five months of growth based on only one of the short experiments (values are averages over 4 rhizotrons; letters denoting significant differences)

Treatment	S	P	K	Ca	Mg	Mn	Al
Control	0.43b	0.40b	0.42b	0.59	0.42ab	0.73	0.02
$CaCO_3$	0.56a	0.57a	0.53a	0.64	0.51a	0.79	0.02
$CaSO_4$	0.33b	0.43b	0.37c	0.51	0.32b	0.71	0.03

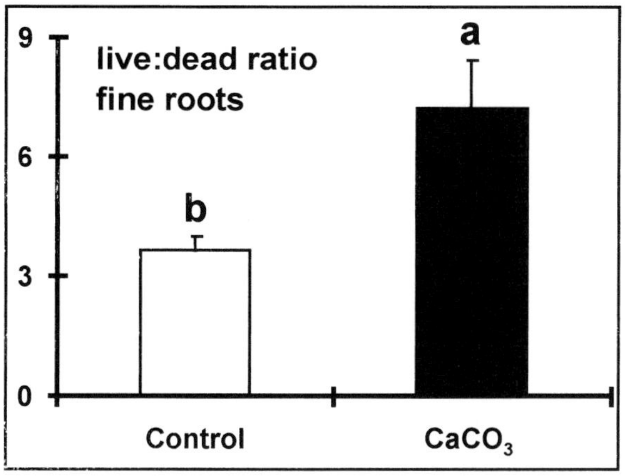

Figure 1. Live:dead ratio of fine roots (<2 mm dia.) on a weight basis (after 13 months); the letters indicate a significant treatment effect at $p<0.05$).

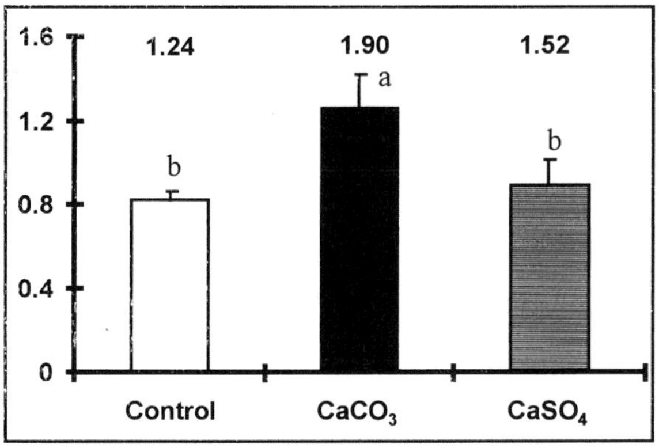

Figure 2. Shoot:root ratios summed for the three short experiments (standardized values); values indicated in the figure are the non-standardized means of the three experiments (the letters indicate significant treatment effects at ($p<0.05$).

non-standardized values were not significant when regrouping the three short experiments, but in the last experiment (with acorns), liming had a significantly higher shoot:root ratio than the gypsum and the control treatment. Since variation between the experiments was considerable, perhaps related to the fact that the latter experiment was carried out with acorns, a standardization was applied by dividing the values by the mean of each experiment (Figure 2). On the basis of these standardized values, liming significantly ($p<0.05$) increased the shoot:root ratio.

The soil

Figure 3 presents the treatment effects for the bulk soil. Liming significantly enhanced exchangeable Ca concentration, CEC, base saturation and Ca/Al ratio in the A-layer, while significantly decreasing exchangeable Al and H^+. Gypsum significantly increased exchangeable Ca, base saturation and Ca/Al ratio in both A and B-layers. In the A-layer, gypsum significantly increased CEC and decreased exchangeable Mn, Mg and Al, while in the B-layer exchangeable Mn and Mg were significantly increased. Concentrations of Na (0.16 and 0.17 $cmol_c\ kg^{-1}$ in A and B, respectively) and K (0.32 and 0.25 $cmol_c\ kg^{-1}$, respectively) were not affected by the treatments. Effects on soil pH were limited to the A-layer, with gypsum showing a significantly lower pH-H_2O relative to both the other treatments, and for pH-KCl, significant increases for both applications ($CaCO_3$>$CaSO_4$>control). These results reflect the higher solubility of the gypsum product as compared to the lime product, resulting in a greater shock reaction and reactivity extending to greater soil depth, displacing elements like Mn, Mg and Ca from the A to the B-layers. The effects of the treatments on the concentrations in the rhizosphere were similar, but weaker than those in the bulk soil.

The differences between bulk and rhizosphere soil are presented in Table 7. As can be seen (and also in Figure 3) the liming and gypsum treatments affected overall concentrations in bulk and rhizosphere soil, but did not have that much effect on the gradient between bulk and rhizosphere, meaning that these gradients were comparable for all three treatments for most elements. In general, most exchangeable elements were more concentrated in the rhizosphere than in the bulk soil, regardless of treatment or soil layer (not significant everywhere). The most important exceptions to this are lower values of exchangeable Mn in the rhizosphere (significant only for the control) and Ca (significant only for liming), as well as the Ca/Al ratio (significant for both lime and gypsum treatment), in the A-layer as compared to the bulk soil. The CEC did not differ significantly between bulk and rhizosphere soil, neither in the A nor in the B-layer. Base saturation was significantly lower in the rhizosphere in the A-layer after liming and higher in the B layer for the control and the liming treatment, as compared to the bulk soil. So, the liming and gypsum treatments mainly affected overall soil chemistry of the bulk soil and thereby the rhizosphere soil chemistry, whereas effects on the gradient between bulk and rhizosphere were limited to some effects on base saturation, exchangeable Ca and Ca/Al ratio in the A-layer. The significantly lower values of Ca in the liming treatment in the A-layer, suggest a relatively high uptake of Ca

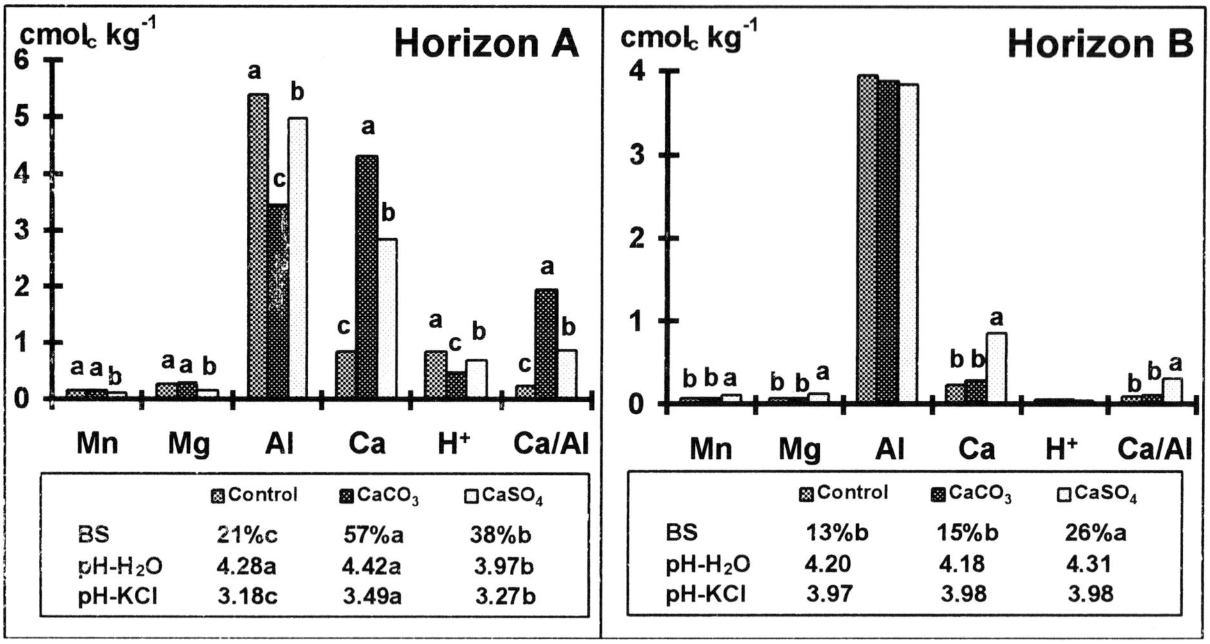

Figure 3. Effects of lime and gypsum applications on the bulk soil in the A and B-layers (letters indicating significant treatment effects at least at $p<0.05$; Na and K not presented).

here as compared to the other two treatments. This is confirmed by the observation of higher Ca tissue concentrations and contents (Tables 4 and 5).

The soil solution composition is presented in Table 8. Generally the gypsum application resulted in the highest solution concentrations (Mn, Mg, Ca, Al, NH_4^+) in both soil layers. The difference in the form of nitrogen is remarkable: a significant increase of NO_3^- by liming as compared to the control treatment (A-layer) and an overall decrease due to the gypsum treatment in both layers as compared to the other treatments. Conversely, gypsum significantly increased NH_4^+ concentration in both layers. Again the higher solubility (leading to higher solution concentrations to a greater depth more quickly) of gypsum as compared to the lime product, is very clear. Whereas $CaCO_3$ had more effect on pH and Ca/Al ratio in the A-layer, the opposite was true in the B-layer. S concentrations increased logically due to $CaSO_4$ application. Na (0.27 and 0.26 mmol$_c$ l^{-1} for A and B-layers, respectively) and K (0.43 and 0.35, respectively) were not significantly affected by the treatments.

In Figure 4 effects of the treatments on the rhizosphere pH are shown, based on comparisons from measured pH values at the rhizoplane with values obtained in the bulk soil at 1 cm from the root. It can be seen that the pH at the root tip is higher (sign. at $p<0.05$) than the pH of bulk soil close to the tip for the control and liming treatment. Furthermore, liming affected the extent of the zone where these differences occur along the axis of the root. The increase of the pH in the rhizosphere was significant ($p<0.05$) even at 2 cm from the root tip. The gypsum treatment had no significant effect on the rhizosphere pH at all. These observations suggest a different uptake balance of cations and anions according to the treatments, perhaps mpre favourable uptake conditions (over a greater part of the root length) after liming.

Discussion

Plant growth

In the present work liming consistently and significantly stimulated seedling growth throughout the experiments. Considering the very low levels of Ca^{2+} and Mg^{2+} and low levels of K^+ in the soil used for the experiments, an addition of Ca, a limiting nutrient, was expected to increase shoot growth. Liming, especially when it also affects the N availability, is known to stimulate seedling growth (Belkacem and Nys, 1995; Gobran et al., 1993; Hüttl and Zöttl et al., 1993). In the present study, root growth also appeared

Table 7. Effect of lime and gypsum applications on exchangeable elements in bulk and rhizosphere soil: the rhizospheric gradient is based only on the short experiments (different letters denoting significant differences at $p < 0.05$ between bulk and rhizosphere soil)

		K		Ca		Mg		Na		Mn		H^+		Al		Ca/Al	
		$cmol_c\ kg^{-1}$														(mol/mol)	
A-layer (short experiments only)																	
Control	bulk	0.31b	*0.01*	0.83	*0.03*	0.27	*0.01*	0.06b	*0.01*	0.16a	*0.00*	0.85	*0.04*	5.40b	*0.11*	0.23a	*0.01*
Control	rhizo	0.41a	*0.03*	0.83	*0.10*	0.35	*0.06*	0.28a	*0.07*	0.13b	*0.01*	1.03	*0.13*	6.06a	*0.19*	0.17b	*0.02*
$CaCO_3$	bulk	0.29b	*0.01*	4.32a	*0.18*	0.30	*0.01*	0.09b	*0.01*	0.17	*0.02*	0.47b	*0.04*	3.44b	*0.15*	1.94a	*0.15*
$CaCO_3$	rhizo	0.38a	*0.04*	3.02b	*0.26*	0.35	*0.05*	0.25a	*0.04*	0.13	*0.01*	0.79a	*0.13*	4.95a	*0.22*	0.92b	*0.14*
$CaSO_4$	bulk	0.27b	*0.01*	2.84	*0.15*	0.17b	*0.02*	0.06b	*0.01*	0.11	*0.01*	0.68	*0.04*	4.97b	*0.15*	0.86a	*0.03*
$CaSO_4$	rhizo	0.40a	*0.04*	2.62	*0.15*	0.27a	*0.05*	0.18a	*0.03*	0.12	*0.01*	0.82	*0.16*	5.44a	*0.16*	0.66b	*0.03*
B-layer (short experiments only)																	
Control	bulk	0.17b	*0.01*	0.23	*0.04*	0.07b	*0.01*	0.07b	*0.02*	0.07b	*0.01*	0.06	*0.02*	3.95	*0.08*	0.09	*0.02*
Control	rhizo	0.36a	*0.05*	0.34	*0.07*	0.20a	*0.05*	0.34a	*0.08*	0.11a	*0.01*	0.08	*0.03*	3.96	*0.10*	0.07	*0.01*
$CaCO_3$	bulk	0.17b	*0.01*	0.29b	*0.02*	0.08b	*0.01*	0.10b	*0.02*	0.07b	*0.01*	0.06	*0.02*	3.87	*0.05*	0.11	*0.01*
$CaCO_3$	rhizo	0.32a	*0.03*	0.48a	*0.06*	0.20a	*0.04*	0.27a	*0.04*	0.14a	*0.02*	0.08	*0.03*	3.91	*0.12*	0.14	*0.02*
$CaSO_4$	bulk	0.20b	*0.01*	0.85	*0.23*	0.13b	*0.01*	0.08b	*0.01*	0.10b	*0.01*	0.04	*0.02*	3.84	*0.14*	0.31	*0.06*
$CaSO_4$	rhizo	0.33a	*0.03*	0.86	*0.06*	0.27a	*0.05*	0.18a	*0.03*	0.14a	*0.01*	0.08	*0.03*	3.81	*0.13*	0.29	*0.02*

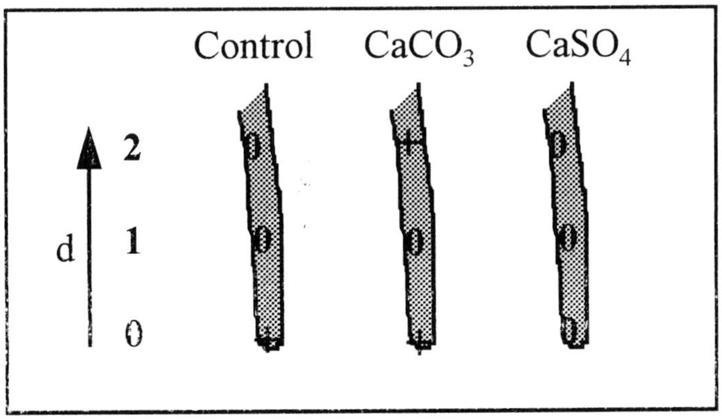

Figure 4. Effects of liming and gypsum applications on the pH at the rhizoplane of fine roots of the B horizon as a function of the distance 'd' from the root apex in cm (presented only up to 2 cm from the apex). Also indicated are pH values that were significantly higher at $p<0.05$ at the rhizoplane as compared to the bulk soil at that particular point (+) or not significantly different (0) between the rhizoplane and bulk soil.

to be stimulated by lime (Table 3), but less so than the stimulation of shoot growth. Effects on fine root length were more consistent than those on fine root weight, so that specific root length tended to be increased by liming. Due to this higher stimulation of the shoot compared to the root, shoot:root ratio increased as a result of liming. This is similar to the observations by Gobran et al. (1993) and Tagliavini et al. (1995), but must be considered in relation to the lime-induced stimulation of N-mineralization occurring on the same soil (Belkacem and Nys, 1995, 1997), which is known to increase shoot:root ratio (Beck, 1996; Seth et al. 1996).

In contrast to the stimulation of growth by liming, gypsum did not stimulate shoot nor root growth clearly, in spite of the similar doses of Ca added. Gypsum even seems to result in lower survival in the experiment where seedlings were grown directly from acorns, or to decrease (not-significantly) above- and below-ground seedling growth in the longest experi-

Table 8. Effects of treatments on soil solution based only on the short experiments (standard errors in italic and significant differences indicated by different letters)

	n	pH	Ca		Mg		Mn		Al		NO_3^-		NH_4^+		Ca/Al	
			mmol$_c$ l^{-1}												(mol/mol)	
A-layer																
Control	18	3.85b	0.31c	*0.04*	0.19b	*0.03*	0.07b	*0.01*	0.29b	*0.01*	1.76b	*0.35*	2.68b	*0.27*	1.55c	*0.17*
$CaCO_3$	17	4.12a	1.51b	*0.08*	0.32b	*0.02*	0.11b	*0.01*	0.20b	*0.01*	3.18a	*0.14*	2.75b	*0.14*	11.4a	*0.51*
$CaSO_4$	16	3.46c	4.73a	*0.31*	0.69a	*0.08*	0.34a	*0.03*	1.07a	*0.11*	0.48c	*0.07*	4.06a	*0.24*	7.38b	*0.56*
B-layer																
Control	17	3.98b	0.99b	*0.07*	0.53b	*0.03*	0.46b	*0.02*	0.88	*0.08*	3.83a	*0.05*	1.51b	*0.19*	1.80b	*0.11*
$CaCO_3$	13	3.97b	1.25b	*0.06*	0.57b	*0.03*	0.46b	*0.03*	1.20	*0.13*	3.94a	*0.03*	1.33b	*0.18*	1.70b	*0.12*
$CaSO_4$	15	4.06a	1.99a	*0.23*	0.87a	*0.06*	0.56a	*0.03*	1.10	*0.10*	3.36b	*0.23*	2.34a	*0.25*	2.97a	*0.41*

ment (13 months). This would suggest a more hostile rooting environment, perhaps leading to a higher mortality of roots and a lower growth over time. Unfortunately, the amount of dead roots was not assessed in the gypsum application, so this could not be verified in this rhizotron experiment. In the field trial on the same substrate, fine root mortality did not significantly differ between the gypsum and the liming treatment (Bakker, 1999) and both had a lower mortality than the control. Other studies indicated that ammonium-sulphate increased fine root necromass (Majdi and Rosengren-Brinck, 1994) or lowered fine root vitality (Clemensson-Lindell and Persson, 1995), and that low live/dead root ratios are a result of high Al concentrations and root damage (Gobran et al., 1993; Persson et al., 1995). The observed higher levels of NH_4^+ (Table 8) and SO_4^{2-} (by means of the application of gypsum) in the soil solution for the gypsum treatment in this work, agrees with these studies.

The observed lime-induced stimulation of live/dead ratios of fine roots (Figure 1), should thus be interpreted in terms of better health and higher survival. Such an increase in survival can be accompanied by lower fine root production, as observed after N-free fertilization (Persson et al., 1995), but higher survival can also be the result of nutrient rich patches, where root longevity is higher (Robinson, 1994). One may speculate about the origin of this apparently higher mortality in the control treatments relative to the liming treatment. Higher mortality of fine roots could be an adaptation to high Al toxicity, leaving behind large amounts of Al in the dying roots. The absence of growth stimulation by $CaSO_4$ in this work is in contrast with the work of Carvalho and Van Raij (1997), who reported equal or slightly higher growth stimulation by $CaSO_4$ than $CaCO_3$ of shoot and root growth for maize as compared to the control. This difference could be related to the fact that maize, a plant species, has more difficulty in growing in an acid subsoil as compared to oak seedlings, so that the beneficial effect of $CaSO_4$ (higher Ca activity, Al complexation by sulfate) is more important in their work.

The observed concentrations of nutrients in foliage (Keltjens and Van Loenen, 1989; Ljungström and Nihlgård, 1995) and fine roots (Majdi and Rosengren-Brinck, 1994; Persson et al., 1995; Yin et al., 1991) in this study are in the range of values commonly reported for other tree species, despite the fact that part of the fine root values in these studies are based on coniferous species. Only the foliar levels of Ca, Mg and P are somewhat low, especially for gypsum in the case of Mg, and P (Bonneau, 1995). The significant increases of Ca in fine roots and foliage by liming and of S in roots by gypsum, were expected. Such increases of foliar levels of Ca after lime and Mg in the case of dolomitic lime, are commonly observed (Ljungström and Nihlgård, 1995; Van Praag et al., 1997). The fact that gypsum did not improve Ca, but increased N while decreasing Mg nutrition, was more unexpected. This probably reflects the higher concentrations of Al^{3+} and NH_4^+ in the soil solution or at the adsorption sites of the root apoplasm, reducing uptake of Ca^{2+}, Mg^{2+} and K^+ (Keltjens and Van Loenen, 1989; Keltjens and Van Ulden, 1987; Kölling et al., 1997; Ljungström and Nihlgård, 1997; Majdi and Rosengren-Brinck, 1994).

Liming significantly enhanced total uptake of S, P, Mg, Ca and K, and percentage allocation of S, P and K to the shoot. In contrast, gypsum only stimulated total

S uptake, but not percentage allocation to the shoot. This higher allocation to the shoot by liming confirms other observations on increased allocation of nutrients such as P to the shoot (Gobran et al., 1993), or a general increase in biomass (carbon) allocation to the shoot when nutrient availability increases (Ericsson, 1995). In the present study, despite a significantly higher proportion of biomass allocated to the roots in the control and gypsum treatment as compared to the liming treatment, uptake (total content) was higher in the liming treatment, except for S in the gypsum treatment. On the contrary, liming significantly increased uptake of nutrients (Table 5) as compared to gypsum and in particular the control treatment. Plants with a fertilizer supply in the immediate locality in nutrient limiting soils, have been shown to have very high uptake rates (Robinson, 1994). Apparently, availability and accessibility of most cations in sufficient quantities in the liming treatment, permits the seedlings to allocate more of their biomass to shoot growth. Since the gypsum treatment added the same amounts of Ca to the soil, an explanation for the absence of a positive effect on growth, shoot:root and uptake, has to be sought in the difference in bulk and rhizosphere chemistry of soil and solution, impeding uptake and reducing root function.

The soil

The effects of liming and gypsum on the bulk soil and bulk soil solution are comparable to those more generally observed (Belkacem and Nys, 1995, 1997). In the soil, the slight increase of CEC is thought to be related to the pH dependent character of the CEC (Bakker et al., 1999). The lower concentrations of exchangeable Mn, Mg and Al in the A-layer compared with a significant increase of Mn and Mg in the B-layer, supports the evidence of displacement of these cations by gypsum. The method used for separating rhizosphere from bulk soil may be criticized as being rather approximate, and reducing the existing differences between bulk and rhizosphere soil. Nevertheless, the concentration effect of most elements in the rhizosphere is in accordance with the observations by Guan (1997). Such a concentration of nutrients in the rhizosphere is related to higher concentrations of exudates, bacteria and organic matter, but also depends on the rate of uptake and release, the mobility of the ions in the soil and the rate of conversion between available and unavailable forms (Darrah, 1993). In the liming treatment, Ca concentration was significantly lower in the rhizosphere, suggesting that uptake was higher than diffusion plus mass flow to the roots.

The solution was affected much more by gypsum than liming as compared to the control. Although Ca concentrations also increased and should have increased Ca availability even more than the liming treatment, gypsum had a negative influence on the soil solution chemistry. First, for Mg and Ca, a shift of the exchangeable ions from the A into the B-layer can be observed, indicating potential leaching losses, and, second, the increase of solution concentrations in both layers of Mn^{2+}, NH_4^+ and Al^{3+}, increases the risk of phytotoxic levels of these ions (Keltjens and Van Ulden, 1987; Keltjens and Van Loenen, 1989). Aluminium toxicity, but also high acidity, are known to have a detrimental effect on root elongation and can lead to deficiencies of nutrients like Mg as a result of impaired uptake (Marschner, 1991). A treatment with ammonium sulphate in Scandinavia (Majdi and Rosengren-Brinck, 1994) also resulted in high NH_4^+ concentrations and a reduced Mg and Mn uptake. This antagonism between Mg and NH_4^+ is confirmed by the higher N and lower Mg levels in the foliage of the gypsum treatment in this study.

The effects on root pH confirm the earlier observations on solution, seedling nutrition and plant growth. The liming treatment has the largest effect on root-induced pH increase and the length of the zone where this increase occurs, is larger for the liming treatment. As ammonium and nitrate weigh heavily in the cation/anion uptake balance, this pH increase means a higher NO_3^-/NH_4^+ uptake ratio, or that more OH^-/HCO_3^- ions are excreted into the soil (Darrah, 1993; Gijsman, 1990; Keltjens and Van Ulden, 1987). Gijsman (1990) concluded that the proportion of N taken up as NO_3^- was very important for root growth and the function of Douglas-fir. When the proportion N–NO_3^- is greater than 65%, there was favourable root growth due to carboxylate and OH^- excretion, between 20 and 65% alkalization only occurred at the root tip, and below 20% physological disturbances occurred (Gijsman, 1990). This beneficial effect of NO_3^- uptake compared to NH_4^+ uptake was also reported by Seith et al. (1996), who found increased uptake of the cations Ca^{2+}, Mg^{2+} and Mn^{2+} with higher NO_3^- uptake.

In the present work, fine root length was only moderately stimulated by liming, but uptake rate was greatly improved. If mass flow is sufficient to fulfil the needs for uptake of the seedlings, an increased fine root system is not expected to increase uptake. For nutrients that are more generally present in insufficient quantities (or having a limited mobility) in the soil solution such as P, K, NH_4^+ (Yin et al., 1991), an increase of the fine root system, may be effective for increasing uptake. This suggests that the uptake conditions at the uptake sites of the root (root tips) have improved considerably in the liming treatment, as total amount taken up of nearly all elements was significantly increased, much more than the small increase in fine root length can explain.

Conclusions

Soil chemistry and development of the oak seedlings were affected differently by liming and gypsum treatments relative to the control. Although both liming and gypsum added the same quantities of Ca to the Ca poor soil, liming was much more effective than gypsum in improving soil and plant fertility status and seedling growth as compared to the control. Whereas liming had more gradual effects, gypsum had a larger and deeper reaching effect on soil and soil solution. It is concluded that the stimulation of seedling growth and nutrition, the higher immobilization and percentage allocation of biomass and nutrients to the shoot by liming, are the result of an improved root environment and function. Nutrient availability is improved and Al toxicity alleviated as a result of liming. Gypsum improved nutrient availability to some extent, but also increased aluminium and ammonium activity, inhibiting the improvement of root function and seedling development.

Acknowledgements

This research was supported by grants from INRA Paris and the Dutch Ministry of Agriculture, Nature Management and Fisheries, Directorate of Science and Knowledge Dissemination. The authors also acknowledge the valuable comments of two anonymous referees about an earlier version of this manuscript.

References

Bakker M R 1999 The effect of lime and gypsum applications on a sessile oak (*Quercus petraea* (M.) Liebl.) stand at La Croix-Scaille (French Ardennes). II. Fine root dynamics. Plant Soil 206, 109–121.

Bakker M R, Nys C and Picard J-F 1999 The effect of liming and gypsum applications on a sessile oak (*Quercus petraea* (M.) Liebl.) stand at La Croix-Scaille (French Ardennes). I. Site characteristics soil chemistry and aerial biomass. Plant Soil 206, 99–108.

Belkacem S and Nys C 1995 Consequences of liming and gypsum top-dressing on nitrogen and carbon dynamics in acid forest soils with different humus forms. Plant Soil 173, 79–88.

Belkacem S and Nys C 1997 Effects of liming and gypsum regimes on chemical characteristics of an acid forest soil and its leachates. Ann. Sci. For. 54, 169–180.

Beck E H 1996 Regulation of shoot/root ratio by cytokinins from roots in *Urtica dioica*: Opinion. Plant Soil 185, 3–12.

Bonneau M 1995 Fertilisation des forêts dans les pays tempérés. Théorie, bases du diagnostic, conseils pratiques, réalisations expérimentales. ENGREF, Nancy. 367 p.

Carvalho M C S and Van Raij B 1997 Calcium sulphate, phosphogypsum and calcium carbonate in the amelioration of acid subsoils for root growth. Plant Soil 192, 37–48.

Clemensson-Lindell A and Persson H 1995 Fine-root vitality in a Norway spruce stand subjected to various nutrient supplies. Plant Soil 168/169, 167–172.

Darrah P R 1993 The rhizosphere and plant nutrition: a quantitative approach. Plant Soil 155/156, 1–20.

Ericsson T 1995 Growth and shoot:root ratio of seedlings in relation to nutrient availability. Plant Soil 168/169, 205–214.

Espiau P and Pedro G 1980 Caractérisation du complexe d'échange des sols acides. Le taux d'acidité d'échange et sa signification pédogénètique sous climat tempéré. Ann. Agron. 31, 363–383.

Gijsman A J 1990 Rhizosphere pH along different root zones of Douglas-fir (*Pseudotsuga menziesii*), as affected by source of nitrogen. Plant Soil 124, 161–167.

Gobran G R, Fenn L B, Persson H and Al Windi I 1993 Nutrition response of Norway spruce and willow to varying levels of calcium and aluminium. Fert. Res. 34, 181–189.

Göransson A and Eldhuset T D 1987 Effects of aluminum on growth and nutrient uptake of *Betula pendula* seedlings. Physiol. Plant. 69, 193–199.

Guan X 1997 Nutrient availability in forest soils. Rhizospheric and sequential leaching studies. PhD Thesis. Acta Universitatis Agriculturae Sueciae Silvestria 36. Uppsala, Sweden.

Hendriks L and Jungk A 1981 Erfassung der Mineralstoffverteilung in Wurzelnähe durch getrennte Analyse von Rhizo- und Restboden. Z. Pflanzenernähr. Bodenk. 144, 276–282.

Hüttl R F and Zöttl H W 1993 Liming as a mitigation tool in Germany's declining forests – reviewing results from former and recent trials. For. Ecol. Manage. 61, 325–338.

Keltjens W G and Van Loenen E 1989 Effects of aluminium and mineral nutrition on growth and chemical composition of hydroponically grown seedlings of five different forest tree species. Plant Soil 119, 39–50.

Keltjens W G and Van Ulden P S R 1987 Effects of Al on nitrogen (NH_4^+ and NO_3^-) uptake, nitrate reductase activity and proton release in two sorghum cultivars differing in Al tolerance. Plant Soil 104, 227–234.

Kölling C, Pauli B, Häberle K-H and Rehfuess K E 1997 Magnesium deficiency in young Norway spruce (*Picea abies* [L.] Karst.) trees induced by NH_4NO_3 application. Plant Soil 195, 283–291.

Ljungström M and Nihlgård B 1995 Effects of lime and phosphate additions on nutrient status and growth of beech (*Fagus sylvatica* L.) seedlings. For. Ecol. Manage. 74, 133–148.

Majdi H and Rosengren-Brinck U 1994 Effects of ammonium sulphate application on the rhizosphere, fine-root and needle chemistry in a *Picea abies* (L.) Karst. stand. Plant Soil 162, 71–80.

Marschner H 1991 Mechanisms of adaptation of plant roots to acid soils. Plant Soil 134, 1–20

Murach D and Schunemann E 1985 Reaktion der Feinwurzeln von Fichten auf Kalkungsmassnahmen. Allgemeine Forstzeitschrift 40, 1151–1154.

Olsthoorn A F M, Keltjens W G, Van Baren B and Hopman M C G 1991 Influence of ammonium on fine root development and rhizosphere pH of Douglas-fir seedlings in sand. Plant Soil 133, 75–81.

Persson H and Ahlström K 1994 The effects of alkalizing compounds on fine-root growth in a Norway spruce stand in southwest Sweden. J. Environ. Sci. Health A29(4), 803–820.

Persson H, Majdi H and Clemensson-Lindell A 1995 Effects of acid deposition on tree roots. Ecol. Bull. 44, 158–167.

Praag H J van, Weissen F, Dreze P and Cogneau M 1997 Effects of aluminium on calcium and magnesium uptake and translocation by root segments of whole seedlings of Norway spruce (*Picea abies* Karst.). Plant Soil 189, 267–273.

Robinson D 1994 The responses of plants to non-uniform supplies of nutrients. New Phytol. 127, 635–674.

Rost-Siebert K 1983 Aluminium-Toxizität und -Toleranz an Keimpflanzen von Fichte (*Picea abies* Karst.) und Buche (*Fagus sylvatica* L.). Allgemeine Forstzeitschrift 38, 686–689.

Schuler G and Zwick N 1992 Die Beeinflussung von Feinwurzelmasse und -vitalität eines Kiefernbestandes (*Pinus sylvestris* L.) mit unterständiger Buche (*Fagus sylvatica* L.) durch pflanzenverfügbare Elemente, sowie Veränderungen nach unterschiedlichen Kalkungsmaßnahmen. Mitt. Forst. Versuchsanstalt Rheinland-Pfalz 21, 69–98.

Seith B, George E, Marschner H, Wallenda T, Schaeffer C, Einig W, Wingler A and Hampp R 1996 Effects of varied soil nitrogen supply on Norway spruce (*Picea abies* [L.] Karst.). I. Shoot and root growth and nutrient uptake. Plant Soil 184, 291–298.

Staaf H, Persson T and Bertills U (Eds) 1996 Skogsmarkskalkning. Resultat och slutsatser från Naturvårdsverkets försöksverksamhet. Naturvärdsverket rapport 4559. Tryck Gotab, Stockholm.

Tagliavini M, Masia A and Quartieri M 1995 Bulk soil pH and rhizosphere pH of peach trees in calcareous and alkaline soils as affected by the form of nitrogen fertilizers. Plant Soil 176, 263–271.

Tennant D 1975 A test of a modified line intersect method of estimating root length. J. Ecol. 63, 995–1001.

Trüby P 1989 Eine Titrationsmethode zur simultanen Bestimmung von H^+ und Aluminium in NH_4Cl-Bodenextrakten. Z. Pflanzenernaehr. Bodenk. 152, 297–300.

Unistat Ltd 1995 Unistat version 4.0 for Windows.

Vogt K A and Persson H 1991 Measuring growth and development of roots. *In* Techniques and Approaches in Forest Tree Ecophysiology. Eds J P Lassoie and T M Hinckley. pp 477–502. CRC Press, Boca Raton.

Yin X, Perry J A and Dixon R K 1991 Temporal changes in nutrient concentrations and contents of fine roots in a Quercus forest. For. Ecol. Manage. 44, 175–184.

A. Stokes (ed.), *The Supporting Roots of Trees and Woody Plants: Form, Function and Physiology*, 419–426.

A new approach to the investigation on the tonogenic groups of root cell walls

N.R. Meychik* and I.P. Yermakov
Plant Physiology Department, Biology Faculty, Moscow State University, 119899 Moscow, Russia

Key words: anion, cation exchange, cell walls, ionogenic groups, pK_a, plant roots

Abstract

Acid-base properties of wheat, lupin, pea root cell walls were investigated. The roots of etiolated and green plants of different age were analysed by the potentiometric method. The ion exchange capacity of root cell walls (S_i) was estimated at various pH values (pH_i 2 to pH_i 12) and constant ion strength of the solution (10 mM). To analyse polysigmoid curves $pH_i = f(S_i)$, Gregor's equation was used. It was shown that Gregor's model fits fairly well the experimental data. The total quantities of cation-exchange (S_t^{cat}) and anion-exchange (S_t^{an}) groups were determined in the root cell walls. It was shown that the quantity of anion exchange groups is varied through a small range (60–185 μmol/g dry wt.) in plant species tested, and that the S_t^{cat} differs widely from 550 to 1300 μmol/g dry wt. For leguininous plants the quantity of acidic groups (fixed anions) is nearly twice as large as that for cereals. It was found that in seedlings as well as in plants, there are 3 cation-exchange groups and one anion-exchange group in root cell walls. The quantity of functional groups of each type (S^j) was estimated, and the corresponding values of n^j and pK_a^j were calculated. It can be assumed that the groups with the $pK_a^1 \approx 3.2$ are amine groups, the ones with $PK_a^2 \approx 5$ are groups of galacturonic acid, the ones with $pK_a \approx 7.5$ are the carboxyl groups of the second species, and the ones with $pK_a^4 \approx 40$ are the phenolic groups. The values of dissociation constants (pK_a^j) and S^j indicate that the root cell walls of wheat, lupin and pea are identical in qualitative structure of ionogenic groups but vary in the quantity of each ionogenic group. It was demonstrated that the summarized quantity of carboxyl groups ($S^2 + S^3$) should be connected directly with the pH gradient in the extracellar space at the membrane surface. The gradient arises from ion-exchange reactions between cations of an outer medium and protons of the ionized carboxyl groups of the cell walls. The results suggest that, S_t^{cat} and S_t^{an} allow the quantitative estimation of ion exchange properties of the cell walls. The resulting parameters (S^j, pK_a^j and n^j) allow prediction of changes in an ionic composition of a medium that bathes the cell membrane, during the first step of mineral nutrition uptake.

Introduction

In higher plants, the apoplast occupies 5% or less of the root cortex's tissue volume (Vakhmistrov, 1967). Structural and ionic characteristics of this small compartment are important for the physiology of the plant because (a) they determine an ionic composition of a medium that bathes the cell membrane, (b) they control an extracellar transport of solutes and (c) they affect mechanical and osmotic phenomena involved in cell growth (Grignon and Sentenac, 1991). As is known, plant cell walls contain high concentrations of uronic acids. All measurements of ionic and electrical properties of the cell walls indicate that the main charges are those of dissociated weak acids, with pK_a values similar to that of polygalacturonic acid (about 3) (Morvan et al., 1979; Starrach, 1985; Richter and Dainty, 1989). The simple calculations with this pK_a value, according to classical Henderson-Hasselbach equation, have shown that the concentration of carboxyl ionogenic groups in plant cell walls must aspire to maximum value at pH 4 ~5, because the degree of dissociation > 99%. Consequently, the

* E-mail: NatalyMeychik@hotmail.com
IPY@1.plantphys.bio.msu.ru

Plant and Soil is the original source of publication of this article. It is recommended that this article is cited as: *Plant and Soil* **217:** 257–264, 1999.

cation exchange capacity at pH 4 $\sim$ 5 should approximate to the maximum magnitude. However, the all known results of potentiometric titration of isolated cell walls indicate that increasing pH in the medium higher than 5, results in increasing the quantity of dissociated acid groups up to pH value 10–11 (Morvan et al., 1979; Richter and Dainty, 1989).

By chemical analyses it has been established that there are several types of ionogenic groups in the composition of different polymer components of the cell walls. There are an amine, a carboxyl, a phenolic, and, very likely, a phosphate groups, that can take part in ion-exchange reactions at certain conditions. But in previous investigations of the acid-base equilibrium in the cell walls, the calculations of the dissociation constant were performed just for the carboxyl groups of polygalacturonic acids, in spite of the fact, that the total ion-exchange capacity was 1.5 – 2 times higher than the concentration of polygalacturonic acidic groups (Morvan et al., 1979; Richter and Dainty, 1989). Authors of these studies have assumed that at pH>7 another class of weak acid groups is ionized, but this statement was not taken into account in the calculations of pK_a value. Consequently, the calculations of fixed charge concentration at corresponding values of medium pH, using the early obtained constant of dissociation, could be invalid. Probably this fact leads to the distinctions between the predicted and the observed magnitudes of cation exchange capacity of the cell walls (Grignon and Sentenac, 1991).

Simultaneously, the experimental and theoretical investigations on physicochemical properties of synthetic polyftmctional cross-linked cation-exchangers (Leykin et al., 1978; Meychik et al., 1989; Shataeva et al., 1979) allow us to raise the problem of an adequate description of acid-base properties of cell wall ion-exchanger. In our study these properties of plant cell walls were investigated over a wide range of pH (2 to 12). The quantity of ionogenic groups of each type and the corresponding pK_a values were estimated for root cell walls.

Materials and methods

Plant material and sample preparation

Roots were obtained from several plants grown in different conditions:

a) 10-days-old lupin seedlings grown in water for 8 days, (*Lupinus albus* L.),
b) etiolated 10-days-old wheat seedlings grown m water for 8 days, (*Triticum aestivum* L.),
c) 38-days-old wheat plants grown in 0.5 N Pryanishnikov medium,
d) 35-days-old pea plants grown in 0.5 N Pryanishnikov medium, (*Pisum sativum* L.).

Plants were raised from seeds at 20–22 °C and 14 h light/10 h dark cycles. Excised roots were fixed by a high temperature (100 °C, 20 minutes), were dried at 50 °C to the constant weight and were stored in glass bags at room temperature until needed.

Isolation of the cell walls was done, as previously indicated (Richter and Dainty, 1989) but several modifications were applied. Each sample of dried roots was placed in ion-exchange column (volume ~0.25 l) and washed in following sequence: 10 mM KOH (volume ~0.5 l), distilled water (volume ~2 l), 10 mM HCl (volume ~0.5 l). All solutions were prepared on distilled water. Finally the sample was washed with distilled water until the absence of chloride-ion in the column eluate. After the washing procedure the sample was dried at 50 °C to the constant weight. This method (named 'standardization') is usually applied for polyflinctional synthetic ion-exchange materials following the process of synthesis and is designed for removal low-molecular components from ion-exchangers and for transfonnation of weakly acidic and basic groups to H-form and 'free' amine form, respectively (Leykin et al., 1978).

Special attention was paid to prove that the sample is chemically stable. To reveal it the standardized sample (~0.1 g) was placed in a flask with a stopper, and 25 ml of 10 mM KOH solution was added. The sample remained in the solution for 24 hours. Before and after equilibration the KOH concentration was determined by titration with 5 mM HCl and an acid-base indicator. Ion-exchange capacity of the cell walls was calculated with the formula:

$$S^{cat(an)} = (C^{in} - C^{eq.})^{*} V/g, \tag{1}$$

where $S^{cat(an)}$ = cation (anion) exchange capacity of isolated cell walls, μmol/g of dry weight (dry wt.); C^{in} and $C^{eq.}$ = initial and final KOH (HCl) concentrations in the solution, mM; V = volume of the solution, ml; g = dry weight of the sample, g.

The whole procedure (standardization and S^{cat} determination) was repeated three times to be absolutely sure in the chemical stability of the sample. It was found that the differences between the observed means of S^{cat} were less than 3% (date not shown).

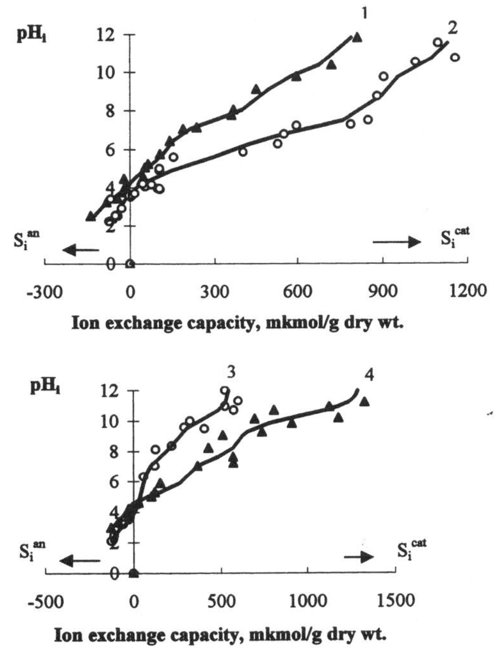

Figure 1. Potentiometric curves of titration of the cell walls. On the abscissa cation exchange capacity (S_i^{cat}, positive means) and anion exchange one (S_i^{an}, negative means) are plotted. S_i^{cat} and S_i^{an} are expressed in inicromole per gram dry weight. The points are experimental data. The lines join the calculated points obtained from equation (3) with parameters of Table 2. 1 - 38-days-old wheat plants; 2 - 10-days-old lupin seedlings; 3 - 10-days-old etiolated wheat seedlings; 4 - 35-days-old pea plants.

Potentiometric titration method. The standardized samples (0.0800 $\pm$ 0.0001 g) were placed in several glass flasks (volume $\sim$ 100 ml of each) with glass stoppers and allowed to equilibrate in a range of KOH or HCl solutions from 0 to 10 mM. The ion strength of the solution in each flask was kept constant (10 mM) by adding KCl salt. The total volume of the solution was 25 ml in each flask. After 24 hours the samples were discarded, and in the solution of each flask the final KOH or HCl concentration was determined by titration with 5 mM HCl or KOH and an acid-base indicator. In each solution the initial (pH_i^{in}) and final or equilibrium (pH_i) pH values were measured by pH-meter. For corresponding pH_i means the calculations of $S_i^{cat(an)}$ values were performed by equation (1).

Results

The experimental curves $pH_i = f(S_i)$ are shown in Figure 1. The dependencies have complicated polysigmoidal character that points to the existence of several different types of ionogenic groups in polymeric cell wall structure of wheat, lupin and pea roots. The maxima of cation-exchange (S_t^{cat}) and anion-exchange (S_i^{an}) capacities are reached at app. pH>10,8 and pH< 3, respectively (Table 1). It means that all ionogenic groups are occupied by K^+ or Cl^-, and S_t^{cat} and S_t^{an} express the total quantities of acidic and basic groups in root cell wall polymeric structure. These groups could be involved in ion exchange reactions at corresponding medium pH. Table 1 shows that the quantity of anion exchange groups is varied through a small range (60–185 μmol/g dry wt.) in plant species tested. S_t^{cat} differs widely (550–1300 μmol/g dry wt.): for legunimous plants the quantity of acidic groups (fixed anions) is nearly twice as large as that for cereals. The dependence S_t^{cat} on the physiological state of roots (compared to etiolated seedlings and 38-days-old plants of wheat) suggests the possibility of applying this parameter to the quantitative characterization of root cell walls.

The experimental curves (Figure 1) were subdivided into j monosigmoidal fragments according to differential curves $(dS_i/dpH_i) = f(pH_i)$. Figure 2 shows that the differential curves have four peaks and several minima (inflexion points). Each minimum approximates to the initial or final points of ionization of the specific group in root cell walls (Leykin et al., 1978; Meychik et al., 1989). This means that for all studied plant species there are four ranges, each by itself represents the titration of the specific ionogenic group of the cell walls. The difference of the ion-exchange capacity between the initial and final points of the titration of the specific ionogenic group approximates to the quantity of ionogenic groups of j-th type (S^j). We emphasize that the indicated method of approximate extent determination of monosigmoidal fragments is unique for polyfunctional ion-exchangers. Other well-known methods of quantitative determination of ionogenic groups in the cell walls are based on chemical destruction of polymeric cell wall structure, for example, the extraction of wall polyphenols, hydrolysis of wall protein (Richter and Dainty, 1989), but it is common knowledge that the extent to which such reactions proceed is always much less than 100%.

By these means, the character of differential curves testifies that there are four ion exchange groups in root cell walls in seedlings as well as in plants. According to the differential curves the S^j values were determined (Table 2).

Table 1. Dependence maxima cation exchange capacity (S_t^{cat}) and anion exchange capacity (S_t^{an}) on plant species. S_t^{cat} and S_t^{an} are the total quantities of acidic and basic groups, respectively, in root cell wall polymeric structure. These groups could be involved in ion-exchange reactions at corresponding means of pH of an outer medium. S_t^{cat} and S_t^{an} are expressed in micromole per gram dry weight (dry wt.) of root cell walls. Means ± SE (3 replicates)

The plant species	S_t^{cat} (μmol/g dry wt.)	S_t^{an} (μmol/g dry wt.)
38-days-old wheat plants	800±30	140±14
10-days-old wheat seedlings	550±25	140±11
10-days-old lupin seedlings	1150±40	60±11
35-days-old pea plants	1300±50	185±20

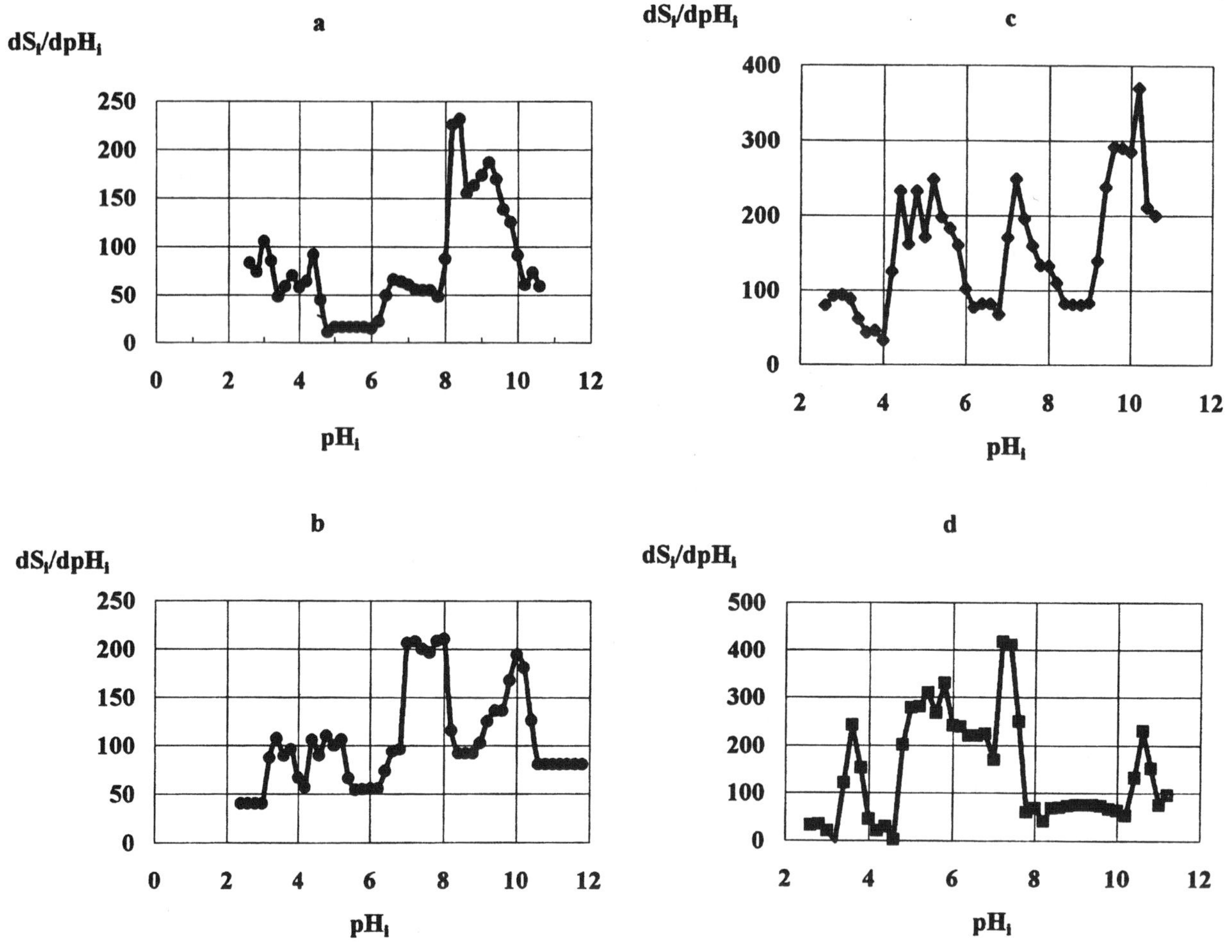

Figure 2. The differential curves obtained from the experimental data of Figure 1. The differential curves have four peaks and several minima. Each minimum approximates to the initial or final points of ionization of the specific group in root cell walls: a - 10-days-old etiolated seedlings of wheat; b - 38-days-old wheat plants; c - 10-days-old lupin seedlings; d - 35-days-old pea plants.

Table 2. The computed parameters of Gregor's equation (2) for each ionogenic group of the cell walls of different plants. Dependence pH_i on $\log[\alpha_i/(1-\alpha_i)]$ is a straight line with y-intercept, the value of which yields $pK_a{}^j$ and with a slope, the value of which yields n^j. The $r^j{}_{corr.}$ is coefficient of correlation. The S^j is the quantity of ionogenic groups of the j-th type

j	$pK_a{}^j$	n^j	$r^j{}_{corr.}$	S^j (μmol/g dry wt.)
		38-days-old wheat plants		
1	3,37	–0,90	0,992	140
2	5,12	0,99	0,989	140
3	7,53	0,93	0,991	330
4	10,11	1,10	0,995	330
		10-days-old wheat seedlings		
1	3,01	–0,91	0,997	140
2	4,52	1,12	0,972	60
3	7,65	1,02	0,964	180
4	10,09	1,04	0,999	310
		10-days-old lupin seedlings		
1	3,28	–1,13	0,961	60
2	5,17	1,16	0,962	400
3	7,07	1,00	0,988	500
4	10,31	1,09	0,984	250
		35-days-old pea plants		
1	3,34	–1,21	0,983	185
2	5,32	0,82	0,999	310
3	7,59	0,91	0,992	310
4	10,38	1,07	0,991	680

To calculate the pK_a value for each ionogenic group, Gregor's equation was used (Gregor et al., 1954; Shataeva et al., 1979). The equation appears as:

$$pH = pK_a + n * log_{10}[\alpha/(1-\alpha)], \qquad (2)$$

where pK_a = apparent constant of dissociation of ionogenic group in cross-linked polymer structure, α = degree of dissociation of the group, α represents the ratio of amount of ionized groups to the total amount of these groups (ionized and unionized), n = constant, depending on structure of cross-linked cell wall polymer. The equation (2) is empirical, but it is usefully employed for description of acid-base processes going on in synthetic polyfunctional ion-exchanger structures (Leykin et al., 1978; Meychik et al., 1989).

Plotting the external pH as function of $\log_{10}[\alpha/(1-\alpha)]$ should thus give a straight line with an intercept on the y axis, the value of which yields pK_a, and with a slope, the value of which yields n constant of the equation (2). The data of Table 2 indicate that such dependence is really the straight line for all types of

Table 3. Parameters of regression analysis for dependencies $S_i{}^{ex} = S_i{}^{cal.} * b + a$, where $S_i{}^{ex}$. is the experimental ion-exchange capacity at pH_i, μmol/g dry wt.; $S_i{}^{cal.}$ is the calculated ion-exchange capacity according to the equation 3 at pH_i, μmol/g dry wt.; $r_{corr.}$ is correlation coefficient

The plant species	b	a	$r_{corr.}$
38-days-old wheat plants	0,96	14,4	0,992
10-days-old wheat seedlings	0,98	4,80	0,996
10-days-old lupin seedlings	0,99	4,78	0,999
35-days-old pea plants	0,95	31,1	0,971

ionogenic groups and for all studied plant species, as evidenced by the correlation coefficients, the values of which are more than 0.97 in all instance. According to the determined S_j value, $\alpha_i = S_i^j / S^j$ at the corresponding experimental pH_i and the equation (2), $pK_a{}^j$ and n^j were calculated for each ionogenic group (Table 2).

Using the derived means of the parameters ($pK_a{}^j$ S^j and n^j), experimental data were tested to conform the equation (Leykin et al., 1978):

$$S_i^{cal.} = S_t^{cat} - \sum_{J,i=1}^{k,m} S^j/[1 + exp10((pK_a^j - pH_i)/n^j)], \qquad (3)$$

where $S_i{}^{cal.}$ = calculated ion-exchange capacity at equilibrium mean of pH_i, μmol/g dry wt.; $S_t{}^{cat}$ = total cation exchange capacity, μmol/g dry wt. (Table 1); S^j = quantity of ionogenic groups of the j-th type, μmol/g dry wt.; $pK_a{}^j$ and n^j = constant of dissociation and constant of equation (2) for the ionogenic groups of the j-th type; k = quantity of experimental points on a curve ($k \geq 23$); m = quantity of types of ionogenic groups.

Fitting the satisfaction of the experimental data to the equation 3 was tested by the following equation of regression analysis:

$$S_i^{ex} = S_i^{cal.} * b + a, \qquad (4)$$

where $S_i{}^{ex}$ experimental ion-exchange capacity at pH_i, μmol/g dry wt.; $S_i{}^{cal.}$ = calculated ion-exchange capacity according to the equation 3 at pH_i, μmol/g dry wt. The calculations show that Gregor's equation fits fairly well the experimental data, as demonstrate the parameters of the equation 4 (Table 3) and the computed curves (Figure 1).

The constant n^j in Gregor's equation (2) defines the deviation of studied systems from an ideal. For all studied plants species and for all ionogenic groups the mean of n^j approximates 1 (Table 2). This fact provides the additional argument for Gregor's model application in the description of acid-base properties of plant root cell walls.

Discussion

It is known that an amine, a carboxyl and a phenolic groups enter into the composition of plant cell walls (Luttge and Higinbotham, 1984). In the pH range covered, these groups are able to dissociate. Our results suggest that there are four ionogenic groups in the polymeric structure of the cell walls, three of them are cation exchange ones, and the forth one exchanges anions. Based on (a) the obtained $pK_a{}^j$ values, (b) the data about chemical composition of plant cell walls and (c) the pK_a values for low-molecular amine, carboxyl and phenolic compounds, we attempt to correlate our potentiometric titration data with the species of ionogenic groups known to exist in plant cell walls. Very likely the region of potentiometric curves at pH 9 to 11, in which the groups with $pK_a{}^4 \sim 10.3$ is ionized, is associated with titration of phenolic groups of root cell walls, because the pK_a of phenol equals 9.98 (Albert and Serzhent, 1964). For low-molecular amines the pK_a is equal to 1 – 3.5 (Albert and Serzhent, 1964), it means that these compounds are weak base. Our experiments have shown that there is no proton release during titration at the region of pH = 2 – 3.8, but there is an acid uptake according to the reaction:

$$R - NH_2 + H^+ + Cl^- \rightarrow [R - NH_3]^+ Cl^-,$$

where R = polymer chain. We assume that groups with $pK_a{}^1 \sim 3.2$ are the amine groups of the cell walls.

It is known that the pK_a of galacturonic acid is 3.42 (Richter and Dainty, 1989). But also it is well known that there is considerable difference between pK_a of low molecular and polymeric carboxyl-contained acids. For example, the pK_a of acrylic acid equals 4.26 (Albert and Serzhent, 1964), the pK_a of polyacrylic acid –4.86 (Shataeva et al., 1979) and the pK_a of ionogenic groups of acrylic acid in the cross-linked polymer structure is equal to 5 – 7.5, depending on the type and quantity of crosslinked agent (Shataeva et al., 1979). Hence, it can be assumed that ionogenic groups with $pK_a{}^2 \sim 5$ (Table 2) are groups of galacturonic acid in cross-linked polymeric structure of root cell walls. It should be noted that the $pK_a{}^2$ value of these groups, obtained in our study ($pK_a{}^2 \sim 5$), differs significantly from that previously derived (about 3). In our opinion it is due to the fact that authors of this investigation have not taken into consideration all kinds of groups in the cell wall structure (Morvan et al., 1979; Starrach et al., 198S; Richter and Dainty, 1989). Consequently, the extent between mitial and final-points of the titration of the groups (i.e. the quantity of ioinogenic groups of each type) could be determined as an error.

Table 4. Phosphorous content in isolated root cell walls of different plant species. The data were obtained by digestion the samples in sulphuric acid with Se catalyst following by use Loury's method of spectrophotometric phosphorous determination. Phosphorous content is expressed in micromole per gram dry weight and percent. Means ± SE (3 replicates)

The plant species	Phosphorous content	
	μmol/g dry wt.	%
38-days-old wheat plants	2,01±0,10	0,0063
10-days-old wheat seedlings	8,95±0,13	0,028
10-days-old lupin seedlings	10,8±0,13	0,033
35-days-old pea plants	22,0±1,00	0,065

It is known that in cross-linked polymer structure the cation exchange groups with $pK_a{}^3 \sim 7.5$ could be either a carboxyl or a phosphate (Shataeva et al., 1979). We used Loury assay to verify the presence of the phosphate groups in root cell wall structure. Table 4 indicates that isolated cell walls contain small amount of phosphorus, in all instances P content is less than 5% of quantity as ionogenic groups with $pK_a{}^3 \sim 7.5$. It figures out that ionogenic groups with $pK_a{}^3 \sim 7.5$ are the carboxyl groups of the second species (different from the carboxyl groups of galacturonic acid).

The root cell walls of wheat, lupin and pea are identical in qualitative structure of ionogenic groups but vary in quantity of each type of the groups, as demonstrate the values of dissociation constants ($pK_a{}^j$) and S^j (Table 2). The total quantity of carboxyl groups ($S^2 + S^3$) is 1.5 – 2 times higher for the lupin and the pea cell walls than for the wheat. For the root cell walls of 38-days-old wheat this parameter is nearly twice larger than for etiolated seedlings. The quantity of carboxyl groups should be connected directly with the pH gradient in the extracellar space at the membrane surface. The gradient arises from ion

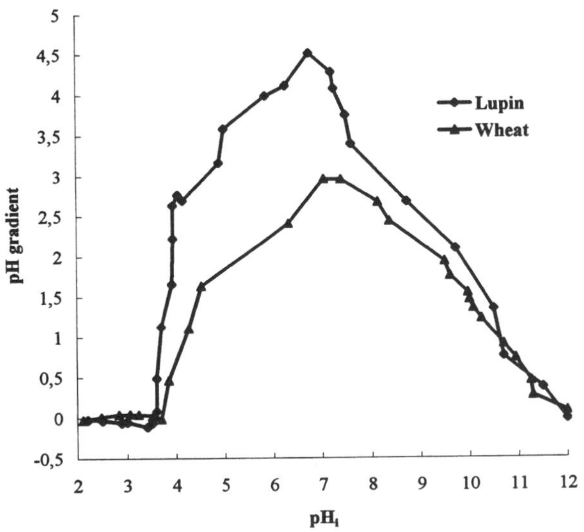

Figure 3. Dependencies of pH gradient means on equilibrated pH. In each point gradient pH was obtained as variance between initial (pH_i^{in}) and final or equilibrium (pH_i) pH means of the solution. pH_i^{in} and pH_i were obtained before and after equilibrium of the cell wall sample with the solution.

exchange reactions between cations of an outer medium and the protons of ionized carboxyl groups of the cell walls. Figure 3 demonstrates that the greater is the quantity of the ionized carboxyl groups, the higher is the pH gradient. For isolated cell walls of the lupin ΔpH is higher than for the wheat at all pH means of the bathing solution (compare the data of Figure 3 and the data of Table 2). Both for the lupin and for the wheat the maximum ΔpH is reached at the same pH means (pH$\sim$7), but for the lupine walls the maximum ΔpH value is $\sim$4.5 pH units, whereas for wheat ΔpH value is $\sim$3 pH units.

The ionization degree of weak acid and base depends only on a medium pH and pK_a value. The later parameter, as is known, is the constant for any acid or base. It means that at particular pH the ionization degree depends only on the nature of acid or base (Albert and Serzhent, 1964). Using this line of reasoning and the data of Table 2, we calculate α_i means for each type of ionogenic groups at corresponding pH_i, according to the formula:

$$\alpha_i = 1/\{1 + 1/10\hat{[}(pH_i - pK_a^j)/n^j]\} \tag{5}$$

Dependencies α_i = (pH_i) (Figure 4) are identical for all studied samples because $pK_a{}^j$ and n^j differ little. The curves on Figure 4 represent the state of the amine, carboxyl and phenolic groups, arranged in polymeric cell wall structure of the lupin, the pea and the wheat. For example, at pH = 6 the carboxyl groups with $pK_a{}^2 \sim 5.2$ are 90% ionized, whereas the all carboxyl groups with $pK_a{}^3 \sim 7.5$ are unionized; at pH = 8 both groups are ionized by 100% and 75%, respectively. It should be noted that the amine and phenolic groups are always closed (are unionized) at physiological conditions (pH = 4 to 8) and so, do not take part in ion exchange reactions with cations of an outer solution (see the Figure 4).

Our results testify that in normal conditions (pH = 4 to 8; Grignon and Sentenac, 1991) the root cell walls have the properties of cross-linked cation-exchanger, but not anion-exchanger (Figure 4). Consequently, once the solution has passed the cell wall phase, its pH is diminished by some value, the concentration of cations is decreased by the same value, and the concentration of anions is unaffected.

For physiological conditions the nutritive medium pH ranges between a weakly acidic and a weakly basic regions. It emphasizes the role of different ionogenic groups during the first step of cation mineral nutrition uptake. The adsorption of cations, at least in the epidermis, is related to the existence of carboxyl groups in the cell walls. The local pH value of the apoplast is the derivative of pH and a composition of the medium, of ion exchange and diffusion processes, of protons transfer through the cell membrane. Probably, the variation of local apoplast pH, which (variation) is associated with cation exchange reactions in cell walls, (a) stimulates the participation of other species of ionogenic groups (for example, anion-exchange groups) in exchangeable processes and (b) is necessary for transporting of mineral solutes into the cell. These anion-exchangeable groups could enter into composition of other structures, for example the plasmalemma.

Finally, the results suggest that, $S_t{}^{cat}$ and $S_t{}^{an}$ allow the quantitative estimation of ion exchange properties of the cell walls. The resulting parameters (S^j, $pK_a{}^j$ and n^j) allow prediction of changes in an ionic composition of a medium that bathes the cell membrane, during the first step of mineral nutrition uptake. We conclude that the high cation exchange capacity of the cell walls and H^+/cation exchange in the apoplast plays the most important role in the physiology of ion movement.

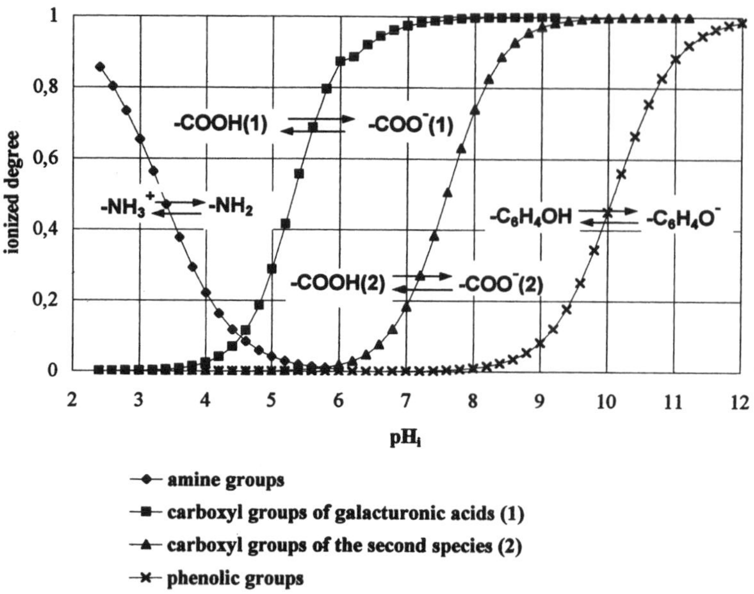

Figure 4. Dependencies of ionized degree of different ionogenic groups on outer medium pH_i. The lines join the calculated points obtained from equation 5 with parameters of Table 2.

Acknowledgements

This work is supported by the Russian Foundation of Fundamental Research (Grant 98-04-48867) and by S. Peterburg's Program 'Russian Universities'.

References

Albert A and Serzhent E 1964 Constants of ionization of acids and basis. Chemistry, Leningrad, USSR. 179p.

Gregor H P, Luttinger L D and Loeble E M 1954 Titration polyacrylic acid with quaternary ammonium basses. J. Amer. Chem. Soc. 76, 5879–5880.

Grignon C and Sentenac H 1991 pH and ionic conditions in the apoplast. Annu. Rev. Plant Physiol. 42, 103–128.

Leykin Y A, Meychik N R and Solovyov V K 1978 Acid-base equilibrium of polyamfolitov with pyridine and phosphate groups. Russian J. Physical Chemistry. 52, 1420–1424.

Luttge U and Higinbotham N 1984 Transport in plants. Sprnger-Verlag, New York, Heidel-berg Berlin. 408 p.

Meychik N R, Leykin Y A, Kosaeva A E and Galitskay N B 1989 The study of acidbase equilibrium and sorptional properties of nitrogen-, hydroxyl-containing ion-exchanger. Russian J. Physical Chemistry. 63, 540–542.

Morvan C, Demarty M and Thellier M 1979 Titration of isolated cell walls of *Lemna minor* L. Plant Physiol. 63, 1117–1122.

Richter C and Dainty 1 1989 Ion behavior in plant cell walls. I. Characterization of the Sphagnum russowii cell wall ion exchanger. Can. J. Bot. 67, 451–459.

Shataeva L A, Kuznetsova N N and Elkin G E 1979 Carboxyl ion exchangers in biology. Science, Leningrad, USSR. 286 p.

Starrach N, Flach D and Mayer W-E 1985 Activity of fixed negative charges of isolated extensor cell walls of the laminar pulvinus of primary leaves of Phaseolus. J. Plant Physiol. 120, 441–455.

Vakhmistrov D B 1967 Localization of the free space in barely roots. Russian Plant Physiol. 14, 397–404.

Developments in Plant and Soil Sciences

1. J. Monteith and C. Webb (eds.): *Soil Water and Nitrogen in Mediterranean-type Environments.* 1981 ISBN 90-247-2406-6
2. J.C. Brogan (ed.): *Nitrogen Losses and Surface Run-off from Landspreading of Manures.* 1981 ISBN 90-247-2471-6
3. J.D. Bewley (ed.): *Nitrogen and Carbon Metabolism.* 1981 ISBN 90-247-2472-4
4. R. Brouwer, I. Gašparíková, J. Kolek and B.C. Loughman (eds.): *Structure and Function of Plant Roots.* 1981 ISBN 90-247-2510-0
5. Y.R. Dommergues and H.G. Diem (eds.): *Microbiology of Tropical Soils and Plant Productivity.* 1982 ISBN 90-247-2624-7
6. G.P. Robertson, R. Herrara and T. Rosswall (eds.): *Nitrogen Cycling in Ecosystems of Latin America and the Caribbean.* 1982 ISBN 90-247-2719-7
7. D. Atkinson, K.K.S. Bhat, M.P. Coutts, P.A. Mason and D.J. Read (eds.): *Tree Root Systems and Their Mycorrhizas.* 1983 ISBN 90-247-2821-5
8. M.R. Sarić and B.C. Loughman (eds.): *Genetic Aspects of Plant Nutrition.* 1983 ISBN 90-247-2822-3
9. J.R. Freney and J.R. Simpson (eds.): *Gaseous Loss of Nitrogen from Plant-Soil Systems.* 1983 ISBN 90-247-2820-7
10. United Nations Economic Commission for Europe (ed.): *Efficient Use of Fertilizers in Agriculture.* 1983 ISBN 90-247-2866-5
11. J. Tinsley and J.F. Darbyshire (eds.): *Biological Processes and Soil Fertility.* 1984 ISBN 90-247-2902-5
12. A.D.L. Akkermans, D. Baker, K. Huss-Danell and J.D. Tjepkema (eds.): Frankia *Symbioses.* 1984 ISBN 90-247-2967-X
13. W.S. Silver and E.C. Schröder (eds.): *Practical Application of* Azolla *for Rice Production.* 1984 ISBN 90-247-3068-6
14. P.G.L. Vlek (ed.): *Micronutrients in Tropical Food Crop Production.* 1985 ISBN 90-247-3085-6
15. T.P. Hignett (ed.): *Fertilizer Manual.* 1985 ISBN 90-247-3122-4
16. D. Vaughan and R.E. Malcolm (eds.): *Soil Organic Matter and Biological Activity.* 1985 ISBN 90-247-3154-2
17. D. Pasternak and A. San Pietro (eds.): *Biosalinity in Action.* Bioproduction with Saline Water. 1985 ISBN 90-247-3159-3
18. M. Lalonde, C. Camiré and J.O. Dawson (eds.): Frankia *and Actinorhizal Plants.* 1985 ISBN 90-247-3214-X
19. H. Lambers, J.J. Neeteson and I. Stulen (eds.): *Fundamental, Ecological and Agricultural Aspects of Nitrogen Metabolism in Higher Plants.* 1986 ISBN 90-247-3258-1
20. M.B. Jackson (ed.): *New Root Formation in Plants and Cuttings.* 1986 ISBN 90-247-3260-3
21. F.A. Skinner and P. Uomala (eds.): *Nitrogen Fixation with Non-Legumes* (Proceedings of the 3rd Symposium, Helsinki, 1984). 1986 ISBN 90-247-3283-2
22. A. Alexander (ed.): *Foliar Fertilization.* 1986 ISBN 90-247-3288-3
23. H.G. v.d. Meer, J.C. Ryden and G.C. Ennik (eds.): *Nitrogen Fluxes in Intensive Grassland Systems.* 1986 ISBN 90-247-3309-X
24. A.U. Mokwunye and P.L.G. Vlek (eds.): *Management of Nitrogen and Phosphorus Fertilizers in Sub-Saharan Africa.* 1986 ISBN 90-247-3312-X

Developments in Plant and Soil Sciences

25. Y. Chen and Y. Avnimelech (eds.): *The Role of Organic Matter in Modern Agriculture.* 1986 ISBN 90-247-3360-X
26. S.K. De Datta and W.H. Patrick Jr. (eds.): *Nitrogen Economy of Flooded Rice Soils.* 1986 ISBN 90-247-3361-8
27. W.H. Gabelman and B.C. Loughman (eds.): *Genetic Aspects of Plant Mineral Nutrition.* 1987 ISBN 90-247-3494-0
28. A. van Diest (ed.): *Plant and Soil: Interfaces and Interactions.* 1987 ISBN 90-247-3535-1
29. United Nations Economic Commission for Europe and FAO (eds.): *The Utilization of Secondary and Trace Elements in Agriculture.* 1987 ISBN 90-247-3546-7
30. H.G. v.d. Meer, R.J. Unwin, T.A. van Dijk and G.C. Ennik (eds.): *Animal Manure on Grassland and Fodder Crops.* Fertilizer or Waste? 1987 ISBN 90-247-3568-8
31. N.J. Barrow: *Reactions with Variable-Charge Soils.* 1987 ISBN 90-247-3589-0
32. D.P. Beck and L.A. Materon (eds.): *Nitrogen Fixation by Legumes in Mediterranean Agriculture.* 1988 ISBN 90-247-3624-2
33. R.D. Graham, R.J. Hannam and N.C. Uren (eds.): *Manganese in Soils and Plants.* 1988 ISBN 90-247-3758-3
34. J.G. Torrey and J.L. Winship (eds.): *Applications of Continuous and Steady-State Methods to Root Biology.* 1989 ISBN 0-7923-0024-6
35. F.A. Skinner, R.M. Boddey and I. Fendrik (eds.): *Nitrogen Fixation with Non-Legumes* (Proceedings of the 4th Symposium, Rio de Janeiro, 1987). 1989 ISBN 0-7923-0059-9
36. B.C. Loughman, O. Gašparíková and J. Kolek (eds.): *Structural and Functional Aspects of Transport in Roots.* 1989 ISBN 0-7923-0060-2; Pb 0-7923-0061-0
37. P. Plancquaert and R. Haggar (eds.): *Legumes in Farming Systems.* 1990 ISBN 0-7923-0134-X
38. A.E. Osman, M.M. Ibrahim and M.A. Jones (eds.): *The Role of Legumes in the Farming Systems of the Mediterranean Areas.* 1990 ISBN 0-7923-0419-5
39. M. Clarholm and L. Bergström (eds.): *Ecology of Arable Land – Perspectives and Challenges.* 1989 ISBN 0-7923-0424-1
40. J. Vos, C.D. van Loon and G.J. Bollen (eds.): *Effects of Crop Rotation on Potato Production in the Temperate Zones.* 1989 ISBN 0-7923-0495-0
41. M.L. van Beusichem (ed.): *Plant Nutrition – Physiology and Applications.* 1990 ISBN 0-7923-0740-2
42. N. El Bassam, M. Dambroth and B.C. Loughman (eds.): *Genetic Aspects of Plant Mineral Nutrition.* 1990 ISBN 0-7923-0785-2
43. Y. Chen and Y. Hadar (eds.): *Iron Nutrition and Interactions in Plants.* 1991 ISBN 0-7923-1095-0
44. J.J.R. Groot, P. de Willigen and E.L.J. Verberne (eds.): *Nitrogen Turnover in the Soil-Crop System.* 1991 ISBN 0-7923-1107-8
45. R.J. Wright, V.C. Baligar and R.P. Murrmann (eds.): *Plant-Soil Interactions at Low pH.* 1991 ISBN 0-7923-1105-1
46. J. Kolek and V. Kozinka (eds.): *Physiology of the Plant Root System.* 1992 ISBN 0-7923-1205-8
47. A.U. Mokwunye (ed.): *Alleviating Soil Fertility Constraints to Increased Crop Production in West Africa.* 1991 ISBN 0-7923-1221-X; Pb 0-7923-1222-8

Developments in Plant and Soil Sciences

48. M. Polsinelli, R. Materassi and M. Vincenzini (eds.): *Nitrogen Fixation* (Proceedings of the 5th Symposium, Florence, 1990). 1991 ISBN 0-7923-1410-7
49. J.K. Ladha, T. George and B.B. Bohlool (eds.): *Biological Nitrogen Fixation for Sustainable Agriculture.* 1992 ISBN 0-7923-1774-2
50. P.J. Randall, E. Delhaze, R.A. Richards and R. Munns (eds.): *Genetic Aspects of Plant Mineral Nutrition.* 1993 ISBN 0-7923-2118-9
51. K.S. Kumarasinghe and D.L. Eskew (eds.): *Isotopic Studies of Azolla and Nitrogen Fertilization of Rice.* 1993 ISBN 0-7923-2274-6
52. F.A. Bliss and G. Hardarson (eds.): *Enhancement of Biological Nitrogen Fixation of Common Baen in Latin America.* 1993 ISBN 0-7923-2451-X
53. M.A.C. Fragoso and M.L. van Beusichem (eds.): *Optimization of Plant Nutrition.* 1993 ISBN 0-7923-2519-2
54. N.J. Barrow (ed.) *Plant Nutrition – From Genetic Engineering to Field Practice.* 1993 ISBN 0-7923-2540-0
55. A.D. Robson (ed.): *Zinc in Soils and Plants.* 1993 ISBN 0-7923-2631-8
56. A.D. Robson, L.K. Abbott and N. Malajczuk (eds.): *Management of Mycorrhizas in Agriculture, Horticulture and Forestry.* 1994 ISBN 0-7923-2700-4
57. P.H. Graham, M.J. Sadowsky and C.P. Vance (eds.): *Symbiotic Nitrogen Fixation.* 1994 ISBN 0-7923-2781-0
58. F. Baluška, M. Čiamporová, O. Gašparíkova and P.W. Barlow (eds.): *Structure and Function of Roots.* 1995 ISBN 0-7923-2832-9
59. J. Abadía (ed.): *Iron Nutrition in Soils and Plants.* 1995 ISBN 0-7923-2900-7
60. P.S. Curtis, E.G. O'Neill, J.A. Teeri, D.R. Zak and K.S. Pregitzer (eds.): *Belowground Responses to Rising Atmospheric* CO_2. Implications for Plants, Soil Biota, and Ecosystem Processes. 1994 ISBN 0-7923-2901-5
61. P.C. Struik, W.J. Vredenberg, J.A. Renkema and J.E. Parlevliet (eds.): *Plant Production on the Threshold of a New Century.* 1994 ISBN 0-7923-2903-1
62. L.O. Nilsson, R.F. Hüttl and U.T. Johansson (eds.): *Nutrient Uptake and Cycling in Forest Ecosystems* (CEC/IUFRO Symposium, Halmstad, Sweden, 1993). 1995 ISBN 0-7923-3030-7
63. H.P. Collins, G.P. Robertson and M.J. Klug (eds.): *The Significance and Regulation of Soil Biodiversity* (International Symposium, East Lansing, 1993). 1995 ISBN 0-7923-3138-9
64. R.A. Date, N.J. Grundon, G.E. Rayment and M.E. Probert (eds.): *Plant-Soil Interactions at Low pH: Principles and Management* (3rd International Symposium, Brisbane, 1993). 1995 ISBN 0-7923-3198-2
65. J.K. Ladha and M.B. Peoples (eds.): *Management of Biological Nitrogen Fixation for the Development of More Productive and Sustainable Agricultural Systems.* (15th Congress of Soil Science, Acapulco, Mexico, 1994) 1995 ISBN 0-7923-3413-2; Pb 0-7923-3414-0
66. C. Rodriguez-Barrueco (ed.): *Fertilizers and Environment.* 1996 ISBN 0-7923-3729-8
67. B. Buerkert, B.E. Allison and M. von Oppen (eds.): *Wind Erosion in Niger.* Implications and Control Measures in a Millet-based Farming System. 1996 ISBN 0-7923-3885-5

Developments in Plant and Soil Sciences

68. O. van Cleemput, G. Hofman and A. Vermoesen (eds.): *Progress in Nitrogen Cycling.* 1996 ISBN 0-7923-3962-2
69. N. Ahmad (ed.): *Nitrogen Economy in Tropical Soils.* 1996 ISBN 0-7923-4094-9
70. M. Rahman, A. Kumar Podder, C. van Hove, Z.N. Tahmida Begum, T. Heulin and A. Hartmann (eds.): *Biological Nitrogen Fixation Associated with Rice Production.* 1996 ISBN 0-7923-4197-X
71. D. Rosen, E. Tel-Or, Y. Hadar and Y. Chen (eds.): *Modern Agriculture and the Environment.* 1997 ISBN 0-7923-4295-X
72. G.H. Elkan and R.G. Upchurch (eds.): *Current Issues in Symbiotic Nitrogen Fixation.* 1997 ISBN 0-7923-4367-0
73. H.M. Anderson, P.W. Barlow, D.T. Clarkson, M.B. Jackson and P.R. Shewry (eds.): *Plant Roots – From Cells to Systems.* 1997 ISBN 0-7923-4369-7
74. Zhu Zhao-liang, Wen Qi-xiao and J.R. Freney (eds.): *Nitrogen in Soils of China.* 1997 ISBN 0-7923-4372-7
75. J.K. Ladha, F.J. de Bruijn and K.A. Malik (eds.): *Opportunities for Biological Nitrogen Fixation in Rice and Other Non-Legumes* (2nd Working Group Meeting, Faisalabad, Pakistan, 1996). 1997 ISBN 0-7923-4514-2
76. R.W. Bell and B. Rerkasem (eds.): *Boron in Soils and Plants* (International Symposium, Chiang Mai, Thailand, 1997). 1997 ISBN 0-7923-4705-6
77. B. Dell, P.H. Brown and R. Bell (eds.): *Boron in Soils and Plants: Reviews.* 1997 ISBN 0-7923-4729-3
78. T. Ando, K. Fujita, T. Mae, H. Matsumoto, S. Mori and J. Sekiya (eds.): *Plant Nutrition for Sustainable Food Production and Environment.* 1997 ISBN 0-7923-4796-X
79. K.A. Malik, M.S. Mirza and J.K. Ladha (eds.): *Nitrogen Fixation with Non-Legumes.* 1998 ISBN 0-7923-4873-7
80. P.A. Finke, J. Bouma and M.R. Hoosbeek (eds.): *Soil and Water Quality at Different Scales.* 1998 ISBN 0-7923-4994-6
81. V. Balasubramanian, J.K. Ladha and G.L. Denning (eds.): *Resource Management in Rice Systems: Nutrients.* 1999 ISBN 0-7923-5199-1
82. J.E. Box, Jr. (ed.): *Root Demographics and Their Efficiencies in Sustainable Agriculture, Grasslands and Forest Ecosystems.* 1998 ISBN 0-7923-5230-0
83. G. Hardarson and W. Broughton (eds.): *Molecular Microbial Ecology of the Soil.* Results from an FAO/IAEA Co-ordinated Research Programme, 1992–1996. 1998 ISBN 0-7923-5252-1
84. C. Kirda, P. Moutonnet, C. Hera and D.R. Nielsen (eds.): *Crop Yield Response to Deficit Irrigation.* Report of an FAO/IAEA Co-ordinated Research Program by using Nuclear Techniques. 1999 ISBN 0-7923-5299-8
85. M.J. McLaughlin and B.R. Singh (eds.): *Cadmium in Soils and Plants.* 1999 ISBN 0-7923-5843-0
86. D. Anac and P. Martin-Prével (eds.): *Improved Crop Quality by Nutrient Management.* 2000 ISBN 0-7923-5850-3
87. A. Stokes (ed.): *The Supporting Roots of Trees and Woody Plants: Form, Function and Physiology.* 2000 ISBN 0-7923-5964-X
88. P.A. Finke, M.F.P. Bierkens and P. De Willigen: *Upscaling and Downscaling Methods for Environmental Research.* 2000 ISBN 0-7923-6339-6

Developments in Plant and Soil Sciences

89. C.J. Staiger, F. Baluska, D. Volkmann and P. Barlow (eds.): *Action: A Dynamic Framework for Multiple Plant Cell Functions*. 2000 ISBN 0-7923-6412-0

Kluwer Academic Publishers – Dordrecht / Boston / London